Electrical Noise:
Fundamentals & Sources

OTHER IEEE PRESS BOOKS

Electrical Noise:
Fundamentals & Sources

Edited by

Madhu S. Gupta

Assistant Professor of Electrical Engineering
Massachusetts Institute of Technology

A volume in the IEEE PRESS Selected Reprint Series.

IEEE
PRESS

The Institute of Electrical and Electronics Engineers, Inc. New York

IEEE International Standard Book Numbers: Clothbound: 0-87942-085-5
Paperbound: 0-87942-086-3

Library of Congress Catalog Card Number 76-57816

Sole Worldwide Distributor (Exclusive of the IEEE):

JOHN WILEY & SONS, INC.
605 Third Ave.
New York, NY 10016

Wiley Order Numbers: Clothbound: 0-471-03116-6
Paperbound: 0-471-03117-8

Contents

Introduction

Scope of This Volume

Noise, the subject of this volume, is an overused word. In addition to its obvious meaning as audible sound ("acoustic noise"), it has been used to denote fluctuations in fluid flow ("hydrodynamic noise"), meteorological variations (observed as atmospherics and as "scintillation noise"), certain types of errors in computation (such as "roundoff noise"), the difference between a quantity and its representation (such as the "quantization noise" in analog-to-digital conversion), and almost all kinds of fluctuations, errors, or deviations from normal expectation. In this volume, the term noise will refer to electrical noise, which manifests itself as a deviation of an electrical current or voltage from its expected value. The scope of the term is further restricted to fluctuation noise (sometimes called internal or random noise) arising in electronic devices and networks. This is to distinguish it from interference (or external noise or pickup), which may have a solar or galactic origin ("cosmic noise"), a natural terrestrial origin ("atmospheric noise"), or a man-made source ("static" or unwanted signal).

The omnipresence and the inevitability of electronic fluctuations lend the study of noise a fundamental character and a wide range of applicability. There are a number of approaches to the study of noise: from the design of systems that are immune to noise, to the investigation of microscopic processes related to noise. On this scale, extending from phenomenological to physical viewpoints, the emphasis here is towards the physical end. The "physical theory" of noise in the present volume focuses on the generation of noise and excludes what may be called the circuit theory of noise, the signal theory of noise, and the system theory of noise. These other subjects will hopefully be the foci for other similar volumes.

Need for this Compilation

The subject of noise in electronic circuits, devices, and systems cannot be called "mature" despite the vast literature on this subject that has accumulated in the past few decades. This statement is supported by at least three different types of evidence. One of the hallmarks of a mature subject is the existence of a set of unifying principles that transcend the individual applications of those principles to narrowly defined problems. A few such principles, like Nyquist's theorem, have indeed emerged in this field, but there is much that has to be rederived when applied to a new situation. To quote a specific example as an illustration, a theory of IMPATT oscillator noise had to be redeveloped [1] rather than being considered a re-application of the earlier work done in the context of lasers [2]. Second, in most mature disciplines, the basic principles and techniques filter down from research papers to textbooks over a period of time, and become so well established that they are taken for granted and used without quoting the original reference, or any reference at all. It appears that this has not taken place to any significant degree in the literature of noise. Nyquist and Schottky still remain two of the most quoted authors in noise literature. Third, unlike a well-established discipline, the vocabulary in the field of noise is still fluid, and even some common terms, such as "shot noise" and "multiplicative noise," are used in more than one sense.

The principal reason for this situation, in the Editor's opinion, is the dearth of the sources of consolidated and compiled information. Much of the research on noise has been carried out in the context of a particular device or system, often by people whose primary interest is in the specific class of devices or system. The transferability of principles and techniques across these boundaries is therefore a slow and uneven process, increasing the need to rely on the literature of the field to provide for this transfer. However, the number and scope of books written in this field is limited (they are listed in Part II), retarding the establishment of a body of facts and principles that can be called common knowledge.

It appears desirable to make another attempt to condense some of the literature on electrical noise into book form. A relatively more expeditious way of doing this is through a compilation. A much more important advantage of a compilation, however, is the fact that each author is writing in a narrower field in which he is well versed; single-author books with wide scopes are sometimes written around one chapter (or group of chapters) closest to the author's heart, the remainder serving as packing material. The scope of a compilation can be chosen to be wider without the hazard of stepping into unknown territory. Undoubtedly, there are also the disadvantages of nonuniform levels of treatment, varying notation, and some unavoidable duplication. The present compilation is intended to help in filling this proverbial "gap."

Purpose of this Compilation

"Knowledge is of two kinds," wrote Samuel Johnson. "We know a subject ourselves or we know where to find information upon it." The present compilation is intended to serve as a source of both kinds of knowledge, but at somewhat different levels. The first kind of knowledge is provided by the reprinted papers and the second kind by the bibliographies.

The primary purpose of this volume is to bring together a set of tutorial review papers on the physical theory and generation of noise. Engineers have always lamented the shortage of good review papers. Prudence suggests that review papers being a scarce resource, we should attempt to get more mileage out of them. Hopefully, their wider dissemination through this reprinting will help towards that end.

The second, equally important, purpose of this volume is to compile some bibliographies on individual topics lying within the scope of the present subject. Whereas the reviews of tutorial nature are most useful to those relatively new to the field, extensive bibliographies are useful to those who have a deeper interest in the subject, and who must dig further into the original literature. The bibliographies are therefore a step beyond the reviews.

PLAN OF THIS COMPILATION

This volume is divided into five parts. The first is devoted to the history of the subject of electrical noise. The second part is an elementary introduction to the three components of the required background in the field: physical mechanisms, mathematical methods, and applications. The third part is intended to introduce the principal noise-generating processes. The fourth part is organized around some of the types of electron devices in which noise phenomena have been studied. Finally, the fifth part provides an introduction to the more commonly used noise generators.

The major content of these parts is a collection of 22 papers on electrical noise, appearing in journals over a period of almost three decades. The criteria on the basis of which these papers were selected are stated below; however, the single most important characteristic common to these papers is their "review" nature. Despite this fact, most of these papers contain a list of references which is "short" (compared to the literature of their field). This deficiency is made up by supplementing the papers with more extensive bibliographies on individual subjects.

Every part is itself divided into three sections. The editorial comments in each part are intended to delineate the subject matter of the part, place the reprinted papers into perspective, and point out the goals and directions of the papers listed in the bibliographies. This is followed by the reprinted papers and then the bibliographies.

THE REPRINTED PAPERS

Compilations of scientific papers by different authors, previously published at different times, are a recent addition to the engineering literature (although one of the best known earlier compilations is in the field of noise [3]). The objectives, and hence the criteria for selecting the reprinted papers, differ considerably from compilation to compilation. Some contain "classic" papers, which introduced ideas that greatly influenced the subsequent development of their subject, while others are organized around a specific application or device. The present volume is primarily a collection of review papers on the physical aspects of the subject of electrical noise.

Most papers included in this compilation are review papers, either explicitly or implicitly. There are several reasons for this choice—as contrasted with original research papers, the review papers are written to have a wider appeal, they have a broader viewpoint and more even-handed treatment of individual research contributions, they do not leave the reader bogged down in details like algebraic manipulations or equipment design, they contain sizable bibliographies for readers wishing further or related information, they do not get obsolete as rapidly, and they have a high tutorial value. Admittedly, a single review paper may not have all of these listed virtues. The term "review paper" is generally applied to several different kinds of papers, differing from each other in their scope, level, and detail. Several different types of review papers are reprinted here, including tutorial as well as research reviews, comprehensive as well as selective reviews, and critical as well as annotated literature reviews.

In addition to this principal requirement, the selection of reprinted papers was guided by several other factors also. First, only the papers in English were considered (in particular, this excluded some very comprehensive survey articles in German). A paper translated from Russian into English is also included. Second, where several review papers on a single subject were available, the paper with a broader coverage, more basic viewpoint, more recent publication date, or more tutorial approach was selected. Third, some of the papers were included here to cover topics on which no review papers are available.

THE BIBLIOGRAPHIES

None of the bibliographies appearing in this volume should be considered exhaustive. In fact, the approximately 1200 references included in these bibliographies were selected from a collection of almost twice as many references. The need for this selection arose partly from limitations of space and partly from striking a utility-size compromise. There is a law of diminishing returns in the preparation of technical bibliographies [4]: as a bibliography becomes more nearly complete, its size increases rapidly and the average relevance of individual items decreases rapidly.

The selection of papers for the bibliographies is based on several of the same criteria stated earlier. The bibliographies are confined to papers in English (and a very few in some European languages); very old or otherwise difficult to find references were rejected in favor of the more recent and accessible ones; duplicative or very similar papers by the same author(s) were avoided; papers in trade magazines or manufacturers' literature were discriminated against in favor of those in scientific journals; journal articles were preferred over conference papers (a list of noise-related conferences is contained in Part II); and the boundaries of the topic were defined rigidly in order to keep the length of bibliographies reasonable.

The bibliography so generated on a single topic is arranged alphabetically by the authors, and for identical author(s), by the date of publication. A very lengthy bibliography, unclassified by subject, is known to be difficult to use. The longest bibliography included here on a single topic contains approximately 300 items in it, and appears to represent the upper limit beyond which further subclassification would be necessary.

Attempts have been made to keep the bibliographies readable (by line indentation), useful (by reference selection), and error-free. The Editor apologizes for significant omissions resulting either from selection or from ignorance, and for citation errors which appear to be as inevitable as electrical noise itself.

The Editorial Comments

Each of the five parts is preceded with brief editorial comments. It is not the purpose of these comments to summarize the contents of either the reprinted papers or the papers listed in the subject bibliographies of the part. That would result in a lengthy book and would duplicate the purpose of the reprinted review papers which themselves are a summary of the original literature. The objective of the introductory comments is to introduce the literature, not the subject matter. Hopefully, the comments will aid the reader in understanding the terminology, goals, approach, or reasoning behind the papers reprinted in this volume or included in the bibliographies following the comments.

References

[1] H. A. Haus, H. Statz, and R. A. Pucel, "Optimum noise measure of IMPATT diodes," *IEEE Trans. Microwave Theory Techn.*, vol. MTT-19, pp. 801–813, Oct. 1971.

[2] H. A. Haus, "Amplitude noise in laser oscillators," *IEEE J. Quantum Electron.*, vol. QE-1, pp. 179–180, July 1965.

[3] N. Wax, Ed., *Selected Papers on Noise and Stochastic Processes.* New York: Dover, 1954.

[4] M. S. Gupta, "Preparation of a technical bibliography," *IEEE Trans. Professional Commun.*, vol. PC-16, pp. 7–10, Mar. 1973.

Part I
Historical Development

THE SUBJECT

"Far too many physicians and scientists, as well as laymen, look upon the history of discovery as an entertaining pastime, a tiresome academic exercise, or merely the record of egotistical aspiration" [1]. However, it is possible to point out several reasons why the history of a scientific subject should be studied. In this volume, the inclusion of a part on historical introduction to the subject was motivated by only one of these reasons. An awareness of the historical development of a subject is helpful in understanding the terminology, motivation, and procedures of present work on that subject. This benefit coincides with the objectives of this compilation.

The origin of the study of noise or, for that matter, of most subjects, cannot be precisely specified in time because it can be traced back almost ad-infinitum. The study of electrical fluctuations is usually traced back to the study of Brownian motion, although the fluctuations studied by Brown were not electrical. Furthermore, it is commonplace to find that a major discovery had already been made or anticipated by someone who was ahead of his times or did not get sufficient exposure. Thus we find that Brownian motion had been observed by Needham and von Gleichen before Brown, and that the Nyquist formula for thermal noise had been given earlier by Lorentz and by Slingelandt. For these reasons, this historical introduction to our subject should be viewed as an aid to awareness rather than as the ultimate resolution of questions about origin and priority.

THE REPRINTED PAPERS

The four papers included in this part deal with four complementary areas in the study of electrical noise, although there is some inevitable overlap between them. MacDonald's paper is the oldest among the papers reprinted in this compilation and is devoted primarily to the physical origin of noise. Johnson (after whom thermal noise is often called Johnson noise) directs his history primarily towards the performance of electron tubes, the principal active device for over three decades. These two together describe the field until the 1930's.

The earliest and the principal application of electronic devices was in the communication of information. The next two papers are related primarily to the developments that arose from recognizing noise limitations in communication. Lebenbaum's paper is devoted to the emergence of a measure of sensitivity for communication receivers. On the theoretical side, the cognizance that messages and noise should be treated with probabilistic concepts initially led to advances in two related directions [2]: statistical theory of communication and information theory. Ragazzini and Chang's paper summarizes the class of problems to which the statistical theory of communication was initially addressed.

THE BIBLIOGRAPHY

The bibliography on noise by Chessin, included here, covers much of the literature on the subject, and on related areas, until about 1954. Along with the references cited by the four reprinted papers, particularly MacDonald's, a fair picture of the early history of the subject can be inferred. The major limitation of the bibliography is that it is confined to the literature in English (and some European languages). One convenient source of early Soviet works is a bibliography prepared by Green [3]. Another compilation of early references worth mentioning is Stumpers' bibliography on information theory [4].

REFERENCES

[1] I. H. Page, "A sense of the history of discovery," *Science*, vol. 186, p. 1161, Dec. 27, 1974.
[2] Y. W. Lee, *Statistical Theory of Communication*. New York: Wiley, 1966, p. 2.
[3] P. E. Green, Jr., "A bibliography of Soviet literature on noise, correlation, and information theory," *IRE Trans. Inform. Theory*, vol. IT-2, pp. 91–94, June 1956.
[4] F. L. Stumpers, "A bibliography of information theory. Communication theory—cybernetics," *IRE Trans. Inform. Theory*, vol. PGIT-2, pp. 1–60, Nov. 1953. Supplements in: *IRE Trans. Inform. Theory*, vol. IT-1, pp. 31–47, Sept. 1955; vol. IT-3, pp. 150–166, June 1957; vol. IT-6, pp. 25–51, Mar. 1960.

The Brownian Movement and Spontaneous Fluctuations of Electricity

D. K. C. MacDONALD, M.A., Ph.D.

Clarendon Laboratory, Oxford

In 1827 the biologist ROBERT BROWN*[1] was studying under his microscope the pollen grains, some 0·0002 inch in length, of the plant *Clarckia pulchella*.

'While examining the form of these particles immersed in water, I observed many of them very evidently in motion. These motions were such as to satisfy me . . . that they arose neither from currents in the fluid, nor from its gradual evaporation, but belonged to the particle itself.'

He examined particles of a very wide range of substances, even including a fragment from the Sphinx, but found the motion to persist for every substance examined. The ramifications of this phenomenon, the *Brownian Movement*, extend to such fields as the opalescence observed near the critical point of a fluid,[4, 5] with its closely related problem of the blue of the sky, the limitation to the range of modern radar and television, and statistical theories of the motion and distribution of the stars.[6]

* Brown was not the first in fact to observe the motion ; NEEDHAM[2] and VON GLEICHEN[3] previously noticed it, but Brown appears to have been the first to examine it at all systematically. His own conclusion however was that he had demonstrated the existence of an elementary form of life in all organic and inorganic matter.

A fundamental feature of the Brownian movement be it in a mechanical, electrical or thermal application, lies in the fact that it sets a lower limit to any coherent stimulus (or signal) which we may hope to observe in a system, since any such small effect tends to be 'masked' or overlaid by the random spontaneous fluctuation of the medium. Viewed from another aspect it enables us to draw certain definite conclusions about the nature of a medium (*e.g.* a liquid in which a particle is floating, or an electric current passing through a valve) and how it behaves.

Much speculation and erroneous theory was put forward during the seventy years following Brown's discovery to account for the phenomenon. Suggestions that the effect arose from electrical forces,[7] evaporization[8, 9] or mechanical shocks[10] were examined and disposed of ; it was found to persist unchanged after the specimen had been kept in the dark for a week,[11] or after an hour's heating,[12] and electromagnetic fields were found to have no effect. Thus it became clear that the effect was entirely fundamental, and A. EINSTEIN,[13] in a series of classical papers, was the first to provide a clear analysis of the problem as arising from continuous

Reprinted with permission from *Res. Appl. in Industry*, vol. 1, pp. 194–203, Feb. 1948. Butterworth & Co., London, UK.

and random molecular bombardments and to apply the analysis to a number of problems. It is of interest to note that he derived an expression for the random flux of electric charge across a section of a conductor, although no method of detecting such an effect was then available. M. VON SMOLUCHOWSKI[14] and M. P. LANGEVIN[15] also made valuable contributions to the theory at this time, showing in particular how the motion progressed from any given initial conditions. It is helpful to note that VON NÄGELI[16] in 1879 had considered the possibility of molecular bombardment but had concluded because the impulse due to one collision was so minute, that this could not be the cause ; for he opined that since all directions in space are equally likely the cumulative effect of many random collisions could only be of the same magnitude. The error is a common one and arises essentially from implicitly regarding a random process as made up of regularly alternating favourable and unfavourable events ; such a process is, however, a highly ordered one and it is those very ' runs ' of favourable (or unfavourable) events, which we sometimes regard as against the ' laws of chance ', which characterize a random process and give rise to the relatively large fluctuations observed.* The essentials are well illustrated by the fundamental ' random-walk ' problem, as first posed by KARL PEARSON[17] in 1905. A man (presumably very drunk) takes steps of equal length, l, from a starting point O one after the other in successively random directions. Where is he likely to be after n such steps ? LORD RAYLEIGH[18] answered the problem immediately where n is large ; the probability that he is at a distance between

r and $r+dr$ from his starting point is $p(r)dr = \dfrac{2r}{nl^2}\varepsilon^{-\frac{r^2}{nl^2}}dr.$

His average† distance is therefore $\displaystyle\int_0^\infty r.p(r)dr = \dfrac{\sqrt{\pi}}{2}.\sqrt{nl},$

and thus increases with the square root of the time for which he continues the walk ; this is illustrated in *Figure 1* omitting the early stages of the process.‡

A glance at these sketches enables us to see clearly the point of Pearson's comment in his reply[20]

' The lesson of Lord Rayleigh's solution is that

* This is a simplified statement to emphasize the essential argument.

† *i.e.* the average observed over a large number of similar experiments, each one starting afresh ; any particular experiment or ' trial ' may yield a result very different from this average, constituting in fact a *fluctuation* from the mean.

‡ *i.e.* the first five steps or so, where the process degenerates from a coherent to a random system. This case has been considered recently in connection with electrical transmission problems by M. SLACK.[19]

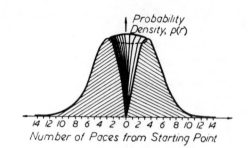

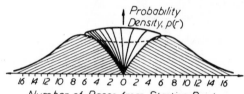

Figure 1. Probability surface of random walk; (a) after 18 random paces; (b) after 72 random paces

in open country the most probable place to find a drunken man . . . is somewhere near his starting point.'

On the other hand, of course, when dealing with molecular processes, the number n of events during a reasonable period of observation will be extremely large, and so even $\sqrt{n}$ will be a very large number.

E. VON SCHWEIDLER[21] and N. R. CAMPBELL[22] (1908–10) considered the problem of fluctuations in relation to radioactive decay and the photoelectric effect ; Campbell derived two fundamental theorems enabling one to calculate the average response, and the mean square fluctuation, due to a random sequence of similar events. If these occur at an average rate $\cdot\mathcal{N}$ per unit time, if an individual stimulus occurring at zero time results in a response $f(t)$ at time t (generally, of course, falling off to zero for large t), and if the responses are linearly additive, then denoting the overall response of the system by θ we have

$$\bar{\theta} = \mathcal{N}\int_0^\infty f(t)dt \qquad \dots(1)$$

$$\overline{(\theta-\bar\theta)^2} = \mathcal{N}\int_0^\infty f(t)^2dt \qquad \dots(2)$$

These formulae, and extensions to higher order deviations,[23] can be applied to a large number of problems, as will be exemplified later.

For the next few years the main interest lay in experimental verification and extension of Einstein's and Smoluchowski's work in mechanical systems. TH. SVEDBERG,[24] A. WESTGREN,[25] R. FÜRTH,[26] J. PERRIN[27] and co-workers examined the average density fluctuations in colloid solutions (*i.e.* liquids with microscopic particles), the distribution of fluctuations and how the fluctuation at any instant is affected by the preceding state (fluctuation 'velocity' or correlation).[28],[29]   In general, experiment and theory showed excellent agreement ; in particular, mutually consistent values of Avogadro's number or Boltzmann's constant, *k*, were obtained from the experiments.   We may note, for example, that Einstein's theory predicts that the mean square random displacement, $\overline{\Delta r^2}$, of a spherical particle radius *a*, in a fluid of viscosity η, at temperature *T*, in time *t* is given by

$$\overline{\Delta r^2} = \frac{kT}{\pi \eta a} . t \qquad \dots (3)$$

One ought certainly also to remember VON SMOLUCHOWSKI'S[30] valuable contributions at this time to the interpretation of phenomenological thermodynamics on the basis of the kinetic-molecular theory ; for the second law of thermodynamics in one form tells us that a closed system moves monotonically to a state of equilibrium from which there can be no self-directed reversal.   Clearly the phenomenon of spontaneous fluctuations contradicts any such law in a rigid form.   Smoluchowski emphasized and interpreted the significance of fluctuations and their 'trend' (correlation) in thermodynamical considerations.

THE 'SHOT' EFFECT

A new phase opened in 1918 when W. SCHOTTKY[31] considering the great increase in electrical amplification then available, mainly resulting from the

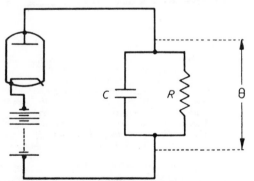

Figure 2. Circuit to illustrate application of Campbell's theorem to the shot effect

impetus of the 1914–18 war, pointed out that the current in a thermionic valve, being composed of individual electrons, should exhibit fluctuations. This is called the 'schrot' or 'shot' effect. If we may assume each electron to leave the emitter randomly and independently and to pass to the

anode independently of its fellows then Campbell's theorem (equation 2) may be applied directly. Considering the circuit of *Figure 2* we may set $f(t) = \frac{e}{C} \varepsilon^{-\frac{t}{RC}}$ (where *e* is the electronic charge), whence the voltage θ observed is given by :

$$\theta = \frac{Ne}{C} \int_0^\infty \varepsilon^{-\frac{t}{RC}} dt = \frac{Ne}{C}(RC) = IR \qquad \dots (4)$$

where $I(=Ne)$ is the mean current and

$$\overline{(\theta - \theta)^2} = \frac{Ne^2}{C^2} \int_0^\infty \varepsilon^{-\frac{2t}{RC}} dt = \frac{IeR}{2C} \qquad \dots (5)$$

Equation 5, or similar expressions for other external circuits, should enable an estimate of *e* to be made by measuring the shot effect. It is perhaps better in view of the number of possible disturbing factors, as T. C. FRY[32] later pointed out, to regard measurements of this sort rather as a proof of the correctness (or otherwise) of the premises than a fundamental method of determining the electrical quantum.

In the same paper, Schottky also rediscussed the thermal fluctuations of electricity in passive conductors and derived more useful expressions for this effect than Einstein, enabling a direct estimate of the 'noise power' in the first tuned circuit to be made.   The first experimental test of the shot effect was undertaken by C. A. HARTMANN in 1921.[33]   This proved very disappointing, the magnitude being 10^2–10^3 smaller than expected.   However, J. B. JOHNSON[34] of Bell Telephone Laboratories shortly after pointed out an error in the evaluation of an integral in Schottky's formula when applied to Hartmann's circuit which, when corrected, restored agreement with theory in order of magnitude. R. FÜRTH[35] also pointed out that the detection method (aural by telephone) is not immediately suitable for comparing the intensity of a fluctuating signal with a standard tone.   He was able to eliminate a further error in magnitude and a spurious frequency-dependence.   Many other investigators[36],[37],[38] have carried out more refined investigations, and excellent agreement with theory has been finally shown when the conditions underlying the theory are satisfied.

'SPACE-CHARGE REDUCTION' AND 'FLICKER EFFECT'

A. W. HULL and N. H. WILLIAMS[37] first reported that the presence of space-charge (*i.e.* when the valve departs from the temperature-limited region) reduces the observed shot-effect very considerably.*   It is

* They mention that the effect was, however, first discovered and brought to their notice by W. L. CARLSON of the General Electric Co.

customary now to express this reduction formally by the insertion of a factor $\Gamma^2(<1)$ in the shot-effect equation. This effect might be anticipated since the electron-transits from cathode to anode are no longer independent of one another. From the engineering point of view, this phenomenon is most

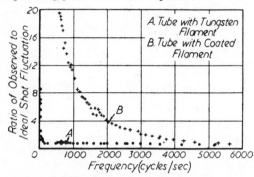

Figure 3.   Experiment to illustrate existence of flicker effect.   (After J. B. Johnson)

valuable, since normal amplifying valves by their very nature—resting in the control of the space-charge by variation of the external potentials—must operate in the space-charge region. Without this noise reduction, the range of modern radar and television would be severely curtailed. The full interpretation and understanding of the effect have engaged attention ever since its discovery. Shortly after this, J. B. JOHNSON[36] confirmed the phenomenon and reported the discovery of another. We may note that in discussing the space-charge reduction effect he said '. . . space-charge is . . . present and makes the electrons pass in a sequence more orderly than the purely chaotic one.' Such a dogmatic statement of cause, although certainly pardonable, is rather typical of the pioneer work and has undoubtedly contributed to later confusion of thought. The new discovery was the fact that at low frequencies the fluctuation magnitude increases from the standard value as the frequency is reduced (*Figure 3*). As is also clear the effect became noticeable first at higher frequencies with oxide-coated emitters than with tungsten filaments. Johnson himself suggested that the phenomenon arose as a result of some secular, relatively gross, variation of the emitter; he says

'The electrical emission at any time depends on the condition of the cathode surface . . . continual change due to evaporation, diffusion, chemical action, structural rearrangements and gas ion bombardment.'

One might liken the true shot effect to the macroscopically 'smooth' boiling of a liquid when a piece of porcelain pot is present to provide a boiling nucleus; the Johnson, or 'flicker', effect (as W. SCHOTTKY[39] later called it) might be likened

to the explosive action which can occur when this precaution is not observed. Schottky analysed Johnson's results and proposed for the flicker effect a formula of the type

$$\overline{(i-I)^2} \propto \frac{I^2}{\omega^2 + a^2} \qquad \ldots(6)$$

where i is the instantaneous valve current and I the mean current, $\omega = 2\pi f$, and $1/a$ is a time constant characteristic of the process. This he derived on the basis that the source is the random variation in density of foreign atoms or molecules on the surface of the cathode. The predicted dependence on I agreed quite well with Johnson's data; the frequency variation however, proved less satisfactory; E. B. MOULLIN,[40] for example, pointed out in 1938 that Johnson's results indicated a variation with $\omega^{-1.2}$. The flicker effect is naturally of great significance in the design of low frequency amplifiers required for example in physiological studies of nerve-potentials[41, 42] (*Figure 4*).

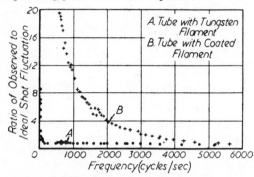

Figure 4a

Figure 4b

Figure 4.   Action potentials from the Sural Nerve (N. Saphenus min) in the rabbit by stimulating the hair by brushing. (a) Strong stimulus clearly visible through 'background-noise'.   (b) Weak stimulus practically obscured by 'background-noise'.   (Courtesy of B. Frankenhaeuser and G. Weddell, Dept. of Human Anatomy, University of Oxford)

SEMI-CONDUCTORS

In Hull and Williams' paper[37] another effect was observed; in estimating the contribution of battery voltage fluctuations in their amplifier, valves were temporarily replaced by passive resistances. When these were wire-wound, no fluctuation* was observed; but '. . . an India-ink resistance (standard grid-leak) gave a fluctuation of the same order of magnitude as the shot-effect.' This effect, dependent also on frequency and current, has been observed in

* *i.e.* none comparable with the shot-effect.

various semi-conductors,[43] carbon-granule microphones[44] and very thin metallic films.[*45] Rather recently it has been recognized that this effect[†] and the 'flicker' effect are closely connected; G. G. MACFARLANE[46] in particular has re-analysed the problems. He regards both phenomena as arising from the diffusion (due to concentration gradients) of clusters of alien atoms (*e.g.* excess barium atoms in the case of a barium oxide emitter) on to the contact surface—or the cathode surface in the case of a valve. This cluster becomes ionized on the (contact) surface and subsequently leaves the surface under the influence of the electric field. He is led to a formula

$$\overline{(i-I)^2} \propto \frac{I^{x+1}}{\omega^x} \qquad \ldots (7)$$

(where $1 < x < 2$), which may be compared with equation 6. This appears to give considerably improved agreement with experiment. The subject of 'low-frequency noise' is at present under active research and further interesting findings are to be expected.

THERMAL FLUCTUATIONS—NYQUIST'S THEOREM

We must, however, now return to 1927-8 when J. B. JOHNSON[47] first reported quantitative and systematic observation[‡] of the fundamental thermal fluctuation of electricity in conductors predicted by Einstein. Johnson examined a wide range of resistance elements, including copper, platinum, *Advance* wire and various aqueous electrolytic solutions. His results showed excellent agreement with a theoretical formula derived by H. NYQUIST in a companion paper.[48] This expresses the thermal fluctuation in the most convenient form for calculation. If a passive resistance of magnitude $R(f)$ at frequency f, at temperature T, be regarded as a constant-voltage or constant-current (fluctuation) generator (by the application of Helmholtz's [Thévenin's] theorem)[§] then within a frequency-range Δf the relevant formulae are

$$\overline{(v-V)^2} = 4R(f)kT\Delta f \qquad \ldots (8a)$$

(where v is the instantaneous voltage, and V the mean voltage)

$$\overline{(i-I)^2} = \frac{4}{R(f)}kT\Delta f \qquad \ldots (8b)$$

* Where number of 'charge-carriers' is abnormally low.

† At any rate in the class of semi-conductors.

‡ He says : '. . . It had been known for some time among amplifier technicians that the "noise" increases as the input resistance is made larger'.

§ See Appendix.

We may note here that the corresponding expression of the formula for the shot-effect in a valve is

$$\overline{(i-I)^2} = 2eI(\Gamma^2)\Delta f \qquad \ldots (9)$$

Nyquist also pointed out that if classical statistics were not applicable then (8a) must be replaced by

$$\overline{(v-V)^2} = 4R(f)\frac{hf}{\varepsilon^{hf/kT}-1}\Delta f \qquad \ldots (10)$$

(where h is Planck's constant).

One therefore sees that formulae 8 applicable to most situations are directly analogous to the Rayleigh-Jeans law; R. E. BURGESS[49] has developed this argument further in relation to the fluctuations generated in an aerial of radiation resistance R. Burgess has in fact proved Nyquist's theorem starting from the black body radiation law and has in a sense echoed Nyquist's suggestion that the law might also be proved by considering the interaction of the Brownian movement of a gas with a telephone diaphragm which could be arranged to present an electrical impedance R. Nyquist himself to derive the law used the idealized mechanism of a loss-free transmission line connecting two resistances of equal value R, and difficulties have since been expressed in the understanding of his proof.[50] The writer believes that a perhaps more satisfactory formulation of Nyquist's proof consists in a direct consideration of a semi-infinite loss-free transmission line of characteristic impedance R.

Measurements of thermal noise yield an experimental value of Boltzmann's constant; Johnson himself obtained fair agreement to about 8 per cent and later E. K. SANDEMANN and L. H. BEDFORD[51] improved this agreement, while E. B. MOULLIN and H. D. M. ELLIS[52] in 1932 after many months of work obtained the value $1\cdot361 \times 10^{-23}$ joules/°K.

It is also noteworthy that at the same time EUGEN KAPPLER[53] following earlier work by GERLACH and LEHRER[54] and EGGERS,[55] made a detailed and accurate examination of the mechanical Brownian movement of a torsion balance (a small mirror suspended on a fine quartz fibre). He recorded the fluctuations directly on a moving photographic film, after a high degree of light beam amplification (*Figures 5*). In view of the long period (~ 30 sec) of the system his experiments lasted over several hours and to avoid extraneous mechanical shocks had to be conducted at night. He determined the average deviation and distribution, obtaining an excellent value for Avogadro's number ($60\cdot59 \times 10^{23}$), and also, following Smoluchowski, examined the correlation of the fluctuation.

THEORIES OF SPACE-CHARGE REDUCTION

In 1930, F. B. LLEWELLYN[56] offered a theory of space-charge reduction of valve noise. If in a diode

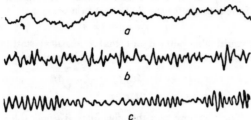

Figure 5. Spontaneous fluctuation of a torsion balance under varying degrees of air pressure; (a) 76 cm Hg ; (b) 10^{-3} mm Hg ; (c) 10^{-4} mm Hg. (After Kappler)

we let $\mathcal{J}$ be the emission, I the valve current, and V_a the anode voltage, then we may write

$$I = I(\mathcal{J}, V_a) \qquad \ldots (11)$$

if we are prepared to regard I as a single-valued function of the independent variables $\mathcal{J}$, V_a and

$$\delta I = \frac{\partial I}{\partial \mathcal{J}} \delta \mathcal{J} + \frac{\partial I}{\partial V_a} \delta V_a \qquad \ldots (12)$$

Clearly the second term merely represents the effect of the anode circuit (composed of the external impedance and that of the valve itself) in modifying the observable magnitude of the fluctuation.* The first term signifies the fundamental modification due to the existence of the space-charge, and Llewellyn's original analysis may be curtailed by the application of Helmholtz's constant-current theorem, enabling us to set $\delta V_a = 0$. It thus appears that

$$\overline{(\delta I)^2} \equiv \overline{(i-I)^2} = \left(\frac{\partial I}{\partial \mathcal{J}}\right)^2 \cdot \overline{(\delta \mathcal{J})^2} \qquad \ldots (13)$$

But clearly Schottky's law (equation 9 with $\Gamma^2 = 1$) must apply to the emission, whence

$$\overline{(\delta \mathcal{J})^2} = 2e\mathcal{J}\Delta f \qquad \ldots (14)$$

therefore

$$\overline{(i-I)^2} = 2e\left(\frac{\partial I}{\partial \mathcal{J}}\right)^2 \mathcal{J}\Delta f \qquad \ldots (15)$$

Comparing with equation 9 we have

$$\Gamma^2 = \frac{\mathcal{J}}{I}\left(\frac{\partial I}{\partial \mathcal{J}}\right)^2 \qquad \ldots (16)$$

Now as the valve becomes highly space-charge limited it is clear from equation 16 that $\Gamma^2 \to 0$; this theoretical result was in disagreement with experiment (in fact, it is now known that under retarding-field conditions $\Gamma^2 = 1$). Llewellyn then suggested that this arose because we must now *add* to the 'shot-noise' the 'thermal-noise' generated by the differential—('slope')—resistance, R_a.

The writer believes that this was one of the most unfortunate proposals in the history of the space-charge effect ; it seems entirely obvious that *a priori*

* See Appendix.

we may (attempt to) evaluate the noise *either* by regarding the valve as a single element, into whose detailed mode of operation we do not inquire and by applying some general theorem if we consider the system to be eligible (*e.g.* the second law of thermo-dynamics if we believe the valve to be in a state of thermal equilibrium) ; *or* we may examine the fluctuation in a relatively more detailed manner* as did Llewellyn. If, however, either method yields an unsatisfactory result we surely must not try to invoke the other approach *in addition*.

In fact, the writer has shown[57] on the basis of a theorem due to J. M. WHITTAKER[58] that the correct formula on Llewellyn's premises is

$$\overline{(i-I)^2} = 2e\left(\frac{\partial I}{\partial \mathcal{J}} \cdot \frac{\mathcal{J}}{I}\right)^2 I\Delta f$$

whence $\Gamma^2 = \left(\frac{\partial I}{\partial \mathcal{J}} \cdot \frac{\mathcal{J}}{I}\right)^2 \qquad \ldots (17)$

This formula now behaves correctly in the temperature-limited and retarding-field region, and it can further be shown that it is in agreement with a formula later derived by W. SCHOTTKY[62] after a rather lengthy discussion on the mechanism of the shot-effect. If we go further and set

$$I = \mathcal{J}\varepsilon^{\frac{eV_m}{kT}} \qquad \ldots (18)$$

where V_m is the internal barrier potential due to space-charge and T the cathode temperature, then we may show readily from equation 17 that

$$\Gamma^2 = \left(1 + \frac{\mathcal{J}e}{kT} \cdot \frac{\partial V_m}{\partial \mathcal{J}}\right)^2 \qquad \ldots (19)$$

Although correct in the two extreme cases, the formula proves unsatisfactory in the intermediate space-charge region ; this arises essentially because we have employed a single-valued analysis. Due, however, to the fact that the emission (and hence also the valve current) is composed of electrons whose velocities are distributed over a range (given, of course, by the Maxwell-Boltzmann law *on the average*) the problem cannot properly be treated on this basis.† The underlying mechanism of space-charge

* This procedure is exactly paralleled in demonstrations of Nyquist's theorem by discussing the detailed conduction mechanism (using a classical model, or otherwise).[59-61]

† It is, however, of interest to note that one may show that : $1 + \frac{\mathcal{J}e}{kT} \cdot \frac{\partial V_m}{\partial \mathcal{J}} = \frac{kT}{IeR_a}$ (*cf.* equation 20) in the space-charge limited region, and the writer believes that a formula $\Gamma^2 \sim \left(1 + \frac{\mathcal{J}e}{kT} \cdot \frac{\partial V_m}{\partial \mathcal{J}}\right)$ is valid over the whole range of valve operation, including the awkward transition to temperature limitation.

reduction is, nevertheless, clearly demonstrated in equation 19; the term $\frac{\mathcal{J}e}{kT}\cdot\frac{\partial V_m}{\partial \mathcal{J}}$ being essentially negative indicates the macroscopic 'compensating action' which arises as a result of the existence of the space-charge barrier in the valve. If one could 'fix' the barrier potential—in the same way that one fixes, in principle, the external potential when applying Thévenin's theorem—then there would simply be no 'space-charge reduction'. Essentially, what this means is that the electron-gas in the valve only interacts effectively through the mean Coulomb potential; any effect of true *smoothing* due to the space-charge cloud is quite negligible.* D. O. NORTH[63] *et al.*, W. SCHOTTKY and E. SPENKE[64] and F. B. LLEWELLYN and A. J. RACK[65] all more or less simultaneously treated the problem taking into account the emission-velocity fluctuations and averaging the barrier-voltage variations over the velocity classes, and achieved finally satisfactory agreement with experiment. This yields the result that for considerable space-charge limitations

$$\Gamma^2 = \frac{2k(\theta T)}{IeR_a} \qquad \ldots\ldots(20)$$

(where T = cathode temperature

R_a = anode slope resistance),

the factor $\theta \approx 0.64$ falling to 0.5 at the onset of retarding-field conditions. This means that using equations 20 and 9 we may conveniently write

$$\overline{(i-I)^2} = \frac{4}{R_a}k(\theta T)\Delta f \qquad \ldots\ldots(21)$$

and regard the fluctuations *alternatively* (from a formal standpoint) as thermal in character at an apparent temperature $\sim 0.64T$. It is not possible here to discuss further in detail the significance of this result; the reader will find material for discussion in the references.[66-70] It seems clear to the writer that any *a priori* treatment of the thermionic valve as a whole as a thermodynamic system in equilibrium is invalid.

The retarding-field region was first attacked by F. C. WILLIAMS[71] and more recently R. FÜRTH and the author[72] have made a study of this region. We

* We note that any real *smoothing* (*i.e.* alteration of the fluctuation rate or 'auto-correlation') as has been proposed in the past[105] would exhibit itself as a more or less rapid decay in the frequency-spectrum; this relationship is demonstrated analytically in the Wiener-Khintchine theorem.[106]

observe in this case that the fluctuation may equally be expressed as

$$\left.\begin{array}{l}\overline{(i-I)^2} = 2eI\Delta f \\[4pt] \qquad = \frac{1}{2}\left(\frac{4}{R_a}kT\Delta f\right)\end{array}\right\} \qquad \ldots\ldots(22)$$

since from equation 18 setting $V_m = V_a$ we may show immediately

$$\frac{1}{R_a} = \frac{eI}{kT} \qquad \ldots\ldots(23)$$

In so far as we may regard a valve under retarding-field conditions as one 'half' of a system in thermal equilibrium[66, 68] the unification of 'shot' and 'thermal' fluctuation is evident in this region.

'PARTITION NOISE'

A further most important source of fluctuation occurs in valves with a positive (screen) grid as well as the anode; in this case some of the beneficial compensating action of the space-charge barrier may be destroyed as a result of the random division of the electron current between screen (I_{g+}) and anode (I_a) ('partition noise'). In this case it may be shown, under reasonable hypotheses,[73, 74, 75, 76] that the appropriate formula for the fluctuation in the anode current is now

$$\overline{(i-I)^2} = 2eI_a\frac{(\Gamma^2 I_a + I_{g+})}{I_a + I_{g+}}\Delta f \qquad \ldots\ldots(24)$$

Satisfactory agreement with experiment is found.

So far we have considered only the magnitude of electrical fluctuations; considerable attention has also been paid to the statistical behaviour thereof. Valuable theoretical contributions have been made by many workers.[77-83] The correlation, and effect of passage through rectifiers, non-linear networks, *etc*, have been studied together with many other related problems. In view of the relative scarcity of published experimental work, R. FÜRTH and D. K. C. MAC-DONALD[84] have recently presented a study of the correlation and distribution of magnitude of electrical fluctuations. *Figures 6a, 6b* are taken from this work.

NOISE FACTOR

The problem of specifying the 'goodness-factor' as regards noise for a complete receiver has led to the specification of a Noise Figure or Noise Factor and its evaluation for various complexes.[85-90] The Noise Figure measures the deterioration of the ratio of signal and noise power available from a given aerial system to that available at the 'output' of the receiver; this output point is generally taken to be the last intermediate frequency stage to avoid the complication of the non-linearity of the detector.

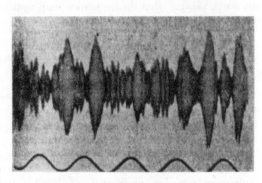

Figure 6 a

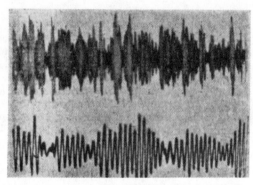

Figure 6 b

Figure 6. (a) *Shot fluctuation record. Dominant frequency 255 kc/s, nominal bandwidth 1·2 kc/s, timing wave 250 c/s. (b) Fluctuation record; upper part—as generated in last intermediate-frequency stage; lower part—after rectification and passage through 1 kc/s filter of bandwidth 100 c/s. Upper and lower traces recorded simultaneously*

The drive towards ultra high frequencies during the recent war naturally produced intensive research into the problems of improving the signal-noise ratio in ultra high frequency receivers and elements. Some of this work is now being published,[91·93] and as an example of particular interest we may note the special travelling-wave pre-amplifier for centimetre waves designed by R. KOMPFNER.[94]

PHOTO-CELLS

The fluctuations in other electrical elements are not without interest. Work on photoelectric cells has been undertaken by B. A. KINGSBURY[95] and R. FÜRTH and D. K. C. MACDONALD[96] confirming in the saturated region the validity of the simple 'shot' formula, and it appears that the formula will also hold in the .retarding-field region ; further work is in progress at present. One problem of

considerable interest in this connection is the possibility of detecting radiation fluctuations by such cells, or related instruments, *e.g.* ultra violet counters.[97]   E. VON SCHWEIDLER's[21] original discussion of fluctuations in radioactive decay is of course germane. It appears however that in the case of the photoelectric cell the quantum efficiency is so low that the electron emission would be essentially random even were the incident photon beam to be perfectly ordered ; in the case of a counter the efficiency fluctuations appear to be so great as entirely to mask any effect of the incident beam. Statistical fluctuations are also important in such problems as the determination of significant coincidences in Geiger-Müller counters.[98]

Inevitably, many problems have been left untouched in this survey ; the writer will be satisfied if some indication has been given here of the fascinating variety of applications of the Brownian movement.

APPENDIX

The 'constant-voltage' theorem was first propounded by H. HELMHOLTZ[99] and later by L. THÉVENIN[100] and others. The precise origin of the analogous 'constant-current' theorem appears to be in some doubt. The former, in its most general form, states that if the open circuit voltage across any two terminals of a linear network be V (which need not necessarily be a constant potential in time) and the impedance measured across these two terminals with all internal e.m.f.'s supposed suppressed be Z, then the network (*Figure A.1*) will under all conditions *as regards any externally measurable effects* behave as an impedance Z in series with an e.m.f. V (*Figure A.2*). The 'constant-current' corollary states that if the current measured in a short circuit link placed externally across the two terminals be I, then the network will similarly behave as an impedance Z in parallel with a current generator I (*Figure A.3*).

Thus, for example, in the case of a space-charge limited valve (equation 9), the equivalent representation would be as in *Figure A.4b* and the actual fluctuation voltage measured in a small band width Δf near a frequency f_o would be given by

$$\overline{(v-V)^2}_{\Delta f} = \frac{2eI\Gamma^2 \Delta f}{\left(\dfrac{R+R_a}{RR_a}\right)^2 + 4\pi^2 f_o^2 C^2} \quad \ldots (\text{A.1})$$

or, the *total* fluctuation

$$\overline{(v-V)^2} = 2eI\Gamma^2 \int_0^\infty \frac{df}{\left(\dfrac{R+R_a}{RR_a}\right)^2 + 4\pi^2 f^2 C^2} \quad \ldots (\text{A.2})$$

For the case of ideal temperature limitation $\Gamma^2 = 1$; ·

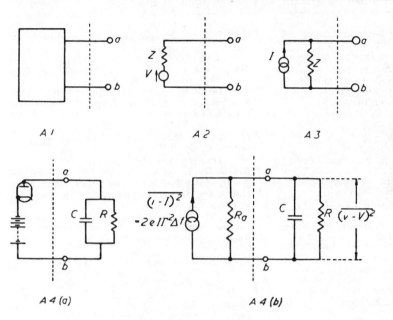

A 1   A 2   A 3

A 4 (a)   A 4 (b)

$R_a \to \infty$, and it may readily be verified that equation A.2 then yields

$$\overline{(v-V)^2} = \frac{eIR}{2C} \qquad \dots (A.3)$$

in agreement with equation 5 in the text.

The applicability of Helmholtz's theorem to electrical fluctuation problems has been discussed,[101-103] but if one is willing to accept the hypothesis underlying Campbell's theorem that individual stimuli give rise to responses characteristic of macroscopic experiments then there can be no further scruples. E. N. ROWLAND[103] attempted to specify single·electron events in a perhaps more plausible manner, but his attempt was unsuccessful; R. FÜRTH[104] maintains that there can be no doubt on the applicability of 'large charge' behaviour. On the other hand, J. M. WHITTAKER[102] sets out to show on fundamental grounds that the valve differential resistance can be treated essentially as set out above but, in the author's view, nullified the validity of his argument by *assuming* 'large charge' behaviour of individual electrons; to quote:

' When the incident current has its average value the effect of a single electron at time τ is to diminish the anode current by an amount $-\dfrac{1}{CR_a}\varepsilon^{-k(t-\tau)}$.'

The whole essence of the fact to be proved, that R_a may be regarded as an 'ordinary' resistance in parallel with the external circuit, appears to lie in just such an assumption, making further proof unnecessary.

REFERENCES

[1] BROWN, R. *Philos. Mag.* 4 (1828) 161
— *Ann. Phys. Chem.* 14 (1828) 294
[2] NEEDHAM. *Nouvelles observations microscopiques.* Paris, 1750
[3] VON GLEICHEN. *Das Neueste aus dem Reiche der Pflanzen oder mikroskopische Untersuchungen. . . .* (1764) 30
[4] KAMMERLINGH ONNES, H. and KEESOM, W. H. *Communication from Phys. Lab. Leiden. No. 104.* (1908)
LORD RAYLEIGH. *Philos. Mag.* 41 (1871) 107, 274
— *Not. Proc. roy. Instn.* Feb 25 (1910)
[5] FÜRTH, R. *Wiener Ber.* 124 (1915) 577
[6] CHANDRASEKHAR, S. *Rev. mod. Phys.* 15 (1943) 68
[7] JEVONS, S. *Proc. Manchr Soc.* 9 (1869) 78
[8] WIENER, CHR. *Ann. Phys. Chem.* 118 (1863) 79
[9] GOUY. *C.R. Acad. Sci., Paris.* 109 (1889) 102
[10] EXNER, F. *Ann. Phys.* 2 (1900) 843
[11] MEADE BACHE. *Proc. Amer. phil. Soc.* 33 (1894) 163
— *Chem. News.* 71 (1895) 47, 83, 96, 107
[12] MALTEZOS. *Ann. Chim. (Phys.), Paris.* 1 (1894) 559
— *C.R. Acad. Sci., Paris.* 121 (1895) 303
[13] EINSTEIN, A. *Ann. Phys.* 17 (1905) 549
— *ibid.* 19 (1906) 289, 371
— *ibid.* 34 (1911) 591
— *Z. Elektrochem.* 13 (1907) 41
— *ibid.* 14 (1908) 235
FÜRTH, R. *Investigation on the Theory of the Brownian Movement.* London, 1926

[14] VON SMOLUCHOWSKI, M. *Ann. Phys.* 21 (1906) 756
[15] LANGEVIN, M. P. *C.R. Acad. Sci., Paris.* 146 (1908) 530
[16] VON NÄGELI. *Münch. Sitzungsber. Math. phys.* 9 (1879) 389
[17] PEARSON, K. *Nature, Lond.* 72 (1905) 294
[18] LORD RAYLEIGH. *Nature, Lond.* 72 (1905) 318
[19] SLACK, M. *J. Instn elec. Engrs.* 93 (1946) 76
[20] PEARSON, K. *Nature, Lond.* 72 (1905) 342
[21] VON SCHWEIDLER, E. *Uber Schwankungen der radioaktiven Umwandlung.* *Congr. int. Radiol. Ionis., Liége.* (1905)
— *Phys. Z.* 11 (1910) 614
— *ibid.* 14 (1913) 198
[22] CAMPBELL, N. R. *Proc. Camb. phil. Soc.* 15 (1909) 117, 310
— *ibid.* 15 (1910) 513
— *Z. Phys.* 11 (1910) 826
[23] RIVLIN, R. S. *Philos. Mag.* 36 (1945) 688
[24] SVEDBERG, TH. *Die Existenz der Molekule.* Leipzig, 1912
— *Z. phys. Chem.* 73 (1911) 547
— *ibid.* 77 (1911) 145
[25] WESTGREN, A. *Arkiv Mat. Ast., och Fys.* 11 (1916) 8, 14
[26] FÜRTH, R. *Ann. Phys.* 53 (1917) 177
— *Schwankungserscheinungen in der Physik.* Brunswick, 1920
[27] PERRIN, J. *Atoms.* London, 1916

[28] DE HAAS-LORENTZ, G. L. *Die Brownsche Bewegung und einige verwandte Erscheinungen.* Brunswick, 1913
[29] CHANDRASEKHAR, S. Stochastic Problems in Physics and Astronomy. *Rev. mod. Phys.* 15 (1943) 1
[30] VON SMOLUCHOWSKI, M. *Phys. Z.* 13 (1912) 1069
[31] SCHOTTKY, W. *Ann. Phys.* 57 (1918) 541
[32] FRY, T. C. *J. Franklin Inst.* 199 (1925) 203
[33] HARTMANN, C. A. *Ann. Phys.* 65 (1921) 51
[34] JOHNSON, J. B. *ibid.* 67 (1922) 154
[35] FÜRTH, R. *Phys. Z.* 23 (1922) 354
[36] JOHNSON, J. B. *Phys. Rev.* 26 (1925) 71
[37] WILLIAMS, N. H. and HULL, A. W. *Phys. Rev.* 25 (1925) 147
— and VINCENT, H. B. *ibid.* 28 (1926) 1250
— and HUXFORD, W. S. *ibid.* 33 (1929) 773
[38] WILLIAMS, F. C. *J. Instn elec. Engrs.* 79 (1936) 350
[39] SCHOTTKY, W. *Phys. Rev.* 28 (1926) 74
[40] MOULLIN, E. B. *Spontaneous Fluctuations of Voltage,* Chap. VI, p. 165. Oxford, 1938
[41] ADRIAN, E. D. and BRONK, D. W. *J. Physiol.* 66 (1928) 81 xii
[42] MATTHEWS, B. H. C. *J. Physiol.* 81 (1934) 28 P
— *ibid.* 93 (1938) 25 P
— and ADRIAN, E. D. *ibid.* 81 (1934) 440
[43] HARRIS, E. J., ABSON, W. and ROBERTS, W. L. In preparation. (1947)
[44] CHRISTENSEN and PEARSON. *Bell. Syst. tech. J.* 15 (1936) 197
[45] BERNAMONT, J. *C.R. Acad. Sci., Paris.* 198 (1934) 2144
— *Ann. Chim. (Phys.), Paris.* 7 (1937) 71
[46] MACFARLANE, G. G. *Proc. phys. Soc., Lond.* 59 (1947) 366
[47] JOHNSON, J. B. *Nature, Lond.* 119 (1927) 50
— *Phys. Rev.* 29 (1927) 367
— *ibid.* 32 (1928) 97
[48] NYQUIST, H. *Phys. Rev.* 29 (1927) 614
— *ibid.* 32 (1928) 110
[49] BURGESS, R. E. *Proc. phys. Soc., Lond.* 53 (1941) 293
[50] MOULLIN, E. B. Ref. [40], p. 34 *et seq*
[51] SANDEMAN, E. K. and BEDFORD, L. H. *Philos. Mag.* 7 (1929) 774
[52] MOULLIN, E. B. and ELLIS, H. D. M. *Proc. Camb. phil. Soc.* 28 (1932) 386
[53] KAPPLER, E. *Ann. Phys.* 11 (1931) 233
— *ibid.* 15 (1932) 545, 550
[54] GERLACH, W. and LEHRER, E. *Naturwissenschaften.* 15 (1927) 15
[55] EGGERS. *Ann. Phys.* 7 (1930) 833
[56] LLEWELLYN, F. B. *Proc. Inst. rad. Engrs.* 18 (1930) 243
[57] MACDONALD, D. K. C. *Thesis, Univ. of Edinb., p. 19
[58] WHITTAKER, J. M. *Proc. Camb. phil. Soc.* 34 (1938) 158
[59] BERNAMONT, J. *Ann. Chim. (Phys.) Paris.* 7 (1937) 71
[60] BELL, D. A. *J. Instn. elec. Engrs.* 82 (1938) 522
[61] BAKKER, C. J. and HELLER, G. *Physica (The Hague)* 6 (1939) 262
[62] SCHOTTKY, W. *Die Telefunken-Röhre.* 8 (1936) 175
[63] NORTH, D. O., *et al.* *R.C.A. Rev.* 4 : 269, 441 ; 5 : 106, 244, 371, 505 ; 6 : 114. (1940–41)
[64] SCHOTTKY, W. and SPENKE, E. *Wiss. Veröff. aus dem Siem. Werk.* 16 (1937) 1
[65] RACK, A. J. *Bell Syst. tech. J.* 17 (1938) 592
[66] SCHOTTKY, W. The relation between shot and thermal fluctuations in electron valves. *Z. Phys.* 104 (1936) 248
[67] WILLIAMS, F. C. *J. Instn elec. Engrs.* 78 (1936) 326, MOULLIN, E. B. *Spontaneous Fluctuations of Voltage,* App., p. 223. London, 1938
[68] NORTH, D. O. *R.C.A. Rev.* 4 (1940) 466

[69] BELL, D. A. *J. Inst. elec. Engrs.* 89 (1942) 207
— *ibid.* 93 (1946) 37
— *Proc. phys. Soc., Lond.* 59 (1947) 403
[70] MACDONALD, D. K. C. *Thesis, Univ. Edinb., pp. 27–34
— *Proc. phys. Soc., Lond.* 59 (1947) 407
[71] WILLIAMS, F. C. *J. Inst. elec. Engrs.* 78 (1936) 326
[72] MACDONALD, D. K. C. and FÜRTH, R. *Nature, Lond.* 157 (1946) 841
— *Proc. Phys. Soc., Lond.* 59 (1947) 375
[73] ZIEGLER, M. *Phil. Tech. Rev.* 2 (1937) 329
[74] NORTH, D. O. *R.C.A. Rev.* 5 (1940) 244
[75] SCHOTTKY, W. *Ann. Phys.* 32 (1938) 195
[76] BAKKER, C. J. and VAN DER POL, B. *Document A.G. No. 67.* Union Rad. Scient. Int. Venice, Sept. 1938
[77] UHLENBECK, G. E. and ORNSTEIN, L. S. *Phys. Rev.* 36 (1930) 823
WANG, M. C. and UHLENBECK, G. E. *Rev. mod. Phys.* 17 (1945) 323
[78] RICE, S. O. *Bell Syst. tech. J.* 23 (1944) 282
— *ibid.* 24 (1945) 46
[79] FRÄNZ, K. *Elekt. Nachr.-Tech.* 17 (Oct 1940)
— *Hockfrequenztech. u. Elektroakust.* 57 (1941) 146
— *Elekt. Nachr.-Tech.* 19 (1942) 166
[80] NORTH, D. O. *The Modification of Noise by Certain Non-linear Devices.* Read before *Amer. Inst. rad. Engrs,* 28 Jan 1944. (Not yet published)
[81] BURGESS, R. E. Several papers, as yet unpublished, on the passage of noise through non-linear devices *etc*
[82] CAMPBELL, N. R. and FRANCIS, V. J. *J. Instn elec. Engrs.* 93 (1946) 45
— *Philos. Mag.* 37 (1946) 289. Includes considerable reference to work of M. Bondi
[83] FRANCIS, V. J. and JAMES, E. G. *Wireless Engr.* 23 (1946) 16
[84] FÜRTH, R. and MACDONALD, D. K. C. *Nature, Lond.* 157 (1946) 807
— *Proc. phys. Soc., Lond.* 59 (1947) 388
[85] FRIIS, H. T. *Proc. Inst. rad. Engrs.* 32 (1944) 419
[86] FRANZ, K. *Elekt. Nachr.-Tech.* 16 (1939) 92
[87] HEROLD, E. W. *R.C.A. Rev.* 6 (1942) 302
[88] NORTH, D. O. *R.C.A. Rev.* 6 (1942) 332
[89] BURGESS, R. E. *Wireless Engr.* 20 (1943) 66
— *ibid.* 22 (1945) 56 ; 23 (1946) 217
— *Proc. phys. Soc., Lond.* 58 (1946) 313
[90] MACDONALD, D. K. C. *Philos. Mag.* 35 (1944) 386
[91] ROBERTS, S. *Proc. Inst. rad. Engrs.* 35 (1947) 257
[92] BREAZALE, W. M. *Proc. Inst. rad. Engrs.* 35 (1947) 31
[93] NORTON, K. A. and OMBERG, A. C. The Maximum Range of a Radar Set. *Proc. Inst. rad. Engrs.* 35 (1947) 4
[94] KOMPFNER, R. *Proc. Inst. rad. Engrs.* 35 (1947) 124
[95] KINGSBURY, B. A. *Phys. Rev.* 38 (1931) 1458
[96] FÜRTH, R. and MACDONALD, D. K. C. *Nature, Lond.* 159 (1947) 608
[97] KOLIN, A. *Ann. Phys.* 21 (1933) 813
[98] FEATHER, N. *Proc. Camb. phil. Soc.* 39 (1943) 84
[99] HELMHOLTZ, H. *Ann. Phys. Chem.* 89 (1853) 222
[100] THÉVENIN, L. *C.R. Acad. Sci., Paris.* 97 (1883) 159
[101] MOULLIN, E. B. Ref. [40], p. 70 *et seq*
[102] WHITTAKER, J. M. *Proc. Camb. phil. Soc.* 34 (1938) 158
[103] ROWLAND, E. N. *Proc. Camb. phil. Soc.* 32 (1936) 580
— *ibid.* 33 (1937) 344
[104] FÜRTH, R. *Phys. Z.* 23 (1922) 354
[105] THATCHER, E. W. and WILLIAMS, N. G. *Phys. Rev.* 39 (1932) 474
[106] WIENER, N. *Acta. Math., Stockh.* 55 (1930) 117
KHINTCHINE. *Math. Ann.* 109 (1934) 604

Electronic noise:

the first two decades

Most of the basic knowledge in the field of electronic noise was gained in the 20-year period following World War I. A considerable amount of the vacuum-tube data obtained was later translated into the semiconductor language

*John Bertrand Johnson**

You could hear a pin drop. (English saying)

You could hear the grass grow. (German version)

Such harmony is in immortal souls;
But whilst this muddy vestment of decay
Doth grossly close it in we cannot hear it.

(Merchant of Venice)

Fifty-two years ago the classical paper on noise in amplifiers was written by Dr. Walther Schottky.[31] The high-vacuum thermionic amplifier could then be called about six years old. Its development had taken place along nearly parallel lines in several countries, including Germany, mostly under rules of strict secrecy. It seems now almost incredible that out of the Germany of those years, faced with military defeat and economic collapse, could come a scientific paper of the quality and technical importance of this paper of Schottky's.

The amplifiers developed at the Siemens-Halske Works no doubt had the same kind of faults as those produced at other laboratories—poor welds, mechanical resonances, unstable cathodes, inadequate pumping, etc. These faults could distort the signals applied to the amplifiers and, since thermionic amplifiers were then being installed in commercial and military telephone systems, the faults became a technical liability. At this time I was employed in the Engineering Department of the Western Electric Company, the Engineering and Supply Division for the Bell System. I was assigned to study some of the

* John Bertrand Johnson (F) died at the age of 83 on November 27, 1970, the day he completed work on this manuscript. Dr. Johnson's obituary appears on page 107 of the January issue of IEEE SPECTRUM.

many projects on vacuum-tube research, came early in touch with Schottky's work, and have some memories of the work that went on. With this as my background, the Editor of IEEE SPECTRUM asked me to write this article on the study of amplifier noise as I saw it develop during about the first two decades of its progress into a rather broad scientific field.

'Wärmeeffekt' and 'Schroteffekt'

In the 1918 paper, Dr. Schottky evidently assumes that the grosser current fluctuations produced by faulty tube structures such as those just enumerated have been, or can be, eliminated, and he is left with two sources of noise that are of a much more fundamental nature. One he calls the "Wärmeeffekt," in English now commonly named "thermal noise." This is a fluctuating voltage generated by electric current flowing through a resistance in the input circuit of an amplifier, not in the amplifier itself. The motion of charge is a spontaneous and random flow of the electric charge in the conductor in response to the heat motion of its molecules. The voltage between the ends of the conductor varies and is impressed upon the input to the amplifier as a fluctuating noise. This flow of energy between molecules and electric current involves not the charge of the electron but rather the rate of flow of power between charge and momentum. It involves the Boltzmann constant k times the absolute temperature T of the system, and a power flow of at least 10^{-17} watt to be audible in a telephone. Schottky believed any other noise source would be much stronger than this.

And here, for the sake of history, we may digress a bit. In estimating the total of noise that is going to be contributed by the "Schroteffekt," the integration of a certain expression is needed that has come to be called the Schottky equation. Schottky performed this integration

Reprinted from *IEEE Spectrum*, vol. 8, pp. 42–46, Feb. 1971.

and got the result $2\pi/r^2$, where r is a damping factor of the circuit, $r = R/L\omega$.

My recollection is that because of some postal delay the 1918 paper did not get to the United States until about 1920. On reading it, I became suspicious of the integration, but in the then-available tables of integration could find no solution for the Schottky equation. I asked my friend, Dr. L. A. MacColl, mathematician, for assistance. He suggested splitting the Schottky expression into four complex factors, integrating each separately and then recombining them for the final result, $2/r$. When, after much labor on my part, this was done, MacColl again looked at the equation and said this was a case for the method of poles and residues and, without putting pencil to paper, read off the correct result. This was impressive, but evidently the method had not yet penetrated down to physicists and engineers, and the more cumbersome method was left in. The method of residues was evidently used later by Fry[5] and by Hull and Williams,[11] but before them several other methods had also been suggested. We correctors, however, chose to abide by Schottky's word that the thermal-effect noise is much smaller than the shot noise, and recognition of the technical importance of the thermal noise was delayed by about a decade. This probably did not matter much, because it was a busy decade spent on other phases of the project.

In the case of the "thermal noise," as we shall call it, the electric charge is in effect held in long bags with walls relatively impervious to electrons at low temperature. The mass transport of charge along the bag, or wires, under the influence of the heat motion, sets up the potential differences that generate the fluctuating output of the amplifier.

When now one end of the conductor, the "cathode" of the tube, is heated to incandescence, electrons can be emitted from the cathode surface to travel across the vacuum toward the anode. The electrons are emitted at random times, independent of each other, and they travel at different velocities, depending on initial velocity and voltage distribution for electron passage. In the case of a small electron emission, a small nearly steady flow of current results, with a superimposed smaller alternating current whose amplitude can be calculated from statistical theory. This small current flowing through the amplifier generates the "Schroteffekt," or shot effect, in the amplifier.

The first experimental work on identifying and measuring the shot effect was done in Schottky's laboratory and published by C. A. Hartmann in 1921. This seemed like a well-designed set of tests, but was a little ahead of its time in the new art. After corrections, it left little doubt of the existence of the shot effect.

The next step came with the publication of three papers in 1925. T. C. Fry[5] covered parts of the theory that he wanted put on a firmer mathematical basis. Through Fry,

the work of Hull and Williams[11] at General Electric and Johnson[14] at Western Electric–Bell Laboratories became known to the participants, which may have given added impetus to the efforts. At GE, the first application of Hull's screen-grid tube in the amplifier increased the accuracy of the GE work to such a point that the value of the charge of the electron found by the shot effect came out close to that of the oil-drop method. The work of Johnson at lower frequencies revealed the existence of the "flicker effect," which could be many times greater than the shot effect, as well as the effect of space charge in reducing the magnitude of both shot effect and flicker effect by large factors (also recognized by Hull and Williams).

FIGURE 1. The effect of space charge on fluctuation noise. Three tubes have filaments composed of tungsten, thoriated tungsten, and barium oxide. $\overline{E^2}$ is the mean-square noise voltage across the output measuring device expressed in arbitrary units. The variation in space current was obtained by changing the cathode temperature, the plate voltage remaining constant. (Copyright 1934, The American Telephone and Telegraph Co.; reprinted by permission)

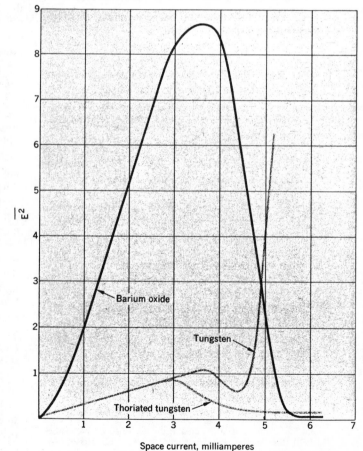

Space current, milliamperes

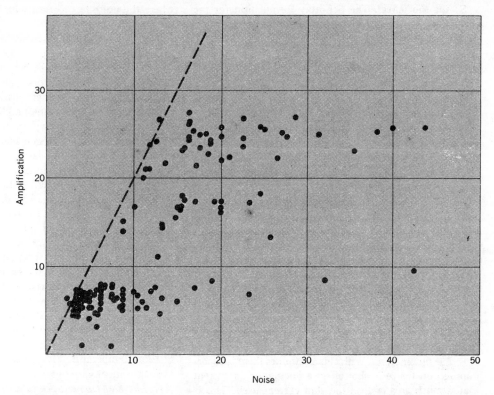

FIGURE 2. Amplification as a function of noise in three-electrode tubes; noise in arbitrary units; each point represents a tube. (Physical Review, 1925, reprinted by permission)

Each of these phenomenons will be discussed in connection with Fig. 1, which is reproduced from the 1934 paper by Pearson.[27]

By the early 1930s, the shot effect had been fairly well established for thermionic diodes, simple amplifiers, and photoelectric tubes.

A typical event that took place during the shot-effect work will be described here. We were visited by Sir J. J. Thomson, and the shot effect was demonstrated to him. Our explanation of it may not have been satisfactory, for as he left the room, the discoverer of the electron, with a forbearing smile and a gentle shake of the head, muttered, "Oh, no, no, no!"

Toward the end of the shot-noise work, a rough exploratory test was made. About 100 triode tubes of various kinds were picked out at random and tested for gain and noise in a circuit of fixed voltage, frequency range, etc. A resistance of 500 kΩ was connected across the input of the tube under test, with the output of the tube resistance–capacitance coupled to the amplifier. For each tube, the observed noise was plotted against the separately measured amplification of the tube, as in Fig. 13 of the 1925 article, here reproduced as Fig. 2. There is one point for each tube and these points are scattered over the right-hand side of the diagram. On the left, the point distribution stops abruptly along a straight sloping line. This suggests that along this line the noise pulses that the amplifier responds to have been amplified by the tube under test by its gain factor, from incoming pulses of more nearly constant value. Could this be the thermal effect predicted by Schottky?

A few simple tests, such as varying the electrical value of the input resistor, its temperature, its size, its material, soon answered the question in the affirmative. The results were discussed with Dr. H. Nyquist, who in a matter of a month or so came up with the famous formula for the effect, based essentially on the thermodynamics of a telephone line, and covering almost all one needs to know about the thermal noise.

The two effects: A and B (or T and S?)

We have, then, two different sources of electrical noise obeying statistical laws. Both have the properties in common that the noise can be described as a power dissipated by the noise source at a point of the amplifier circuit, and that for frequencies above certain values the noise power is constant up to very high frequencies. For thermal noise this constant power extends also to low values, while for shot noise there are many exceptions and variations.

T: the thermal effect. By the Nyquist[23] formulation, the thermal effect may be expressed as a voltage applied by the source to the input point of the amplifier at the high-impedance grid-leak resistor:

$$\text{Thermal formulation } \overline{V_T{}^2} = 4kTR \qquad (1)$$

Here $\overline{V_T{}^2}$ is the mean-square noise fluctuation per unit bandwidth as measured by a thermocouple voltmeter; R is the resistance of the input circuit; T is temperature in degrees Kelvin; k is Boltzmann's constant, 1.38×10^{-23} joules/degree K. This can also be written

$$W_T = 4kT \text{ watts per unit resistance} \qquad (2)$$

per unit bandwidth. The total for any case is then obtained by linear integration over the resistance and bandwidth range.

There is not much more to be done with this formulation except to consider the slope resistance of the tube, which will be done later.

S: the shot effect. The Schottky formulation for the shot effect per unit bandwidth may similarly be written

$$\overline{J_S{}^2} = 2ei \tag{3}$$

or

$$W_S = 2eiR_1 \tag{4}$$

where the charge on the electron $e = 1.602 \times 10^{-19}$ coulomb; i = dc space current, in amperes, flowing in space from cathode to anode (negative); R_1 = total resistance between cathode and anode, including that internal to the tube (function of frequency); and W_S = power per cycle dissipated in R_1.

This formulation was found by the early workers to hold under some carefully controlled conditions, including choice of cathode materials, freedom from space-charge effects, and choice of frequency band. When these conditions were judiciously selected, the experiments yielded, for instance, very nearly the correct value for the charge on the electron, as was shown in the tests of the 1920s. More complicated effects were also observed; they were subjected to a rather concentrated theoretical attack in the 1930s and will briefly be described in the following paragraphs.

1. *The flicker effect.* With some cathodes there is superimposed on the pure shot effect a fluctuation in current that is much greater than the shot current itself. This is illustrated in Fig. 1. The linear portion of the curve, obtained from tubes having filaments of tungsten and thoriated tungsten, gives the values the pure shot noise should have. The noise data were recorded as the temperature of the cathode was raised, the plate voltage of the diode being supplied by a fixed battery through a constant resistance. A measure of the cathode temperature is given by the indicated total current, in milliamperes. In the barium oxide tube, the noise increased more rapidly and reached a maximum value approximately ten times that of pure shot noise. The reason for this excess noise was surmised by Johnson to be fluctuations in the work function of the cathode surface due to particle migration, and was discussed at length by Schottky, who called it "Fackeleffekt."

2. *Space-charge depression.* Still in Fig. 1, after passing through a maximum, the noise in all three of the tubes decreases toward values eventually far below the theoretical shot value, at first thought to be effectively zero. This is an important feature, for it is in this low noise range that thermionic devices can be used as amplifiers. Schottky ascribes this noise depression to the smoothing effect of a dense space-charge layer near the cathode—between cathode and grid in a triode, for instance—and he works out a plausible theory for it.

3. *Frequency and flicker effect.* With fixed operating conditions, except for the natural frequency of a narrow-band circuit that the device works into, the noise output depends on this frequency. Normally the noise varies with this frequency f as

$$\overline{J_S{}^2} = f^{-n} \tag{5}$$

where n may lie in the range 1.2–0.9, depending on the material and condition of the cathode. For very pure materials, this increase in noise may be unobservable at frequencies above a few thousand hertz. Oxide cathodes, and perhaps all cathodes, show the effect down to very low frequencies, such as perhaps one cycle per month, where the noise has merged with the natural drift of the device.

The f^{-n} law has been discussed theoretically by Schottky and others.

4. *Ionic effects.* Ions may be generated from gas in the device, or from the electrodes, either by photoelectric or collision processes. The ion current would normally be small and make only a small addition to the dc electron current. But if, say, a heavy positive ion becomes trapped in the negative potential well that is created by the electron space charge, then a large pulse of electrons may be released through the potential minimum to make a noise pulse. This effect was described by Johnson[14] and studied by Ballantine[1] and others. The sharp rise of the noise at high currents as depicted for the tungsten tube in Fig. 1 is a result of ions emitted from its filament.

5. *Thermal noise in plate current.* A curious situation developed in about 1930. Llewellyn[21] suggested that the internal resistance R_0 of the thermionic device is really in parallel with the external resistance R_1, the parallel combination taken as the thermal noise source of the output circuit. Llewellyn suggested that this slope or differential resistance should be considered at the cathode temperature in combining it with the external resistance at room temperature. The result seemed to give reasonable agreement with observations.

6. *The half-temperature rule.* In making more careful measurements, however, Pearson[28] concluded that the temperature of the slope resistance should be half of the absolute temperature of the cathode in order to get agreement with Eq. (1) or (2). There seems at first to be no physical basis for this peculiar situation, but further experiments seemed to agree. Some found it hard to believe that there could be such a coupling between a stream of electrons and their source (the cathode). The most careful calculation of the effect, based on certain assumptions, was made by Rack[29], who found that over a considerable part of the mid-temperature range the value of the temperature should be taken as $0.644T$ instead of $0.500T$. The most plausible explanation of the effect is probably presented by Schottky[33], who arrived at about 0.500 for the factor, but his presentation has to do with a certain rectification of the noise signal in the output circuit of the device and is not easy to repeat here.

Another facet of the $\frac{1}{2}T$ rule is that for very small currents the noise can be derived from either the Schottky equation ei or the Nyquist equation kT. This is in the region where the current to the anode is too small to set up appreciable space charge, because of too low a cathode temperature. This seems to have been first noticed by F. C. Williams[42], but was also discussed by Schottky and others. The temperature must again be taken as $\frac{1}{2}T$, but tubes are probably not often used in these regions, except possibly for logarithmic response.

Rating of tubes

We have, then, two fundamental sources of noise in an electronic circuit: thermal noise, which can be calculated from the input parameters; and shot noise, which is modified by various device parameters and can, in some cases, be calculated, or can be measured for individual devices. In the device, the effects of these sources are added into a noise-power spectrum. In a diode, this is fairly simple, but in a grid-controlled tube it is more complicated since each electrode must be considered.

Up to 1940, the period of this review, the method proposed by Johnson for grid-controlled tubes was followed closely by others. This technique involved short-circuiting the input, setting the other parameters at some operating condition, and measuring the noise at the output of the device in this condition. This was then considered the noise figure introduced by the device itself, and it could be expressed in terms of a resistance at the input that would give the same amount of thermal noise. This would normally be a few hundred to a few thousand ohms.

Several measurements of this kind will be referred to but no details will be given here because methods may have changed and, moreover, because most of the tubes tested are now obsolete.

Tests on a few U.S. tube types were made by Pearson,[27] whereas Moullin and Ellis[22] reported tests on some British tubes. Spenke[38] studied some German tubes, on which he presented extended and careful discussions. Probably the most extensive and detailed discussion and measurements on U.S.-made tubes for our period were reported by Thompson, North, and Harris.[41]

I would like to acknowledge some debts: for the short early days of my participation, three friends, long departed: Hendrick van der Bijl, Oliver Buckley, and Harold Arnold, for technical guidance and management support; for aid in the preparation of this manuscript, the staffs of Bell Telephone Laboratories and of Thomas A. Edison Industries; for love, cooperation, and understanding: in the early times, Clara, and in the latter days, Ruth.

BIBLIOGRAPHY

The following list of references is appended for readers who want to go a little further into the early history of our subject than is done in this brief review. It should help establish the approximate sequence of the important steps made in the first two decades. The list does not pretend to be complete, and it contains many items that are not specifically referred to in the main text.

By 1940 this basic work had been about completed, and from there on the work on noise took different directions. First, there was the highly mathematical study of how to extract a weak signal from a background of noise. Then came the transistor and the translation of the vacuum-tube data into the semiconductor language. This opened up new fields of applications, such as low-temperature work, rocketry, and space research. A few recent references may open the door to these fields.

1. Ballantine, S., "Fluctuation noise due to collision ionization in electronic amplifier tubes," *Physics*, vol. 4, pp. 294–306, Sept. 1933.

2. Campbell, N., "Discontinuities in light emission," *Proc. Cambridge Phil. Soc.*, vol. 15, pp. 310–328, 1910.

3. De La Garza, A., "Tracking and telemetry," *Space Aeronaut.*, vol. 52, pp. 195–198, July 1969.

4. Ellis, H. D., and Moullin, E. B., "Measurement of Boltzmann's constant by means of the fluctuations of electron pressure in a conductor," *Proc. Cambridge Phil. Soc.*, vol. 28, pp. 386–402, July 30, 1932.

5. Fry, T. C., "The theory of the Schroteffekt," *J. Franklin Inst.*, vol. 99, pp. 203–320, 1925.

6. Fürth, R., "Die Bestimmung der Elektronenladung aus dem Schroteffekt an Glühkathodenröhren," *Phys. Z.*, vol. 23, pp. 354–362, 1922.

7. Fürth, R., "Die Bestimmung der Elektronenladung aus dem Schroteffekt an Glühkathodenröhren. Discussion," *Phys. Z.*, vol. 23, p. 438, 1922.

8. Hartmann, C. A., "Die Bestimmung der Elektrischen Elementarquantums aus dem Schroteffekt," *Ann. Physik.*, vol. 65, pp. 51–78, 1921.

9. Hartmann, C. A., "The present position of the Schrot-effect, problem," *Phys. Z.*, vol. 23, pp. 436–438, 1922.

10. Hull, A. W., "Measurements of high-frequency amplification with shielded-grid pliotrons," *Phys. Rev.*, vol. 27, pp. 439–454, Apr. 1926.

11. Hull, A. W., and Williams, N. H., "Determination of elementary charge E from measurements of shot effect," *Phys. Rev.*, vol. 25, pp. 147–173, 1925.

12. Ising, G., "Natürliche Empfanglichkeitsgrenze der Waage," *Ann. Physik*, vol. 8, pp. 905–928, 1931.

13. Johnson, J. B., "Bemerkung zur Bestimmung des Elektrischen Elementarquantums aus dem Schroteffekt," *Ann. Physik*, vol. 67, pp. 154–156, 1922.

14. Johnson, J. B., "The Schottky effect in low-frequency circuits," *Phys. Rev.*, vol. 26, pp. 71–85, 1925.

15. Johnson, J. B., "Thermal agitation of electricity in conductors," *Phys. Rev.*, vol. 32, pp. 97–109, 1928.

16. Johnson, J. B., and Llewellyn, F. B., "Limits to amplification," *Elec. Eng.*, vol. 53, pp. 1449–1454, Nov. 1934.

17. Kingsbury, B. A., "The shot effect in photoelectric currents," *Phys. Rev.*, vol. 38, pp. 1458–1476, Oct. 15, 1931.

18. Kozanowski, H. N., and Williams, N. H., "Shot effect of the emission from oxide cathodes," *Phys. Rev.*, vol. 36, pp. 1314–1329, Oct. 15, 1930.

19. Lee, D. H., and Nicolet, M.-A., "Thermal noise in double injection," *Phys. Rev.*, vol. 184, pp. 806–808, Aug. 15, 1969.

20. Letzter, S., and Webster, N., "Noise in amplifiers," *IEEE Spectrum*, vol. 7, pp. 67–75, Aug. 1970.

21. Llewellyn, F. B., "A Study of noise in vacuum tubes and attached circuits," *Proc. IRE*, vol. 18, pp. 243–265, Feb. 1930.

22. Moullin, E. B., and Ellis, H. D. M., "Spontaneous background noise in amplifiers due to thermal agitation and shot effects," *J. IEE* (London), vol. 74, pp. 323–356, 1934.

23. Nyquist, H., "Thermal agitation of electric charge in conductors," *Phys. Rev.*, vol. 32, pp. 110–113, 1928.

24. v. Orbán, F., "Schroteffekt und Wärmegeräusch im Photozellenverstärker," *Z. Tech. Phys.*, vol. 13, pp. 420–424, 1932.

25. v. Orbán, F., "Schroteffekt und Wärmegeräusch im Photozellenverstärker. II," *Z. Tech. Phys.*, vol. 14, pp. 137–143, 1933.

26. Ornstein, L. S., and Burger, H. C., "Zur Theorie des Schroteffektes," *Ann. Physik*, vol. 70, no. 8, pp. 622–624, 1923.

27. Pearson, G. L., "Fluctuation noise in vacuum tubes," *Bell System Tech. J.*, vol. 13, pp. 634–653, 1934.

28. Pearson, G. L., "Shot effect and thermal agitation in an electron current limited by space charge," *Physics*, vol. 6, pp. 6–9, Jan. 1935.

29. Rack, A. J., "Effect of space charge and transit time on the shot noise in diodes," *Bell System Tech. J.*, vol. 17, no. 4, pp. 592–619, 1938.

30. Rothe, H., and Klein, W., "Multi-grid tubes," *Telefunkenröhren*, p. 174, Nov. 1936.

31. Schottky, W., "Über spontane Stromschwankungen in verschiedenen Elektrizitätsleitern," *Ann. Phys.*, vol. 57, pp. 541–567, 1918.

32. Schottky, W., "Small-shot effect and flicker effect," *Phys. Rev.*, vol. 28, pp. 74–103, July 1926.

33. Schottky, W., "Schroteffekt unde Raumladungschwelle," *Telefunkenröhren*, pp. 175–195, Nov. 1936.

34. Schottky, W., "Raumladungsswächung beim Schroteffekt und Funkeleffekt," *Physica*, vol. 4, no. 2, pp. 175–180, 1937.

35. Schottky, W., "Raumladungsswächung des Schroteffektes, Part 1. Theoretische Grundlagen und Hauptergebnisse," *Wiss. Veröffent. Siemens-Werken*, vol. 16, no. 2, pp. 1–18, 1937.

36. Schottky, W., "Zusammenhänge zwischen korpuskularen und thermische Schwankungen in Elektronenröhren," *Z. Physik*, vol. 104, pp. 248–274, 1937.

37. Schottky, W., "Zur Theorie des Elektronenrauschens in Mehrgitterröhren," *Ann. Phys.*, vol. 32, pp. 195–204, May 1938.

38. Spenke, E., "Raumladungsschwächung des Schroteffektes. II. Durchführung der Theorie für ebenen Anordnungen," *Wiss. Veröffentl. Siemens-Werken*, vol. 16, no. 2, pp. 19–41, 1937.

39. Thatcher, E. W., and Williams, N. H., "Shot effect in space-charge-limited currents," *Phys. Rev.*, vol. 39, pp. 474–496, Feb. 1, 1932.

40. Thompson, B. J., and North, D. O., "Shot effect in tubes," *Electronics*, p. 31, Nov. 1936.

41. Thompson, B. J., North, D. O., and Harris, W. A., "Fluctuations in space-charge-limited currents at moderately high frequencies," *RCA Rev.*, pp. 269–285, Jan. 1940; pp. 441–472, Apr. 1940; pp. 106–124, July 1940; pp. 244–260, Oct. 1940; pp. 371–388, Jan. 1941; pp. 505–524, Apr. 1941; pp. 114–124, July 1941.

42. Williams, F. C., "Fluctuation voltage in diodes and in multi-electrode valves," *J. IEE* (London), vol. 79, pp. 349–360, 1936.

43. Williams, F. C., "Fluctuation noise in vacuum tubes which are not temperature-limited," *J. IEE* (London), vol. 78, pp. 326–332, 1936.

44. van der Ziel, A., "Noise in double-injection space-charge-limited solid-state diodes," *IEEE Trans. Microwave Theory and Techniques*, vol. MTT-16, p. 308, May 1968.

Some Perspectives on Receiver Noise Performance and Its Measurement

microwave engineers take it for granted that they can define receiver sensitivity and measure it, but can they--and how did they arrive at present definitions and measurement technology?

MATTHEW T. LEBENBAUM   *Applied Electronics Div., AIL div. of Cutler Hammer*

It isn't enough to say that "this is a good receiver" or "that is a sensitive receiver". The question is "how good" or "how sensitive". The need to establish a quantitative measure of a performance parameter rather than a qualitative one is a real one if there is to be progress towards optimizing a design. The question of "how much better" or "how much more sensitive", is one receiver than another must be answered if a rational choice is to be made. And in the absolute sense, one must have a quantitative measure if the feasibility of an overall system is to be determined at all.

Before trying to characterize that performance parameter of a receiver loosely termed sensitivity, one has to recognize what the ultimate limitation is to the ability of a receiver to deliver a usable signal at its output. In this enlightened day and age, it seems almost childish to ask the question, for the idea of "noise," "noise temperature," "galactic noise," etc., — are all so much a part of our modern technology that we cannot remember a time when it was otherwise.

But even as late as 1938, the IRE Standards defined the sensitivity of a radio receiver as "that characteristic which determines the minimum strength of signal input capable of causing a desired value of signal output". This is simply a definition of gain! Implicit in this is the fact that, in most applications prior to that period, internally generated noise in the receiver was not the limiting factor in receiving weak signal, but rather externally generated noise.

Until some time in the '30's, the limitation to the reliable reception of weak signals was "interference," either man-made or natural phenomena. In the vicinity of urban areas, arc noise from street lighting or trolley cars and from coronal discharges on high-voltage transmission lines produced an uncomfortable noise background in LF and HF receivers, and electrical storms did the same. Jansky's famous discovery of natural electromagnetic radiation from our own Galaxy in 1933 was made as the result of a study of interference to long-distance communications. It was not until the late '30's, when the higher frequencies began to be exploited, that attention began to be focussed on the dual problems of how to reduce the noise generated internally in a receiver, and how to characterize the noise performance of the resulting design.

Anyone who truly desires to obtain an insight into the problems of low-noise receiver design and its measurement could well spend some leisure time in rereading the early articles of E. W. Herold,[1] D. O. North[2] and H. W. Friis[3].

Not only are they informative (and instructive even today) but they remind us that it is possible to discuss technical matters with style. Much of the material of that period in US journals was produced at either RCA or Bell Telephone Laboratories, and this has had one unfortunate result. North coined the term "noise factor" and Friis used "noise figure" and this usage has never been resolved. In official IEEE Standards and in articles by Standards committees, the terms are used interchangeably, and in the extreme of fence-straddling, are combined as "noise figure (factor)". Fortunately, as technology has advanced and noise figures (factors) have been reduced, the new term, "noise temperature" has come into use, and to my knowledge, there has been no other term used as a noun in the various forms of "effective noise temperature," "operating noise temperature," etc.

Although, as Friis points out, not only was the "absolute sensitivity in short-wave radio receivers" determined by thermal-agitation noise, but he had shown this experimentally as early as 1928. But it was not until 1942 that there was sufficient usage of the "short-wave" bands that it became imperative that some standardization of terms be accomplished, and that we should decide exactly what it was that we were trying to measure. It is unfortunately true that the pressures produced by international conflict produce major advances in many fields. World War II was the pressure cooker that speeded major advances in medicine, chemistry, physics, engineering, and other fields. It is only a personal opinion, but a strongly held one of mine, that the tremendous emphasis placed on the problem of detection of aircraft because of the gathering storm clouds of conflict in Europe (block that metaphor!) underlies the present electronics age, which some pronounce that we live in.

And as we moved into these "short-wave" bands, the receiver engineer began to focus on the problems of absolute sensitivity and its measurement. Rapidly, the concept of noise figure* as a measure of the quality of a radio receiver gained favor, as it was realized that the absolute sensitivity of a receiver in the UHF and microwave bands was, to a large extent, governed by the noise generated

---

* For some long forgotten and probably inconsequential reason, the writer has used the term noise figure rather than noise factor (although he had read North before Friis) and will use this terminology until he can talk about really low-noise receivers in terms of noise temperature.

Reprinted with permission from *Microwave J.*, vol. 16, pp. 31–32, 34, Jan. 1973.

internally in the first few stages of the receiver, rather than being controlled by man-made interference, or "atmospherics".

It must be realized that this measure, the noise figure, is a restrictive one. It is a measure of the noise that the receiver adds to whatever signal and noise enter the system from a prescribed antenna, and only describes the performance of the linear portions of the receiver prior to detection. It is obvious that the detection and post detection processing of the incoming signal can have a significant effect on the utilization of the signal, but this is a separate problem. Noise figure is a useful, but restricted, figure of merit. The object of good receiver design is to establish the best signal-to-noise ratio at the output for a given signal input. This is the ultimate measure of quality in a receiver. But all other factors being equal (bandwidth, dynamic range, transient response, demodulation techniques, etc.) the receiver with the better noise figure is the better one. And so it is important to measure the noise figure, or the noise temperature, of a system accurately.

A few definitions are essential to any meaningful discussion of measurement methods. The noise figure, as Friis[3] first defined it, modified by the following discussions of Friis and North,[4] was standardized in 1952[5] and modified in 1957[6]. The reader should refer to the standards or to literature on the subject, for exact definitions. It is not the purpose of this article to provide a working collection of formulae, but hopefully, to give a little perspective of the present as it derived from the past. Probably the most complete detailed discussion that collects all of the definitions and formulae in a single place is the little book by Mumford and Scheibe[7]. As an example of some of the confusion that has existed in the past, they list nine definitions of noise figure that have been used over the years, in spite of IEEE standardization.

For our purposes, I should like only to define noise figure and noise temperature in basic terms. The noise figure F of the network is defined as the ratio of the available signal-to-noise ratio at the input terminals (when the temperature of the input termination is 290 K and the bandwidth is limited by the receiver) to the available signal-to-noise ratio at its output terminals. It is a measure of the degradation of the signal-to-noise ratio available at the input terminals as it passes through the linear portions of the receiver. It is important to realize that, since the temperature at the input termination is specified as 290 K, F is essentially independent of antenna characteristics. The difference between the noise figures of two receivers, therefore, is *not* a direct measure of the change in performance of an *operating* "real-life" system when the one receiver is substituted for another.

The output noise from a receiver is proportional to the sum of the noise generated in the source and that added by the receiver. Since the noise generated by thermal means in a resistor is proportional to the absolute temperature of the resistor, it has become convenient to express noise powers in terms of the temperature to which a resistor would have to be heated or cooled to produce the same output noise as the source or receiver. And so the terms antenna temperature or effective receiver temperature are derived. The noise power available from a resistor at 290 K in a 1-MHz bandwidth would be 4×10^{-15} watts. How much more convenient to use 290 K instead of

4×10^{-15} watts. This alone would be sufficient reason for substituting temperature for power (and temperatures can be added, just as powers can). There are other reasons for using the noise temperature concept, as mentioned later, when receiver noise temperatures are low and antenna temperatures deviate from 290 K.

In the early days of microwave-receiver design, when one was struggling to get down from a 20 dB noise figure to 10 dB, most work was being done in war-time research labs, either Government or industrial. Getting equipment designed, built, and into operational use was more pressing than arguing about an odd half of a dB, and so measuring techniques were crude and more a check on whether one was in the right ballpark than not. A standard signal generator, double the output power, and measure the bandwidth was the usual method. Measuring noise bandwidth rather than the 3-dB bandwidth was a refinement that was observed in the breach. Perhaps some who were more careful will shudder at the above, and perhaps it is unfair to those who cared. Maybe it is only because the writer worked with wide-tuning range search receivers and often was happy to settle for 20-dB noise figures (or more — remember the AN/APR-1 and AN/APR-4); his friends down the road at Radiation Lab who worked with radar were far more concerned with sensitivity and worked at a fixed frequency where they could optimize performance and achieve noise figures approaching the 10-dB mark; all we could claim was a wide-tuning range and some signal output.

But post-war developments changed all that. Contracts were based on competitive bids, and specifications always asked for just a little bit more than was easily available. Then measurement accuracy became more and more important, particularly when final delivery had to be made. Greene's classic "The Art of Noisemanship"[8] is still readable today — it was defined as the art of measuring noise figure *independent* of device performance, and was broken into two sections, one for the seller and one for the buyer. Although tongue-in-check, it still contains an excellent cautionary list (by quoting the opposite) of what to avoid in noise figure measurement.

The biggest single advance in noise measurement was the invention of the gas-discharge noise source by W. W. Mumford[9] in 1949. Temperature-limited diodes, whose output was low and whose accuracy was questionable at frequencies greater than a few hundred megahertz, suddenly were replaced by a noise source that could be, and was, broadbanded to cover a full waveguide band in a single unit. And it had an output high enough to lend itself to relatively accurate measurements over the full range of waveguide bands up to (today) upwards of 40 GHz. Also, ingenious designs incorporated the gas-discharge tube into coaxial designs that went down as low as 100 MHz.

Jokingly, in the late '40's and early '50's, when a noise figure was quoted, one would ask "whose dB's?" Mumford's invention, still in effective use today, made it far more difficult to practice noisemanship. Yet even as noise figures moved under the 10-dB level, a few odd tenths of a dB here and there were forgivable and not too important. In 1955, masers were invented and shortly thereafter, parametric amplifier work started in earnest. Suddenly, the old concepts of measuring receiver performance, although still completely valid, did not quite fill the bill.

Just as gain was expressed in dB's, because it was awkward to talk about a power gain of 126,000 times rather than 51 dB, so noise figures were expressed in dB's because the ratio was large. People like to express ideas in numbers from 1 to 100 and in extremes, from 1 to 1000. So, as noise figures suddenly plunged to small fractions of a dB, the idea of noise temperature became more and more popular.

And this usage was emphasized by the fact that the operating performance of a receiving system was, with low-noise receivers and antennas, a function of both receiver and antenna. As long as most applications used antennas that looked toward the horizon or towards the earth, as search radars, point-to-point or tropospheric scatter communications systems or navigation radars did, the effective temperature of the source was not too different from 290 K as a source temperature, the difference in dB's between the noise figures of two receivers was a reasonable measure of the difference in performance between two systems. But with the advent of low-noise receivers and the increasing importance of radio astronomy in the 1950's and satellite communications in the 1960's, a performance figure that assumed a 290 K antenna was no longer useful as antennas tilted upward and antenna temperatures dropped to the 10-K to 50-K range.

To put this in perspective, take two receivers with noise figures of 0.25 dB (17 K and 3 dB (290 K), not too untypical of cooled and last year's uncooled parametric amplifiers for operation in a ground satellite communication terminal. The difference between the two, as far as noise figure is concerned, is 2.75 dB; significant but some might question whether cooling was worth the cost and complication. But if the antenna points at high angles, an antenna temperature of 25 K is not unreasonable. The system operating temperature, which is $T_a + T_e$, is 42 K in the cooled case and 315 K in the uncooled, and the ratio of the two is 7.5 or 8.75 dB! And cooling the amplifier is looked at in a different light.

It is not within the scope of this article to delve deeply into noise temperature and noise figure, their definitions and subtleties; it is only to point out that the ideas that have evolved with time have evolved because of a need to express in meaningful numbers an important performance parameter of an electronic system. Mumford and Scheibe's book and perhaps the 1962 IRE Standards (62 IRE 7.S2)[10] and the accompanying article by the IRE Subcommittee 7.9 on Noise[11] are required reading to anyone truly wishing to measure a quantity that *is dependent* on device performance. And with the performance that can be expected of modern low-noise receiver front-ends, the measurement of this performance becomes both contractually and electronically of great importance.

Although the accuracy of Mumford's gas discharge noise source is now assured with availability of calibration against NBS standards, there are some disadvantages of using this type of source for measuring modern very low-noise devices, as he himself shows. As a result, almost every procurement specification now calls out the requirement that measurements shall be made using hot/cold body loads with well known absolute temperatures. And sophisticated as the equipment gets, so does the buyer; he says in effect, you must prove to me that your loads really do have "well-known" temperatures. Further he is often as (or more) aware of some of the pitfalls of making accurate noise temperature measurements as the designer is.[12,13]

In a very short time, we have seen receiver performance, in terms of noise temperature, improve from points in the 30,000 to 300,000 K range to as low as 5 K and commonly in the 10 to 100 K range. And this improvement came about, not just "because it was there," but because there was a real need for this performance. Measuring techniques have had to keep pace. But one can paraphrase, each in his own way, W. W. Hansen's remark concerning the theory of microwaves that "the idea of impedance cannot be used as a substitute for thought." Noise temperature is a convenient way to think about receiver performance, but it is not complete in itself. How it is measured requires a thorough understanding of the system to which it is applied and the way in which it is measured.

REFERENCES

1. Herold, E. W., "An Analysis of the Signal-to-Noise Ratio of Ultrahigh Frequency Receivers," *RCA Review*, Vol 6 No 3, January 1942, pp 302-331.
2. North, D. O., "The Absolute Sensitivity of Radio Receivers," *RCA Review*, Vol 6 No 3, January 1942, pp 332-344.
3. Friis, H. T., "Noise Figures of Radio Receivers," *Proc IRE*, Vol 33 No 2, July 1944, pp 419-422.
4. Friis, H. T. and D. O. North, "Discussion on Noise Figures of Radio Receivers," *Proc IRE*, Vol 33 No 2, February 1945, pp 125-127.
5. "Standards on Receivers: Definitions of Terms, 1952," *Proc IRE*, Vol 40 No 12, December 1952, pp 1681-1685.
6. "Standards on Electron Tubes: Definition of Terms, 1957", *Proc IRE*, Vol 45 No 7, July 1957, pp 983-1010.
7. Mumford, W. W. and E. H. Scheibe, *Noise: Performance Factors in Communication Systems,* Horizon House - Microwave, Inc., 1968.
8. Greene, J. C., "The Art of Noisemanship," *Proc IRE*, Vol 49 No 7, July 1961, pp 1223-1224.
9. Mumford, W. W., "A Broadband Microwave Noise Source," *BSTJ*, Vol 28 No 4, October 1949, pp 608-618.
10. "Standards on Electron Tubes" Definition of Terms, 1962," *Proc IEEE*, Vol 51 No 3, March 1963, pp 434-435.
11. Adler, R., R. S. Engelbrecht, S. W. Harrison, H. A. Haus, M. T. Lebenbaum, and W. W. Mumford, "Description of the Noise Performance of Receiving Systems," *Proc IEEE*, Vol 51 No 3, March 1963, pp 436-442.
12. Sleven, R. L., "A Guide to Accurate Noise Measurement," *Microwaves*, July 1964.
13. *Handbook of Microwave Measurements*, Chapter XVII, "Noise Factor," A. J. Hendler and M. T. Lebenbaum, pp 865-887.

Noise and Random Processes*

J. R. RAGAZZINI†, FELLOW, IRE AND S. S. L. CHANG, FELLOW‡, IRE

Summary—Early investigators in the field of communications first realized that the presence of unwanted random noise was an important factor following the discovery that the maximum gain of an amplifier was limited by the discrete nature of currents in electron tubes. Called *shot effect*, this was first explained by W. Schottky and later by many other investigators. Much research on this problem during the second and third decades of the twentieth century finally led to the rigorous formulation of the phenomenon by B. J. Thompson and others in 1940. Concurrently, the problem of spontaneous thermal noise effects in conductors was studied and formulated. By 1940, the situation was developed to an extent that the application of mathematical statistics to explain and solve broader noise problems in systems was inevitable. About this time, the basic contributions of N. Wiener led to an understanding of the optimum linear filtration of signals imbedded in random noise. His work influenced the entire course of development of theory on the optimization of filters designed to abstract a signal out of its noisy environment.

I. INTRODUCTION

THE THEORY of noise and random processes in electronic devices was developed in two stages. The first stage spanning the two decades following 1918 brought about the understanding of the nature and effects of noise in vacuum tubes and circuits. The second stage, initiated by Wiener in the early 1940's, established the theoretical basis for the analysis and synthesis of systems subjected to random signals and noise. His rationale for optimization and design of optimum linear filters subjected to Gaussian signals and noise set the direction for much of the subsequent research on this subject. While a detailed discussion of optimum filtration and detection of signals embedded in random noise is treated in a separate paper on information theory, the subject is discussed briefly to demonstrate its link to the early work on the subject. The literature on noise and random processes is so extensive as to make a complete coverage impractical. Only highlights and significant stepping stones in the development of the theory are described.

II. EARLY WORK ON SOURCES OF NOISE

The first realization that unwanted random noise was a factor to contend with in the field of communications came during World War I when attempts were being made to design high-gain vacuum-tube amplifiers. It was soon found that there was a limit to the number of stages which could be cascaded in the quest for high gain due to an unacceptably high background noise which masked the weak signals being amplified. In his classic paper[1] Schottky first explained one of these effects and formulated the random component in the plate current of a vacuum tube.

Schottky ascribed the random fluctuations in the plate current to the fact that this current is composed not of a continuum but rather of a sequence of discrete increments of charge carried by each electron arriving at the plate at random times. The average rate of charge arrival constitutes the dc component of the plate current on which is superimposed a fluctuation component as each discrete charge arrives. He referred to this phenomenon as *schroteffekt* or *shot-effect* as we call it.

In arriving at his result, Schottky made a number of important simplifying assumptions. First, he assumed that the transit time of an electron from cathode to plate was infinitesimal so that the current pulse produced by each electron could be represented by an impulse or delta function. This immediately postulated that the random component of the current had an infinite frequency spectrum or was "white." This approximation is not a serious limitation except at extremely high frequencies when the transit time is a sensible fraction of a period.

The second and more limiting assumption is that the only force acting on the electron in transit is the electrostatic field between cathode and plate. The forces produced by other electrons in the space charge are neglected. This assumption is valid only when the plate current is temperature limited. The expression for the power spectrum of the random component of the plate current obtained by Schottky was

$$W(f) = 2eI \text{ a}^2/\text{cps}, \tag{1}$$

where e is the charge of the electron in coulombs and I is the dc plate current in amperes.

For the first time a quantitive expression for the random component of plate current became available and estimates of the ratio of signal-to-noise power could be made. But the formula suffers from a very serious defect, namely, that it is applicable only when the plate current is temperature limited. In practice, vacuum tubes were then and are now being operated with plate currents far below temperature saturation and the measured values of random noise in the plate current were much lower than those predicted by the Schottky formula. For many years attempts were made to explain

* Received by the IRE, October 2, 1961; revised manuscript received, December 14, 1961.
† College of Engineering, New York University, N. Y.
‡ Dept. of Elec. Engrg., College of Engineering, New York University, N. Y.

[1] W. Schottky, "Theory of shot effect," *Ann. phys.*, vol. 57, pp. 541–568; December, 1918.

Reprinted from *Proc. IRE*, vol. 50, pp. 1146–1151, May 1962.

this reduction of shot noise,[2-5] but most were empiric and did not get to the basic explanation of the mechanism by which space-charge could affect the result.

A major contribution to the understanding of the source of shot noise in space-charge-limited vacuum tubes was made in a set of papers by Thompson, North, and Harris.[6] Briefly, these investigators showed that a negative compensatory current pulse is generated each time an electron crosses the potential minimum between cathode and plate. Referring to Fig. 1, the region between cathode and plate is divided into an α and β region to the left and right of the potential minimum, respectively. As an electron e passes from the α to the β region, it increases the population of the β region, thereby depressing the potential minimum slightly. This inhibits the transit of a certain amount of charge out of the α region and returns it to the cathode. In effect, this is equivalent to a reverse coherent current pulse, as shown in Fig. 2, where i_n is the forward pulse of current produced by an electron entering the β region and i_c is the negative current pulse. The relation between the two is given by

$$i_c = \lambda i_n, \tag{2}$$

where λ is a constant dependent on the degree of space-charge saturation of the plate current. This modification results in a power spectrum for the random component of the plate current given by

$$W(f) = 2\Gamma^2 eI \text{ a}^2/\text{cps}, \tag{3}$$

where Γ^2 is $(1-\lambda^2)$. For zero space charge, Γ becomes unity and the power spectrum reduces to Schottky's result.

Thompson, North, and Harris computed Γ based on theoretical considerations and proved the result by experiment. A typical plot of Γ vs cathode temperature is shown in Fig. 3. It is seen that the power spectrum of the shot noise in a vacuum tube can be as little as 4 to 5 per cent of the values given by the Schottky formula. The authors extended the theory to cover multi-electrode vacuum tubes. This work essentially completed the basic theory of random noise in vacuum tubes insofar as random noise was concerned.

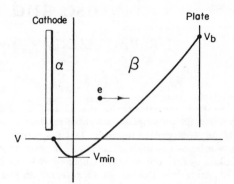

Fig. 1—Potential variation in space-charge-limited diode showing α and β regions.

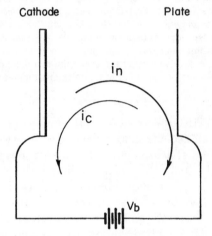

Fig. 2—Noise and compensating pulse produced in space-charge-limited diode.

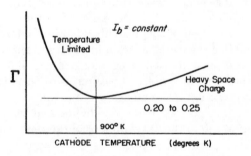

Fig. 3—Typical variation of Γ in a space-charge-limited diode with oxide-coated cathode.

In addition to fluctuation effects produced by vacuum tubes, it was found that random noise signals were generated in metallic resistors made of homogeneous materials. These effects were found to be temperature dependent and are known as thermal noise. A number of basic contributions to the understanding of thermal noise were made in the 1920's and 1930's[4,7,8] among which

[2] T. C. Fry, "Theory of shot effect," *J. Franklin Inst.;* 1925. Also, "Probability and Its Engineering Uses," D. Van Nostrand, Co., Inc., Princeton, N. J.; 1928.

[3] F. B. Llewellyn, "Study of noise in vacuum tubes and attached circuits," Proc. IRE, vol. 18, pp. 243–265; February, 1930.

[4] E. B. Moullin and H. D. M. Ellis, "The spontaneous background noise in amplifiers due to thermal agitation and shot effects," *J. IEE*, vol. 74, pp. 323–356; April, 1934.

[5] A. J. Rack, "Effect of space charge and transit time on shot noise in diodes," *Bell Sys. Tech. J.*, vol. 17, pp. 592–619; October, 1938.

[6] B. J. Thompson, D. O. North, and W. A. Harris, "Fluctuations in space charge-limited currents at moderately high frequencies," *RCA Rev.*, part 1 vol. 4, pp. 269–285; January, 1940; pt. 2, vol. 5, pp. 106–125; July, 1940; pt. 3, pp. 244–260, October, 1940; pt. 4, vol. 5, pp. 374–388; January, 1941; pt. 5, pp. 505–524; April, 1941.

[7] J. B. Johnson, "Thermal agitation of electricity in conductors," *Phys. Rev.*, vol. 32, 2nd ser., pp. 97–109; July, 1928.

[8] H. Nyquist, "Thermal agitation of electricity in conductors," *Phys. Rev.*, vol. 32, 2nd ser., pp. 110–113; July, 1928.

was the outstanding paper by Johnson in 1925. The source of thermal conductor noise was traced to the random excitation of the electron gas in the conductor in consequence of its existence in an environment of thermally-agitated molecules. The effect is similar to the Brownian movement of particles suspended in a liquid in which the thermally-agitated molecules of the liquid collide with the suspended particle and impart to it a certain amount of energy. Since the particle is cohesive, collision with any one of its molecules sets the entire particle in motion thereby resulting in random movements observable under a microscope.

The basic principle used to explain Brownian movement is that of equipartition of energy. The cohesive molecular group constituting the particle has imparted to it by collision an equal share of energy per degree of freedom just as the individual free molecules in the liquid have themselves equal shares of thermal energy. In the case of Brownian movement, the particle has three degrees of freedom thereby giving three independent energy storage modes, each of which stores a mean energy given by $\frac{1}{2}KT$ where K is the Boltzman constant and T the absolute temperature.

Analagously, in the case of the electron gas in a conductor, the same cohesive property that exists in the particle can be postulated. If one electron is set in motion by a collision with a thermally-agitated molecule, the entire electron gas in the circuit is set in motion by the suddenly changing magnetic field generated by the accelerating electron. As a result, the principle of equipartition of energy can be applied to this phenomenon and the electron gas derives a share of energy equal to $\frac{1}{2}KT$ for each energy-storage mode such as a condenser or inductor in the circuit. Based on considerations like these, the power spectrum of equivalent generator associated with each resistor is given by

$$W(f) = 4KTR \text{ v}^2/\text{cps} \tag{4}$$

where K, the Boltzman constant, is 1.372×10^{-32} joules/°K, T is absolute temperature in °K, and R is the resistance in ohms. The correctness of the hypothesis and the resultant formula for the power spectrum has been proved experimentally. Thus, all resistances in a circuit can be represented by a resistance and a white noise generator in series as shown in Fig. 4. Resistance components which are not homogeneous, such as composition resistors, generally produce unpredictable noise effects several times greater than homogeneous resistors, due to "crazy contact" effects. As in the case of tube noise, it is fair to assume that thermal noise spectra are white since the electron pulses produced by molecular collision are extremely short in duration.

Thus, by the late 1930's there was good understanding of the mechanisms producing random noise in electronic circuits and it was possible to relate these sources to the performance of a system. The link which made this possible is the fact that the power spectrum is related to

the Fourier spectrum of the random signal by the following:[9]

$$W(f) = \lim_{T \to \infty} \frac{2}{T} \mid G(j\omega) \mid^2, \tag{5}$$

where $G(j\omega)$ is the Fourier transform of the random signal $g(t)$, and $\omega = 2\pi f$. Thus, if such a random signal is applied to a linear filter as shown in Fig. 5, it follows readily from the fact that the Fourier transform of the output is the product of the Fourier transform of the input and the transfer function of the filter that the relation between the power spectrum at the input and the output is given by

$$W_y(f) = \mid H(j\omega) \mid^2 W_x(f), \tag{6}$$

where $H(j\omega)$ is the transfer function of the filter or the Fourier transform of its impulsive response. Coupled with the fact that the power spectrum $W_x(f)$ is related to the mean-square value of the signal by the integral

$$\overline{x^2} = \int_0^{+\infty} W_x(f)df, \tag{7}$$

the net effect of noise signals generated spontaneously in electronic circuits can be calculated.

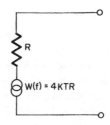

Fig. 4—Equivalent circuit of resistor including thermal-agitation noise-voltage generator.

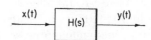

Fig. 5—Block diagram of linear filter with random input and output.

For convenience in analysis, the quantity spectral density $\Phi_{xx}(j\omega)$ is often used instead of power spectrum. Spectral density is an even function which extends from negative to positive infinite frequency and has values at each frequency half those of the power spectrum. Thus, the mean-square value of a signal is given by

$$\overline{x^2} = \frac{1}{2\pi} \int_{-\infty}^{\infty} \Phi_{xx}(j\omega)d\omega. \tag{8}$$

It should be noted here that noise sources so far considered are quite correctly assumed to generate white

[9] S. O. Rice, "Mathematical analysis of random noise," *Bell Sys. Tech. J.*, vol. 23, pp. 282–332, 1944; and vol. 24, pp. 46–156, 1945.

noise spectra based on an understanding of the originating mechanisms. Until the 1940's a certain amount of optimization by adjustment of circuit parameters to obtain optimum signal-to-noise ratios was possible by straightforward analysis. It was later that fundamental work was done in optimum filtering of signals in the presence of noise which is not necessarily white as described in subsequent sections.

III. More General Statistical Tools

The techniques which were used in expressing and analyzing the effects of random noise produced by vacuum tubes and resistors were, in fact, applicable to general problems. For instance, electronic equipments became part of more comprehensive systems such as radar-tracking systems, fire-control systems, navigation systems, etc., during the late 1930's and early 1940's. Analysis and optimization of such systems were urgently required, especially under the stresses of World War II when extension of system capabilities became a matter of national survival.

In most cases involving more comprehensive systems, the mechanisms generating random signals cannot be clearly specified or postulated as they can for shot effect and thermal noise. As a result, the statistical properties of random noise must be determined by experiment from time records of typical situations. A direct determination of power spectrum from these data is not as convenient as through the indirect approach using the autocorrelation function. As far back as 1930[10] Wiener showed a simple interrelationship between the autocorrelation function and the power spectrum, a relation which was to prove most useful in experimental as well as analytical operations. This expression is known as the Wiener-Khintchine relation in recognition of later independent work done by Khintchine.[11]

In order for the Wiener-Khintchine relation to apply, a number of assumptions must be made regarding the random signal. First, the random process of which a particular time record is a member must be stationary; that is, its statistical properties are independent of the origin of time. Secondly, the random process must satisfy the ergodic hypothesis. This implies that the stationary random time functions, an ensemble of which constitutes the random process, possess the same properties whether averaged over time on any one function or over an ensemble of functions at any given time. This means that such characteristics as the autocorrelation function can be obtained by integrating over time as well as over the ensemble. Within these limitations, the autocorrelation function of a random time function $x(t)$ is given[12] by

$$\phi_{xx}(\tau) = \lim_{T \to \infty} \frac{1}{T} \int_{-T/2}^{T/2} x(t)x(t - \tau)dt. \qquad (9)$$

The function is a measure of the dependence of the values the time function at any one time to another time displaced τ seconds. For instance, white noise is completely uncorrelated, the correlation function being zero for all finite values of displacement τ. Its mathematical expression is the impulse-response function $\delta(\tau)$ multiplied by a constant.

The Wiener-Khintchine relation states that the spectral density $\Phi_{xx}(j\omega)$ is the Fourier transform of the autocorrelation function[13]

$$\Phi_{xx}(j\omega) = \int_{-\infty}^{\infty} \phi_{xx}(\tau)e^{-j\omega\tau}d\tau, \qquad (10)$$

and conversely, the autocorrelation function is the inverse Fourier transform of the power spectrum

$$\phi_{xx}(\tau) = \frac{1}{2\pi} \int_{-\infty}^{\infty} \Phi_{xx}(j\omega)e^{j\omega\tau}d\omega. \qquad (11)$$

Similar expressions exist for the relation between two different random functions, $x(t)$ and $y(t)$:

$$\phi_{xy}(\tau) = \lim_{T \to \infty} \frac{1}{T} \int_{-T/2}^{T/2} x(t)y(t + \tau)dt \qquad (12)$$

and

$$\Phi_{xy}(j\omega) = \int_{-\infty}^{+\infty} \phi_{xy}(\tau)e^{-j\omega\tau}d\tau \qquad (13)$$

$$\phi_{xy}(\tau) = \frac{1}{2\pi} \int_{-\infty}^{+\infty} \Phi_{xy}(j\omega)e^{j\omega\tau}d\omega, \qquad (14)$$

where $\phi_{xy}(\tau)$ is known as the crosscorrelation function and $\Phi_{xy}(j\omega)$ the cross spectral density.

Eq. (6) can also be put in a more general form. Let y_1 and y_2 be the outputs from filters $H_1(j\omega)$ and $H_2(j\omega)$ with x_1 and x_2 as the respective inputs. Then

$$\Phi_{y_1y_2}(j\omega) = H_1(-j\omega)H_2(j\omega)\Phi_{x_1x_2}(j\omega). \qquad (15)$$

In addition to its application in the analysis of systems in which random signals and noise exist, the Wiener-Khintchine relation is very useful for the experimental determination of pertinent properties of the signal. For instance, a computation and plot of the autocorrelation function can be made directly from the experimental records of a random signal. Usually, the procedure is to fit a convenient analytical expression to the experimental plot of the autocorrelation function and from this the power spectrum of the signal can be obtained. In cases where complex systems such as radar tracking systems must be optimized, information of this type is usually obtained in this manner.

[10] N. Wiener, "Generalized harmonic analysis," *Acta Math.*, vol. 55, pp. 117–258; 1930

[11] A. Khintchine, "Korrelationtheorie der stationaischen prozesse," *Ann. Math.*, vol. 109, pp. 604–615; 1934.

[12] H. M. James, N. B. Nichols, and R. S. Phillips, "Theory of Servomechanisms," McGraw-Hill Book Co., Inc., New York, N. Y.; pp. 271–273; 1947.

[13] *Ibid.*, p. 283.

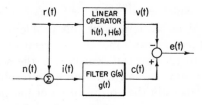

Fig. 6—Block diagram showing criterion for optimum filter.

In most systems which are of interest to the engineer and the designer, random processes are assumed to be Gaussian or at least assumed to be satisfactorily approximated by such a distribution. As discussed previously, a random signal or noise is usually generated by a large number of independent events (the shot-effect noise is generated by thermoinic emission of individual electrons, etc.). It follows from the central limit theorem in probability theory that, not only is the amplitude of such a signal normally distributed, but the joint distribution function of various amplitudes at various times is also a multivariate normal distribution function. A random signal with this property is called Gaussian. A significant implication of the above is that all statistical properties can be determined from a knowledge of the autocorrelation function or, equivalently, the power spectrum. Furthermore, it can be shown that Gaussian signals remain Gaussian after passing through linear filters. In practical problems, therefore, the measurement of the autocorrelation and its conversion into the power spectrum by Fourier transformation gives all the necessary statistical parameters needed for optimization of systems using procedures which were developed in the World War II years and later.

An outstanding piece of work in this regard is due to Rice.[9] Rice gave an estimate of the closeness to Gaussian distribution of a signal of the shot-effect type; and assuming the distribution to be Gaussian, he derived equations for a number of significant relations, for instance: the distribution of the total power of a random signal, the number of zero crossings and maxima and minima per second, and the power spectrum of the output signal from a nonlinear device.

IV. The Concept of Optimization

There are two types of problems in communication and control in which random signals play an essential role, namely: a) processing and separation of signal from noise, and b) detection of weak signals embedded in noise. Fig. 6 illustrates a problem of type a). The desired output $v(t)$ is the result of a specified linear operation on a signal $r(t)$. The available input signal $r(t)$ is contaminated with random noise $n(t)$. The problem is to design a filter for $i(t)$ so that the output $c(t)$ is closest to $v(t)$ in some well-defined sense.

A major contribution to this problem was made by Wiener[14,15] who, in 1942, worked out an optimization theory of such significance that its influence was to dominate at least the next two decades. He introduced a criterion for closeness of the signal filtered signal $c(t)$ to the ideal signal $v(t)$ on the basis of a least squares differ-

ence between the two. Perhaps both mathematical expediency and universality of Gaussian random noise (and signal) conspired to influence him and other pioneer investigators to select the least square error criterion as they did.

The criterion for optimization in the case of random signal and noise is expressed as a cost $k(t)$ defined as

$$k(t) = \lim_{T \to \infty} \frac{1}{T} \int_{-T/2}^{+T/2} [e(t)]^2 dt. \tag{16}$$

The design of the so-called optimum filter $G(s)$ is dictated by a minimization of this cost function subject to the conditions that the filter be linear and physically realizable. The latter requires that impulsive response $g(t)$ be zero for negative time. By taking advantage of the fact that the noise and signal are Gaussian, and of the Wiener-Khintchine relation, Wiener solved the problem for the optimum linear filter.

Many modifications of his procedure[16,17] have appeared but, basically, the Wiener approach including the least-square-error criterion has not been altered fundamentally. Booton[18] has extended Wiener's theory to the case where both the signal $r(t)$ and the noise $n(t)$ are random but nonstationary. The difficulty with his result is that it resulted in no general analytic solution, although later investigators[19] obtained a solution which can be implemented only by the use of computers.

In the signal detection problem, many of the methods and criteria used by Wiener influenced the analysis of the problem of estimating the presence or nonpresence of a signal such as a pulse in the presence of random noise. A least-squares-error criterion was used by North[20] to maximize the predicted signal-to-noise ratio. Referring to Fig. 7, the impulse-response function of the predetection filter is denoted as $g(t)$. The problem is to decide whether a filtered output is due to noise alone $a_0(t)$, or due to signal plus noise $a_1(t)$. North solved this

[14] J. H. Laning and R. H. Battin, "Random Processes in Automatic Control," McGraw-Hill Book Co., Inc., New York, N. Y.; 1956.

[15] N. Wiener "The Extrapolation, Interpolation, and Smoothing of Stationary Time Series," Mass. Inst. Tech., Cambridge, Rept. of the Services 19, Res. Project DIC-6037; February, 1942. (Published by John Wiley and Sons, Inc., New York, N. Y.; 1949.)

[16] H. W. Bode and C. E. Shanon, "A simplified derivation of linear least square smoothing and prediction theory," Proc. IRE, vol. 38, pp. 417–425; April, 1950.

[17] L. A. Zadeh and J. R. Ragazzini, "An extension of Wiener's theory of prediction," *J. Appl. Phys.*, vol. 21, pp. 645–655; July, 1950.

[18] R. C. Booton, Jr., "An optimization theory for time-varying linear systems with nonstationary statistical inputs," Proc. IRE, vol. 40, pp. 977–981; August, 1952.

[19] K. S. Miller and L. A. Zadeh, "Solution of an integral equation occurring in the theories of prediction and detection," IRE Trans. on Information Theory, vol. IT-2, pp. 72–75; June, 1956.

[20] D. O. North, "Analysis of the Factors Which Determine Signal to Noise Discrimination in Radar," RCA Labs., Princeton, N. J., Rept. No. PIR-6C; June, 1943.

problem for white noise and later investigators[21-26] further enlarged on variations of this problem. The concept of autocorrelation and crosscorrelation function, its relation to the spectral density and the selection of a least-mean-square-error criterion have resulted in significant advances in the optimization of linear filters to which are applied Gaussian noise and/or signal.

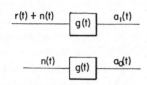

Fig. 7—Block diagram illustrating the decision to be made in an optimum detection problem.

[21] L. A. Zadeh, and J. R. Ragazzini, "Optimum filters for the detection of signals in noise," Proc. IRE, vol. 40, pp. 1223–1231; October, 1952.

[22] J. Neyman and E. S. Pearson, "On the problem of the most efficient tests of statistical hypothesis," *Phil. Trans. Royal Soc. London*, vol. A231, pp. 289–338; 1933.

[23] J. L. Lawson and G. E. Uhlenbeck, "Threshold Signals," McGraw-Hill Book Co., Inc., New York, N. Y., ch. 7; 1950.

[24] M. Schwartz, "Statistical Approach to the Automatic Search Problem," Ph.D. dissertation, Harvard University, Cambridge, Mass., 1951.

[25] P. M. Woodward and I. L. Davies, "A theory of radar information," *Phil. Mag.*, vol. 41, series 7, pp. 1001–1031; October, 1950. See also Proc. IRE, vol. 39, pp. 1521–1524, December, 1951; and *J. IEE*, vol. 99, pp. 37–49; March, 1952.

[26] D. Middleton and D. Van Meter, "Detection and extraction of signals in noise from the point of view of statistical decision theory," *J. Soc. Industrial and Appl. Math.*, vol. 3, pp. 192–253; December, 1955; and vol. 4, pp. 86–119; June, 1956.

V. Future Trends

As a result of intensive research in the past few decades, the field of noise and random processes is rapidly becoming mature. It is very unlikely that major contributions such as those made by Wiener are to be expected in the future. However, there are a few discernible trends which are noteworthy.

The understanding of the sources of noise in conductors, semiconductors and electron tubes is fairly well understood so that no important developments should be expected in these areas. On the other hand, in the field of optimization, developments on the concept, analysis and synthesis of adaptive systems which are responsive either to varying inputs or to environment are to be expected. The general subject of nonlinear optimum systems is still in a somewhat undeveloped state. The nonlinearity may already be present in an existing part of the system or may be deliberately introduced in the filter because the inputs are not Gaussian. Many special cases have been worked out, but results of a general nature are very few.

With the availability of fast computers of high capacity, it is expected that their employment as active in-line and off-line devices to implement optimum systems will increase. However, the underlying theory which postulates the employment of such computers in optimum systems is incomplete, and it is expected that research along these lines will be intensified.

A Bibliography on Noise

P. L. CHESSIN†

THIS BIBLIOGRAPHY is an attempt to collect a reasonably complete set of references pertaining primarily to fluctuation type noise, characteristic of shot and thermal noise, up to the year 1954. Of necessity these references touch upon such diverse fields as electrical engineering, aerodynamics, quantum mechanics, mathematical statistics, and pure mathematics. Although the material has been divided into a number of broad categories, it should be pointed out that many references could appear with almost equal validity under more than one heading. The classification is somewhat arbitrary and the reader is advised to consider related topics in all cases. Within each category the items are arranged chronologically with respect to their appearance in print. The attempt has been made to survey as accurately as possible the field in English, and to include as many pertinent and accessible foreign-language articles as practicable.

Because noise is the relentless and undesirable associate of intelligence transmitted or received by acoustical and electronic systems, it is essential for any proper theory of communication to provide suitable knowledge and methods for studying the physical properties of noise and its interaction with a desired signal. Not only is a successful technique of measurement required to control or lessen the noise, but also adequate information and theory is requisite to guide experiments and to interpret data. The purpose, therefore, of this bibliography is to present a coherent source of information references for the case of noise belonging primarily to the fluctuation type (random normal processes), rather than impulsive noise such as atmospheric and solar static, the cataloguing of which would have been prohibitive due to the enormous number of publications of observations.

Classification Chart

1. Source Works: basic texts, monographs, and auxiliary bibliographies.
2. Internal Noise Sources: tube noise, circuit element noise, etc.
3. External Noise: signals received in the presence of noise, control systems with statistical inputs.
4. Noise Generation and Measurements.
5. Impulsive Type Noise: atmospheric and solar noise.
6. Modulation and Noise: noise and PM, PCM, PTM, AM, and FM systems.
7. Radar Applications: jitter, clutter, glint, scintillation, etc.
8. Noise, Communication and Filtering: data smoothing, Wiener theory, etc.
9. Statistical Theory: general studies of random noise, generalized harmonic analysis, etc.
10. Author Index: alphabetical listing.
11. Appendix: publisher listing.

In past years, since Schottky*[2-1] first noted and correctly interpreted shot effect, the subject of fluctuation noise was quite mysterious even to the best of engineers and scientists. In 1928, however, Johnson[2-3] and Nyquist[2-4] cleared up the subject of thermal agitation noise in circuits. From about 1935, fluctuation noise in the plate circuit of amplifier tubes has been thoroughly worked out[2-15], and, with some further work on mixer noise and converter noise[3-4,3-7], it is possible to use accurate qualitative data on tube noise for receiver calculations. At high frequencies, however, input noise must be considered; this too has been evaluated[2-20,2-19]. The signal-to-noise ratio of a radio receiver at ultra-high frequencies is primarily dependent on noise relations for tubes and circuits[2-29], but also one must consider the input signal and its transmission to and through the receiver. Early published works[4-3,2-12] on signal-to-noise ratio did not include induced input noise and hence was not strictly applicable at uhf. Until North's paper[3-8] on the quantity called "noise factor", there was no widely accepted basis on which to compare qualitatively experimental or analytical results. The extension of the analysis to include induced noise was made[3-7], and the interpretation of results in terms of noise factor has become common. As a result it was possible to discuss signal-to-noise ratio at uhf with considerable clarity[2-29,2-38].

Since the paper by Armstrong[6-1] drew attention to the possibilities of frequency modulating regarding the reduction of noise, a considerable quantity of work in this direction has been published. The earlier theoretical treatment of signal and noise, however, had been confined in the main to the case where the noise energy is small as compared to the signal energy. Now rigorous treatments valid for all signal-to-noise energy ratios may be found. These treatments are usually developed by methods which Rice[9-12] and Fränz[4-8, 4-10] applied to similar problems and which are based on the Fourier Spectrum of the noise[1-1, 1-5, 1-8, 1-12, 9-2, 9-9, 9-17].

In the past fifteen years a great deal of study has been aimed at the mathematical analysis of random noise in communication systems[9-46]. Theoretical contributions have been made by many people, among whom are Rice,

† Westinghouse Electric Corp., Air Arm Division, Friendship International Airport, Baltimore 27, Md. *Univ. of Maryland, College Park, Md.*

* *Editorial note*—All superscripts refer to bibliographical data which follows, in this manner: In superscript [2-1], e.g., "2" designates section 2 (Internal Noise Sources), and "-1" refers to the first author listed, etc.

Reprinted from *IRE Trans. Inform. Theory*, vol. IT-1, pp. 15–31, Sept. 1955.

mentioned earlier,[9-27] and Middleton[3-19, 3-25, 3-32, 3-46, 3-58, 7-12, 9-59]. The former's papers give a unified view of the fundamental methods of noise analysis for both linear and nonlinear circuits. The latter has solved a good number of nonlinear problems of practical importance.

Since the mathematics is quite complicated[9-67, 9-68], a need for developing experimental techniques to determine the important statistical characteristics of random noise in circuits, and, incidentally, to check existing theoretical results and to solve problems not susceptible to theoretical analysis was felt in many quarters. The method usually consists of finding the autocorrelation function[9-85] and by a Fourier cosine transformation determining the power density spectrum[9-72, 9-74]. At MIT's Research Laboratory of Electronics, for example, there are available an electronic digital correlator[9-41] and other devices, such as the electronic differential analyzer[9-28] and the delay line filter[8-37]. The last two machines accomplish the transformation of the correlation experimentally obtained by the first. Some work by Knudtzon[8-26] was done in this direction where, however, experimental studies were limited to the case of linear circuits. On the other hand, Weinberg and Kraft[9-86] have extended the scope of investigation via the correlation technique to include nonlinear devices.

The determination of an optimum design for an automatically tracking radar-controlled system is hampered by the difficult problem of analyzing the character of the radar noise. The engineer who wishes to minimize the effect of the noise in the tracking loop, for example, may use the Wiener "Optimum Synthesis" technique[9-6] to determine the characteristics of an optimum linear control system. The equations which are based on the criterion of minimizing the rms error on a statistical basis are discussed in[1-6]. The synthesis of such a system requires the aforementioned autocorrelation function and power-density spectra of the input message and noise characteristics, in order that the power spectra may be placed into the design equations to obtain the optimum system. Phillips and Weiss[3-16] concerned themselves with the best smoothing of positional data for gunnery prediction for noise spectra of arbitrary forms, while Floyd, Zadeh, and others[8-22, 8-32, 8-39] extended Wiener's theory[1-9] to nonstationary time series, and obtained a more general solution to this problem.

1. Source Works

1-1   Paley, R. E. A. C., and Wiener, N., *Fourier Transforms in the Complex Domain.* AMS Colloq. Publ., Vol. 19, 1934, reprinted 1954.

1-2   Cramer, H., *Random Variables and Probability Distributions.* London, Cambridge University Press, 1937.

1-3   Moulin, E. B., *Spontaneous Fluctuations of Voltage.* New York, Oxford, 1938.

1-4   Kendall, M. G., *Advanced Theory of Statistics.* (2 v.), London, Griffin, 1943.

1-5   Cramer, H., *Mathematical Methods of Statistics.* Princeton, N. J., Princeton University Press, 1946.

1-6   James, H. M., Nichols, N. B., and Phillips, R. S., *Theory of Servomechanisms.* New York, McGraw-Hill, 1947.

1-7   Thomas, H. A., and Burgess, R. E., "Survey of Existing Information and Data on Radio Noise over the Frequency Range 1–30 Mc/s." *Special Report #15*, Dept. Sci. & Indust. Res., London, 1947. (Contains a 186-item bibliography.)

1-8   Goldman, S., *Frequency Analysis, Modulation, and Noise.* New York, McGraw-Hill, 1948.

1-9   Wiener, N., *Extrapolation, Interpolation and Smoothing of Stationary Time Series.* New York, Tech. Press and John Wiley, 1949.

1-10   *Symposium on Noise Reduction, Part III.* North American Aviation Aerophysics Laboratory Report #AL-930, Feb., 1949 (contains an exhaustive bibliography).

1-11   Lawson, J. L., and Uhlenbeck, G. E., *Threshold Signals.* New York, McGraw-Hill, 1950.

1-12   Goldman, S., *Information Theory.* New York, Prentice-Hall, 1953.

1-13   Black, H. S., *Modulation Theory.* New York, Van Nostrand, 1953.

1-14   Blanc-Lapierre, A., and Fortet, R., *Theory of Aleatory Functions. Applications to Diverse Phenomena of Fluctuation.* Paris, Masson, 1953.

1-15   Bell, D. A., *Information Theory—and its Engineering Applications.* New York, Pitman Publ. Co., 1953.

2. Internal Noise Sources

2-1   Schottky, W., "Spontaneous Current Fluctuations in Various Conductors." *Ann. Phys.*, Vol. 57 (December 20, 1918), pp. 541–567.

2-2   Hull, A. W., and Williams, N. H., "Determination of Elementary Charge E from Measurements of Shot Effect." *Phys. Rev.*, Vol. 25 (1925), p. 147.

2-3   Johnson, J. B., "Thermal Agitation of Electricity in Conductors." *Phys. Rev.*, Vol. 32 (July, 1928), pp. 97–109.

2-4   Nyquist, H., "Thermal Agitation of Electric Charge in Conductors." *Phys. Rev.*, Vol. 32 (July, 1928), pp. 110–113.

2-5   Williams, N. H., and Huxford, W. S., "Determination of the Charge of Positive Thermions from Measurements of Shot Effect." *Phys. Rev.*, Vol. 33 (1929), p. 773.

2-6   Ballantine, S., "Fluctuation Noise in Radio Receivers." Proc. IRE, Vol. 18 (August, 1930), p. 1377.

2-7   Case, N. P., "Receiver Design for Minimum Fluctuation Noise." Proc. IRE, Vol. 19 (June, 1931), pp. 963–971.

2-8   Wilbur, D. A., *Thermal Agitation of Electricity in Conductors.* Dissertation, University of Michigan, 1932.

2-9   McNally, J. O., "Analysis and Reduction of Output Disturbances Resulting from the AC Operation of the Heaters of Indirectly Heated Cathode Triodes." Proc. IRE, Vol. 20 (August, 1932), pp. 1263–1284.

2-10   Moullin, E. B., "Measurement of Shot Voltage Used to Deduce the Magnitude of Secondary Thermionic Emission." *Proc. Roy. Soc.*, Vol. 147 (1934), p. 109.

2-11   Williams, F. C., "Fluctuation Voltage in Diodes and in Multi-electrode Valves." *Jour. IEE* (London), Vol. 79 (1936), p. 349.

2-12   Williams, F. C., "Thermal Fluctuations in Complex Networks." *Jour. IEE* (London), Vol. 81 (December, 1937), pp. 751–760.

2-13   Shockley, W., and Pierce, J. R., "A Theory of Noise for Electron Multipliers." Proc. IRE, Vol. 26 (March, 1938), pp. 321–333.

2-14   Rack, A. J., "Effect of Space Charge and Transit Time on the Shot Noise in Diodes." *Bell Sys. Tech. Jour.*, Vol. 17 (October, 1938), pp. 592–619.

2-15   Thompson, B. J., North, D. O., and Harris, W. A., "Fluctuations in Space-Charge-Limited Currents at Moderately High Frequencies." *RCA Rev.*, Vol. 4 (January, 1940), pp. 269–285; Vol. 5 (July,) pp. 106–124; (January, 1941), pp. 371–388; (April) pp. 505–524; Vol. 6 (July, 1941), pp. 114–124.

2-16   North, D. O., "Fluctuations in Space-Charge-Limited Currents at Moderately High Frequencies, Part II, Diodes and Negative-grid Triodes." *RCA Rev.*, Vol. 4 (April, 1940), pp. 441–472.

2-17   North, D. O., "Fluctuations in Space-Charge-Limited Currents at Moderately High Frequencies." *RCA Rev.*, Vol. 5 (October, 1940), pp. 244–260.

2-18   Wald, M., "Noise Suppression by Means of Amplitude Limiters." *Wire. Engr.*, Vol. 17 (October, 1940), pp. 432–438.

2-19   Bakker, C. J., "Fluctuations and Electron Inertia." *Physica*, Vol. 8 (January, 1941), pp. 23–43.

2-20   North, D. O., and Ferris, W. R., "Fluctuations Induced in Vacuum-tube Grids at High Frequencies." Proc. IRE, Vol. 29 (February, 1941), pp. 49–50.

2-21   Bell, D. A., "Measurement of Shot and Thermal Noise." *Wire. Engr.*, Vol. 18 (March, 1941), pp. 95–98.

2-22   Harris, W. A., "Fluctuations in Vacuum Tube Amplifiers and Input Systems." *RCA Rev.*, Vol. 5 (April, 1941), p. 4.

2-23   Burgess, R. E., "Receiver Input Circuits, Design Considerations for Optimum Signal/Noise Ratio." *Wire. Engr.*, Vol. 20 (1943), p. 66.

2-24   Schiff, L. I., *Noise in Crystal Rectifiers*. Report #126, National Defense Research Council, Univ. of Penna., March 10, 1943.

2-25   Rice, S. O., "Filtered Thermal Noise-Fluctuation Energy as a Function of Interval Length." *Jour. Acous. Soc. Am.*, Vol. 14 (April, 1943), pp. 216–227.

2-26   Weisskopf, V. F., *On the Theory of Noise in Conductors, Semi-conductors, and Crystal Rectifiers*. Report #14-133, National Defense Research Council, May 12, 1943.

2-27   Beers, Y., *Noise from Local Oscillators*. Report #304, MIT Rad. Lab., June 8, 1943.

2-28   Smith, R. N., *Crystal Noise as a Function of DC Bias*. Report #14-167, National Defense Research Council, Purdue Univ., June 25, 1943.

2-29   Herold, E. W., and Malter, L., "Some Aspects of Radio Reception at Ultra-High Frequency—Part II, Admittances and Fluctuation Noise of Tubes and Circuits; Part III, Signal/Noise Ratio of Radio Receivers." Proc. IRE, Vol. 31 (September, 1943), pp. 491–500, pp. 501–510.

2-30   Lawson, A. W., Miller, P. H., and Stephens, D. E., *Noise in Silicon Rectifiers at Low Temperatures*. Report #14-189, National Defense Research Council, Univ. of Penna., October 1, 1943.

2-31   Bennett, W. R., "Response of Linear Rectifier to Signal and Noise." *Bell Sys. Tech. Jour.*, Vol. 23 (January, 1944), pp. 97–113.

2-32   Bennett, W. R., "Response of Linear Rectifier to Signal and Noise." *Jour. Acous. Soc. Am.*, Vol. 15 (January, 1944), pp. 164–170.

2-33   Pierce, J. R., *Noise Calculations for Reflex Oscillators*. Report #MM-44-140-4, Bell. Tele. Lab., Jan. 29, 1944.

2-34   Blanc-Lapierre, A., "Photo-Counters and Poisson's Law." *Compt. Rend. Acad. Sci.* (Paris), Vol. 218 (January 31, 1944), pp. 188–190.

2-35   Blanc-Lapierre, A., "Shot Effect and Fluctuations at the Output of a Linear Amplifier." *Compt. Rend. Acad. Sci.* (Paris), Vol. 218 (February 14, 1944), pp. 272–274.

2-36   Brill, E. R., *The Improvement in Minimum Detectable Signal in Noise Through the Use of the Long-After-Glow CR Tube and Through Photographic Integration*. Report #411-84, Office of Scientific and Research Development, February, 1944.

2-37   Miller, P. H., Lewis, M. N., Schiff, L. I., and Stephens, D. E., *Noise Spectrum of Silicon Rectifiers*. Report #14-256, National Defense Research Council, Univ. of Penna., March 20, 1944.

2-38   Dishal, M., "Theoretical Gain and Signal-to-Noise Ratio of the Grounded-Grid Amplifier at Ultra-High Frequencies." Proc. IRE, Vol. 32 (May, 1944), pp. 276–284.

2-39   Jones, M. C., "Grounded-Grid Radio-Frequency Voltage Amplifiers." Proc. IRE, Vol. 32 (July, 1944), pp. 276–284.

2-40   Kalmus, H. P., "Some Notes on Superregeneration with Particular Emphasis on its Possibilities for FM." Proc. IRE, Vol. 32 (October, 1944), pp. 591–600.

2-41   Sziklai, G. C., and Schroeder, A. C., "Cathode-Coupled Wide-Band Amplifiers." Proc. IRE, Vol. 33 (October, 1945), pp. 701–708.

2-42   Kuper, J. B. H., and Waltz, M. C., *Measurements on Noise from Reflex Oscillators*. Report #872, MIT Radiation Laboratory, December 21, 1945.

2-43   Knipp, J. K., *Theory of Noise from the Reflex Oscillator*. Report #873, MIT Radiation Laboratory, January 10, 1946.

2-44   Schremp, E. J., "A Generalization of Nyquist's Thermal Noise Theorem." *Phys. Rev.*, Vol. 69 (March 1, 1946), p. 255.

2-45   Burgess, R. E., "Fluctuation Noise in a Receiving Aerial." *Proc. Phys. Soc.* (*London*), Vol. 58 (May 1, 1946), pp. 313–321.

2-46   Beard, E. G., "Random Noise in Radio Receivers." *Philips Tech. Commun.* (Australia), (July, 1946), pp. 14–22.

2-47   Blanc-Lapierre, A., "Study of Fluctuations Produced by the Shot Effect in Amplifiers." *Rev. Sci.* (*Paris*), Vol. 84 (June/July, 1946), pp. 75–94.

2-48   Kac, M., and Siegert, A. J. F., "Note on the Theory of Noise in Receivers with Square Law Detectors." *Phys. Rev.*, Vol. 70 (September, 1946), p. 449.

2-49   Strutt, M. J. O., "Noise Figure Reduction in Mixer Stages." Proc. IRE, Vol. 34 (December, 1946), pp. 942–950.

2-50   Bennett, W. R., "The Biased Ideal Rectifier." *Bell Sys. Tech. Jour.*, Vol. 26 (January, 1947), pp. 139–169.

2-51   Miller, P. H., Jr., "Noise Spectrum of Crystal Rectifiers." Proc. IRE, Vol. 35 (March, 1947), pp. 252–257.

2-52   Roberts, S., "Some Considerations Governing Noise Measurements on Crystal Mixers." Proc. IRE, Vol. 35 (March, 1947), pp. 257–265.

2-53   Kac, M., and Siegert, A. J. F., "Theory of Noise in Radio Receivers with Square Law Detectors." *Jour. Appl. Phys.* Vol. 18 (April, 1947), pp. 383–397.

2-54   Macfarlane, G. G., "A Theory of Flicker Noise in Valves and Impurity Semi-conductors." *Proc. Phys. Soc.*, Vol. 59 (May 1, 1947), pp. 366–375, 403–408.

2-55   MacDonald, D. K. C., and Fürth, R., "Spontaneous Fluctuations of Electricity in Thermionic Valves under Retarding Field Conditions." *Proc. Phys. Soc*, Vol. 59 (May 1, 1947), pp. 375–388, 403–408.

2-56   Fürth, R., and MacDonald, D. K. C., "Statistical Analysis of Spontaneous Electrical Fluctuations." *Proc. Phys. Soc.*, Vol. 59 (May 1, 1947), pp. 388–408; see also: *Nature* (London), Vol. 157 (June 15, 1946), p. 807.

2-57   van der Ziel, A., and Versnel, A., "Total Emission Noise in Diodes." *Nature*, Vol. 159 (May 10, 1947), pp. 640–641.

2-58   Roos, W., "Background Noise In Amplifiers." *Tech. Mitt. Schweiz. Telegr.-Teleph. Verw.*, Vol. 25 (August 1, 1947), pp. 143–147.

2-59   Schremp, E. J., *Noise in Linear Networks*. Atti del Congresso internazionale della Radio (Rome), (September/October, 1947), pp. 402–411, (in English).

2-60   Goodman, B., "How Sensitive is your Receiver?" *QST*, Vol. 31 (September, 1947), pp. 13–21.

2-61   Rauch, L. L., "Fluctuation Noise in Pulse-Height Multiplex Radio Links." Proc. IRE, Vol. 35 (November, 1947), pp. 1192–1198.

2-62   Peterson, L. C., "Space-Charge and Transit-Time Effects on Signal and Noise in Microwave Tetrodes." Proc. IRE, Vol. 35 (November, 1947), pp. 1264–1272.

2-63   Fürth, R., "A Unified Theory of Spontaneous Electrical Fluctuations in Thermionic Valves." *Nature* (London), Vol. 160 (December 13, 1947), pp. 832–833.

2-64   Pierce, J. R., "Noise in Resistances and Electron Streams." *Bell Sys. Tech. Jour.*, Vol. 27 (January, 1948), pp. 158–174.

2-65   MacDonald, D. K. C., "Spontaneous Fluctuations." *Report Progress in Physics*, Vol. 12, 1948-1949, pp. 56–81.

2-66   van der Ziel, A., and Versnel, A., "Induced Grid Noise and Total-Emission Noise." *Phil. Res. Rep.*, Vol. 3 (February, 1948), pp. 13–23.

2-67   Fürth, R., "On the Theory of Electrical Fluctuations." *Proc. Roy. Soc.*, Vol. 192 (March 18, 1948), pp. 593–615.

2-68   Campbell, N. R., Francis, V. J., and James, E. G., "Valve Noise and Transit Time." *Wire. Engr.*, Vol. 25 (May, 1948), pp. 148–157.

2-69   Harris, E. J., "Circuit and Current Noise." *Electronics Engr.*, Vol. 20 (May, 1948), pp. 145–148.

2–70 Wallman, H., Macnee, A. B., and Gadsen, C. P., "A Low-Noise Amplifier." Proc. IRE, Vol. 36 (June, 1948), pp. 700–708.

2–71 van der Ziel, A., and Versnel, A., "The Noise Factor of Grounded-Grid Valves." Phil. Res. Rep., Vol. 3 (August, 1948), pp. 255–270.

2–72 Harris, W. A., "Some Notes on Noise Theory and Its Application to Input Circuit Design." RCA Rev., Vol. 9 (September, 1948), pp. 406–418.

2–73 MacDonald, D. K. C., "On the Theory of Electrical Fluctuations." Proc. Roy. Soc., Vol. 195 (December 7, 1948), pp. 225–230.

2–74 Callendar, M. V., "Thermal Noise Output in AM Receivers." Wire. Engr., Vol. 25 (December, 1948), pp. 395–399.

2–75 Freeman, J. J., "Noise Spectrum of a Diode with a Retarding Field." Jour. Res. Nat. Bur. Stand., Vol. 42 (January, 1949), pp. 75–88.

2–76 Lavoo, N. T., "Measured Noise Characteristics at Long Transit Angles." Proc. IRE, Vol. 37 (April, 1949), pp. 383–386.

2–77 Fraser, D. B., "Noise Spectrum of Temperature-limited Diodes." Wire. Engr., Vol. 26 (April, 1949), pp. 129–132.

2–78 Krumhansl, J. A., and Beyer, R. T., "Barkhausen Noise and Magnetic Amplifiers: Part I—Theory of Magnetic Amplifiers." Jour. Appl. Phys., Vol. 20 (May, 1949), pp. 432–436.

2–79 Sulzer, P. G., "Noise Figures for Receiver Input Circuits." Tele-Tech, Vol. 8 (May, 1949), pp. 40–42, 57.

2–80 Krumhansl, J. A., and Beyer, R. T., "Barkhausen Noise and Magnetic Amplifiers: Part II—Analysis of the Noise." Jour. Appl. Phys., Vol. 20 (June, 1949), pp. 582–586.

2–81 Slinkman, R. W., "Temperature-limited Noise Diode Design." Sylvania Technologist, Vol. 2 (October, 1949), pp. 6–8.

2–82 Ashcroft, H., and Hurst, C., "Transit Time Correction Factor for Cylindrical Noise Diodes." Proc. Phys. Soc., Vol. 62 (October 1, 1949), pp. 639–646.

2–83 Mooers, T., "Low-Frequency Noise in Transistors." Proc. NEC (Chicago), Vol. 5 (1949), pp. 17–22.

2–84 Paolini, E., and Canegallo, G., "Resistor Noise." Alta. Frequenza, Vol. 18 (December, 1949), pp. 254–267 (in Italian, with English, French, and German summaries), (French edition: Radio Tech. Dig. (Franç), Vol. 4 (April, 1950), pp. 91–107.

2–85 Meltzer, B., "A Note on the Identity of Thermal Noise and Shot Noise." Phil. Mag., Vol. 40 (December, 1949), pp. 1224–1226.

2–86 Richardson, J. M., "The Linear Theory of Fluctuations Arising from Diffusional Mechanisms—An Attempt at a Theory of Contact Noise." Bell Sys. Tech. Jour., Vol. 29 (January, 1950), pp. 117–141.

2–87 du Pre, F. K., "Suggestion Regarding the Spectral Density of Flicker Effect Noise." Phys. Rev., Vol. 77 (January 1, 1950), p. 615.

2–88 Hettlich, A., "Dimensions and Noise of Resistors." Frequenz, Vol. 4 (January, 1950), pp. 14–25.

2–89 Bell, R. L., "Induced Noise and Noise Factor." Nature (London), Vol. 165 (March 18, 1950), pp. 443–444. (See also "Induced Grid Noise," Wire. Engr., Vol. 27 (March, 1950), pp. 86–94; and also Proc. IRE, Vol. 39 (September, 1951), pp. 1059–1063.

2–90 Knol, K. S., and Diemer, G., "Theory and Experiments on Electrical Fluctuations and Damping of Double-Cathode Valves." Phil. Res. Rep., Vol. 5 (April, 1950), pp. 131–152.

2–91 van der Ziel, A., "On the Noise Spectra of Semi-Conductor Noise and of Flicker Effect." Physica, 's Grav., Vol. 16 (April, 1950), pp. 359–372 (in English).

2–92 Schultz, M. A., "Linear Amplifiers." Proc. IRE, Vol. 38 (May, 1950), pp. 475–485.

2–93 van der Ziel, A., "Note on Total Emission Damping and Total Emission Noise." Proc. IRE, Vol. 38 (May, 1950), p. 562.

2–94 van der Ziel, A., "Thermal Noise at High Frequencies." Jour. Appl. Phys., Vol. 21 (May, 1950), pp. 399–401.

2–95 van der Ziel, A., "Noise Suppression in Triode Amplifiers." Canad. Jour. Res., Vol. 28 (June, 1950), pp. 189–198.

2–96 Boyer, R. F., "Random Noise in Dielectric Materials." Jour. Appl. Phys., Vol. 21 (June, 1950), pp. 469–477.

2–97 Chavasse, P., and Lehmann, R., "Background Noise and the Use of White Noise in Acoustics." Ann. Télécommun., Vol. 5 (June, 1950), pp. 229–236.

2–98 Böer, W., "Limiting the Noise Spectrum in Oscillatory Circuits." Ann. Phys. (Lpz), Vol. 8 (July 31, 1950), pp. 87–92.

2–99 Singleton, H. E., Theory of Nonlinear Transducers. Tech. Report #160, MIT Research Laboratory of Electronics, August 12, 1950.

2–100 Moses, R. C., "Germanium-Diode Impulse-Noise Limiters." Sylvania Tech., Vol. 3 (October, 1950), pp. 1–5.

2–101 Macfarlane, G. G., "Theory of Contact Noise in Semi-conductors." Proc. Phys. Soc. (London), Vol. 63b (October 1, 1950), pp. 807–814.

2–102 Castellini, N. R., "Noise Figure of the Magnetic Amplifier." Proc. NEC (Chicago), Vol. 6 (1950), pp. 52–58.

2–103 Lehmann, G., "Background Noise in Amplifiers." Rev. gén. eléct., Vol. 59 (December, 1950), pp. 543–553.

2–104 Harris, I. A., "A Note on Induced Grid Noise and Noise Factor." Jour. Brit. IRE, Vol. 10 (December, 1950), pp. 396–400.

2–105 Petritz, R. L., and Siegert, A. J. F., "On the Theory of Noise in Semi-Conductors." Phys. Rev., Vol. 79 (1950).

2–106 Cutler, C. C., and Quate, C. F., "Experimental Verification of Space Charge and Transit Time Reduction of Noise in Electron Beams." Phys. Rev., Vol. 80 (December 15, 1950), pp. 875–878.

2–107 Kleen, W., "Receiver Noise." Fernmeldetech. Z., Vol. 4 (January/February/April, 1951), pp. 19–25, 56–63, 182.

2–108 Agdur, B. M., and Åsdal, G. G. L., "Noise Measurements on a Travelling-Wave Tube." Acta. polytech. (Stockholm), No. 86 (1951).

2–109 Watanabe, Y., and Honda, N., "Theory of Noise in the Transistor." Sci. Rep. Res. Inst. Tohoku. Univ., Ser B, Vol. 1–2 (March, 1951), pp. 313–325.

2–110 Callen, H. C., and Welton, T. A., "On the Theory of Thermal Fluctuations or Generalized Noise." Phys. Rev., Vol. 82 (1951), p. 296.

2–111 Miller, D. A., and Slusser, E. A., "Expediting TV Receiver Noise Calculations." TV Engr., Vol. 2 (April, 1951), pp. 12, 13, 35.

2–112 van der Ziel, A., "Induced Grid Noise in Triodes." Wire. Engr., Vol. 28 (July, 1951), pp. 226–227.

2–113 Robinson, F. N. H., and Kompfner, R., "Noise in Traveling-Wave Tubes." Proc. IRE, Vol. 39 (August, 1951), pp. 918–926.

2–114 van der Ziel, A., "Noise Suppression in Triode Amplifiers." Canad. Jour. Tech., Vol. 29 (Dec., 1951), pp. 540–553.

2–115 Schooley, A. H., and George, S. F., "Input vs Output S/N Characteristics of Linear, Parabolic, and Semi-Cubical Detectors." Proc. NEC (Chicago), Vol. 7 (1951), pp. 151–161; (see also Tele-Tech, Vol. 11 (July, 1952), pp. 60–63, 75).

2–116 Bladier, B., "Background Noise in Amplifiers—Its Reduction—Its Application in Physiology." Acustica, Vol. 2, No. 1 (1952), pp. 23–34, (in French).

2–117 Solomon, S. S., "Thermal and Shot Fluctuations in Electrical Conductors and Vacuum Tubes." Jour. Appl. Phys., Vol. 23 (January, 1952), pp. 109–112.

2–118 Watkins, D. A., "Traveling-Wave Tube Noise Figure." Proc. IRE, Vol. 40 (January, 1952), pp. 65–70.

2–119 Blaquière, A., "The Effect of Background Noise on the Frequency of Valve Oscillations. Ultimate Accuracy of Electronic Clocks." Compt. Rend. Acad. Sci. (Paris), Vol. 234 (January 21, 1952), pp. 419–421.

2–120 Blaquière, A., "The Effect of Background Noise on the Amplitude of (Valve-) Maintained Oscillators." Compt. Rend. Acad. Sci. (Paris), Vol. 234 (February 11, 1952), pp. 710–712.

2–121 Blaquière, A., "Effect of Background Noise on the Frequency of Valve Oscillators." Compt. Rend. Acad. Sci. (Paris), Vol. 234 (March 10, 1952), pp. 1140–1142; (see also "Extension of Nyquist's Theory to the Case of Nonlinear Characteristics." Vol. 233 (July 30, 1951), pp. 345–347).

2-122 Parzen, P., "Effect of Thermal-Velocity Spread on the Noise Figure in Traveling-Wave Tubes." *Jour. Appl. Phys.*, Vol. 23 (April, 1952), pp. 394–406.

2-123 Tokahashi, H., "Generalized Theory of Thermal Fluctuations." *Jour. Phys. Soc. Japan*, Vol. 7 (1952), pp. 439–446.

2-124 Kleen, W., and Ruppel, W., "Calculation of the Noise Figure of the Travelling-Wave: Part I." *Arch. elekt. übertragung*, Vol. 6 (May, 1952), pp. 187–194.

2-125 Tucker, D. G., "Linear Rectifiers and Limiters." *Wire. Engr.*, Vol. 29 (May, 1952), pp. 128–137.

2-126 Watkins, D. A., "The Effect of Velocity Distribution in a Modulated Electron Stream." *Jour. Appl. Phys.*, Vol. 23 (May, 1952), pp. 568–573.

2-127 Tucker, D. G., "The Synchrodyne and Coherent Detector." *Wire. Engr.*, Vol. 29 (July, 1952), pp. 184–188.

2-128 Petritz, R. L., "A Theory of Contact Noise." *Phys. Rev.*, Vol. 87 (August 1, 1952), pp. 535–536.

2-129 Tomlinson, T. B., "Space Charge Reduction of Low Frequency Fluctuations in Thermionic Emitters." *Jour. Appl. Phys.*, Vol. 23 (August, 1952), pp. 894–899.

2-130 Montgomery, H. C., "Electrical Noise in Semiconductors." *Bell Sys. Tech. Jour.*, Vol. 31 (September, 1952), pp. 950–975.

2-131 Gratama, S., "Noise in Receivers and Amplifiers." *Tijdschr. ned. Radiogenoot*, Vol. 17 (September/November, 1952), pp. 207–247 (59 references).

2-132 Strutt, M. J. O., "Methods of Reducing the Ratio of Noise to Signal at the Output Terminals of Amplifiers." *Elektrotech. Z.*, Vol. 73 (October 11, 1952), pp. 649–653.

2-133 Bell, D. A., "Current Noise in Semiconductors. A Reexamination of Bernamont's Data." *Phil. Mag.*, Vol. 43 (October, 1952), pp. 1017–1111.

2-134 Prof Group on Electron Devices, "On the Shot Effect of *p-n* Junctions." TRANS. IRE, *PGED-1* (November, 1952), pp. 20–24.

2-135 Freeman, J. J., "On the Relation between the Conductance and the Noise Power Spectrum of Certain Electronic Streams." *Jour. Appl. Phys.*, Vol. 23 (November, 1952), pp. 1223–1225.

2-136 Petritz, R. L., "On the Theory of Noise in *p-n* Junctions and Related Devices." *Proc. IRE*, Vol. 40 (November, 1952), pp. 1440–1456.

2-137 Keonjian, E., and Schaffner, J. S., "An Experimental Investigation of Transistor Noise." PROC. IRE, Vol. 40 (November, 1952), pp. 1456–1460.

2-138 Montgomery, H. C., "Transistor Noise in Circuit Applications." PROC. IRE, Vol. 40 (Nov., 1952), pp. 1461–1471.

2-139 Amakusu, K., "Excess Thermal Noise due to Current Flow Through (B$_a$S$_r$O) Oxide Coating." *Jour. Appl. Phys.*, Vol. 23 (December, 1952), pp. 1330–1332.

2-140 Pierce, J. R., "A New Method of Calculating Noise in Electron Streams." PROC. IRE, Vol. 40 (December, 1952), pp. 1675–1680.

2-141 Lindemann, W. W., and van der Ziel, A., "On the Flicker Noise caused by an Interface Layer." *Jour. Appl. Phys.*, Vol. 23 (December, 1952), pp. 1410–1411.

2-142 Davenport, W. B., *Signal-to-Noise Ratios in Bandpass-Limiters.* Report #234, MIT Research Laboratory of Electronics, 1952.

2-143 Bell, D. A., "Current Noise in Semiconductors." *Wire. Engr.*, Vol. 30 (January, 1953), pp. 23–24.

2-144 Blaquière, A., "Effect of Background Noise on the Frequency of Tube Oscillators. Ultimate Accuracy of Electronic Clocks." *Ann. Radioélect.*, Vol. 8 (January, 1953), pp. 36–80.

2-145 Rowe, H. E., "Noise Analysis of a Single-Velocity Electron Gun of Finite Cross Section in an Infinite Magnetic Field." TRANS. IRE (January, 1953), pp. 36–46.

2-146 Berktay, H. O., "Noise in Interface Resistance." *Wire. Engr.*, Vol. 30 (February, 1953), pp. 48–49.

2-147 Mattson, R. H., and van der Ziel, A., "Shot Noise in Germanium Filaments." *Jour. Appl. Phys.*, Vol. 24 (February, 1953), pp. 222–223.

2-148 Rollin, B. V., and Templeton, I. M., "Noise in Semiconductors at Very Low Frequencies." *Proc. Phys. Soc.* (London), Vol. 66 (March 1, 1953), pp. 259–261.

2-149 Fremlin, J. H., "Noise in Thermionic Valves." *Proc. IEE* (London), Vol. 100 (March, 1953), pp. 91–92.

2-150 Evans, J. H., "The Noise Factor of Centimetric Superheterodyne Receivers." *Electronics Engr.*, Vol. 25 (March, 1953), pp. 98–100.

2-151 Preston, G. W., and Gardner, R. K., "The Noise Performance of Triggered Pulse Generators." *1953 IRE Conv. Rec.*, Part 8, Inform. Theory, pp. 96–100.

2-152 Burgess, R. E., "Contact Noise in Semiconductors." *Proc. Phys. Soc.*, Vol. 66 (April 1, 1953), pp. 334–335.

2-153 George, T. S., and Urkowitz, H., "Fluctuation Noise in Microwave Superregenerative Amplifier." PROC. IRE, Vol. 41 (April, 1953), pp. 516–521.

2-154 Blaquière, A., "Power Spectrum of a Nonlinear Oscillator Perturbed by Noise." *Ann. Radioélect.*, Vol. 8 (April, 1953), pp. 153–179.

2-155 Suhl, H., "Theory of Magnetic Effects on the Noise in a Germanium Filament." *Bell Sys. Tech. Jour.*, Vol. 32 (May, 1953), pp. 647–664.

2-156 Tomlinson, T. B., "Temperature Dependence of Low Frequency Fluctuations in Thermionic Emitters." *Jour. Appl. Phys.*, Vol. 24 (May, 1953), pp. 611–615.

2-157 Preston, G. W., "The Equivalence of Optimum Transducers and Sufficient and Most Effective Statistics." *Jour. Appl. Phys.*, Vol. 24 (July, 1953), pp. 841–844.

2-158 Strum, P. B., "Some Aspects of Mixer Crystal Performance." PROC. IRE, Vol. 41 (July, 1953), pp. 875–889.

2-159 Inuishi, Y., and Tsung-che, Y., "Influence of Impurity Atoms on Flicker Noise." *Jour. Phys. Soc.* (Japan), Vol. 8 (July/August, 1953), pp. 565–567.

2-160 van der Ziel, A., "Note on the Shot Effect in Semi-Conductors and Flicker Effect in Cathodes." *Physica*, Vol. 19 (August, 1953), pp. 742–744.

2-161 Tomlinson, T. B., and Price, W. L., "Theory of the Flicker Effect." *Jour. Appl. Phys.*, Vol. 24 (August, 1953), pp. 1063–1065.

2-162 Middleton, D., Gottschalk, W. M., and Wiesner, J. B., "Noise in CW Magnetrons." *Jour. Appl. Phys.*, Vol. 24 (August, 1953), pp. 1065–1066.

2-163 Templeton, I. M., and MacDonald, D. K. C., "The Electrical Conductivity and Current Noise of Carbon Resistors." *Proc. Phys. Soc.*, Vol. 66 (August 1, 1953), pp. 680–687.

2-164 Karr, P. R., "Effective Circuit Bandwidth for Noise with a Power-Law Spectrum." *Jour. Res. Nat. Bur. Stand.*, Vol. 51 (August, 1953), pp. 93–94.

2-165 van der Ziel, A., "A Simpler Explanation for the Observed Shot Effect in Germanium Filaments." *Jour. Appl. Phys.*, Vol. 24 (August, 1953), p. 1063.

2-166 Peter, R. W., and Rultz, J. A., "Influence of Secondary Electrons on Noise Factor and Stability of Traveling-Wave Tubes." *RCA Rev.*, Vol. 14 (Sept., 1953), pp. 441–452.

2-167 Montgomery, H. C., and Clark, M. A., "Shot Noise in Junction Transistors." *Jour. Appl. Phys.*, Vol. 24 (October, 1953), pp. 1337–1338.

2-168 Houlding, N., "Noise Factor of Conventional VHF Amplifiers." *Wire. Engr.*, Vol. 30 (November/December, 1953), pp. 281–290, 299–306.

2-169 Blaquière, A., "Effect of Valve Noise on Oscillators." *Compt. Rend. Acad. Sci.* (Paris), Vol. 237 (November 23, 1953), pp. 1316–1318.

2-170 Watkins, D. A., "Low-noise Traveling-Wave Tubes for X-Band." PROC. IRE, Vol. 41 (December, 1953), pp. 1741–1746.

2-171 Robinson, F. N. H., "Microwave Shot Noise in Electron Beams and the Minimum Noise Factor of Traveling Wave Tubes and Klystrons." *Jour. Brit. IRE*, Vol. 14 (February, 1954), pp. 79–87.

2-172 Machulup, S., "Noise in Semiconductors: Spectrum of a Two-parameter Random Signal." *Jour. Appl. Phys.*, Vol. 25 (March, 1954), pp. 341–343.

2-173 Slocum, A., and Shive, J. N., "Shot Dependence of *p-n* Junction Phototransistor Noise." *Jour. Appl. Phys.*, Vol. 25 (March, 1954), p. 406.

2-174 Handley, P. A., and Welch, P., "Valve Noise Produced by Electrode Movement." PROC. IRE, Vol. 42 (March, 1954), pp. 565–573.

2–175   Nozick, S., and Winkler, S., "Noise Limitation on Storage Tube Operation." 1954 IRE Convention Record—Part 3.

2–176   Talpey, T. E., and Macnee, A. B., "The Nature of the Uncorrelated Component of Induced Grid Noise." 1954 IRE Convention Record—Part 4.

2–177   Carlisle, R. W., and Pearson, H. A., "A Simple Transistor-Noise and Gain-Test Set." 1954 IRE Convention Record—Part 10.

3. External Noise

3–1   Beverage, H. H., and Peterson, H. O., "Diversity Receiving System of RCA Communications, Inc. for Radiotelegraphy." Proc. IRE, Vol. 19 (April, 1931), pp. 531–562.

3–2   Jansky, K. G., "Minimum Noise Levels Obtained on Short-Wave Radio Receiving Systems." Proc. IRE, Vol. 25 (December, 1937), pp. 1517–1531; Vol. 26 (April, 1938), p. 400.

3–3   Landon, V. D., and Reid, J. D., "A New Antenna System for Noise Reduction." Proc. IRE, Vol. 27 (March, 1939), pp. 188–192.

3–4   Herold, E. W., "Superheterodyne Converter System Considerations in Television Receivers." RCA Rev., Vol. 4 (January, 1940), pp. 324–337.

3–5   Slater, J. C., Noise and the Reception of Pulses. Report #115, MIT Radiation Laboratory, February 13, 1941.

3–6   Hansen, W. W., Coincidence Method of Noise Reduction. Report #119, MIT Radiation Laboratory, August 25, 1941.

3–7   Herold, E. W., "Analysis of S/N Ratio of UHF Receivers." RCA Rev., Vol. 6 (January, 1942), pp. 302–331.

3–8   North, D. O., "The Absolute Sensitivity of Radio Receivers." RCA Rev., Vol. 6 (January, 1942), pp. 332–343.

3–9   Herold, E. W., "The Operation of Frequency Converters and Mixers." Proc. IRE, Vol. 30 (February, 1942), pp. 84–103.

3–10   Gray, F., Noise in a Detected Carrier Pulse. Report #MM 42-130-94, Bell Tele. Laboratory, December 10, 1942.

3–11   Goudsmit, S. A., and Weiss, P. R., Statistics of Circuit Noise. Report #43-20, MIT Radiation Laboratory, January 29, 1943.

3–12   Jordon, W. H., Action of Linear Detector on Signals in the Presence of Noise. Report #61-23, MIT Radiation Laboratory, July 6, 1943.

3–13   Van Vleck, J. H., The Spectrum of Clipped Noise. Report #51, Radio Research Laboratory, July 21, 1943.

3–14   Tiberio, U., "On the Evaluation of the S/N Ratio in Oscillographic Receivers." Alta. Freq., Vol. 12 (July–September, 1943), pp. 316–323 (in Italian, with English, French, and German summaries).

3–15   Siegert, A. J. F., On the Fluctuations in Signals Returned by Many Independent Scatterers. Report #465, MIT Radiation Laboratory, November 12, 1943.

3–16   Phillips, R. S., and Weiss, P. R., Theoretical Calculation on Best Smoothing of Position Data for Gunnery Prediction. Report #532, MIT Radiation Laboratory, February 16, 1944.

3–17   Cobine, J. D., Curry, J. R., Gallagher, C. J., and Ruthberg, S., Video Transformers for Noise Voltage. Report OEMsr-411-244, Radio Research Laboratory, October 30, 1945.

3–18   Bunimovich, V. I., "Effect of the Fluctuations (Noise) and Signal Voltages on a Non-Linear System." Jour. Phys. (USSR), Vol. 10, No. 1 (1946), pp. 35–48.

3–19   Middleton, D., "The Response of Biased Saturated Linear and Quadratic Rectifiers to Random Noise." Jour. Appl. Phys., Vol. 17 (October, 1946), pp. 778–801.

3–20   Allen, E. W., Jr., "Very-High-Frequency and Ultra-High-Frequency Signal Ranges as Limited by Noise and Co-channel Interference." Proc. IRE, Vol. 35 (February, 1947), pp. 128–152.

3–21   Middleton, D., "An Approximate Theory of Eddy-Current Loss in Transformer Cores Excited by Sine Wave or by Random Noise." Proc. IRE, Vol. 35 (March, 1947), pp. 270–281.

3–22   Jelonek, F., "Noise Problems in Pulse Communication." Jour. IEE (London), part IIIA, Vol. 94 (1947), pp. 533–545.

3–23   Cunningham, W. J., Goffard, S. J., and Licklider, J. C. R., "The Influence of Amplitude Limiting and Frequency Selectivity upon the Performance of Radio Receivers in Noise." Proc. IRE, Vol. 35 (October, 1947), pp. 1021–1026.

3–24   Cobine, J. D., Curry, J. R., Gallagher, C. J., and Ruthberg, S., "High-Frequency Excitation of Iron Cores." Proc. IRE, Vol. 35 (October, 1947), pp. 1060–1067.

3–25   Middleton, D., "Some General Results in the Theory of Noise Through Non-linear Devices." Quart. Appl. Math., Vol. 5 (January, 1948), pp. 445–498.

3–26   Laplume, J., "On the Number of Signals Discernible in the Presence of Random Noise in a Transmission System with Limited Passband." Comp. Rend. Acad. Sci. (Paris), Vol. 226 (April, 1948), pp. 1348–1349.

3–27   Ross, A. E., "A Type of Nonlinear Random Distortion (Noise) in Pulse Communication." Phys. Rev., Vol. 73 (May, 1948), p. 1126.

3–28   Middleton, D., "Spurious Signals Caused by Noise in Triggered Circuits." Jour. Appl. Phys., Vol. 19 (September, 1948), pp. 817–830.

3–29   Lehan, F. W., Noise Effects in FM-FM Telemetering. Memo No. 4-47, Jet Propulsion Lab., February 14, 1949.

3–30   Macfarlane, G. G., "On the Energy-Spectrum of an Almost Periodic Succession of Pulses." Proc. IRE, Vol. 37 (October, 1949), pp. 1139–1143.

3–31   Lebenbaum, M. T., "Design Factors in Low-Noise Figure Input Circuits." Proc. IRE, Vol. 38 (January, 1950), pp. 75–80; (see also p. 539 (May, 1950) for correction of table i; also Hudson, A. C., "Note on Low-Noise Figure Input Circuits." (June, 1950), pp. 684–685.

3–32   Middleton, D., On the Distribution of Energy in Noise and Signal Modulated Waves. Tech. Report #99, Cruft Lab., March 1, 1950.

3–33   Marshall, J., "Low-Noise FM Front End." Rad. and Electr., Vol. 21 (June, 1950), pp. 58–63.

3–34   Oswald, J., "Signals with Limited Spectra and their Transformations." Câbles et Trans. (Paris), Vol. 4 (July, 1950), pp. 197–215 (detailed mathematical treatment).

3–35   Dwork, B. M., "Detection of a Pulse Superimposed on Fluctuation Noise." Proc. IRE, Vol. 38 (July, 1950), pp. 771–774.

3–36   Middleton, D., "The Effect of a Video Filter on the Detection of Pulsed Signals in Noise." Jour. Appl. Phys., Vol. 21 (August, 1950), pp. 734–740.

3–37   Harrington, J. V., and Rogers, T. F., "Signal-to-Noise Improvement through Integration in a Storage Tube." Proc. IRE, Vol. 38 (October, 1950), pp. 1197–1203.

3–38   Wilmotte, R. M., "Interference Caused by More than One Signal." Proc. IRE, Vol. 38 (October, 1950), pp. 1145–1150.

3–39   Hanse, H., The Optimization and Analysis of Systems for the Detection of Pulsed Signals in Random Noise. MIT Doctoral Dissertation, January, 1951.

3–40   Bieber, C. F., and Clark, H. L., Signal-to-Noise Ratio as a Function of Optical Systems Parameters. Report #3803, United States Naval Research Laboratory, January 9, 1951.

3–41   Jones, S., "The Determination of the Best Response of a Servo When Noise is Present with the Error Signal." TRE Jour. (England) (January, 1951), pp. 1–31.

3–42   Wulfsberg, K. N., Signal-to-Noise Improvement through Integration in a High-Q Filter. Report #ATI 93917, USAF Research Laboratory, January, 1951.

3–43   Oswald, J., "Random Signals with Limited Spectra." Câbles et Trans. (Paris), Vol. 5 (April, 1951), pp. 158–177 (theoretical treatment).

3–44   Lingard, A., Noise and Varying Parameter in Fixed-Gun Fighter Fire Control. MIT EE Doctoral Thesis, May 1951; Instru. Lab. Report #T-7.

3–45   Burgess, R. E., "The Rectification and Observation of Signals in the Presence of Noise." Phil. Mag., Vol. 42 (May, 1951), pp. 475–503.

3–46   Middleton, D., "The Distribution of Energy in Randomly Modulated Waves." Phil. Mag., Vol. 42 (July, 1951), pp. 689–787.

3-47 Lerner, R. M., "The Effect of Noise on the Frequency Stability of a Linear Oscillator." *Proc. NEC* (Chicago), Vol. 7 (1951), pp. 275–286.

3-48 Middleton, D., "On the Distribution of Energy in Noise and Signal-Modulated Waves: Part I—Amplitude Modulation." *Quart. Appl. Math.*, Vol. 9 (January, 1952), pp. 337–354; (see also Cruft Lab. Tech. Report #99, March, 1950; same title).

3-49 Booton, R. C., Jr., and Siefert, W. W., *The Effect of Noise on Limited Acceleration Homing Missiles.* Report #60, MIT Dyn. Anal. and Control Laboratory, 1952.

3-50 Booton, R. C., Jr., *Nonlinear Control Systems with Statistical Inputs.* Report #R-61, MIT Dyn. Anal. and Control Laboratory, March, 1952.

3-51 Smith, O. J. M., "Separating Information from Noise." TRANS. IRE (Circuit Theory Group) (1952), pp. 81–100.

3-52 Middleton, D., "On the Distribution of Energy in Noise- and Signal-Modulated Waves: Part II—Simultaneous Amplitude and Angle Modulation." *Quart. Appl. Math.*, Vol. 10 (April, 1952), pp. 35–36.

3-53 Mellen, R. H., "Thermal-Noise Limit in the Detection of Underwater Acoustic Signals." *Jour. Acous. Soc. Am.*, Vol. 24 (September, 1952), pp. 478–480.

3-54 Jones, S., *The Determination of the Best Form of Response for a Servo when an Extraneous Random Disturbance is Present with the Error Signal.* Automatic & Manual Control. New York, Acad. Press, 1952, pp. 139–147, 147–157.

3-55 Fano, R. M., *Communication in the Presence of Additive Gaussian Noise.* Proc. Lond. Symp. (1952), and New York, Acad. Press.

3-56 Johnson, R. A., and Middleton, D., *Measurement of Auto- and Cross-Correlation Functions of Modulated Carrier and Noise Following a Non-Linear Device.* Proc. Lond. Symp. (1952), and New York, Academic Press.

3-57 Meyer-Eppler, W., *Exhaustion Methods of Selecting Signals from Noisy Backgrounds.* Proc. Lond. Symp. (1952), and New York, Acad. Press.

3-58 Middleton, D., *Statistical Criteria for the Detection of Pulsed Carriers in Noise.* Elec. Res. Div., USAF Cambridge Research Center, November, 1952.

3-59 Reich, E., and Swerling, P., "The Detection of a Sine Wave in Gaussian Noise." *Jour. Appl. Phys.*, Vol. 24 (March, 1953), pp. 289–296.

3-60 Reiger, S., "Error Probabilities of Binary Data Transmission Systems in the Presence of Random Noise," 1953 IRE Convention Record, Pt. 8—Info. Theory, pp. 72–79.

3-61 Middleton, D., "Statistical Criteria for the Detection of Pulsed Carriers in Noise. Part I." *Jour. Appl. Phys.*, Vol. 24 (April, 1953), pp. 371–378.

3-62 Middleton, D., "Statistical Criteria for the Detection of Pulsed Carriers in Noise. Part II." *Jour. Appl. Phys.*, Vol. 24 (April, 1953), pp. 379–391.

3-63 McCombie, C. W., "Fluctuation Theory in Physical Measurements." *Rep. Prog. Phys.*, Vol. 16 (1953), pp. 266–320.

3-64 Tucker, D. G., and Griffiths, J. W. R., "Detection of Pulse Signals in Noise." *Wire. Engr.*, Vol. 30 (November, 1953), pp. 264–273.

3-65 Booton, R. C., Jr., Seifert, W. W., and Mathews, M. V., *Nonlinear Servomechanisms with Random Inputs.* Report #70, MIT Dyn. Anal. & Control Laboratory, 1953.

3-66 Wilmotte, R. M., "Reception of an FM Signal in the Presence of a Stronger Signal in the Same Frequency Band, and other Associated Results." *Proc. IEE* (London), Vol. 101, Part III (March, 1954), pp. 69–75.

3-67 Mazelsky, B., "Extension of Power Spectral Methods of Generalized Harmonic Analysis to Determine Non-Gaussian Probability Functions of Random Input Disturbances and Output Responses of Linear Systems." *Jour. Aero. Sci.*, Vol. 21 (March, 1954), pp. 145–153 (see also, Tick, L. J., letters to *Editor*, same *Journal*).

3-68 Rochefort, J. S., "Matched Filters for Detecting Pulsed Signals in Noise." 1954 IRE Convention Record, Part 4.

3-69 Lampard, D. G., "The Minimum Detectable Change in the Mean Noise-Input Power to a Radio Receiver." *Proc. IEE* (London), part III, Vol. 101 (March, 1954), pp. 111–113; (see also IEE Monograph, November, 1953, *Proc. IEE*, Part IV (February, 1954)).

4. NOISE GENERATION AND MEASUREMENT

4-1 Brown, R., Englund, C. R., and Friis, H. T., "Radio Transmission Measurements." PROC. IRE, Vol. 11 (April, 1923), pp. 115–152.

4-2 Espenschied, L., "Methods for Measuring Interfering Noises." PROC. IRE, Vol. 19 (November, 1931), pp. 1951–1955.

4-3 Llewellyn, F. B., "A Rapid Method of Estimating the S/N Ratio of a High Gain Receiver." PROC. IRE, Vol. 19 (March, 1931), pp. 416–420.

4-4 Alger, P. L., "Progress in Noise Measurements." *Elec. Engr.*, Vol. 52 (November, 1933), p. 741.

4-5 Peterson, H. O., "A Method of Measuring Noise Levels on Short-Wave Radio-Telegraph Circuits." PROC. IRE, Vol. 23 (February, 1935), pp. 128–132.

4-6 Jansky, K. G., "An Experimental Investigation of the Characteristics of Certain Types of Noise." PROC. IRE, Vol. 27 (December, 1939), pp. 763–769.

4-7 Burrill, C. M., "Progress in the Development of Instruments for Measuring Radio Noise." PROC. IRE, Vol. 29 (August, 1941), pp. 433–442.

4-8 Fränz, K., "Contribution to the Calculation of the Relations between Signal Strength and Noise Strength at the Last Stage of Receivers." *Elek. Nach. Tech.*, Vol. 17 (1940), pp. 215–230; also, Vol. 19 (1942), pp. 285–287.

4-9 Ragazzini, J. R., "The Effect of Fluctuation Voltages on the Linear Detector." PROC. IRE, Vol. 30 (June, 1942), pp. 277–287.

4-10 Fränz, K., and Vellat, T., "The Influence of Carrier Frequency on the Noise Following Amplitude Limiters and Linear Rectifiers." *Elek. Nach. Tech.*, Vol. 20 (1943), pp. 183–189.

4-11 MacDonald, D. K. C., "A Note on Two Definitions of Noise Figure in Radio Receivers." *Phil. Mag.*, Vol. 35 (1944), pp. 386–395.

4-12 Friis, H. T., "Noise Figures of Radio Receivers." PROC. IRE, Vol. 32 (July, 1944), pp. 419–422; also, Vol. 33 (February, 1945), pp. 125–127.

4-13 Smith, J. E., "Theoretical Signal-to-Noise Ratios." *Electronics*, Vol. 19 (June, 1946), pp. 150–152.

4-14 Pettit, J. M., "Specification and Measurement of Receiver Sensitivity at the Higher Frequencies." PROC. IRE, Vol. 35 (March, 1947), p. 302.

4-15 Bell, R. L., "Linearity Range of Noise-Measuring Amplifiers." *Wire. Engr.*, Vol. 24 (April, 1947), pp. 119–122.

4-16 Cobine, J. D., and Curry, J. R., "Electrical Noise Generators." PROC. IRE, Vol. 35 (Sept., 1947), pp. 875–880.

4-17 van der Ziel, A., "Method of Measurement of Noise Ratios and Noise Factors." *Philips Res. Rep.*, Vol. 2 (October, 1947), pp. 321–330.

4-18 Spratt, H. G. M., "Noise and its Measurement." *Elec. Rev.* (London), Vol. 144 (April 8, 1949), pp. 565–567.

4-19 Noday, G., "Technique of Noise Measurement for UHF Receivers." *Ann. Radioélect.*, Vol. 4 (July, 1949), pp. 257–260.

4-20 Eisengrein, R. H., *Measurement and Analysis of Noise in a Fire-Control System.* MIT Servomech Laboratory, (October, 1949).

4-21 Fowler, C. J., and Nicholson, F. T., *Progress Report No. 14 of Investigation of the Measurement of Noise.* Univ. of Penna., NOBSR #39269, October, 1949.

4-22 Jastrum, P. S., and McCouch, G. P., "A Video-Frequency Noise-Spectrum Analyzer." PROC. IRE, Vol. 37 (October, 1949), pp. 1127–1133.

4-23 Mumford, W. W., "A Broad-band Microwave Noise Source." *Bell Sys. Tech. Jour.*, Vol. 28 (October, 1949), pp. 608–618.

4-24 Meister, H., "A Thermal-Noise Generator for Low-Frequency Tests." *Tech. mitt. schweiz. Telegr.-Teleph. verw.*, Vol. 28 (August 1, 1950), pp. 320–324 (in French and German).

4-25 Harwood, H. D., and Shorter, D. E. L., "A High-Level Noise Source for the Audio-Frequency Band." *Jour. Sci. Instr.*, Vol. 27 (September, 1950), pp. 250–251.

4-26 Boff, A. F., "A Set for Noise and Distortion Tests on Carrier and Broadcast Systems." *Marconi Rev.*, Vol. 13 (3rd Quarter 1950), pp. 110–118.

4-27 Smith, R. A., "The Relative Advantages of Coherent and Incoherent Detectors, A Study of their Output Noise Spectra under Various Conditions." *Proc. IEE* (London), Vol. 98, part IV (1951), pp. 43–54.

4-28 Burgess, R. E., "The Measurement of Fluctuation Noise by Diode and Anode-Band Voltmeters." *Proc. Phys. Soc.*, Vol. 64 (June 1, 1951), pp. 508–518.

4-29 Bennett, R. R., and Fulton, A. S., "The Generation and Measurement of Low-Frequency Random Noise." *Jour. Appl. Phys.*, Vol. 22 (September, 1951), pp. 1187–1191.

4-30 Cheek, R. C., and Moynihan, J. D., "A Study of Carrier-Frequency Noise on Power Lines: Parts 1 and 2." *Trans. AIEE*, Vol. 70 (1951), pp. 1127–1133, 1325–1334.

4-31 Peterson, A. P. G., "A Generator of Electrical Noise." *Gen. Rad. Rev.*, Vol. 26 (December, 1951), pp. 1–9.

4-32 Chinnock, E. L., "A Portable, Direct-Reading Microwave Noise Generator." *Proc. IRE*, Vol. 40 (February, 1952), pp. 160–164.

4-33 Nordby, K. S., "Amplifier Noise, Particularly in the Audio-Frequency Range." *Tech. mitt. schweiz. Telegr-.Teleph. verw.*, Vol. 30 (June 1, 1952), pp. 185–197.

4-34 Hamer, E. G., "Noise Performance of VHF Receivers." *Electronics Engr.*, Vol. 25 (February, 1953), pp. 68–71.

4-35 Deutsch, R., and Hance, H. V., "A Note on Receivers for Use in the Studies of Signal Statistics." 1953 IRE Convention Record, Part 8—Information Theory, pp. 7–13.

4-36 "Standards on Modulation Systems: Definition of terms, 1953." *Proc. IRE*, Vol. 41 (May, 1953), pp. 612–615; (see also "Standards on Electron Devices: Methods of Measuring Noise." *Proc. IRE*, Vol. 41 (July, 1953), pp. 890–896).

4-37 Chavasse, P., and Pimonow, L., "Comparison of Microphone Calibrations using Pure Tones or using a Noise Source." *Ann. Télécommun.*, Vol. 8 (August/September, 1953), pp. 267–270.

4-38 Bess, L., "Possible Mechanism for 1/f Noise Generation in Semiconductor Filaments." *Phys. Rev.*, Vol. 91 (September, 1953), p. 1569.

4-39 Sherman, S., and Lakatos, A., *Noise Generation for Analog Simulation.* Symp. III, Simul. and Comput. Techs., BuAer and USNADC, October 12, 1953, pp. 155–192.

4-40 Bennett, R. R., "Analogue Computing Applied to Noise Studies." *Proc. IRE*, Vol. 41 (October, 1953), pp. 1509–1513.

4-41 Houlding, N., "Valve and Receiver Noise Measurement at UHF." *Wire. Engr.*, Vol. 31 (January, 1954), pp. 15–26.

4-42 Bernstein, R., Bickel, H., and Brookner, E., "A Generator of Uniformly Distributed Random Noise." 1954 IRE Convention Record—Part 10.

4-43 Winter, D. F., "A Gaussian Noise Generator for Frequencies down to 0.001 Cycles per Second." 1954 IRE Convention Record—Part 4.

5. Impulsive Type Noise

5-1 Landon, V. D., "Impulse Noise in FM Reception." *Electronics*, Vol. 14 (February, 1941), pp. 26–30, 73–76.

5-2 Smith, D. B., and Bradley, W. E., "The Theory of Impulsive Noise in Ideal FM Receivers." *Proc. IRE*, Vol. 34 (October, 1946), pp. 743–751.

5-3 Tellier, J. C., "An Analysis of the Behavior of a Limiter-Discriminator FM Detector in the Presence of Impulse Noise." *Proc. NEC* (Chicago), Vol. 3 (1947), pp. 680–696.

5-4 Stumpers, F. L. H. M., "On the Calculation of Impulse-Noise Transients in Frequency-Modulation Receivers." *Philips Res. Rep.*, Vol. 2 (December, 1947), pp. 468–474.

5-5 Haeff, A. V., "On the Origin of Solar Radio Noise." *Phys. Rev.*, Vol. 75 (May 15, 1949), pp. 1546–1551.

5-6 Pawsey, J. L., and Yabsley, D. E., "Solar Radio-Frequency Radiation of Thermal Origin." *Aust. Jour. Sci. Res.*, Ser. A, Vol. 2 (June, 1949), pp. 198–213.

5-7 Rogers, D. C., "Suppressing Impulse Noise." *Wireless World*, Vol. 55 (November, 1949), pp. 489–492.

5-8 Gerson, N. C., "Noise Levels in the American Sub-Arctic." *Proc. IRE*, Vol. 38 (August, 1950), pp. 905–916.

5-9 Wallace, P. R., "Interpretation of the Fluctuating Echo from Randomly Distributed Scatterers: Part II." *Canad. Jour. Phys.*, Vol. 31 (September, 1953), pp. 995–1009.

5-10 Lapin, S. P., and Suran, J. J., "Transient Response of Selective Networks and Impulsive Noise in Narrow Band FM Receivers." 1954 IRE Convention Record—Part 8.

6. Modulation and Noise

6-1 Armstrong, E. H., "A Method of Reducing Disturbances in Radio-Signalling by a Method of Frequency Modulation." *Proc. IRE*, Vol. 24 (May, 1936), pp. 689–740.

6-2 Crosby, M. G., "Frequency Modulation Noise Characteristics." *Proc. IRE*, Vol. 25 (April, 1937), pp. 472–514.

6-3 Landon, V. D., "Noise in Frequency Modulation Receivers." *Wireless World*, Vol. 47 (June, 1941), pp. 156–158.

6-4 Goldman, S., "Noise and Interference in Frequency Modulation." *Electronics*, Vol. 14 (August, 1941), pp. 37–42.

6-5 Sorbacher, R. I., and Edson, W. A., "Tubes Employing Velocity Modulation." *Proc. IRE*, Vol. 31 (August, 1943), p. 439.

6-6 Chireix, H., "Determination of Noise Power and S/N Ratio for the Case of Simplex or Multiplex Radio Transmission on Ultra-Short Waves by (A) Amplitude- or Duration-Modulated Pulses; (B) Frequency-Modulated Pulses." *Ann. Radioeléct.*, Vol. 1 (July, 1945), pp. 55–64.

6-7 Marchand, N., "Interference in Frequency Modulation." *Commun.*, Vol. 26 (February, 1946), pp. 38–40, 42.

6-8 Ashby, R. M., Martin, F. W., and Lawson, J. L., *Modulation of Radar Signals from Airplanes.* Report #914, MIT Rad. Laboratory, March 28, 1946.

6-9 Roddam, R., "Noise and Pulse Modulation." *Wire. World*, Vol. 52 (October, 1946), pp. 327–329.

6-10 Gladwin, A. S., "Energy Distribution in the Spectrum of a Frequency Modulated Wave: Part 2." *Phil. Mag.*, Vol. 38 (April, 1947), pp. 229–251.

6-11 Moskowitz, S., and Grieg, D. D., "Noise-suppression Characteristics of Pulse Modulation." *Elec. Commun.* (London), Vol. 24 (June, 1947), pp. 271–272; (see also Vol. 26 (1949), pp. 46–51).

6-12 Fitch, E., "The Spectrum of Modulated Pulses." *Jour. IEE* (London), Part IIA, Vol. 94 (1947), pp. 556–564.

6-13 Nicholson, M. G., "Comparison of Amplitude and Frequency Modulation." *Wire. Engr.*, Vol. 24 (July, 1947), pp. 197–208.

6-14 Nowotny, W., "Noise Reduction with Bandwidth in the Principal Modulation Systems." *Elektrotech. und Maschin.*, Vol. 64 (July/August, 1947), pp. 116–125.

6-15 Libois, L. J., "Signal-to-Noise Ratio in Different Methods of Radio Transmission. Spectrum of Pulse Modulation." *Onde Élec.*, Vol. 27 (November, 1947), pp. 411–425.

6-16 Goldman, S., "Some Fundamental Considerations Concerning Noise Reduction and Range in Radar and Communication." *Proc. NEC* (Chicago), Vol. 3 (1947), p. 191; (full paper, *Proc. IRE*, Vol. 36 (May, 1948), pp. 584–594).

6-17 Moskowitz, S., and Grieg, D. D., "Noise-suppression Characteristics of Pulse-Time Modulation." *Proc. IRE*, Vol. 36 (April, 1948), pp. 446–450.

6-18 Middleton, D., *Rectification of Sinusoidally Modulated Carrier in Random Noise.* Tech. Report #45, Cruft Laboratory, July, 1948 (see 6–28).

6-19 Stumpers, F. L. H. M., "Noise in a Pulse-Frequency-Modulation System." *Philips Res. Rep.*, Vol. 3 (August, 1948), pp. 241–254.

6-20 Moss, S. H., "Frequency Analysis of Modulated Pulses." *Phil. Mag.*, Vol. 39 (September, 1948), pp. 663–691.

6-21 Stumpers, F. L. H. M., "Theory of Frequency-Modulation Noise." *Proc. IRE*, Vol. 36 (September, 1948), pp. 1081–1092.

6-22 Håård, B., "Signal-to-Noise Ratios in Pulse Modulation Systems." *Ericsson Tech.*, No. 47 (1948) (in English).

6-23 Middleton, D., "Rectification of a Sinusoidally Modulated Carrier in the Presence of Noise." *Proc. IRE*, Vol. 36 (December, 1948), pp. 1467–1477.

6-24 Bennett, W. R., "Noise in PCM Systems." *Bell Tele. Lab. Rec.*, Vol. 26 (December, 1948), pp. 495–499.

6-25 Fubini, E. G., and Johnson, D. C., "Signal-to-Noise Ratio in AM Receivers." *Proc. IRE*, Vol. 36 (December, 1948), pp. 1461–1466; (see also Airborne Instr. Labs., Inc. Report #159-1, March, 1948).

6-26 Blachman, N. M., "The Demodulation of a FM Carrier and Random Noise by a Limiter and Discriminator." *Jour. Appl. Phys.*, Vol. 20 (January, 1949), pp. 38–47.

6-27 Hall, A. C., and Seifert, W. W., *Noise in Nonlinear Servo Systems*. Report #AL-930, North Amer. Avia., February, 1949, pp. 85–113.

6-28 Clavier, A. G., Panter, P. F., and Dite, W., "Signal-to-Noise Ratio Improvement in a PCM System." PROC. IRE, Vol. 37 (April, 1949), pp. 355–359.

6-29 Middleton, D., "The Spectrum of Frequency-Modulated Waves after Reception in Random Noise: Parts 1 and 2." *Quart. Appl. Math.*, Vol. 7 (July, 1949), pp. 129–174; Vol. 8 (April, 1950), pp. 59–80; (see also Cruft Lab. Tech. Reports #33, March 8, 1948, and 62, Nov., 1948).

6-30 Middleton, D., "On the Theoretical Signal-to-Noise Ratios in FM Receivers: A Comparison with Amplitude Modulation." *Jour. Appl. Phys.*, Vol. 20 (July, 1949), p. 724; (corrections to Vol. 20 (April, 1949), pp. 334–351). (See also Cruft Lab. Tech. Report #38, June, 1948.)

6-31 Runge, W., "Comparison of Signal/Noise Ratios of Modulation Arrangements." *Arch. Elek. Übertragung.*, Vol. 3 (August, 1949), pp. 155–159.

6-32 Kettel, E., "Noise Factor for Communication by Code Modulation." *Arch. Elek. Übertragung*, Vol. 3 (August, 1949), pp. 161–164.

6-33 Panter, P. F., and Dite, W., "Signal-to-Noise Improvement in a Pulse-Count-Modulation System." *Elec. Commun.*, Vol. 26 (September, 1949), pp. 257–262.

6-34 Blachman, N. M., "The Demodulation of a Frequency-Modulated Carrier and Random Noise by a Discriminator." *Jour. Appl. Phys.*, Vol. 20 (October, 1949), pp. 976–983; (see also Cruft Lab. Tech. Report #31, March 5, 1948).

6-35 Kretzmer, E. R., "An Application of Auto Correlation Analysis." *Jour. Math. Phys.*, Vol. 29 (October, 1950), pp. 179–190.

6-36 Libois, L. J., "Background Noise and Distortions in Code Modulation." *Câbles et Trans.* (Paris), Vol. 6 (January, 1952), pp. 65–79.

6-37 Bittel, H., "Characteristics of Noises and Noise Voltages." *Z. angew. Phys.*, Vol. 4 (April, 1952), pp. 137–146.

6-38 Stumpers, F. L. H. M., "Signal/Noise Ratio for Various Modulation Systems." *Tidjschr. ned. Radiogenoot.*, Vol. 17 (September/November, 1952), pp. 249–260.

6-39 Bordewijk, J. L., "Intermodulation Noise." *Tidjschr. ned Radiogenoot.*, Vol. 17 (September/November, 1952), pp. 261–279.

6-40 Goldman, S., *Information Theory of Noise Reduction in Various Modulation Systems*. Proc. Lond. Symp., New York, Academic Press, 1952.

6-41 Abbott, W. R., "Separating Signal from Noise in a PCM System." *ORDB Symp. on Info. Theory Appl. to Guided Missile Problems* (February 2, 1953), pp. 237–239.

6-42 Page, R. M., "Comparison of Modulation Methods." 1953 IRE Convention Record, Part 8—Info. Theory, pp. 15–20, 22–25.

6-43 George, S. F., "Appendices to 'Comparison of Modulation Methods.'" 1953 IRE Convention Record, Part 8—Info. Theory, pp. 20–25.

6-44 Fontanellaz, G., "The Reception of Weak Amplitude-Modulated Signals with Linear Detection." *Tech. mitt. schweiz. Telegr.-Teleph. verw.*, Vol. 31 (July 1, 1953), pp. 177–181 (in German).

6-45 Blachman, N. M., "A Comparison of the Informational Capacities of Amplitude- and Phase-Modulation Communication Systems." PROC. IRE, Vol. 41 (June, 1953), pp. 748–759.

6-46 van de Weg, H., "Quantizing Noise of a Single Integration Delta Modulation System with an *N*-digit Code." *Philips Res. Rep.*, Vol. 8 (October, 1953), pp. 367–385.

6-47 Youla, D. C., "The Use of the Method of Maximum Likelihood in Estimating Continuous-Modulated Intelligence which has been Corrupted by Noise." TRANS. IRE—Information Theory, (March, 1954), pp. 90–105.

6-48 Deutsch, R., "Detection of Modulated Noise-like Signals." TRANS. IRE—Information Theory (March, 1954), pp. 106–122.

7. RADAR APPLICATIONS

7-1 Bruce, E., "Development in Short-Wave Directive Antennas." PROC. IRE, Vol. 19 (August, 1931), pp. 1406–1434.

7-2 Goudsmit, S. A., *Comparison between Signal and Noise*. Report #43-21, MIT Radiation Laboratory, January 29, 1943.

7-3 Norton, K. A., *Maximum Range of a Radar Set*. US Signal Corps ORG-P-9-1, February, 1943.

7-4 North, D. O., *An Analysis of the Factors which Determine Signal-Noise Discrimination in Pulsed Carrier Systems (Radar)* RCA Report #PTR-6-C, June, 1943.

7-5 Eaton, T. T., and Wolff, I., *An Experimental Investigation of the Factors which Determine S/N Discrimination in Pulse Radar Systems—in Particular the Effect of Integration*. RCA Report #PTR-7-C, June, 1943.

7-6 Sutro, P. J., *The Theoretical Effect of Integration on the Visibility of Weak Signals through Noise*. OSRD Report #411-77, February 4, 1944.

7-7 Bloch, F., Hammermesh, M., and Philips, M., *Return Cross Sections from Random Oriented Resonant Half-Wavelength Chaff*. Report #411-TM-127, Radio Research Laboratory, June 19, 1944.

7-8 Middleton, D., and Sutro, P. J., "Analysis of a Possible A/J System Against Window." Report #411-128, Radiation Research Laboratory, December 9, 1944.

7-9 Lewis, W. B., "Radar Receivers." *Jour. IEE* (London), Vol. 93, No. 1 (1946), pp. 272–279.

7-10 Haeff, A. V., "Minimum Detectable Radar Signal and its Dependence upon Parameters of Radar Systems." PROC. IRE, Vol. 34 (November, 1946), pp. 857–861.

7-11 Berkner, L. V., "Naval Airborne Radar." PROC. IRE, Vol. 34 (September, 1946), pp. 671–707.

7-12 Van Vleck, J., and Middleton, D., "A Theoretical Comparison of the Visual, Aural, and Meter Reception of Pulsed Signals in the Presence of Noise." *Jour. Appl. Phys.*, Vol. 17 (November, 1946), pp. 940–971.

7-13 DeLano, R. H., *A Summary of Elementary Probability Relations and of Some Theory Relating Specifically to the Analysis of Random Noise*. Hughes Aircraft Co. Tech. Memo. #140, March, 1947.

7-14 Busignies, H., and Dishal, M., "Relation between Bandwidth, Speed of Indication, and Signal/Noise Ratio in Radio Navigation and Direction Finding." *Elec. Commun.*, Vol. 24 (June, 1947), pp. 264–265.

7-15 DeLano, R. H., *Signal/Noise Relations in a Conventional Simple Phase Detector with Noise-free Reference*. Hughes Aircraft Co. Tech. Memo. #166, August 4, 1947.

7-16 Norton, K. A., and Omberg, A. C., "The Maximum Range of a Radar Set." PROC. IRE, Vol. 35 (January, 1947), pp. 4–24; (September, 1947), pp. 927–931.

7-17 DeLano, R. H., *Reduction of Noise Fluctuation by Integration*. Hughes Aircraft Co. Tech. Memo, #182, October 16, 1947.

7-18 Marcum, J. I., *A Statistical Theory of Target Detection by Pulsed Radar*. RAND Report #RA 15061, December 1, 1947 (mathematical appendix: #R-113, July, 1948).

7-19 M. W. Kellogg Co., *System and Radar Noise Analysis of Two-Beam Command Guidance for MX-800*. ORDB Report #129, December 1, 1947.

7-20 Goldman, S., *Some Fundamental Considerations Concerning Noise Reduction and Range in Radar and Communications*. MIT Tech. Report #32, December 15, 1947 (see also PROC. IRE, Vol. 36 (May, 1948), pp. 584–594: 6–16.

7-21 Levy, M., "Signal/Noise Ratio in Radar." *Wire. Engr.*, Vol. 24 (December, 1947), pp. 349–352; Vol. 25 (March, 1948), pp. 97–98; Vol. 25 (July, 1948), pp. 236–237.

7-22 Meade, J. E., *Radar Tracking Noise as it is Affected by Lobing Methods*. Report #AL-930, North Am. Avia., February, 1949, pp. 31–41.

7-23 Budenbom, H. T., *Monopulse Radar Noise*. Report #AL-930, North Am. Avia., February, 1949, pp. 43–66.

7-24 Hastings, A. E., and Meade, J. E., *Improvement of Radar Tracking*. USNRL Report #3424, February 24, 1949.

7-25 Goulder, M. E., *Noise Problems of FM/CW Radar as Applied to Guided Missiles*. Report #AL-930, North Am. Avia., February, 1949, pp. 249–259.

7–26 Busignies, H., and Dishal, M., "Some Relations between Speed of Indication, Bandwidth, and Signal-to-Noise Ratio in Radio Navigation and Direction Finding." *Elect. Commun.*, Vol. 26 (September, 1949), pp. 228–242; [see also: Proc. IRE, Vol. 37 (May, 1949), pp. 478–488; (September, 1949), p. 1096].

7–27 Hastings, A. E., *Methods of Obtaining Amplitude-Frequency Spectra.* USNRL Report #3466, May 16, 1949.

7–28 Hastings, A. E., *Amplitude Fluctuation in Radar Echo Pulses.* USNRL Report #3487, June 21, 1949.

7–29 DeLano, R. H., "Signal-to-Noise Ratios of Linear Detectors." Proc. IRE, Vol. 37 (October, 1949), pp. 1120–1126.

7–30 Woodward, P. M., "Theory of Radar Information." *Proc. Lond. Symp.* (1950); also Trans. IRE—Information Theory (February, 1953), pp. 108–113, 182–186.

7–31 Muchmore, R. B., *Calculation of Range of Automatic Tracking Radar. Advanced Radar Target Seeker.* Tech. Memo. #193, Hughes Aircraft Co., 1950.

7–32 DeLano, R. H., *Angular Scintillation of Radar Targets.* Tech. Memo. #233, Hughes Aircraft Co., April 24, 1950.

7–33 Muchmore, R. B., *et al.*, *Guidance and Control-1-Summary Report.* Tech. Memo. #235, Hughes Res. and Dev. Lab., July 1, 1950.

7–34 Cunningham, W. J., May, J. C., and Skalnik, J. G., "Integration-Noise Reducer for Radar." *Electronics*, Vol. 23 (September, 1950), pp. 76–78.

7–35 Muchmore, R. B., *Aircraft Scintillation Studies.* Symp. in Reflect. Studies, ORDB, September 27, 1950.

7–36 Woodward, P. M., and Davies, I. L., "A Theory of Radar Information." *Phil. Mag.*, Vol. 41 (October, 1950), pp. 1001–1017.

7–37 DeLano, R. H., *Angular Scintillation of Radar Targets with Monopulse and Interferometer Target Seekers.* Tech. Memo. #257, Hughes Aircraft, November 1, 1950.

7–38 Meade, J. E., Hastings, A. E., and Gerwin, H. L., *Noise in Tracking Radar.* USNRL Report #3759, November 15, 1950.

7–39 Brockner, C. E., "Angular Jitter in Conventional Conical-Scan, Automatic-Tracking Radar Systems." Proc. IRE, Vol. 39 (January, 1951), pp. 51–55.

7–40 Kaplan, S. M., and McFall, R. W., "The Statistical Properties of Noise Applied To Radar Range Performance." Proc. IRE, Vol. 39 (January, 1951), pp. 56–60; (see also Vol. 40 (April, 1952), pp. 487–489).

7–41 Davies, I. L., *Theory of Radar Information—Existence Probabilities for Signals in Noise.* Tech. Note #109, Telecommun. Res. Est. (England), February 9, 1951.

7–42 Ross, A. W., "Visibility of Radar Echoes." *Wire. Engr.*, Vol. 28 (March, 1951), pp. 79–92.

7–43 Schwartz, M., *Statistical Approach to the Automatic Search Problem.* Doctoral Dissertation, Harvard, June, 1951.

7–44 Ruina, J., *The Effects of Noise on Range Tracking Systems.* Report #R-247-51, Microwave Res. Inst., June 13, 1951.

7–45 Gabler, R. T., *An Estimate of 'Glint' Noise from Radar Tracking Data.* RAND Report #683, September 10, 1951.

7–46 Spencer, R. E., "The Detection of Pulse Signals near the Noise Threshold." *Jour. Brit. IRE*, Vol. 11 (October, 1951), pp. 435–454.

7–47 Angel, J. B., *Errors in Angle Radar Systems Caused by Complex Targets.* Report #77, MIT Dyn. Anal. and Contr. Lab., October, 1951.

7–48 Schwartz, L. S., "Second-Detector Signal-to-Noise Improvement." *Proc. NEC* (Chicago), Vol. 7 (1951), pp. 141–150; (see also *Tele-Tech.*, Vol. 11 (October, 1952), pp. 56, 107).

7–49 Woodward, P. M., "Information Theory and the Design of Radar Receivers." Proc. IRE, Vol. 39 (December, 1951), pp. 1521–1524.

7–50 Hastings, A. E., Meade, J. E., and Gerwin, H. L., *Noise in Tracking Radars Part II—Distribution Functions and Further Power Spectra.* USNRL Report #3929, January 16, 1952 (see also 7–38).

7–51 Muchmore, R. B., *Theoretical Scintillation Spectra.* Tech. Memo. #271, Hughes Aircraft, March 1, 1952.

7–52 Davies, I. L., "On Determining the Presence of Signals in Noise." *Proc. IEE* (London), Part III, Vol. 99 (March, 1952), pp. 45–51.

7–53 Bennett, R. R., and Mathews, W. E., *Analytical Determination of Miss Distances for Linear Homing Navigation Systems*, Tech. Memo. #260, Hughes Air., March 31, 1952.

7–54 George, T. S., "Fluctuations of Ground Clutter in Airborne Radar Equipment." *Proc. IEE* (London), Part IV, Vol. 99 (April, 1952), pp. 92–99.

7–55 George, T. S., "Fluctuations of Ground Clutter Return in Airborne Radar." *Proc. IEE* (London), Part III, Vol. 99 (May, 1952), pp. 160–161 (summary only).

7–56 Tiberio, U., "Echo and Noise in a Radar System Subjected to Deliberate Interference." *Alta. Freq.*, Vol. 21 (June, 1952), pp. 137–151.

7–57 Spetner, L. M., "Some Notes on the Detected Spectrum of Radar Target Noise." Report #CF-1872, Johns Hopkins Appl. Phys. Lab., September 30, 1952.

7–58 Middleton, D., *Statistical Methods for the Detection of Pulsed Radar in Noise.* Proc. Lond. Symp., New York, Academic Press, 1952.

7–59 Graham, J. W., "Analysis of Angular Scintillation of Radar Echos." (1952).

7–60 Muchmore, R. B., *Review of Scintillation Measurements.* Tech. Memo. #272, Hughes Aircraft, December 1, 1952.

7–61 Carlton, A. G., *Optimum Guidance Filtering.* Symp. Info. Theory Applic. to Guided Missiles Problems, ORDB, pp. 1–5, February 2, 1953.

7–62 Davis, H., "Radar Problems and Information Theory." 1953 IRE Convention Record, Part 8—Info. Theory, pp. 39–47.

7–63 Bennett, R. R., *Analytical and Simulation Techniques Applicable to the Problem of Noise in Missile Guidance Systems.* Symp. Info. Theory Applic. to Guided Missile Prob., ORDB, pp. 77–87, February 2, 1953.

7–64 Seifert, W. W., *Simulation Techniques in a Multiple-Target Study Involving Random Signals.* Symp. Info. Theory Applic. to Guided Missile Prob., ORDB, pp. 101–112, February 2, 1953.

7–65 Zimmerman, H. J., and Armstrong, D. B., *Aircraft Echo Characteristics.* Symp. Info. Theory Applic. to Guided Missile Prob., ORDB, pp. 309–318, February 2, 1953.

7–66 Meade, J. E., *Target Noise Characteristics.* Symp. Info. Theory Applic. to Guided Missile Prob., ORDB, pp. 339–352, February 2, 1953.

7–67 DeLano, R. H., *A Theory of Target Glint or Angular Scintillation in Radar Tracking.* Symp. Info. Theory Applic. to Guided Missile Prob., ORDB, February 2, 1953, pp. 353–390; (see also 1953 IRE Conv. Rec., Part I, pp. 13–19, March, 1953).

7–68 Wright, J. R., *Aircraft Target Noise Studies.* Symp. Info. Theory Applic. to Guided Missile Prob., ORDB, pp. 391–398, February 2, 1953.

7–69 Steward, M. H., *Scintillation Noise in Interceptor Radars.* Symp. Info. Theory Applic. to Guided Missile Prob., ORDB, pp. 401–412, February 2, 1953.

7–70 Muchmore, R. B., *Survey of Radar Scintillation Measurements.* Symp. Info. Theory Applic. to Guided Missile Prob. ORDB, pp. 443–452, February 2, 1953.

7–71 Peterson, W. W., and Birdsall, T. G., *The Theory of Signal Detectability, Parts I & II.* Report #13, U. Mich. Engr. Res. Inst., June, 1953.

7–72 Loeb, J., "The London Conference, Sept. 1952: Part 2—Radar and Television." *Onde Élect.*, Vol. 33 (July, 1953), pp. 478–481 (reviews some papers).

7–73 Stone, W. M., "On the Statistical Theory of Detection of a Randomly Modulated Carrier." *Jour. Appl. Phys.*, Vol. 24 (July, 1953), pp. 935–939.

7–74 Middleton, D., "Noise in CW Magnetrons." *Jour. Appl. Phys.*, Vol. 24 (August, 1953), p. 1065.

7–75 Urkowitz, H., "Filters for Detection of Small Radar Signals in Clutter." *Jour. Appl. Phys.*, Vol. 24 (August, 1953), pp. 1024–1031.

7–76 Floyd, G. F., *Radar Tracking System.* Symp. on Airborne Electronics, IRE-ORDB, August 18, 1953.

7–77 Illman, R. W., *Missile Guidance and Control.* Symp. on Airborne Electronics, IRE-ORDB, August 18, 1953.

7–78 DeLano, R. H., "A Theory of Target Glint or Angular Scintillation in Radar Tracking." Proc. IRE, Vol. 41 (December, 1953), pp. 1778–1784 (see 7–67).

7-79 Booton, R. C., Jr., "Nonlinear Control Systems with Random Inputs." TRANS. IRE (Circuit Theory), (March, 1954), pp. 9–18.

7-80 Davis, R. C., "Detectability of Random Signals in the Presence of Noise." TRANS. IRE (Information Theory) (March, 1954), pp. 52–62.

7-81 Slepian, D., "Estimation of Signal Parameters in the Presence of Noise." TRANS. IRE (Information Theory) (March, 1954), pp. 68–89.

8. NOISE, COMMUNICATION, AND FILTERING

8-1 Nyquist, H., "Certain Factors Affecting Telegraph Speed." *Bell Sys. Tech. Jour.*, Vol. 3 (April, 1924), p. 324.

8-2 Nyquist, H., "Certain Topics in Telegraph Transmission Theory." *AIEE Trans.*, Vol. 47 (April, 1928), p. 617.

8-3 Hartley, R. V. L., "The Transmission of Information." *Bell Sys. Tech. Jour.*, Vol. 7 (July, 1928), pp. 535–564.

8-4 Francis, V. J., and James, E. G., "Rectification of Signal and Noise." *Wire. Engr.*, Vol. 25 (1946), p. 16.

8-5 Stumpers, F. L. H. M., "On a Non-Linear Noise Problem." *Philips Res. Rep.*, Vol. 2 (August, 1947), pp. 241–259.

8-6 Blanc-Lapierre, A., and Fortet, R., "Extension of the Method of Filters and of Nonstationary Uncertain Functions." *Compt. Rend. Acad. Sci.* (Paris), Vol. 222 (1947), p. 1270.

8-7 Eggleston, P., and Karmack, W. C., "Bandwidth vs Noise in Communication Systems." *Electronics*, Vol. 21 (June, 1948), pp. 72–75.

8-8 Fink, D. G., "Bandwidth vs Noise in Communication Systems." *Electronics*, Vol. 21 (June, 1948), pp. 72–75.

8-9 Grant, E. F., *An Exposition of Norbert Wiener's Theory for the Prediction and Filtering of a Single Stationary Time Series.* Report #1136, Cambridge Field Sta., May, 1948.

8-10 Earp, C. W., "Relationship between Rate of Transmission of Information, Frequency Bandwidth, and Signal-to-Noise Ratio." *Elec. Commun.* (London), Vol. 25 (June, 1948), pp. 178–195.

8-11 Shannon, C. E., "A Mathematical Theory of Communication." *Bell Sys. Tech. Jour.*, Vol. 27 (July, October, 1948), pp. 379–423, 623–656.

8-12 Blackman, R. B., Bode, H. W., and Shannon, C. E., *Data Smoothing and Prediction in Fire-Control Systems.* ORDB Report #13 MGC 12/1, August 15, 1948, pp. 71–160.

8-13 Bubb, F. W., *Theory of Best Automatic Control Systems for Prediction and Smoothing.* GE Report TR-55377, August 30, 1948.

8-14 Cohen, R., *Some Analytical and Practical Aspects of Wiener's Theory of Prediction.* Tech. Report #69, MIT Res. Lab. of Electronics, 1948.

8-15 Dickey, F. R., Emslie, A. G., and Stockman, H., *Extraction of Weak Signals from Noise by Integration.* Cambridge Res. Lab. Report #E 5038 (September, 1948); (see also *Proc. NEC* (Chicago), Vol. 4 (1948), pp. 102–120.

8-16 Shannon, C. E., "Communication in the Presence of Noise." PROC. IRE, Vol. 37 (January, 1949), pp. 10–21.

8-17 Bubb, F. W., *Optimum Design of Linear Predictors with Regard to Noise Reduction.* Rept. AL-930, North Am. Avia., February, 1949, pp. 67–83.

8-18 Enequist, L. N., *On Some Integral Equations Arising in the Study of Noise in Linear Systems.* Rept. AL-930, North Am. Avia., February, 1949, pp. 231–248, Part 1.

8-19 Jordon, W. B., *Introduction to Basic Theory of Noise and Information.* Rept. AL-930, North Am. Avia., February, 1949, pp. 1–20, Part 1.

8-20 Landon, V. D., *Comparison of Modulation Types with Regard to Noise for Radio Communication Systems.* Rept. AL-930, North Am. Avia., February, 1949, pp. 23–29, Part 1.

8-21 Fano, R. M., *The Transmission of Information.* Tech. Rept. #65, MIT Res. Lab. of Electronics, March, 1949.

8-22 Zadeh, L., and Floyd, G. F., *An Extension of Wiener's Theory to Non-Stationary Time Series.* Rept. 8390, M. W. Kellogg Co., April, 1949.

8-23 Lee, Y. W., "Filtering and Prediction." (Lecture 5-AIEE-IRE) Theory of Communication Series, N,Y., May 9, 1949.

8-24 Tuller, W. G., "Theoretical Limitations on the Rate of Transmission of Information." PROC. IRE, Vol. 37 (May, 1949), pp. 468–478.

8-25 Laemmel, A. E, *General Theory of Communication.* Rept. R-208-49, Microwave Res. Inst., July, 1949.

8-26 Knudtzon, N., *Experimental Study of Statistical Characsritics of Filtered Random Noise.* Tech. Report #115, MIT Res. Lab. of Electronics, July, 1949.

8-27 Lee, Y. W., and Stutt, C. A., *Statistical Prediction of Noise.* Tech. Rept. #129, MIT Res. Lab. of Electronics, July 12, 1949; (see also *Proc. NEC*, Vol. 5 (1949), pp. 342–365.

8-28 Feldman, C. B., and Bennett, W. R., "Bandwidth and Transmission Performance." *Bell Sys. Tech. Jour.*, Vol. 28 (July, 1949), pp. 490–595.

8-29 Wiesner, J. B., "Statistical Theory of Communication." *Proc. NEC* (Chicago), Vol. 5 (1949), pp. 334–341.

8-30 Cherry, E. C., "A History of the Theory of Information." Proc. Lond. Symp., 1950; (see also TRANS. IRE (Information Theory), (February, 1953), pp. 22–43, 167–168).

8-31 Bode, H. W., and Shannon, C. E., "A Simplified Derivation of Linear Least Square Smoothing and Prediction Theory." PROC. IRE, Vol. 38 (April, 1950), pp. 417–425.

8-32 Zadeh, L. A., and Ragazzini, J. R., "An Extension of Wiener's Theory of Prediction." *Jour. Appl. Phys.*, Vol. 21 (July, 1950), pp. 645–655.

8-33 Lee, Y. W., Cheatham, Jr., T. P., and Wiesner, J. B., "Application of Correlation Analysis to the Detection of Periodic Signals in Noise." PROC. IRE, Vol. 38 (October, 1950), pp. 1165–1171; (see also PROC. IRE, Vol. 39 (September, 1951), pp. 1094–1096).

8-34 Costas, J. P., *Interference Filtering.* Tech. Rept. #185, MIT Res. Lab. of Electronics, March, 1951.

8-35 Malatesta, S., "Basis of Information Theory." *Alta. Frequenza*, Vol. 29 (June/August, 1951), pp. 128–159.

8-36 Booton, R. C., Jr., "An Optimization Theory for Time-varying Linear Systems with Non-stationary Statistical Inputs." PROC. IRE, Vol. 40 (August, 1952), pp. 977–981; (see also MIT, Dynamic Anal. Contr. Lab., Rept. #72, July, 1951).

8-37 Stutt, C. A., *Experimental Study of Optimum Filters.* Tech. Rept. #182, MIT Res. Lab. of Electronics, 1951.

8-38 Sullivan, A. W., and Barney, J. M., "A Non-Linear Statistical Filter." *Proc. NEC* (Chicago), Vol. 7 (1951), pp. 85–91.

8-39 Zadeh, L. A., and Miller, K. S., "Generalized Ideal Filters." *Jour. Appl. Phys.*, Vol. 23 (February, 1952), pp. 223–228.

8-40 Icole, J., and Oudin, J., "Time Analysis and Filtering." *Ann. Télécommun.*, Vol. 7 (February, 1952), pp. 99–108.

8-41 Woodward, P. M., and Davies, I. L., "Information Theory and Inverse Probability in Telecommunication." *Proc. IEE* (London), Vol. 99 (March, 1952), p. 37.

8-42 Davenport, W. B., Jr., Johnson, R. A., and Middleton, D., "Statistical Errors in Measurements on Random Time Functions." *Jour. Appl. Phys.*, Vol. 23 (April, 1952), pp. 377–388.

8-43 Jacot, H., "Theory of the Prediction and Filtering of Stationary Time Series According to Norbert Wiener." *Ann. Télécommun.*, Vol. 7 (May/August, 1952), pp. 241–249, 297–303, 325–335.

8-44 Malatesta, S., "Contribution to the Statistical Study of Communications." *Alta. Frequenza*, Vol. 21 (August/October, 1952), pp. 163–198.

8-45 Davis, R. C., "On the Theory of Prediction of Nonstationary Stochastic Processes." *Jour. Appl. Phys.*, Vol. 23 (September, 1952), pp. 1047–1053.

8-46 Fromageot, A., "Concept of Entropy in the Calculus of Probability." *Ann. Télécommun.*, Vol. 7 (September, 1952), pp. 388–396.

8-47 Zadeh, L. A., and Ragazzini, J. R., "Optimum Filters for the Detection of Signals in Noise." PROC. IRE, Vol. 40 (October, 1952), pp. 1223–1231.

8-48 Slattery, T. G., "The Detection of a Sine Wave in the Presence of Noise by the Use of a Non-Linear Filter." PROC. IRE, Vol. 40 (October, 1952), pp. 1232–1236.

8-49 Arthur, G. R., "A Note on the Approach of Narrow Band Noise After a Non-Linear Device to a Normal Probability Density." *Jour. Appl. Phys.*, Vol. 23 (October, 1952), pp. 1143–1144.

8-50 Pike, E. W., "A New Approach to Optimum Filtering." *Proc. NEC* (Chicago), Vol. 8 (1952), pp. 407–418.

8-51 DiToro, M. J., *Inverse Probability Applied to Pulse Communication Through a Randomly Time Variable and Noisy Transmission Medium.* Symp. on Information Theory Applications to Guided Missile Problems, Res. and Dev. Board, pp. 215–236, February 2, 1953.

8-52 Young, G. O., and Gold, B., "Noise Problems of Theoretical and Practical Interest." 1953 IRE Convention Record, Part 8—(Information Theory), pp. 2–6.

8-53 Zadeh, L. A., "Optimum Non-Linear Filters for the Extraction and Detection of Signals." 1953 IRE Convention Record, Part 8—(Information Theory), pp. 57–65.

8-54 Lehan, F. W., and Parks, R. J., "Optimum Demodulation." 1953 IRE Convention Record, Part 8—(Information Theory), pp. 101–103.

8-55 Berkowitz, R. S., "Optimum Linear Sloping and Filtering Networks." Proc. IRE, Vol. 41 (April, 1953), pp. 532–537.

8-56 Bell, D. A., "Wiener's Theory of Linear Filtering." *Wireless Enar.*, Vol. 30 (June, 1953), pp. 136–142.

8-57 Fromageot, A., "The London Conference, September, 1952: Part 1—Applications of the Theory of Information." *Onde Élect.*, Vol. 33 (July, 1953), pp. 473–478.

8-58 Young, G. O., and Gold, B., "Effect of Limiting on the Information Content of Noisy Signals." 1954 IRE Convention Record, Part 4.

8-59 White, W. D., "Information Losses in Regenerative Pulse-Code Systems." 1954 IRE Convention Record, Part 4.

8-60 Stateman, N. H., and Ritterman, M. B., "Theoretical Improvement in S/N of TV Signals by Equivalent Comb Filter Technique." 1954 IRE Convention Record, Part 4.

8-61 DiToro, M. J., "Time Varying Quasi-Linear Method of Speech Noise Suppression." 1954 IRE Convention Record, Part 4.

9. Statistical Theory

9-1 Wiener, N., "Harmonic Analysis of Irregular Motion." *Jour. Math. and Phys.*, Vol. 5 (1926), pp. 99–189.

9-2 Wiener, N., "Generalized Harmonic Analysis." *Acta Math.* (Stockholm), Vol. 55 (1930), pp. 117–258.

9-3 Landon, V. D., "A Study of the Characteristics of Noise." Proc. IRE, Vol. 24 (November, 1936), pp. 1514–1521.

9-4 Landon, V. D., "Distribution of Amplitude with Time in Fluctuation Noise." Proc. IRE, Vol. 29 (February, 1941), pp. 50–55; (see also Vol. 30 (September, 1942), pp. 425–429; (November, 1942), p. 526).

9-5 Wiener, N., *Response of Non-Linear Device to Noise.* Rept. #129, MIT Rad. Lab., April 6, 1942.

9-6 Wiener, N., *Statistical Method of Prediction in Fire Control.* Rept. #59, Nat. Def. Council, December 1, 1942.

9-7 Chandrasekhar, S., "Stochastic Problems in Physics & Astronomy." *Rev. Mod. Phys.*, Vol. 15 (Jan., 1943), pp. 1–89.

9-8 Hausz, W., *A Mathematical Treatment of Signal-to-Noise Ratio.* EMT #540, Gen. Elect. Co., May 22, 1943.

9-9 Uhlenbeck, G. E., *Theory of Random Processes.* Rept. #454, MIT Rad. Lab., October 15, 1943.

9-10 Hurwitz, H., and Kac, M., "Statistical Analysis of Certain Types of Random Functions." *Jour. Math. Phys.*, Vol. 23 (January, 1944), p. 195; (see also *Ann. Math. Statist.*, Vol. 15 (1944), p. 173).

9-11 Emslie, A. G., *Coherent Integration.* Rept. #103-5, MIT Rad. Lab., May 16, 1944.

9-12 Rice, S. O., "Mathematical Analysis of Random Noise." *Bell Sys. Tech. Jour.*, Vol. 23 (July, 1944), pp. 282–333; also Vol. 24 (January, 1945), pp. 46–157.

9-13 Wang, M. C., and Uhlenbeck, G. E., "On the Theory of the Brownian Motion." *Rev. Mod. Phys.*, Vol. 17 (1945), p. 323.

9-14 Bartlett, M. S., "On the Theoretical Specification and Sampling Properties of Autocorrelated Time Series." *Jour. Roy. Statist. Soc.*, Vol. 8 (January, 1946), pp. 27–41.

9-15 Stumpers, F. L. H. M., "Some Investigations on Oscillations with Frequency Modulation." *Philips* (Res. Lab.) *Tech. Rev.*, Vol. 8 (September, 1946), pp. 287–288.

9-16 Siegert, A. J. F., "On the First Passage Time Problem for a One-Dimensional Markoffian Gaussian Random Function." *Phys. Rev.*, Vol. 70 (September, 1946), p. 449.

9-17 Rice, S. O., "Fourier Series in Random Processes." *Phys. Rev.*, Vol. 69 (1946), pp. 676.

9-18 Domb, C., "The Resultant of a Large Number of Events of Random Phase." *Proc. Camb. Phil. Soc.*, Vol. 42 (October, 1946), pp. 245–299.

9-19 Hilton, P. J., "The Effect of Fluctuation Noise Interference on Pulse Distortion." *Phil. Mag.*, Vol. 37 (October, 1946), pp. 685–693.

9-20 Blanc-Lapierre, A., and Fortet, R., "On the Structure of Strictly Stationary Uncertain Functions with Totally Discontinuous Spectrum." *Compt. Rend. Acad. Sci.* (Paris), Vol. 222 (1946), p. 1155.

9-21 Cunningham, L. B., and Hynd, W. R., "Random Processes in Problems of Air Warfare." (Supplement), *Jour. Roy. Statist. Soc.*, Vol. 8 (1946), No. 1.

9-22 Vann, J. D., *A Statistical Approach to an Investigation of the Value of Integrators in Reducing the Minimum Detectable Signal.* Tech. Rept. #55315, G. E. Co., May 29, 1947.

9-23 Schwartz, L. S., *Statistical Methods in the Design and Development of Electronic Systems.* Rept. R-3111, US Nav. Res. Lab., July, 1947.

9-24 Crawford, J. E., *Reduction of Noise Fluctuation by Integration.* Gen. Elect. Co., 1947.

9-25 Blanc-Lapierre, A., "Notes on the 'Energy' Properties of Random Functions." *Compt. Rend. Acad. Sci.* (Paris), Vol. 225 (November 24, 1947), p. 982.

9-26 Bennett, W. R., "Distribution of the Sum of Randomly Phased Components." *Quart. Appl. Math.*, Vol. 5 (January, 1948), pp. 385–393.

9-27 Rice, S. O., "Statistical Properties of a Sine Wave Plus Random Noise." *Bell Sys. Tech. Jour.*, Vol. 27 (January, 1948), pp. 109–157.

9-28 Macnee, A. B., *An Electronic Differential Analyzer.* Tech. Rept. #90, MIT Res. Lab. of Electronics, 1948.

9-29 Wold, H. A., "On Prediction in Stationary Time Series." *Ann. Math. Statist.*, Vol. 19 (1948), pp. 558–567.

9-30 MacDonald, D. K. C., "Some Statistical Properties of Random Noise." *Proc. Camb. Phil. Soc.*, Vol. 45 (1949), p. 368; (see also *Phil. Mag.*, Vol. 40 (1949), p. 561).

9-31 Lee, Y. W., *Communication Applications of Correlation Analysis.* Symp. Appli. Autocorr. Analy. to Phys. Problems, ONR, pp. 4–23, June 3, 1949.

9-32 Middleton, D., *Noise and Non-Linear Communication Problems.* Symp. Appli. Autocorr. Analy. to Phys. Problems, ONR, pp. 24–46, June 3, 1949.

9-33 Rudnick, P., *A System for Recording & Analyzing Random Processes.* Symp. Appli. Autocorr. Analy., pp. 68–73, 1949.

9-34 Tukey, J. W., *The Sampling Theory of Power Spectrum Estimates.* Symp. Appli. Autocorr. Analy. to Phys. Problems, ONR, pp. 47–67, June 13, 1949.

9-35 Lee, Y. W., Cheatham, T. P., and Wiesner, J. B., *The Application of Correlation Functions in the Detection of Small Signals in Noise.* Tech. Rept. #141, MIT Res. Lab. of Electronics, October, 1949.

9-36 Rice, S. O., "Communication in the Presence of Noise—Probability of Error for Two Encoding Schemes." *Bell Sys. Tech. Jour.* Vol. 29 (January, 1950) pp. 60–93.

9-37 Costas, J. P., *Periodic Sampling of Stationary Time Series.* Tech. Rept. #156, MIT Res. Lab. Elect., May 16, 1950.

9-38 Lee, Y. W., and Wiesner, J. B., "Correlation Functions and Communication Applications." *Electronics*, Vol. 23 (June, 1950), pp. 86–92.

9-39 Corum, C. E., *Signal-to-Noise Improvement by Pulse Integration.* Rept. 3699, US Naval Res. Lab., July, 1950.

9-40 Stumpers, F. L. H. M., "On a First-Passage-Time Problem." *Philips Res. Rept.*, Vol. 5 (August, 1950), pp. 270–281.

9-41 Singleton, H. E., *A Digital Electronic Correlator.* Tech. Rept. #152, MIT Res. Lab. of Electronics, 1950.

9-42 MacDonald, D. K. C., "The Statistical Analysis of Electrical Noise." *Phil. Mag.*, Vol. 41 (August, 1950), pp. 814–181, 863.

9-43 Blanc-Lapierre, A., "Some Statistical Functions Useful for the Study of Background Noise." *Compt. Rend. Acad. Sci.* (Paris), Vol. 231 (September 18, 1950), pp. 566–567.

9-44 Lee, Y. W., *Detection of Periodic Signals in Random Noise by the Method of Correlation.* Tech. Rept. #157, MIT Res. Lab. Elect., 1950.

9–45 Lee, Y. W., *Application of Statistical Methods to Communication Problems.* Tech. Rept. #181, MIT Res. Lab. of Electronics, September, 1950.

9–46 Armstrong, D. B., *A Survey of the Theory of Random Noise; Its Behavior in Electronic Circuits.* Seminar Paper, Dept. of E. E., MIT, 1950.

9–47 MacDonald, D. K. C., "Fluctuations and the Theory of Noise." *Proc. Lond. Symp.* (1950), pp. 114–120, 187–189; (see also TRANS. IRE, Info. Theory (Feb., 1953)).

9–48 Bell, D. A., "The Autocorrelation Function." *Wireless Engr.,* Vol. 28 (January, 1951), pp. 31–32.

9–49 Callen, H. C., and Welton, T. A., "Irreversibility and Generalized Noise." *Phys. Rev.,* Vol. 83 (Jan., 1951) pp. 34–40.

9–50 Weinberg, L., and Kraft, L. G., *Experimental Study of Non-Linear Devices by Correlation Methods.* Tech. Rept. #178, MIT Res. Lab. of Electronics, January 20, 1951.

9–51 Siegert, A. J. F., "On the First-Passage-Time Probability Problem." *Phys. Rev.,* Vol. 81 (Feb. 15, 1951), pp. 617–623.

9–52 Fano, R. M., *Signal-to-Noise Ratio in Correlation Detection.* Tech. Rept. #186, MIT Res. Lab. Elect., February, 1951.

9–53 Cheatham, T. P., Jr., *An Electronic Correlator.* Tech. Rept. #122, MIT Res. Lab. of Electronics, March 28, 1951.

9–54 Doizelet, H., "The Electronic Correlator." *Rad. Tech. Dig.* (France), Vol. 5, No. 4 (1951), pp. 187–200.

9–55 Bunimovich, V. I., "Voltage Peaks of Fluctuation Noise." *Zh. Tekh. Fiz.,* Vol. 21 (June, 1951), pp. 625–636.

9–56 Bunimovich, V. I., "Observed Groups of Peaks of Electrical Fluctuations." *Zh. Tekh. Fiz.,* Vol. 21 (June, 1951), pp. 637–646.

9–57 Hunt, F. V., "Perturbation and Correlation Methods for Enhancing the Space Resolution of Directional Receivers." PROC. IRE, Vol. 39 (July, 1951), p. 840.

9–58 Lucas, J. L., and Raymond, R. C., "The Power Spectrum of a Narrow-Band Noise Passed Through a Non-Linear Impedance Element." *Jour. Appl. Phys.,* Vol. 22 (September, 1951), pp. 1211–1213.

9–59 Middleton, D., "On the Theory of Random Noise, Phenomenological Models: Parts 1 and 2." *Jour. Appl. Phys.,* Vol. 22 (September, 1951), pp. 1143–1163; (see also Vol. 22 (November, 1951), p. 1326).

9–60 O'Neill, R., Thaxton, H. M., and Cutler, S., "The Autocorrelation Function From the Ergodic Hypothesis." *Proc. NEC* (Chicago), Vol. 7 (1951), pp. 74–77.

9–61 Costas, J. P., "Synchronous Detection of Amplitude-Modulated Signals." *Proc. NEC* (Chi.), Vol. 7 (1951), pp. 121–129; (also *Tele-Tech,* Vol. 11 (July, 1952), pp. 55–57, 119).

9–62 Page, C. H., "Instantaneous Power Spectra." *Jour. Appl. Phys.,* Vol. 23 (January, 1952), pp. 103–106.

9–63 Norton, K. A., Schultz, E. L., and Yarbrough, H., "The Probability Distribution of the Phase of the Resultant Vector Sum of a Constant Vector Plus a Rayleigh Distributed Vector. *Jour. Appl. Phys.,* Vol. 23 (January, 1952), pp. 137–141.

9–64 Hoff, R. S., and Johnson, R. C., "A Statistical Approach to the Measurement of Atmospheric Noise." PROC. IRE, Vol. 40 (February, 1952), pp. 185–187.

9–65 Francis, V. J., "Random Noise. Rate of Occurrence of Peaks." *Wireless Engr.,* Vol. 29 (Feb., 1952), pp. 37–40.

9–66 Bussgang, J. J., *Crosscorrelation Functions of Amplitude-Distorted Gaussian Signals.* Tech. Rept. #216, MIT Res. Lab. of Electronics, 1952.

9–67 Emerson, R. C., *Applications of the Kac-Siegert Method for Finding Output Probability Densities for Receivers with Square-Law Detectors,* RAND Rept. P-294, April, 1952.

9–68 Siegert, A. J. F., "On the Evaluation of Noise Samples." *Jour. Appl. Phys.,* Vol. 23 (1952), pp. 737–742.

9–69 Robin, L., "Autocorrelation Function & Power-Density Spectrum of Clipped Thermal Noise-Filtering of Simple Periodic Signals in Such Noise." *Ann. Télécommun.,* Vol. 7 (September, 1952), pp. 375–387.

9–70 Casimir, H. B. G., "Symposium on Noise. General Introduction." *Tijdschrift Ned Radiogenoot.,* Vol. 17 (September/November, 1952), pp. 199–206 (in Dutch).

9–71 Blaquière, A., "Power Spectrum of a Non-Linear Oscillator with a Frequency/Amplitude Law, Perturbed by Noise." *Compt. Rend. Acad. Sci.* (Paris), Vol. 235 (November 12, 1952), pp. 1201–1203.

9–72 Pélegrin, M. J., *Application of the Statistical Technique to the Servo-Mechanism Field;* "Automatic and Manual Control." New York, Academic Press, pp. 123–137, 147–157, 1952.

9–73 Booton, R. C., Jr., *Application of Nonstationary Statistical Theory to Guided-Missile Design.* Symp. on Information Theory Applications to Guided Missile Problems, ORDB pp. 89–100, February 2, 1953.

9–74 Golay, M. J. E., "Correlation Versus Linear Transforms." PROC. IRE, Vol. 41 (February, 1953), pp. 268–271; (September, 1953), p. 1187.

9–75 Page, R. M., Brodzinsky, A., and Zirm, R. R., "A Microwave Correlator." PROC. IRE, Vol. 41 (January, 1953), pp. 128–131.

9–76 Rudnick, P., "The Detection of Weak Signals by Correlation Methods." *Jour. Appl. Phys.,* Vol. 24 (February, 1953), pp. 128–131.

9–77 Slade, J. J., Jr., Fich, S., and Molony, D. A., "Detection of Information by Moments." 1953 IRE Convention Record, Part 8—(Information Theory), pp. 66–71.

9–78 Arthur, G. R., "The Statistical Properties of the Output of a Frequency Sensitive Device." 1953 IRE Convention Record, Part 8—(Information Theory), pp. 80–90.

9–79 Booton, R. C., Jr., *The Analysis of Non-Linear Control Systems with Random Inputs.* Symp. on Non-Linear Circuit Analysis, Bklyn. Poly. Inst., pp. 369–391, April 23, 1953.

9–80 Duncan, D. B., "Response of Linear Time-Dependent Systems to Random Inputs." *Jour. Appl. Phys.,* Vol. 24 (May, 1953), pp. 609–611.

9–81 Weber, J., "Quantum Theory of a Damped Electrical Oscillator and Noise." *Phys. Rev.,* Vol. 90 (June 1, 1953), pp. 977–982.

9–82 Marshall, J. S., and Hitschfeld, W., "Interpretation of the Fluctuating Echo from Randomly Distributed Scatters: Part I," *Cand. Jour. Phys.,* Vol. 31 (September, 1953), pp. 962–994.

9–83 Emerson, R. C., "First Probability Densities for Receivers with Square-Law Detectors." *Jour. Appl. Phys.,* Vol. 24 (September, 1953), pp. 1168–1176.

9–84 Duncan, D. B., "Response of Undamped Systems to Noise." *Jour. Appl. Phys.,* Vol. 24 (September, 1953), pp. 1252–1253.

9–85 Bennett, W. R., "The Correlatograph." *Bell Sys. Tech. Jour.,* Vol. 32 (September, 1953), pp. 1173–1185.

9–86 Weinberg, L., and Kraft, L. G., "Measurements of Detector Output Spectra by Correlation Methods." PROC. IRE, Vol. 41 (September, 1953), pp. 1157–1166.

9–87 Pfeffer, I., and Favreau, R. R., *Application of Computer Techniques to Evaluation of Complex Statistical Functions.* Symp. III on Simulation and Computing Techniques, BuAer & US NADC, pp. 251–272, October 14, 1953.

9–88 Meyer-Eppler, W., "Correlation and Autocorrelation in Communication Engineering." *Arch. Elek. Übertragung,* Vol. 7 (October/November, 1953), pp. 501–504, 531–536, (46 references).

9–89 Sugar, G. R., "Estimation of Correlation Coefficients from Scatter Diagrams." *Jour. Appl. Phys.,* Vol. 25 (March, 1954), pp. 354–358.

9–90 Siegert, A. J. F., "Passage of Stationary Processes Through Linear and Non-Linear Devices." TRANS. IRE—Information Theory (March, 1954), pp. 4–25.

9–91 Middleton, D., "Statistical Theory of Signal Detection." TRANS. IRE—Info. Theory, (March, 1954), pp. 26–51.

9–92 Gold, B., and Young, G. O., "Response of Linear Systems to Non-Gaussian Noise." TRANS. IRE—Information Theory (March, 1954), pp. 63–67.

9–93 Smith, O. J. M., "Statistically Almost Optimum Network Design." TRANS. IRE—Information Theory (March, 1954), p. 123.

9–94 Keilson, J., "A Suggested Modification of Noise Theory." *Quart. Appl. Math.,* Vol. 12 (April, 1954), pp. 71–76.

9–95 Orr, L. W., "Wide-Band Amplitude Distribution Analysis of Voltage Sources." 1954 IRE Convention Record, Part 10.

9–96 Huggins, W. H., "Network Approximation in the Time Domain." Rept. #E5048A, AF Res. Lab.

9–97 *Statistical Theory of Noise.* Report #102, Res. Div., Philco Radio Corp.

10. Author Index

Johnson, D. C., 6–25
Johnson, J. B., 2–3
Johnson, R. A., 3–56, 8–42
Johnson, R. C., 9–64
Jones, M. C., 2–39
Jones, S., 3–41, 3–54
Jordon, W. B., 8–19
Jordon, W. H., 3–12

K

Kac, M., 2–48, 2–53, 9–10
Kalmus, H. P., 2–40
Kaplan, S. M., 7–40
Karmack, W. C., 8–7
Karr, P. R., 2–164
Keilson, J., 9–94
Kellogg, M. W. Co., 7–19
Kendall, M. G., 1–4
Keonjian, E., 2–137
Kettel, E., 6–32
Kleen, W., 2–107, 2–124
Knipp, J. K., 2–43
Knol, K. S., 2–90
Knudtzon, N., 8–26
Kompfner, R., 2–113
Kraft, L. G., 9–50, 9–86
Kretzmer, E. R., 6–35
Krumhansl, J. A., 2–78, 2–80
Kuper, J. B. H., 2–42

L

Laemmel, A. E., 8–25
Lakatos, A., 4–39
Lampard, D. G., 3–69
Landon, V. D., 3–3, 5–1, 6–3, 8–20, 9–3, 9–4
Lapin, S. P., 5–10
Laplume, J., 3–26
Lavoo, N. T., 2–76
Lawson, J. L., 1–11, 2–30, 6–8
Lebenbaum, M. T., 3–31
Lee, Y. W., 8–32, 8–27, 8–33, 9–31, 9–35, 9–38, 9–44, 9–45
Lehan, F. W., 3–29, 8–54
Lehmann, G., 2–103
Lehmann, R., 2–97
Lerner, R. M., 3–47
Levy, M., 7–21
Lewis, M. N., 2–37
Lewis, W. B., 7–9
Libois, L. J., 6–15, 6–36
Licklider, J. C. R., 3–23
Lindemann, W. W., 2–141
Lingard, A., 3–44
Llewellyn, F. B., 4–3
Loeb, J., 7–72
Lucas, J. L., 9–58

M

MacDonald, D. K. C., 2–55, 2–56, 2–65, 2–73, 2–163, 4–11, 9–30, 9–42, 9–47
Macfarlane, G. G., 2–54, 2–101, 3–30
Machulup, S., 2–172
Macnee, A. B., 2–70, 2–176, 9–28
Malatesta, S., 8–35, 8–44
Malter, L., 2–29
Marchand, N., 6–7
Marcum, J. I., 7–18
Marshall, J., 3–33, 9–82
Martin, F. W., 6–8
Mathews, M. V., 3–65
Mathews, W. E., 7–53
Mattson, R. H., 2–147
May, J. C., 7–34
Mazelsky, B., 3–67
McCombie, C. W., 3–63
McCouch, G. P., 4–42
McFall, R. W., 7–40
McNally, J. O., 2–9

Meade, J. E., 7–22, 7–24, 7–38, 7–50, 7–66
Meister, H., 4–24
Mellen, R. H., 3–53
Meltzer, B., 2–85
Meyer-Eppler, W., 3–57, 9–88
Middleton, D., 2–162, 3–19, 3–21, 3–25, 3–28, 3–32, 3–36, 3–46, 3–48, 3–52, 3–56, 3–58, 3–61, 3–62, 6–18, 6–23, 6–29, 6–30, 7–8, 7–12, 7–58, 7–74, 8–42, 9–32, 9–59, 9–91
Miller, D. A., 2–111
Miller, K. S., 8–39
Miller, Jr., P. H., 2–30, 2–37, 2–51
Molony, D. A., 9–75
Montgomery, H. C., 2–130, 2–138, 2–167
Mooers, T., 2–83
Moses, R. C., 2–100
Moskowitz, S., 6–11, 6–17
Moss, S. H., 6–20
Moullin, E. B., 1–3, 2–10
Moynihan, J. D., 4–30
Muchmore, R. B., 7–31, 7–33, 7–35, 7–51, 7–60, 7–70
Mumford, W. W., 4–23

N

Nichols, N. B., 1–6
Nicholson, F. T., 4–21
Nicholson, M. G., 6–13
Noday, G., 4–19
Nordby, K. S., 4–33
North American Avia. Aerophys. Lab., 1–10
North, D. O., 2–15, 2–16, 2–17, 2–20, 3–8, 7–4
Norton, K. A., 7–3, 7–16, 9–63
Nowotny, W., 6–14
Nozick, S., 2–175
Nyquist, H., 2–4, 8–1, 8–2

O

Omberg, A. C., 7–16
O'Neill, R., 9–60
Orr, L. W., 9–95
Oswald, J., 3–34, 3–43
Oudin, J., 8–40

P

Page, C. H., 9–62
Page, R. M., 6–42, 9–75
Paley, R. E. A. C., 1–1
Panter, P. F., 6–28, 6–33
Paolini, E., 2–84
Parks, R. J., 8–54
Parzen, P., 2–122
Pawsey, J. L., 5–6
Pearson, H. A., 2–177
Pélegrin, M. J., 9–72
Peter, R. W., 2–166
Peterson, A. P. G., 4–31
Peterson, H. O., 3–1, 4–5
Peterson, L. C., 2–62
Peterson, W. W., 7–71
Petritz, R. L., 2–105, 2–128, 2–136
Pettit, J. M., 4–14
Pfeffer, I., 9–87
Philco Rad. Co., 9–83
Philips, M., 7–7
Phillips, R. S., 1–6, 3–16
Pierce, J. R., 2–13, 2–33, 2–64, 2–140
Pike, E. W., 8–50
Pimonow, L., 4–37
Preston, G. W., 2–145, 2–151, 2–157
Price, W. L., 2–161

Q

Quate, C. F., 2–106

R

Rack, A. J., 2–14

Ragazzini, J. R., 4–9, 8–32, 8–47
Rauch, L. L., 2–61
Raymond, R. C., 9–58
Reich, E., 3–59
Reid, J. D., 3–3
Reiger, S., 3–60
Rice, S. O., 2–25, 9–12, 9–17, 9–27, 9–36
Richardson, J. M., 2–86
Ritterman, M. B., 8–60
Roberts, S., 2–52
Robin, L., 9–69
Robinson, F. N. H., 2–113, 2–171
Rochefort, J. S., 3–68
Roddam, T., 6–9
Rogers, D. C., 5–7
Rogers, T. F., 3–37
Rollin, B. V., 2–148
Roos, W., 2–58
Ross, A. E., 3–27
Ross, A. W., 7–42
Rowe, H. E., 2–145
Rudnick, P., 9–33, 9–76
Ruina, J., 7–44
Rultz, J. A., 2–166
Runge, W., 6–31
Ruppel, W., 2–124
Ruthberg, S., 3–17, 3–24

S

Schaeffner, J. S., 2–134
Schiff, L. I., 2–24, 2–37
Schooley, A. H., 2–115
Schottky, W., 2–1
Schremp, E. J., 2–44, 2–59
Schroeder, A. C., 2–41
Schultz, E. L., 9–63
Schultz, M. A., 2–92
Schwartz, L. S., 7–48, 9–23
Schwartz, M., 7–43
Seifert, W. W., 3–49, 3–65, 6–27, 7–64
Shannon, C. E., 8–11, 8–12, 8–16, 8–31
Sherman, S., 4–39
Shive, J. N., 2–173
Shockley, W., 2–13
Shorter, D. E. L., 4–25
Siegert, A. J. F., 2–48, 2–53, 2–105, 3–15, 9–16, 9–51, 9–68, 9–90
Singleton, H. E., 2–99, 9–41
Skalnik, J. G., 7–34
Slade, J. J., Jr., 9–77
Slater, J. C., 3–5
Slattery, T. G., 8–48
Slepian, D., 7–81
Slinkman, R. W., 2–81
Slocum, A., 2–173
Slusser, E. A., 2–111
Smith, D. B., 5–2
Smith, J. E., 4–13
Smith, O. J. M., 3–51, 9–93
Smith, R. A., 4–27
Smith, R. N., 2–28
Solomon, S. S., 2–117
Sorbacher, R. I., 6–5
Spencer, R. E., 7–46
Spetner, L. M., 7–57
Spratt, H. G. M., 4–18
Stateman, N. H., 8–60
Stephens, D. E., 2–30, 2–37
Steward, M. H., 7–69
Stockman, H., 8–15
Stone, W. M., 7–73
Strum, P. B., 2–158
Strutt, M. J. O., 2–49, 2–131
Stumpers, F. L. H. M., 5–4, 6–19, 6–21, 6–38, 8–5, 9–15, 9–40
Stutt, C. A., 8–27, 8–37
Sugar, G. R., 9–89

Suhl, H., 2–155
Sullivan, A. W., 8–38
Sulzer, P. G., 2–79
Suran, J. J., 5–10
Sutro, P. T., 7–6, 7–8
Swerling, P., 3–59
Sziklai, G. C., 2–41

T

Talpey, T. E., 2–176
Tellier, J. C., 5–3
Templeton, I. M., 2–148, 2–163
Thaxton, H. M., 9–60
Thomas, H. A., 1–7
Thompson, B. J., 2–15
Tiberio, U., 3–14, 7–56
Tick, L. J., 3–67
Tokahashi, H., 2–123
Tomlinson, T. B., 2–129, 2–156, 2–161
Tsung-Che, Y., 2–159
Tucker, D. G., 2–125, 2–127, 3–64
Tukey, J. W., 9–34
Tuller, W. G., 8–24

U

Uhlenbeck, G. E., 1–11, 9–9, 9–13
Urkowitz, H., 2–153, 7–75

V

van der Ziel, A., 2–57, 2–66, 2–71, 2–91, 2–93, 2–94, 2–95, 2–112, 2–114, 2–147, 2–160, 2–165, 4–17
van de Weg, H., 6–46
Vann, J. O., 9–22
Van Vleck, J., 3–13, 7–12
Vellat, T., 4–10
Versnel, A., 2–57, 2–66, 2–71

W

Wald, M., 2–18
Wallace, P. R., 5–9
Wallman, H., 2–70
Waltz, M. C., 2–42
Wang, M. C., 9–13
Watanabe, Y., 2–109
Watkins, D. A., 2–118, 2–126, 2–170
Weber, J., 9–81
Weinberg, L., 9–50, 9–86
Weiss, P. R., 3–11, 3–16
Weisskopf, V. F., 2–26
Welch, P., 2–174
Welton, T. A., 2–110, 9–49
White, W. D., 8–59
Wiener, N., 1–1, 1–9, 9–1, 9–2, 9–5, 9–6
Wiesner, J. B., 2–162, 8–29, 8–33, 9–35 9–38
Wilbur, D. A., 2–8
Williams, F. C., 2–11, 2–12
Williams, N. H., 2–2, 2–5
Wilmotte, R. M., 3–38, 3–66
Winkler, S., 2–175
Winter, D. F., 4–43
Wold, H. A., 9–29
Wolff, I., 7–5
Woodward, P. M., 7–30, 7–36, 7–49, 8–41
Wright, J. R., 7–68
Wulfsberg, K. N., 3–42

Y

Yabsley, D. E., 5–6
Yarbrough, H., 9–63
Youla, D. C., 6–47
Young, G. O., 8–52, 8–58, 9–92

Z

Zadeh, L. A., 8–22, 8–32, 8–39, 8–47, 8–53
Zimmerman, H. J., 7–65
Zirm, R. R., 9–75

11. Appendix

A

Academic Press,
 New York, N. Y.
Acta Mathematica,
 Stockholm, Sweden
Acta Polytechnica,
 Stockholm, Sweden
Acustica,
 Paris, France
AIEE Journal,
 New York, N. Y.
AIEE, Transactions,
 New York, N. Y.
Air Force Research Laboratory,
 Cambridge, Mass.
Alta Frequenza,
 Milan, Italy
American Mathematics Society,
 Providence, R. I.
Annales de Radioélectricite,
 Paris, France
Annales de Télécommunication,
 Paris, France
Annalen der Physik,
 Leipzig, Germany
Annals of Mathematical Statistics,
 Baltimore, Md.
Applied Physics Laboratory,
 Silver Spring, Md.
Archiv für Elektrotechnik (Übertragung),
 Berlin, Germany
Atti del Congresso internazionale della
 Radio,
 Rome, Italy
Australian Journal of Applied Science,
 Melbourne, Australia

B

Bell Laboratories Record,
 New York, N. Y.
Bell System Technical Journal,
 New York, N. Y.

C

Câble set Transmission,
 Paris, France
Cambridge University Press,
 London, England
Canadian Journal of Research,
 Ottawa, Canada
Communications,
 New York, N. Y.
Comptes Rendus de l'Academie des Sciences
 Paris, France
Cruft Laboratory,
 Boston, Mass.

D

Dynamic Analysis & Control Laboratory,
 Boston, Mass.

E

Electrical Communications,
 London, England
Electrical Engineering,
 New York, N. Y.
Electrical Reviews,
 London, England
Electronic Engineering,
 London, England
Electronics,
 New York, N. Y.
Elektrische Nachrichten-Technik,
 Berlin, Germany

Elektrotechnische Zeitschrift,
 Berlin, Germany
Ericsson Technics,
 Stockholm, Sweden

F

Fernmeldetechnische Zeitschrift,
 Brunswick, Germany
Frequenz,
 Berlin, Germany

G

General Electric Reviews,
 Schenectady, N. Y.
Griffin Press,
 London, England

H

Hughes Aircraft Company,
 Culver City, Calif.

J

Jet Propulsion Laboratory,
 Pasadena, Calif.
Journal, Acoustical Society of America,
 New York, N. Y.
Journal, British Institution of Radio
 Engineers,
 London, England
Journal, Institution of Electrical Engineers,
 London, England
Journal of Applied Physics,
 New York, N. Y.
Journal of Mathematics and Physics,
 Cambridge, Mass.
Journal of Research,
 National Bureau of Standards,
 Washington, D. C.
Journal of Scientific Instruments,
 London, England
Journal of the Aeronautical Sciences,
 New York, N. Y.
Journal of the Physical Society of Japan,
 Tokyo, Japan
Journal of the Royal Statistical Society,
 London, England

K

M. W. Kellogg Company,
 New York, N. Y.

M

Marconi Review,
 London, England
McGraw-Hill Publishing Company,
 New York, N. Y.
Microwave Research Institute,
 Brooklyn, N. Y.

N

National Defense Research Council,
 Washington, D. C.
Nature,
 London, England
North American Aviation Company,
 Downey, Calif.

O

Office of Scientific and Research Develop-
 ment,
 Washington, D. C.
Onde Eléctrique,
 Paris, France
Oxford Press,
 New York, N. Y.

P

Philco Radio Company,
Philadelphia, Pa.
Philips Research Reports,
Eindhoven, Holland
Philips Technical Review,
Eindhoven, Holland
Philosophical Magazine,
London, England
Physica,
The Hague, Holland
Physical Review,
New York, N. Y.
Prentice-Hall Publishing Company,
New York, N. Y.
Princeton University Press,
Princeton, N. J.
Proceedings, Cambridge Philosophical
Society,
London, England
Proceedings, Institute of Radio Engineers,
New York, N. Y.
Proceedings, Institution of Electrical
Engineers,
London, England
Proceedings, National Electronics Confer-
ence,
Chicago, Ill.
Proceedings, Physical Society,
London, England
Proceedings, the Royal Society,
London, England

Q

Quarterly of Applied Mathematics,
Providence, R. I.

QST,
West Hartford, Conn.

R

Radiation Laboratory,
Boston, Mass.
Radio-Electronic Engineering,
Chicago, Ill.
Radio Research Laboratory,
Boston, Mass.
RCA Review,
Princeton, N. J.
Reports of the Physical Society on Progress
in Physics,
London, England
Research Laboratory of Electronics,
Boston, Mass.
Review of Scientific Instruments,
New York, N. Y.
Reviews of Modern Physics,
New York, N. Y.
Revue Générale de l'Eléctricité,
Paris, France
Revues Scientifiques,
Paris, France

S

Science Reports of the Research Institute of
Tohoku University,
Japan
Servomechanism Laboratory,
Boston, Mass.
Sylvania Technologist,
Kew Gardens, N. Y.

T

Telecommunication Research
Establishment,
England
Tele-Tech,
New York, N. Y.
Tijdschrift van het Nederlandsch Radio-
genootschap,
Baarn, Holland
Transactions of the IRE,
New York, N. Y.
TV Engineering,
New York, N. Y.

U

United States Naval Research Laboratory,
Washington, D. C.
University of Michigan,
Ann Arbor, Mich.

V

Van Nostrand Publishing Company,
New York, N. Y.

W

Wireless Engineers,
London, England
Wireless World,
London, England

Z

Zeitschrift für angewandte Physik,
Berlin, Germany
Zhurnal Tekhnoskoi Fiziki,
Leningrad, U.S.S.R.

Part II
General Introduction

The Subject

The purpose of this part is to introduce the subject of electrical noise at an elementary level as a background preparation for subsequent material. An adequate preparation should have three ingredients: 1) a knowledge of the various basic physical processes generating noise to provide an overview of the field; 2) familiarity with mathematical methods used in noise studies to provide some understanding of analytical details; and 3) an awareness of potential applications to provide motivation for, and examples of, noise studies. These three topics outline the contents of the present part.

The subject of the physical origin of noise is taken up for detailed discussion in the next part, while the subject of noise applications is introduced in one of the reprinted papers. Therefore, only the mathematical methods required in the study of noise in electronic circuits, devices, and systems are briefly considered in this section.

The theory of probability and stochastic processes is one of the major branches of mathematics, and it would be presumptuous to expect to review this entire field here in a couple of short survey articles. The class of problems to which this branch of mathematics is addressed is large and varied. However, not all of these problems are of direct interest to electronic engineers concerned with circuits, devices, and systems (at least at present). As Snell [1] points out in his short survey of stochastic processes, "... although many of these problems arose naturally from applications, an even greater number arose because a mathematician was looking for an interesting problem to solve or from his desire to give a kind of mathematical completeness to a subject already studied. It is not possible to separate out the part which really means something in real life."

All that is possible in a short review paper is to point out some of the directions wherein the progress of work has led to results of use to electronic engineers in past. The present scope is further narrowed (and simplified) by confining the interest to the study of fluctuation phenomena in electronic devices and circuits, and excluding other areas of interest to electronic engineers, such as statistical theory of communication, information theory, reliability theory, electromagnetic interference, and system theory of stochastic systems.

At the elementary level, the required mathematical apparatus consists of basic ideas from probability theory, various averages characterizing a random variable, functions and transformations of random variables, and, in particular, the second-order properties (autocovariance and autocorrelation) and

spectral analysis which are of great importance in engineering work. More advanced work requires the use of results from the theory of stochastic processes. A stochastic (or random) process is a mathematical model of a sequence of experiments with random outcomes. Stochastic processes can be classified in a number of ways, usually depending upon the properties of the distribution function, accounting for the various adjectives like discrete, linear, Markov, Gaussian, stationary, etc., which are used to qualify stochastic processes. The determination of the distribution function of a stochastic process (usually by finding a differential equation satisfied by the distribution function) or one of its moments, and the determination of the properties of the various types of stochastic processes constitute the most important class of problems from the present point of view. Of course, the topics of interest are different for applications of the theory of stochastic processes in other areas like system theory and information theory [2].

The Reprinted Papers

The physical origin of noise is the subject of the first two papers reprinted in this part. Both of these are two decades old, but they are still useful for a first look at noise mechanisms. The two papers together cover most of the important noise mechanisms, although they do not explicitly deal with the more recently studied mechanisms like Barkhausen noise in magnetic materials, quantum noise in optoelectric devices, flux-flow noise in superconducting devices, intervalley scattering noise in hot-electron devices, etc. The first paper, by Pierce, emphasizes noise mechanisms of primary importance in electron tubes, while the second paper, by Burgess, is devoted to those noise sources which are of interest in semiconductor materials and devices. There is, of course, an inevitable overlap between the two.

The basic probabilistic ideas and the theory of stochastic processes are reviewed in the next two papers. Bennett's paper on the mathematical treatment of noise problems is an elementary and lengthy survey of the subject. It should therefore be useful both as an initial tutorial introduction to the subject as well as for review purposes. By contrast, Lin's survey of stochastic processes is a highly condensed outline of the subject. It can therefore serve primarily as a "road-map" and as a resource listing. Both Bennett and Lin have written books [3], [4] on the subject of these papers, which enlarge upon the papers and provide many examples of the applications of mathematical methods discussed here. A more in-depth introduction to the mathematical methods can be found in several other short reviews appearing elsewhere [1], [5] and in the books listed in one of the bibliographies in this part.

The paper on noise applications contains short summaries of a wide variety of applications, but is by no means exhaustive. Some of the applications mentioned therein are well established, such as the use of noise for the identification, characterization, and recognition of linear systems, a subject to which entire books have been devoted [6], particularly in connection with nuclear reactors. Some other applications are just beginning to be used, such as the study of biological membranes through noise, around which a sizable literature has already accumulated.

The Bibliographies

Two bibliographies are included in this part, one on noise in electronic circuits and devices, and the other on probability theory and stochastic processes. Both of these bibliographies are devoted to sources for general introduction. Thus the bibliography on noise includes books, compilations, conference proceedings, and standards related to noise, but not individual papers. The papers will be included in subject bibliographies on specific subjects in other parts. Similarly, the bibliography on mathematical methods contains selected references only to general books and some advanced treatises, but not publications reporting research results. In both cases, the purpose has been to provide a list of resource materials for background reading, for which state-of-the-art literature is neither convenient nor efficient. The bibliographies also include some items from related subject areas, such as the statistical theory of communication.

References

[1] J. L. Snell, "Stochastic processes," in *Handbook of Mathematical Psychology*, vol. III, R. D. Lee *et al.*, Eds. New York: Wiley, 1965, pp. 411–485.
[2] E. Wong, "Recent progress in stochastic processes—A survey," *IEEE Trans. Inform. Theory*, vol. IT-19, pp. 262–275, May 1973.
[3] W. R. Bennett, *Electrical Noise*. New York: McGraw-Hill, 1960.
[4] Y. K. M. Lin, *Probabilistic Theory of Structural Dynamics*. New York: McGraw-Hill, 1967.
[5] J. C. Samuels, "Elements of stochastic processes," in *Continuum Physics*, vol. 1, A. C. Eringen, Ed. New York: Academic, 1971, pp. 605–663.
[6] R. W. Harris and T. J. Ledwidge, *Introduction to Noise Analysis*. New York: Academic, 1974.

Two Tutorial Papers on Noise

Previous tutorial papers have been largely survey papers or papers reviewing progress in certain special fields. The Tutorial Papers Subcommittee of the Committee on Education has concluded that there is need for another type of paper which will help the engineer to keep abreast of developments in basic theory. Engineers some years out of college tend to lose the ability to read with profit even important papers in their own fields of interest, because new concepts have crept in with which they are only vaguely familiar and new methods have been adopted for handling basic problems. Such new concepts and methods usually first appear in the literature in scattered fragments, and a considerable time may elapse before a consecutive account is available which the nonspecialist engineer can study effectively. It is hoped that the new Tutorial Papers can shorten the gap between the evolution of new and useful ideas and their application to solving the everyday practical problems of the radio engineer.

The subject of noise is encountered in every branch of radio engineering and is of the highest importance. But is is an elusive subject and can be handled only by statistical methods with which the average engineer is unfamiliar. The subject is difficult to learn because new kinds of reasoning must be employed. The following two papers on noise, therefore, are not light, easy reading, but their study will make possible a real understanding of a large group of basic problems.

The project of writing a tutorial paper on noise as a part of the Tutorial Papers program was planned in detail at a meeting on January 26, 1954, attended by W. R. Bennett, F. K. Bowers, W. R. Davenport, B. McMillan, J. R. Pierce, S. O. Rice, C. E. Shannon, and D. Slepian. The original proposal was to obtain a number of individual contributions from different authors on selected topics. What has now emerged divides the subject into two main parts.

The first paper, by J. R. Pierce, deals primarily with descriptions of the various physical phenomena which lead to noise, and the second paper, by W. R. Bennett, explains mathematical techniques which have been developed for quantitative evaluations. The distinction is one of emphasis rather than complete separation, for it is necessary to include calculations in the physical description and to refer to physical models in the mathematical treatment.

The inclusion of topics within these main divisions is representative rather than exhaustive. A comprehensive coverage of the field appears to be inconsistent with a reasonable length of text. The second article in particular has grown bulkier than intended and still neglects a number of important analytic phases of the subject. Among the topics not even mentioned are Smoothing and Prediction of Noise, Likelihood Criteria for Detection of Signals in Noise, and Statistics in a Finite Time Interval. We do not claim that one who reads these articles will thereby be prepared to read the papers in the literature on the topics not included, but we think he will be helped. The topics treated are naturally the ones with which the authors are most familiar. An attempt is made to establish a background of fundamental concepts which are applied to progressively more difficult situations as the subject is developed. It is hoped that the readers who so wish will be able to apply these principles to similar problems on their own and that those who do not wish to derive formulas themselves will be aided toward a more effective use of results available in the more complete sources.

W. N. Tuttle, *Chairman*
W. R. Bennett
Tutorial Papers Subcommittee

Physical Sources of Noise*

J. R. PIERCE†, FELLOW, IRE

Summary—Johnson noise is caused by the fluctuation of charges or polarizable molecules in lossy materials. Its magnitude can best be calculated by means of statistical mechanics, which tells us that each degree of freedom must have associated with it an energy of $\frac{1}{2} kT$. This leads to the usual expressions for Johnson noise, including the fact that the available thermal noise power, from a resistor, a lossy network, a lossy dielectric, or an antenna is always kTB. In the case of a resistor, network, or dielectric, T is the temperature of the lossy material. In the case of an antenna, T is the average temperature of the environment which the antenna "sees." Consistent with the power kTB available from an antenna, there is a particular density of radiation in space which is a function of temperature and frequency. Noise figure is defined in terms of Johnson noise.

Shot noise, due to the discrete nature of electron flow, is generally distinct from Johnson noise, although in some electron devices the expression for the noise in the electron flow has the same form as that for Johnson noise. When the noise in electron flow is greater or less than pure shot noise, the motions of the electrons must be in some degree correlated. In an electron stream of low noise, the random interception of a fraction of the electron flow can reduce the correlation and increase the noise. Johnson noise and shot noise have a flat frequency spectrum. Some active devices, such as vacuum tubes and transistors, have a 1/f frequency spectrum at low frequencies.

* Original manuscript received by the IRE, January 17, 1956; revised manuscript received, February 3, 1956.
† Bell Telephone Labs., Inc., Murray Hill, N.J.

Reprinted from *Proc. IRE*, vol. 44, pp. 601–608, May 1956.

Introduction

MANY SORTS of electric signals are called noise. In the early days of radio we were most familiar with the crash and crackle of static. Later, we encountered the rasp of ignition noise and the hiss of the thermal and shot noise generated in radio circuits themselves. In the end, many engineers have come to regard any interfering signal of a more or less unpredictable nature as noise. An interfering sine wave of absolutely constant frequency is predictable, and it can be filtered out with negligible loss of channel capacity. Crosstalk from one telephone channel to another is not predictable, at least, not in the same way as a sine wave is, and it cannot be eliminated, at least not by practical means. The engineer is apt to consider it as so much noise.

The study of noise began with the consideration of certain physical sources of noise and the sorts of noise that they generate. At first, only very simple properties of the noise signals so generated were understood and described. As the art has progressed, a mathematical theory of noise has grown up. This theory is a part of the general field of statistics, and it deals with signals which have an unpredictable, a statistical, a random element. Some understanding of this theory is essential in dealing with engineering problems involving noise.

The theory of noise as it is usually presented is not valid for all signals or phenomena which the engineer may identify as noise, though it may be adequate in a practical sense in dealing with signals (multichannel crosstalk, for instance) which do not meet the mathematician's strict requirements of randomness. The theory of noise is best adapted to handling signals which originate in truly random processes, such as the emission of electrons from a photo-surface or a hot cathode, or the thermal agitation of charges in a resistor. When a cathode emits electrons at so slow a rate that we observe their effects in a circuit as separate pulses, we have *impulse noise*, and the theory of noise has something to say about this. When electrons are randomly emitted so rapidly that the pulses they produce in the circuit overlap, the statistics of large numbers applies, and the theory of noise tells us a great deal that must be true of a large class of noise signals, despite differences in the exact nature of their sources.

The statistical theory of noise will be treated in a paper by W. R. Bennett.[1] In this paper, as an introduction to noise, we consider important physical sources which produce noise of just the type to which the statistical theory applies exactly.

Johnson Noise

The first source of noise which we consider is Johnson noise, the thermal noise from a resistor. The engineering fact is that a resistor of resistance R acts as a noise generator. The resistor as a noise generator can be described as a zero-impedance voltage generator in series with the resistance of the resistor. The mean square voltage $\overline{V^2}$ (V is the open-circuit voltage of the generator) is given by

$$\overline{V^2} = 4kTRB \qquad (1)$$

This generator is shown in Fig. 1.

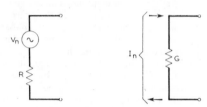

Fig. 1—The thermal power source associated with a resistance or conductance can be represented as a series voltage source or a shunt current source.

Also, as shown in Fig. 1, we can as easily describe the resistor as a noise generator as a mean square impressed current from an infinite-impedance source, $\overline{I^2}$, (I is the short-circuit current), in shunt with the conductance G of the resistor.

$$\overline{I^2} = 4kTGB \qquad (2)$$

In (1) and (2) B is the bandwidth in which the noise voltage or current lies, k is Boltzmann's constant, and T is the absolute temperature in degrees Kelvin, which is the temperature in degrees centigrade plus 273°.

Expressions (1) and (2) also apply to the noise from any complex passive circuit which is all at one temperature. A passive circuit is one to which no power is supplied except by heating it; that is, no dc current is supplied, or no regular mechanical motion. In case of such circuits, R is the resistive component of the impedance, or G is the conductive component of the admittance. Because R and G vary with frequency f, B is taken as a small incremental bandwidth df and $\overline{V^2}$ or $\overline{I^2}$ must be evaluated by integration.

What is the source of Johnson noise? In an ordinary resistor, it is a summation of the effects of the very short current pulses of many electrons as they travel between collisions, each pulse individually having a flat spectrum. In this case the noise is a manifestation of the Brownian movement of the electrons in the resistor. In a resistor consisting of two opposed, close-spaced, hot, electron-emitting cathodes, it is a result of the current pulses of randomly-emitted electrons passing from one cathode to the other. In a lossy dielectric it is the result of random thermal excitations of polarizable molecules, forming little fluctuating dipoles. The case of noise in the radiation resistance of an antenna, which will be discussed later, is quite involved.

For any particular sort of resistor, it should be possible to trace out the source and calculate the magnitude of the Johnson noise, and indeed, this approach has been used. However, there is something very general about

---

[1] "Methods of solving noise problems," Proc. IRE, pp. 609–637; this issue.

an expression which applies to so wide a variety of physical systems. It turns out that there is a very general way of deriving the expression for Johnson noise; this is afforded by thermodynamics and statistical mechanics.

Consider a network containing many resistors. If we heat one hotter than the rest, energy tends to flow from the hot resistor to the cooler resistors. Johnson noise is such energy flowing as electric power. Even when the resistors are all at the same temperature, power will flow back and forth between them through the connecting network, always so that on the average a resistor receives just as much power as it sends out. We may note that if we know the resistance of one portion of the network, its internal impedance, and the impedance of the rest of the network acting as a load, by using (1) we can calculate the power flow. But really, it is from a knowledge of the general behavior of systems in equilibrium, that is, all at the same temperature, that we can derive (1) and (2).

Statistical mechanics tells us how much energy must on the average be associated with each *degree of freedom* of a system when the system is in thermal equilibrium. In an electrical network of inductors, capacitors, and resistors in which there are no inductors connected directly in series nor capacitors connected directly in shunt, the number of degrees of freedom is the number of inductors plus the number of capacitors. In setting up a signal on the network, we are free to specify arbitrary currents in all the inductors and arbitrary voltages across all capacitors. If we know these currents and voltages at any instant, we can (disregarding Johnson noise, which acts as a power source) calculate all future voltages and currents.

Classical statistical mechanics says that in a system (in our case, such an electrical network) which is in equilibrium (all at the same temperature) there is on the average an energy

$$\tfrac{1}{2}kT \text{ joules}$$

associated with each degree of freedom:

$$k = 1.380 \times 10^{-23} \text{ joule/degree.}$$

According to quantum mechanics, the energy is less than this at high frequencies; Nyquist and others have used the quantum-mechanical expression to get the correct result. However, even up to tens of thousands of megacycles the classical expression is accurate.

Let us consider two simple circuits as particular examples, to see how things work out. In these circuits an inductance L is in series with a resistance R at temperature T, and a capacitance C is in shunt with a conductance G at a temperature T, as shown in Fig. 2.

Let $\overline{I^2}$ be the total mean square noise current in the inductor. We can write

$$\tfrac{1}{2}L\overline{I^2} = \tfrac{1}{2}kT.$$

On the left we have the average power in the inductance. On the right we have the average value this must have according to statistical mechanics. Accordingly,

$$\overline{I^2} = \frac{kT}{L} \cdot \tag{3}$$

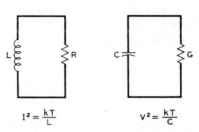

$$I^2 = \frac{kT}{L} \qquad V^2 = \frac{kT}{C}$$

Fig. 2—The total Johnson noise current squared in an inductance in series with a resistance, and the total Johnson noise voltage squared across a capacitance in shunt with a conductance, are independent of the resistance or the conductance.

This must be true regardless of the value of R. If R is low we have a narrow-band circuit; if R is high we have a broad-band circuit. As the noise current is made up of various frequency components, more low-frequency current must flow in the narrow-band case when R is small than in the broad-band case when R is large, if the total mean squared current is to be the same in both cases.

In a similar way, in the case of the capacitance C and the conductance G in shunt we easily find that

$$\tfrac{1}{2}C\overline{V^2} = \tfrac{1}{2}kT$$

$$\overline{V^2} = \frac{kT}{C} \cdot \tag{4}$$

If the conductance is small, we have a narrow-band circuit with high low-frequency noise components. If the conductance is large we have less low-frequency noise but more bandwidth.

Relations (3) and (4) of course apply to capacitors and inductors not merely in the simple circuits we have considered, but to capacitors and inductors anywhere in all circuits, no matter how complicated they may be. In any case, we see that the noise voltage or current squared is proportional to the temperature T.

Among the circuits to which (3) and (4) apply is the resonant circuit shown in Fig. 3, which consists of an inductance L, a capacitance C and a conductance G in shunt. This circuit is characterized by a resonant frequency f_0, a Q, and an impedance the square of whose magnitude is $|Z|^2$, given by

$$f_0 = \frac{1}{2\pi\sqrt{LC}}$$

$$Q = \sqrt{C/L}/G$$

$$|Z|^2 = \frac{(L/C)Q^2}{Q^2\left(\dfrac{f}{f_0} - \dfrac{f_0}{f}\right)^2 + 1} \cdot$$

We can regard this circuit as excited by the impressed shunt Johnson noise current I. If we make the Q very high, so that the bandwidth of the circuit is very narrow, we can find out something about the spectral distribution of the impressed noise current.

Because the mean square value of a number of voltages or currents of different frequencies is the sum of the mean square values of the different frequency components, we will assume that

$$d\overline{I^2} = F(f)df. \tag{5}$$

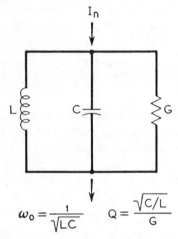

$$\omega_0 = \frac{1}{\sqrt{LC}} \qquad Q = \frac{\sqrt{C/L}}{G}$$

Fig. 3—We can use a tuned circuit to explore the frequency spectrum of Johnson noise.

That is, each frequency range df contributes some part $d\overline{I^2}$ of the total mean square current. In (3), $F(f)$ is a function of frequency.

Let us now calculate the total mean square noise voltage across the capacitance and set it equal to the value given by (2). This gives

$$\frac{kT}{C} = \frac{L}{C} Q^2 \int_0^\infty \frac{F(f)df}{Q^2\left(\dfrac{f}{f_0} - \dfrac{f_0}{f}\right)^2 + 1}.$$

Now suppose we make Q very large, so that the band is very narrow. Then $F(f)$ will not vary much over the band, and we can take it outside the integral and give f the value f_0. With some rearrangement and substitutions we obtain

$$kT = \frac{F(f_0)}{2\pi G} \int_0^\infty \frac{Q\,d(f/f_0)}{Q^2(f/f_0 - f_0/f)^2 + 1}.$$

The value of the integral is found to be independent of Q and f_0, so that $F(f_0)$ is a constant, independent of frequency. Let us call the integral I. The substitution

$$\frac{f}{f_0} = e^x$$

leads to

$$I = \int_{-\infty}^\infty \frac{Qe^x dx}{4Q^2 \sinh^2 x + 1}.$$

The value is unchanged if we replace e^x by e^{-x}, so that

$$I = \int_{-\infty}^\infty \frac{Q \cosh x\, dx}{4(Q \sinh x)^2 + 1}$$

$$= \frac{1}{2} \int_{-\infty}^\infty \frac{\frac{1}{2} du}{u^2 + (\frac{1}{2})^2}.$$

By Pierce 480

$$I = \frac{\pi}{2}$$

$$F(f_0) = 4kTG$$

$$d\overline{I^2} = 4kTG\,df.$$

This is often written in the form (2), where

$$\overline{I^2} = 4kTGB. \tag{2}$$

Here B is bandwidth and $\overline{I^2}$ is understood to be the part of the mean square impressed noise current lying in the bandwidth B.

The spectrum of Johnson noise is flat; it is called *white* noise.

We can evaluate the series noise voltage much as we evaluated the shunt noise current, and we obtain (1):

$$\overline{V^2} = 4kTRB \tag{1}$$

We can of course obtain (1) directly from (2) by asking what current flows when the resistance is shorted; this is

$$\overline{I^2} = \frac{\overline{V^2}}{R^2} = \frac{4kTB}{R} = 4kTGB.$$

What happens if we connect two resistances in series or two conductances in parallel? In a given frequency range, the voltages or currents produced by different resistances are uncorrelated; they have random phases, and the mean square of the sum of the separate voltages or currents is equal to the sum of the mean square voltages or currents of the separate resistors. This is in accord with (1) and (2), taking R and G as total resistance and conductance.

As we have noted, for a complex impedance the series noise voltage generator at any frequency can be calculated from the resistive component R of the impedance, and the shunt noise current generator from the conductive component G of the admittance.

Relations (3) and (4) tell us something which is sometimes useful in connection with networks. From (1) or (2) we can calculate the mean square thermal noise current or voltage lying in a narrow frequency range for any network, simply by associating with the resistance or conductance series voltage generators or impressed currents according to these relations.

There is another way of expressing the information in (1) or (2). We may ask, what is the thermal noise power P available from a resistor? We will draw off the maximum power if we supply a matched load of the same resistance. Thus, the available noise power in the bandwidth B can be obtained by calculating the noise power flowing into a resistance R from a source with an internal resistance R and an open-circuited voltage given by (1). This power is

$$P = \frac{1}{2} \frac{\overline{V^2}}{(2R)^2}$$

$$P = kTB. \tag{6}$$

Thus, the Johnson noise power available from a resistor is independent of the resistance.

Let us assume that a resistance R is matched to a waveguide W, and the waveguide is terminated in an antenna, perhaps a horn antenna, as shown in Fig. 4. A power P given by (6) is radiated from the antenna in a beam; to the beam we assign some nominal solid angle ψ steradians. Suppose that the antenna is surrounded by a perfectly absorbing box D (a "black box") at the same temperature as the resistance R. Since there can be no average power flow between two objects at the same temperature, a power P given by (6) will flow back from the box D to the resistance R. This power must be independent of the size and the directivity of the antenna.

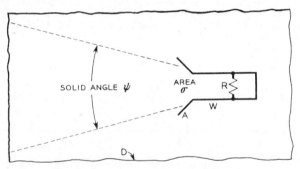

Fig. 4—An antenna A in a black box D.

No matter what direction we point the antenna in the box, the power received will be the same. The box is full of electromagnetic power flow in all directions. The antenna receives power over a solid angle ψ, and over the area σ of the antenna. At a given wavelength, ψ is inversely proportional to σ; that is, the bigger the antenna the narrower the beam, so the product of $\psi\sigma$ remains constant as the antenna size is varied.

We know, however, that for an antenna of given size, the shorter the wavelength, or the higher the frequency, the narrower the beam. If the antenna is to receive a power P per unit bandwidth, the radiation flux for unit solid angle and unit area must be proportional to the square of the frequency.

Consider a little area $d\sigma$ receiving radiation from a direction θ with respect to the normal to the area, over a solid angle $d\psi$. (See Fig. 5.) The area of $d\sigma$ seen from the direction of radiation is

$$\cos \theta d\sigma.$$

From this and the previous argument, we see that the power dP received must be

$$dP = Af^2 \cos \theta d\psi d\sigma,$$

where A is a constant. When the constant is evaluated, we find

$$dP = \frac{f^2 df}{c^2} kT \cos \theta d\psi d\sigma. \tag{7}$$

Here c is the velocity of light. This is the power in one polarization; there is an equal power in the other polarization which is not accepted by the antenna. It is also the power radiated by an absorbing surface in one polarization in a direction θ to the normal, over a solid angle $d\psi$ and over an area $d\sigma$.

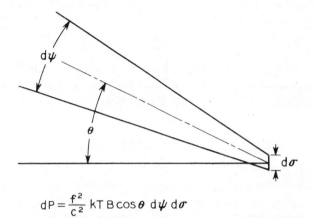

$$dP = \frac{f^2}{c^2} kT B \cos\theta \, d\psi \, d\sigma$$

Fig. 5—The power radiated or absorbed in one polarization at a frequency f in a direction making an angle θ with the normal, in a bandwidth B, a solid angle d_4, over an area $d\sigma$.

Why cannot one get the total power radiated per unit area from a solid body by integrating twice dP as given by (7) with respect to $d\psi$ and df? The answer is, because (7) is incorrect at high frequencies, and gives an infinite answer; this is the "ultra-violet catastrophe" of pre-quantum physics.

We can, however, use (7) to calculate the power received in a given bandwidth by an antenna from some hot body of narrow extent, such as the sun.

Relation (7) is accurate at microwave frequencies, and it is consistent with the fact that the total noise power received from the walls of a nonreflecting enclosure held at temperature T must be kTB. Ordinarily, however, what surrounds an antenna is neither perfectly absorbing nor at a uniform temperature. In this case we may use an average temperature T_A in order to calculate how much Johnson noise an antenna will receive.

If we used the antenna in question as a transmitting antenna, part of the power might be absorbed directly by trees in the beam, part might be reflected from the ground into the sky, to be absorbed ultimately by cosmic matter, and some might go directly to the sky. Suppose a fraction p_n of the total radiated power is either directly, or after reflection, or due to imperfect reflection, ultimately absorbed by matter of temperature T_n. Then, the Johnson noise power received when the antenna is used for receiving instead of transmitting will be correctly given by (6), where as T we use an average temperature T_A given by

$$T_A = \sum_n p_n T_n. \qquad (8)$$

Johnson noise serves as a reference for the noisiness of radio receivers and amplifiers. Suppose, for instance, that the input of an amplifier is connected to an antenna pointed at an absorbing body of temperature T, or, simply to a transmission line terminated at the other end in a resistance of temperature T. Let the ratio of the power output to the available power at the antenna or resistive source be N at some reference frequency f. Then, if the amplifier introduced no noise but merely amplified Johnson noise, the power output P_j over a narrow bandwidth B would be

$$P_j = NkTB.$$

Actually, the amplifier will generate noise. Let the total output noise power be P_t. Then the noise factor or noise figure NF of the amplifier is defined as

$$NF = \frac{P_t}{P_j} = \frac{P_t}{NkTB} . \qquad (9)$$

This defines noise figure for some narrow band B at a specific frequency. Under such circumstances, P_t is found to be proportional to B, and the noise figure is independent of B and varies somewhat with frequency, as P_t/B and N vary with frequency.

Shot Noise and Other Noise in Electron Tubes

Electricity is not a smooth fluid; it comes in little pellets, that is, electrons. The flow of electrons in a vacuum tube is accompanied by a noise of the same nature as the patter of rain on a roof. Schottky, who first investigated this phenomenon, called it the *Schroteffekt* (from shot); it is now usually called simply *shot noise*.

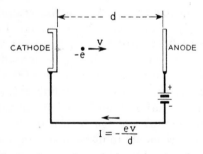

Fig. 6—An electron leaves the cathode at time t_1 and travels with increasing speed between cathode and anode and strikes the anode at time t_2.

Consider an electron of charge $-e$ moving with a velocity v between parallel plane conductors a distance d apart, for instance, a cathode and an anode, as shown in Fig. 6. If the cathode and anode are connected by a wire of zero impedance, a current I flows to the wire, and I is given by

$$I = - \frac{ev}{d} .$$

The current starts when the electron leaves the cathode at time t_1 and ends when the electron reaches the anode at time t_2. If the electron leaves the cathode with zero velocity and is accelerated by a dc field between the cathode and the anode, the current pulse, that is, I plotted against time, will look something as shown in Fig. 7. The total charge Q flowing as a result of the passage of the electron is

$$Q = \int_{t_1}^{t_2} I dt = \int_{t_1}^{t_2} - \frac{ev}{d} dt = \int_0^d - \frac{edx}{d}$$

$$Q = - e.$$

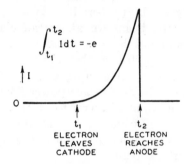

Fig. 7—The current flow in the case of Fig. 6 starts at time t_1, increases as the electron is accelerated between cathode and anode, and stops when the electron strikes the anode. The total transfer of charge between cathode and anode is the electronic charge, $-e$.

That is, a charge equal to the charge of the electron flows into the circuit in the form of a short pulse during the transit of the electron between cathode and anode.

Suppose that the cathode and anode form part of a shunt resonant circuit of inductance L, capacitance C and conductance G, as shown in Fig. 3. If the current pulse due to the passage of the electron is short compared with the period of the circuit, we can regard the transfer of charge as instantaneous and say that the passage of the electron from cathode to anode is equivalent to putting a charge $-e$ on the capacitance C. If we do this we give to the circuit an energy W:

$$W = \frac{1}{2} \frac{e^2}{C} .$$

We see that the energy given to the circuit is dependent only on C, and is not dependent on the resonant frequency. The energy is gradually dissipated in the conductance G as the circuit oscillates with exponentially decaying amplitude following the pulse.

If we put N charges on a circuit at random times, over an interval T, the waves initiated have random phases, and the powers add, so the total energy is NW:

$$NW = \frac{1}{2} N \frac{e^2}{C} .$$

The rate P at which power is dissipated in the circuit is thus

$$P = \frac{NW}{T} = \frac{1}{2} \frac{N}{t} \frac{e^2}{C} .$$

The number N of electrons which flow in a time T is related to the average electron current I_0:

$$N = \frac{IT}{e},$$

so that

$$P = \frac{1}{2}\frac{eI_0}{C}. \tag{10}$$

Thus, the power in the circuit is dependent on C only and is independent of the resonant frequency.

As before, we will assume that the impressed noise current I is such that

$$\overline{I^2} = F(f)df. \tag{11}$$

This current flows into an impedance Z:

$$Z = \frac{1}{G}\frac{(1-jQ)\left(\dfrac{f}{f_0}-\dfrac{f_0}{f}\right)}{1+Q^2\left(\dfrac{f}{f_0}-\dfrac{f_0}{f}\right)^2}. \tag{12}$$

By equating the power calculated from (11) and (12) with the power as given by (10) we obtain

$$\frac{1}{2}\frac{eI_0}{C} = \frac{1}{G}\int_0^\infty \frac{F(f)df}{1+Q^2\left(\dfrac{f}{f_0}-\dfrac{f_0}{f}\right)^2}.$$

As before, we can assume that Q is very high, so that the circuit impedance is appreciable over a very narrow frequency range only, and take $F(f)$ outside of the integral. This gives, with some rearrangement and substitutions

$$eI_0 = \frac{F(f_0)}{\pi}\int_0^\infty \frac{Qd(f/f_0)}{1+Q^2\left(\dfrac{f}{f_0}-\dfrac{f_0}{f}\right)^2}.$$

This is the same integral we encountered before. Its value is independent of Q and f_0. Thus, we find that $F(f_0)$ is a constant, so that the noise current per unit bandwidth is independent of frequency.

$$F(f_0) = 2eI_0$$

$$\overline{I^2} = 2eI_0B. \tag{13}$$

Like Johnson noise, shot noise has a flat spectrum.[2] This is really what we should expect of a random collection of very short pulses, each of which has a flat spectrum.

In a given frequency range, a current of electrons can have either more or less noise than pure shot noise as given by (13). For the noise to be either greater or less than shot noise, the times of emission of various electrons must be correlated in some degree.

Suppose, for instance, that the I_0/e electrons which leave the cathode each second were emitted at evenly spaced times e/I_0 seconds apart. The flow would have no ac component at all at frequencies below I_0/e. The only frequencies present would be I_0/e and its harmonics.

On the other hand, suppose that the electrons left n at a time but that the bunches of n electrons left at uncorrelated times. This is equivalent to saying that the current is carried by randomly emitted charges of charge ne, and from (13) the noise current would be

$$\overline{I^2} = 2(ne)I_0B.$$

Random processes other than the random emission of electrons can give rise to noise. Suppose, for instance, that electrons enter an electron multiplier periodically, so that there is no noise in the electron stream at ordinary frequencies. If the electron multiplier gave out exactly the same number N of electrons each time one went in, there would be no noise in the output. It doesn't, however, for sometimes an electron produces more electrons and sometimes fewer. This randomness gives rise to noise in the output current of the multiplier.

Similarly, noise can be introduced into an initially noiseless electron flow if the electrons randomly hit or miss the wires of a grid, with a certain average interception of current. Such noise is called *partition noise* or *interception noise*. If a small fraction only of the current is intercepted, the added noise is roughly equal to shot noise for the intercepted current.

At high frequencies and long transit times the excitation of a circuit may depend on the velocity of the entering electrons. In such a case the random variation of velocity of emission from one electron to another, associated with the Maxwellian velocity distribution of electrons leaving a cathode, can give rise to noise. This fluctuation in velocity is also responsible for what is commonly called the *modified* or *reduced* shot noise in space-charge-limited flow of electrons from a cathode.

A very simple theory of noise in space-charge-limited diodes and triodes at frequencies low enough so that transit time is not important predicts that the noise can be represented by an impressed noise current in the plate circuit I^2, given by

$$\overline{I^2} = (.644)4kT_cgB. \tag{14}$$

Here T_c is the temperature of the cathode and g is the conductance of the diode or the transconductance of the triode. More elaborate theories lead to a factor which, in various circumstances, may be a little greater or a little less than .644.

Expression (14) is of the same form as that for Johnson noise, but the expression cannot be derived in a manner analogous to that which we have used in treating Johnson noise, for a conducting diode or triode

---

[2] At frequencies for which the transit time from cathode to plate is comparable with the period, the noise induced in the plate circuit is a function of frequency.

cannot be in thermal equilibrium. It is interesting to note that when the plate of a diode is very negative, so that there is no potential minimum between cathode and plate, the impressed noise current is given by

$$\overline{I^2} = (.5)4kT_cgB$$
$$= 2eI_0B.$$

That is, the noise is shot noise, but it can also be expressed in a form analogous to that for Johnson noise, and like expression (14).

At moderate frequencies, noise in triodes and diodes agrees fairly well with (14), being perhaps a little higher. At frequencies high enough so that transit time is important, noise associated with the electronic loading of the grid circuit becomes important. At low audio frequencies and below, *flicker noise* appears. This typically but not always has a $1/f$ spectrum (discussed in the following section). Flicker noise is very variable from tube to tube. It has been ascribed to fluctuations in the work function of the cathode surface.

A detailed treatment of the generation of noise through random processes in active devices is very complicated and would be out of place here.

Noise with a $1/f$ Spectrum

It is clear that different Johnson noise spectra are associated with circuits for which the impedance varies in different manners with frequency. However, Johnson noise is in a sense inherently white noise in that the fundamental relation between the noise source—the resistance or conductance—and the amount of noise per unit bandwidth is independent of frequency. Shot noise is in the same sense inherently white noise too, although it can give rise to different spectra in circuits in different transfer admittances or in tubes of different transit times. Presumably it must be, because in a close-spaced diode formed of opposed cathodes at the same temperature and with no average current, shot noise and Johnson noise are two names for the same thing.

Some important sorts of noise are generated *only* in nonequilibrium systems, for instance, in systems in which dc current flows. Among these are *contact noise*, such as is produced in a carbon microphone, and the noise produced in carbon resistors and in silicon and germanium diodes and transistors. Both contact noise and transistor noise have a spectrum such that the noise power per unit bandwidth varies nearly as $1/f$ over a large frequency range, though it may be constant at high frequencies and at very low frequencies.

A noise made up of a random sequence of short pulses, or impulses, as are Johnson noise and shot noise, has a flat spectrum. A noise made up of a random sequence of step functions would have a $1/f^2$ spectrum. This is because a step is the integral of an impulse, and the amplitude of any frequency component of the step is $1/2\pi f$ times that for the impulse. If the amplitude varies as $1/f$, the power will vary as $1/f^2$.

One could obtain a $1/f$ spectrum down to any given frequency by a proper mixture of pulses of various lengths. The longer pulses might actually be a result of a correlation between short pulses, so that pulses tended to occur in groups or bursts over a period longer than that of one pulse; this could give, in effect, a "long" pulse.

Transistor noise has been attributed to the trapping of the holes or electrons (*carriers*) which form the current flow. The trapping and subsequent release of a charge carrier is equivalent to a rectangular pulse in the current. The effect may be strengthened by the charge of the trapped carrier modulating the flow of other charges. By assuming a particular distribution of trapping times, or pulse lengths, a $1/f$ spectrum can be obtained. The matter of noise in semiconductors is by no means thoroughly understood, and somewhat different mechanisms have been suggested.

Actually, the power spectrum cannot vary as $1/f$ right down to $f=0$, for this would imply an infinite noise power. Measurements do show a $1/f$ spectrum down to frequencies as low as 10^{-4} cycles per second.

Bibliography

This is not a comprehensive bibliography. It is intended to direct the reader to some important early papers and to later papers or books in which more detailed treatments and some further references are given.

Johnson Noise

Johnson, J. B., "Thermal Agitation of Electricity in Conductors." *Physical Review*, Vol. 32 (July 1928), pp. 97–109.
Nyquist, H., "Thermal Agitation of Electric Charge in Conductors." *Physical Review*, Vol. 32 (July 1928), pp. 110–113.
Spenke, E., "Zur korpuskularen Behandlungsweise des Thermischen Rauschens elektrischer Widerstände." *Wissenschaftliche Veröffentlichungen aus den Siemens-Werken*, Vol. 18 No. 2 (February 16, 1939), pp. 54–72.

Shot Noise

Schottky, W., "Spontaneous Current Fluctuations in Various Conductors." *Annalen der Physik*, Vol. 57 No. 23 (1918), pp. 541–567.

Noise in Tubes and Multipliers

Pierce, J. R., "Noise in Resistances and Electron Streams." *Bell System Technical Journal*, Vol. 27 (January, 1948), pp. 158–174.
Beck, A. H. W., *Thermionic Valves*. Cambridge, Cambridge University Press, 1953.

Transistor Noise

Shockley, W., *Electrons and Holes in Semiconductors*. New York, D. Van Nostrand Co., 1950.
MacFarlane, G. G., "A Theory of Contact Noise in Semiconductors." *Proceedings of the Physical Society*, Vol. 63 (October 1, 1950), pp. 807–813.
Bess, Leon, "A Possible Mechanism for 1/f Noise Generation in Semiconductor Filaments." *Physical Review*, Vol. 91 (September 15, 1953), p. 1569.

General

van der Ziel, A., *Noise*. New York Prentice-Hall, 1954.
MacDonald, D. K. C., "Brownian Movement and Spontaneous Fluctuations of Electricity." *Research* (London), Vol. 1 (February, 1948), pp. 194–203.

Electronic fluctuations in semiconductors*

By Prof. R. E. Burgess, B.Sc., A.M.I.E.E.,† Radio Research Station, Slough, Bucks.

Electrical fluctuations which occur in semiconductors, particularly on the passage of current, are of considerable fundamental and technological significance. In many systems the limiting sensitivity is determined almost entirely by the noise of a semiconductor device. The present paper reviews the various types of spontaneous fluctuation which arise and comparison is made with the corresponding processes in a vacuum tube.

Particular attention is paid to modulation noise which is conspicuous in semiconductors at low frequencies, especially when current-carrying contacts or barriers occur.

Electrical fluctuations in semiconductors have attracted considerable attention over the last fifteen years, for two main reasons; because many radio and electronic devices depend essentially for their operation on the properties of a semiconductor and because the noise exhibited by these devices when carrying current can be considerably in excess of the thermal and shot noise components. Usually the spectral density of the noise decreases with increasing frequency and for this reason it is particularly a limitation at the lower frequencies; for instance, the transistor cannot compete with the vacuum tube at audio frequency in respect of signal/noise ratio, and this limitation in the performance of a new device has been responsible for increased attention to the general problem.

Empirically it has been observed that devices which are accurately reproducible in their macroscopic parameters (current-voltage-temperature characteristics) may exhibit noise levels which differ greatly from one sample to another. It is accordingly believed that the fluctuation processes may be markedly structure-sensitive and depend upon certain types of imperfection to a much greater extent than the macroscopic characteristics. It is also known experimentally that there is a particular tendency for noise to be generated in transition regions such as occur between a metal electrode and a semiconductor or between semiconductors having opposing types of conductivity. Since one or both of these types of transition region are inevitably present in practical devices it is particularly important that contact and junction noise should be understood.

There is little doubt that further improvements will result from more refined techniques of preparing semiconducting materials so as to eliminate chemical and physical imperfections and from advances in methods of fabrication so as to control the conditions existing at the surfaces and at electrode contacts.

In the present paper the various sources of noise in semiconductors are considered. Where possible the individual processes are compared with those which are analogous and familiar in the vacuum tube. We shall not be greatly concerned with *purely* spatial fluctuations, i.e. variations from point to point of the properties of the semiconductor as might arise from a random distribution of impurities; so long as such a distribution is invariant in time it would not of itself provoke temporal fluctuations which are our concern here.

It is appropriate to consider some of the terminology used to describe the fluctuations and for this purpose we regard the fluctuating variable x in which we are interested as a statistically stationary time series having certain well defined attributes; mean value $\bar{x}$, variance σ^2, autocorrelation coefficient $r(t)$, and the spectral density $S(f)$.

The spectral density may be defined as follows: if a fluctuating voltage be transmitted by a narrow band pass filter centred on frequency f and having unity transmission over a bandwidth Δf, the mean square voltage at the output would be given by $S(f)$. Δf when averaged over a time long compared with $1/\Delta f$. If the variable under examination contained a d.c. component $\bar{x}$ or a discrete sine-wave component of a certain frequency f_0, these would give rise to spectral lines at $f = 0$ or f_0 and would correspond to delta function terms in $S(f)$. In the fluctuations we are considering there may in general be a non-zero mean value of the quantity but not usually any discrete frequencies. The relations between the autocorrelation coefficient and the spectral density of the continuum may therefore be written

$$\sigma^2 r(t) = \int_0^\infty S(f) \cos (2\pi ft)df \tag{1}$$

$$S(f) = 4\sigma^2 \int_0^\infty r(t) \cos (2\pi ft)dt \tag{2}$$

The latter relation is frequently of convenience in the calculation of the spectral density of processes in which the statistical behaviour in time is readily formulated.

In more physical terms, the variables whose fluctuations are of interest include: current, voltage, resistance or conductance, number of carriers, number of ions, number of occupied traps, rates of generation and recombination of carriers, temperature, and height of a potential barrier.

A phenomenological description of the noise current and voltage of a two-terminal device connected in a circuit may be attempted in terms of the "fluctuating characteristic" concept. It is worthwhile to examine this in a little detail since the question of representation of noise generators in non-linear systems is involved. Consider a simple circuit having a battery of e.m.f. E in series with a source resistance S and a noisy element of resistance $R(t)$ which will be assumed for generality to have a non-linear current-voltage characteristic. Let it be possible to speak of an instantaneous characteristic, so that the characteristic at time t in general differs from the mean characteristic (as observed macroscopically). The fluctuations will always be regarded as sufficiently small for first-order effects only to be important. On account of the non-linearity the d.c. resistance R and the a.c. differential resistance ρ differ. The following relations then hold for the fluctuations of voltage current and resistance:

$$V = -S\Delta I = \frac{SE}{(S+R)^2}\Delta R = \frac{SI}{S+R}\Delta R \tag{3}$$

* Based on a lecture given before the Electronics Group of The Institute of Physics in London on 18 September, 1954.

† Now at Department of Physics, University of British Columbia, Vancouver.

whereas the equivalent noise generators are

$$e = -\rho i = \frac{S+\rho}{S+R} I \Delta R \qquad (4)$$

It is important to note that the noise e.m.f. e is not in general equal to $I\Delta R$. Equality only obtains when the noisy element R is linear ($R = \rho$) or when the impedance of the external circuit is very large compared with both R and ρ. Similarly the noise current generator i is not in general equal to $-I\Delta R/R$ unless R is linear or the impedance of the external circuit is very small compared with both R and ρ. It is also clear that thermal noise, which arises purely spontaneously in the absence of any current flow from an external circuit cannot be represented in terms of a fluctuating characteristic.

The types of noise which will be considered in order are: thermal noise, shot noise, partition noise, avalanche noise, and modulation noise. Much of the discussion is a review of previous work but some original contributions are included which it is intended to treat more fully elsewhere.

THERMAL NOISE

In any two-terminal passive element in thermodynamical equilibrium at absolute temperature T the fluctuation e.m.f. appearing at the terminals has the spectral density

$$S_e(f) = 4kTR(f) \qquad (5)$$

where $R(f)$ is the real part of the impedance presented at the terminals at frequency f and k is Boltzmann's constant; this result, due to Nyquist is valid so long as $hf \ll kT$, or $f \ll 2 \times 10^{10} T \sec^{-1}$. This equation relates the electrical manifestation of the thermal energy of the current carriers (electrons, holes, ions) to the atoms of the material as a whole through the dissipative mechanism represented by R. Its proof rests on purely thermodynamical arguments and is therefore independent of mechanisms of conduction. However, proofs based on particular models of conduction-electron distribution and scattering have been given[1, 2] and lead to the same result.

The noise current spectral density is given by

$$S_i(f) = 4kTG(f) \qquad (6)$$

where $G(f)$ is the real part of the admittance at frequency f.

Although the thermal noise of a semiconductor or a semiconductor device is a well-defined quantity when equilibrium is established, this does not correspond to the practical conditions in which the devices normally operate. Usually there will be current flowing and thus additional types of noise can arise. Nevertheless, if the applied electrical field is everywhere such that the velocity distribution of the current carriers is unchanged apart from the linear superposition of a drift velocity in the direction of the field it is reasonable to assume that to a first approximation the thermal noise component is still given by Nyquist's equation. Physically we are saying that the thermal noise originates from the random discontinuities in the motions of the carriers due to their collisions with the atoms and its value is intimately connected with the appearance of the term kT in the energy distribution function for the carriers. In silicon and germanium at room temperature it seems that the superposition of drift and thermal velocities is valid for electric fields such that the drift velocity does not exceed 3×10^6 cm/s.[3, 4]

SHOT NOISE

The term "shot noise" was originally applied to the fluctuations of current in a saturated vacuum diode due to the randomness of electron emission from the cathode. If the electrons are emitted at an average rate $\nu(=I/e)$ per second and each electron travels to the anode without interaction with the other electrons, the spectral density of the current is given by

$$S_i(f) = 2\nu|F(f)|^2 = 2(I/e)|F(f)|^2 \qquad (7)$$

where $F(f)$ is the Fourier transform of the current impulse due to the passage of a single electron:

$$F(f) = \int_0^\tau i(t) \exp(-j\omega t)dt \qquad (8)$$

At low frequencies such that the electron transit time τ is small compared with $1/\omega$, the transform $F(f) \simeq e$ and the spectral density assumes the simple form $2eI$. The concept of randomness of rate of emission implies that the process is determined by a stationary Poisson distribution. The random distribution of emission *velocities* can for some purposes be regarded as contributing additional thermal noise.[5]

If the vacuum diode be replaced by a uniform semiconductor a number of important differences arise. In general there are two types of current carrier—electrons and holes; these carriers will undergo many collisions during their drift motion with the electric field and furthermore they may be generated and may recombine at points throughout the bulk and on the surface of the semiconductor as well as at the electrodes. Thus the current impulses due to individual transits may vary widely in their duration and their integrated value $F(0) < e$ since it is only equal to the electronic charge for a carrier which passes completely from one electrode to the other. We here separate the carrier motion into the superposition of a drift motion and thermal agitation and to the latter we ascribe thermal noise given by Nyquist's equation to which must be added the shot noise due to drift which we now calculate.

Consider the case of an n-type semiconductor rod of cross section A, length l and with uniform electric field E. If the electron mobility is μ and the mean lifetime τ we can assert that if N electrons are in motion at any given moment, the number of those still in motion at a time t later is, on the average

$$N \exp(-t/\tau)(1 - E\mu t/l) \quad \text{for } t \leqslant l/E\mu \qquad (9)$$

and zero for greater values of t.

Thus

$$\overline{N_0 N_t} - (\overline{N})^2 = [\overline{N^2} - (\overline{N})^2] \exp(-t/\tau)(1 - E\mu t/l) \qquad (10)$$

and since the instantaneous drift current is $Ne\mu E/l$ the spectral density of the shot noise is

$$S(f) =$$

$$4\left(\frac{e\mu E}{l}\right)^2 [\overline{N^2} - (\overline{N})^2] \int_0^{l/E\mu} \exp(-t/\tau)(1 - E\mu t/l) \cos(\omega t)dt \qquad (11)$$

$$= 2eI\left[\frac{\overline{N^2} - (\overline{N})^2}{\overline{N}}\right] W\left(\omega\tau, \frac{\omega l}{E\mu}\right) \qquad (12)$$

where $W = \dfrac{2\theta^2/\phi^2}{1 + \theta^2} \times$

$$\left\{ \frac{\phi}{\theta} - \frac{1 - \theta^2}{1 + \theta^2} + \frac{\exp(-\phi/\theta)}{1 + \theta^2}[(1 - \theta^2)\cos\phi - 2\theta\sin\phi] \right\}$$ (13)

in which $\theta \equiv \omega\tau$ and $\phi = \omega l/E\mu$. Davydov and Gurevich[6] also considered this particular case but derived an erroneous result. A number of special cases may now be considered:

(i) If the frequency is sufficiently low for both θ and ϕ to be small compared with unity

$$W = \frac{2\theta^2}{\phi^2}\left(\frac{\phi}{\theta} - 1 + \exp\frac{-\phi}{\theta}\right)$$ (14)

$$\simeq 1 \text{ if } l \ll E\mu\tau$$

as expected since we now have all the carriers traversing the semiconductor and full shot noise results.

(ii) If $l \gg E\mu\tau$ so that the life-path along the field is negligible compared with the length of the specimen

$$W = \frac{2\theta/\phi}{1 + \theta^2} = \frac{2E\mu\tau}{l}\frac{1}{1 + \omega^2\tau^2}$$ (15)

in which the shot effect is reduced by both a path factor and a frequency term.

(iii) If $l \ll E\mu\tau$

$$W = 2(1 - \cos\phi)/\phi^2$$ (16)

which corresponds to electrons crossing the distance l with uniform drift velocity $E\mu$ and is analogous to the noise induced by an electron beam in a klystron gap.

The ratio of the variance of N to its mean value which occurs in the spectral density depends upon the mechanism of carrier generation and recombination or trapping.[7] It is not unity as is commonly assumed.

Another important instance of shot noise arises in the other extreme from a uniform semiconductor, namely the motion of carriers across a high-field transition region, e.g. at a metallic contact or at a p–n junction. Normally the carrier velocities in such a region would be of the order of 10^7 cm/s and the width of the region would lie in the range 10^{-5} to 10^{-3} cm so that the transit time would be negligible except at the highest microwave frequencies. Furthermore it may be readily shown that since the change in quasi-Fermi level for the carriers across the transition region is very nearly equal to the applied voltage, the effect of each electron transit is effectively to induce a current impulse $e\delta(t)$, and thus full shot noise may be attributed to the flow.

In a semiconductor diode *at zero bias* there is no net current across the barrier layer, but since the noise induced by carrier transits in independent of the direction of transit one may consider the noise to equal twice the shot noise attributable to the flow I_0 in either direction. Then it is clear that since the noise must also be given by Nyquist's formula the zero-voltage conductance of the diode is

$$G_0 = eI_0/kT$$ (17)

If a biased semiconductor diode be illuminated a photo-current in the reverse direction is added to whatever current is already flowing due to the bias. Experimental data[8] may be interpreted to show that the increase of noise on illuminating a reverse-biased diode is equal to the shot noise associated with the increase of current *at a constant bias voltage.* This shot noise behaviour occurs even at low

frequencies where the dark current may exhibit appreciable modulation noise. This result is very significant from the theoretical point of view, and is discussed elsewhere.[9]

The zero-bias condition is of importance in the video-detector type of receiver in which the low-level signals (usually received over a broad band at microwaves) are directly rectified and passed to a video amplifier. The r.m.s. noise current is inversely proportional to the square root of the crystal conductance, while the signal current is proportional to the applied signal power, the factor of proportionality being $e/2kT$ for an ideal detector. In this system there is no theoretical noise problem and the practical problem is to make the rectification coefficient approach the ideal as nearly as possible.

In a microwave superheterodyne receiver using a crystal diode mixer the noise appearing at the intermediate frequency is approximately given by the shot noise which one would attribute to the mean current flow through the diode. In good diodes at 30 or 45 Mc/s this is fairly closely true although in poor diodes there is a vestige of the modulation noise spectrum. The experimental work of Miller[10] and Nicoll[11] with both d.c. and a.c. excitation of silicon diodes illustrates the relative importance of shot and modulation noise at different frequencies.

If a diode has a reverse characteristic of the form

$$I = I_0[1 - \exp(-eV/kT)]$$ (18)

where I_0 in general increases slowly with the reverse voltage V, and if the h.f. noise be attributed wholly to the sum of the shot noises of two components of current:

$$S_i(f) = 2eI_0[1 + \exp(-eV/kT)]$$ (19)

then this may be expressed also in the following manner:

$$S_i(f) = 4kTG + 2eI\left(1 - \frac{2kT}{eI_0}\frac{dI}{dV}\right)$$ (20)

showing that the noise could be represented as the sum of thermal noise and a shot noise term which is somewhat reduced below the full value of the effect of non-saturation of the reverse current.

PARTITION NOISE

If a current *in vacuo* or in a semiconductor has the possibility of dividing between two or more electrodes, the process of partition will, because it is a statistical process, influence the noise in the resulting components of the current.

Let us first consider the case of a vacuum tube in which a cathode current I_0 having mean square noise $2eI_0F^2$ per unit bandwidth (F^2 = space-charge reduction factor) can divide between two electrodes (1) and (2). Then if I_1 and I_2 are the mean currents to these electrodes ($I_1 + I_2 = I_0$) the probabilities that an electron arrives at one or the other are I_1/I_0 and I_2/I_0 respectively. The noise in the current to electrode (1) will have the component $i_{01}^2 = 2eI_0F^2(I_1/I_0)^2$ due to the original cathode noise and a partition noise component given by $i_{p1}^2 = 2eI_1 . I_2/I_0$ as determined from the variance of a binomial distribution. Thus the total noise in electrode (1) is

$$\overline{(i_{01} + i_{p1})^2} = \overline{i_1^2} = 2eI_1(I_1F^2 + I_2)/I_0$$ (21)

The noise in electrode (2) may be similarly expressed with $i_{p2} = -i_{p1}$:

$$\overline{(i_{02} + i_{p2})^2} = \overline{i_2^2} = 2eI_2(I_2F^2 + I_1)/I_0$$ (22)

As might be expected, if the cathode current exhibits full shot noise ($F^2 = 1$) the currents I_1 and I_2 will likewise do so since partition leaves unaffected the randomness of the original electron stream.

The noises of the two currents are correlated since they have equal and opposite partition components (electrons which do not go to one electrode go to the other) and also have related fractions of the initial cathode current noise. Thus

$$\overline{i_1 i_2} = \overline{(i_{01} + i_{p1})(i_{02} + i_{p2})} = -2e(I_1 I_2/I_0)(1 - F^2) \quad (23)$$

Hence negative correlation always exists between the current fluctuations at the two electrodes. This correlation must be taken into account when circuital or electronic coupling exists between the two electrodes. Only when $F^2 = 1$ does the correlation become zero and the noises in (1) and (2) can be regarded as completely independent.

Now an analogous situation arises in an NPN junction transistor. The cathode current corresponds to the electron stream passing from the emitter into the base, the recombination of injected carriers in the base region corresponds to the current passing to electrode (1) and the current carried to the collector to that at electrode (2). Thus if the emitter be assumed to have unity electron injection efficiency we identify α with the ratio I_2/I_0. Since for normal current densities in semiconductors space charge effects are negligible we put $F^2 = 1$. The presence of a non-zero resistance in the base circuit results in a common impedance carrying the base current which will influence both the impedance and the noise appearing at the other electrodes. A further feature which does not arise in the vacuum tube is that not all of the collector current I_0 arises from carriers transmitted from the emitter across the base; a relatively small component I_{c0} is due to electrons thermally generated in the base region diffusing across to the collector junction. This component generates shot noise in the base-collector circuit independently of the current contribution from the emitter. On the basis of the shot noise of the emitter current and of I_{c0} and the partition noise associated with current division between base and collector together with the thermal noise of the base resistance it is possible to construct a reasonably satisfactory model for the noise behaviour of a junction transistor at frequencies sufficiently high that the $1/f$ component is unimportant yet not so high that the phase angle of the current transmission factor α is significant.[12]

AVALANCHE NOISE

When a barrier region is subjected to reverse bias the electric field may reach the order of 10^5 V/cm or greater, and at these fields there occur phenomena which cause a rapid increase of current and eventual breakdown; it has furthermore been observed that the current is "noisy" in this region, becoming increasingly impulsive as breakdown is approached.

One possible phenomenon is analogous to field emission and consists of the excitation of electrons from the valence band to the conduction band across the forbidden gap. It was first discussed by Zener in connection with dielectric breakdown in crystals. Similar excitation may also occur between impurity atoms and the valence or conduction band.

The second type of phenomenon is the electron multiplication process which arises from collision ionization of neutral atoms by electrons which have acquired appreciable energy from the field between collisions. This process is analogous to the electron avalanches which can occur in electrical discharges in gases, and the theory for the ionization rate per unit path length has been given by Wolff.[13]

McKay[14] has studied the processes of reverse voltage breakdown in silicon and germanium p–n junctions while Tomura and Abiko[15] have investigated selenium rectifiers. McKay concluded that electron and hole multiplication is the important process in silicon and germanium junctions while only in narrow germanium junctions is the Zener emission likely to be of significance.

As a first approximation the noise of electron avalanches may be treated from the point of view of the shot effect and it is somewhat analogous to the noise of a secondary emission multiplier. If n is the number of electrons in any one avalanche, the current it carries is nev where v is the average frequency of such avalanches while the low frequency Fourier transform of the current impulse $F(0) = ne$. Thus the shot noise spectral density at frequencies low compared with the reciprocal of the transit time is given by:

$$S = 2vn^2 e^2 = 2eI(\overline{n^2}/\bar{n}) \quad (24)$$

This is the formula used by Haworth and Bozorth[16] in their study of electron multiplication in Pyrex, and they inferred values of $\overline{n^2}/\bar{n}$ ranging from 10 to 10^6 as the electric field was increased by a factor of 4.

Noise measurements[15] at 600 kc/s on reverse-biased selenium rectifiers were interpreted as corresponding to values of $\overline{n^2}/\bar{n}$ ranging from unity (corresponding to normal shot noise without electron multiplication) up to 100.

In silicon junctions McKay[14] observed that at the onset of breakdown there appears a distinctive form of impulsive noise consisting of a random sequence of rectangular current pulses of variable duration but of constant amplitude. The spectral density of such a pulse train would be expected to be of the form $S_0/(1 + \omega^2 \tau^2)$ where $1/\tau$ is the sum of the transition probabilities corresponding to the on-and-off processes.[17] It is possible that the inevitable inhomogeneity of the semiconductor in the neighbourhood of the junction gives rise to small regions (or "weak spots") in which breakdown occurs for lower applied voltages than elsewhere and this localized breakdown will switch from an "off" to an "on" condition and back again, triggered by random fluctuations.

MODULATION NOISE

The term modulation noise is applied to those processes in which some controlling parameter (such as temperature, barrier height or rate of generation of carriers) undergoes fluctuations and thereby influences the flow of current on a collective or semi-macroscopic scale. To illustrate this concept by means of an analogy, consider a vacuum triode having a floating grid capable of capturing electrons which can leak away slowly through a high-resistance path to the cathode. Then in addition to the normal shot effect of the electron stream to the anode there will be grosser fluctuations of current due to modulation of the anode current by the fluctuations of grid potential; these fluctuations are an example of modulation noise and they have the typical characteristics of being able to exceed greatly the shot effect and having a spectral density which increases with decreasing frequency.

Similarly the flicker effect in a vacuum tube is due to the slow random variations in emissivity of small patches of the cathode surface; here the modulation is effectively one of a barrier height, i.e. of the work function of localized regions of the cathode.

In both these examples involving tubes the physical

mechanism involves the random arrival and departure of electrons or ions at a controlling region (the grid or the cathode surface). Similar effects can occur in a semiconductor due to the motion of carriers and the diffusion of ions, causing a variation in the effectiveness of potential barriers to influence the flow of current. This implies that co-operative effects are occurring and that the motion of carriers is no longer truly random.

Experimentally[10,11,18–22] the appearance of modulation noise becomes most marked at the lower frequencies (i.e. below about 10 kc/s) since it has a spectral density which increases with decreasing frequency.

These low frequency studies have been made in systems ranging from the simplest, namely bars of single-crystal germanium with potential probes, to the complicated in the form of point-contact diodes and transistors. In all these cases there is evidence of the presence of a component having a $1/f^n$ variation where n is usually in the range $1 \cdot 0$ to $1 \cdot 1$ together with components having the form appropriate to processes with a well-defined relaxation time τ:

$$S = \frac{A}{f^n} + \frac{B}{1 + \omega^2\tau^2} \tag{25}$$

The dependence of A and B on current (or voltage) and temperature has not been established but there is evidence that they are not in general proportional to I^2 as is suggested by unsophisticated theories. Not all experimenters have used the same noise generator representation. The preferable spectral density is that of the mean square current generator $S_i(f)$; sometimes the e.m.f. density $S_e(f)$ is used and sometimes the available power density $S_p(f)$.

These various forms are related by

$$S_e(f) = |Z(f)^2|S_i(f) \tag{26}$$

$$S_p(f) = S_e(f)/4R(f) = S_i(f)/4G(f) \tag{27}$$

where Z, R and G are the impedance, resistance and conductance of the device at frequency f. An alternative to the available power spectrum is that of the equivalent noise temperature ratio defined by

$$t(f) = S_e(f)/4kTR(f) = S_i(f)/4kTG(f) \tag{28}$$

These representations lead to spectral densities which vary with frequency, applied voltage and temperature. Thus when testing any specific model of noise generation it is important to choose the appropriate form in order that such phrases as "a $1/f$ component" or "temperature-independent noise" shall be meaningful.

In the simplest model for modulation noise the current I depends upon a barrier height ϕ by a factor of the form $\exp(-e\phi/kT)$. Thus if ϕ fluctuates due to some controlling parameter x fluctuating, the current fluctuations are given by

$$\overline{(\Delta I)^2} = \left(\frac{e}{kT}\right)^2 I^2 \left(\frac{d\phi}{dx}\right)^2 \overline{(\Delta x)^2} \tag{29}$$

Now only if $d\phi/dx$ is independent of I (or the applied voltage V) will the noise spectral density be proportional to I^2. In many physical systems, however, $d\phi/dx$ will depend on I or V; for example in a metal-semiconductor contact the barrier height is influenced by image force or by the surface state space charge. Each of these controls depends upon both the applied voltage and the ionic density in the barrier layer. Hence if the ionic density fluctuates so will the barrier height but to an extent which depends upon the field at the contact and thus on the applied voltage. The measurements of Hyde[19] on point-contact germanium rectifiers may be interpreted to suggest that the modulation noise current spectral density varies more nearly as I^2V rather than as I^2 and this type of dependence can be understood in terms of the model just discussed.

In bulk semiconductors such as the germanium filaments which have been examined experimentally in some detail, the mechanism for the modulation noise appears to be a variation of the rate at which carriers (or carrier pairs) are generated and recombine. Surface conditions have a marked effect suggesting that the modulation centres are primarily associated with superficial physical or chemical imperfections whose electrical state changes either by transitions in the state of ionization or by surface migration. The observed effect of magnetic fields on filament noise confirms at least qualitatively that surface sources and sinks for minority carriers are largely responsible for filament noise.[20,23]

A very general analysis of the noise due to diffusional mechanisms has been given by Richardson.[24] It is particularly applicable to the diffusion of ions in and out of a barrier region; the "coupling" between the current and the spatial location of the ions is determined completely generally by a vector function and the diffusion of the ions may take place in the volume or over the surface. Various specific models which lead to convergent spectra do not exhibit a $1/f$ behaviour over a wide range of frequency. This type of behaviour can be obtained by a rather special diffusional model but ensuring convergence at $f = o$ and $f = \infty$ requires the introduction of suitable boundary conditions for the coupling function.

A different approach to the $1/f$ type of spectrum is to consider the distribution in energy of modulation processes, and this treatment is now presented by considering electronic transitions between a modulation centre and a continuous band.

Consider a process in which a single modulation centre makes alternate random transitions between two states 1 and 2 (e.g. unionized and ionized) and let p_{12} and p_{21} be the transition probabilities per unit time. If such a transition causes a change A in the current flow in the system the relevant spectrum is that of a random rectangular wave. The resulting spectral density of the current may be derived from Machlup's analysis[17] (which was applied to a different problem) and is found to be

$$S(f) = \frac{4A^2\nu(\tau)\tau^2}{1 + \omega^2\tau^2} \tag{30}$$

where $1/\tau = p_{12} + p_{21}$ and $1/\nu(\tau) = 1/p_{12} + 1/p_{21}$.

If the transitions involve an energy difference E, the probabilities will be both proportional to $\exp(-E/kT)$; the ratio of the probabilities p_{12}/p_{21} is clearly the ratio r of the average time spent in state 2 to that spent in state 1, and is independent of E. If now E can have values lying between E_1 and E_2 the probability distribution of τ is proportional to $\exp(-E/kT)$ and thus to $1/\tau$ whence

$$P(\tau) = \frac{kT}{(E_2 - E_1)\tau} = \frac{1}{\tau \ln(\tau_2/\tau_1)} \tag{31}$$

and

$$\nu(\tau) = \frac{r}{(1 + r)^2\tau} \tag{32}$$

Thus the composite spectrum is

$$S(f) = 4A^2 \frac{kT}{E_2 - E_1} \frac{r}{(1 + r)^2} \int_{\tau_1}^{\tau_2} \frac{d\tau}{1 + \omega^2 \tau^2} \qquad (33)$$

$$= 4A^2 \frac{kT}{E_2 - E_1} \frac{r}{(1 + r)^2} \frac{\tan^{-1}\omega\tau_2 - \tan^{-1}\omega\tau_1}{\omega} \qquad (34)$$

where τ_1 and τ_2 correspond to the extreme cases of the allowable transitions with energy differences E_1 and E_2.

Thus for frequencies such that $1/\tau_1 \gg \omega \gg 1/\tau_2$ we have a $1/f$ spectrum. At extreme low frequencies S becomes constant while at high frequencies it behaves as $1/f^2$. The $1/f$ behaviour at intermediate frequencies depends on the Boltzmann factor in the transition probabilities, and since this is a general attribute it is not surprising that the $1/f$ spectral component has been found so ubiquitously in a wide variety of types of current noise. Note that it is *not* dependent on any assumed distribution of energy levels of centres. Since in fact the spectrum departs at the extremes from the $1/f$ form (and this could occur within the experimental range of frequencies) it is proposed that the generic form of equation (34) should be termed the "arctan" type of spectrum. Nothing generally can be asserted about the temperature since, in addition to the factor T, the parameters A and r will also depend on temperature.

The model just discussed may be illustrated by an example. Let the modulation centre be a hole trap which is either neutral (state 1) or negative (state 2) due to transitions between the centre and the valence band. Then we identify E_1 with the difference in energy level between the centre and the top of the valence band and $E_2 - E_1$ with the width of the valence band. The ratio r is simply $b \exp(-E_0/kT)$ where E_0 is the height of the trapping level above the Fermi level, and b is a spin factor which will be $\frac{1}{2}$ or 2. The factor $r/(1 + r)^2$ will be small if $|E_0|$ is large compared with kT and consequently only levels near the Fermi level will experience appreciable fluctuations. In this connection it should be noted that the total mean square current fluctuation integrated over all frequencies is

$$\overline{(\Delta I)^2} = \int_0^\infty S(f)df = A^2 r/(1 + r)^2 \qquad (35)$$

which is, as expected, the variance of a quantity which can assume the values 0 or A with probabilities $1/(1 + r)$ and $r/(1 + r)$ respectively.

The mean square fluctuations in the number of carriers or ions in a semiconductor can also be calculated either by a purely statistical argument using the Fokker–Planck approach, or in the case where thermodynamical equilibrium applies by the application of the Einstein free-energy method. These methods have been discussed in a recent paper[7] and shown to be consistent; particularly simple asymptotic results may be derived when the numbers involved are large as is usually the case.

If a variety of modulation centre levels occur, the spectrum will contain as many terms each having the form of equation (34). But so long as ω is between but remote from $1/\tau_1$ or $1/\tau_2$ for each of these levels the $1/f$ type of spectrum is obtained for each individual centre as well as for the aggregate. The distribution of relaxation times necessary for this type of spectrum arises simply from the existence of continuous bands (conduction or valence) to which transitions can be made, be it only from a single bound level or several. However as mentioned above the greatest weight will be given to those levels which lie near the Fermi level.

ACKNOWLEDGEMENTS

The work described above was carried out as part of the programme of the Radio Research Board. This paper is published by permission of the Director of Radio Research of the Department of Scientific and Industrial Research.

REFERENCES

(1) BAKKER, C. J., and HELLER, G. *Physica*, **6**, p. 262 (1939).
(2) BERNAMONT, J. *Ann. Phys.* [Paris], **7**, p. 71 (1937).
(3) RYDER, E. J. *Phys. Rev.*, **90**, p. 766 (1953).
(4) SHOCKLEY, W. *Bell. Syst. Tech. J.*, **30**, p. 990 (1951).
(5) FREEMAN, J. J. *J. Appl. Phys.*, **23**, p. 1223 (1952).
(6) DAVYDOV, B., and GUREVICH, B. *J. Phys. USSR*, **7**, p. 138 (1943).
(7) BURGESS, R. E. *Physica*, **20**, p. 1007 (1954).
(8) SLOCUM, A., and SHIVE, J. N. *J. Appl. Phys.*, **25**, p. 406 (1954).
(9) BURGESS, R. E. *Proc. Phys. Soc.* [London] (in course of publication).
(10) MILLER, P. H. *Proc. Instn Radio Engrs*, **35**, p. 252 (1947).
(11) NICOLL, G. *Proc. Instn Elect. Engrs*, **101** (Pt. III), p. 317 (1954).
(12) VAN DER ZIEL, A. *J. Appl. Phys.*, **25**, p. 815 (1954).
(13) WOLFF, P. A. *Phys. Rev.*, **95**, p. 1415 (1954).
(14) McKAY, K. G. *Phys. Rev.*, **94**, p. 877 (1954).
(15) TOMURA, M., and ABIKO, Y. *J. Phys. Soc. Japan*, **7**, p. 524 (1952).
(16) HAWORTH, F. E., and BOZORTH, R. M. *Physics*, **5**, p. 15 (1934).
(17) MACHLUP, S. *J. Appl. Phys.*, **25**, p. 341 (1954).
(18) BROPHY, J. *J. Appl. Phys.*, **25**, p. 222 (1954).
(19) HYDE, F. J. *Proc. Phys. Soc.* [London] B, **66**, p. 1017 (1953).
(20) MONTGOMERY, H. C. *Bell Syst. Tech. J.*, **31**, p. 950 (1952).
(21) VAN VLIET, K. M., and others. *Physica*, **20**, p. 481 (1954).
(22) VAN DER ZIEL, A. *Noise* (New York: Prentice-Hall, Inc., 1954).
(23) SUHL, H. *Bell Syst. Tech. J.*, **32**, p. 647 (1953).
(24) RICHARDSON, J. M. *Bell Syst. Tech. J.*, **29**, p. 117 (1950).

Methods of Solving Noise Problems[*]

W. R. BENNETT[†], SENIOR MEMBER, IRE

Summary—A tutorial exposition is given of various analytical concepts and techniques of proved value in calculating the response of electrical systems to noise waves. The relevant probability theory is reviewed with illustrative examples. Topics from statistics discussed include probability density, moments, stationary and ergodic processes, characteristic functions, semi-invariants, the central limit theorem, the Gaussian process, correlation, and power spectra. It is shown how the theory can be applied to cases of noise and signal subjected to such operations as filtering, rectification, periodic sampling, envelope detection, phase detection, and frequency detection.

SIGNIFICANCE OF PROBABILITY DENSITY

NOISE-LIKE phenomena have the common property that we do not specify precisely what magnitudes are observed at what times. The reasons for our failure to give such complete descriptive data are various—we may not know enough, we may consider the system to be so complicated that complete use of all possible knowledge is not practical, or we may decide that all we really need to know is furnished by a smoothed-out picture of the true state of affairs. Whatever our motives, by refraining from precise specification, we resign ourselves to a statistical description and we must get used to the appropriate mathematical tools.

The first concept we introduce is that of the probability density function. Even though the numerical outcome of a noise measurement may vary widely when what we choose to call the same experiment is repeated many times, we can distinguish relative likelihood of obtaining various values. Suppose a single property of the noise, as for instance the instantaneous voltage or current, is under measurement. Let x represent a typical measured value. Imagine each x to define a point at the corresponding distance from a fixed reference point on a straight line. Then if we divide this line into small equal intervals of length Δx, and count the number of points in each interval, we approximate a density function of x, which we shall call $p(x)$. We proceed as if there existed such a function defined by

Symbolically

$$\text{Prob } (x_1 < x < x_2) = \int_{x_1}^{x_2} p(x)dx. \tag{2}$$

A second important function called the distribution function $P(x)$ is defined as the probability that the value is less than some specified x. It is given by

$$P(x) = \int_{-\infty}^{x} p(x)dx. \tag{3}$$

It follows that

$$p(x) = \frac{d}{dx} P(x) = P'(x) \tag{4}$$

for such values of x for which $P(x)$ has a derivative.

Note that not all functions are suitable for probability density or distribution functions. To be meaningful, $p(x)$ cannot be negative or imaginary. Also if it is certain that every measurement must yield some real value, we must have

$$\int_{-\infty}^{\infty} p(x)dx = 1 \tag{5}$$

signifying a certainty that the value of x lies somewhere between $-\infty$ and ∞. From the non-negative character of $p(x)$, the value of $P(x)$ cannot decrease with increasing x, and from (5)

$$P(-\infty) = 0, \qquad P(\infty) = 1. \tag{6}$$

An instructive use of the probability density function occurs in the calculation of averages. Suppose we wish to know what average value of some specified function of the quantity x would be obtained from a large number of measurements. The function might for example be the value of x raised to some power, it might be the exponential or sine of x, and of course it might in the simplest nontrivial case be proportional to x itself. In

$$p(x) = \mathop{\text{Lim}}_{\substack{\Delta x=0 \\ N=\infty}} \frac{(\text{Number of Values in Range } \Delta x \text{ at } x)/\Delta x}{\text{Total Number of Values} = N}. \tag{1}$$

We call $p(x)$ the probability density function. As with other densities with which we are familiar, we obtain the probability that a particular measured value lies in an interval of infinitesimal length dx centered at x by multiplication of $p(x)$ by dx. Then if we wish to find the probability that the value is in a specified larger range, say x_1 to x_2, we integrate $p(x)$ through this range.

general let $F(x)$ represent the function of x which is to be averaged and suppose data from many measurements are available. Assume that we calculate the average of $F(x)$ in the usual way by dividing the sum of the values obtained for $F(x)$ by the number of observations. By considering fine divisions of width dx on the x scale, and noting that for a large number of trials a fraction $p(x)dx$ of the total observations belong to an interval of length dx containing the value of x and hence give the obser-

[*] Original manuscript received by the IRE, January 17, 1956.
[†] Bell Telephone Labs., Inc., Murray Hill, N. J.

Reprinted from *Proc. IRE*, vol. 44, pp. 609–638, May 1956.

vation $F(x)$, we reason that in the limit as the number of measured values becomes very large

$$\text{av } F(x) = \int_{-\infty}^{\infty} F(x)p(x)dx. \qquad (7)$$

Here we have used the abbreviation "av $F(x)$" to mean the average value of $F(x)$ approached by a very large number of trials. Other designations in use include "$<F(x)>$" and "mathematical expectation of $F(x)$," abbreviated as $E[F(x)]$.

The positive integer powers constitute a particular set of functions with averages of outstanding interest. We define the average of x^n as the nth moment of the distribution, and write

$$m_n = \text{av } x^n = \int_{-\infty}^{\infty} x^n p(x)dx. \qquad (8)$$

We note that $m_0 = 1$. Of the others, our attention is most often concentrated on m_1, which is the ordinary arithmetical mean, and m_2, which is the mean square. To the electrical engineer dealing with the case in which x is a voltage or current, m_1 represents the constant or dc component of the process and m_2 multiplied by the conductance or resistance respectively gives the mean power. Frequently it is found convenient to study the data with the average subtracted from all values. This concentrates attention on the ac component of the phenomenon. The corresponding power averages are called central moments and are defined by

$$\mu_n = \text{av } (x - m_1)^n = \int_{-\infty}^{\infty} (x - m_1)^n p(x)dx. \qquad (9)$$

Evidently $\mu_1 = 0$. The most important central moment is μ_2, which is defined as the variance of the distribution. From the definitions,

$$
\begin{aligned}
\mu_2 &= \int_{-\infty}^{\infty} (x^2 - 2m_1 x + m_1^2)p(x)dx \\
&= \int_{-\infty}^{\infty} x^2 p(x)dx - 2m_1 \int_{-\infty}^{\infty} xp(x)dx + m_1^2 \int_{-\infty}^{\infty} p(x) \\
&= m_2 - 2m_1^2 + m_1^2 = m_2 - m_1^2 \\
&= \text{av } x^2 - (\text{av } x)^2. \qquad (10)
\end{aligned}
$$

The square root of the variance is called the standard deviation, σ, that is

$$\sigma = (\mu_2)^{1/2}, \quad \text{and} \quad \sigma^2 + m_1^2 = m_2. \qquad (11)$$

In electrical language, the standard deviation is the root-mean-square or rms value of the ac component. The variance is the mean square and when multiplied by the conductance or resistance as appropriate gives the mean power represented by the ac component. In noise theory, it is usually convenient to express results in terms of the variance directly rather than complicate the formulas by introducing conductance or resistance to convert to power values. The variance is sometimes referred to as ac power in a one-ohm circuit.

Up to this point we have spoken somewhat blithely about dc and ac components in a process without revealing our plan of relating actual measurements to the statistical quantities we have been discussing. There are two basic physical programs for doing this which we should now explain. One is based on a series of measurements on one system throughout a very long time and the other calls for a simultaneous set of measurements on a large number of similar systems. In the first case, which we might call the extension-in-time program, we evaluate probability density and distribution functions by observations over a considerable period of time. For example, we could install a set of level recorders at the output of the system and note the fractions of time in which output values are found in sufficiently fine sub-intervals of the range of variation. To measure average values we use integrating meters, such as may be approximated by voltmeters, ammeters, or wattmeters with long time constants, to measure average voltage, current, or power. The averaging time is prescribed roughly as long enough so that if it were made longer the result would not be changed. Since we must conclude our experiment some time in order to use the data, we cannot extend our measurements throughout all time. A question then arises as to whether we would get the same statistical results if we repeated the entire experiment an hour, a day, or a week later. Such processes as would give the same probability density function and consequently all the other statistical parameters derived therefrom no matter where we localize our observation interval in time are called stationary. A truly stationary process for all future and past time is difficult to imagine, but over the elapsed time in which we retain interest in our apparatus the concept of stationarity is often an appropriate one.

The other program which we might call an extension in space but which is more often called the ensemble method furnishes a more elegant basis for mathematical analysis than the first mentioned. We suppose that a large number of systems are available which differ only in ways of which we are ignorant or which we choose to ignore. Identical measuring instruments are inserted in each system and a set of instantaneous values are read on all meters at the same time. From these data all the statistical functions and parameters are determined by arithmetic calculation. The question of how long a time over which to continue our observations in the first method is replaced by the question of how many systems should we measure in the second. Here again we use enough so that if we added more the results would not be significantly changed.

A stationary process from the ensemble point of view is now defined as one in which the statistics measured at any two distinct instants of time are the same. We thereby avoid one difficulty of the extension-in-time program in that an infinite time is required there to complete the measurements, and no time is left to take a new run. However it could be argued that we are likewise in difficulty when we call for an infinite number of measurements on distinct systems at one time. In any practical case we must be able to get our data from a finite number of observations. A more important difference from the theoretical viewpoint is that in some cases of importance the limits exist over the ensemble whereas no limit may be approached for the corresponding averages over a long time.

A matter of considerable theoretical interest now arises in that we can not say in general that the two methods we have described for obtaining statistics of noise processes will yield the same results even when the process is stationary in accordance with both definitions. We therefore introduce a new term by defining an *ergodic* process as one in which the statistics over a long time interval for any one system are the same as the statistics over the ensemble of systems at any one instant of time. An ergodic process is always stationary, but a stationary process can be nonergodic. Examples of the latter tend to be somewhat artificial. To illustrate assume that a certain model of amplifier is placed in quantity production. The tubes for the amplifier are obtained from two different manufacturing lots and the tubes of one lot are definitely noisier than the others. The two lots of tubes are thoroughly mixed. Then the average noise power from any one amplifier tends to be either high or low depending on whether it is equipped with tubes from the noisy or quiet lot. The noise power averaged over the ensemble of amplifiers however is intermediate between the noisy and quiet species.

Illustrative Examples of Probability Density

The Uniform or Rectangular Distribution

In Fig. 1(a), the variable has no preference with respect to any part of the range α to $\alpha + \beta$, but it is never outside this range. Then

$$p(x) = \begin{cases} 0, & x < \alpha \text{ and } x > \alpha + \beta \\ k, & \alpha < x < \alpha + \beta. \end{cases} \quad (12)$$

The constant k is determined from the condition

$$\int_{\alpha}^{\alpha+\beta} k\,dx = 1, \text{ whence } k = 1/\beta. \quad (13)$$

The average value is given by

$$m_1 = \int_{\alpha}^{\alpha+\beta} \frac{x\,dx}{\beta} = \frac{1}{2\beta}\left[x^2\right]_{\alpha}^{\alpha+\beta} = \alpha + \frac{\beta}{2} \quad (14)$$

as could have been determined by inspection. The mean square is

$$m_2 = \int_{\alpha}^{\alpha+\beta} \frac{x^2\,dx}{\beta} = \frac{1}{3\beta}\left[x^3\right]_{\alpha}^{\alpha+\beta} = \alpha^2 + \alpha\beta + \beta^2/3. \quad (15)$$

The variance is

$$\mu_2 = m_2 - m_1^2 = \beta^2/12. \quad (16)$$

The standard deviation is

$$\sigma = \frac{\beta}{2\sqrt{3}}. \quad (17)$$

The distribution function [Fig. 1(b)] is

$$P(x) = \begin{cases} 0, & x < \alpha \\ (x - \alpha)/\beta, & \alpha < x < \alpha + \beta \\ 1, & x > \alpha + \beta \end{cases}. \quad (18)$$

These results are immediately applicable to the problem of "quantizing noise" arising from the errors in conversion of a signal in analog form to a digital approximation. Suppose the range of signal magnitudes is divided into uniform intervals of width Δ. All magnitudes falling within each interval are equated to a single value at midinterval. If the intervals are made small, the magnitudes falling within any particular one are uniformly distributed within the interval and the errors, which are the differences between the original value and the midinterval approximation, are uniformly distributed between $-\Delta/2$ and $\Delta/2$. Therefore the errors have a uniform distribution in which $\alpha = -\Delta/2$ and $\beta = \Delta$. From the above results, then the dc component of the error is zero and the rms error is $\Delta/2\sqrt{3}$. These are fundamental relations for digital-to-analog conversion in which the quantizing steps are all the same size. It is also applicable in straight numerical computation as a means of estimating the error from rounding off at a certain number of decimal places.

A rectangular distribution is often assumed for the phase of a sine wave with respect to a particular origin of time or with respect to the phase of some other sine wave. What we mean by such an assumption is that no information is available to make any particular phase angle more likely than any other. If, as is often the case, angles differing by multiples of 2π can not be distinguished, it is convenient to take a fixed interval of 2π radians as defining the complete range of possible angles. In Fig. 1(a) this would mean $\beta = 2\pi$ and α arbitrary. We may then wish to calculate the statistics of some function of the angle.

In general, if we are given the probability density function $q(\theta)$ of a variable θ, and we wish to calculate the probability density function $p(x)$ for $x = F(\theta)$, it is convenient to invert the functional relationship and write $\theta = f(x)$. Then by direct substitution,

$$q(\theta)d\theta = q[f(x)]f'(x)dx \quad (19)$$

where $f'(x) = df(x)/dx$. Hence if $f(x)$ is a single valued function of x,

$$p(x) = q[f(x)]f'(x). \quad (20)$$

If $f(x)$ is multiple valued, the expression on the right is summed over all the values. We shall apply these relations in the next example.

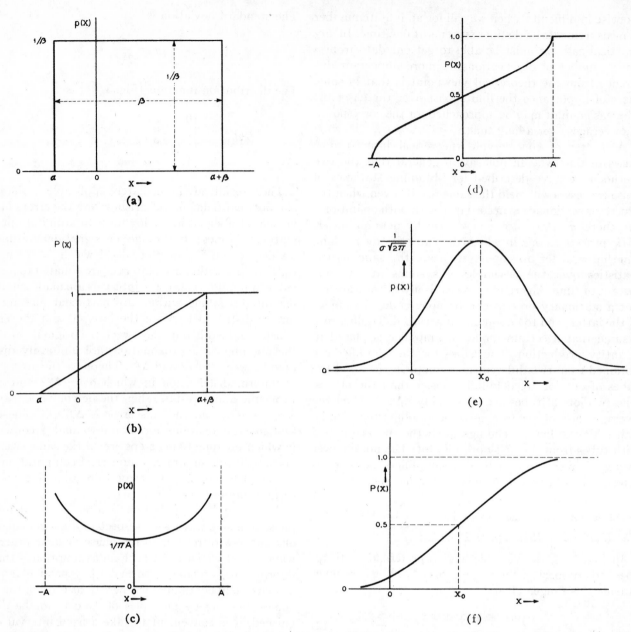

Fig. 1—(a) Probability density function for rectangular distribution, (b) Distribution function for rectangular case. (c) Probability density function for sinusoidal distribution. (d) Distribution function for sinusoidal case. (e) Probability density function for Gaussian distribution. (f) Distribution function for Gaussian case.

The Sinusoidal Distribution

Let

$$x = A \sin \theta \qquad (21)$$

where θ has a uniform distribution in range $-\pi/2$ to $3\pi/2$. Then the values of x are distributed like those of the ordinates of a sine wave with amplitude A. To find the probability density function for x, we write $\alpha = -\pi/2, \beta = 2\pi$, and

$$q(\theta) = \begin{cases} 0, \ \theta < -\pi/2 \text{ and } \theta > 3\pi/2 \\ \dfrac{1}{2\pi}, \ -\pi/2 < \theta < 3\pi/2. \end{cases} \qquad (22)$$

Then solving (21) for θ as a function of x, we find

$$\theta = f(x) = \arcsin \frac{x}{A} \qquad (23)$$

$$f'(x) = (A^2 - x^2)^{-1/2}. \qquad (24)$$

We note that each value of x in the range $-A$ to A corresponds to two values of θ and hence that we must multiply the right hand side of (20) by two to obtain $p(x)$. Then

$$p(x) = \begin{pmatrix} 0, \ x < -A \text{ and } x > A \\ (A^2 - x^2)^{-1/2}/\pi, \ -A < x < A \end{pmatrix}. \qquad (25)$$

Fig. 1(c) shows the graph of this function.

The average value over the ensemble is given by

$$m_1 = \int_{-A}^{A} \frac{x\,dx}{\pi(A^2 - x^2)^{1/2}} = 0. \tag{26}$$

Likewise the mean square is

$$m_2 = \int_{-A}^{A} \frac{x^2\,dx}{\pi(A^2 - x^2)^{1/2}} = \frac{A^2}{2}. \tag{27}$$

Since $m_1 = 0$, m_2 is also the variance. The distribution function, Fig. 1(d), is

$$P(x) = \int_{-A}^{x} \frac{dx}{\pi(A^2 - x^2)^{1/2}} = \frac{1}{\pi}\left(\frac{\pi}{2} + \arcsin\frac{x}{A}\right). \tag{28}$$

This example shows that it is possible for the probability density to become infinite provided the integral over a finite interval including the singularity exists.

The sinusoidal distribution is appropriate for the noise produced by cw interference of constant amplitude and random phase. What we have illustrated above is the ensemble approach which assumes an infinite number of possible sources of form $A \sin(\omega t + \theta)$ with A fixed and θ distributed uniformly throughout the range α to $\alpha + 2\pi$. We averaged over the ensemble with t fixed, noting that adding the constant ωt to θ does not change the distribution except for a trivial shift in α. We can also work out time averages over a single function of time belonging to the ensemble. Let

$$E(t) = A \sin(\omega t + \theta) \tag{29}$$

with A, ω, and θ fixed. To evaluate the mean square over a long time we write

$$\begin{aligned}
\overline{E}^2 &= \operatorname*{Lim}_{T\to\infty} \frac{1}{T}\int_{t_1}^{t_1+T} E^2(t)\,dt \\
&= \operatorname*{Lim}_{T\to\infty} \frac{1}{T}\int_{t_1}^{t_1+T} \frac{A^2}{2}[1 + \cos 2(\omega t + \theta)]\,dt \\
&= \frac{A^2 T}{2T} = \frac{A^2}{2}.
\end{aligned} \tag{30}$$

Since the process is stationary the result does not depend on t_1. The fact that we obtained the same result as by the ensemble method is illustrative of the ergodicity of the process.

The Gaussian Distribution

One of the most important distributions in noise theory is that in which the probability density is proportional to the exponential of a negative quadratic function of the values of the variable. The general form for one variable is

$$p(x) = \frac{1}{\sigma\sqrt{2\pi}} e^{-(x-x_0)^2/2\sigma^2}. \tag{31}$$

The parameters have been adjusted here to give the required condition

$$\int_{-\infty}^{\infty} p(x)\,dx = 1, \tag{32}$$

and also

$$m_1 = \int_{-\infty}^{\infty} xp(x)\,dx = x_0 \tag{33}$$

$$m_2 = \int_{-\infty}^{\infty} x^2 p(x)\,dx = x_0^2 + \sigma^2 \tag{34}$$

$$\mu_2 = m_2 - m_1^2 = \sigma^2. \tag{35}$$

That is x_0 is the mean, σ^2 is the variance, and σ is the standard deviation. This distribution is called normal or Gaussian. An important theorem in statistics called the central limit theorem shows under very general assumptions that the distribution of the sum of an indefinitely large number of other independently distributed quantities must approach the Gaussian distribution, no matter what the individual distributions may be. We shall not prove this theorem here, but shall indicate later by examples the sort of mechanism by which it operates. The phenomenon called Johnson noise, resistance noise, or thermal noise associated with thermal agitation of electrons in conductors gives rise to voltages and currents with Gaussian distribution. It may be thought of as the sum of a very large number of practically independent pulses. When the receiver is linear, there is no preference between positive and negative values, and the mean value x_0, which represents the dc component, is zero. The parameter σ is then the rms value.

Figs. 1(e) and 1(f) show the nature of the Gaussian functions for $p(x)$ and $P(x)$. The latter may be expressed by

$$P(x) = \frac{1}{2}\left[1 + \operatorname{erf}\frac{(x - x_0)}{(\sqrt{2}\sigma)}\right],$$

where

$$\operatorname{erf} z = \frac{2}{\sqrt{\pi}}\int_{0}^{z} e^{-\lambda^2}\,d\lambda. \tag{36}$$

The theory we have given so far is sufficient to solve a number of interesting noise problems. Suppose we are asked to calculate the dc component in the output of a half-wave linear rectifier when the input is resistance noise. The probability density function of the input wave with the negative lobes cut off is given by

$$p(x) = \frac{1}{\sigma\sqrt{2\pi}} e^{-x^2/2\sigma^2} + \frac{1}{2}\delta(x), \qquad x \geqq 0 \tag{37}$$

where $\delta(x)$ is defined by

$$\delta(x) = 0 \qquad x \neq 0 \tag{38}$$

$$\int_{0-\epsilon}^{0+\epsilon} \delta(x)\,dx = 1, \qquad \epsilon > 0. \tag{39}$$

The δ function is introduced as a convenient symbol for the fact that the suppressed negative lobes transfer all their accumulated probability to the value $x = 0$.

Then if a_1 is the conductance of the rectifier during the conducting phase, the dc component is

$$
\begin{aligned}
\text{av } (a_1 x) &= a_1 \int_{-\infty}^{\infty} x p(x) dx \\
&= a_1 \int_{0}^{\infty} \frac{x}{\sigma \sqrt{2\pi}} e^{-x^2/2\sigma^2} dx + a_1 \int_{-\infty}^{\infty} x \delta(x) dx \\
&= \frac{a_1}{\sigma \sqrt{2\pi}} \left[-\sigma^2 e^{-x^2/2\sigma^2} \right]_{0}^{\infty} = \frac{a_1 \sigma}{\sqrt{2\pi}}.
\end{aligned}
\tag{40}
$$

Rectification with other functional laws can be treated in a similar manner.

A curious property of the Gaussian distribution is that no matter how large a value we may consider there is a finite probability of it being exceeded in an observation. The probability diminishes so rapidly at very large values however that for practical purposes peaks exceeding some reasonable value can be regarded as impossible.

DISTRIBUTION OF SUMS—THE CHARACTERISTIC FUNCTION

There are many physical situations in which we wish to relate the statistics of the measured sum of a number of components to the statistics of the individual contributions. We have already mentioned one example— the representation of thermal noise as the sum of a large number of practically independent pulses. Another case is that of atomic radiation considered as the sum of a large number of independent sine waves of limited duration. An important case to which we shall devote much attention in this paper is that of the sum of a wanted signal and an undesired noise component. The fundamental relation, given $p_1(x)$ and $p_2(y)$ as the probability density function of two *independent* quantities x and y, and that $z = x + y$, is

$$
p(z) = \int_{-\infty}^{\infty} p_1(x) p_2(z - x) dx.
\tag{41}
$$

That is, in order to obtain the value z for a given x, we must have y in the range $(z - x)dx$. This is a compound event in which the individual events are independent and hence the probability is found by multiplying the separate probabilities. The total probability is found by integrating over all x.

Integrals of the form (41) are commonly known as convolution integrals. The communication engineer has become familiar with them in relation to Fourier analysis. As an example, suppose that $g_1(\lambda)$ is the amplitude of a wave at the time λ, and $g_2(t - \lambda)$ is the response of a linear network at time $t - \lambda$ after it has been excited by a unit impulse. Then the response of the network to $g_1(t)$ is given as a function of time t by

$$
g(t) = \int_{-\infty}^{\infty} g_1(\lambda) g_2(t - \lambda) d\lambda.
\tag{42}
$$

A basic theorem on Fourier transforms states that if $g_1(t)$, $g_2(t)$ have the Fourier transforms $S_1(\omega)$, $S_2(\omega)$ respectively; that is, if

$$
S_1(\omega) = \int_{-\infty}^{\infty} g_1(t) e^{-i\omega t} dt
\tag{43}
$$

and

$$
S_2(\omega) = \int_{-\infty}^{\infty} g_2(t) e^{-i\omega t} dt
\tag{44}
$$

then the Fourier transform of $g(t)$ is $S_1(\omega)S_2(\omega)$. That is

$$
S_1(\omega) S_2(\omega) = \int_{-\infty}^{\infty} g(t) e^{-i\omega t} dt
\tag{45}
$$

and by inversion

$$
g(t) = \frac{1}{2\pi} \int_{-\infty}^{\infty} S_1(\omega) S_2(\omega) e^{i\omega t} d\omega.
\tag{46}
$$

This means that we can calculate $g(t)$ by calculating the inverse Fourier transform of the product of the Fourier transforms of $g_1(t)$ and $g_2(t)$. (Note that the sign of i may be changed throughout without changing the validity.)

We can apply these same results of Fourier analysis to (41). We must then conclude that the probability density function for the sum of two independent variables is the inverse Fourier transform of the product of the Fourier transforms of the individual probability density functions. It therefore appears that an important function to be defined is the Fourier transform of a probability density. In statistics this function is called the *characteristic function* (ch.f.). It is a function of a dummy variable ξ, and is in fact seen to be the average value of $e^{i\xi x}$. That is, the ch.f. corresponding to the variable x is defined by

$$
C_x(\xi) = \text{av } e^{i\xi x} = \int_{-\infty}^{\infty} e^{i\xi x} p(x) dx.
\tag{47}
$$

If we write $C_x(\xi)$ for the ch.f. of x, $C_y(\xi)$ for the ch.f. of y, and $C_z(\xi)$ for the ch.f. of $z = x + y$, then

$$
C_z(\xi) = C_x(\xi) C_y(\xi)
\tag{48}
$$

and $p_2(z)$, the probability density function of the sum $z = x + y$ is

$$
p_2(z) = \frac{1}{2\pi} \int_{-\infty}^{\infty} C_x(\xi) C_y(\xi) e^{-i\xi z} d\xi.
\tag{49}
$$

Extending the process now to the sum of n variables, let $z_n = x_1 + x_2 + \cdots + x_n$. Then

$$
C_{z_n}(\xi) = C_{x1}(\xi) C_{x2}(\xi) \cdots C_{x_n}(\xi)
\tag{50}
$$

$$
p(z_n) = \frac{1}{2\pi} \int_{-\infty}^{\infty} C_{x1}(\xi) C_{x2}(\xi) \cdots C_{x_n}(\xi) e^{-i\xi z} d\xi.
\tag{51}
$$

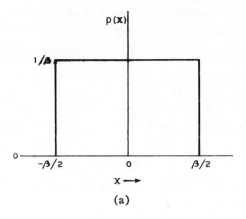

(a)

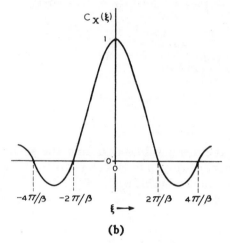

(b)

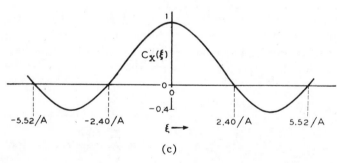

(c)

Fig. 2—(a) Symmetrical uniform probability density function. (b) Characteristic function of symmetrical uniform case. (c) Characteristic function of sinusoidal case.

We illustrate by calculating the characteristic functions for the previous examples and show how they may be used to calculate distributions of corresponding sums.

Consider first the symmetrical uniform distribution with $\alpha = -\beta/2$, Fig. 2(a).

Then

$$C_x(\xi) = \int_{-\beta/2}^{\beta/2} e^{i\xi x}\, \frac{dx}{\beta} = \frac{2 \sin \beta\xi/2}{\beta\xi}. \tag{52}$$

This function is plotted in Fig. 2(b). The probability density function of the sum of n uniformly distributed variables in the same range is

$$p_n(x) = \frac{1}{2\pi} \int_{-\infty}^{\infty} \frac{\sin^n \beta\xi/2}{(\beta\xi/2)^n}\, e^{-ix\xi} d\xi. \tag{53}$$

It will be noted that as n is made very large the integrand shrinks rapidly as we move away from the origin and that only the values very near $\xi = 0$ contribute much. When $\xi = 0$, the integrand becomes unity. In the neighborhood of $\xi = 0$, the integrand is of form

$$\frac{\left[\dfrac{\beta\xi}{2} - \dfrac{(\beta\xi)^3}{48} + \cdots\right]^n e^{-ix\xi}}{(\beta\xi/2)^n}$$

$$= \left[1 - \frac{(\beta\xi)^2}{24} + \cdots\right]^n e^{-ix\xi}$$

$$= \left[1 - \frac{n}{24}(\beta\xi)^2 + \cdots\right] e^{-ix\xi}. \tag{54}$$

We may approximate the integral by replacement with a function which has the same initial terms in its power series expansion and drops rapidly to zero away from the origin. Such a function is

$$e^{-n(\beta\xi)^2/24 - ix\xi}.$$

Then for large n,

$$p_n(x) = \frac{1}{2\pi} \int_{-\infty}^{\infty} e^{-n\beta^2\xi^2/24} \cos x\xi\, d\xi \tag{55}$$

or

$$p_n(x) = \frac{1}{\pi} \sqrt{\frac{6\pi}{n\beta^2}}\, e^{-6x^2/n\beta^2}$$

$$= \frac{1}{\sigma_n \sqrt{2\pi}}\, e^{-x^2/2\sigma_n^2} \tag{56}$$

where

$$\sigma_n = \beta \sqrt{\frac{n}{12}} = \sigma_1 \sqrt{n}. \tag{57}$$

Thus we tend toward a Gaussian distribution with $\sqrt{n}$ times the standard deviation of the original or n times the variance. That is, the mean power values in independent uniform noise sources add directly, and if enough are added a Gaussian distribution is approached. Since each component represents an independent occurrence, the Gaussian limit could have been deduced from the central limit theorem previously mentioned. Our calculation may be regarded as a verification for this particular case. Since the range of each component is $-\beta/2$ to $\beta/2$, the range of the sum cannot exceed $-n\beta/2$ to $n\beta/2$. The Gaussian approximation, which gives nonzero probability for all finite ordinates, should only be applied to values well within the range of possible sums. As n becomes indefinitely large, the excluded range recedes away from any values of physical interest.

An engineering application of the above result may be made to the case of resultant quantizing noise when a number of alternate analog-to-digital and digital-to-analog conversions are made on a signal. We can also identify the sum as noise caused by superimposing independently occurring triangular pulses of equal height $\beta/2$, random sign, and random times of occurrence.

As a second example, consider the sum of a number of independent sinusoidal distributions such as might be obtained by combining the outputs of nonsynchronous oscillators. From (25), a component of peak amplitude A has the ch.f.

$$C_x(\xi) = \frac{1}{\pi} \int_{-A}^{A} \frac{e^{i\xi x}dx}{(A^2 - x^2)^{1/2}}$$

$$= J_0(A\xi). \tag{58}$$

where J_0 is the zero order Bessel function of the first kind. The graph of $C_x(\xi)$ is shown in Fig. 2(c). Hence the probability density function for the sum of n independent sinusoidal sources with peak values A_1, A_2, $\cdots$, A_n is

$$p_n(x) = \frac{1}{2\pi} \int_{-\infty}^{\infty} J_0(A_1\xi)J_0(A_2\xi) \cdots J_1(A_n\xi)e^{-ix\xi}d\xi. \tag{59}$$

The limiting case when n is large may be studied by the same method as used for the sum of n uniform distributions. We substitute the first two terms of the power series for each Bessel function, namely

$$J_0(x) = 1 - \frac{x^2}{4} \text{ for } x \text{ small}, \tag{60}$$

and note that for ξ small,

$$J_0(A_1\xi)J_0(A_2\xi) \cdots J_0(A_n\xi),$$

$$= 1 - \frac{A_1^2 + A_2^2 + \cdots + A_n^2}{4} \xi^2. \tag{61}$$

Replacing this expression by $\exp\left[-(A_1^2 + A_2^2 + \cdots + A_n^2)\xi^2/4\right]$ which has the same first two terms in its power series expansion, we then deduce that for n large

$$p_n(x) \sim \frac{1}{2\pi} \int_{-\infty}^{\infty} e^{-ix\xi - (A_1^2 + A_2^2 + \cdots + A_n^2)\xi^2/4}d\xi$$

$$= \frac{1}{\sigma\sqrt{2\pi}} e^{-x^2/2\sigma^2} \tag{62}$$

where

$$\sigma^2 = \tfrac{1}{2}(A_1^2 + A_2^2 + \cdots + A_n^2). \tag{63}$$

This shows that the resultant distribution approaches the Gaussian form with standard deviation equal to the square root of the sum of the mean squares of each individual distribution. This is another example of the central limit theorem. By taking the A's unequal we have illustrated that the component distributions do not have to be the same to produce the Gaussian limit.

In the case of the Gaussian distribution, the ch.f. is

$$C_x(\xi) = \frac{1}{\sigma\sqrt{2\pi}} \int_{-\infty}^{\infty} e^{-(x-x_0)^2/2\sigma^2 + ix\xi}dx$$

$$= \frac{1}{\sigma\sqrt{2\pi}} \int_{-\infty}^{\infty} e^{-u^2/2 + i\xi(\sigma u + x_0)}\sigma du$$

$$= e^{i\xi x_0 - \sigma^2\xi^2/2}. \tag{64}$$

This is an example of the self-reciprocal property of the Gaussian function with respect to Fourier transformation. Since the product of Gaussian functions is also Gaussian, the distribution of the sum of Gaussian variables must remain Gaussian. This is certainly to be expected from the central limit theorem, for if the addition of other distributions tends toward the Gaussian, the Gaussian form must represent an equilibrium-case preserving its form when more like itself are added.

We calculate for the probability density function of the sum of n Gaussian variables having mean values a_1, a_2, $\cdots$, a_n and standard deviation σ_1, σ_2, $\cdots$, σ_n

$$p_n(t) = \frac{1}{2\pi} \int_{-\infty}^{\infty} e^{-\xi^2/2(\sigma_1^2 + \sigma_2^2 + \cdots \sigma_n^2) + i\xi(a_1 + a_2 + \cdots + a_n) - ix\xi}d\xi$$

$$= \frac{1}{\sigma\sqrt{2\pi}} e^{-(x-x_0)^2/2\sigma} \tag{65}$$

where

$$x_0 = a_1 + a_2 + \cdots + a_n \tag{66}$$

$$\sigma^2 = \sigma_1^2 + \sigma_2^2 + \cdots + \sigma_n^2. \tag{67}$$

That is, the mean values add, and the squares of the variances add. Any number of Gaussian distributions may be combined in this way.

In many practical problems our ultimate goal is the determination of the moments, particularly the first and second, which give the dc voltage or current and the ac power. The probability density function is in these cases merely a means to an end, and we should be on the alert for the possibility of simpler methods of getting the moments. The characteristic function furnishes one such possibility, for we observe that

$$C_x(\xi) = \text{av } e^{i\xi x} = \text{av} \sum_{r=0}^{\infty} \frac{(i\xi x)^r}{r!}$$

$$= \sum_{r=0}^{\infty} \frac{(i\xi)^r}{r!} \text{ av } x^r = \sum_{r=0}^{\infty} \frac{(i\xi)^r}{r!} m_r. \tag{68}$$

This equation shows that m_r, which is the rth moment or average rth power, may also be defined as

$$m_r = \begin{array}{l} r!/i^r \text{ times coefficient of } \xi^r \text{ in the expansion of} \\ C_x(\xi) \text{ in powers of } \xi. \end{array} \tag{69}$$

We thus see that we can compute moments from the characteristic function directly. For example, to obtain the moments of the sum of n variables, we may calculate the ch.f. of each, multiply the ch.f.'s together, and expand the product in a power series in ξ. The moments are then given by (69). Calculation of the Fourier transform of the resultant ch.f. with a subsequent integration for the typical moment is thereby avoided.

If we combine (68) with (48), letting the moments of x, y, and z be represented respectively by m_{xr}, m_{yr}, m_{zr}, we note that

$$\sum_{r=0}^{\infty} \frac{(i\xi)^r}{r!} m_{zr} = \sum_{r=0}^{\infty} \sum_{s=0}^{\infty} \frac{(i\xi)^{r+s}}{r!s!} m_{xr}m_{ys}. \tag{70}$$

Now if we equate coefficients of the first and second powers of ξ on the two sides of the equation, we verify that

$$m_{z1} = m_{x1} + m_{y1} \qquad (71)$$

$$m_{z2} - m_{z1}^2 = m_{x2} - m_{x1}^2 + m_{y2} - m_{y1}^2. \qquad (72)$$

Eq. (71) states that the mean value of the sum is equal to the sum of the means. This is elementary and is in fact true of dependent as well as independent processes. In electrical engineering terminology, the equivalent statement is that dc components may be added algebraically. Eq. (72) says that the variance of the sum of *independent* components is equal to the sum of the variances of the individual components. It is important to observe that independence was assumed in our derivation, for this law does not hold for dependent processes. In electrical engineering language, the equivalent statement is that the total power from independent components is the sum of the power in the individual components. The statement does not hold for example when two sine waves of the same frequency and fixed phase are added. In the latter case if the two amplitudes are equal, the power in the sum may have any value from zero when the waves are oppositely phased to four times the power of each component when phases coincide.

It is sometimes convenient to make use of a complete set of parameters which like the mean and variance are additive when distributions of sums of independent quantities are calculated. Such a set is furnished by the semi-invariants. The rth semi-invariant s_r is defined as the coefficient of $(i\xi)^r/r!$ in the power series expansion of $\ln C_x(\xi)$. That is, noting from (47) that $C_x(0)=1$ and hence $\ln C_x(0) = s_0 = 0$, we write

$$\ln C_x(\xi) = \sum_{r=1}^{\infty} \frac{(i\xi)^r}{r!} s_r \qquad (73)$$

Taking exponentials of both sides and recalling (68), we find

$$C_x(\xi) = e^{\sum_{r=1}^{\infty}(i\xi)^r s_r/r!} = \sum_{r=0}^{\infty} \frac{(i\xi)^r}{r!} m_r. \qquad (74)$$

This equation may be used to obtain either the semi-invariants in terms of the moments, or vice versa. The results are

$$\left.\begin{array}{l} s_1 = m_1 \\ s_2 = m_2 - m_1^2 \\ s_3 = m_3 - 3m_1m_2 + 2m_1^2, \text{ etc.} \end{array}\right\} \qquad (75)$$

and

$$\left.\begin{array}{l} m_1 = s_1 \\ m_2 = s_2 + s_1^2 \\ m_3 = s_3 + 3s_1s_2 + s_1^2, \text{ etc.} \end{array}\right\}. \qquad (76)$$

Now since the ch.f. $C_n(\xi)$ for the distribution of $x_1 + x_2 + \cdots + x_n$ is

$$C_n(\xi) = C_{x_1}(\xi) C_{x_2}(\xi) \cdots C_{x_n}(\xi), \qquad (77)$$

it follows that

$$\ln C_n(\xi) = \ln C_{x_1}(\xi) + \ln C_{x_2}(\xi) + \cdots + \ln C_{x_n}(\xi) \qquad (78)$$

so from (73) if we let $s_r^{(k)}$ represent the rth semi-invariant of the kth distribution, and s_{nr} represent the rth semi-invariant of the sum,

$$\sum_{r=1}^{\infty} \frac{(i\xi)^r}{r!} s_{nr} = \sum_{r=1}^{\infty} \frac{(i\xi)^r}{r!} s_r^{(1)} + \sum_{r=1}^{\infty} \frac{(i\xi)^r}{r!} s_r^{(2)} + \cdots$$

$$+ \sum_{r=1}^{\infty} \frac{(i\xi)^r}{r!} s_r^{(n)}$$

$$= \sum_{r=1}^{\infty} \frac{(i\xi)^r}{r!} [s_r^{(1)} + s_r^{(2)} + \cdots + s_r^{(n)}]. \qquad (79)$$

Therefore

$$s_{nr} = s_r^{(1)} + s_r^{(2)} + \cdots + s_r^{(n)}. \qquad (80)$$

In other words as previously implied, the rth semi-invariant of the distribution of the sum of any number of independent variables is equal to the sum of the rth semi-invariants of the individual distributions. The first and second semi-invariants are identical with the mean and variance respectively.

A practical calculating procedure to determine the moments for the sum is then:

1) Calculate individual moments.
2) Calculate individual semi-invariants from moments by (75).
3) Add semi-invariants to obtain semi-invariants of sum.
4) Calculate moments of sum by (76).

An interesting example of the use of semi-invariants to determine the moments of the sum of n quantities whose logarithms have Gaussian distributions has been given by Holbrook and Dixon [2].

Two-Dimensional Probability Theory

Up to now we have dealt with probability density functions of a single variable. We are able to obtain statistical information as to relative numbers of occurrence of different magnitudes in this way, but we cannot deduce anything about the time scale over which we would expect to observe such a representative set of values. This defect must be remedied because all our engineering data are obtained with the circuits which are frequency selective—that is, they do not respond equally to all impressed waves, but are sensitive also to the rapidity with which changes are made. The communication engineers' statement of this is that circuits respond appreciably only to those disturbances which fall within their transmission bands.

Our one-variable theory dealing with values isolated from each other in time was demonstrated to be capable of solving one problem in frequency selectivity—that of distinguishing between dc and ac. That is, by taking averages over a long time we can resolve a noise wave into one component which is constant and therefore contains only the frequency zero and a second component which is variable with zero mean and hence can be said to contain only frequencies greater than zero. This resolution is entirely too crude to be adequate for any but the simplest situations.

The next step is to study the statistics of pairs of values of voltage or current separated by specified instants of time. It turns out that these additional statistics are sufficient for practically all the problems of engineering importance. We accordingly digress to state some needed facts about probability relations concerning two coordinates x and y which may be dependent on each other—that is, specifying the value of one affects the statistics of the other. Such distributions are called bivariate. It is convenient to consider first two generalized variables x and y. Later we shall specialize to the case in which x and y represent values of the same voltage or current wave at two different instants of time.

The necessary techniques are an easy generalization of what has gone before. We define $p(x, y)$, the probability density function of two coordinates, as to be that function which when multiplied by the infinitesimal area $dxdy$ gives the probability that the value of the first coordinate is in the range x to $x+dx$ and the value of the second coordinate is simultaneously within the range y to $y+dy$. Then as before the probability that the first coordinate falls somewhere in the finite range x_1 to x_2 while the second falls somewhere in the range y_1 to y_2 is expressed by

Prob $(x_1 < x < x_2, y_1 < y < y_2)$

$$= \int_{x_1}^{x_2} \int_{y_1}^{y_2} p(x, y)dxdy. \quad (81)$$

As before we may express the probability that the values are less than specified ones in terms of the probability density function. The complete distribution function $P(x, y)$ is now

$$P(x, y) = \int_{-\infty}^{x} \int_{-\infty}^{y} p(x, y)dxdy. \quad (82)$$

What are called marginal distribution functions define probabilities of x without regard to y and vice versa. There are two of these

$$P_y(x) = \int_{-\infty}^{x} dx \int_{-\infty}^{\infty} p(x, y)dy = \int_{-\infty}^{x} p_y(x)dx \quad (83)$$

$$P_x(y) = \int_{-\infty}^{y} dy \int_{-\infty}^{\infty} p(x, y)dx = \int_{-\infty}^{y} p_x(y)dy. \quad (84)$$

The functions $p_y(x)$, $p_x(y)$ defined by

$$p_y(x) = \int_{-\infty}^{\infty} p(x, y)dy \quad (85)$$

$$p_x(y) = \int_{-\infty}^{\infty} p(x, y)dx \quad (86)$$

may be called marginal probability density functions.

Evidently when the indicated derivatives exist

$$p(x, y) = \frac{\partial^2 P(x, y)}{\partial_x \partial_y} \quad (87)$$

$$p_y(x) = dP_y(x)/dx \quad (88)$$

$$p_x(y) = dP_x(y)/dy. \quad (89)$$

We must likewise have

$$P(\infty, \infty) = \int_{-\infty}^{\infty} \int_{-\infty}^{\infty} p(x, y)dxdy = 1 \quad (90)$$

$$P_y(\infty) = \int_{-\infty}^{\infty} p_y(x)dx = 1 \quad (91)$$

$$P_x(\infty) = \int_{-\infty}^{\infty} p_x(y)dy = 1. \quad (92)$$

Averages are computed in the same way as before except that we now have two integrations to perform

$$\text{av } F(x, y) = \int_{-\infty}^{\infty} \int_{-\infty}^{\infty} F(x, y)p(x, y)dxdy. \quad (93)$$

For functions of one of the coordinates only, we may use

$$\text{av } F(x) = \int_{-\infty}^{\infty} F(x)p_y(x)dx \quad (94)$$

$$\text{av } F(y) = \int_{-\infty}^{\infty} F(y)p_x(y)dy. \quad (95)$$

In the special case in which the values of x and y are independent, $p(x, y)$ can be written as $p_y(x)p_x(y)$; *i.e.*, as the product of a function of x only and one of y only.

The moments now become doubly infinite in number and require a double subscript. We write

$$m_{jk} = \text{av } x^j x^k = \int_{-\infty}^{\infty} \int_{-\infty}^{\infty} x^j y^k p(x, y)dxdy. \quad (96)$$

We see that $m_{00} = 1$ and that the means of x and y separately are

$$m_{10} = x_0 = \text{av } x = \int_{-\infty}^{\infty} \int_{-\infty}^{\infty} xp(x, y)dxdy$$

$$= \int_{-\infty}^{\infty} xp_y(x)dx \quad (97)$$

$$m_{01} = y_0 = \text{av } y = \int_{-\infty}^{\infty} \int_{-\infty}^{\infty} yp(x, y)dxdy$$

$$= \int_{-\infty}^{\infty} yp_x(y)dy. \quad (98)$$

As in the one-variable case, we often wish to subtract out the mean values, and correspondingly we define the central moments as

$$\mu_{jk} = \text{av } [(x - x_0)^j(y - y_0)^k]$$

$$= \int_{-\infty}^{\infty} \int_{-\infty}^{\infty} (x - x_0)^j(y - y_0)^k p(x, y)dxdy. \quad (99)$$

The most important central moments are the second order ones μ_{20}, μ_{11}, and μ_{02}. Since

$$\mu_{20} = \text{av } (x - x_0)^2 = \int_{-\infty}^{\infty} x^2 p_y(x)dx$$

$$= \text{av } x^2 - (\text{av } x)^2 \qquad (100)$$

$$\mu_{02} = \text{av } (y - y_0)^2 = \int_{-\infty}^{\infty} y^2 p_x(y)dy$$

$$= \text{av } y^2 - (\text{av } y)^2 \qquad (101)$$

we see that they are simply the variances of x and y separately. The quantity

$$\mu_{11} = \text{av } [(x - x_0)(y - y_0)]$$

$$= \int_{-\infty}^{\infty}\int_{-\infty}^{\infty} (xy - x_0 y - y_0 x + x_0 y_0)p(x, y)dxdy$$

$$= \text{av } (xy) - x_0 y_0 - y_0 x_0 + x_0 y_0$$

$$= \text{av } (xy) - (\text{av } x)(\text{av } y) \qquad (102)$$

is defined as the covariance of the two quantities. It is the most significant contribution of two-dimensional theory over the one-dimensional case and will enable us to define the power spectra, or density of mean power along the frequency scale, for our random processes. We remark that when two quantities are independent, their covariance vanishes and their average product is equal to the product of their individual averages. The converse is not true in general but is true for jointly Gaussian processes, which we shall discuss next.

In the two-dimensional jointly Gaussian distribution, the probability density function is proportional to the exponential of a negative quadratic function in x and y. By assuming the most general function of this type and determining the constants by evaluating x_0, y_0, μ_{20}, μ_{02}, μ_{11} and also equating the integral to unity, we find

$$p(x, y) = \frac{1}{2\pi A} e^{[-\mu_{02}(x-x_0)^2 + \mu_{20}(y-y_0)^2 - 2\mu_{11}(x-x_0)(y-y_0)]/2A^2} \quad (103)$$

where

$$A^2 = \mu_{20}\mu_{02} - \mu_{11}^2. \qquad (104)$$

Fig. 3 shows the surface defined by (103). It is symmetrical about an axis which is displaced and rotated with respect to the coordinate axes. By the central limit theorem, this is the form of probability density function approached for the values of x and y when x is the sum of the independent values $x_1, x_2, \cdots x_n$, y is the sum of the independent values $y_1, y_2, \cdots y_n$, the probability density function $p_k(x_k, y_k)$ holds for the values x_k, y_k, and n is large.

The characteristic function of a bivariate distribution is a double Fourier transform which we write as

$$C_{xy}(\xi_1, \xi_2) = \text{av } e^{i(\xi_1 x + \xi_2 y)}$$

$$= \int_{-\infty}^{\infty}\int_{-\infty}^{\infty} e^{i(\xi_1 x + \xi_2 y)} p(x, y)dxdy. \quad (105)$$

The distribution function of the sum of bivariates may be calculated from the ch.f. as in the single variable case. That is, if

$$\begin{pmatrix} X = x_1 + x_2 \\ Y = y_1 + y_2 \end{pmatrix} \qquad (106)$$

and the ch.f.'s of (x_1, y_1) and x_2, y_2 are given, then

$$C_{XY}(\xi_1, \xi_2) = C_{x_1 y_1}(\xi_1, \xi_2)C_{x_2 y_2}(\xi_1, \xi_2) \qquad (107)$$

and

$$p(X, Y) = \frac{1}{(2\pi)^2}\int_{-\infty}^{\infty}\int_{-\infty}^{\infty} e^{-i(\xi_1 X + \xi_2 Y)}C_{XY}(\xi_1, \xi_2)d\xi_1 d\xi_2. \quad (108)$$

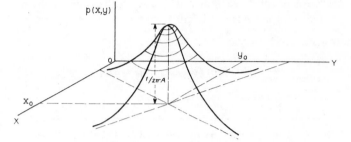

Fig. 3—Two-dimensional Gaussian probability density function.

The bivariate moments can be calculated by expanding the ch.f. in a power series in x and y by analogy with the single variable case. From (105) we obtain

$$C_{xy}(\xi_1, \xi_2) = \text{av } (e^{i\xi_1 x} e^{i\xi_2 y})$$

$$= \text{av } \sum_{j=0}^{\infty}\sum_{k=0}^{\infty} \frac{(i\xi_1 x)^i (i\xi_2 y)^k}{j!k!}$$

$$= \sum_{j=0}^{\infty}\sum_{k=0}^{\infty} \frac{i^{j+k}\xi_1^j \xi_2^k}{j!k!} \text{ av } x^i y^k$$

$$= \sum_{j=0}^{\infty}\sum_{k=0}^{\infty} \frac{i^{j+k}\xi_1^j \xi_2^k}{j!k!} m_{jk} \qquad (109)$$

Therefore m_{jk} is $j! \, k!/i^{j+k}$ times the coefficient of $\xi_1^j \xi_2^k$ in the power series expansion of $C_{xy}(\xi_1, \xi_2)$.

As in the one-dimensional case, the ch.f. of the bivariate Gaussian distribution is a bivariate Gaussian function of the transformation variables ξ_1 and ξ_2. To simplify the results we set x_0 and $y_0 = 0$ which can always be accomplished by a change of variable. Then the Gaussian ch.f. may be written

$$C_{xy}(\xi_1, \xi_2) = \frac{1}{2\pi A}\int_{-\infty}^{\infty}\int_{-\infty}^{\infty} e^{i\xi_1 x + i\xi_2 y - (\mu_{02}x^2 + \mu_{20}y^2 - 2\mu_{11}xy)/2A^2}dxdy$$

$$= e^{-(\mu_{20}\xi_1^2 + \mu_{02}\xi_2^2 + 2\mu_{11}\xi_1\xi_2)/2}. \qquad (110)$$

We may expand this in a power series in ξ_1 and ξ_2 by using the exponential series

$$C_{xy}(\xi_1, \xi_2) = 1 - \frac{\mu_{20}\xi_1^2 + \mu_{02}\xi_2^2 + 2\mu_{11}\xi_1\xi_2}{2}$$

$$+ \frac{(\mu_{20}\xi_1^2 + \mu_{20}\xi_2^2 + 2\mu_{11}\xi_1\xi_2)^2}{2!2^2} - \cdots . \quad (111)$$

Now invoking the rule derived in (109), we note, for example, that the coefficient of ξ_1^2 is $-\mu_{20}/2$ and multiplying this value by $2! \, 0!/i^{2+0} = -2$ gives μ_{20}. In a similar manner it may be checked that μ_{02} and μ_{11} are actually the specific moments stated.

The characteristic function of a multivariate Gaussian distribution of n dependent quantities is the natural generalization of (110). It may be written for the case in which the average of each variable is zero as

$$C_{x_1 x_2 \cdots x_n} = \text{av} \left(e^{i(\xi_1 x_1 + \xi_2 x_2 + \cdots + \xi_n x_n)} \right)$$

$$= \exp\left[-\frac{1}{2} \sum_{j=1}^{n} \sum_{k=1}^{n} \lambda_{jk} \xi_j \xi_k \right]. \qquad (112)$$

Here λ_{jk} is the covariance of x_j and x_k. The generalized central moments for this multivariate case are given by

$$\mu_{j_1 j_2 \cdots j_n} = \text{av} \left(x_1^{i_1} x_2^{i_2} \cdots x_n^{i_n} \right)$$

$$= \frac{j_1! j_2! \cdots j_n!}{i^{i_1+i_2+\cdots+i_n}} \text{Coeff. of } \xi_1^{i_1} \xi_2^{i_2} \cdots \xi_n^{i_n} \text{ in } C_{x_1 x_2 \cdots x_n}. \quad (113)$$

The multivariate probability density function is not so simply expressed in terms of second order moments because the Fourier transform of (112) introduces a quadratic form reciprocal to that having the coefficients λ_{jk}. The result may be expressed as

$$p(x_1, x_2, \cdots x_n) = \frac{\exp\left[-\frac{1}{2\lambda} \sum_{j=1}^{n} \sum_{k=1}^{n} \Lambda_{jk} x_j x_k \right]}{(2\pi)^{n/2} \lambda^{1/2}} \qquad (114)$$

where λ is the nth order determinant with typical element λ_{jk} and Λ_{jk} is the cofactor of λ_{jk} in λ. Note that $\lambda_{jk} = \lambda_{kj}$ and $\Lambda_{jk} = \Lambda_{kj}$ throughout.

The central limit theorem for a multivariate distribution states under very general conditions that if each $x_1, x_2 \cdots x_n$ is itself the sum of a large number of independent quantities with arbitrary distributions the probability density function $p(x_1, x_2, \cdots x_n)$ approaches the n-dimensional Gaussian form.

AUTOCORRELATION AND POWER SPECTRA

An immediate application of bivariate statistics may be made to the case in which the two values represent measurements made on a single noise source at two specified instants of time. In the case of the extension-in-time program, we consider a single noise wave $v(t)$ and let $x = v(t_1)$, $y = v(t_1 + \tau)$. If the noise wave represents a stationary process, the statistics of x and y do not depend on t_1 but only on the time difference τ. In the case of the ensemble representation we make observations at the instants t_1 and $t_1 + \tau$ on a large number of similar noise sources. We thereby obtain an ensemble of paired values, which we may write

$$(x, y) = (v(t_1), v(t_1 + \tau)). \qquad (115)$$

If the ensemble $[v(t)]$ is stationary, the statistics of the ensemble (x, y) do not depend on t_1 but only on the time shift τ. Finally if the noise process is ergodic, the ensemble statistics and time statistics coincide.

In the extension-in-time program for stationary processes it is customary to define the average product of $v(t)$ and $v(t+\tau)$ as the autocorrelation function, which we shall designate here as $R_v(\tau)$. Then

$$R_v(\tau) = \lim_{T \to \infty} \frac{1}{T} \int_0^T v(t) v(t + \tau) dt. \qquad (116)$$

If the process is ergodic, the autocorrelation can also be computed by averaging over the ensemble, thus

$$R_v(\tau) = \text{av} \, (xy) = \int_{-\infty}^{\infty} \int_{-\infty}^{\infty} xy p(x, y) dx dy. \quad (117)$$

We recall that the covariance of the jointly distributed quantities x and y has been previously defined by (102) as av $[x - \text{av} \, x) \, (y - \text{av} \, y)]$. The autocovariance of $v(t)$ may accordingly be defined as the covariance of x and y when $x = v(t)$ and $y = v(t+\tau)$. It is equal to the autocorrelation in the case of ac ergodic noise sources. For ergodic sources in general, we have the relation that the autocovariance is equal to the autocorrelation minus the square of the average. The autocovariance of the sum of two independent processes is the sum of the autocovariances of the individual processes as may be seen by writing out the expression for the required average and setting the covariances of one process with the other equal to zero.

It is possible to proceed further by defining the third order probability density function $p(x, y, z)$ where $x = v(t)$, $y = x(t + \tau_1)$, $z = v(t + \tau_1 + \tau_2)$ and the third order autocorrelation function

$$\psi(\tau_1, \tau_2) = \text{av} \, (xyz) \qquad (118)$$

Similar generalization can be extended to any order. Second order statistics are however sufficient for most engineering problems and in one important case, that of the Gaussian process, the higher order statistics are completely determined when the second order is known.

THE POWER SPECTRUM

It has proved to be of inestimably great value in communication engineering to resolve waves into sinusoidal components of specified frequencies. The central application may be summarized thus: if a wave $v(t)$, which to be definite we shall say is a voltage, can be represented in the form

$$v(t) = \int_{-\infty}^{\infty} F(\omega) e^{i t \omega} \frac{d\omega}{2\pi} \qquad (119)$$

and if a linear system has the steady state transmission function $Y(\omega)$, which for definiteness, we shall say means that a voltage $V_0 e^{i\omega t}$ produces current $I_0 e^{i\omega t}$, with $I_0 = Y(\omega) V_0$, then the current $I(t)$ in response to $v(t)$ is

$$I(t) = \int_{-\infty}^{\infty} Y(\omega) F(\omega) e^{i t \omega} \frac{d\omega}{2\pi} \qquad (120)$$

The formal rule for calculating $F(\omega)$ from $v(t)$ is

$$F(\omega) = \int_{-\infty}^{\infty} v(t)e^{-i\omega t}dt \qquad (121)$$

$F(\omega)d\omega/2\pi$ is interpreted physically as the amplitude (in complex form) of an infinitesimal component of $v(t)$, containing frequencies in the range $d\omega$ at ω.

This is the conventional Fourier integral theory as applied to linear systems and it is highly satisfactory when the operations indicated can be given numerical meaning. In case of a noise wave it is inadequate because the integral (121) does not converge for a typical $v(t)$ from a noise ensemble. The difficulty can be remedied by basing our calculations on mean square voltages in infinitesimal frequency intervals instead of first powers.

Suppose we consider a finite interval of time from $t=0$ to $t=T$. If we replace the values of $v(t)$ outside this interval by zero, we can write

$$F(\omega) = \int_{0}^{T} v(t)e^{-i\omega t}dt \qquad (122)$$

which can be calculated even for a noise wave because the limits are finite.

Then for $0 < t < T$,

$$v(t) = \int_{-\infty}^{\infty} F(\omega)e^{i\omega t}\frac{d\omega}{2\pi}. \qquad (123)$$

The total energy represented by $v(t)$ in the interval 0 to T is proportional to

$$\begin{aligned}
\int_{0}^{T} v^2(t)dt &= \int_{0}^{T} v(t)dt \int_{-\infty}^{\infty} F(\omega)e^{i\omega t}\frac{d\omega}{2\pi} \\
&= \int_{-\infty}^{\infty} F(\omega)\frac{d\omega}{2\pi} \int_{0}^{T} v(t)e^{i\omega t}dt \\
&= \int_{-\infty}^{\infty} F(\omega)F(-\omega)\frac{d\omega}{2\pi} = \int_{-\infty}^{\infty} \frac{|F(\omega)|^2}{2\pi}d\omega. \quad (124)
\end{aligned}$$

The average power S_T in the voltage wave in the time interval from 0 to T is obtained (again for unit resistance) by dividing by T. Hence

$$S_T = \frac{1}{T}\int_{0}^{T} v^2(t)dt = \int_{-\infty}^{\infty} \frac{|F(\omega)|^2}{2\pi T}d\omega. \qquad (125)$$

Let us define the function $W_v(\omega, T)$ by

$$W_v(\omega, T) = \frac{|F(\omega)|^2}{2\pi T}. \qquad (126)$$

We may then write (125) in the form

$$S_T = \int_{-\infty}^{\infty} W_v(\omega, T)d\omega. \qquad (127)$$

This expression for average power over a finite but arbitrarily long time interval T is in the form of an integral of a density function of frequency. We can interpret this as meaning that an infinitesimal frequency interval $d\omega$ at ω contributes average power $W_v(\omega, T)d\omega$ to the total. It is tempting to assume now that T approaches infinity and define the corresponding limit of $W_v(\omega, T)$ as the spectral density of $v(t)$. This has in effect been implied in much of the literature, and for most problems the results thereby obtained are correct. The procedure is not without its hazards, however, and we must caution the reader that a straightforward limiting process applied to (126) in specific cases will almost always fail to converge.

There are two ways of escaping from this seeming mathematical quandary. First, if we prefer to continue with a single function of time, we examine a finite segment of frequency interval by defining the function

$$\Omega(\omega, T, \Delta\omega) = \int_{\omega-\Delta\omega/2}^{\omega+\Delta\omega/2} W_v(\omega, T)d\omega. \qquad (128)$$

Now if $\Delta\omega$ is held fixed and T made indefinitely large a limit is actually approached, defining the function

$$\Omega(\omega, \Delta\omega) = \lim_{T\to\infty} \Omega(\omega, T, \Delta\omega). \qquad (129)$$

We may now take the limit of the ratio of this quantity to $\Delta\omega$ as $\Delta\omega$ approaches zero and thereby define the spectral density by

$$W_v(\omega) = \lim_{\Delta\omega\to 0} \frac{\Omega(\omega, \Delta\omega)}{\Delta\omega}. \qquad (130)$$

The subscript v is inserted to designate the power spectrum of $v(t)$.

The second procedure is based on the ensemble approach. We first average $W_v(\omega, T)$ over the ensemble of possible noise functions and then take the limit as T approaches infinity. In symbols, we define the spectral density as

$$W_v(\omega) = \lim_{T\to\infty}\left[\text{av}\,\frac{|F(\omega)|^2}{2\pi T}\right]. \qquad (131)$$

When the calculation is made in this way, the mathematical difficulties disappear, except when the ensemble contains periodic components, as will be discussed below. If the process is ergodic, this definition suffices for the spectral density of a single function from the ensemble as well. Note that we have defined the spectral density in terms of ω expressed in radians per second. We shall adopt the convention here that $w_v(f) = 2\pi W_v(\omega)$ is the density when the frequency is expressed in cps.

The ensemble approach is the preferred one in modern analytical work. Experimental work in spectral analysis on the other hand with its long tradition going back to the earliest work in optics, has been almost entirely based on the first procedure outlined above. The bandwidth $\Delta\omega$ is the width of the resolving filter and in practice it must always be finite. The averaging interval T is not infinite, but is sufficiently large to make the effect of any further increment negligible. In general, the smaller $\Delta\omega$ the longer T should be. In practice there is very little difficulty in fitting $\Delta\omega$ and T to the scale of the experiment. Fig. 4 indicates the type of apparatus suitable for the experimental determination of $W_v(\omega)$ from the definition (130). We note that for mathematical convenience we have defined a spectral density function for both positive and negative frequencies. It is clear from (126) that the function so defined must be an even function of frequency since $F(\omega)$ and $F(-\omega)$ are conjugate. In physical apparatus the mean power contributions from intervals $d\omega$ at $-\omega$ and ω add to give a single contribution equal to their sum at ω. Hence the measured outputs correspond to $2W_v(\omega)\Delta\omega$.

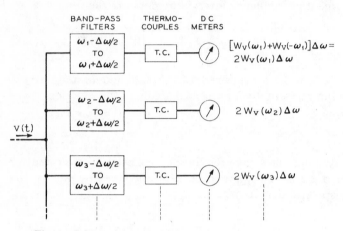

Fig. 4—Schematic circuit for measuring power spectrum.

If $v(t)$ contains one or more purely sinusoidal components, we tend to approach indefinitely large spectral densities at the corresponding frequencies. There is no analytical difficulty here if we deal with the integral of the spectral density up to frequency w, for the value of this integral can change abruptly as we pass the frequency of the sinuosidal component. A convenient symbolic representation is obtained by writing $P_0\delta(f-f_0)$ for the spectral density of a component having mean power P_0 and frequency f_0. The usual rules for calculating with δ functions are then applied.

A great convenience obtained by introduction of the power spectrum is the associated application of steady state network theory to evaluate effects of transmission through linear systems. Thus going back to (119) and (120) we note that if $v(t)$ has the power spectrum $w_v(f)$, the power spectrum of $I(t)$ is

$$w_I(f) = |Y(\omega)|^2 w_v(f). \qquad (132)$$

The analytical simplifications thereby achieved are indispensable to orderly progress.

The power spectrum has been defined in terms of squaring and averaging operations on the time function. It is accordingly a second-order statistic and can be related to the other second-order statistic we have

defined as the autocorrelation. We develop this relation in the next section.

RELATION BETWEEN AUTOCORRELATION AND THE POWER SPECTRUM

Let us write $R_v(\tau, T)$ for the approximate autocorrelation evaluated by averaging over a finite interval T. Then

$$R_v(\tau, T) = \frac{1}{T}\int_0^T v(t)v(t+\tau)dt. \qquad (133)$$

Substituting the expression (123) for $v(t+\tau)$, we then find

$$R_v(\tau, T) = \frac{1}{2\pi T}\int_0^T v(t)dt \int_{-\infty}^{\infty} F(\omega)e^{i\omega(t+\tau)}d\omega$$
$$= \frac{1}{2\pi T}\int_{-\infty}^{\infty} F(\omega)e^{i\omega\tau}d\omega \int_0^T v(t)e^{i\omega t}dt. \qquad (134)$$

The last integral may be evaluated from (122) giving

$$R_v(\tau, T) = \frac{1}{2\pi T}\int_{-\infty}^{\infty} F(\omega)e^{i\omega\tau}F(-\omega)d\omega$$
$$= \int_{-\infty}^{\infty} \frac{|F(\omega)|^2}{2\pi T} e^{i\omega\tau}d\omega$$
$$= \int_{-\infty}^{\infty} W_v(\omega, T)e^{i\omega\tau}d\omega. \qquad (135)$$

This is seen to be a generalization of (127). The difficulties with the limits are similar to those previously described in defining the spectral density and need not be discussed again. With similar reservations, we pass to the limit for T large and write

$$R_v(\tau) = \int_{-\infty}^{\infty} W_v(\omega)e^{i\omega\tau}d\omega. \qquad (136)$$

We note that

$$R_v(0) = \int_{-\infty}^{\infty} W_v(\omega)d\omega = \lim_{T\to\infty} S_T. \qquad (137)$$

That is, the autocorrelation for zero time shift is equal to the average power over all time. This is the maximum value which $R_v(\tau)$ can have since $W_v(\omega)$ is always positive and the term $e^{i\omega\tau}$ with absolute value unity can not increase the integrand of (136) over that of (137) at any value of ω. Thus the autocorrelation function is the Fourier transform of the power spectrum. Also by the inverse Fourier transform theorem

$$W_v(\omega) = \frac{1}{2\pi}\int_{-\infty}^{\infty} R_v(\tau)e^{-i\omega\tau}d\tau. \qquad (138)$$

Hence we may compute the power spectrum from the autocorrelation function and when the noise source is ergodic from the autocovariance of the ensemble. It is in fact convenient from the mathematical point of view to define the ac power spectrum of a stationary ensemble of functions as the Fourier transform of the autocovariance. It is desirable from the computational standpoint to deal with the dc term separately.

The power spectrum, like the autocorrelation from which it can be linearly calculated, is an additive property for independent noises. This is consistent with our familiar procedure of adding average power from incoherent sources to obtain the average total power in their sum.

We shall now give some illustrations. Consider an ergodic Gaussian noise source, $v(t)$, with mean value zero. The probability density function $p(x, y)$ with $x = v(t)$, $y = v(t+\tau)$ is obtained by setting $x_0 = y_0 = 0$ in (103), giving

$$p(x, y) = \frac{1}{2\pi\sqrt{\mu_{20}{}^2 - \mu_{11}{}^2}}$$
$$\cdot \exp\left[-\frac{\mu_{20}{}^2(x^2 + y^2) - 2\mu_{11}xy}{2(\mu_{20}{}^2 - \mu_{11}{}^2)}\right]. \quad (139)$$

Here $\mu_{20} = \mu_{02}$ is the variance σ^2 or in this case, the mean square value of $v(t)$, and μ_{11}, is the autocovariance or autocorrelation $R_v(\tau)$. The power spectrum $W_v(\omega)$ is given by (137). Similarly, if we are given an ac Gaussian noise source with power spectrum $W_v(\omega)$, we obtain the probability density function $p(x, y)$ by inserting in (139)

$$\mu_{11} = \int_{-\infty}^{\infty} W_v(\omega)e^{-i\omega\tau}df = R_v(\tau) \quad (140)$$

$$\mu_{20} = [\mu_{11}]_{\tau=0} = R_v(0) = \sigma^2. \quad (141)$$

Let us complete a problem started in the second section—the half-wave rectification of a given band of Gaussian noise. Before we were able to calculate only the average ac power in the rectifier output. Now we can calculate the complete power spectrum. To do so, let the output voltage of the rectifier be $u(t)$ and let

$$\left.\begin{array}{c} \xi = u(t) \\ \eta = u(t + \tau) \end{array}\right\}. \quad (142)$$

Then ξ is equal to x when x is positive, but is zero when x is negative. Likewise η is equal to y when y is positive and is zero when y is negative. The probability density function $q(\xi, \eta)$ can then be expressed in terms of $p(x, y)$ by

$$\left.\begin{array}{l} q(\xi, \eta) = p(\xi, \eta) \text{ when } \xi > 0 \text{ and } \eta > 0 \\ q(\xi, \eta) = 0 \text{ when } \xi < 0 \text{ or when } \eta < 0 \end{array}\right\}. \quad (143)$$

These expressions do not include the origin at which all the probability associated with negative values of either x or y is condensed. Since ξ and η are both zero here there is no contribution to the autocorrelation ensemble average and we write for the autocorrelation function $R_u(\tau)$ of the half-wave rectifier output

$$R_u(\tau) = \int_0^\infty \int_0^\infty \xi\eta\, p(\xi, \eta)d\xi d\eta$$

$$= \frac{1}{2\pi(\mu_{20}{}^2 - \mu_{11}{}^2)} \int_0^\infty \int_0^\infty \xi\eta$$

$$\cdot \exp\left[-\frac{\mu_{20}{}^2(\xi^2 + \eta^2) - 2\mu_{11}\xi\eta}{2(\mu_{20}{}^2 - \mu_{11}{}^2)}\right]d\xi d\eta$$

$$= \frac{1}{2\pi}\left[(\mu_{20}{}^2 - \mu_{11}{}^2)^{1/2} + \mu_{11} \arccos\frac{-\mu_{11}}{\mu_{20}}\right]$$

$$= \frac{1}{2\pi}\left\{[R_v{}^2(0) - R_v{}^2(\tau)]^{1/2}\right.$$

$$\left. + R_v(\tau) \arccos\frac{-R_v(\tau)}{R_v(0)}\right\}. \quad (144)$$

The autocorrelation of the rectified output is thus expressed in terms of the autocorrelation of the Gaussian noise input. The power spectrum $W_u(\omega)$ of the rectified output follows immediately by use of the Fourier transform relationship (138) replacing the subscript v by u throughout. There is however a computational difficulty in that the $R_u(\tau)$ defined by (144) gives a divergent integral when substituted in (138). This can be seen from the fact that $R_v(\tau)$ approaches zero as τ goes to $\pm\infty$, and hence $R_u(\tau)$ approaches the constant $R_v{}^2(0)/2\pi$. To resolve this difficulty, we observe that the autocorrelation function of a constant voltage E is by (116).

$$R_E(\tau) = \lim_{T \to \infty} \frac{1}{T}\int_{t+T}^{T} E^2 dt = E^2 \quad (145)$$

and hence that the presence of an additive constant R_0 in an autocorrelation means that there is a dc component E_0 in the wave satisfying

$$E_0 = R_0{}^{1/2} \quad (146)$$

and hence having mean power $E_0{}^2 = R_0$. We therefore subtract a dc component of power R_0 to obtain the ac component of the wave. The autocorrelation of the ac component of $u(t)$ is therefore

$$R_{uac}(\tau) = R_u(\tau) - \frac{R_v{}^2(0)}{2\pi}. \quad (147)$$

Since we have already calculated the dc component for half-wave rectified Gaussian noise by a more elementary procedure, we can check this part of our more sophisticated calculation. Our earlier result was given by (40) in the form $\sigma/(2\pi)^{1/2}$ when we set the rectifier constant a_1 equal to unity. In the present solution we substitute

$$R_0 = \frac{R_v{}^2(0)}{2\pi} = \frac{\sigma^2}{2\pi} \quad (148)$$

and hence

$$E_0 = R_0^{1/2} = \sigma/(2\pi)^{1/2} \qquad (149)$$

as before.

But we now may go on and calculate the remainder of the spectrum of the rectified output. The result depends of course on the power spectrum of the input. Some interesting facts can be deduced from (144) by a power series expansion without specifying the spectrum. The following result is obtained

$$
\begin{aligned}
R_u(\tau) = & \frac{\mu_{20}}{2\pi} + \frac{\mu_{11}}{4} + \frac{\mu_{11}^2}{4\pi\mu_{20}} + \frac{\mu_{11}^4}{24\pi\mu_{20}^3}\left[\frac{1}{2} + \frac{3^2}{5\cdot 2\cdot 3!}\frac{(\mu_{11})^2}{(\mu_{20})}\right. \\
& \left. + \frac{3^2\cdot 5^2}{5\cdot 7\cdot 2^2\cdot 4!}\frac{(\mu_{11})^4}{(\mu_{20})} + \frac{3^2\cdot 5^2\cdot 7^2}{5\cdot 7\cdot 9\cdot 2^3 5!}\frac{(\mu_{11})^6}{(\mu_{20})} + \cdots \right] \\
= & \frac{R_v^2(0)}{2\pi} + \frac{R_v(\tau)}{4} + \frac{R_v^2(\tau)}{4\pi R_v(0)} + \frac{R_v^4(\tau)}{24\pi R_v^3(0)} \\
& \cdot \left\{\frac{1}{2} + \frac{3^2}{5\cdot 2\cdot 3!}\left[\frac{R_v(\tau)}{R_v(0)}\right]^2 + \cdots\right\}.
\end{aligned} \qquad (150)
$$

The first term of the series after the constant is one-fourth the autocorrelation function of $v(t)$, the input wave. The Fourier transform of this term in accordance with (138) is therefore one-fourth of the original power spectrum. This verifies a familiar fact that a half-wave linear rectifier with a multifrequency sine wave input delivers among other things a set of fundamental components of half amplitude and hence one quarter power. The next term is proportional to the square of $R_v(\tau)$. This is the kind of behavior found in autocorrelation when the output voltage is proportional to the square of the input. It is well-known that in such cases we may resolve the output into components with frequencies which are sums and differences of the frequencies present in the original wave. The Fourier transform of this term therefore leads to a sum and difference spectrum. Similarly the higher power terms represent contributions from correspondingly higher orders of modulation. It should be noted that we have taken the linear rectifier merely as a typical example and that the procedure is applicable to the response of nonlinear devices in general. For further applications see [3].

We have shown earlier in (62) how a Gaussian process could be approached as the limiting form of the sum of a large number of sinusoidal distributions. This suggests a method of constructing Gaussian ensembles with specified spectral density functions. First, divide the frequency range into small intervals of width $\Delta\omega$. Then assume sinusoidal components of random phase, frequencies equal to each midinterval value, and amplitudes proportional to the square root of the spectral density function. The ensemble of sums of these sinusoidal components approaches the required Gaussian ensemble as the division points are taken infinitesimally close together. Specifically, the ensemble

$$v(t) = 2\sum_{n=1}^{N}\sqrt{W_v(\omega_1+n\Delta\omega)\Delta\omega}\cos\left[(\omega+n\Delta\omega)t+\theta_n\right] \qquad (151)$$

with $N = (\omega_2-\omega_1)/\Delta\omega$ and the θ_n's uniformly and independently distributed approaches a Gaussian ensemble with spectral density $W_v(\omega)$ in the range $\omega_1 < |\omega| < \omega_2$. The mean square of the typical component is one-half the square of its amplitude, or $2W_v(\omega_1+n\Delta\omega)\Delta\omega$, which represents a contribution of $W_v(\omega_1+n\Delta\omega)\Delta\omega$ centered around the frequency $\omega_1+n\Delta\omega$ and a like contribution centered around the frequency $-\omega_1-n\Delta\omega$.

We may use this model to solve a wide variety of noise problems by first evaluating the response of the system under study to the wave (151) with N finite and then taking the limit of our solution as N goes to infinity. It is to be noted that the solution thus obtained is for a Gaussian noise source. Compared to modern techniques, the approximation of a noise source by a sum of sine waves is crude, but it is nevertheless effective in situations where the algebra can be kept under control. It forms a transition between the kinds of waves transmitted in multichannel frequency division systems and the waves produced by Gaussian noise. When the number of channels in such systems is large, it has been found helpful to use noise waves in both experiment and analysis as a simulation of average loading conditions [4, 18].

A disadvantage of the sine-wave series as a model of noise is the cumbersomeness of the calculations in complicated situations. For an example of moderate complexity see the author's paper [5] on linear rectification of a sine wave superimposed on noise. For an analysis of some considerably more intricate situations, we cite a paper by Lewin [6], in which the question of coherence between various terms obtained in the computing process becomes an issue. It seems that it is very hard to insure analytical rigor in such calculations. We shall return later however to the use of the sine-wave series as a very valuable adjunct in explaining the properties of narrow band noise.

We next illustrate some typical spectral density functions and their associated autocorrelations.

Band Limited White

In this case, the input spectrum is spread uniformly over a specified range of frequencies and vanishes outside. In theory, it is obtained by passing white noise through an ideal filter. Since neither white noise nor ideal filters exist it is only an approximation to reality. We shall continue for computational simplicity the model containing both positive and negative frequencies and write for this case,

$$W(\omega) = \begin{cases} W_0, & \omega_1 < |\omega| < \omega_2 \\ 0, & |\omega| < \omega_1 \text{ and } |\omega| > \omega_2. \end{cases} \qquad (152)$$

The plot of this spectrum is shown in Fig. 5(a). The mean total power is equal to the area of the two rectangles, giving

$$P = 2W_0(\omega_2 - \omega_1). \qquad (153)$$

The autocorrelation function $R(\tau)$ is found to be

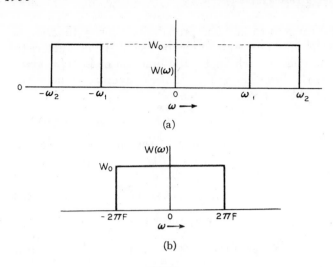

(a)

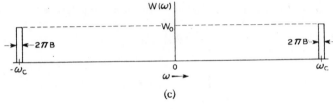

(b)

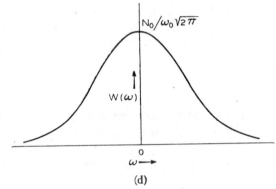

(c)

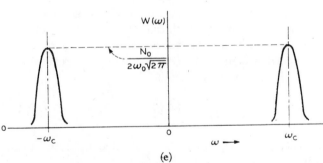

(d)

(e)

Fig. 5—(a) Power spectrum of band-limited white noise. (b) Power spectrum of low pass white noise. (c) Power spectrum of narrow band white noise. (d) Gaussian low pass power spectrum. (e) Gaussian band pass power spectrum.

$$R(\tau) = \left[\int_{-\omega_2}^{-\omega_1} + \int_{\omega_1}^{\omega_2} \right] W_0 e^{i\omega\tau} d\omega$$

$$= \frac{4W_0}{\tau} \sin \frac{(\omega_2 - \omega_1)}{2} \tau \cos \frac{(\omega_2 + \omega_1)}{2} \tau. \quad (154)$$

Substitution of this term in (144) leads to rather involved integrations, but evaluation has at least been reduced to an explicit computational procedure. From here on it is a matter of ingenuity in evaluating integrals or effective employment of machine techniques.

Special cases of the band-limited white noise of considerable importance are
The Low Pass Case:

$$\omega_1 = 0, \quad \omega_2 = 2\pi F, \quad R(\tau) = \frac{2W_0 \sin 2\pi F\tau}{\tau}, \quad (155)$$

[See Fig. 5(b).]
and
The Narrow Band Case:

$$\omega_1 \gg \omega_2 - \omega_1.$$

Here we would write for convenience,

$$\frac{\omega_1 + \omega_2}{2} = \omega_c, \quad \omega_2 - \omega_1 = 2\pi B$$

$$R(\tau) = \frac{4W_0}{\tau} \sin \pi B\tau \cos \omega_c\tau. \quad (156)$$

[See Fig. 5(c).]

In many problems, the exact shape of the spectrum within the narrow range to which it is confined is not critical. We shall devote a later section to an intensive treatment of the narrow band case with arbitrary spectral shape.

The Gaussian Spectrum

A Gaussian low pass spectrum has spectral density [Fig. 5(d)]

$$W(\omega) = \frac{N_0}{\omega_0(2\pi)^{1/2}} e^{-\omega^2/2\omega_0^2}. \quad (157)$$

N_0 is the mean total power. At $\omega = \omega_0$, the power per unit bandwidth is down one-quarter neper or 2.17 db relative to the density at zero frequency. The spectrum is of considerable importance because it simulates the gradual kind of cutoff which is more realistic in physical networks than the abrupt transition of the ideal filter. Specifically a Gaussian frequency selectivity is approached by an RC-coupled amplifier of many stages. We must not confuse the Gaussian frequency spectrum with the Gaussian probability density function previously discussed. Both are Gaussian functions, but the variables are different. A computational advantage of the Gaussian spectrum is that its Fourier transform is also Gaussian and hence the autocorrelation is Gaussian. Furthermore a Gaussian function raised to a power remains Gaussian so that a series expansion such as (150) in power of the autocorrelation can be Fourier-transformed term by term to yield a corresponding series of Gaussian spectral densities.

The autocorrelation function for the Gaussian low pass spectrum is

$$R(\tau) = \frac{N_0}{\omega_0(2\pi)^{1/2}} \int_{-\infty}^{\infty} e^{-\omega^2/2\omega_0^2} e^{i\omega\tau} d\omega$$

$$= N_0 e^{-(\omega_0\tau)^2/2}. \quad (158)$$

The Gaussian band pass spectrum is also important.

It is

$$W(\omega) = \frac{N_0}{2\omega_0(2\pi)^{1/2}} \left[e^{-(\omega-\omega_c)^2/2\omega_0^2} + e^{-(\omega+\omega_c)^2/2\omega_0^2} \right] \quad (159)$$

$$R(\tau) = N_0 e^{-(\omega_0\tau)^2/2} \cos \omega_1\tau. \quad (160)$$

[See Fig. 5(e).]

The narrow band case in which $\omega_0 \ll \omega_c$ is the one most frequently of interest. Here the effect of the tails crossing over from positive to negative and negative to positive frequencies is negligible, and we obtain a good approximation for a sharply-tuned multi-stage amplifier response to white noise. This is the network-theory analog of the central limit theorem on probability density functions. The Gaussian band pass case will be treated along with other narrow band distributions in a later section.

In general, the power spectrum, like the autocorrelation function, is representative of second order statistics and does not give a complete statistical description of the process. An important exception is the Gaussian process (referring now to Gaussian probability function), which yields statistics of all orders in terms of second order only. We assume here and later that the average value of the ensemble has been subtracted out, so that when we refer to Gaussian noise we mean "ac Gaussian Noise." For this case the autocorrelation $R_v(\tau)$ is sufficient to determine all orders of statistics. Since the autocorrelation is calculated from the spectral density $W(\omega)$ and vice versa, we can equally well relate all orders of statistics to $W(\omega)$.

SIGNAL AND NOISE

The study of noise itself is interesting and important, but a still more useful analysis deals with combinations of noise and signal waves. By "signal" we mean some information-bearing component of a wave which we desire to isolate. We regard the undesirable remainder as noise. The separation may be of varying degrees ranging all the way from freeing a high fidelity musical program from any trace of a noise background to answering with yes or no the question: is there or is there not a signal present in this wave? As we approach the latter type of problem the use of statistics rather than an exact description becomes more and more the proper approach.

The signal may take on many forms, but whatever it is, we usually fall back on Fourier analysis to resolve it into sinusoidal components, and to simplify the problem we usually begin the analysis in terms of one sinusoidal wave accompanied by noise. An important ensemble for study is that of the sine wave plus Gaussian noise. We may write this ensemble as

$$[y(t)] = [x(t) + A \cos (\omega_0 t + \theta)] \quad (161)$$

where $[x(t)]$ is a Gaussian ensemble and $[A \cos (\omega_0 t + \theta)] = [s(t)]$ is an independent sinusoidal ensemble defined with A and ω_0 constant and θ uniformly distributed in

the range 0 to 2π. The probability density function of $[y(t)]$ may be obtained by the use of the characteristic functions. The characteristic function of $[x(t)]$ is given by (64). We shall assume that the average value of the ensemble is zero and hence

$$C_x(\xi) = e^{-\sigma^2\xi^2/2}. \quad (162)$$

The characteristic function of the sinusoidal ensemble has been found in (58). We write it in our present notation as

$$C_s(\xi) = J_0(A\xi). \quad (163)$$

It follows that the ch.f. of $[y(t)]$ is

$$C_y(\xi) = C_x(\xi)C_s(\xi) = J_0(A\xi)e^{-\sigma^2\xi^2/2}. \quad (164)$$

Hence the probability density function for $[y(t)]$ is

$$p_y(z) = \frac{1}{2\pi} \int_{-\infty}^{\infty} J_0(A\xi) e^{iz\xi - \sigma^2\xi^2/2} d\xi$$

$$= \frac{1}{\pi} \int_0^{\infty} J_0(A\xi) e^{-\sigma^2\xi^2/2} \cos z\xi d\xi. \quad (165)$$

It seems we are forced to leave this function in integral form. It is sometimes convenient to replace $J_0(A\xi)$ by an integral representation

$$J_0(A\xi) = \frac{1}{\pi} \int_{-\pi/2}^{\pi/2} e^{-iA\xi \sin \theta} d\theta \quad (166)$$

and perform the integration with respect to ξ. The equivalent form for $P_y(z)$ thereby obtained is

$$p_y(z) = \frac{1}{\pi(2\pi)^{1/2}} \int_{-\pi/2}^{\pi/2} e^{-(z-A \sin \theta)^2/2\sigma^2} d\theta. \quad (167)$$

The probability density function is sufficient to solve the problem of calculating the dc output when a sine wave plus Gaussian noise is applied to a rectifier. If the rectifier is linear half wave with forward conductance α, the dc component y_0 is

$$y_0 = \alpha \int_0^{\infty} z p_y(z) dz. \quad (168)$$

The resulting integral can be evaluated in closed form to give

$$y_0 = \alpha(\sigma/2\pi)^{1/2} e^{-A^2/4\sigma}$$

$$\cdot \left\{ I_0(A^2/4\sigma) + \frac{A^2}{2\sigma} \left[I_0(A^2/4\sigma) + I_1(A^2/4\sigma) \right] \right\}. \quad (169)$$

We may calculate the spectrum of the rectified output if we first determine the autocorrelation function of $y(t)$. The latter can be computed from the two-dimensional probability density function of $y(t)y(t+\tau)$, and this in turn can be found by combining bivariate characteristic functions of the component ensembles. The two-dimensional ch.f. of the Gaussian ensemble has previously been given by (110). The ch.f. of the two-dimensional sinusoidal ensemble is

$$C_s(\xi_1, \xi_2)$$

$$= \text{av } (e^{i\xi_{1s}(t)}e^{i\xi_{2s}(t+\tau)})$$

$$= \text{av exp } (i\xi_1 A \cos (\omega_0 t + \theta) + i\xi_2 A \cos [\omega_0(t + \tau) + \theta])$$

$$= \frac{1}{2\pi} \int_{-\pi+\alpha}^{\pi+\alpha} e^{iA(\xi_1 \cos (\omega_0 t+\theta)+\xi_2 \cos [\omega_0(t+\tau)+\theta])}d\theta$$

$$= \frac{1}{2\pi} \int_{-\pi+\alpha}^{\pi+\alpha} d\theta e^{iA\sqrt{(\xi_1+\xi_2 \cos \omega_0\tau)^2+\xi_2^2 \sin^2 \omega_0\tau} \cos (\omega_0 t+\theta+\beta)} \quad (170)$$

where α is arbitrary and

$$\tan \beta = \frac{\xi_2 \sin \omega_0\tau}{\xi_1 + \xi_2 \cos \omega_0\tau}. \quad (171)$$

By changing variable to $\phi = \theta + \beta + \omega_0 t$, we observe that

$$C_s(\xi_1, \xi_2) = J_0(A\sqrt{\xi_1^2 + \xi_2^2 + 2\xi_1\xi_2 \cos \omega_0\tau}). \quad (172)$$

Writing $C_x(\xi_1, \xi_2)$ for the two-dimensional ch.f. of the Gaussian ensemble, we have from (110)

$$C_x(\xi_1, \xi_2) = e^{[-\sigma^2(\xi_1^2+\xi_2^2)+2R_x(\tau)]/2} \quad (173)$$

and thence from (107), the two-dimensional ch.f. of the $y(t)$ ensemble is

$$C_y(\xi_1, \xi_2) = C_s(\xi_1, \xi_2)C_x(\xi_1, \xi_2). \quad (174)$$

If $y_1 = y(t)$ and $y_2 = y(t+\tau)$, the required two-dimensional probability density function is

$$p(y_1, y_2) = \frac{1}{4\pi^2} \int_{-\infty}^{\infty} \int_{-\infty}^{\infty} C_y(\xi_1, \xi_2)e^{-i(y_1\xi_1+y_2\xi_2)}dy_1 dy_2. \quad (175)$$

The autocorrelation function of a rectifier output when sine wave and noise form the input can now be computed by the method used in (144). The spectral density follows from the Fourier transform of the autocorrelation. The results are not simple, but this is not a simple problem—the remarkable thing is that a solution can be computed at all. Variations in the procedure are possible and the complications may be changed if not actually reduced by employing some of them. For example, we can compute the autocorrelation function of the rectified output directly from the characteristic function of the input in the important case in which the rectifying characteristic is itself expressible by a Fourier integral. The latter representation may not be possible when the path of integration is along the real axis, but becomes so when a properly chosen path in the complex plane is used. It is also useful to note that if we are interested only in the discrete components of the output spectrum, we may take the limit of the autocorrelation function as the shift time τ approaches infinity. This removes the contribution of the continuous part of the spectrum. For further treatment of these problems see papers by Rice and Middleton, [3] and [7].

DISCRETE SAMPLING OF A NOISE SOURCE— INTRODUCTION TO TIME SERIES

In many cases the measured noise data consist of observations made at discrete instants of time. This is a familiar situation in statistical work of all kinds. The instants of observation usually have a uniform spacing. The name "time series" is commonly given to such a collection of measured values at specified instants. In our treatment so far we have tacitly assumed that the functions of time we are studying could be observed continuously. The question then arises as to how we might apply our results to cases wherein only regularly spaced samples are available. The "sampling theorem" gives us an important leverage on those cases in which the power spectrum of the noise source vanishes outside a finite range. Physically we can only approximate such a condition, but the approximation is often sufficiently good to make the mathematical model useful. The theorem states that if a noise wave contains no frequencies of absolute value equal to or greater than W cps it can be completely reconstructed from its samples taken at instants $1/2W$ apart [8]. Specifically if $x(t)$ represents the noise wave, we may write the identity

$$x(t) = \sum_{n=-\infty}^{\infty} x\left(\frac{n}{2W}\right) \frac{\sin \pi(2Wt - n)}{\pi(2Wt - n)}. \quad (176)$$

The right-hand side is the response of an ideal low pass filter cutting off at $f = \pm W$ to impulses proportional to the sampled values. The samples and the continuous functions $x(t)$ thus give completely equivalent data, since we can either by experiment or calculation recover the complete $x(t)$ by passing the samples through an ideal low pass filter.

Now consider what would happen if we sample a wave for which the spectral density does not vanish for frequencies exceeding W. Assume

$$x(t) = \sum_{n=-\infty}^{\infty} y\left(\frac{n}{2W}\right) \frac{\sin \pi(2Wt - n)}{\pi(2Wt - n)} \quad (177)$$

where $y(t)$ is not band-limited. This is the ensemble of functions produced by passing the y-samples through an ideal filter cutting off at W. The function $x(t)$, is, of course, band-limited because it is a sum of ideal filter responses. Also by setting $t = n/2w$, we see that

$$y\left(\frac{n}{2W}\right) = x\left(\frac{n}{2W}\right). \quad (178)$$

That is, the values of $y(n/2W)$ are actually samples of the band-limited function $x(t)$, as well as of the unlimited function $y(t)$. A difference is that $x(t)$ can be generated from these samples, while $y(t)$ in general cannot. We ask the question as to what relation there is between the statistics of $x(t)$ and $y(t)$.

We may compute autocorrelation of ensemble $x(t)$ by averaging $x(t) x(t+\tau)$ over the y-ensemble with t fixed.

$$R_x(\tau) = \text{av } x(t)x(t + \tau)$$

$$= \sum_{m=-\infty}^{\infty} \sum_{n=-\infty}^{\infty} \text{av } y\left(\frac{m}{2W}\right)y\left(\frac{n}{2W}\right)\frac{\sin \pi(2Wt - m)}{\pi(2Wt - m)}$$

$$\cdot \frac{\sin \pi[2W(t + \tau) - n]}{\pi[2W(t + \tau) - n]}. \quad (179)$$

We may write

$$\text{av } y\left(\frac{m}{2W}\right)y\left(\frac{n}{2W}\right) = R_y\left(\frac{n-m}{2W}\right) \quad (180)$$

where $R_y(\tau)$ is the autocorrelation function of the y-ensemble, or if y is ergodic, the autocorrelation function of any $y(t)$. The summation in (179) appears to depend on the value of t but this is an illusion which can be dispelled by means of the following useful identity

$$\sum_{m=-\infty}^{\infty}\sum_{n=-\infty}^{\infty} \phi(n-m) \frac{\sin \pi(x-m)\sin \pi(x-n+a)}{\pi^2(x-m)(x-n+a)}$$

$$= \sum_{j=-\infty}^{\infty} \phi(j) \frac{\sin (j-a)\pi}{(j-a)\pi}. \quad (181)$$

The following derivation of this formula due to D. Slepian is much simpler than the process by which I obtained it. Let $g(x)$ be any function of x band-limited to the range $-1/2 < f < 1/2$. Then

$$g(x) = \sum_{m=-\infty}^{\infty} g(m) \frac{\sin \pi(x-m)}{\pi(x-m)}. \quad (182)$$

An example of such a band-limited function is

$$g_1(x) = \frac{\sin \pi(x-y)}{\pi(x-y)} = \frac{\sin \pi(y-x)}{\pi(y-x)} \quad (183)$$

where y is a constant. The Fourier transform of $g_1(x)$ is equal to $\exp(-iyf)$ in the range $-1/2 < f < 1/2$ and is zero outside this range. Substituting (183) in (182) gives the identity

$$\frac{\sin \pi(x-y)}{\pi(x-y)} = \sum_{m=-\infty}^{\infty} \frac{\sin \pi(y-m)}{\pi(y-m)} \frac{\sin \pi(x-m)}{\pi(x-m)}. \quad (184)$$

Assign to y the value

$$y = x - j + a$$

where j is zero or any positive or negative integer and a is a constant. Then

$$\frac{\sin \pi(j-a)}{\pi(j-a)} = \sum_{m=-\infty}^{\infty} \frac{\sin \pi(x-j+a-m)}{\pi(x-j+a-m)}$$

$$\cdot \frac{\sin \pi(x-m)}{\pi(x-m)}. \quad (185)$$

Multiply both sides by $\phi(j)$ and sum over all values of j

$$\sum_{j=-\infty}^{\infty} \phi(j) \frac{\sin \pi(j-a)}{\pi(j-a)}$$

$$= \sum_{j=-\infty}^{\infty} \sum_{m=-\infty}^{\infty} \phi(j) \frac{\sin \pi(x-j+a-m)}{\pi(x-j+a-m)} \frac{\sin \pi(x-m)}{\pi(x-m)}.$$

Substitution of $j = n - m$ in the double summation now gives (181).

Applying (181) to (179) and (180), we obtain

$$R_x(\tau) = \sum_{n=-\infty}^{\infty} R_y\left(\frac{n}{2W}\right) \frac{\sin \pi(n+2W\tau)}{\pi(n+2W\tau)}. \quad (186)$$

The results of this section are expressed in cps instead of radians as used hitherto. We shall use a lower case w to represent spectral density as a function of f in cps; $w(f) = 2\pi W(\omega)$. The power spectrum $w_x(f)$ of the band-limited ensemble is then, by (138),

$$w_x(f) = \int_{-\infty}^{\infty} e^{-i2\pi f\tau} R_x(\tau)d\tau$$

$$= \begin{cases} 0, & |f| > W \\ \dfrac{1}{2W} \displaystyle\sum_{n=-\infty}^{\infty} R_y\left(\frac{n}{2W}\right) e^{in\pi f/W}, & |f| < W. \end{cases} \quad (187)$$

We next replace R_y by its expression in terms of the power spectrum of y with the result that for $|f| < W$,

$$w_x(f) = \frac{1}{2W} \sum_{n=-\infty}^{\infty} e^{in\pi f/W} \int_{-\infty}^{\infty} w_y(\lambda)e^{in\pi\lambda/W}d\lambda. \quad (188)$$

Further simplification requires in effect an inversion of the order of summation and integration, but as the expression stands a divergent series would block progress. The difficulty could be avoided by judicious manipulation of δ functions, but fortunately we have available an impeccable theorem from analysis to produce the desired form of result immediately. This is Poisson's Summation Formula [9]:

$$\sum_{n=-\infty}^{\infty} \int_{-\infty}^{\infty} \phi(z)e^{inz}dz = 2\pi \sum_{n=-\infty}^{\infty} \phi(2n\pi). \quad (189)$$

In this case we set $\pi(f+\lambda)/W = z$ and find that

$$2\pi\phi(z) = w_y\left(\frac{zW}{\pi} - f\right). \quad (190)$$

Therefore for $|f| < W$,

$$w_x(f) = \sum_{n=-\infty}^{\infty} w_y(f + 2nW). \quad (191)$$

The interpretation of this result is clear. The contribution to the spectrum at the typical frequency f within the filter band is the sum of spectral densities occurring f cps above and below each harmonic of the sampling frequency $2W$. In Fig. 6 all the small full line

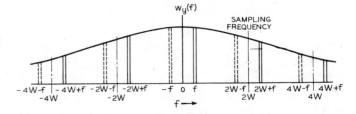

Fig. 6—Power spectrum obtained by low sampling rate.

rectangular slices shown contribute to the spectrum at f. The dashed rectangles give an equal contribution to $-f$. In the language of the communication engineer, the sampling process is equivalent to modulating the height of periodic short pulses by the noise. This may also be expressed as modulating the amplitude of all harmonics (including the zeroth) of the sampling frequency, since a very sharp pulse contains all harmonics with nearly the same amplitude. The result is that each harmonic beats down into the low pass filter band all the noise in the range W cps above and below it. There is no way of telling from the output which part of the original spectrum of $y(t)$ produces the individual contributions to the spectrum of $x(t)$. This phenomenon has been called "sideband overlap" in telephony. J. W. Tukey has contributed the name "aliasing." We remark that the result (191) can be obtained far more easily [10] by the sine-wave composition treatment of noise based on (141) than by the method used here; in fact a practical communication engineer would probably regard the result as intuitively obvious. However, as pointed out before, the sine-wave model must lead to Gaussian noise and is hence less general than our treatment here.

In communication engineering it is possible to limit the input band by filtering before sampling and thereby avoid multiple contributions to the desired output band. In statistical data such as obtained on prices, crops, weather, and the like, the input circuit is not accessible for filtering, and it may be difficult to separate the aliased from the original spectrum.

The total power represented by the band-limited noise is

$$\sigma_x^2 = \psi_x(0) = \sum_{n=-\infty}^{\infty} \psi_y\left(\frac{n}{2W}\right)\frac{\sin n\pi}{n\pi} = \psi_y(0) = \sigma_y^2. \quad (192)$$

Hence the mean total power of the limited and unlimited ensembles is the same. If the samples of the $y(t)$-ensemble are mutually independent

and

$$\left.\begin{array}{c} \psi_y\left(\dfrac{n}{2W}\right) = 0 \text{ when } n \neq 0 \\[2ex] \psi_x(\tau) = \psi_y(0)\,\dfrac{\sin 2\pi W\tau}{2\pi W\tau} \\[2ex] w_x(f) = \begin{cases} 0, & |f| > W \\ \sigma_y^2/2W, & |f| < W \end{cases} \end{array}\right\}. \quad (193)$$

This case thus gives a constant spectral density over the range $-W$ to W and zero spectral density outside. This is what we previously called band-limited white noise, (152).

In the important case in which the ensemble $y(t)$ is Gaussian, the values $y(n/2W)$ are Gaussian variates. If we then consider values of $x(t)$ at times other than the sampling instants we obtain values related to Gaussian variates by a linear transformation. It follows that values of $x(t)$ for any t, or set of t values, are also Gaussian variates and that their statistics are completely determined by either the autocovariance or power spectrum of the ensemble $x(t)$ as determined by (186) and (191). This convenient property may be summarized in the statement that the noise produced by filtered samples of Gaussian noise is also Gaussian. We can not make a **corresponding** statement about most other kinds of noise. The response of a low pass filter to samples having uniform or sinusoidal distribution, for example, does not have uniform or sinusoidal distribution respectively when the complete wave form of the filter output is considered.

CROSSCORRELATION AND CROSS-POWER SPECTRA

We next consider the statistics of two or more dependent noise sources. These might, for example, be one noise wave applied as input to a network and the resulting noise wave received as output from the network. Another important example consists of the amplitude and phase variations of a narrow band noise source.

As in the single noise source case, second order statistics are sufficient for most purposes. Corresponding to the autocorrelation function $R_x(\tau)$ for the ensemble of single functions $[x(t)]$, we define the crosscorrelation functions $R_{xy}(\tau)$ and $R_{yx}(\tau)$ for the ensemble of paired functions $[x(t), y(t)]$ by

$$R_{xy}(\tau) = \text{av}\,[x(t)y(t + \tau)]$$
$$= \int_{-\infty}^{\infty}\int_{-\infty}^{\infty} x_1 y_2 p(x_1,\,y_2)dx_1 dy_2 \quad (194)$$

where $x_1 = x(t)$, $y_2 = y(t+\tau)$.

$$R_{yx}(\tau) = \text{av}\,[y(t)x(t + \tau)] = R_{xy}(-\tau). \quad (195)$$

Corresponding to the power spectrum $W_x(\omega)$, we define the cross-power spectra

$$W_{xy}(\omega) = \frac{1}{2\pi}\int_{-\infty}^{\infty} R_{xy}(\tau)e^{-i\omega\tau}d\tau$$
$$= \lim_{T=\infty}\frac{F_x(-\omega)F_y(\omega)}{2\pi T} \quad (196)$$

$$W_{yx}(\omega) = \frac{2}{2\pi}\int_{-\infty}^{\infty} R_{yx}(\tau)e^{-i\omega\tau}d\tau$$
$$= \lim_{T\to\infty}\frac{F_y(-\omega)F_x(\omega)}{2\pi T} = W_{xy}^*(\omega). \quad (197)$$

In the above, $F_x(\omega)$ and $F_y(\omega)$ are defined by

$$F_x(\omega) = \int_0^T x(t)e^{-i\omega t}dt \quad (198)$$

$$F_y(\omega) = \int_0^T y(t)e^{-i\omega t}dt. \quad (199)$$

The asterisk means "conjugate of." We shall not enter into another discussion of the limiting processes indicated above as the arguments are entirely similar to those discussed in the definition of the spectral density of a single source. The measurement of cross spectra has been discussed by Barnes and Krendel [11, 12].

The cross-power spectra are seen to be conjugate complex numbers. The real parts are even functions of frequency and the imaginary parts are odd functions of frequency. The crosscorrelations can be calculated from the cross-power spectra by inverting the Fourier transforms, thus

$$R_{xy}(\tau) = \int_{-\infty}^{\infty} W_{xy}(\omega)e^{i\tau\omega}d\omega \qquad (200)$$

$$R_{yx}(\tau) = \int_{-\infty}^{\infty} W_{yx}(\omega)e^{i\tau\omega}d\omega. \qquad (201)$$

If we write

$$W_{xy}(\omega) = U_{xy}(\omega) + iV_{xy}(\omega) \qquad (202)$$

where U_{xy} and V_{xy} are real, then

$$U_{xy}(-\omega) = U_{xy}(\omega) = U_{yx}(\omega) \qquad (203)$$

$$V_{xy}(-\omega) = -V_{xy}(\omega) = V_{yx}(\omega) \qquad (204)$$

and

$$R_{xy}(\tau) = 2\int_{0}^{\infty} U_{xy}(\omega)\cos\tau\omega d\omega$$

$$-2\int_{0}^{\infty} V_{xy}(\omega)\sin\tau\omega d\omega. \qquad (205)$$

Given any two noise ensembles $[x(t)]$ and $[y(t)]$ with specified power spectra $W_x(\omega)$, $W_y(\omega)$ respectively and cross-power spectrum $W_{xy}(\omega) = U_{xy}(\omega) + iV_{xy}(\omega)$, it is possible to resolve either ensemble into the sum of three components which have cross spectra purely real, purely imaginary, and zero respectively relative to the other. Specifically, if we take $x(t)$ as the reference, we may write

$$[y(t)] = [\alpha(t)] + [\beta(t)] + [\gamma(t)] \qquad (206)$$

where

$$W_{x\alpha} = U_{xy}, \quad W_{x\beta} = iV_{xy}, \quad W_{xy} = 0. \qquad (207)$$

The resolution is easily affected from the limiting definition (196). We first note that if $W_x = 0$, then $W_\alpha = W_\beta = W_\gamma = 0$. When W_x is not zero then

$$\left.\begin{array}{ll} W_\alpha = U_{xy}^2/W_x, & W_{\alpha\beta} = iU_{xy}V_{xy}/W_x \\ W_\beta = V_{xy}^2/W_x, & W_{\alpha\gamma} = 0 \\ W_\gamma = W_y - |W_{xy}|^2/W_x, & W_{\beta\gamma} = 0. \end{array}\right\} \qquad (208)$$

In accordance with the above resolution, we may define some particular types of paired noise sources with interesting properties. *Incoherent* noise ensembles $[x(t)]$ and $[y(t)]$ have the property that $W_{xy} = 0$ for all frequencies. The converse is not necessarily true, for in general there might be higher order cross spectra and corresponding higher order crosscorrelations which do not vanish. For the important case of jointly Gaussian noise pairs however, the vanishing of W_{xy} is sufficient to insure that all orders of cross spectra vanish and hence that $W_{xy} = 0$ is a necessary and sufficient condition for incoherence. For other sources, we might define $W_{xy} = 0$ as equivalent to "second order incoherence." The word "incoherence" as used here is equivalent to "independence" as used in general statistical terminology, and is introduced here because of the similarity to the generally accepted usage of the term in optics.

If W_{xy} does not vanish at some ω, the sources are *partially coherent*. If furthermore $|W_{xy}|^2 = W_x W_y$, the conherence is total in the second order sense for non-Gaussian sources and totally coherent without further qualification in the case of jointly Gaussian pairs. If in addition to this condition, we also have W_{xy} equal to a purely real quantity U_{xy} at all frequencies, the noise sources will be said to be colinear. Here two conditions can be distinguished. When $U_{xy}(\omega)$ is positive the phases are additive, and if $U_{xy}(\omega)$ is negative the phases are subtractive. A null spectrum can be obtained from the sum of two colinear noise sources at any frequency by relative amplitude adjustment and choice of sign. Eq. (205) shows that the crosscorrelation of colinear noise ensembles is an even function of τ.

If in the case of total coherence, the value of $W_{xy}(\omega)$ has a pure imaginary value iV_{xy} at all frequencies, the sources will be said to be in *quadrature*, indicating that a ninety-degree phase shift exists between components at the same frequencies. A positive value of V_{xy} indicates that the x-component lags the corresponding y-component by ninety degrees, while a negative sign indicates a corresponding leading phase angle. From (205) we see that the cross-correlation function of quadrature noise sources is an odd function of τ and vanishes for $\tau = 0$. Simultaneous samples of Gaussian quadrature noise pairs are therefore independent.

A particular virtue of the cross spectrum in analysis is the aid it furnishes in calculating the effect of linear operations on the noise. From the definition, we see that passing the wave $x(t)$ through the transfer admittance $Y_1(\omega)$ and the wave $y(t)$ through the transfer admittance $Y_2(\omega)$ gives two resulting waves, $\xi(t)$, $\eta(t)$, with cross spectra

$$\left.\begin{array}{l} W_{\xi\eta} = Y_1(-\omega)Y_2(\omega)W_{xy}(\omega) \\ W_{\eta\xi} = Y_1(\omega)Y_2(-\omega)W_{yx}(\omega) \end{array}\right\}. \qquad (209)$$

The cross correlations of ξ and η are then calculable by Fourier transforming the cross spectra.

If $\xi(t)$ represents the input and $\eta(t)$ the output of a network with transfer admittance $Y(\omega)$, when a single noise source $x(t)$ is applied, $Y_1(\omega) = 1$, $Y_2(\omega) = Y(\omega)$, $W_{xy}(\omega) = W_x(\omega)$ and

$$W_{\eta\xi} = Y(\omega)W_x(\omega). \qquad (210)$$

If $[x(t)]$ is a white noise source, $W_x(\omega)$ is a constant, and it follows that the cross spectrum is in fact proportional to the complex transfer admittance of the network.

Cross spectra of derivatives and integrals of noise waves are readily obtained from the original cross spectra since they are linear operations describable in terms of admittance functions. Thus if we represent differentiations with respect to time by a dot above the symbol,

$$W_{\dot{x}y}(\omega) = -i\omega W_{xy}(\omega) \qquad (211)$$

$$W_{x\dot{y}}(\omega) = i\omega W_{xy}(\omega). \qquad (212)$$

Then from (200),

$$R_{\dot{x}y}(\tau) = \frac{1}{2\pi} \int_{-\infty}^{\infty} W_{\dot{x}y}(\omega) e^{i\tau\omega} d\omega$$

$$= -\frac{1}{2\pi} \int_{-\infty}^{\infty} i\omega W_{xy}(\omega) e^{i\tau\omega} d\omega = -R_{xy}'(\tau) \quad (213)$$

where the prime represents differentiation with respect to τ. Similarly,

$$R_{x\dot{y}}(\tau) = R_{xy}'(\tau) = -R_{\dot{x}y}(\tau). \quad (214)$$

Cross correlations of higher derivatives can be evaluated from these; *e.g.*,

$$R_{xy}''(\tau) = -\frac{d}{d\tau} \left[R_{\dot{x}y}(\tau) \right] = R_{xy}''(\tau), \text{ etc.} \quad (215)$$

We also note that

$$W_{x\dot{x}}(\omega) = i\omega W_x(\omega),$$

and

$$\left| W_{x\dot{x}}^2 \right| = W_x W_{\dot{x}}. \quad (216)$$

Since $W_x(\omega)$ is real, we see that a noise wave and its derivative form a pair of quadrature noise sources. Also,

$$R_{x\dot{x}}(\tau) = R_x'(\tau). \quad (217)$$

Likewise a noise source and its second derivative are colinear since

$$W_{xx}'' = \omega^2 W_x(\omega) \quad (218)$$

and

$$W_{xx}''^2 = W_x W_x''. \quad (219)$$

Relations between a noise source and its integral may be evaluated in a similar way dividing by $i\omega$ instead of multiplying. If the integral is taken over a fine time, say from $-T$ to 0, the transfer factor becomes

$$Y(\omega) = \frac{1 - e^{-iT\omega}}{i\omega} . \quad (220)$$

When the integration is taken over all time we may justify discarding the term in $e^{-iT\omega}$, particularly if the integral is subject to further linear operations.

It is interesting and useful to extend the model based on randomly phased sine waves, (151), to include cross-correlated Gaussian ensembles. Given the jointly Gaussian ensemble $[x(t), y(t)]$ with self-power spectra $W_x(\omega)$, $W_y(\omega)$ and cross-power spectrum $W_{xy}(\omega) = U_{xy}(\omega) + iV_{xy}(\omega)$ we can approximate $x(t)$ say by

$$x_N(t) = 2 \sum_{n=1}^{N} \sqrt{W_x(\omega_n)\Delta\omega} \cos(\omega_n t + \theta_n) \quad (221)$$

as in (151), when N is large, $\Delta\omega$ is small but not zero, $\omega_n = \omega_1 + n\Delta\omega$, and θ_n is random. Then $y(t)$ may be approximated by

$$y_N(t) = 2 \sum_{n=1}^{N} \left[U_{xy}(\omega_n) \sqrt{W_x^{-1}(\omega_n)\Delta\omega} \cos(\omega_n t + \theta_n) \right.$$

$$- V_{xy}(\omega_n) \sqrt{W_x^{-1}(\omega_n)\Delta\omega} \sin(\omega_n t + \theta_n)$$

$$+ \sqrt{[W_y(\omega_n) - |W_{xy}(\omega_n)|^2/W_x(\omega_n)]\Delta\omega}$$

$$\left. \cdot \cos(\omega_n t + \psi_n) \right]. \quad (222)$$

where the angles ψ_n are random with respect to each other and with respect to θ_n.

Envelope, Phase, and Instantaneous Frequency—Evaluation of Noise in AM, PM, and FM Systems

The case in which the width of the frequency interval occupied by the noise is small compared with the mid-frequency of the interval is the next important communication problem we shall consider. The systems commonly designated as radio frequency (rf) and intermediate frequency (if) generally have this narrow band property. The waves passed by such systems consist of oscillations in the high frequency range. The amplitude and frequency of these oscillations can change with time only at rates comparable with the bandwidth.

If $v(t)$ is a representative wave of a narrow band noise ensemble with spectrum centered at frequency ω_c radians per second, it is convenient to seek a resolution of form

$$v(t) = x(t) \cos \omega_c t - y(t) \sin \omega_c t \quad (223)$$

where $x(t)$ and $y(t)$ are slowly varying functions of time relative to oscillations of frequency ω_c.

Such resolution could be performed physically by multiplying the wave by $2 \cos \omega_c t$ to determine $x(t)$ and by $-2 \sin \omega_c t$ to determine $y(t)$. The results of such multiplications are

$$\left.\begin{array}{l} 2v(t) \cos \omega_c t = x(t) + x(t) \cos 2\omega_c t - y(t) \sin 2\omega_c t \\ -2v(t) \sin \omega_c t = y(t) - y(t) \cos 2\omega_c t - x(t) \sin 2\omega_c t \end{array}\right\} \quad (224)$$

The only low frequency terms in the two cases are contained in $x(t)$ and $y(t)$ respectively and they are separable from the high prequency terms by a low pass filter. This is the principle of homodyne or local carrier detection of rf waves. It is to be noted that (224) can be used to give added precision to the requirement on relative magnitudes of the frequencies involved. It follows by expressing products of sine and cosine terms by sines and cosines of sums and differences that if $x(t)$ and $y(t)$ have zero spectral density at frequencies equal to or greater than ω_c then the highest frequency in $x(t)$ and $y(t)$ will be less than the lowest frequency in the other terms and separation is possible. The resolution is thus physically significant when the original noise bandwidth of $v(t)$ is less than $2\omega_c$.

In the above representation we note that the frequency ω_c can be chosen arbitrarily to a considerable extent since to make the resolution physically significant we require only that the spectral density be negligibly small at frequencies exceeding $2\omega_c$. In the important case in which a signal is present in a narrow band of noise, it becomes convenient to use the signal as a reference to make the resolution unique. Specifically if the signal is a sine wave, we designate the signal as

$$s(t) = A \cos \omega_c t \qquad (225)$$

and study the ensemble

$$u(t) = v(t) + s(t)$$
$$= [x(t) + A] \cos \omega_c t - y(t) \sin \omega_c t. \qquad (226)$$

A further important significance now appears when we rewrite the equation in the equivalent form

$$u(t) = \rho(t) \cos [\omega_c t + \phi(t)] \qquad (227)$$

where

$$\rho^2(t) = [x(t) + A]^2 + y^2(t) \qquad (228)$$

$$\tan \phi(t) = \frac{y(t)}{x(t) + A}. \qquad (229)$$

The function $\rho(t)$ with positive sign is commonly called the envelope of the rf wave and gives the low frequency response of an amplitude detector. The function $\phi(t)$ determined as to quadrant by (229) but ambiguous to within any multiple of 2π is called the phase of the wave and gives the low frequency response to a phase detector. Still another important function is the time rate of change of phase

$$\dot{\phi}(t) = \frac{d}{dt} \phi(t) = \frac{[x(t) + A]\dot{y}(t) - \dot{y}(t)\dot{x}(t)}{[x(t) + A]^2 + y^2(t)} \qquad (230)$$

which is the *instantaneous frequency displacement* from ω_c. $\dot{\phi}(t)$ is by accepted convention the ac component of the response of an ideal frequency detector to the wave. It is uniquely determined by (230).

A model leading to an appropriate mathematical description of a narrow band Gaussian ensemble is obtained as in (151) by a finite approximation to $v(t)$ of form

$$v_N(t) = 2 \sum_{n=1}^{N} \sqrt{W_v(\omega_1 + n\Delta\omega)\Delta\omega} \cos [(\omega_1 + n\Delta\omega)t + \theta_n] \quad (231)$$

where the narrow band is assumed to extend from ω_1 to $\omega_1 + \Omega$ and $\Omega = N\Delta\omega$. The θ_n's are uniformly and independently distributed. By adopting this model our discussion will be confined in this section to Gaussian narrow noise sources. Let

$$\omega_c = \omega_1 + \Omega/2 \qquad (232)$$

and replace $\omega_1 + n\Delta\omega$ in the argument of the cosine by $(\omega_1 + n\Delta\omega + \omega_c) - \omega_c$. We then find that we can write

$$v_N(t) = x_N(t) \cos \omega_c t - y_N(t) \sin \omega_c t \qquad (233)$$

where, for large N

$$x_N(t) = 2 \sum_{n=0}^{N} \sqrt{W_v(\omega_1 + n\Delta\omega)\Delta\omega}$$
$$\cdot \cos [(\omega_1 - \omega_c + n\Delta\omega)t + \theta_n]$$
$$y_N(t) = 2 \sum_{n=0}^{N} \sqrt{W_v(\omega_1 + n\Delta\omega)\Delta\omega}$$
$$\cdot \sin [(\omega_1 - \omega_c + n\Delta\omega)t + \theta_n] \qquad (234)$$

The substitution $n = m + N/2$ results in the equations

$$x_N(t) = 2 \sum_{m=-N/2}^{N/2} \sqrt{W_v(\omega_1 + n\Delta\omega)\Delta\omega}$$
$$\cdot \cos (m\Delta\omega t + \theta_{m+N/2})$$
$$y_N(t) = 2 \sum_{m=-N/2}^{N/2} \sqrt{W_v(\omega_1 + n\Delta\omega)\Delta\omega}$$
$$\cdot \sin (m\Delta\omega t + \theta_{m+N/2}) \qquad (235)$$

Now by letting $\Delta\omega$ become small and N large with the θ's independent, we deduce from the central limit theorem that the ensemble $[x_N(t), y_N(t)]$ approaches the jointly Gaussian ensemble $[x(t), y(t)]$. Furthermore the power spectra of $x(t)$ and $y(t)$ are equal and are given by

$$W_x(\omega) = W_y(\omega) = W_v(\omega_c + \omega) + W_v(\omega_c - \omega). \quad (236)$$

The two terms of (236) come from combining the power contributions of the corresponding positive and negative values of n in (235). Physically they are the upper and lower sideband contributions to the low frequency spectrum. We remark that the computational procedure can be carried out without regard to any limitation on bandwidth of the original noise but that the physical significance of the resolution is lost when the low frequency spectra cannot be separately observed.

It is seen by comparison with (222) that in the limit $x(t)$ and $y(t)$ constitute noise sources in quadrature with

$$W_{xy}(\omega) = iV_{xy}(\omega), \qquad (237)$$

$$V_{xy}(\omega) = W_v(\omega_c - \omega) - W_v(\omega_c + \omega). \qquad (238)$$

The cross-correlation $R_{xy}(\tau)$ is accordingly found from (205), and is

$$R_{xy}(\tau) = -\frac{1}{\pi} \int_0^\infty [W_v(\omega_c - \omega) - W_v(\omega_c + \omega)] \sin \tau\omega d\omega. \quad (239)$$

If the spectral density of $v(t)$ is symmetrical about ω_c, $W_v(\omega_c - \omega) = W_v(\omega_c + \omega)$ and $V_{xy}(\omega) = 0$. For this case then, the cross correlation vanishes, and since we are dealing with jointly Gaussian sources, $x(t)$ and $y(t+\tau)$ constitute independent ensembles. The auto-correlations of x and y are equal and are given by

$$R_x(\tau) = R_y(\tau)$$
$$= \frac{1}{\pi} \int_0^\infty [W_v(\omega_c - \omega) + W_v(\omega_c + \omega)] \cos \tau\omega d\omega. \quad (240)$$

We note that even in the unsymmetrical case the cross correlation vanishes when $\tau = 0$. Since the ensemble is jointly Gaussian, this means that $x(t)$ and $y(t)$ are independent. The probability density function of x and y evaluated at the same instant of time is therefore the product of single-variable Gaussian functions in x and y, that is,

$$p(x, y) = \frac{1}{2\pi\sigma^2} e^{-(x^2+y^2)/2\sigma^2} \tag{241}$$

where

$$\sigma^2 = R_x(0) = R_y(0) = R_v(0) \tag{242}$$

is the variance of $x(t)$, $y(t)$, and $v(t)$. We note that in the absence of signal (228) and (229) may be also written

$$\left.\begin{array}{l} x(t) = \rho(t) \cos \phi(t) \\ y(t) = \rho(t) \sin \phi(t) \end{array}\right\} \tag{243}$$

and hence transforming from noise components in quadrature to envelope and phase noise is like transforming from rectangular to polar coordinates. Hence if $q(\rho, \phi)$ is the probability density function with respect to envelope and phase of the noise wave,

$$p(x, y)dxdy = p(\rho \cos \phi), (\rho \sin \phi)\rho d\rho d\phi$$
$$= q(\rho, \phi)d\rho d\phi \tag{244}$$

and

$$q(\rho, \phi) = \frac{\rho}{2\pi\sigma^2} e^{-\rho^2/2\sigma^2}. \tag{245}$$

The probability density function of the noise envelope alone is obtained by averaging over all phases and is

$$q_1(\rho) = \int_0^{2\pi} q(\rho, \phi)d\phi = \frac{\rho}{\sigma^2} e^{-\rho^2/2\sigma^2}. \tag{246}$$

This is known as the Rayleigh distribution. Much confusion has been caused in discussions when one participant had in mind this distribution which is appropriate to the envelope of narrow band Gaussian noise while the other was thinking of the previously introduced distribution of wide band Gaussian noise. The probability density function of ϕ is

$$q_\phi(\phi) = \int_0^\infty q(\rho, \phi)d\rho = \frac{1}{2\pi}. \tag{247}$$

The values of $\phi(t)$ are thus found to have a rectangular distribution.

We may obtain the probability density function for the combination of the sine wave signal and noise by substituting

$$\xi(t) = x(t) + A \tag{248}$$

in (241). The resulting probability density function in ξ and y is

$$p(x, y) = r(\xi, y) = \frac{1}{2\pi\sigma^2} e^{-[(\xi-A)^2+y^2]/2\sigma}. \tag{249}$$

Now transform to polar coordinates by

$$\left.\begin{array}{l} \xi = \rho \cos \phi \\ y = \rho \sin \phi \end{array}\right\} \tag{250}$$

where ρ now represents the envelope and ϕ the phase of the signal plus noise. Then the joint probability density function $q(\rho, \phi)$ is calculated to be

$$q(\rho, \phi) = \frac{\rho}{2\pi\sigma^2} e^{-(\rho^2+A^2-2A\rho \cos \phi)/2\sigma}. \tag{251}$$

The probability density function of the envelope itself is

$$q_\rho(\rho) = \int_0^{2\pi} q(\rho, \phi)d\phi = \frac{\rho}{\sigma^2} I_0\left(\frac{A\rho}{\sigma^2}\right) e^{-(\rho^2+A^2)/2\sigma^2} \tag{252}$$

while the probability density function of the phase is

$$q_\phi(\phi) = \int_0^\infty q(\rho, \phi)d\rho = \frac{e^{-A^2/2\sigma^2}}{2\pi} + \frac{A \cos \phi}{2\sigma\sqrt{2\pi}}$$
$$\cdot \left[1 + \mathrm{erf}\left(\frac{A \cos \phi}{\sqrt{2}\sigma}\right)\right] e^{-A^2 \sin \phi^2 1/2\sigma^2}. \tag{253}$$

Eqs. (252) and (253) reduce to (246) and (247) respectively when $A = 0$.

Application of (252)—Probability of Error in Carrier Telegraph Signal Received Through Gaussian Noise

Suppose a telegraph message is sent by on-off bursts of a sine wave and is received by an envelope detector. Also suppose that Gaussian noise is picked up at the receiver input. We wish to calculate the probability of error in the received signal. The result will depend on how the decisions are made at the detector output. We consider the simple case in which a threshold indicator operates at the midpoint of each signaling interval. If the envelope is greater than some fixed threshold value ρ_0, the signal is judged to be a mark; otherwise it is called a space. We note that we have two kinds of envelope ensembles to choose between—1) the mark ensemble which has the probability density function $q_\rho(\rho)$ of (252) for sine wave plus noise and 2) the space ensemble which has the probability density function $q_\rho(\rho)$ of (246) for envelope of noise alone. We designate the latter function here as $q_{\rho 0}(\rho)$.

There are two kinds of error which can occur:

Type I) send "SPACE" receive "MARK"
Type II) send "MARK" receive "SPACE."

Let p_s represent the probability of sending a space and p_M the probability of sending a mark. Then the probability p_I of an error of Type I is p_s multiplied by the probability that the envelope value from noise alone exceeds ρ_0. Therefore

$$P_I = p_s \int_{\rho_0}^{\infty} q_{\rho_0}(\rho) d\rho = p_s \int_{\rho_0}^{\infty} \frac{\rho e^{-\rho^2/2\sigma^2}}{\sigma^2} d\sigma \quad (254)$$

$$= p_s e^{-\rho_0^2/2\sigma^2}.$$

The probability p_{II} of a Type II error is p_M multiplied by the probability that noise superimposed on the sine wave depresses the envelope below ρ_0. Therefore

$$p_{II} = p_M \int_0^{\rho_0} q_\rho(\rho) d\rho$$

$$= p_M \int_0^{\rho_0} \frac{I_0(A\rho/\sigma^2)}{\sigma^2} \rho e^{-(\rho^2+A^2)/2\sigma^2} d\rho$$

$$= p_M e^{-(A^2+\rho_0^2)/2\sigma^2} \sum_{n=1}^{\infty} \left(\frac{\rho_0}{A}\right)^n I_n(A^2/\sigma^2). \quad (255)$$

The evaluation of the integral in terms of a sum is accomplished by successive integration by parts making use of the indefinite integral

$$\int z^n I_{n-1}(az) dz = z^n I_n(az)/a. \quad (256)$$

The two types of error are not equally probable in general, but can be made so for a specific signal-to-noise ratio by a proper choice of the threshold level ρ_0. Determination of the proper value requires the solution of a transcendental equation. If $p_M = p_s$ and $\rho_0 = A/2$ the probability of Type II errors is smaller than Type I. The two types of errors can be made equally probable here by setting the threshold level at more than half the peak carrier amplitude. We reproduce the results of such a computation in Fig. 7 opposite. The corresponding curve for baseband pulse transmission is shown for comparison. In the latter case the two kinds of error are equally probable for threshold at half the pulse peak. The total probability of error is given by (36) with $x_0 = 0$ and $x =$ half the pulse peak. The power scale of the abscissa applies to a rectangular baseband pulse. The curve would be shifted 3 db to the left for a sinusoidal baseband pulse.

A more potent probability density function from which the spectra of envelope and phase can be obtained is the four-dimensional one in $x(t)$, $y(t)$, $x(t+\tau)$, $y(t+\tau)$. Here the fundamental simplifying property is that in a jointly Gaussian process all multivariate distributions obtained by linear operations on the functions are also jointly Gaussian. Hence if we write

$$\begin{aligned} x_1 &= x(t) & x_2 &= x(t+\tau) \\ x_3 &= y(t) = y_1 & x_4 &= y(t+\tau) = y_2 \end{aligned} \quad (257)$$

we refer back to (112)–(114) and note that the probability density function can be entirely constructed from

$$C_{x_1 x_2 x_3 x_4} = \exp\left[-\frac{1}{2} \sum_{j=1}^{4} \sum_{k=1}^{4} \lambda_{jk} \xi_j \xi_k\right] \quad (258)$$

where λ_{jk}, is the typical crosscorrelation or average of $x_j x_k$. In our case

$$\left.\begin{aligned} \lambda_{11} &= \lambda_{22} = \lambda_{33} = \lambda_{44} = R_x(0) = R_y(0) = R_v(0) = \sigma^2 \\ \lambda_{12} &= \lambda_{34} = R_x(\tau) = R_y(\tau) = R_I \\ \lambda_{13} &= \lambda_{24} = R_{xy}(0) = 0 \\ \lambda_{14} &= -\lambda_{23} = R_{xy}(\tau) = R_Q \end{aligned}\right\} \quad (259)$$

$$\lambda = \begin{vmatrix} \sigma^2 & R_I & 0 & R_Q \\ R_I & \sigma^2 & -R_Q & 0 \\ 0 & -R_Q & \sigma^2 & R_I \\ R_Q & 0 & R_I & \sigma^2 \end{vmatrix} = (\sigma^4 - R_I^2 - R_Q^2).^2 \quad (260)$$

Let

$$A_0 = \sigma^4 - R_I^2 - R_Q^2. \quad (261)$$

Then

$$A_0^2 = \lambda \quad (262)$$

$$\Lambda_{11} = \Lambda_{22} = \Lambda_{33} = \Lambda_{44} = -\sigma^2 A_0$$

$$\Lambda_{12} = \Lambda_{34} = -R_I A_0$$

$$\Lambda_{14} = -\Lambda_{23} = -R_Q A_0$$

$$\Lambda_{13} = \Lambda_{24} = 0 \quad (263)$$

$$p(x_1, x_2, y_1, y_2)$$

$$= \frac{1}{4\pi^2 A_0} \exp\left(-\frac{1}{2A_0}\left[\sigma^2(x_1^2 + x_2^2 + y_1^2 + y_2^2)\right.\right.$$

$$\left.\left. - 2R_I(x_1 x_2 + y_1 y_2) - 2R_Q(x_1 y_2 - x_2 y_1)\right]\right). \quad (264)$$

This probability density function is sufficient to calculate the autocorrelation and hence the power spectrum of the envelope and phase of the noise ensemble. It is also adequate to obtain the same quantities when a sine wave signal is added to the noise. The latter includes the former as a special case. To calculate the autocorrelation of the envelope of signal and noise, we compute the ensemble average of $\rho(t)\rho(t+\tau)$, with $\rho(t)$ defined by (239) and (240). The above formulas are sufficient for this. We have

$$R_\rho(\tau) = \text{av } [(x_1+A)^2 + y_1^2]^{1/2}[(x_2+A)^2 + y_2^2]^{1/2}$$

$$= \int_{-\infty}^{\infty}\int_{-\infty}^{\infty}\int_{-\infty}^{\infty}\int_{-\infty}^{\infty} [(x_1+A)^2 + y_1^2]^{1/2}$$

$$\cdot [(x_2+A)^2 + y_2^2]^{1/2} p(x_1, x_2, y_1, y_2) dx_1 dx_2 dy_1 dy_2. \quad (265)$$

A double transformation to polar coordinates is indicated to evaluate this integral.

The autocorrelation of the phase is likewise

$$R_\phi(\tau) = \text{av}\left(\arctan\frac{y_1}{x_1+A}\right)\left(\arctan\frac{y_2}{x_2+A}\right)$$

$$= \int_{-\infty}^{\infty}\int_{-\infty}^{\infty}\int_{-\infty}^{\infty}\int_{-\infty}^{\infty} \arctan\frac{y_1}{x_1+A}$$

$$\cdot \arctan\frac{y_2}{x_2+A} p(x_1, x_2, y_1, y_2) dx_1 dx_2 dy_1 dy_2. \quad (266)$$

Here again we would introduce polar coordinates.

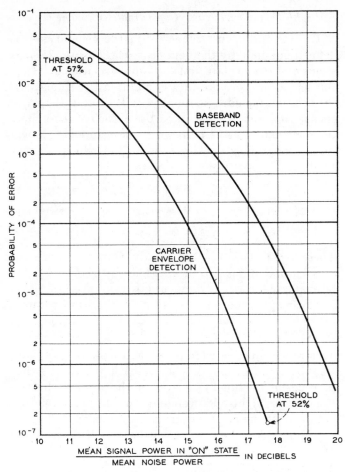

Fig. 7(a)—Probability of error from Gaussian noise in detection of random on-off pulses.

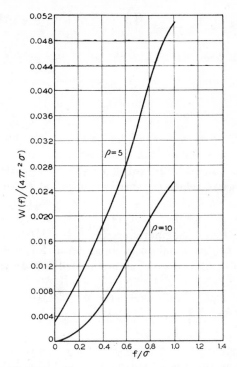

Fig. 7(b)—Power spectrum of instantaneous frequency of sine wave plus noise.

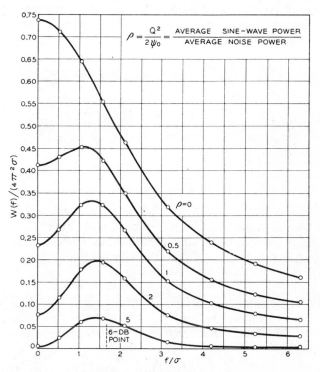

Fig. 7(c)—Power spectrum of instantaneous frequency of sine wave plus noise.

The formula for $R_\phi(\tau)$ should be accompanied by a reminder that by our definition the phase has been confined to a single interval of width 2π. This is all right for a static case in which angles are ambiguous by any additive multiple of 2π, but it may not be good for a phase which varies continuously with time. In the latter case when we reach the edge of our basic 2π interval and the slope does not change sign, continuity requires us to cross the boundary into the next interval rather than shift abruptly by 2π to stay within one interval. For this reason, we would use (266) only for the case in which the phase excursion can be wholly confined to a single range of width 2π. In such cases which might be approximated by low index phase or frequency modulation with no dc signal component, and with signal large compared to noise, we could use (266) not only for the phase, but for the instantaneous frequency as well, since

$$W_{\dot\phi}(\omega) = |i\omega|^2 W_\phi(\omega) = \omega^2 W_\phi(\omega) \qquad (267)$$

Signal-to-Noise Ratio in FM with Strong Carrier

Before proceeding to a more comprehensive look at the fm case, let us test the theory developed so far by working through the familiar case in which the amplitude of the sine wave is large compared to the rms value of the noise. The principal term in the numerator of

(230) for the instantaneous frequency is then $A\dot{y}(t)$ and in the denominator, A^2. Hence we have approximation

$$\dot\phi(t) = \frac{\dot y(t)}{A}. \qquad (268)$$

The spectral density of the instantaneous frequency is then

$$W_{\dot{\phi}}(\omega) = \frac{|i\omega|^2}{A^2} W_y(\omega) = \frac{\omega^2[W_v(\omega_c+\omega)+W_v(\omega_c-\omega)]}{2W_c} \quad (269)$$

where

$$W_c = \frac{A^2}{2}. \quad (270)$$

W_c is the mean square of the carrier wave. This is the familiar formula for the mean square frequency error spectrum caused by perturbing a sine wave with narrow band noise of relatively small mean power. If the power spectrum of the noise is flat throughout the range of interest, we obtain the type of error spectrum often referred to as triangular since the rms value in a narrow band is proportional to ω. In general, for the case of small noise, the rms of $\dot{\phi}$ in a narrow band $\Delta\omega$ at ω is

$$\dot{\phi}_{\text{rms}} = \omega \sqrt{\frac{W_v(\omega_c+\omega)+W_v(\omega_c-\omega)}{2W_c}} \Delta\omega. \quad (271)$$

The signal-to-noise ratio S^2/N^2 in the output of an fm receiver is commonly expressed as the ratio of the mean square frequency deviation caused by modulating the carrier frequency with a full load sine wave test signal to the mean square frequency deviation produced by noise when the carrier frequency is constant. The denominator is therefore given by

$$N^2 = \left[\int_{-\omega_2}^{-\omega_1} + \int_{\omega_1}^{\omega_2}\right] W_{\dot{\phi}}(\omega)d\omega = 2\int_{\omega_1}^{\omega_2} W_{\dot{\phi}}(\omega)d\omega \quad (272)$$

where f_1 and f_2 are the lower and upper cutoff frequencies of the filter following the frequency detector. The numerator is

$$S^2 = \frac{\omega_s^2}{2} \quad (273)$$

where ω_s is the peak frequency deviation caused by the test fm signal. If the output filter is essentially low pass with cutoff at ω_a, and the noise band is flat,

$$W_v(\omega_c+\omega) = W_v(\omega_c-\omega) = W_v \quad (274)$$

$$N^2 = \int_0^{\omega_a} \frac{4\omega^2 W_v}{2W_c} d\omega = \frac{2W_v\omega_a^3}{3W_c} \quad (275)$$

$$\frac{S^2}{N^2} = \frac{3W_c}{4W_v\omega_a}\left(\frac{\omega_s}{\omega_a}\right)^2 = \frac{3}{2}\kappa^2\frac{W_c}{W_a} \quad (276)$$

where $\kappa = \omega_s/\omega_a$ is the ratio of test signal peak frequency deviation to width of accepted signal output band, and $W_a = 2W_v\omega_a$ is the equivalent noise power accepted by the receiver input in a band of width equal to the accepted output band. This formula is well-known in the fm art and its derivation is given here only as a warm-up exercise for a really tough problem—fm with no restriction on the noise power relative to signal power.

FM with Signal and Noise Comparable in Strength

The instantaneous frequency in the general case is defined uniquely by the nonlinear relationship (257). It seems best to calculate first the autocorrelation and then the spectrum. We note that in addition to $x(t)$ and $y(t)$ the relation includes $\dot{x}(t)$ and $\dot{y}(t)$. To obtain the autocorrelation, we must also make use of each of these functions evaluated at $t+\tau$, so we require an eight-dimensional probability density function, in which we now shall set

$$x_1 = x(t), \qquad x_2 = x(t+\tau) \qquad x_3 = \dot{x}_1 = \dot{x}(t),$$
$$x_4 = \dot{x}_2 = \dot{x}(t+\tau) \quad x_5 = y_1 = y(t), \qquad x_6 = y_2 = y(t+\tau),$$
$$x_7 = \dot{y}_1 = \dot{y}(t), \qquad x_8 = \dot{y}_2 = \dot{y}(t+\tau). \quad (277)$$

These eight variables are jointly Gaussian since the various operations are linear. The covariances λ_{jk} can be evaluated from our previous observation on the effect of linear operation on cross spectra. Thus for instance,

$$\lambda_{18} = \text{av}\left[x(t)\dot{y}(t+\tau)\right] = R_{x\dot{x}}(\tau) = R_x'(\tau). \quad (278)$$

This representative example should make it clear how all the λ's can be found. The evaluation of the Λ's then proceeds in a manner exactly as in the previous illustration and the eight-fold probability density function thus obtained. If the noise spectrum is symmetrical about ω_c a considerable number of zero values in the determinant appear. The autocorrelation of the instantaneous frequency is then

$$R_{\dot{\phi}}(\tau) = \text{av}\left[\dot{\phi}(t)\dot{\phi}(t+\tau)\right]$$
$$= \int_{-\infty}^{\infty}\cdots\int_{-\infty}^{\infty} \frac{[(x_1+A)\dot{y}_1-y_1\dot{x}_1][(x_2+A)\dot{y}_2-y_2\dot{x}_2]}{[(x_1+A)^2+y_1][(x_2+A)^2+y_2]}$$
$$\cdot p(x_1, x_2, \dot{x}_1, \dot{x}_2, y_1, y_2, \dot{y}_1, \dot{y}_2)$$
$$\cdot dx_1d\dot{x}_2d\dot{x}_1d\dot{x}_2dy_1dy_2d\dot{y}_1d\dot{y}_2. \quad (279)$$

A prescription for the calculation is thereby achieved. The spectrum of the instantaneous frequency is

$$W_{\dot{\phi}}(\omega) = \frac{1}{2\pi}\int_{-\infty}^{\infty} R_{\dot{\phi}}(\tau)e^{-i\omega\tau}d\tau. \quad (280)$$

The spectrum of the phase is obtained by noting that the admittance function needed to produce it from the frequency is $1/i\omega$ corresponding to integration in the time domain and hence

$$W_{\phi}(\omega) = W_{\dot{\phi}}(\omega)/\omega^2. \quad (281)$$

Fortunately for the engineer it should not often be necessary to work through calculations of this type. Altruistic investigators have done the work for representative problems and their results are available in tabular and graphical form. We should at this point then merely cite the appropriate references and give helpful explanatory hints on how to use them effectively. It is possible to understand how the results were obtained and to use them intelligently without checking through the details of the evaluations.

S. O. Rice [13] has computed the power spectrum of ϕ for various ratios of sine wave power to noise power for the case in which the sine wave frequency is at the center of a narrow Gaussian band of noise. This is the noise spectrum illustrated in Fig. 5(e). The dc component is not shown and is not usually of interest. Rice uses the symbol ρ for the ratio of mean sine wave power to mean total noise power. In our notation, $\rho = A^2/2 N_0$, since the mean square of $A \cos \omega_c t$ is $A^2/2$ and N_0 in (159) is the total power represented by the noise spectrum. The frequency abscissa is f/σ where σ is the frequency displacement from the center of the band at which the spectral density is down one-quarter neper or **2.17** db. In our notation, $\sigma = \omega_0/2\pi = f_0$. The ordinate is labeled $W(f)/4\pi^2\sigma$ and includes the contributions at f and $-f$. In our notation, this is $W_\phi(f)/2\pi^2 f_0 = 2W_{\dot\phi}(\omega)/\omega_0$. We reproduce Rice's published curves here plus a supplementary curve computed by him since the publication of his paper.

We shall illustrate the use of these curves starting with typical data available to the engineer. In the case of an fm receiver, the ratio ρ can be calculated from the signal power at the input and the noise figure of the rf- and if-part of the receiver. The noise figure is defined as the ratio [16] of the signal-to-noise ratio at the input to the available signal-to-noise ratio at the output. If the receiver itself did not contribute any noise the noise figure would be unity. In the case in which the noise accompanying the input signal is thermal noise only, we have in effect a noise source with flat spectrum of density 204 db below one watt in a band of width one cps at room temperature. A noise figure of F in effect increases the spectral density of the equivalent source to 204–10 log F below one watt per cps. If the mean carrier input power is $W_c = A^2/2$, the appropriate value of ρ to calculate the spectral density of the mean square frequency of signal and noise when the carrier wave is at constant frequency is W_c/N_0, where N_0 is the mean power of that part of the input which the selective circuits of rf- and if-branches accept. If we assume the rf- and if-branches are adequately described by a Gaussian band pass curve, the input spectrum of interest is that of Fig. 5(e). We add the contributions at corresponding positive and negative frequencies and write

$$2\left(\frac{N_0}{2f_0\sqrt{2\pi}}\right) = 10^{-20.4}F \text{ watts/cps} \qquad (282)$$

which corresponds to a mean total of N_0 watts of noise power with a band pass spectrum down 2.17 db at $f_0 = \omega_0/2\pi$ cps away from midband and with maximum density at midband equal to thermal noise density multiplied by the noise figure. Then

$$N_0 = f_0 F 10^{-20.4}\sqrt{2\pi}. \qquad (283)$$

If we wish to base our calculations on any other loss point—say f_k cps from midband (or ω_k radians per

second) with transmission k db down, we may use the Gaussian relations,

$$2.17 f_k^2/f_0^2 = k \text{ db} \qquad (284)$$

from which

$$f_0 = f_k\sqrt{\frac{2.17}{k}} = \frac{1.473 f_k}{\sqrt{k}} \qquad (285)$$

$$N_0 = 10^{-20.4}F f_k\sqrt{\frac{4.34\pi}{k}}. \qquad (286)$$

The ρ curve we select then is given by

$$\rho = \frac{W_c}{N_0} = \frac{10^{20.4}W_c}{F f_k}\sqrt{\frac{k}{4.34\pi}}. \qquad (287)$$

If we represent the abscissa of the typical curve as x_ρ and the ordinate by y_ρ, then

$$x_\rho = \frac{f}{f_0} = \frac{\omega}{\omega_0}, \qquad y_\rho = 2W_{\dot\phi}(\omega)/\omega_0. \qquad (288)$$

The mean square frequency error in the detector output at frequency $\omega = \omega_0 x_\rho$ is $2W_{\dot\phi}(\omega) = \omega_0 y_\rho$. The total noise output from the receiver is expressed as in (272) by

$$N^2 = 2\int_{\omega_1}^{\omega_2} W_{\dot\phi}(\omega)d\omega = 2\int_{\omega_1/\omega_0}^{\omega_2/\omega_0} \frac{\omega_0 y_\rho}{2}\cdot\omega_0 dx_\rho$$

$$= \frac{2.17\omega_k^2}{k}\int_{\omega_1/\omega_0}^{\omega_2/\omega_0} y_\rho dx_{\rho_0} \qquad (289)$$

Finally from (273),

$$\frac{S^2}{N^2} = \frac{k}{4.34}\left(\frac{\omega_s}{\omega_k}\right)^2\left[\int_{\omega_1/\omega_0}^{\omega_2/\omega_0} y_\rho dx_\rho\right]^{-1}. \qquad (290)$$

Corresponding results for a flat band of input noise have been given by F. L. H. M. Stumpers. Stumpers uses an entirely different analytical approach from that given here. The more general problem in which the signal wave is frequency modulated has been treated by D. Middleton [15] by an extension of the methods here described. A discussion of various aspects of the problem is also given in Chapter 13 of vol. 24 of the Radiation Lab. Series [17].

When the signal wave is frequency modulated by a band of noise, the autocorrelation of the signal becomes the two-dimensional characteristic function of the noise. This helpful computing aid has been applied by Bennett, Curtis, and Rice to the problem of evaluating interchannel interference in FM transmission of carrier groups [18].

ACKNOWLEDGMENT

Preparation of this paper was aided by a preliminary draft by B. McMillan and D. Slapian dealing with the more strictly mathematical features of noise theory. The writer also wishes to thank J. R. Pierce for advice and encouragement.

BIBLIOGRAPHY

[1] Cramer, H., *Mathematical Methods of Statistics*. Princeton, New Jersey, Princeton University Press, 1951.
[2] Holbrook, B. D., and Dixon, J. T., "Load Rating Theory for Multi-Channel Amplifier," *Bell System Technical Journal*, Vol. 18 (October, 1939), pp. 624–644.
[3] Rice, S. O., "Mathematical Analysis of Random Noise," *Bell System Technical Journal*, Vol. 23 (July, 1944), pp. 282–332; Vol. 24 (January, 1945), pp. 46–156.
[4] Bennett, W. R., "Cross-Modulation in Multichannel Amplifiers," *Bell System Technical Journal*, Vol. 19 (October, 1940), pp. 587–610.
[5] Bennett, W. R., "Response of a Linear Rectifier to Signal and Noise," *Journal of the Acoustical Society of America*, Vol. 15 (January, 1944), pp. 164–172.
[6] Lewin, L., "Interference in Multi-Channel Circuits," *Wireless Engineer*, Vol. 27 (December, 1950), pp. 294–304.
[7] Middleton, D., "Some General Results in the Theory of Noise Through Non-Linear Devices," *Quarterly of Applied Mathematics*, Vol. 5 (January, 1948), pp. 445–498.
[8] Whittaker, J. M., *Interpolatory Function Theory*, No. 33, *Cambridge Tracts in Mathematics and Mathematical Physics*. Cambridge, England, Cambridge University Press, 1935.
[9] Courant, R., and Hilbert, D., *Methoden der Mathematischen Physik*. Berlin, Germany, Springer, Vol. 1, p. 65, 1924.
[10] Bennett, W. R., "Time Division Multiplex Systems," *Bell System Technical Journal*, Vol. 20 (April, 1941), pp. 119–221.
[11] Barnes, G. H., *Data Reduction Equipment for the Analysis of Human Tracking*. Franklin Institute Laboratories for Research and Development, Final Report No. F-2333, September 5, 1952–May 15, 1953.
[12] Krendel, E. S., and Barnes, G. H., *Interim Report on Human Frequency Response Studies*. WADC Technical Report 54-370, June, 1954.
[13] Rice, S. O., "Properties of a Sine Wave Plus Random Noise," *Bell System Technical Journal*, Vol. 27 (January, 1948), pp. 109–157.
[14] Stumpers, F. L. H. M., "Theory of Frequency Modulation Noise," PROCEEDINGS OF THE IRE, Vol. 36 (September, 1948), pp. 1081–1092.
[15] Middleton, D., "The Spectrum of Frequency-Modulated Waves After Reception in Random Noise—Part I," *Quarterly of Applied Mathematics*, Vol. 7 (July, 1949), pp. 129–173; Part II, Vol. 8 (April, 1950), pp. 59–80.
[16] Friis, H. T., "Noise Figures of Radio Receivers," PROCEEDINGS OF THE IRE, Vol. 32 (July, 1944), pp. 419–422.
[17] Lawson, J. L., and Uhlenbeck, G. E., *Threshold Signals*. New York, New York, McGraw-Hill Book Company, Inc., Chapter 13, 1950.
[18] Bennett, W. R., Curtis, H. E., and Rice, S. O., "Interchannel Interference in FM and PM Systems Under Noise Loading Conditions," *Bell System Technical Journal*, Vol. 34 (May, 1955), pp. 601–636.

Random Processes

Y. K. LIN

PROFESSOR OF AERONAUTICAL AND ASTRONAUTICAL ENGINEERING
UNIVERSITY OF ILLINOIS

Scope

The present paper is a brief survey of the modern theory of random processes addressed to workers in applied mechanics who desire to use random processes in their analyses. The writer realizes that it is not possible, nor is it essential, to compile an exhaustive bibliography. Therefore, references are made only to a selected few. For a rather comprehensive bibliography up to about 1959, the reader is directed to [Ref. 1]. Furthermore, since our emphasis is the theory itself, contributions to various applications are generally not cited, even though they may be quite significant.

Characterization of a Random Process

A random process is an indexed family of random variables. Therefore, to explain the concept of a random process, it is necessary to begin with the concept of random variables.

Briefly, a random variable is an uncertain number whose value is governed by a certain rule of chance. Let X be a random variable. A complete specification of X is given by its probability distribution function $F_X(x)$ which gives the probability of the event $X \leqslant x$, abbreviated as Prob $[X \leqslant x]$. When several random variables are involved, they may be characterized by their joint probability distribution function. For example, the joint probability distribution function $F_{X_1 X_2}(x_1, x_2)$ represents the probability of the event $X_1 \leqslant x_1$ and $X_2 \leqslant x_2$, abbreviated as Prob $[X_1 \leqslant x_1$ and $X_2 \leqslant x_2]$, etc. The derivatives of probability distributions when they exist, are called the probability densities.

Two types of random processes are possible. The first is a family of random variables with a discrete index. Denoted by X_n, the index n ranges over a set of discrete values, say the integers. By a choice of fixed index values, for example, by letting $n = 1, 2$, etc., one obtains random variables X_1, X_2, etc. The second type of random process is associated with a continuous index. De-

noted by $X(t)$, the index t varies over all values in an interval or a number of intervals. By fixing t values at $t = t_1$, t_2, etc., one is concerned with random variables $X(t_1)$, $X(t_2)$, etc. In applied mechanics, the first type of random processes is seldom used. Thus we shall limit our following discussion to continuously valued indices. A continuously indexed random process is also called a random function. Its index is often referred to as the parameter.

As a family of random variables, a random process is characterized by the law of chance governing the behavior of the members of the family. In ascending degrees of completeness, a random process is described by the corresponding probability distribution functions of ascending orders. These distribution functions and their meanings are listed below:

$$F_X(x_1, t_1) = \text{Prob } [X(t_1) \leqslant x_1]$$

$$F_X(x_1, t_1; x_2, t_2) = \text{Prob } [X(t_1) \leqslant x_1 \text{ and } X(t_2) \leqslant x_2]$$

$$\cdots \cdots \cdots \cdots \cdots \cdots \cdots$$

A mathematical difficulty arises, since the parameter t may assume an uncountably infinite number of values and $X(t)$ actually represents a family of uncountable random variables. The fundamental question of whether a random function possesses a unique set of probability distribution functions was answered by the celebrated extension theorem of Kolmogorov [Ref. 2], which was proved in the context of measure theory. It was primarily due to this fundamental contribution that Kolmogorov has been credited with having laid the foundation of the modern theory of random process.

Another scheme to overcome the difficulty of uncountable random variables is to extend the set of probability distribution functions to a probability distribution functional. The value of an nth order distribution function $F_X(x_1, t_1; x_2, t_2; \ldots; x_n, t_n)$ depends on the choice of n pairs (x, t) values. These pairs may be represented by n points in the x-t plane. As n increases and neighboring t values become infinitely close, these points will

Reprinted with permission from *Appl. Mechanics Rev.*, vol. 22, pp. 825–831, Dec. 1969.

trace a curve which is, in fact, the trace of a function $x(t)$. A distribution functional, denoted by $F_X[X(t)]$, may be defined as a limiting distribution function in the sense described above. Of course, the value of this functional depends on the choice of $x(t)$ function over the entire domain of t.

When this method is applied to random processes, it is more convenient to work with the characteristic functional which is the Fourier-Stieltjes functional transform of the probability distribution functional. The use of characteristic functional was first discussed by Bochner in 1947 [Ref. 3].

In many applications, one is concerned only with some statistical properties, not necessarily the detailed probabilistic structure, of a random process. The first order statistical property is given by the mean function $E[X(t)]$, and the second order property is described by the correlation function $E[X(t_1) X(t_2)]$ where $E[\quad]$ denotes an average taken over the entire ensemble of possible realizations of the random process and weighted by the probability distribution. Sometimes, instead of the correlation function, it may be more convenient to work with the covariance function, which is computed from $E[X(t_1) X(t_2)] - E[X(t_1)] E[X(t_2)]$. An important role is played by correlation functions in the study of turbulence of fluid since their first introduction by G. I. Taylor in 1921 [Ref. 4].

Stationary and Ergodic Processes

The adjective "stationary" is used to describe a random process whose probabilistic structure is unchanged upon a shift of the parametric origin; that is, $X(t)$ and $X(t-a)$ are governed by the same set of probability distribution functions. The term "weakly stationary" is used to indicate a broader class of random processes where stationarity is valid for the first and the second order distribution functions, but may not hold for higher order distributions. From the above definitions, it follows that the mean function of a random process which is at least weakly stationary reduces to a constant, and the correlation function $E[X(t_1) X(t_2)]$ of such a process depends only on $\tau = t_1 - t_2$.

In the class of stationary random processes, there is a subclass of ergodic processes. The unique feature of an ergodic process is that the ensemble average of a function of the random process is almost always equal to the corresponding average obtained from a single realization and taken over the range of parametric values. Wiener [Ref. 5] has shown that certain stationary processes which are related to the Brownian motion process are ergodic. This result was further extended by Doob [Ref. 6]. Itô [Ref. 7] has proved the ergodicity of a stationary Gaussian process whose spectral density is continuous.

From an engineering standpoint, one may be interested merely in the substitution of a parametric mean for an ensemble mean, or a parametric correlation function for an ensemble correlation function. For these limited purposes, it is sufficient to know that the ergo-

dicity on the level of the first moment is guaranteed if the process is weakly stationary and if its covariance function is integrable on τ ($= t_1 - t_2$) from 0 to ∞. See, for example, [Ref. 8]. The conditions for ergodicity on the level of correlation function can be similarly established; these conditions involve a higher order of stationarity and the integrability of a function which is related to the fourth and the second moment functions of the process.

Integration and Differentiation of a Random Process

In practical applications, it is often necessary to apply differentiation and/or integration operations to a random process. Since a random process is a family of random variables, these operations must be interpreted in terms of the convergence of a sequence of random variables. There are four different ways in which a random sequence can converge to a limit: convergence with probability one, convergence in probability, convergence in distribution, and convergence in L_r. The most convenient one to use is the L_2 convergence, also known as convergence in the mean or the mean square convergence, which is defined as

$$\lim_{n \to \infty} E[\,|X_n - X|^2\,] = 0$$

Either the convergence with probability one or the mean square convergence implies the other two modes of convergence. The calculus of random functions based on the concept of L_2 convergence has been developed through the work of Slutsky, Cramér, Loéve, and others. A rigorous exposition of this subject can be found in [Ref. 9] and a very readable account in [Ref. 10]. The L_2 calculus has gained increasing popularity in recent years, since the criterion for the existence of an integral or a derivative is simple and easily verified and since the operation of an L_2 limit and the operation of an ensemble average may be interchanged. It is interesting to remark that by use of the commutability of L_2 limit and ensemble average, one can write down immediately a set of input-output relationships for randomly excited linear systems in terms of statistical properties. See, for example, [Ref. 11].

Spectral Analysis

Under suitable conditions a random process may be represented by a Fourier-Stieltjes integral as follows

$$X(t) = \int e^{i\omega t} dZ(\omega)$$

where $Z(\omega)$ is an orthogonal random process with a parameter ω; that is, $E\left\{[Z(\omega_4) - Z(\omega_3)]\,[Z(\omega_2) - Z(\omega_1)]\right\} = 0$ whenever $\omega_4 > \omega_3 \geqslant \omega_2 > \omega_1$. If, in addition, $X(t)$ is at least weakly stationary, then its correlation function $E[X(t_1) X(t_2)]$ depends only on

$\tau = t_1 - t_2$. Denoted by $R_{XX}(\tau)$, such a correlation function also has a Fourier-Stieltjes integral representation

$$R_{XX}(\tau) = \int e^{i\omega\tau} dF(\omega)$$

where it can be shown that $dF(\omega) = E\left\{|dZ(\omega)|^2\right\}$. The above expression is known as the Wiener-Khintchine relation [Refs. 12, 13], and $F(\omega)$ is called the spectral distribution function of the random process. The derivative of $F(\omega)$, when it exists, is called the spectral density.

When a random process is nonstationary, the correlation function can no longer be expressed as a function of $t_1 - t_2$. In this case, it is sometimes convenient to deal with the generalized spectral density which is the double Fourier transform of the correlation function [Refs. 14, 15]. Recently Priestley [Ref. 16] introduced the concept of evolutionary spectral density, which is considered especially useful in dealing with mildly nonstationary processes.

Markov Processes

Among the most widely studied random processes are the Markov processes, which are defined on the basis of conditional probability. A random process $X(t)$ is said to be Markovian if, on the condition that $X(t_1) = x_1$, $X(t_2) = x_2, \ldots, X(t_n) = x_n$ where $t_1 < t_2 < \ldots < t_n$, the probabilistic structure of $X(t)$ for $t > t_n$ is only dependent on x_n (that is, independent of all the other conditional values $x_1, x_2, \ldots, x_{n-1}$). A sufficient condition for a random process to be Markovian is that it has independent increments; that is, $X(t_4) - X(t_3)$ and $X(t_2) - X(t_1)$ are independent random variables whenever $t_4 > t_3 \geqslant t_2 > t_1$.

The simplest example of a Markov process is the Wiener process, which is a mathematical model for Brownian motion of a free particle. Although Einstein [Ref. 17] and Smoluchowski [Ref. 18] had previously analyzed Brownian motion from a probabilistic standpoint, it was Wiener [Ref. 19] who gave the first rigorous mathematical treatment of the problem. The Ornstein-Uhlenbeck [Ref. 20] process is also a mathematical model for Brownian motion, but it is obtained by including viscous resistance in the equation of motion for the particle. In this sense the O-U process is a generalization of Wiener's process. Further extensions to the case of Brownian motion in a potential field require a Markov vector model in a phase plane (or phase space). Such extensions were made by Chandrasekar [Ref. 22], Wang and Uhlenbeck [Ref. 23], Kramers [Ref. 24], and Pontryagin, Andronov and Witt [Ref. 25]. The case of nonstationary disturbances was considered by Lin [Ref. 26]. The method of attack used by these authors is to solve a parabolic differential equation called the Fokker-Planck equation. This equation governs the diffusion of probability measure essentially under the assumption that the disturbances causing the changes (i.e., the increments) in the random vector are small, infinitely frequent, and statistically independent at any two instants

of time, an assumption which justifies the random vector being Markovian. The solutions obtained in [Ref. 24] and [Ref. 25] apply to nonlinear systems. Further extensions to other nonlinear cases have been made by Barrett [Ref. 27], Chuang and Kazda [Ref. 28], Ariaratnam [Ref. 29], Caughey [Refs. 30 and 31], and others. Caughey's solutions were written for N degrees of freedom nonlinear systems; however, with suitable interpretations [Ref. 32], these solutions also apply to the case of infinitely many degrees of freedom. Furthermore, these results are analogous to the principle of equi-partitioned energy in the kinetic theory of gases; therefore, it is not surprising to find that additional restriction must be placed on the relation between generalized excitations and generalized dampings in order to ensure that the same amount of average energy is induced in each generalized coordinate. All the solutions for nonlinear cases mentioned above are incomplete in the sense that only asymptotic forms (i.e., the stationary solutions) have been obtained.

When random disturbances play the role of parametric excitations, the question of stability of the system arises. Again, if the disturbances are small but very frequent, and if they can be reasonably idealized as being statistically independent at any two instants of time which are sufficiently close in the scale of the time constant of the system, then the use of a Markov model is in order. The Fokker-Planck method also has been applied to such problems: [Refs. 33, 34, and 35].

In the mathematical literature, the Fokker-Planck equation is generally known as the Kolmogorov forward equation in recognition of his fundamental work [Ref. 36]. In this paper, Kolmogorov obtained another equation also governing the diffusion of probability of a Markov process, known as the Kolmogorov backward equation. The forward and backward equations are adjoint equations. The existence and uniqueness of a solution to these equations were discussed by Feller [Ref. 37].

Gaussian Processes

There are several equivalent definitions of a Gaussian process, one of which is based on the concept of jointly distributed Gaussian random variables. To explain this definition we must begin with the definition of Gaussian random variables.

Random variables $X_1, X_2, \ldots, X_n$ are said to be jointly Gaussian distributed if their joint probability density is given by

$$p(x_1, x_2, \ldots, x_n) = (2\pi)^{-n/2} |S|^{-1/2}$$

$$\exp\left[-(2|S|)^{-1} \sum_{j=1}^{n} \sum_{k=1}^{n} \alpha_{jk}(x_j - \mu_j)(x_k - \mu_k)\right]$$

where $\mu_j = E[X_j]$, where $|S|$ is the determinant of a matrix S whose (j, k) element is $E[(X_j - \mu_j)(X_k - \mu_k)]$, and where α_{jk} is the cofactor of the (j, k) element of matrix S.

A random process $X(t)$ is a Gaussian process if by any choice of n parametric values $t_1, t_2, \ldots, t_n$ the resulting n random variables $X(t_1), X(t_2), \ldots, X(t_n)$ are jointly distributed Gaussian random variables.

Another definition of a Gaussian process is based on the following unique form of its characteristic functional:

$$M_X[\theta(t)] = \exp \left[i \int \mu_X(t) \theta(t) \, dt \right.$$
$$\left. - \frac{1}{2} \iint K_{XX}(t_1, t_2) \theta(t_1) \theta(t_2) \, dt_1 \, dt_2 \right]$$

where $\mu_X(t)$ is the mean function and $K_{XX}(t_1, t_2)$ is the covariance function of $X(t)$.

It is clear from either of the above definitions that a Gaussian random process is completely characterized by its mean function and its covariance function.

The important role played by Gaussian random processes in practical applications is due partly to their remarkably simple mathematical properties. In particular, any linear operation on a Gaussian process results in another Gaussian process, a conclusion which can easily be reached by use of the mean square calculus. The modeling of a physical phenomenon by a Gaussian process is justified if the phenomenon is a consequence of numerous small independent causes of random nature. The mathematical statement of this fact is contained in the central limit theorem, the various versions of which have been the subjects of numerous studies. In view of this theorem it is not surprising to find that the Wiener process [Ref. 19], the O-U process [Ref. 20], and each component of the Markov vector investigated by Wang and Uhlenbeck [Ref. 23] are also Gaussian processes. It is interesting to remark, however, that the components of a Markov vector corresponding to nonlinear system response, such as those discussed in [Refs. 24, 25, 27-31], are not Gaussian processes.

Point Processes

The name "point process" is used to represent a system of points which are distributed randomly in an interval. In what follows we shall regard such an interval as a time interval. Numerous studies on point processes may be found under the title queueing theory. See, for example, [Ref. 38].

The simplest point process is the Poisson system of points, where the arrival of every random point on the interval is independent of the others. Such a point process is completely characterized by the expected number of arrivals per unit time. For the treatment of the more general case where the rates of arrival of random points are correlated at different instants of time, the method of regeneration, the method of product densities [Ref. 39], and the method of generating functional [Ref. 40] have been developed.

One useful class of random process may be constructed from superposition of random pulses, each of which

begins at a random point [Ref. 41]. In particular, when all pulses have the same deterministic shape and have independent and identically distributed random amplitudes, the statistical properties of the constructed random process can be computed from those of the point process and the random amplitudes. The well-known theorem of Campbell [Ref. 42] gives simple relations of this type corresponding to the Poisson system of points. The special case of Poisson arriving impulses is called a shot noise. Upon further specialization to a constant arrival rate, a shot noise is reduced to a white noise. We remark that since white noise is defined merely on the basis of the second order statistical property, this model represents but one example of white noise, not the only white noise.

Threshold Crossings, Peaks and Envelope of a Random Process

Sometimes, it is necessary to obtain statistical information about the event when a random process passes through a given threshold. This problem, generally known as the threshold crossing problem, was first investigated by S. O. Rice [Ref. 43], who obtained a simple expression for the expected number of threshold crossings per unit time of a stationary random process. Rice's result, however, is equally valid for the nonstationary case for which the crossing rate is, of course, time dependent. The axis crossing problem is the special case where the threshold is at the level zero. Therefore, papers in this general area may also be found under the title of zero crossing or axis crossing. Reference [44] contains a useful bibliography of some such contributions up to about 1962. The determination of more detailed statistical properties of threshold crossings is much more complicated. The problem of random interval between crossings was considered in [Refs. 46-47].

The event of the first threshold crossing is known as the first excursion or first passage. Given the initial value of a random process, one is concerned with the random time at which the first excursion occurs. Exact solutions for the probability distribution [Ref. 48] and the ensemble average [Ref. 41] have been obtained only for a one-dimensional Markov process by use of the Kolmogorov backward equation. For other random processes, various approximations have been attempted, [Refs. 49-53]. Some computer simulation results were reported in [Ref. 54].

The occurrence of a peak in a random process $X(t)$ is the same as the event of a zero crossing of its derivative $X'(t)$ from above. Therefore an expression for the expected number of peaks per unit time can be derived in a similar manner as that used to obtain the expected crossing rate [Ref. 43]. For a smooth random process the probability is zero for any two peaks to occur simultaneously. Then the expected rate of peak occurrence multiplied by a short time interval is approximately equal to the probability for one peak to occur within this short interval. In such a case, the expected rate of peaks which occur above a certain level divided by the expected rate of peaks of all magnitudes is equal to one minus the probability distribution function for the peak magnitude (or height) conditional on the occurrence of a peak. To avoid the ambiguity of having a condition with zero probability, this probability distribution function is often defined in the sense of "horizontal window" of Kac and Slepian [Refs. 55-56].

One problem closely related to the distribution of peak magnitudes is the distribution of rises and falls in a random process. This problem was investigated by J. R. Rice and F. P. Beer [Ref. 57].

The envelope of a random process, if suitably defined, is also a random process. The most widely known definition is the S. O. Rice definition for the envelope of a narrow-band random process. The term "narrow-band" is often used to describe a weakly stationary random process whose spectral density is negligible except within a small interval in the frequency domain. A typical realization of a narrow-band random process has the appearance of a sinusoid with slowly varying amplitude and phase. Thus nearly all the peaks of a narrow band process are positive, and nearly every up-crossing over the axis is followed by a peak. For the case of a narrow band Gaussian process, Rice has shown that the first-order probability distribution of the envelope is the Rayleigh distribution, which, incidentally, coincides with the peak magnitude distribution.

Rice's definition of an envelope is difficult to apply when the original random process is not Gaussian. Motivated by the need for a simple definition of an envelope which is suitable for the response of nonlinear system to random excitations, Crandall has suggested the use of the total energy in the system as a basis for such a definition [Ref. 58]. According to this definition, the magnitude of the envelope at an instant of time is equal to the would-be deflection if both the kinetic and potential energies at this instant are converted entirely to potential energy. Crandall has shown that for a narrow band Gaussian process, his definition gives the same result as that from Rice's definition in the limit as the bandwidth is reduced to zero.

One statistical quantity closely related to threshold crossings, peaks and envelope is the "average clump size," a name coined by Lyon [Ref. 59] to represent the extent of clustering of high level peaks in a random process. Lyon suggested that the average clump size could be approximated by the ratio of the expected up-crossing rate of the original process and the expected up-crossing rate of the envelope. However, at high thresholds, Lyon's approximation may lead to a meaningless result of an average clump size smaller than one. To avoid this inconsistency an alternative approximate procedure to compute the average clump size was suggested recently by Racicot [Ref. 60].

Concluding Remarks

In concluding this survey, it is considered appropriate to cite a number of useful textbooks and monographs. For general reference, Parzen [Ref. 61], Papoulis [Ref. 62], Takács [Ref. 63], Bartlet [Ref. 64], Blanc-Lapierre and Fortet [Ref. 65], and Cox and Miller [Ref. 66] are suggested. These books contain practical examples, and they are less abstract than some other books on random processes. Yaglom [Ref. 8] and Cramer and Leadbetter [Ref. 67] are excellent texts on stationary random processes. Bharucha-Reid [Ref. 68] and Dynkin [Ref. 69] specialize in Markov processes. For measurement and analysis of random data, Blackman and Tukey [Ref. 70] and Bendat and Piersol [Ref. 71] are suitable references.

References

[1] Wold, H. O. A., editor, "Bibliography on Time Series and Stochastic Processes," The MIT Press, Cambridge, Mass., (1965).

[2] Kolmogorov, A. N., "Grundbegriffe der Wahrscheinlichkeitsrechnung" Springer, Berlin, (1933).

[3] Bochner, S., "Stochastic processes," *Ann. of Math.* **48**, 1014-1061 (1947).

[4] Taylor, G. I., "Diffusion by continuous movement," *Proc. London Math. Soc.* **20** (2), 196-212 (1921).

[5] Wiener, N., "The homogeneous chaos," *Amer. J. Math.* **60**, 897-936 (1938).

[6] Doob, J. L., "Stochastic Processes," Wiley, New York, (1953); AMR **8** (1955), Rev. 1261.

[7] Itô, K., "On the ergodicity of a certain stationary process," *Proc. Imp. Acad. Japan* **20**, 54-55 (1944).

[8] Yaglom, A. M., "An introduction to the theory of stationary random functions," English translation by Silverman, R. A., Prentice-Hall, Englewood Cliffs, N. J. (1962); AMR **16** (1963), Rev. 4377.

[9] Loeve, M., "Probability Theory," 3d ed., Van Nostrand, Princeton, N. J. (1963).

[10] Moyal, J. E., "Stochastic processes and statistical physics," *J. Roy. Statist. Soc.* **B11**, 150-210 (1949).

[11] Lin, Y. K., "Probabilistic Theory of Structural Dynamics," McGraw-Hill, New York, N. Y. (1967); AMR **21** (1968), Rev. 1699.

[12] Wiener, N., "Generalized harmonic analysis," *Acta Math.* **55**, 117-258 (1930).

[13] Khintchine, A., "Korrelations theorie der stationaren stochastischen Prozesse," *Math. Ann.* **109**, 604-615 (1934).

[14] Lampard, D. G., "Generalization of the Wiener-Khintchine theorem to nonstationary processes," *J. Appl. Phys.* **25**, 802-803 (1954).

[15] Bendat, J. S., Enochson, L. D., Klein, G. H., and Piersol, A. G., "Advanced concepts of stochastic processes and statistics for flight vehicle vibration estimation and measurement," ASD-TDR-62-973, Thompson Ramo Wooldridge, Inc. (1962).

[16] Priestley, M. B., "Evolutionary spectra and nonstationary processes," *J. Royal Statist. Soc.* **27(B)**, 204-237 (1965).

[17] Einstein, A., "Uber die von der molekularkinetischen Theorie der Wärme geforderte Bewegung von in ruhenden Flüssigkeiten suspendierich Teilchen," *Ann. Physik* **17**, 549-560 (1905). Collected in "Investigations on the Theory of Brownian Movement," Dover, New York, (1956); AMR **9** (1956), Rev. 3372.

[18] v. Smoluchowski, M., "Drei Vortrage uber Diffussion, Brownsche Bewegung und Koagulation von Kolloidteilchen," *Physik Zeits* **17**, 557-585 (1916).

[19] Wiener, N., "Differential space," *J. Math. and Physics* **2**, 131-174 (1923).

[20] Uhlenbeck, G. E., and Ornstein, L. S., "On the theory of the Brownian motion," *Phys. Rev.* **36**, 823-841, (1930). Collected in [21]; AMR **8** (1955), Rev. 324.

[21] Wax, N., editor, "Selected Papers on Noise and Stochastic Processes," Dover, New York, (1955); AMR **8** (1955), Rev. 324.

[22] Chandrasekhar, S., "Stochastic problems in physics and astronomy," *Rev. of Mod. Phys.* **15**, 1-89 (1943). Collected in [21]; AMR **8** (1955), Rev. 324.

[23] Wang, M. C., and Uhlenbeck, G. E., "On the theory of the Brownian motion II," *Rev. of Mod. Phys.* **17**, 323-342 (1945). Collected in [21]; AMR **8** (1955), Rev. 324.

[24] Kramers, H. A., "Brownian motion in a field of force and the diffusion model of chemical reactions," *Physica* **7**, 284-304 (1940).

[25] Pontryagin, L., Andronov, A., and Witt, A., "On the Statistical Investigation of Dynamical Systems," *Zh. Eksperim. i Teor. Fiz.* **3**, 165-180 (1933).

[26] Lin, Y. K., "Nonstationary shot noise," *J. Acoust. Soc. Amer.* **36**, 82-84 (1964); AMR **17** (1964), Rev. 3619.

[27] Barrett, J. F., "Application of Kolmogorov's equations to randomly disturbed automatic control systems," Proc. 1st. Intern. Congr. Intern. Federation Autom. Control, Moscow (1960).

[28] Chuang, K., and Kazda, L. F., "A study of nonlinear systems with random inputs," *Trans. AIEE* **78**, Part II, 100-105 (1959).

[29] Ariaratnam, S. T., "Random vibrations of nonlinear suspensions," *J. Mech. Eng. Sci.* **2**, 195-201 (1960); AMR **14** (1961), Rev. 3520.

[30] Caughey, T. K., "Derivation and application of the Fokker-Planck equation to discrete nonlinear dynamic systems subjected to white random excitation," *J. Acoust. Soc. Amer.* **35**, 1683-1692 (1963).

[31] Caughey, T. K., "On the response of a class of nonlinear oscillators to stochastic excitation," Proc. Colloq. Intern. du Centre National de la Recherche Scientifique, No. 148, pp. 393-402, Marseille, September, (1964).

[32] Lin, Y. K., "Response of linear and nonlinear continuous structures subject to random excitation and the problem of high level excursions," Paper presented at Intern. Conf. on Structural Safety and Reliability, Washington, D. C., April, 1969 (Proceedings to be published by Pergamon Press).

[33] Caughey, T. K., and Dienes, J. K., "The behavior of linear systems with random parametric excitation," *J. Math. and Phys.* **41**, 300-318 (1962).

[34] Bogdanoff, J. L., and Kozin, F., "Moments of the output of linear random systems," *J. Acoust. Soc. Amer.* **34**, 1063-1066 (1962); AMR **16** (1963), Rev. 671.

[35] Ariartnam, S. T., "Dynamic stability of a column under random loading," in "Dynamic stability of structures," Pergamon Press, New York (1966).

[36] Kolmogorov, A., "Uber die analytischen Methoden in der Wahrscheinlichkeitsrechnung," *Math. Ann.* **104**, 415-458 (1931).

[37] Feller, W., "The parabolic differential equations and the associated semigroups of transformations," *Ann. Math.* **55**, 468-519 (1952).

[38] Doig, A., "Bibliography on the theory of queues," *Biometrika* **44**, 490-514 (1957).

[39] Ramakrishnan, A., "Stochastic processes associated with random divisions of a line," *Proc. Camb. Phil. Soc.* **49**, 473-485 (1953).

[40] Kuznetsov, P. I., Stratonovich, R. L., and Tikhonov, V. I., "On the mathematical theory of correlated random points," *Izv. Akad. Nauk SSSR, Ser. Mat.* **20** (2) 167-178 (1956).

[41] Stratonovich, R. L., "Topics in the theory of random noise," Vol. 1, English translation by Silverman, R. A., Gordon and Breach, New York (1963).

[42] Campbell, N., "The study of discontinuous phenomena," *Proc. Camb. Phil. Soc.* **15**, 117-136. "Discontinuities in light emission," *Proc. Camb. Phil. Soc.* **15**, 310-328, (1909).

[43] Rice, S. O., "Mathematical analysis of random noise," *Bell System Tech. J.* **23**, 282-332 (1944), **24**, 46-156 (1945). Collected in [21]; AMR **8** (1955), Rev. 324.

[44] Slepian, D., "The one-sided barrier problem for Gaussian Noise," *Bell System Tech. J.* **41**, 463-501 (1962).

[45] McFadden, J. A., "The axis-crossing intervals of random functions II," *IRE Trans. on Information Theory* **4**, 14-24 (1958).

[46] Longuet-Higgins, M. S., "The distribution of intervals between zeros of a stationary random function," *Phil. Trans. Roy. Soc. (London), A.* **254**, 557-599 (1962).

[47] Stratonovich, R. L., "Topics in the theory of random noise," Vol. 2, English translation by Silverman, R. A., Gordon and Breach, New York (1967).

[48] Darling, D. A., and Siegert, A. J. F., "The first passage problem for a continuous Markov process," *Ann. Math. Statist.* **24**, 624-632 (1953).

[49] Coleman, J. J., "Reliability of aircraft structures in resisting chance failure," *Operations Research* **7**, 639-645 (1959); AMR **13** (1960), Rev. 5138.

[50] Rosenblueth, E., and Bustamante, J. I., "Distribution of structural response to earthquakes," *J. Eng. Mech. Division, ASCE* **88** (EM3), 75-106 (1962).

[51] Gray, A. H., Jr., "First-passage time in a random vibrational system," *J. Appl. Mech.* **33**, 187-191 (1966).

[52] Shinozuka, M., and Yang, J. N., "On the bound of first excursion probability," Technical Report 32-1304, Jet Propulsion Lab., Pasadena, Calif., Aug. (1968).

[53] Lin, Y. K., "On first-excursion failure of randomly excited structures," Proc. AIAA Structural Dynamics and Aeroelasticity Specialist Conference, New Orleans, La., April 16-17 (1969).

[54] Crandall, S. H., Chandiramani, K. L., and Cook, R. G., "Some first passage problems in random vibration," *J. Appl. Mech.* **33**, 532-538 (1966); AMR **20** (1967), Rev. 2402.

[55] Kac, M., and Slepian, D., "Large excursions of Gaussian processes," *Ann. Math. Statist.* **30**, 1215-1228 (1959).

[56] Leadbetter, M. R., "Extreme value theory and stochastic processes," Paper presented at Intern. Conf. on Structural Safety and Reliability, Washington, D. C. (1969) (Proceedings to be published by Pergamon Press).

[57] Rice, J. R., and Beer, F. P., "On the distribution of rises and falls in a continuous random process," *Trans. ASME, J. Basic Eng.* **87** D, 398-404 (1965); AMR **19** (1966), Rev. 670.

[58] Crandall, S. H., "The envelope of random vibration of a lightly damped nonlinear oscillator," *Zagadnienia drgan nieliniowych* **5**, 120-130 (1964).

[59] Lyon, R. H., "On the vibration statistics of a randomly excited hard-spring oscillator II," *J. Acoust. Soc. Amer.* **33**, 1395-1403 (1961).

[60] Racicot, R. L., "Random vibration analysis—application to wind loaded structures," Ph.D. Thesis, Division of Solid Mechanics, Structures, and Mechanical Design, Case Western Reserve University (1969).

[61] Parzen, E., "Stochastic processes," Holden-Day, San Francisco (1962).

[62] Papoulis, A., "Probability, random variables, and stochastic processes," McGraw-Hill, New York (1965).

[63] Takács, L., "Stochastic processes: problems and solutions," Methnen, London (1960).

[64] Bartlet, M. S., "An introduction to stochastic processes," Cambridge University Press (1966).

[65] Blanc-Lapierre, A., and Fortet, R., "Theory of random functions," English translation by J. Gani, Gordon and Breach, New York (1965).

[66] Cox, D. R., and Miller, H. D., "The Theory of stochastic processes," Wiley (1965).

[67] Cramér, H., and Leadbetter, M. R., "Stationary and related stochastic processes," Wiley (1967).

[68] Bharucha-Reid, A. T., "Elements of the theory of Markov processes and their applications," McGraw-Hill, New York (1960).

[69] Dynkin, E. B., "Markov processes," English translation by Fabius, J., Greenberg, V., Maitra, A., and Majone, G., Academic Press, New York (1965).

[70] Blackman, R. B., and Tukey, J. W., "The measurement of power spectra from the point of view of communication engineering," Dover, New York (1959).

[71] Bendat, J. S., and Piersol, A. G., "Measurement and analysis of random data," Wiley, New York (1966); AMR **20** (1967), Rev. 7589.

Applications of Electrical Noise

MADHU-SUDAN GUPTA, MEMBER, IEEE

Abstract—The omnipresent noise in electronic circuits and devices is generally considered undesirable. This paper describes some of the applications to which it has been put. Short descriptions of a wide variety of applications are given together with references for further details. The applications fall in four categories: applications in which noise is used as a broad-band random signal; measurements in which the random noise is used as a test signal; measurements in which noise is used as a probe into microscopic phenomena; and the applications where noise is a conceptual or theoretical tool. Many examples of applications in each of these categories are given. Some of the applications included are only of historical interest now, and a few are, as yet, only proposals.

I. INTRODUCTION

THE TITLE of this article might appear to be surprising, if not self-contradictory. For a long time, electrical engineers have studied electrical noise (or fluctuation phenomena) in circuits and devices much as physicians study a disease: not out of any affection for the subject but with a desire to eliminate it. In fact, noise is sometimes defined as "any undesirable signal." A typical chapter entitled "Noise" in an electrical engineering text would usually begin by giving such reasons for studying noise as these: "Noise contaminates the signal, setting a limit on the minimum signal strength required for proper communication; therefore, the reliability of communication could be improved, power requirement could be reduced, or larger distances could be reached by reducing noise;" or, "Noise limits the accuracy of measurements, and should be minimized to improve precision." Such statements, although true, do not fully reveal the uses to which a study of noise could be put. In this article, we propose to treat noise as a tool rather than a nuisance.

To the academically oriented, the existence of noise is reason enough to study it; it just *might* turn out to be useful. Most work on noise, however, has been motivated by more immediate applications, reduction of noise being only one of them, although a major one. This article is devoted to some of the other applications. Not very infrequently, the study of noise, or the use of noise as a tool in the study of something else, has led to significant advancements, an outstanding example being the work of Gunn [1]. He was measuring noise in semiconductor materials, while studying the properties of hot electrons, when he discovered what is now called Gunn effect. This triggered a great deal of research work because the effect has applications in making transferred-electron-type microwave oscillators and amplifiers. Of course, one should not rely on serendipity; there are plenty of other reasons for studying noise. This article will present some of them.

Most electrical engineers are aware that random-noise generators are standard laboratory instruments, so there must be

some use for noise, but the variety of applications is often not appreciated. Furthermore, the words "applications of noise" are not synonymous with "applications of noise sources." A much broader point of view is taken here to include any use to which the study or measurement of noise can be put.

The purpose of this article is threefold. First, it is intended to generate interest in the study of noise for reasons other than as a performance limitation. Second, it is aimed toward motivating the readers to apply random-noise techniques in their own work by providing, in summary form, the case histories of past successful applications of noise. Finally, it is hoped that the article would induce some fresh thinking on how noise may be employed for other new and useful purposes.

II. TYPES OF APPLICATIONS OF NOISE

For ease of study and systematization, an attempt is made in this section to classify the types of situations in which noise can be made to serve a useful purpose. One possible method of listing these situations is according to the area of application. Thus noise finds applications in biomedical engineering, circuit theory, communication systems, computers, electroacoustics, geosciences, instrumentation, physical electronics, reliability engineering, and other fields. A second method is to classify the applications by *how* noise is useful rather than *where*. This second approach will be followed here because it is more fundamental and illuminating.

There are several possible characteristics of electrical noise which make it useful in the applications discussed here. Thus there are applications based upon the fact that a noise signal can be broad-band, may arrive from a known or desired direction in space, can have a very small amplitude, or may be uncorrelated between nonoverlapping frequency bands. Very often, however, a single application is based on the existence of several of these characteristics, making it difficult to classify the applications according to these characteristics unambiguously. The noise applications are, therefore, somewhat arbitrarily clustered here into the following four categories, depending upon how the noise is employed.

A. As a Broad-Band Random Signal

A random signal can have some properties which are desirable at times, e.g., it can be incoherent and broad-band, it can be used to establish the presence or the direction of location of its source, it can simulate a random quantity, and it can be used to generate another random quantity. These properties lead to the application of random noise in electronic countermeasures, microwave heating by noise, simulation of random quantities, stochastic computing, and generation of random numbers.

B. As a Test Signal

There are many instances of measurements in which one needs a broad-band signal with precisely known properties like

Manuscript received October 24, 1974; revised March 13, 1975. This work was supported by the Joint Services Electronics Program.

The author is with the Research Laboratory of Electronics and the Department of Electrical Engineering and Computer Science, Massachusetts Institute of Technology, Cambridge, Mass. 02139.

Reprinted from *Proc. IEEE*, vol. 63, pp. 996–1010, July 1975.

amplitude probability density, rms value, or autocorrelation function. Electrical random-noise generators are a convenient source of such signals, having statistical parameters which are either known in advance or can be easily manipulated. This explains the use of noise in such techniques (some of them very accurate) as measurement of impulse response, insertion loss, and linearity and intermodulation of communication equipment, as well as in noise-modulated distance-measuring radar.

C. As a Probe into Microscopic Phenomena

The use of cosmic radio noise to glean information about the sources of this noise is well known in radio astronomy. Similarly, the fact that the electrical noise is caused by the motion (or emission, or recombination, or ionizing collision, etc.) of individual carriers suggests the possibility that such microscopic processes may be studied through the fluctuations. In particular, noise measurements can be used for estimating the physical parameters which are related to these microscopic processes. This forms the basis for the application of noise in the determination of parameters like carrier lifetime in semiconductors, fundamental physical constants, and device constants, as well as in testing semiconductors for uniformity and for estimating the reliability of semiconductor devices.

D. As a Conceptual Tool

While noise has been the motivating cause for the development of new disciplines like information theory and statistical theory of communication, certain other fields, notably circuit theory, have also benefited from the study of noise. In addition to being a subject of interest in itself, noise is also useful as a vehicle for theoretical investigations and modeling of other physical systems. For example, the thermal noise of a resistor serves as a model for fluctuations in any linear dissipative system in thermodynamic equilibrium.

Specific examples of these applications are discussed in the remainder of this article.

III. USE OF NOISE AS A BROAD-BAND RANDOM SIGNAL

A. Microwave Heating by Noise

The use of microwave power for heating in industrial processes and in microwave ovens is well known. It has been suggested that high-level microwave noise sources, like crossed-field noise generators, could be used in these applications of microwaves, especially when the load is of a nonresonant nature [2]. In such applications, the wide-band noise sources have the advantage over the resonant microwave sources (like magnetrons) that they are relatively insensitive to the load VSWR. It should be pointed out that the noise power output, available from crossed-field noise generators, is approximately the same as that available from crossed-field amplifiers (although at a lower efficiency at present). Therefore, the use of noise is not limited to lower power microwave heating applications.

B. Simulation of Random Phenomena

Noise generators have a number of practical applications in the simulation of random quantities in system evaluation and testing. One of the things that a noise signal can simulate is, of course, noise itself. This simple observation accounts for the use of noise sources in such applications as radar simulation. It is often necessary to carry out, under realistic conditions,

the testing of new radar and sonar systems, and the training of personnel working with these systems, without actually placing the system in field use. This task is carried out by designing electronic simulators which generate signals resembling those encountered by the system in actual operating environment. Noise generators are an essential part of such simulators. For example, in one radar simulator [3] used for training, the random fading of signals is duplicated by modulating the signal by low-frequency random noise. The modulating signal itself is generated by cross correlating two narrow-band random-noise signals. In another radar simulator [4], the probability of detection of a target is measured by introducing noise into the videodetector. The random characteristics of noise typically observed in normal pulsed radar receivers are reproduced by adding the output of a noise source to a pulse train and rectifying it by a full-wave rectifier before applying it to the video detector. Such measurements of the visual detectability of signals in the presence of noise are not confined to radar simulation; they are also used in situations like seismic detection, where the noisy signals are recorded on charts [5]. Similar studies of the intelligibility of audio signals in the presence of noise are carried out in the evaluation of speech communication systems and in physioacoustics [6]. All such measurements may be likened to the measurement of noise figure, in that they determine the performance of a system in the presence of noise.

Noise signals are also used for simulating random vibrations in mechanical systems; the combination of a random-noise generator and a shake table is widely used to test the response of mechanical structures to random vibrations [7].

C. Source Detection and Location Through Noise Measurement

There are many well-known applications in which an object is detected or located through the measurement of the noise emitted by it. At infrared frequencies, for example, the emissions are used for the surveillance and tracking of targets, the monitoring of temperature and chemical composition, and other military and industrial applications. At microwave and radio frequencies, the noise is received to detect and identify the radio stars and atmospheric processes. At audiofrequencies, the acoustic noise from boiling liquids and vibrations of components in nuclear power reactors are picked up by sensors for fault-location or failure-detection. The literature devoted to these established technologies is very extensive [8], [9] and no attempt will be made to summarize it here. The following brief discussion is restricted to the basic principles of noise-source detection and location.

If the noise source of interest is known to move on a path, its presence can be detected by a receiver located near the path, even in the presence of a strong noise background. A practical example of this is in the detection of a passing ship by an omnidirectional hydrophone [10], [11]. (While the noise in these applications is acoustic and not electrical, the technique is included here because of its possible generality.) The effect of the passing noise source is to increase the received noise power. This increase can be identified by comparing the noise power measured over a short averaging time with a long-time average of the noise power. If the background noise level is stationary, the long-term averaging time can be increased for estimating the background noise to any desired degree of precision, so that the detectability approaches its maximum value determined by the receiver noise [10]. For nonstationary background noise, the detectability is lower and depends upon the correlation time of the background noise [11].

When the spatial location of a noise source is to be established, it is necessary to use more than one receiver and cross correlate the received signals. The method is based upon the fact that if two received signals $x(t)$ and $y(t)$ come from the same noise source $n(t)$, but with different transit-time delays T_x and T_y, their cross-correlation function is the same as the autocorrelation function of $n(t)$:

$$\phi_{xy}(\tau) = \lim_{T \to \infty} \frac{1}{2T} \int_{-T}^{T} x(t - \tau)\, y(t)\, dt \qquad (1a)$$

$$= \lim_{T \to \infty} \frac{1}{2T} \int_{-T}^{T} n(t - \tau - T_x)\, n(t - T_y)\, dt$$

$$= \phi_{nn}(\tau + T_x - T_y) \qquad (1b)$$

and is, therefore, a maximum for $\tau = T_y - T_x$. Thus the differential transit time can be determined by an experimental measurement of $\phi_{xy}(\tau)$. If the velocity of propagation of signals is determined (possibly with the aid of a separate measurement using a known noise source), the measured value of $T_y - T_x$ yields the locus of possible locations of the unknown noise source.

For a noise source known to be along a straight line (e.g., a straight tube or a wire) two collinear detectors on the two sides of the source are sufficient for locating it [12]. The signals received at the two detectors are cross correlated and the differential delay τ required to make the cross-correlation function $\phi_{12}(\tau)$ maximum is found. A third noise detector may be used to locate a source anywhere in the plane by triangulation [13]. Such techniques have been used for locating noise sources in nuclear reactors (where again the noise is acoustic rather than electrical).

D. Self-Directional Microwave Communication

Directional antennas are used for point-to-point radio communication in order to reduce the required power and the possibility of interference with other users. When the locations of transmitting and receiving antennas are unknown or varying, it is necessary to first scan the field of view and locate the source of a pilot signal, and then to steer the antenna beam in the proper direction. An automatic method of producing properly directed beams has recently been proposed which eliminates the need for a pilot signal, scanning, and steering [14]. It is based upon the ambient noise radiated by a retrodirective antenna array. A retrodirective antenna is defined as an antenna which reradiates a beam in the same direction from which a beam is incident on it. A planar array of antenna elements connected with lines of suitable electrical length can serve as a retrodirective antenna and its retrodirective gain can be increased by inserting amplifiers in the interconnecting lines. Two such antennas which are within each other's field of view will each receive the noise output radiated by the other array, and will amplify and reradiate it towards the other. As the received noise is coherent across the receiving aperture, it is not masked by the locally generated noise. After a sufficient number of return trips around the loop between the two antennas, the signal builds up to a carrier with the

radiated beams of the two antennas focused upon each other and it can then be modulated.

The spectrum of the carrier signal thus established depends upon several factors. If the voltage gain around the loop is less than one, the power spectral density of the signal tends towards a steady state, while if the gain is greater than one, it continues to build up until limiting occurs in the system. The width of the spectrum becomes narrower for larger gain, and the center frequency of the spectrum is simply the inverse of the two-way transit time between the antennas (including any interval delays) or an integral multiple of this fundamental frequency. Finally, the number of round trips required to build up to a given peak power decreases rapidly as the loop gain increases, but only slowly as the number of antenna elements in the array increases.

In many ways, the establishment of carrier in this system is similar to the buildup of oscillations in a multimode oscillator with delay. The noise signal acts as the initial excitation which grows due to feedback. The omnipresence of noise makes this application possible.

E. Applications in Electronic Countermeasures

A well-known application of high-power broad-band noise sources is in "active" jamming of radars [15] and communication equipment [16]. Radar jamming is called active if the jammer radiates a signal at the frequency of operation of the radar system, as distinguished from passive jamming which employs nonradiating devices like chaff. Active jammers themselves are either deception jammers, which radiate false echoes to confuse the radar, or noise jammers, which introduce sufficient noise in the radar receiver to mask the target echo. The noise power required for this purpose can be considerably smaller than that produced by the radar transmitter, because the jamming signal travels only one way from the target to the receiver.

Most modern radars can be tuned rapidly over several hundred megahertz so that a jammer, to be effective, must either distribute its available transmitter power uniformly over the band covered by the radar (called broad-band barrage jamming), or be capable of automatic tuning to match the radar signal frequency (called narrow-band or "spot" jamming). A third possibility is to use swept-spot jamming in which the jammer frequency is swept at a very high rate across the entire band of frequencies to be jammed such that the radar receiver is unable to recover between sweeps. Broad-band jamming is the most reliable of the three schemes but requires the largest amount of power to be successful.

The broad-band jamming signal can be generated either by a noise source centered at carrier frequency or by noise modulating a CW signal. As the noise power radiated by a jammer serves only to increase the background noise level in the receiver, the methods of reducing the effect of jamming are the same as the methods of reducing receiver noise figure. This implies that if a broad-band high-frequency noise source is used to generate the jamming signal, there is nothing that can be done to reduce jamming if the receiver is already a low-noise state-of-the-art receiver. However, if the wide-band noise was generated by frequency-modulating a CW signal by a low-frequency random-noise source, the addition of some signal-processing circuits to the receiver can reduce the effectiveness of jamming. The superiority of a high-frequency noise source stems from the fact that its output in nonoverlapping frequency bands is uncorrelated.

F. Computational Applications

Noise signals have applications in analog computation, digital computation (at least in principle), and in stochastic computation. The simulation of random processes on an analog computer requires a random signal, having prescribed statistical properties, to represent randomly varying variables, initial conditions, or parameters [17]. Noise generators used in such applications must satisfy very strict requirements in respect to stability of output, uniformity of power spectrum, wide bandwidth, and absence of hum or periodic signals. A number of such simulation studies have been made in fields like equipment failure, queuing problems in traffic control, and the nonlinear response of an airplane to random excitations [18], [19].

In digital computers, analog random-noise sources can be used to design a hard-wired source of random numbers [20], although pseudorandom-noise generators may be used in special-purpose machines while the software methods are universally used in general-purpose computers. Random-number tables can also be generated by analog noise sources; the well-known table of random numbers published by the Rand Corporation, Santa Monica, Calif., was prepared in this manner [21]. A random-frequency pulse source, producing approximately 10 000 pulses per second, was gated once per second by a constant-frequency pulse. The resulting pulses were passed through a 5-place binary counter and then converted into a decimal number by a binary-to-decimal converter. The published table is a transformed version obtained by discarding 12 of the 32 states, retaining the last digit of the decimal number, and adding pairs of digits modulo 10. The resulting table had to be tested for statistically significant bias and the circuits refined until a satisfactory table was produced.

A relatively newer application for random noise is in stochastic computers [22], [23], which are low-cost analog computers built with digital components. Stochastic computers use the probability (of switching a digital circuit) as an analog quantity and carry out the operations of multiplication, addition, integration, etc. by using digital integrated circuits that are randomly switched. The heart of such a computer is the generator of clocked random-pulse sequences (CRPS) which represent the analog quantities. In the unipolar system (which is one of the several possible forms of representation), an analog quantity is represented by the average value of a random-pulse train. Two such analog quantities can be multiplied together simply by applying the two pulse trains at the inputs of an AND gate, which results in another pulse train at the output having an average value equal to the product, as shown in Fig. 1. This is because the joint probability of two independent events is equal to the product of the probabilities of the two individual events. Similarly, integration of a quantity can be performed by a digital counter.

Stochastic computers are useful in applications such as process control where cost rather than speed or accuracy is important. The speed of these computers is limited by the width of the pulses, and the accuracy is determined by the limited number of pulses that can be used, so that the average value of the pulse train, which represents the analog variable, is subject to random-variance error.

There are several ways in which the random-pulse trains can be generated to represent an analog variable, for example, by level detection of white random noise using a trigger level proportional to the variable, and subsequent shaping by sampling of clock pulses. Alternatively, sampled flip-flops driven

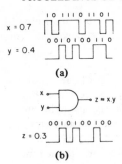

Fig. 1. (a) Unipolar representation of two quantities x and y by the average value of random pulse trains. (b) Multiplication of the analog quantities x and y by the use of an AND gate.

by noise may be used, or pseudorandom noise may be generated using shift registers.

G. Biological and Medical Applications

Noise signals are employed for biomedical purposes in at least three distinct types of applications—simulation, measurement, and therapy. Noise applications in simulation arise from the fact that fluctuations are inherent in biological systems, as they are in electronic systems. For example, the threshold of neurons at the trigger point fluctuates by 10 percent of its resting voltage. Noise signals are thus used in the simulation of the activity of neurons to produce randomness in the voltage waveforms and in the time required to generate the successive pulses [24]. The measurement of the response of biological systems to random stimuli is of interest in physiological and other studies. One system which has been extensively studied in this manner is the human auditory system [25] whose response to noise may consist of signal masking, fatigue, or damage. An interesting medical application of noise is in inducing sleep or anaesthesia [26] and in the suppression of dental pain in a technique called audioanalgesia [25]. A dental patient listens, via earphone, to relaxing music, and switches to filtered random noise on feeling pain, increasing the intensity of noise as necessary to suppress pain. While the physiological and psychological factors responsible for this phenomenon are not well understood, the collected data show that audioanalgesia has "about the same level of effectiveness as morphine" [25].

IV. Use of Noise as a Random Test Signal in Measurements

A. Measurement of Noise

Perhaps the best known of all applications of noise sources is their use in the measurement of the noisiness of electronic devices, circuits, and systems. The purpose of the measurement may be to determine the noisiness of a device, to calibrate the output of a standard noise source, to establish the limitations of a system, to optimize a design, etc., but the ultimate goal is usually the reduction of noise.[1] The literature on the subject of noise measurement is very sizable and only some surveys are referred to here.

The term "noise measurement" has been used for a number of different measurements, only three of which are briefly

---

[1] The critical reader could claim that this is not an "application" of noise in the true sense: the need for such characterization arises due to the problems and limitations created by noise in the first place. Pushing our analogy between noise and disease farther, this is like using virus in a vaccine to help build resistance against the virus, although in a less direct manner.

mentioned here. First, "noise measurement" refers to the measurement of the noise voltage or noise power output of a one-port device or network. Various techniques are used for carrying out the noise spectrum measurement, depending upon the frequency of measurement, the desired accuracy, the impedance level of the one-port, and other such considerations. In general, the noise output is determined by comparing it with the output of a standard noise source [27], [28]. Second, "noise measurement" is commonly used to imply the measurement of the noise figure of a linear two-port. Again there is a number of ways in which this measurement may be carried out. Most of them require that a noise source be connected at the input port and the resulting increase in the available noise power at the output port be measured [29], [30]. Third, "noise measurement" may mean the measurement of the noise modulation of the output of an oscillator or signal generator. The demodulation is then carried out in some suitable manner. The measurement of the demodulated noise power is again facilitated by comparison with standard noise-source output [31].

B. Measurement of Bandwidth

If the transfer function of a linear two-port network or system, such as a bandpass filter, is $H(\omega)$, its effective bandwidth is defined as

$$B_{\text{eff}} = \frac{1}{H_m^2} \int_0^\infty |H(\omega)|^2 \, d\omega \qquad (2)$$

where H_m is the peak value of $|H(\omega)|$. The effective bandwidth is an important parameter in system design and performance calculations. Equation (2) suggests one method of determining B_{eff} by experimental measurement of the frequency response $|H(\omega)|^2$ and graphical integration of response curve. A more direct method is to use white noise (with a frequency range sufficient to encompass the passband) as the input signal so that the noise power spectrum at the output is determined by the frequency response $|H(\omega)|^2$. Measurement of the total output noise power then yields the integral of $|H(\omega)|^2$ and hence the effective bandwidth. An alternative method is to use a "standard" bandpass network. The output noise power from the unknown network is then detected and compared against that from the standard network, so that only the ratio of voltages need be measured experimentally [32].

C. Measurement of Antenna Characteristics

The experimental measurement of antenna characteristics on a test range becomes more difficult and less accurate as the size of the antenna increases. This is because the far-field criterion is more difficult to satisfy, and the possibility of spurious ground reflection is large at low elevation angles. Cosmic radio noise sources are then used for antenna measurement [33], [34] because collectively they have several different desirable properties: they always satisfy the far-field requirement, radiate unpolarized waves over a wide continuous band of frequency, have a very small angular size (essentially a point source), have a fixed position in the sky which is precisely known in many cases, have a steady power output that does not vary over short intervals and is accurately known in many cases, and cover a wide range of elevation angle and power density. The set of cosmic noise sources can, therefore, serve as an ideal test source from radio frequencies to milli-

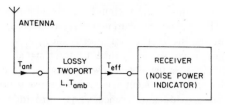

Fig. 2. Radiometric arrangement for measurement of insertion loss of a matched dissipative microwave two-port.

meter waves (although the background radiation from the galaxy is large below 1 GHz). A black disk at a known temperature and located in the far field of the antenna has also been used like cosmic noise sources in a technique called the "artificial moon" method [34].

Several different antenna parameters can be, and have been, measured with the help of cosmic noise sources. Pointing of the electrical axis of an antenna can be carried out if the position of the radio noise source in the sky is accurately known. The true focal point of an antenna reflector can be established because the change in antenna gain is a sensitive indicator of the accuracy of focussing. The radiation pattern of an antenna can be measured accurately (because of the small angular size of noise sources) over a large dynamic range by selecting a sufficiently strong source. If, in addition, the power density of the radiation is separately determined by a small calibrated antenna and the absolute power level can be measured, absolute antenna characteristics like the aperture and the beam efficiency can also be obtained [33].

D. Measurement of Insertion Loss

The measurement of the operating noise temperature of an antenna system has been used for an accurate evaluation of the insertion loss of antenna radome material [35] and of antenna feed components between antenna aperature and receiver input port [36]. In principle, the insertion loss of any well-matched (low VSWR) two-port can be measured by the radiometric method. Fig. 2 shows a simple circuit arrangement in which a lossy two-port is present between the antenna and the noise power indicator. If 1) the connecting transmission lines are lossless, 2) the components are matched to the lines, and 3) the lossy two-port is dissipative (i.e., resistive rather than reactive), then the noise power reaching the indicator expressed as a noise temperature, is

$$T_{\text{eff}} = \frac{T_{\text{ant}}}{L} + \left(1 - \frac{1}{L}\right) T_{\text{amb}} \qquad (3)$$

where T_{ant} is the antenna noise temperature, T_{amb} is the ambient (physical) temperature of the lossy two-port, and L is its loss (the ratio of input to output power). The operating noise temperature T_{eff} of the antenna is measured first with the two-port under test installed in the system, and then with the two-port replaced by one having a known insertion loss. The insertion loss of the two-port under test is then calculated from the measured noise temperatures using the relationship (3) between the insertion loss L and noise temperature T_{eff}.

E. Measurement of Microwave Signal Power

The measurement of microwave power is invariably carried out by converting the power into a suitable quantity, such as the temperature rise of a bolometer or the signal voltage across a rectifier, which can then be measured. In order to relate the measured quantity to the original signal power, it is usually necessary to calibrate the system by using a known amount of power. The power measurement is, therefore, effectively carried out by the method of substitution, in which the unknown signal is compared against a known signal. In practice, the known signal is usually at low frequency, although an RF signal is desirable for higher accuracy. A noise source can also be used to provide a reference signal, provided its output power is precisely known. A resistor is an appropriate source in this application because its thermal noise output can be calculated accurately.

The absolute power level of a CW signal received from a spacecraft has been measured by using a microwave noise signal as the reference [37]. The procedure is similar to the Y factor method of measurement of the noise figure of a linear two-port. In essence, the received output power, which consists of the system noise power and the signal power, is compared with the system noise power alone. The system noise power itself is measured by connecting at the system input a microwave thermal noise source whose output is precisely known. This technique results in a significantly increased accuracy over other methods of measurement.

F. Impulse Response Measurement

The impulse response of a linear two-port network or system can be measured with the help of stationary random noise. The random signal is applied to the system input and the resulting output is cross correlated with the input signal. It is well known [38] that if a random signal $x(t)$ with autocorrelation function $\phi_{xx}(\tau)$ is applied at the input of a linear system having an impulse response $h(t)$, the cross-correlation function between the input and the resulting output $y(t)$ is given by the convolution integral

$$\phi_{xy}(\tau) = \int_{-\infty}^{\infty} h(t)\phi_{xx}(\tau-t)\, dt. \tag{4}$$

This relationship can be used to calculate the impulse response $h(t)$ from a knowledge of ϕ_{xx} and ϕ_{xy}. For causal, lumped, linear, time-invariant systems, this calculation can be carried out algebraically [39].

In most practical situations, the choice of the input signal $x(t)$ is arbitrary. The solution of the integral equation (4) for $h(t)$ is greatly facilitated by the use of white noise as the input signal [38]. If the bandwidth of the input signal is much larger than that of the system under test, $\phi_{xx}(\tau)$ is effectively the impulse function $\delta(\tau)$ and (4) simplifies to

$$\phi_{xy}(\tau) = h(\tau). \tag{5}$$

The impulse response is thus directly measured without the need for involved calculations. Having selected white noise as the input signal, there is still a variety of random signals available to choose from. In particular, the use of a random telegraph signal (which switches randomly between two fixed amplitude levels) has the advantage that the cross correlation is easily determined, and the problems due to a limited dynamic range of the system are avoided [40].

A wide variety of linear systems have been characterized using this method, such as electrochemical electrodes [41] and nuclear reactors [42]. This method of system characterization has two major advantages: 1) the additive random noise, contributed by the system, is eliminated in the process of cross correlation, and 2) the dynamic behavior of the system is measured with a minimum of perturbation or interference with

normal operation. Apart from characterization, the method is also useful as a part of the corrective loop in the design of an adaptive system. As $h(t)$ completely characterizes a linear system, a measurement of $\phi_{xy}(\tau)$ as a function of τ (or for several discrete values of τ) is sufficient for making decisions concerning the performance of a linear system. This measured system characteristic together with a predetermined criterion is usable for the continuous monitoring and readjustment of the system parameters in the design of an adaptive (self-adjusting) system [43].

G. Characterization of Nonlinear Systems

A linear system is completely described by means of its response to sinusoidal signals ("the frequency response"), or to an impulse ("the impulse response"), from which its response to any arbitrary input can be found. For nonlinear systems, the situation is much more complex. In one formulation, a complete description of the system requires a set $\{h_n\}$ of Wiener kernels [44]. The zeroth order kernel h_0 is a constant, the first-order kernel $h_1(\tau)$ is the impulse response of the best linear approximation (in the mean-square error sense) of the system, the second-order kernel $h_2(\tau_1, \tau_2)$ is the nonlinear interaction of the inputs at two time instants τ_1 and τ_2 in past, and the third-order kernel $h_3(\tau_1, \tau_2, \tau_3)$ and other higher order kernels are similarly defined. The experimental determination of Wiener kernels for a nonlinear system is a difficult process that has been attempted only infrequently, and has been carried out by using white Gaussian noise at the input to excite the system [45]. The kernels are then calculated using the expression

$$h_n(\tau_1, \tau_2, \cdots, \tau_n) = \frac{1}{n![\Phi_{xx}(f)]^n}$$
$$\overline{\cdot \left\{ y(t) - \sum_{m=0}^{n-1} G_m[h_m, x(t)] \right\}}$$
$$\overline{\cdot x(t-\tau_1)x(t-\tau_2)\cdots x(t-\tau_n)} \tag{6}$$

where $x(t)$ is the input white Gaussian noise signal with power spectral density $\Phi_{xx}(f)$, resulting in the output $y(t)$, and $\{G_m\}$ is a complete set of orthogonal functionals in terms of which $y(t)$ can be expanded. This technique has been experimentally verified for electronic systems [45] and has been employed for the characterization of biological systems [46].

A very considerable amount of work in the theory of nonlinear systems has, however, been carried out using an "incomplete" description of the system, such as the response of the system to a specific class of excitations. The excitation signal waveform may be a set of sinusoids, an exponential, a staircase, etc. In particular, the response to randomly varying input signals having specified statistical characteristics is a useful description of the system [47], [48]. The effect of the nonlinear system is to transform the statistics of the input noise, and a measure of this transformation serves as a system parameter. For example, if the input and the output random signals of a nonlinear system have an almost Gaussian amplitude probability distribution, an "effective gain" of the system [49] may be defined as a function of the variance of the Gaussian distribution (or the mean-square value of the signal). This is analogous to defining, for a sinusoidal signal, a gain that is dependent upon the signal amplitude. The utility of white Gaussian noise in testing nonlinear systems stems from the fact that it contains a broad range of frequencies and a wide variety of waveforms. It, therefore, represents all possible input signals. It has been used, for example, in the testing and calibration of a nonlinear instrumentation system used for the analysis of television signals [50].

H. Measurement of Linearity and Intermodulation in a Communication Channel

The wide-band characteristic of white noise is useful for measuring the linearity and intermodulation generation in very broad-band communication equipment. Such measurements are important in evaluating the quality of multichannel communication systems. When a large number of telephone channels is to be carried by a coaxial cable or a broad-band radio link, nonlinear distortions existing in the system, if any, will introduce unwanted intermodulation products of the various components of the multiplex signal. The calculation of the intermodulation noise, so introduced, is very difficult because of the large number of channels. Since statistical properties of white noise are similar to those of a complex multichannel signal with a large number of intermittently active channels, white noise is used to simulate such a signal [51]. A method called "noise loading" is commonly used to measure the performance of communication equipment in accordance with the standard specifications. A band-limited Gaussian noise is introduced at the input into the system under test. The noise power in a test channel is measured first with all channels loaded with white noise, and then with all but the test channel loaded with white noise. The ratio of the first to the second measurement is called the noise power ratio from which the channel noise, due to intermodulation, can be calculated [52]. The spectral density of input noise can be shaped to match the signal under actual operating conditions. Instruments are commercially available to carry out this measurement [53].

Several different extensions and variations of the basic noise loading test are possible [54]. For example, the intermodulation can be computed from the spectral density of the output of a nonlinear device with a band of noise applied at the input. It is also possible to determine the order of the intermodulation noise by measuring the effect of input noise power on the intermodulation noise generated. For each decibel change in input power, the intermodulation noise changes 2 dB for a second-order system, 3 dB for a third-order system, etc.

I. Measurement of the Small-Signal Value of Nonlinear Components

In general, the measurement of the value of a circuit element requires that an external signal be applied to the element through a bridge or some other measuring system. The value of a nonlinear element, however, depends upon the magnitude of signal applied to it, so that the measured value is found for a specific signal level, and that level has to be specified. The small-signal value can usually be determined as a limit by extrapolation. A more direct method of measuring this limiting value is by employing the naturally occurring noise signal in the element itself. This method is useful even for reactive circuit elements which have only a small resistive component and hence low thermal noise. For example, Korndorf et al. [55] have measured the inductance of a coil with a ferromagnetic core by connecting a known capacitor across it and measuring the frequency of maximum thermal noise voltage across this resonant circuit by means of a tuned amplifier. The application is, therefore, based upon the broad-band nature of omnipresent noise.

J. Noise-Modulated Distance-Measuring Radar

One of the most interesting applications of noise is in a noise-modulated radar. In a conventional radar, the transmitted energy is periodically modulated in amplitude or frequency. The range of such radars is, therefore, limited by the repetition period of the periodic modulating signal, and there is an ambiguity in target location due to reflections from targets for which the transit time delay is greater than the repetition period. As this ambiguity is inherent in periodic modulating signals, it can be circumvented by using a nonperiodic modulating signal, for instance a noise signal. Still another attractive feature of noise-modulated radars is their noninterference with neighboring radars. With collision-avoidance radars being considered for every automobile, it is absolutely essential that the large number of radars within each other's range be protected from mutual interference. Noise modulation may be examined for this purpose.

The idea of noise-modulated radars has been around for quite some time [56], [57]. In principle, the transmitted energy may be modulated in amplitude, phase, or frequency by a random signal $x(t)$, e.g., Gaussian noise. The modulation of the signal reflected from a target is, therefore, also the same random signal $x(t + T)$, delayed by a period T, neglecting the contamination of the signal due to external noise, Doppler effect, etc. If the modulation of the outgoing carrier is cross correlated with the modulation of the returned signal, the result is the value of the autocorrelation $\phi_{xx}(\tau)$ of the random signal $x(t)$ at $\tau = T$. The range of the target can be deduced from this measurement of $\phi_{xx}(\tau = T)$, provided $\phi_{xx}(\tau)$ is a monotonic function of τ so that its inverse is a single-valued function (i.e., given $\phi_{xx}(\tau)$, one can find a unique τ). According to the Wiener–Khintchine theorem, the autocorrelation function depends upon the power spectrum of the modulating random signal. Therefore, a power spectrum can be chosen which would give a desired autocorrelation function. Merits of the various possible power spectra and modulation methods, and their limitations, have been discussed by Horton [57] who has also described the tests made on an experimental radar.

Several modified versions of the basic noise radar scheme have also been mentioned in the literature; for example, the use of naturally occurring noise from stars in order to eliminate the transmitter entirely [58], and the use of optical frequencies in a noise radar [59], and the use of noise-modulated ultrasonic radar for blood-flow measurement [60].

K. Continuous Monitoring of System Performance

The use of a noise generator for checking the system performance in manufacturing or laboratory is commonly known. The procedure can be extended to in-service monitoring of radar and communication equipment in the field. This technique has now become feasible with the development of solid-state noise sources, which have a smaller power consumption, weight, volume, radio-frequency interference, turnon time, and turnoff time, but higher noise power output and reliability than gas-discharge noise sources. Chasek [61] has described in detail the use of avalanche-diode microwave noise generators in continuous monitoring of gradually deteriorating performance parameters of radar and relay receivers, such as noise figure, distortion, gain, transmission flatness, and gain- and phase-tracking. As a result, the need for retuning or servicing the equipment is recognized before its performance becomes unacceptable. As the noise signal is very small and unrelated to all other signals, the monitoring can be carried out while the equipment is in operation, thus reducing the downtime due to checkups.

L. Applications in Acoustic Measurements

There are several applications of random noise in acoustic measurements. For example, the repeatability of room reverberation measurements is increased by using narrow-band random noise to generate acoustic signal rather than a pure tone because of the flat spectrum of noise in the passband. A test-signal bandwidth of approximately $20/T$ is a good compromise for optimum smoothing without losing too much spectral resolution, where T is the reverberation time [62]. Similarly, the indoor testing of the frequency response of loudspeakers using a narrow-band random noise instead of a single frequency has the advantage of smoothing out the sharp peaks and valleys introduced into the response curves by the room [63]. In another method of loudspeaker testing, broad-band random noise is considered as a voice sample for which the input and output of the loudspeaker are compared to determine the fidelity of reproduction. This method of testing is considered to be more "natural" because it resembles the manner in which a human listener would judge the fidelity [64]. Random noise is also used to obtain the diffused-field response of microphones and for testing close-talking microphones [63].

V. Use of Noise for the Measurement of Physical Parameters

A. Measurement of Impedance and Transducer Characteristics

The use of random noise as a test signal for measuring the characteristics of circuits and systems has already been described. In particular, it was mentioned that the small-signal value of a nonlinear inductor can be measured by measuring its resonant frequency with a known capacitor using the internally generated noise signal. In that application, the noise signal serves only as a small broad-band signal and is not used to measure something which is inherent in the noise generation mechanism. In the application under discussion now, the magnitude of internally generated noise is used as a measure of the quantity which itself is responsible for noise generation, namely the resistive part of the impedance of a two-terminal circuit or device.

According to Nyquist's theorem, any dissipative two-terminal circuit element having an impedance $Z = R + jX$ will, in thermal equilibrium at temperature T, exhibit the mean-square noise voltage and current

$$\overline{e^2} = 4kTBR \tag{7}$$

and

$$\overline{i^2} = 4kTBR/(R^2 + X^2). \tag{8}$$

These relationships can be inverted to obtain R and X from a measurement of noise voltage and current.

$$R = \frac{\overline{e^2}}{4kTB} \tag{9}$$

$$X^2 = \frac{\overline{e^2}}{\overline{i^2}} \left[1 - \frac{\overline{e^2} \cdot \overline{i^2}}{(4kTB)^2} \right]. \tag{10}$$

The sign of the reactance X can be determined by carrying out this measurement at two different frequencies while the band-

width B is best calculated by measuring the noise of a known impedance. The measurements can be made rapidly by designing an automatic system for this purpose [65].

Obviously, this technique can be used not only for measuring impedance but also for other quantities which can be calculated from impedance. For example, the measurement of the sensitivity of a hydrophone as a function of frequency has been carried out in this way. Goncharov [66] has shown that, under certain simplifying assumptions, the current sensitivity of an electroacoustic transducer is proportional to the square root of the real part of its driving-point impedance:

$$E_I(\omega) = a \cdot (\text{Re}\,[Z])^{1/2} \tag{11}$$

where a is approximately independent of frequency. Therefore, the frequency dependence of the sensitivity of a hydrophone can be determined simply by measuring the noise voltage spectrum at its terminals. As another example, this method of impedance measurement has been used for determining the temperature dependence of the dielectric constant of ferromagnetic materials [65].

The advantage of measuring impedance via noise lies in the fact that the signals involved are of very small amplitude and the small-signal value of the impedance of nonlinear devices can be measured in thermal equilibrium. In addition, the environment of the device may sometimes make it desirable to avoid an applied signal. The measurement of hydrophone impedance mentioned previously illustrates this point. When an acoustic transducer is maintained in an enclosure, the reflected waves strongly influence the value of its impedance, so that a point-by-point measurement of this impedance by a bridge circuit must be carried out with very small frequency steps (depending upon the enclosure dimensions) and is very time consuming. A more rapid measurement is possible through the spectral analysis of thermal noise.

B. Measurement of Minority Carrier Lifetime

Most electronic applications of semiconductors are based upon creating deviations in carrier concentrations from their equilibrium values. These "excess" concentrations build up to a steady value in the presence of a steady excitation and decay towards zero when the excitation is removed. The time constant of this exponential decay is, therefore, a parameter of fundamental importance in semiconductor work and is called the minority carrier lifetime. It has been experimentally measured in a variety of ways, some of which are based upon the measurement of noise.

The minority carrier lifetime in semiconductors can be obtained through noise measurements because the current fluctuations arise due to the generation and recombination of carriers which take place at a rate depending upon the carrier lifetime. As a result, the spectrum of current fluctuations depends upon the lifetime τ. Hill and van Vliet [67] showed that if the generation–recombination through surface states is negligible and the number of recombination centers is small compared with the equilibrium concentrations of electrons and holes in a semiconductor sample, the noise current spectrum is given by

$$S_i(\omega) = \frac{K\tau}{(1 + \omega^2 \tau^2)} \quad \text{A}^2/\text{Hz}. \tag{12}$$

This equation has been experimentally verified by measuring the noise spectra for GE samples. The carrier lifetime is, there-

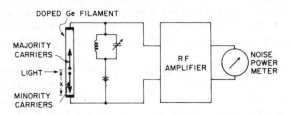

Fig. 3. Setup for measurement of minority carrier lifetime in semiconductors.

fore, directly found from the frequency at which the power spectral density drops to a half of its low-frequency value $K\tau$.

Okazaki and Oki [68] used still another method to measure the lifetime of minority carriers in germanium through the measurement of noise. A schematic diagram of their experimental arrangement is shown in Fig. 3. A source of light liberates hole–electron pairs at a spot on the surface of a germanium filament, and the RF noise in the current, collected at the ends of the filament, is measured over a narrow frequency band. The number of photogenerated minority carriers that recombine before reaching the ohmic contact depends upon the distance x between the illuminated spot and the contact collecting the minority carriers. If it is assumed that the noise power is proportional to the number of minority carriers which recombine in a mean free path, the mean-square value of noise current in a narrow bandwidth is given by

$$\overline{i^2}(x) = a \cdot n \left[1 - e^{-x/\mu E \tau} \right] \tag{13}$$

where a is a constant of proportionality, n is the number of injected minority carriers, μ and τ are their mobility and lifetime, and E is the electric field applied to the sample. The lifetime τ is then determined by measuring the variation of the noise power as a function of spot location x. Okazaki [69] has compared these lifetime measurements with those made using the Haynes–Shockley method to demonstrate the validity of lifetime calculations from noise measurements. Similar measurements have been carried out in silicon also [70].

C. Other Material and Carrier Transport Properties

Several other parameters have been estimated by measurements on noise. The structure and composition of ferromagnetic materials can be tested by the measurement of the spectrum of Barkhausen noise [71]. The capture cross section for the recombination of excess carriers in nickel-doped germanium has been determined by measuring the generation–recombination noise [72]. The measured amplitude distribution of the noise of a resistor has been used to calculate the time-of-flight associated with the Lorentz mean-free path [73]. The mobility, density, and lifetime of hot electrons in the minima of a many-valley semiconductor can be estimated from the noise current spectral density [74]. In all such cases, the property being measured determines (or depends upon the same parameters that determine) the characteristics of the generated noise.

D. Determination of Junction Nonuniformity

It is well known that the charge carriers moving at high speeds in a semiconductor under the influence of a large applied electric field can ionize lattice atoms by impact ionization, a phenomenon called avalanche breakdown. The ratio of the injected to the avalanche-generated current is called the multiplication factor M. The presence of small nonuniformities and inhomogeneities in a transverse plane (perpendicular to

the direction of current flow) in the semiconductor influences the local value of M so that the imperfections in bulk semiconductors or junctions can be detected through the resulting variation of M over the transverse plane. This spatial variation of the local value of M is very conveniently detected through noise measurement.

The avalanche-generated current is noisy because of the randomness of the avalanche process. The mean-square value of this noise current is proportional to the cube of the multiplication factor [75], making the noise current a very sensitive measure of M. In order to detect the variation of M in the transverse plane, it is necessary to create a localized spot of avalanche ionization in this plane and scan the entire cross section of the semiconductor by this spot. Local variations of M due to nonuniformities cause large variations in the mean-square noise current which is monitored. Such an experimental arrangement, utilizing a laser beam to inject primary photocurrent at a spot, has been used to measure the spatial variation of avalanche noise and hence detect nonuniformities in silicon avalanche diodes [76].

E. Measurement of Transistor Parameters

Several different transistor parameters can be obtained by means of noise measurements on transistors. An example of such a parameter is the effective base resistance r_b which is conveniently evaluated by noise measurements (particularly if the base region is inhomogeneous) in at least four different ways:

1) Chenette and van der Ziel's method requiring the measurement of equivalent input noise resistance of the transistor [77];
2) Plumb and Chenette's method involving the minimization of the open circuit emitter flicker noise voltage [78];
3) Gibbon's method requiring transistor noise figure measurement [79]; and
4) Hsu's method of plotting the collector short-circuit noise current against the square of dc collector current [80].

The last two of these methods are briefly described here.

Gibbons proposed the evaluation of r_b by measuring the transistor noise figure as a function of source resistance at low frequencies, where flicker noise is much larger than other types of noise. For large emitter current and small collector voltage, the low-frequency noise figure becomes a minimum when the source resistance is equal to base resistance, which is thus found. The ideality factor n for the emitter junction, which appears in the junction current–voltage characteristic

$$I_E = I_s \left[\exp\left(\frac{q_e V}{nkT}\right) - 1 \right] \tag{14}$$

can also be estimated by this method. For large collector voltage and small emitter current, the low-frequency noise figure becomes a minimum for a source resistance equal to $r_b + nkT/q_e I_E$. This condition yields n if the base resistance r_b is first found, as described previously [79].

In high-gain transistors, flicker noise is not the predominant source of noise, and the base resistance can be measured by a method recently reported by Hsu [80]. For small bias currents, shot noise dominates, and the collector output short-circuit noise current, expressed as equivalent saturated diode current I_{eq}, is proportional to the base resistance and the square of collector current. Therefore, the slope of I_{eq} versus I_c^2 plot gives the base resistance. This technique can be extended to measure the transconductance g_m and the small-

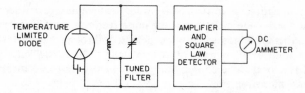

Fig. 4. Hull and Williams' apparatus for measurement of electronic charge.

signal common emitter current gain β as well. g_m can be calculated directly from base resistance and I_{eq}. For the measurement of β, an external resistance R_s is introduced in series with the base and adjusted until I_{eq} is maximum. This value of R_s, together with the base resistance and I_c, can be used to calculate β.

F. Measurement of Physical Constants

The measurement of fundamental constants like Planck's constant, the velocity of light, and the gravitational constant has occupied a central place in the physicists' interest and a sizable amount of the literature of physics. Noise measurements have been used to obtain the values of two basic constants—the electronic charge q_e and Boltzmann's constant k.

Schottky, who predicted the shot effect and derived the relationship called Schottky's theorem

$$\overline{i_n^2} = 2q_e IB \qquad (15)$$

was also the first person to suggest that this relationship could be used for the determination of the electronic charge. Such a measurement was carried out by Hull and Williams in 1925, using the shot noise in a temperature-limited vacuum diode [81], [82]. Fig. 4 shows the essential features of their apparatus. They reported the average electronic charge to be 4.76×10^{-10} ESU, with the measurements having a scatter of within 2 percent of this mean value (1 coulomb = 3×10^9 ESU). Later measurements of q_e from shot noise yielded a more precise value of 4.7972×10^{-10} ESU [83].

The measurement of Boltzmann's constant k by noise was first reported by Johnson [84], who was also the first person to measure the thermal noise of a resistor. His equipment consisted of an amplifier to amplify the thermal noise voltage of a resistor connected at its input, and a thermocouple ammeter at the output to measure the mean-square noise current. Johnson calculated Boltzmann's constant by using Nyquist's theorem

$$\overline{i_n^2} = 4kTB/R \qquad (16)$$

and found an average value of 1.27×10^{16} ergs/C, with a mean deviation of 13 percent. The accuracy of this measurement was later improved considerably by Ellis and Moullin [85].

G. Measurement of Temperature

As early as 1946, Lawson and Long [86], [87] proposed that the thermal noise of a resistor can be measured for an accurate estimation of very low temperatures. The most important advantage of noise thermometry lies in the fact that the thermal noise voltage generated by the resistor is independent of the composition of the resistor, previous thermal or mechanical treatments, the mass and nature of charge carriers, and the environment of the resistor other than temperature. In principle, the idea is very simple. A large resistor is kept in contact with the temperature to be measured and is con-

nected at the input of a low-noise amplifier. The noise voltage at the output of the amplifier is a direct measure of the temperature. The method is limited by the noise contribution of the amplifier, which can be reduced by using a piezoelectric quartz crystal in place of the resistor. The advantage of quartz lies in its high Q; all of the thermal energy in a particular mode is confined to a narrow band of frequencies, and the noise voltage measurement can be made over a very small bandwidth, thus improving the signal-to-noise ratio. Temperatures down to a fraction of a kelvin can be measured in this way [88].

A noise thermometer can provide an absolute temperature standard through the use of Nyquist's theorem stated in (7), and, therefore, does not require any calibration. In practice, however, the temperature measurement is usually carried out by comparing the noise voltage of two known resistors, one at the unknown temperature and the other at a known (e.g., room) temperature [89]. Several modifications of this basic principle of noise thermometry have been employed. In one system used for the measurement of high temperatures (275 K–1275 K) in nuclear reactors, the need for a simultaneous measurement of the value of the resistor was eliminated by measuring noise power rather than noise voltage, thus improving the accuracy of measurement [90]. In another system intended for low temperatures (0.01 K–0.3 K), a superconducting quantum interference device (SQUID) was used to keep the noise contribution of the measurement system low, of the order of a millikelvin [91].

An improvement of the preceding technique has become possible with the development of superconducting weak links (Josephson junctions) which can be used for measuring temperatures of the order of (and possibly below) a millikelvin. It is based upon the fact that the measurement of a frequency can be carried out to a higher accuracy than the measurement of a voltage, so that the voltage fluctuations, due to thermal noise in a resistor, can be measured more precisely if they are first converted into frequency fluctuations. One of the important characteristics of a Josephson junction is that if a voltage V is applied across it, the resulting current alternates at a frequency

$$f = \frac{2q_e V}{h} \qquad (17)$$

where h is Planck's constant, i.e., at approximately 484 GHz per mV. Random fluctuations in V will cause corresponding fluctuations in f, so that the current spectrum will have a finite line width.

In a noise thermometer constructed by Kamper and Zimmerman [92], a small resistor (of the order of 10^{-5} Ω) is biased by a 10-mA constant current source and the voltage across it is applied to a Josephson junction, causing it to oscillate at a frequency of around 40 MHz. The thermal noise voltage across the resistor frequency modulates the oscillations, generating sidebands close to the center frequency (within 1 kHz of it). In order that the thermal noise of resistor be predominant, the resistance value is chosen to be small compared with the resistances of the junction and the current supply. The spectral line width due to thermal noise broadening is given by

$$\Delta f = \frac{16\pi q_e^2 kRT}{h^2} \qquad (18)$$

and is a direct measure of the temperature T. It can be calculated from the variance σ^2 of the number of cycles in a fixed gate time τ_g of a frequency counter, by using the relationship

$$\Delta f = 2\pi \tau_g \sigma^2. \qquad (19)$$

The temperature is thus determined from a digital measurement of frequency.

H. Evaluation of Cathodes in Electron Tubes

The quality of the cathodes of electron tubes is conventionally evaluated in a manufacturing process through the measurement of tube parameters like the transconductance g_m. Experimental studies have shown that the shot noise of an electron tube is a more sensitive indicator of cathode life and activation than its transconductance [93]. In addition, the noise testing has no detrimental effect on the tube. Noise measurements have, therefore, been used for monitoring the aging and for optimizing the activation time of the cathode during the manufacture of electron tubes [94].

I. Prediction of Device Reliability

The reliability of electronic devices has conventionally been measured and specified in terms of statistical measures, like mean time between failures. There is experimental evidence to indicate that noise measurements may be useful as a technique for reliability testing. This technique has three distinct advantages over the conventional lifetime tests—it is nondestructive and does not use up a considerable fraction of the life of the device tested, the lifetime of a specific individual device can be measured rather than an average lifetime for a lot, and measurements do not require a long time. There are several different ways in which the measurement of the noise in a device can be employed to get information concerning the lifetime of the device.

First, the noise measurements can be used for identifying failure-prone devices because the manufacturing defects and the potential instability mechanisms become apparent through their influence on device noise. For example, it has been found that transistors with low $1/f$ noise exhibit longer life spans [95], reverse-biased p-n junction diodes having a noise power spectrum with multiple peaks undergo a more rapid degradation than those with a single-peak spectrum [96], and thin metal films having constrictions in their cross section due to scratches, notches, or pits have higher noise index than uniform films at large current densities [97].

Second, a continuous monitoring of noise can also be used to predict the impending failure of a device. For instance, it has been shown experimentally that the low-frequency $1/f$ noise output of a transistor increases by two or three orders of magnitude shortly before its failure [95]. Lifetime tests have been carried out on large batches of transistors also [98].

Taratuta [99] suggested a third and novel method for predicting lifetime and calculating failure probability of semiconductor devices, based upon the measurement of noise power spectral density of the device. The argument for relating noise spectrum to reliability goes as follows. A major cause of the failure of semiconductor devices is heating due to the random transient processes, which momentarily change the operating characteristics of the device. These transients are, therefore, accompanied by carrier density fluctuations which in turn give rise to noise. The external current through the device can be thought of as containing random pulses of different durations and magnitudes, but only the larger of these pulses are potentially destructive. By making several rather severe assumptions concerning the nature of pulses, it is possible to deduce their distribution from the measured noise spectrum for the device, and thereby to calculate the probability of the appearance of a destructive pulse in the device. Whether this method is practical is open to question; it does require making major assumptions concerning pulse shapes, rate of change of pulse duration, etc., and the calculated failure rate might turn out to be a sensitive function of these assumptions rather than the measured noise power spectrum.

J. Study of Ion Transportation Through Nerve Membranes

As in other systems, the measurement of noise generated in biological systems provides useful information about the physical mechanism involved. For example, the voltage fluctuations across the membrane of a neuron are related to the transport of ions through this permeable membrane and yield information about ion flux [100], [101]. In the resting state, the concentration of sodium, potassium, and chlorine ions inside the cell is different from that in the outside interstitial fluid, generating a potential difference across the cell membrane. An external voltage applied across the membrane causes a partial depolarization and an influx of sodium ions, which disturbance propagates along the nerve fiber. The study of membrane noise is significant not only because it influences the transmission of information within the nervous system but also because it provides a tool for studying the membrane processes on a molecular level.

The measurements of noise voltage across the membrane have shown the presence of 1) shot noise due to the motion of single ions within the membrane, 2) low-frequency $1/f$ noise related to potassium ion transport, 3) high-frequency excess noise due to fluctuations in membrane conductance, 4) burst noise, and 5) relaxation noise [100]. Each source of noise, being the result of different physical mechanisms, is a potential source of information about the various membrane properties. In addition, the membrane impedance may also be obtained from the measurement of thermal noise spectrum [101]. Such noise measurements are important because they provide a means of verifying the validity of the microscopic models of membrane processes. The attempts at interpretation and prediction of the observed noise are helping to improve the model and understanding of the ion transport through nerve membranes.

VI. Uses of Noise as a Conceptual Tool

A. Modeling and Analysis of Stochastic Systems

A number of physical systems such as communication, power, and transportation systems, having nondeterministic features, can be usefully modeled as stochastic systems. Much of the existing body of mathematical formulations and results, developed for dealing with noise in electronic circuits, devices, and systems, is then applicable to new fields. The study of random vibrations in mechanical systems is one such field that has benefited much from analogies to electrical noise and where the borrowed results have been explicitly acknowledged [102]. Another discipline is the investigation of random fluctuations in some parameters of nuclear power reactors like reactor power, temperature, and coolant flow, for the purpose of studying reactor kinetics [42] or monitoring malfunctions [103], [104].

Beyond the mere transfer of mathematical techniques, it is possible to use the concepts and principles developed with electrical noise as guides in working with other physical sys-

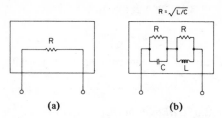

Fig. 5. Slepian–Goldner problem of black-box identification.

tems and for gaining an alternative and perhaps more intuitive viewpoint in other fields. Thermodynamics is such a field where the issues and processes are exemplified by, or lend themselves to an interpretation in terms of, electrical noise. Thus Onsager's relations for irreversible processes can be understood through noise in reciprocal two-port networks [105]. Such treatments have primarily a pedagogical value because the thermodynamic principles were already known and understood before they were interpreted in terms of electrical noise.

A rare but outstanding example of a case where the study of electrical noise preceded, motivated, and guided the development of a thermodynamic principle is the work of Callen and Welton [106] on linear dissipative systems. They observed that Nyquist's theorem (7) on thermal noise was unique in physics because it relates a property of a system in equilibrium (the voltage fluctuations) to a parameter (the electrical resistance) characterizing an irreversible process. This led them to extend the theorem and obtain a relationship between the fluctuations of the generalized forces in a linear dissipative system and the generalized resistance defined for such a system. This general formulation is capable of unifying a number of known results (on the Brownian motion, the Planck's radiation law, and the pressure fluctuations) and predicting new relationships.

B. Noise in Circuit Theory

One of the basic problems in the theory of electrical conduction has been to interpret impedance, which is a macroscopically observable variable, in terms of the motion of charge carriers, i.e., at microscopic level. This relationship between impedance and charge carriers can be described with the help of current fluctuations. For example, the self-inductance of an aggregate of electrons in a metallic conductor can be determined by calculating the current fluctuations [107]. More generally, it has been shown that the real part of the admittance of any linear dissipative system, as calculated from the response of the system to a generalized driving force [108], is identical to the generalized conductance obtained by Callen and Welton from fluctuation-dissipation considerations [106]. Some general theorems in circuit theory have resulted from such works. Thus Richardson [109] has shown that, for a system in equilibrium, a single impedance operator describes its transient response to an applied perturbation as well as its spontaneous transient behaviour starting from a given initial condition in the absence of external perturbation.

Thermal noise is a useful tool in theoretical investigations of circuit elements and models because 1) thermal noise is present in all dissipative systems, and 2) the available noise power depends only upon the temperature of the system. When used in conjunction with the laws of thermodynamics, noise considerations can be used to examine the validity of the circuit model of a physical system or device. Two examples of this application of noise to the verification of circuit models are described here.

The following problem in circuit theory was proposed by Penfield [110] a few years ago as a paradox. He argued that the torque generated by a universal motor (a dc motor with its field coil connected in series with its armature) is proportional to the square of the field current (which is also the armature current). Therefore, the thermal noise current of a resistor at the same temperature as the motor will generate a finite torque and deliver energy in violation of the second law of thermodynamics. This immediately suggests that the described model of the motor is oversimplified and, therefore, invalid for certain operating conditions. The resolution of the paradox indeed resulted in the development of several improved circuit models which do not lead to perpetual motion and hence have a wider range of applicability.

Similar arguments, applied to nonlinear circuit elements, suggest that there exists a relationship between noise and nonlinearity of a system because a diode cannot be used to rectify its own noise and thereby convert random noise into dc current, in violation of thermodynamic laws. Gunn [111] showed on such thermodynamic grounds that a diode with a large nonlinearity must also develop a large amount of electrical noise which imposes restrictions on the type of rectification mechanism involved. This work gave the unexpected result, which was experimentally verified by Gunn, that the equivalent circuit of a diode connected across a resistor at a lower temperature should include a constant current source, which depends upon both the idealized diode model and the load, and drives a current through the diode in the reverse (high resistance) direction.

Penfield has demonstrated the use of thermal noise concepts for proving results in network theory, and in that process has developed an analogy between frequency converters and heat engines [112]. He argues that a sinusoidal signal represents high-grade energy while the thermal noise in a resistor is low-grade energy, and the two are, therefore, analogous to work and heat, respectively, in thermodynamics. Therefore, a sinusoidally pumped three-frequency upconverter, working between two thermal noise sources (resistors) at different frequencies, is equivalent to a heat engine (or refrigerator) operating between two temperatures. Furthermore, the conversion efficiency of the lossless (reactive) upconverter, as predicted by Manley–Rowe equations, is the same as the Carnot efficiency for reversible heat engines, resulting from the second law of thermodynamics. This leads to the result that a sinusoidally pumped three-frequency upconverter obeys Manley–Rowe equations if and only if it is reversible in thermodynamic sense. This result may be generalizable to other physical systems.

Goldner [113], [114] proposed an interesting problem in circuit theory which he solved in a thought experiment using thermal noise. The problem is this: given two black boxes shown in Fig. 5, with ideal R, L, and C elements, how can the boxes be distinguished from each other by terminal measurements alone? The impedance of the boxes being identical, no transient or steady-state experiment would solve the problem. However, the two networks can be made to have different noise properties as follows. If a battery is connected across the terminals of each of the black boxes, the current in box (b) would heat only the resistor in parallel with the capacitor. Starting from uniform temperature for all elements, the two resistors in box (b) will attain different final temperatures. The noise power spectral density at the output terminals of each of the boxes can now be measured. Analysis shows that box (a) will have a larger noise power spectral density so that the two networks can be distinguished.

C. Noise and Quantum Mechanics

The study of noise in quantum electronic devices (in particular, masers and lasers) has led to the development of some analogies between quantum mechanics and the analyses of noisy circuits and systems. Heffner [115] initiated the interest in this direction when he proved the theorem that it is impossible to construct a noiseless linear (phase-preserving) amplifier. The theorem follows directly from the quantum mechanical uncertainty principle, which states that in measuring the two canonically conjugate variables E, the energy of the system, and t, the precise time at which the system possesses this energy, the uncertainties in the measured values (rms deviations from the mean in an ensemble of measurements) are related by

$$\Delta E \, \Delta t \geqq \hbar/2. \tag{20}$$

This relationship can be transformed, by substituting $E = nh\nu$ and $\phi = 2\pi\nu t$ to the form

$$\Delta n \, \Delta \phi \geqq 1/2 \tag{21}$$

where Δn is the uncertainty in the number of quanta in an oscillation, and $\Delta \phi$ is the uncertainty in its phase. If a noiseless linear amplifier precedes the instrument used to measure n and ϕ of a signal, it will increase n (the strength of the signal) without affecting the measurement of ϕ. Thus Δn can be reduced arbitrarily without influencing $\Delta \phi$, in contradiction to the uncertainty principle. This leads to the theorem that a noiseless linear amplifier cannot be constructed. Note that the connection between the uncertainty principle and the theorem is direct and does not require any models, constructs, or frameworks. In fact, the impossibility of a noiseless linear amplifier may be taken as the fundamental principle, and the uncertainty principle be treated as a manifestation of noise.

More recently, Haus [116] demonstrated that an analogy exists between the correlation functions of thermal noise and the commutator brackets of the quantum-mechanical operator amplitudes for a conservative system, because their space-time dependences are identical. This analogy allows one to obtain the commutators for a system by applying to it the fluctuation dissipation theorem (Nyquist's theorem) and solving an analogous thermal noise problem. The importance of this technique lies in unifying the classical and the quantum-mechanical methods of analyzing linear distributed systems in steady state. In the classical method, the steady-state analysis is carried out by assuming a sinusoidal excitation in the system, and the input and output signals can then be viewed as the boundary conditions imposed upon the equations describing the excitation. By contrast, the quantum-mechanical analysis is carried out using the equations of motion, which describe the evolution of the complete universe including the system, starting from an initial excitation specified by imposing the initial conditions. When the system to be analyzed quantum mechanically is an incomplete part of the universe (for example, a system of finite size, exchanging energy with the rest of the universe at its input and output ports, as in the classical method), it becomes necessary to find the commutators for the excitation amplitudes that must be imposed at the boundaries (input and output ports) in order to maintain the proper space-time dependence of commutator brackets. If no excitation was imposed at the boundaries, the initial excitation of the system would decay to a zero value with time. This situation would be similar to that of a noisy system terminated in a noiseless universe (such as loads maintained at absolute zero temperature), whereupon its energy would vanish in time, by leaking out through the ports as noise waves. The Nyquist theorem yields the noise sources that must be connected to the ports of the system to maintain it in steady state, and by analogy, also gives the commutators for excitation amplitudes at system boundaries.

VII. Epilogue

In an article of this nature, it would be difficult to list every proposed or possible application of noise exhaustively, let alone describe each in detail. The applications selected for discussion here have been taken from a variety of areas to emphasize the breadth of possibilities, including some that may never have been actually exploited (for example, Taratuta's method of predicting the reliability of semiconductor devices) and some that do not represent the current practice and state of the art (for example, the measurement of Boltzmann's constant).

The title "applications of electrical noise" has been broadly interpreted here, as is evident from the variety of ways in which noise enters into the different applications discussed. In particular, some of the "applications of noise to x" can be viewed alternatively as "applications of x to noise." For example, the measurement of electronic charge from Schottky's theorem may be thought of as the verification of the theorem (and of the assumption of independent emissions of electrons); such duality of purpose is typical of physical research [117]. Recall that when the nature of X-rays was in question, the use of the diffraction patterns from crystalline lattice was suggested for the estimation of the wavelength of X-rays. The X-rays are now commonly used to determine the structure of crystalline solids.

Discussions of the individual applications have been necessarily short, although a large number of references to the original literature is included for locating the details of specific applications. No attempt has been made here to quote every author describing a given application of noise. Where the literature on an application is extensive, reference is made to a review of the field (as, for example, in the case of membrane noise) or to a few of the earliest and the most recent papers (as in noise thermometry) to indicate the origin as well as the current status of the field. The cited references show that some of the applications of noise are as old as the discovery of electrical fluctuations and many have been known for a long time. In fact, some of the applications are only of historical interest now. At the same time, there are many newer applications attesting to a continued interest in the field.

Acknowledgment

The author would like to thank Prof. R. B. Adler, Prof. H. A. Haus, Prof. R. L. Kyhl, and Prof. P. L. Penfield, Jr., of Massachusetts Institute of Technology, for their helpful comments on this paper, and the anonymous reviewer for pointing out two references.

References

[1] J. B. Gunn, "The Gunn effect," *Int. Sci. Technol.*, no. 46, pp. 43–56, Oct. 1965.

[2] H. L. McDowell and G. K. Farney, "Crossed-Field noise generation devices," in *Microwave Power Engineering*, vol. 1, E. C. Okress, Ed. New York: Academic Press, 1968.

[3] R. G. Hundley, "Simulating tactical radar and sonar," *Electronics*, vol. 36, pp. 25–31, Dec. 13, 1963.

[4] B. E. Williams and R. D. Wilmot, "Analysis of a simplified radar simulator for measuring probability of detection," *IEEE Trans. Instrum. Meas.*, vol. IM-18, pp. 58–62, Mar. 1969.

[5] H. Blatter, "Visual detection of chart-recorded signals in the presence of noise," *IEEE Trans. Geosci. Electron.*, vol. GE-5, pp. 89–93, Dec. 1967.

[6] H. Fletcher, *Speech and Hearing in Communication*. New York: Van Nostrand, 1953.

[7] V. H. Hellmann, "Noise acceleration—a new test method for the assessment of microphony of electron tubes," (in German), *Frequenz*, vol. 13, pp. 83–89, Mar. 1959.

[8] *Proc. IEEE*, (Special issue on infrared technology for remote sensing) vol. 63, Jan. 1975.

[9] *Proc. IEEE*, (Special issue on radio and radar astronomy), vol. 61, Sept. 1973.

[10] F. B. Tuteur, "On the detection of transiting broadband targets in noise of uncertain level," *IEEE Trans. Commun. Technol.*, vol. COM-15, pp. 61–69, Feb. 1967.

[11] ——, "On the detection of a moving noise source in a nonstationary noise background," *J. Acoust. Soc. Amer.*, vol. 44, no. 4, pp. 912–918, 1968.

[12] G. J. Cybula, R. W. Harris, and T. J. Ledwidge, "Location of a boiling noise source by noise analysis techniques," *Proc. Inst. Radio Electron. Eng.* (Australia), vol. 35, pp. 310–316, Oct. 1974.

[13] D. Schwalm, "Some remarks on failure detection in nuclear power reactors by noise measurements," *Atomkernenergie*, vol. 19, pp. 263–270, July 1972.

[14] E. L. Gruenberg, H. P. Raabe, and C. T. Tsitsera, "Self-Directional microwave communication system," *IBM J. Res. & Develop.*, vol. 28, pp. 149–163, Mar. 1974.

[15] R. J. Schlesinger, *Principles of Electronic Warfare*. Englewood Cliffs, N.J.: Prentice-Hall, 1961.

[16] H. Magnuski, "Jamming of communication systems using FM, AM, and SSB modulation," *IRE Trans. Mil. Electron.*, vol. MIL-5, pp. 8–11, Jan. 1961.

[17] G. A. Korn, *Random Process Simulation and Measurement*. New York: McGraw-Hill, 1966.

[18] V. C. Rideout, "Random process studies," in *Computer Handbook*, Sec. 5.9, H. D. Huskey and G. A. Korn, Eds. New York: McGraw-Hill, 1962.

[19] B. Mazelsky and H. B. Amey, Jr., "On the simulation of random excitation for airplane response investigations on analog computers," *J. Aeronaut. Sci.*, vol. 24, pp. 633–649, Sept. 1957.

[20] H. F. Murry, "A general approach for generating natural random variables," *IEEE Trans. Comput.*, vol. C-19, pp. 1210–1213, Dec. 1970.

[21] The Rand Corporation, *A Million Random Digits with 100 000 Normal Deviates*. New York: Free Press, 1955.

[22] S. T. Ribeiro, "Random-pulse machines," *IEEE Trans. Electron. Comput.*, vol. EC-16, pp. 261–276, June 1967.

[23] R. B. Gaines, "Stochastic computer thrives on noise," *Electronics*, vol. 40, pp. 72–79, July 10, 1967.

[24] A. S. French and R. B. Stein, "A flexible neural analog using integrated circuits," *IEEE Trans. Bio-Med. Eng.*, vol. BME-17, pp. 248–253, July 1970.

[25] K. D. Kryter, *The Effects of Noise on Man*. New York: Academic Press, 1970, pp. 526–527.

[26] S. Schuy, "Electronic noise and its use in electro-medicine," in German, *Elektrotech. Maschinenbau*, vol. 84, pp. 449–453, Nov. 1967.

[27] A. van der Ziel, *Noise*. Englewood Cliffs, N.J.: Prentice-Hall, 1954.

[28] D. A. Bell, *Electrical Noise: Fundamentals and Physical Mechanisms*. Princeton, N.J.: Van Nostrand, 1960.

[29] T. Mukaihata, "A survey of techniques in microwave noise measurement," *Instrum. Soc. Amer. Trans.*, vol. 3, pp. 342–352, Oct. 1964.

[30] C. K. S. Miller, W. C. Daywitt, and M. G. Arthur, "Noise standards, measurements, and receiver noise definitions," *Proc. IEEE*, vol. 55, pp. 865–877, June 1967.

[31] B. G. Bosch and W. A. Gambling, "Techniques of microwave noise measurement," *J. Brit. Inst. Radio Eng.*, vol. 21, pp. 503–515, June 1961.

[32] C. R. A. Ammerman, "Direct measurement of bandwidth," *Elec. Eng.*, vol. 69, pp. 207–212, Mar. 1950.

[33] J. W. M. Baars, "The measurement of large antennas with cosmic radio sources," *IEEE Trans. Antennas Propagat.*, vol. AP-21, pp. 461–474, July 1973.

[34] A. D. Kuz'min and A. E. Salomonovich, *Radioastronomical Methods of Antenna Measurement*. New York: Academic Press, 1966.

[35] B. L. Seidel and C. T. Stelzreid, "A radiometric method for measuring the insertion loss of radome materials," *IEEE Trans. Microwave Theory Tech.*, vol. MTT-16, pp. 625–628, Sept. 1968.

[36] C. T. Stelzreid and T. Y. Otoshi, "Radiometric evaluation of antenna-feed component losses," *IEEE Trans. Instrum. Meas.*, vol. IM-18, pp. 172–183, Sept. 1969.

[37] C. T. Stelzreid and M. S. Reid, "Precision power measurements of spacecraft CW signal level with microwave noise standards," *IEEE Trans. Instrum. Meas.*, vol. IM-15, pp. 318–324, Dec. 1966.

[38] Y. W. Lee, *Statistical Theory of Communication*. New York: Wiley, 1960, ch. 13.

[39] J. S. Thorp and M. L. Mintz, "A method of system identification with random inputs," *Proc. IEEE*, vol. 54, pp. 290–291, Feb. 1966.

[40] B. Chatterjee and A. B. Bhattacharyya, "Measurement of an impulse response of a system with a random input," *IEEE Trans. Automat. Contr.*, vol. AC-8, pp. 186–187, Apr. 1963.

[41] M. Ichise, Y. Nagayanagi, and T. Kojima, "Application of pseudo-random signals and cross-correlation techniques in electroanalytical chemistry," *J. Electroanal. Chem. Interfacial Electrochem.*, vol. 49, pp. 187–198, Jan. 10, 1974.

[42] R. E. Uhrig, *Random Noise Techniques in Nuclear Reactor Systems*. New York: Ronald Press, 1970.

[43] G. W. Anderson, J. A. Aseltine, A. R. Mancini, and C. W. Sarture, "A self-adjusting system for optimum dynamic performance," *IRE Nat. Convention Rec.*, pt. 4, pp. 182–190, 1958.

[44] N. Wiener, *Nonlinear Problems in Random Theory*. New York: Technology Press and Wiley, 1958.

[45] Y. W. Lee and M. Schetzen, "Measurement of the Wiener kernels of a nonlinear system by cross-correlation," *Int. J. Contr.*, vol. 2, pp. 237–254, Sept. 1965.

[46] P. Z. Marmarelis and K. I. Naka, "White-Noise analysis of a neuron chain: an application of the Wiener theory," *Science*, vol. 175, pp. 1276–1278, Mar. 17, 1972.

[47] J. F. Barrett and J. F. Coales, "An introduction to the analysis of nonlinear control systems with random inputs," *Proc. Inst. Elec. Eng.*, pt. C, vol. 103, pp. 190–199, 1956.

[48] A. A. Pervozvanskii, *Random Processes in Nonlinear Control Systems*. New York: Academic Press, 1965.

[49] J. C. West, *Analytical Techniques for Non-Linear Control Systems*. Princeton, N. J.: Van Nostrand, 1960.

[50] A. J. Seyler and P. R. Wallace, "Calibration of an instrument for statistical experiments using Gaussian noise as test signal," *Elec. Eng. (Melbourne)*, vol. EE-3, pp. 46–54, Mar. 1967.

[51] E. Peterson, "Gas-Tube noise generator for circuit testing," *Bell Lab. Rec.*, vol. 18, pp. 81–84, Nov. 1939.

[52] P. J. Icenbice and H. E. Fellhaur, "Linearity testing techniques for sideband equipment," *Proc. IRE*, vol. 44, pp. 1775–1782, Dec. 1956.

[53] W. Oliver, "White noise loading of multi-channel communication systems," *Electron. Eng.*, vol. 37, pp. 714–717, Nov. 1965.

[54] Bell Telephone Laboratories Technical Staff, *Transmission Systems for Communications*, Bell Telephone Laboratories, 1970.

[55] S. V. Korndorf *et al.*, "Electrical-Fluctuations resonant method for measuring inductance," *Meas. Tech.*, vol. 16, pp. 1588–1589, Oct. 1973.

[56] R. Bourret, "A proposed technique for the improvement of range determination with noise radar," *Proc. IRE*, vol. 45, pp. 1774, Dec. 1957. (Discussion: G. L. Turin, "A proposed technique for the improvement of range determination with noise radar," *Proc. IRE*, vol. 46, pp. 1757–1758, Oct. 1958.)

[57] B. M. Horton, "Noise-Modulated distance measuring systems," *Proc. IRE*, vol. 47, pp. 821–828, May 1959.

[58] M. P. Grant, G. R. Cooper, and A. K. Kamal, "A class of noise radar systems," *Proc. IEEE*, vol. 51, pp. 1060–1061, July 1963.

[59] H. E. Band, "Noise-Modulated optical radar," *Proc. IEEE*, vol. 52, pp. 306–307, Mar. 1964.

[60] C. P. Jethwa, M. Kaveh, G. R. Cooper, and F. Saggio, "Blood flow measurements using ultrasonic pulsed random signal Doppler system," *IEEE Trans. Sonics Ultrason.*, vol. SU-22, pp. 1–11, Jan. 1975.

[61] N. Chasek, "Avalanche diodes permit in-service measurements of critical parameters in microwave equipment," *Electronics*, vol. 43, pp. 87–91, Jan. 19, 1970.

[62] M. R. Schroeder, "Frequency-Correlation functions of frequency responses in rooms," *J. Acoust. Soc. Amer.*, vol. 34, pp. 1819–1823, Dec. 1962.

[63] L. L. Beranek, *Acoustic Measurements*. New York: Wiley, 1949.

[64] E. Villchur, "A method of testing loudspeakers with random noise input," *J. Audio Eng. Soc.*, vol. 10, pp. 306–309, Oct. 1962.

[65] F. Micheron and L. Godefroy, "Automatic impedance measurements using thermal noise analysis," *Rev. Sci. Instr.*, vol. 43, pp. 1460–1465, Oct. 1972.

[66] K. V. Goncharov, "On the possibility of investigating the frequency sensitivity characteristics of transducers by a spectral

analysis of their thermal noise," *Soviet Phys.—Acoust.*, vol. 5, pp. 120–122, Jan.–Mar. 1959.

[67] J. E. Hill and K. M. van Vliet, "Generation-Recombination noise in intrinsic and near-intrinsic germanium crystals," *J. Appl. Phys.*, vol. 29, pp. 177–182, Feb. 1958.

[68] S. Okazaki and H. Oki, "Measurement of lifetime in Ge from noise," *Phys. Rev.*, vol. 118, pp. 1023–1024, May 15, 1960.

[69] S. Okazaki, "Confirmation of lifetimes by noise and by Haynes–Shockley method," *J. Appl. Phys.*, vol. 32, pp. 712–713, Apr. 1961.

[70] S. Okazaki and M. Hiramatsu, "Spectrum of fluctuations and lifepath in silicon," *Solid-State Electron.*, vol. 8, pp. 401–407, Apr. 1965.

[71] W. Kurzmann and W. Willmann, "Non-destructive testing of structure and composition based on the measurement of Barkhausen noise voltage," (in German), *Technik*, vol. 25, pp. 264–268, Apr. 1970.

[72] F. M. Klaassen, J. Blok, and H. C. Booy, "Generation-Recombination noise of Ni-doped Ge in the temperature range 350–100°K and capture crosssection of Ni in Ge," *Physica (Utreĉht)*, vol. 27, pp. 48–66, Jan. 1961.

[73] B. R. Gossick, "Distribution in energy of Johnson noise pulses," *J. Appl. Phys.*, vol. 21, pp. 847–850, Sept. 1950.

[74] N. G. Ustinov, "Determination of the intravalley mobility of hot electrons from fluctuations of the current," *Sov. Phys.—Semicond.*, vol. 8, pp. 1016–1017, Feb. 1975.

[75] A. S. Tager, "Current fluctuations in a semiconductor (dielectric) under the conditions of impact ionization and avalanche breakdown," *Sov. Phys.—Solid State*, vol. 6, pp. 1919–1925, Feb. 1965.

[76] I. M. Naqvi, C. A. Lee, and G. C. Dalman, "Measurements of multiplication effects on noise in silicon avalanche diodes," *Proc. IEEE*, vol. 56, pp. 2051–2052, Nov. 1968.

[77] E. R. Chenette and A. van der Ziel, "Accurate noise measurements on transistors," *IRE Trans. Electron Devices*, vol. ED-9, pp. 123–128, Mar. 1962.

[78] J. L. Plumb and E. R. Chenette, "Flicker noise in transistors," *IRE Trans. Electron Devices*, vol. ED-10, pp. 304–308, Sept. 1963.

[79] J. F. Gibbons, "Low-Frequency noise figure and its application to the measurement of certain transistor parameters," *IRE Trans. Electron Devices*, vol. ED-9, pp. 308–315, May 1962.

[80] S. T. Hsu, "Noise in high-gain transistors and its application to the measurement of certain transistor parameters," *IEEE Trans. Electron Devices*, vol. ED-18, pp. 425–431, July 1971.

[81] A. W. Hull and N. H. Williams, "Determination of elementary charge from measurement of shot effect," *Phys. Rev.*, 2nd series, vol. 25, pp. 147–173, Feb. 1925.

[82] N. H. Williams and W. S. Huxford, "Determination of the charge of positive thermions from measurements of shot effect," *Phys. Rev.*, 2nd ser., vol. 33, pp. 773–788, May 1929.

[83] L. Stigmark, "A precise determination of the charge of the electron from shot noise," *Ark. Fys.*, vol. 5, pp. 399–426, 1952.

[84] J. B. Johnson, "Thermal agitation of electricity in conductors," *Phys. Rev.*, vol. 32, pp. 97–109, July 1928.

[85] H. D. Ellis and E. B. Moullin, "A measurement of Boltzmann's constant by means of the fluctuations of electron pressure in a conductor," *Proc. Cambridge Phil. Soc.*, vol. 28, pp. 386–402, July 1932.

[86] A. W. Lawson and E. A. Long, "On the possible use of Brownian motion for low temperature thermometry," *Phys. Rev.*, 2nd Series, vol. 70, pp. 220–221, Aug. 1–15, 1946.

[87] J. B. Garrison and A. W. Lawson, "An absolute noise thermometer for high temperatures and high pressures," *Rev. Sci. Instr.*, vol. 20, pp. 785–794, Nov. 1949.

[88] P. M. Endt, "The use of spontaneous voltage fluctuations for the measurement of low temperatures," *Physica (Utrecht)*, vol. 16, pp. 481–485, May 1950.

[89] H. G. Brixy, "Temperature measurement in nuclear reactors by noise thermometry," *Nucl. Instrum. Meth.*, vol. 97, pp. 75–80, Nov. 15, 1971.

[90] C. J. Borkowski and T. V. Blalock, "A new method of Johnson noise thermometry," *Rev. Sci. Instr.*, vol. 45, pp. 151–162, Feb. 1974.

[91] R. P. Gifford, R. A. Webb, and J. C. Wheatley, "Principles and methods of low-frequency electric and magnetic measurements using an RF-biased point-contact superconducting device," *J. Low Temp. Phys.*, vol. 6, pp. 533–610, Mar. 1972.

[92] R. A. Kamper and J. E. Zimmerman, "Noise thermometry with the Josephson effect," *J. Appl. Phys.*, vol. 42, pp. 132–136, Jan. 1971.

[93] W. Dahlke and F. Dlouhy, "A cathode test utilizing noise measurements," *Proc. IRE*, vol. 46, pp. 1639–1645, Sept. 1958.

[94] L. G. Sebestyen, "The evaluation of oxide-cathode quality by shot-noise tests," *J. Brit. Inst. Radio Eng.*, vol. 21, pp. 463–467. May 1961.

[95] A. van der Ziel and H. Tong, "Low-Frequency noise predicts when a transistor will fail," *Electronics*, vol. 28, pp. 95–97, Nov. 28, 1966.

[96] Y. D. Kim and R. P. Misra, "Noise spectral density as a diagnostic tool for reliability of p-n junctions," *IEEE Trans. Rel.*, vol. R-18, pp. 197–200, Nov. 1969.

[97] J. L. Vossen, "Screening of metal film defects by current noise measurements," *Appl. Phys. Lett.*, vol. 23, pp. 287–289, Sept. 15, 1973.

[98] V. S. Pryanikov, "Prediction of lifetime of transistors from their internal noise," (in Russian) *Izv. Vyssh. Ucheb. Zaved. Radioelektron.*, vol. 14, no. 7, pp. 819–821, 1971.

[99] A. S. Taratuta, "Method of prediction of the lifetime of semiconductor devices," *Radio Eng. Electron. Phys. (USSR)*, vol. 10, pp. 1933–1935, Dec. 1965.

[100] A. A. Verveen and L. J. DeFelice, "Membrane noise," in *Progress in Biophysics and Molecular Biology*, vol. 28, A. J. V. Butler and D. Noble, Eds. Oxford, England: Pergamon 1974, pp. 189–265.

[101] C. F. Stevens, "Inferences about membrane properties from electrical noise measurements," *Biophys. J.*, vol. 12, pp. 1028–1047, Aug. 1972.

[102] S. H. Crandall and W. D. Mark, *Random Vibrations in Mechanical Systems.* New York: Academic Press, 1963.

[103] J. A. Thie, *Reactor Noise.* New York: Rowman and Littlefield, 1963.

[104] ——, "Reactor-Noise monitoring for malfunctions," *Reactor Technol.*, vol. 14, pp. 354–365, winter 1971.

[105] H. B. G. Casimir, "Reciprocity theorems and irreversible processes," *Proc. IEEE*, vol. 51, pp. 1570–1573, Nov. 1963.

[106] H. B. Callen and T. A. Welton, "Irreversibility and generalized noise," *Phys. Rev.*, 2nd ser., vol. 83, pp. 34–40, July 1, 1951.

[107] L. Brillouin, "Fluctuations in a conductor," *Helv. Phys. Acta* (in French), vol. 1, suppl. 2, p. 47, 1934.

[108] J. L. Jackson, "A note on 'Irreversibility and generalized noise,'" *Phys. Rev.*, vol. 87, pp. 471–472, Aug. 1, 1952.

[109] J. M. Richardson, "Noise in driven systems," *IRE Trans. Inform. Theory*, vol. IT-1, pp. 62–65, Mar. 1955.

[110] P. Penfield, Jr., "Unresolved paradox in circuit theory," *Proc. IEEE*, vol. 54, pp. 1200–1201, Sept. 1966. (Discussions: vol. 55, pp. 474–477, Mar. 1967; vol. 55, pp. 2047–2048, Nov. 1967; vol. 55, pp. 2173–2174, Dec. 1967; vol. 56, p. 328, Mar. 1968; vol. 56, p. 1145, June 1968; vol. 56, p. 1225, July 1968; vol. 56, pp. 2073–2076, Nov. 1968; vol. 57, p. 711, Apr. 1969.)

[111] J. B. Gunn, "Thermodynamics of nonlinearity and noise in diodes," *J. Appl. Phys.*, vol. 39, pp. 5357–5361, Nov. 1968.

[112] P. Penfield, Jr., "Thermodynamics and the Manley–Rowe equations," *J. Appl. Phys.*, vol. 37, pp. 4629–4630, Dec. 1966.

[113] R. B. Goldner, "An interesting black box problem solved by a noise measurement," *Proc. IRE*, vol. 50, p. 2509, Dec. 1962.

[114] F. S. Macklem, "Dr. Slepian's black box problem," *Proc. IEEE*, vol. 51, p. 1269, Sept. 1963.

[115] H. Heffner, "The fundamental noise limit of linear amplifiers," *Proc. IRE*, vol. 50, pp. 1604–1608, July 1962.

[116] H. A. Haus, "Steady-State quantum analysis of linear systems," *Proc. IEEE*, vol. 58, pp. 1599–1611, Oct. 1970.

[117] D. K. C. MacDonald, "Fluctuations and theory of noise," *IRE Trans. Inform. Theory*, vol. IT-1, pp. 114–120, Feb. 1953.

Bibliography for Part II

```
****************************************************************************
       2-A. NON-SERIAL LITERATURE ON NOISE
****************************************************************************
****************************************************************************
       *** BOOKS ***
****************************************************************************
```

BELL, D. A.
    INFORMATION THEORY AND ITS ENGINEERING APPLICATIONS
        PITMAN, LONDON, 1953.
BELL, D. A.
    STATISTICAL METHODS IN ELECTRICAL ENGINEERING
        CHAPMAN & HALL, LONDON, 1953.
BELL, D. A.
    ELECTRICAL NOISE   FUNDAMENTALS AND PHYSICAL MECHANISMS
        VAN NOSTRAND, LONDON, 1960.
BENDAT, J. S.
    PRINCIPLES AND APPLICATIONS OF RANDOM NOISE THEORY
        WILEY, NEW YORK, 1958.
BENDAT, J. S. AND A. G. PIERSOL
    MEASUREMENT AND ANALYSIS OF RANDOM DATA
        WILEY, NEW YORK, 1966.
BENDAT, J. S. AND A. G. PIERSOL
    RANDOM DATA ANALYSIS AND MEASUREMENT PROCEDURES
        WILEY - INTERSCIENCE, NEW YORK, 971
BENNETT, W. R.
    ELECTRICAL NOISE
        MC GRAW-HILL, NEW YORK, 1960.
BENNETT, W. R.
    INTRODUCTION TO SIGNAL TRANSMISSION
        MC GRAW-HILL, NEW YORK, 1970.
BENNETT, W. R. AND J. R. DAVEY
    DATA TRANSMISSION
        MC GRAW-HILL, NEW YORK, 1965.
BITTEL, H. AND L. STORM
    RAUSCHEN
        SPRINGER-VERLAG, BERLIN, 1971.
BLACHMAN, N. M.
    NOISE AND ITS EFFECT ON COMMUNICATION
        MC GRAW-HILL BOOK CO., INC., NEW YORK, N. Y., 1966.
BLANC-LAPIERRE, A. AND B. PICINBONO
    PROPRIETES STATISTIQUES DU BRUIT DE FOND
        MASSON, PARIS, 1961.
BROWN, W. M. AND C. J. PALERMO
    RANDOM PROCESSES, COMMUNICATIONS AND RADAR
        MC GRAW-HILL BOOK CO., NEW YORK, N.Y., 1969.
BULL, C. S.
    FLUCTUATIONS OF STATIONARY AND NON-STATIONARY ELECTRON CURRENTS
        BUTTERWORTHS, LONDON, 1966.
CONNOR, F. R.
    NOISE
        (INTRODUCTORY TOPICS IN ELECTRONICS & TELECOMMUNICATIONS SERIES).
        HERMAN PUB., BOSTON, MASS. 1974.
DAVENPORT, W. B. AND W. L. ROOT
    AN INTRODUCTION TO THE THEORY OF RANDOM SIGNALS AND NOISE
        MC GRAW-HILL BOOK CO., INC., NEW YORK, N.Y., 1958.
DOWING, J. J.
    MODULATION SYSTEMS AND NOISE
        PRENTICE HALL, INC., ENGLEWOOD CLIFFS, N.J., 1964.
FICCKI, R.
    ELECTRICAL INTERFERENCE
        HAYDEN BOOK CO., NEW YORK AND ILIFFE LTD., LONDON, 1964.
FREEMAN, J. J.
    PRINCIPLES OF NOISE
        JOHN WILEY AND SONS, INC., NEW YORK, N.Y., 1958.
GOLDMAN, S.
    FREQUENCY ANALYSIS, MODULATION AND NOISE
        MC GRAW HILL BOOK CO., INC., NEW YORK, N.Y., 1948.
GRIVET, P. AND A. BLAQUIERE
    LE BRUIT DE FORD (RANDOM NOISE) IN FRENCH.
        MASSON ET CIE, PARIS, 1958.
HALL, C.,
    QUESTIONS AND ANSWERS ABOUT NOISE IN ELECTRONICS,
        HOWARD W. SAMS, INC., 1973.
HAUS, H. A. AND R. B. ADLER
    CIRCUIT THEORY OF LINEAR NOISY NETWORKS
        WILEY, NEW YORK, 1959.
KING, R. A.
    ELECTRICAL NOISE
        CHAPMAN & HALL, LONDON, 1966.
LANGE, F. H.
    CORRELATION TECHNIQUES (TRANSLATED FROM GERMAN BY P. B. JOHNS, ED.)
        VAN NOSTRAND, PRINCETON, 1967.
```
```

LANING, J. H. AND R. H. BATTIN
    RANDOM PROCESSES IN AUTOMATIC CONTROL
        MC GRAW-HILL BOOK CO., INC., NEW YORK, N.Y., 1956.
LAWSON, J. L. AND G. E. UHLENBECK, EDS.
    THRESHOLD SIGNALS, RADIATION LABORATORY SERIES, VOL. 24
        MC GRAW-HILL BOOK CO. INC., NEW YORK, N.Y., 1950.
LEBEDEV, V. L.
    RANDOM PROCESSES IN ELECTRICAL AND MECHANICAL SYSTEMS (TRANSLATED
    FROM RUSSIAN)
        U.S. DEPT. OF COMMERCE, WASHINGTON, D.C., 1961.
LOUISELL, W. H.
    RADIATION AND NOISE IN QUANTUM ELECTRONICS
        MC GRAW-HILL BOOK CO., NEW YORK, N.Y., 1964.
LOUISELL, W. H.
    QUANTUM STATISTICAL PROPERTIES OF RADIATION
        MC GRAW - HILL, NEW YORK, 1973.
MAC DONALD, D. K. C.
    NOISE AND FLUCTUATIONS   AN INTRODUCTION
        JOHN WILEY & SONS, INC., NEW YORK, N.Y., 1962.
MENZEL, D. H.
    THE RADIO NOISE SPECTRUM
        HARVARD UNIVERSITY PRESS, CAMBRIDGE, MASS., 1960.
MITROPOLSKII, A. K.
    CORRELATION EQUATIONS FOR STATISTICAL COMPUTATIONS
    (TRANSLATED FROM RUSSIAN BY E. S. SPIEGELTHAL)
        CONSULTANTS BUREAU, NEW YORK, 1966
MIX, D. F.
    RANDOM SIGNAL ANALYSIS
        ADDISON-WESLEY, READING, MASS., 1969.
MORRISON, R
    GROUNDING AND SHIELDING TECHNIQUES IN INSTRUMENTATION
        JOHN WILEY & SONS, INC., NEW YORK, 196..
MOTCHENBACHER, C. D. AND F. C. FITCHEN
    LOW-NOISE ELECTRONIC DESIGN
        WILEY-INTERSCIENCE, NEW YORK, 1973.
MOULIN, E. B.
    SPONTANEOUS FLUCTUATIONS OF VOLTAGE
        OXFORD UNIVERSITY PRESS, LONDON, 1938.
MUMFORD, W. AND E. SCHEIBE
    NOISE PERFORMANCE FACTORS IN COMMUNICATION SYSTEMS
        HORIZON HOUSE, DEDHAM, MASS., 1968.
OTT, H. W.
    NOISE REDUCTION TECHNIQUES IN ELECTRONIC SYSTEMS
        WILEY-INTERSCIENCE, NEW YORK, 1976.
PANTER, P. F.
    MODULATION, NOISE, AND SPECTRAL ANALYSIS   APPLIED TO INFORMATION
    TRANSMISSION
        MC GRAW-HILL BOOK CO., INC., NEW YORK, N.Y., 1965.
PFEIFER, H.
    ELEKTRONISCHES RAUSCHEN   TEIL 1
        B. G. TEUBNER, LEIPZIG, GERMANY, 1959.
RHEINFELDER, W. A.
    DESIGN OF LOW-NOISE TRANSISTOR INPUT CIRCUITS
        HAYDEN BOOK CO., NEW YORK, N.Y., 1964.
ROBINSON, F. N. H.
    NOISE IN ELECTRICAL CIRCUITS
        OXFORD UNIVERSITY PRESS, LONDON, 1962.
ROBINSON, F. N. H.
    NOISE AND FLUCTUATIONS IN ELECTRONIC DEVICES AND CIRCUITS
        OXFORD UNIVERSITY PRESS, LONDON, 1974.
ROWE, H. E.
    SIGNALS AND NOISE IN COMMUNICATION SYSTEMS
        VAN NOSTRAND REINHOLD, NEW YORK, 1965.
SCHWARTZ, M.
    INFORMATION TRANSMISSION, MODULATION, AND NOISE
        MC GRAW-HILL BOOK CO., NEW YORK, N.Y., 1959.
SCHWARTZ, M., W. R. BENNETT AND S. STEIN
    COMMUNICATION SYSTEMS AND TECHNIQUES
        MC GRAW-HILL, NEW YORK, 1966.
STRATONOVICH, R. L.
    TOPICS IN THE THEORY OF RANDOM NOISE, VOLS. I AND II.
    TRANSLATED FROM RUSSIAN BY R. A. SILVERMAN
        GORDON AND BREACH, NEW YORK, N.Y., 1963 AND 1967.
VAN DER ZIEL, A.
    NOISE
        PRENTICE HALL, ENGELWOOD CLIFFS, N.J., 1954.
VAN DER ZIEL, A.
    FLUCTUATION PHENOMENA IN SEMICONDUCTORS
        BUTTERWORTHS SCIENTIFIC PUBLICATIONS, LONDON, AND ACADEMIC PRESS,
        INC., NEW YORK, N.Y., 1959.
VAN DER ZIEL, A.
    NOISE   SOURCES, CHARACTERIZATION, MEASUREMENT
        PRENTICE HALL, ENGLEWOOD CLIFFS, N.J., 1970.
VAN DER ZIEL, A.
    NOISE IN MEASUREMENTS
        JOHN WILEY, NEW YORK, 1976.

VINCENT, C. H.
    RANDOM PULSE TRAINS   THEIR MEASUREMENT AND STATISTICAL PROPERTIES
    (IEE MONOGRAPH NO. 13)
        PETER PEREGRINUS, LONDON 1973.

****
        *** PAPER AND REPRINT COLLECTIONS ***
****
BURGESS, R. E., ED.
    FLUCTUATION PHENOMENON IN SOLIDS
        ACADEMIC PRESS, NEW YORK, 1965.
KUZNETSOV, P. L., R. L. STRATONOVICH AND V. I. TIKHONOV
    NONLINEAR TRANSFORMATIONS OF STOCHASTIC PROCESSES (A COLLECTION OF
    RUSSIAN PAPERS IN ENGLISH TRANSLATION)
        PERGAMON PRESS, INC., NEW YORK, N.Y., 1965.
SMULLIN, L. D. AND H. A. HAUS, EDS.
    NOISE IN ELECTRON DEVICES
        TECHNOLOGY PRESS OF MIT AND JOHN WILEY & SONS, INC., CAMBRIDGE,
        MASS, 1959.
WAX, N., EDITOR
    SELECTED PAPERS ON NOISE AND STOCHASTIC PROCESSES
        DOVER PUBLICATIONS, INC., NEW YORK, N.Y., 1954.

****
        *** CONFERENCES, SYMPOSIA, AND MEETINGS ***
****

SYMPOSIUM ON NOISE REDUCTION   PART 3, NORTH AMERICAN AVIATION, INC.,
LOS ANGELES, CALIFORNIA, FEBRUARY 1949.
    RECORD PUBLISHED AS -
        AEROPHYSICS LABORATORY REPORT NO. AL-930, FEBRUARY 1950.

SYMPOSIUM ON NOISE, SPONSORED BY TELECOMMUNICATIONS SECTION, KON. INST.
VAN INGENIEURS, MAY 1952.
    PRESENTED PAPERS PUBLISHED IN -
        TIJDSCHRIFT VAN HET NEDERLANDSCH RADIOGENOOTSCHAP (AMSTERDAM),
        VOL. 17, SEPTEMBER-NOVEMBER 1952. IN DUTCH.

1954 SYMPOSIUM ON FLUCTUATION PHENOMENA IN MICROWAVE SOURCES, NEW YORK,
NOVEMBER 1954.
    PROCEEDINGS PUBLISHED AS -
        IRE TRANS. ELECTRON DEVICES, VOL. ED-1, NO. 4, DECEMBER 1954.
    SUMMARY OF IMPORTANT POINTS OF PAPERS BY W. H. HUGGINS
        IRE TRANS. ELECTRON DEVICES, VOL. ED-1, NO. 4, PP. 271-273,
        DECEMBER 1954.

CONFENENCE ON NOISE IN ELECTRONIC DEVICES, BALDOCK, GREAT BRITAIN, 1959
    PROCEEDINGS OF THE CONFERENCE PUBLISHED BY -
        THE INSTITUTE OF PHYSICS AND THE PHYSICAL SOCIETY, LONDON, ENGLAND
        CHAPMAN & HALL, LONDON, AND REINHOLD, NEW YORK, 1959.

SYMPOSIUM ON THE APPLICATION OF LOW NOISE RECEIVERS TO RADAR AND ALLIED
EQUIPMENT
    PROCEEDINGS OF THE SYMPOSIUM PUBLISHED BY
        LINCOLN LABORATORY, MASSACHUSSTTS INSTITUTE OF TECHNOLOGY,
        LEXINGTON, MASS., NOVEMBER 1960.

FIRST SYMPOSIUM ON ENGINEERING APPLICATIONS OF RANDOM FUNCTION THEORY
AND PROBABILITY
    PROCEEDINGS OF THE SYMPOSIUM PUBLISHED AS -
        PROCEEDINGS OF THE FIRST SYMPOSIUM ON ENGINEERING APPLICATIONS OF
        RANDOM FUNCTION THEORY AND PROBABILITY
        BOGDANOFF, J. L. AND F. KOZIN, EDS.
        JOHN WILEY AND SONS, INC., NEW YORK, 1962.

FIFTH AGARD AVIONICS PANEL CONFERENCE ON LOW NOISE ELECTRONICS, OSLO,
NORWAY, JULY 1961.
    PROCEEDINGS OF THE CONFERENCE PUBLISHED AS
        SOLID STATE ELECTRONICS, VOL. 4, OCTOBER 1962.

THIRD ANNUAL FLUCTUATIONS IN SOLIDS SYMPOSIUM, ARMOUR RESEARCH
FOUNDATION, CHICAGO, ILLINOIS, APRIL 1959.
    NO PUBLISHED RECORD.   SUMMARY BY J. J. BROPHY IN
        PHYSICS TODAY, VOL. 12, NO. 9, PP. 30-32, SEPTEMBER 1959.

FOURTH ANNUAL FLUCTUATIONS IN SOLIDS SYMPOSIUM, ARMOUR RESEARCH
FOUNDATION, CHICAGO, ILLINOIS, MAY 1960.
    NO PUBLISHED RECORD.   SUMMARY BY J. J. BROPHY IN
        PHYSICS TODAY, VOL. 13, NO. 10, PP. 38-40, OCTOBER 1960.

FIFTH ANNUAL FLUCTUATIONS IN SOLIDS SYMPOSIUM, ARMOUR RESEARCH
FOUNDATION, CHICAGO, ILLINOIS, MAY 1961.
    NO PUBLISHED RECORD.   SUMMARY BY J. J. BROPHY IN
        PHYSICS TODAY, VOL. 14, NO. 9, PP. 44-47, SEPTEMBER 1961.

SIXTH ANNUAL FLUCTUATIONS IN SOLIDS SYMPOSIUM, UNIVERSITY OF MINNESOTA,
MINNEAPOLIS, MINN., MAY 1962.
    NO PUBLISHED RECORD.   SUMMARY BY J. J. BROPHY IN
        PHYSICS TODAY, VOL. 15, NO. 9, PP. 50-54, SEPTEMBER 1962.

SEVENTH FLUCTUATIONS IN SOLIDS SYMPOSIUM, UNIVERSITY OF MINNESOTA,
MINNEAPOLIS, MINN., MAY 1964.
    NO PUBLISHED RECORD.   SUMMARY BY J. J. BROPHY IN
        PHYSICS TODAY, VOL. 17, NO. 12, PP. 42-48, DECEMBER 1964.

SYMPOSIUM ON THE RESPONSE OF NONLINEAR SYSTEMS TO RANDOM EXCITATION
    PAPERS PUBLISHED IN -
        ACOUSTICAL SOCIETY OF AMERICA JOURNAL, VOL. 35, NO. 11, NOVEMBER
        1963.

NOISE REDUCTION CONFERENCE, LAWERENCE RADIATION LABORATORY, UNIVERSITY
OF CALIFORNIA, LIVERMORE, CALIF., MARCH 1968.
    PROCEEDINGS OF THE CONFERENCE PUBLISHED AS A REPORT BY -
        CLEARINGHOUSE FOR FEDERAL SCIENTIFIC AND TECHNICAL INFORMATION,
        NATIONAL BUREAU OF STANDARDS, U. S. DEPT. OF COMMERCE, SPRINGFIELD,
        VIRGINIA, CONF-680303.

CONFERENCE ON FLUCTUATIONS IN SUPERCONDUCTORS, PACIFIC GROVES, CALIF.,
MAY 1968.
    PROCEEDINGS OF THE CONFERENCE PUBLISHED BY -
        LOW TEMPERATURE PHYSICS DEPARTMENT, STANFORD RESEARCH INSTITUTE,
        MENLO PARK, CALIFORNIA, 94025, MAY 1968.

CONFERENCE ON THE PHYSICAL ASPECTS OF NOISE IN ELECTRONIC DEVICES
UNIVERSITY OF NOTTINGHAM, ENGLAND, SEPTEMBER 1968.
    PROCEEDINGS OF THE CONFERENCE PUBLISHED AS -
    PHYSICAL ASPECTS OF NOISE IN ELECTRONIC DEVICES
        THE INSTITUTE OF PHYSICS AND THE PHYSICAL SOCIETY, LONDON, ENGLAND
        1968.

CONFERENCE ON FLUCTUATION PHENOMENA IN CLASSICAL AND QUANTUM SYSTEMS
CHANIA, CRETE, GREECE, AUGUST 1969.
    PROCEEDINGS OF THE CONFERENCE EDITED AND PUBLISHED AS FOLLOWS -
        E. D. HAIDEMENAKIS, EDITOR
        G. K. HALL, BOSTON, 1971.

WORKSHOP ON FLUCTUATION PHENOMENA IN SINGLE AND DOUBLE INJECTION DEVICES
BASEL, SWITZERLAND, MAY 28-29, 1970.
    PROCEEDINGS OF THE WORKSHOP PUBLISHED IN
        SOLID-STATE ELECTRONICS, VOL. 14, NO. 5, MAY 1971.

INTERNATIONAL CONFERENCE ON NOISE IN ACTIVE SEMICONDUCTOR DEVICES,
TOULOUSE, FRANCE, SEPTEMBER 1971.
    PROCEEDINGS OF THE CONFERENCE PUBLISHED AS -
        LE BRUIT DE FOND DE COMPASANTS ACTIFS SEMICONDUCTEURS, CENTRE
        NATIONAL DE LA RECHERCHE SCIENTIFIQUE, PARIS, 1972.

SYMPOSIUM ON NOISE IN ELECTRONIC MATERIALS AND DEVICES, GAINESVILLE,
    FLORIDA, U.S.A., DECEMBER 10 - 11, 1972.

FOURTH INTERNATIONAL CONFERENCE ON PHYSICAL ASPECTS OF NOISE IN SOLID
STATE DEVICES, NOORDRWIJKERHOUT, THE NETHERLANDS, SEPTEMBER 1975.
    PROGRAM AND CONTRIBUTED PAPERS PUBLISHED BY THE NETHERLANDS' PHYSICAL
    SOCIETY.
    INVITED PAPERS PUBLISHED AS A SPECIAL ISSUE OF PHYSICA.

****
    *** STANDARDS ***
****

IRE STANDARDS COMMITTEE
    IRE STANDARDS ON RECEIVERS   DEFINITION OF TERMS
        PROC. IRE, VOL. 40, NO. 12, PP. 1681-1685, DECEMBER 1952.
IRE STANDARDS COMMITTEE
    STANDARDS ON SOUND RECORDING AND REPRODUCING   METHODS OF MEASUREMENT
    OF NOISE
        PROC. IRE, VOL. 41, NO. 4, PP. 508 - 512, APRIL 1953.
IRE STANDARDS COMMITTEE
    STANDARDS ON ELECTRON DEVICES   METHODS OF MEASURING NOISE
        PROC. IRE, VOL. 41, NO. 7, PP. 890-896, JULY 1953
        PROC. IRE, VOL. 45, NO. 7, PP. 983-1010, JULY 1957.
IRE STANDARDS COMMITTEE
    IRE STANDARDS ON METHODS OF MEASURING NOISE IN LINEAR TWOPORTS, 1959
        PROC. IRE, VOL. 48, NO. 1, PP. 60-68, JANUARY 1960.
IRE SUBCOMMITTEE ON NOISE
    REPRESENTATION OF NOISE IN LINEAR TWOPORTS
        PROC. IRE, VOL. 48, NO. 1, PP. 69-74, JANUARY 1960.
IRE STANDARDS COMMITTEE, 62 IRE 7.52
    IRE STANDARDS ON ELECTRON TUBES   DEFINITION OF TERMS, 1962
        PROC. IRE, VOL. 51, NO. 3, PP. 434-435, MARCH 1963.

IRE SUBCOMMITTEE 7.9 ON NOISE
    DESCRIPTION OF THE NOISE PERFORMANCE OF AMPLIFIERS AND RECEIVING
    SYSTEMS
        PROC. IEEE, VOL. 51, NO. 3, PP. 436-442, MARCH 1963.
IEEE STANDARDS COMMITTEE
    IEEE STANDARDS ON DEFINITIONS, SYMBOLS AND METHODS OF TEST FOR
    SEMICONDUCTOR TUNNEL (ESAKI) DIODES AND BACKWARD DIODES
        IEEE TRANS. ELECTRON DEVICES, VOL. ED-12, NO. 6, PP. 373-386,
        JUNE 1965, SECTION 5 ON NOISE CHARACTERISTICS.
STRAUS, T. M.
    RELATIONSHIP BETWEEN THE NCTA, EIA, AND CCIR DEFINATION OF SIGNAL-
    TO-NOISE RATIO
        IEEE TRANS BROADCASTING, VOL BC-20, NO 3, PP. 36-41, SEPT. 1974.

****
    2-B. BOOKS ON PROBABILITY THEORY AND STOCHASTIC PROCESSES
****
ARLEY, N.
    ON THE THEORY OF STOCHASTIC PROCESSES AND THEIR APPLICATION TO THE
    THEORY OF COSMIC RADIATION
        JOHN WILEY AND SONS, NEW YORK, 1943.
ARNOLD, L.
    STOCHASTIC DIFFERENTIAL EQUATIONS THEORY AND APPLICATIONS
        WILEY INTERSCIENCE, NEW YORK, 1974.
BAILEY, N. T. J.
    THE ELEMENTS OF STOCHASTIC PROCESSES WITH APPLICATIONS TO THE NATURAL
    SCIENCES
        JOHN WILEY, NEW YORK, 1964.
BARBER, M. N. AND B. W. NINHAM
    RANDOM AND RESTRICTED WALKS, THEORY AND APPLICATIONS
        GORDON AND BREACH, NEW YORK, 1970.
BARTLETT, M. S.
    AN INTRODUCTION TO STOCHASTIC PROCESSES
        CAMBRIDGE UNIVERSITY PRESS, CAMBRIDGE, ENGLAND, 1955.
BHARUCHA-REID, A. T.
    ELEMENTS OF THE THEORY OF MARKOV PROCESSES AND THEIR APPLICATIONS
        1960.
BHARUCHA-REID, A. T., ED.
    PROBABILISTIC METHODS IN APPLIED MATHEMATICS
        ACADEMIC PRESS, NEW YORK, 1968.
BHARUCHA-REID, A. T.
    RANDOM INTEGRAL EQUATIONS
        ACADEMIC PRESS, NEW YORK, 1972.
BHAT, U. N.
    ELEMENTS OF APPLIED STOCHASTIC PROCESSES
        JOHN WILEY, NEW YORK, 1972.
BLACKMAN, R. B., AND J. W. TUKEY
    THE MEASUREMENT OF POWER SPECTRA
        DOVER PUBLICATIONS, INC., NEW YORK, 1958.
BLANC-LAPIERRE, A.
    MODELES STATISTIQUES POUR L'ETUDE DE PHENOMENES DE FLUCTUATIONS
        MASSON, PARIS, 1963.
BLANC-LAPIERRE, A. AND R. FORTET
    THEORY OF RANDOM FUNCTIONS, 2 VOLS. (TRANSLATED FROM FRENCH)
        GORDON AND BREACH, NEW YORK, 1967-1968.
BREIMAN, L.
    PROBABILITY AND STOCHASTIC PROCESSES   WITH A VIEW TOWARDS APPLICATIONS
        HOUGHTON MIFFLIN, BOSTON, 1969.
BREIPOHL, A. M.
    PROBABILISTIC SYSTEMS ANALYSIS
        JOHN WILEY & SONS, NEW YORK, 1970.
CASHWELL, E. D. AND C. J. EVERETT
    A PRACTICAL MANUAL ON THE MONTE CARLO METHOD FOR RANDOM WALK PROBLEMS
        PERGAMON PRESS, NEW YORK, 1959.
CHUNG, K. L.
    MARKOV CHAINS WITH STATIONARY TRANSITION PROBABILITIES
        SPRINGER VERLAG, BERLIN, 1960.
CHUNG, K. L.
    ELEMENTARY PROBABILITY THEORY WITH STOCHASTIC PROCESSES
        SPRINGER VERLAG, NEW YORK, 1974.
COX, D. R. AND H. D. MILLER
    THE THEORY OF STOCHASTIC PROCESSES
        WILEY, NEW YORK, 1965.
CRAMER, H.
    MATHEMATICAL METHODS OF STATISTICS
        PRINCETON UNIVERSITY PRESS, PRINCETON, N. J., 1946.
CRAMER, H.
    STATIONARY AND RELATED STOCHASTIC PROCESSES SAMPLE FUNCTION
    PROPERTIES AND THEIR APPLICATIONS
        JOHN WILEY, NEW YORK, 1967.
CRAMER, H.
    RANDOM VARIABLES AND PROBABILITY DISTRIBUTIONS, 3RD EDITION
        CAMBRIDGE UNIVERSITY PRESS, LONDON, 1970.
DAVENPORT, W. B.
    PROBABILITY AND RANDOM PRECESSES

         MC GRAW - HILL BOOK CO., NEW YORK, 1970.
DEUTSCH, R.
    NONLINEAR TRANSFORMATIONS OF RANDOM PROCESSES
        PRENTICE - HALL, ENGLEWOOD CLIFFS, N. J., 1962.
DOOB, J. L.
    STOCHASTIC PROCESSES
        JOHN WILEY, NEW YORK, 1953.
DUBES, R. C.
    THE THEORY OF APPLIED PROBABILITY
        PRENTICE - HALL, ENGLEWOOD CLIFFS, N. J., 1968.
DYNKIN, E. B.
    THEORY OF MARKOV PROCESSES
        PERGAMON PRESS, LONDON, 1960.
EISEN, M.
    INTRODUCTION TO MATHEMATICAL PROBABILITY THEORY
        PRENTICE - HALL, ENGLEWOOD CLIFFS, N. J., 1969.
EPHREMIDES, A., ED.
    RANDOM PROCESSES, PART II POISSON AND JUMP POINT PROCESSES
    (A COLLECTION OF REPRINTS)
        DOWDEN, HUTCHINSON, & ROSS, INC., STROUDSBURG, PENN. AND
        HALSTED PRESS, NEW YORK, 1975.
EPHREMIDES, A. AND J. B. THOMAS, EDITORS
    RANDOM PROCESSES   MULTIPLICITY THEORY AND CANONICAL DECOMPOSITION
    (A COLLECTION OF REPRINTS)
        DOWDEN, HUTCHINSON, & ROSS, INC., STROUDSBURG, PENN.
FELLER, W.
    AN INTRODUCTION TO PROBABILITY THEORY AND ITS APPLICATIONS, VOL. I.
    3RD EDITION
        JOHN WILEY AND SONS, NEW YORK, 1968.
FELLER, W.
    AN INTRODUCTION TO PROBABILITY THEORY AND ITS APPLICATIONS, VOL. II
    2ND EDITION
        JOHN WILEY AND SONS, NEW YORK, 1971.
FREEDMAN, D.
    MARKOV CHAINS
        HOLDEN - DAY, SAN FRANSISCO, 1971.
FREEDMAN, D.
    BROWNIAN MOTION AND DIFFUSION
        HOLDEN - DAY, SAN FRANSISCO, 1971.
GIKHMAN, I. I. AND A. V. SKOROHOD
    INTRODUCTION TO THE THEORY OF RANDOM PROCESSES
        W. B. SAUNDERS CO., PHILADELPHIA, PA., 1969.
GIKHMAN, I. I. AND A. V. SKOROHOD
    STOCHASTIC DIFFERENTIAL EQUATIONS
        SPRINGER VERLAG, BERLIN, 1972.
GIKHMAN, I. I. AND A. V. SKOROHOD
    THE THEORY OF STOCHASTIC PROCESSES, VOL. I AND VOL. II
        SPRINGER VERLAG, BERLIN, 1974.
GIRAULT, M.
    STOCHASTIC PROCESSES
        SPRINGER VERLAG, BERLIN, 1966.
GNEDENKO, B. V.
    THE THEORY OF PROBABILITY
        CHELSEA, NEW YORK, 1962.
HADDAD, A. H., EDITOR
    NONLINEAR SYSTEMS   PROCESSING OF RANDOM SIGNALS - CLASSICAL ANALYSIS
    (A COLLECTION OF REPRINTS)
        DOWDEN, HUTCHINSON, & ROSS, INC., STROUDSBURG, PENN. AND
        HALSTED PRESS, NEW YORK, 1975.
HAMMERSLEY, J. M. AND D. C. HANDSCOMB
    MONTE CARLO METHODS
        METHUEN, LONDON, 1964.
HARRIS, T. E.
    THE THEORY OF BRANCHING PROCESSES
        SPRINGER-VERLAG, BERLIN, 1963.
HOEL, P. G. AND C. J. STONE
    INTRODUCTION TO STOCHASTIC PROCESSES
        HOUGHTON MIFFLIN, BOSTON, 1972.
JEFFREYS, H.
    THEORY OF PROBABILITY
        CLARENDON PRESS, OXFORD, 1961.
JENKINS, G. M., AND D. G. WATTS
    SPECTRAL ANALYSIS AND ITS APPLICATIONS
        HOLDEN-DAY, INC., SAN FRANCISCO, 1968.
KARLIN, S. AND H. M. TAYLOR
    A FIRST COURSE IN STOCHASTIC PROCESSES, 2ND ED.
        ACADEMIC PRESS, NEW YORK, 1975.
KEILSON, J.
    GREEN'S FUNCTION METHODS IN PROBABILITY THEORY
        GRIFFIN, LONDON, 1965.
KEMENY, J. G. AND J. L. SNELL
    FINITE MARKOV CHAINS
        VAN NOSTRAND, NEW YORK, 1960.
KHARKEVICH, A. A.
    SPECTRA AND ANALYSIS
        PLENUM PUBLISHERS, NEW YORK, 1960.

KOLMOGOROV, A. N.
    FOUNDATIONS OF THE THEORY OF PROBABILITY
        CHELSEA, NEW YORK, 1950.
KORN, G. A.
    RANDOM-PROCESS SIMULATION AND MEASUREMENT
        MC GRAW-HILL BOOK CO., NEW YORK, N.Y., 1966.
LEVY, P.
    PROCESSES STOCHASTIQUES ET MOUVEMENT BROWNIEN
        GAUTHIER-VILLARS, PARIS, 1948.
MARTIN, J. C.
    MEASUREMENTS AND CORRELATION FUNCTIONS
        GORDON AND BREACH, NEW YORK, 1970.
MCKEAN, JR., H. P.
    STOCHASTIC INTEGRALS
        ACADEMIC PRESS, NEW YORK, 1969.
MCSHANE, E. J.
    STOCHASTIC CALCULUS AND STOCHASTIC MODELS
        ACADEMIC PRESS, NEW YORK, 1974.
MURTHY, V. K.
    THE GENERAL POINT PROCESS APPLICATIONS TO STRUCTURAL FATIGUE,
    BIOSCIENCE, AND MEDICAL RESEARCH
        ADDISON-WESLEY, READING, MA., 1974.
PAPOULIS, A.
    PROBABILITY, RANDOM VARIABLES, AND STOCHASTIC PROCESSES
        MC GRAW - HILL BOOK CO., NEW YORK, 1965.
PARZEN, E.
    STOCHASTIC PROCESSES
        HOLDEN-DAY, SAN FRANCISCO, 1962.
PERVOZVANSKII, A.A.,
    RANDOM PROCESS IN NONLINEAR CONTROL SYSTEMS,
        ACADEMIC PRESS, NEW YORK, 1965
PFEIFFER, P. E. AND D. A. SCHUM
    INTRODUCTION TO APPLIED PROBABILITY
        ACADEMIC PRESS, NEW YORK, 1973.
PORTER, C.E., EDITOR,
    STATISTICAL THEORIES OF SPECTRA FLUCTUATIONS
    (A COLLECTION OF REPRINTS AND ORIGINAL PAPERS)
        ACADEMIC PRESS, NEW YORK, 1965.
PRABHU, N. U.
    STOCHASTIC PROCESSES BASIC THEORY AND ITS APPLICATIONS
        MACMILLAN, NEW YORK, 1965.
PUGACHEV, V. S.
    THEORY OF RANDOM FUNCTIONS AND ITS APPLICATIONS TO PROBLEMS IN
    AUTOMATIC CONTROL
        PERGAMON PRESS, LONDON, 1965.
ROSENBLATT, M.
    RANDOM PROCESSES
        OXFORD UNIVERSITY PRESS, LONDON, 1962.
    RANDOM PROCESSES, 2ND EDITION
        SPRINGER VERLAG, NEW YORK, 1974.
ROSS, S. M.
    INTRODUCTION TO PROBABILITY MODELS
        ACADEMIC PRESS, NEW YORK, 1972.
ROZANOV, YU. A.
    STATIONARY RANDOM PROCESSES, TRANSLATED BY A. FEINSTEIN
        HOLDEN - DAY, SAN FRANSISCO, 1967.
SCHREIDER, YU. A., EDITOR
    THE MONTE CARLO METHOD THE METHOD OF STATISTICAL TRIALS
        PERGAMON PRESS, NEW YORK, 1966.
SNYDER, D. L.
    RANDOM POINT PROCESSES
        WILEY-INTERSCIENCE, NEW YORK, 1975.
SOONG, T. T.
    RANDOM DIFFERENTIAL EQUATIONS IN SCIENCE AND ENGINEERING
        ACADEMIC PRESS, NEW YORK, 1974.
SPANIER, J. AND E. M. GELBARD
    MONTE CARLO PRINCIPLES AND NEUTRON TRANSPORT PROBLEMS
        ADDISON - WESLEY, READING, MA., 1969.
SPITZER, F. L.
    PRINCIPLES OF RANDOM WALK, 2ND ED.
        SPRINGER VERLAG, NEW YORK, 1976.
SRINIVASAN, S. K.
    STOCHASTIC THEORY AND CASCADE PROCESSES
        AMERICAN ELSEVIER, NEW YORK, N.Y., 1969.
SRINIVASAN, S. K.
    STOCHASTIC POINT PROCESSES AND THEIR APPLICATIONS
        HAFNER, NEW YORK, 1974.
SRINIVASAN, S. K. AND R. VASUDEVAN
    INTRODUCTION TO RANDOM DIFFERENTIAL EQUATIONS AND THEIR APPLICATIONS
        AMERICAN ELSEVIER PUBLISHING CO., NEW YORK, 1971.
STRATONOVICH, R. L.
    CONDITIONAL MARKOV PROCESSES AND THEIR APPLICATION TO THE THEORY OF
    OPTIMAL CONTROL
        AMERICAN ELSEVIER PUBLISHING CO., NEW YORK, 1968.
SULLINS, W.L.
    MATRIX ALGEBRA FOR STATISTICAL APPLICATIONS

INTERSTATE, DANVILLE, ILL., 1973.

SVESHNIKOV, A. A.
  APPLIED METHODS OF THE THEORY OF RANDOM FUNCTIONS
    PERGAMON PRESS, LTD., OXFORD, ENGLAND, 1965.

TAKACS, L.
  STOCHASTIC PROCESSES   PROBLEMS AND SOLUTIONS
    METHUEN, LONDON, 1960.

TAKACS, L.
  COMBINATORIAL METHODS IN THE THEORY OF STOCHASTIC PROCESSES
    JOHN WILEY, NEW YORK, 1967.

THOMAS, J. B.
  INTRODUCTION TO APPLIED PROBABILITY AND RANDOM PROCESSES
    WILEY, NEW YORK, 1971.

THOMASIAN, A. J.
  THE STRUCTURE OF PROBABILITY THEORY WITH APPLICATIONS
    MC GRAW - HILL, NEW YORK, 1969.

TSOKOS, C. P. AND W. J. PADGETT
  RANDOM INTEGRAL EQUATIONS WITH APPLICATIONS TO LIFE SCIENCES AND
  ENGINEERING
    ACADEMIC PRESS, NEW YORK, 1974.

USPENSKY, J. V.
  INTRODUCTION TO MATHEMATICAL PROBABILITY
    MC GRAW - HILL BOOK CO., NEW YORK

VON MISES, R.
  PROBABILITY, STATISTICS, AND TRUTH
    MACMILLAN, NEW YORK, 1957.

WADSWORTH, G. P. AND J. G. BRYAN
  INTRODUCTION TO PROBABILITY AND RANDOM VARIABLES
    MC GRAW - HILL BOOK CO., NEW YORK, 1960.

WIENER, N.
  NONLINEAR PROBLEMS IN RANDOM THEORY
    TECHNOLOGY PRESS OF MIT, CAMBRIDGE, MASS., 1958.

WONG, E.
  STOCHASTIC PROCESSES IN INFORMATIONAL AND DYNAMICAL SYSTEMS
    MC GRAW-HILL, NEW YORK, 1971.

YAGLOM, A. M.
  AN INTRODUCTION TO THE THEORY OF STATIONARY RANDOM FUNCTIONS
  TRANSLATED FROM RUSSIAN BY R. A. SILVERMAN
    PRENTICE-HALL, ENGLEWOOD CLIFFS, N. J., 1962.

YEH, J-C.
  STOCHASTIC PROCESSES AND THE WIENER INTEGRAL
    MARCEL DEKKER, NEW YORK, 1973.

Part III
Physical Theory of Noise

A number of physical mechanisms responsible for the generation of noise were briefly described in the previous part. Some of these are taken up for a more detailed discussion in this part.

The mechanisms of noise generation in devices can be classified in several different ways, for example, on the basis of whether the charge carriers are in thermal equilibrium, and whether the transport of mobile charge carriers is involved. In practice, however, the noise in devices is usually ascribed to one or more of the following sources:

1) thermal noise (or, more generally, diffusion noise),

2) shot noise (including generation-recombination noise),

3) $1/f$ noise (and other excess low-frequency noise, such as burst or popcorn noise),

4) quantum noise,

and others.

where t_n are random points in time with density $\lambda(t)$ and h_n are real functions of time. As the name suggests, $1/f$ noise is defined in terms of the nature of the noise power spectrum, without reference to a specific physical mechanism (indeed, no single mechanism can at present explain all observed $1/f$ noise). Finally, quantum noise is that noise which would be expected on the basis of the quantum mechanical uncertainty principle and is identical with thermal noise or with shot noise for some limiting cases (idealized devices).

Clearly, then, the above list of noise sources does not form an exhaustive set of mutually exclusive noise mechanisms. The set is nevertheless in common use and has become ingrained due to historical rather than logical reasons. The present part is therefore organized along the same lines. For the purpose of physical understanding, a somewhat more systematic classification of noise source is the following (after Chenette and van Vliet [1]):

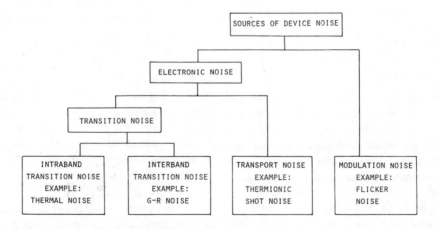

This list is not a classification of noise sources because it does not have a single basis. Thus, thermal noise is defined in terms of a physical process in dissipative materials. The term shot noise, on the other hand, is now the name of a mathematical model, being a special type of random process, expressible as

$$s(t) = \sum_{n=-\infty}^{\infty} h_n(t - t_n) \qquad (1)$$

Of the various kinds of noise sources mentioned above, the present part is confined to thermal, generation-recombination, and $1/f$ noise sources. Some basic principles behind the notion of quantum noise are also discussed in Oliver's paper reprinted in this part. The only principal omission then appears to be "shot noise," and it requires an explanation.

The term "shot noise" has been generalized in its scope several times in the last few decades. In the 1920's, shot noise referred to the fluctuations in the plate current of a vacuum diode, operating in the temperature-limited region of its current voltage characteristic, which arose from the random emission of electrons from the cathode. The definition was soon generalized to other vacuum tubes under different operating conditions, and the noise was then qualified as "modified shot noise" provided its origin could be traced back to the random emission of electrons from the cathode. Still later, the noise in semiconductor devices due to the random generation and recombination of carriers was called shot noise due to its resemblance to the shot noise in electron tubes, the transfer of a carrier from valence to conduction band (or vice versa) being analogous to the emission of a carrier from the cathode.

At the same time, the utility of shot noise as a model for a wide variety of other physical processes led to the mathematical definition of shot noise as a random process:

$$s(t) = \sum_{n=-\infty}^{\infty} h(t - t_n) \tag{2}$$

consisting of terms formed by shifting a given function of time $h(t)$ by random time intervals t_n; the shot noise is called uniform or nonuniform according to whether the density of instants t_n along the time scale is uniform or nonuniform. A further generalization is one in which the function $h(t)$ is not fixed for all n, as in (1).

These generalizations broaden the scope of the term shot noise a great deal. Thus the generation-recombination noise in a bulk semiconductor, the avalanche noise in a semiconductor p-n junction biased into breakdown, the Barkhausen noise in magnetic tapes, and the photon noise in the optical output of light-emitting diodes may all be viewed as "types" of shot noise. Viewed in this manner, shot noise is discussed in this volume in this as well as the next two parts. In particular, the statistics of shot noise are discussed in Oliver's paper in this part, while thermionic shot noise is discussed in Whinnery's paper on microwave tube noise in Part IV.

The Reprinted Papers

The four papers reprinted in this part are devoted to the three topics of thermal noise, generation-recombination noise, and 1/f noise. The first paper is a tutorial one by Oliver, discussing the fundamental concepts of thermal and quantum noise. It also includes brief statements of thermodynamic and quantum mechanical prerequisites for understanding these concepts. In addition, it introduces the elementary ideas of shot noise limitation in optical communication. Another noteworthy tutorial paper on thermal noise is that by Siegman [2], and it could well have been included in this volume. Siegman's discussion is more phenomenological and is applied to electrical networks.

An excellent, although somewhat dated, survey of generation-recombination (g-r) noise in semiconductors and photoconductors is the second reprinted paper by van Vliet, discussing both experimental and theoretical aspects of the subject. A considerable amount of work has been done on the subject since this paper was published (in 1958), including the treatment of g-r noise from a more generalized viewpoint [3].

This paper also includes a summary of the understanding of 1/f noise at the time.

More detailed introduction to 1/f noise is provided in the next two reprinted papers. Hooge has summarized a large body of experimental results on 1/f noise in his paper, although restricting it to a small class of devices. Radeka's paper is directed principally to the influence of 1/f noise in measurements, but it also discusses the questions of models, generation, and properties of 1/f noise. Two earlier reviews [4], [5] by Bell on 1/f noise have a broader scope and may be consulted for a more tutorial introduction to this subject. A survey of the current trends of work on 1/f noise may be found in another very recent review article by Hooge [6].

The Bibliographies

The paper on generation-recombination noise by van Vliet contains a sizable bibliography on its subject. The list of references in the other reprinted papers does not approach an equally comprehensive coverage of the literature. Two bibliographies are therefore included in this part: one on thermal and one on 1/f noise. The latter is supplemented by a third, short bibliography on burst or popcorn noise occurring in some semiconductor devices at low frequency.

The bibliography on thermal noise is confined to papers on the physical theory. Thus it includes various derivations of Nyquist's theorem based on specific microscopic models and generalizations of Nyquist's theorem as a fluctuation-dissipation theorem. It excludes the large number of studies on Brownian motion in various physical systems and on thermal noise in a nonelectronic context. It also excludes most of the sizable literature on the calculation of thermal noise in specific electronic devices (see the next part). The exception is a small selection of papers on noise in devices where thermal noise is the primary source (such as field-effect transistors and double-injection diodes), and is included for illustrative purposes.

Because 1/f noise continues to be an actively pursued area in the physical theory of noise, a sizable bibliography of papers on the subject of 1/f noise is included here. No attempt is made to subclassify this bibliography despite its size because of the large overlap between the possible classifications. The papers span a wide range of activities, including experimental measurements of 1/f noise as well as theoretical efforts towards constructing a model of 1/f noise.

Experimental studies of 1/f noise included in the bibliography are of two different kinds, depending upon their motivation. In the more pragmatic studies, 1/f noise is studied as a limitation to the performance of a useful electronic device in which it is significant, such as crystal rectifiers, MOS transistors, or photodiodes. The results of such ·studies are either in the form of a measure of the noise limitation (signal-to-noise ratio, error-rate, detection threshold, etc.) or in the form of the methods of noise reduction (a set of operating conditions, a surface treatment, or processing methods) for the device. In the more fundamental studies, the purpose is to understand the origin of 1/f noise. The experimental studies usually measure the magnitude of 1/f noise and its dependence upon the various experimental parameters, such as the magnetic field, the dc, or RF bias current magnitude, the temperature, or the dimensions and surface conditions of the device or component in question. The results of such studies are used to localize the origin of $1/f$ noise (e.g., surface rather than the bulk, minority carriers rather than the majority carriers) and to test the theories and models of $1/f$ noise.

The theories and models of $1/f$ noise have been constructed at several levels: from purely mathematical or quantum-mechanical, to phenomenological models in terms of equivalent circuits. The objective in each case has been to explain the existence (and, for more successful theories, the magnitude) of $1/f$ noise power spectrum and its dependence upon experimental parameters such as dc or RF bias current magnitude, temperature, dimensions, or the surface conditions of the device or component in question.

REFERENCES

[1] E. R. Chenette and K. M. van Vliet, "Noise in electronic devices," in *Methods of Experimental Physics*, vol. 2, part B, 2nd ed., L. Marton, Ed. New York: Academic, 1975, pp. 461–500.

[2] A. E. Siegman, "Thermal noise in microwave systems, Part I," *Microwave J.*, vol. 4, pp. 81–90, Mar. 1961.

[3] K. M. van Vliet, "General transport theory of noise in p-n junction-like devices. I. Three dimensional Green's function formulation," *Solid-State Electron.*, vol. 15, pp. 1033–1053, Oct. 1972.

[4] D. A. Bell, *Electrical Noise: Fundamentals and Physical Mechanisms*. London: Van Nostrand, 1960.

[5] D. A. Bell, "Present knowledge of $1/f$ noise," in *Proc. Conf. on Physical Aspects of Noise in Electronic Devices*, The Institute of Physics and the Physical Society, London. Stevenage, England: Peter Peregrinus Ltd., 1968.

[6] F. N. Hooge, "$1/f$ noise," *Physica B & C*, vol. 83, pp. 14–23, May 1976.

Thermal and Quantum Noise

B. M. OLIVER, FELLOW, IEEE

Abstract—One purpose of this article is to develop the theory of black body radiation, thermal noise, and quantum noise from a few basic physical principles. A second purpose is to show how these results apply to certain areas such as antenna theory and ideal receivers. It is hoped that having this related material collected and presented in the language of the electronics engineer will be of tutorial value.

TWO FUNDAMENTAL sources of noise limit the useful sensitivity of any linear amplifier. One is thermal noise, which is black body radiation in a single propagation mode; the other is quantum noise, which is a manifestation of the uncertainty principle. Thermal noise is usually present; quantum noise is unavoidable. Until recently quantum noise was of little interest to electronics engineers since in their devices it was generally negligible compared with thermal noise. Over the frequency ranges and at the temperatures involved, thermal noise could be considered to have a flat spectral distribution. With the advent of masers and lasers this situation is no longer true. Today it is important that we know the limitations imposed by noise at very low temperatures and at very high frequencies.

The original works on thermodynamics, statistical mechanics, black body radiation, and quantum mechanics comprise a substantial fraction of all physics literature. The theories and results in these areas have been well established by rigorous and often beautiful proofs, and have been carefully verified by experiment. In compressing so much physics into one article it is necessary to forego rigor and completeness. The intent here has been to develop understanding of already well established principles, using plausible arguments. The reader who desires to dig deeper will find the effort rewarding. Many of the original papers on fluctuations [1]–[4], on shot noise [5] and on thermal noise [6], [7] make interesting reading. In addition, many recent works [8]–[11] provide comprehensive treatments of matters treated only superficially, or not at all, in this paper.

PART I: BLACK BODY RADIATION AND THERMAL NOISE

The first two laws of thermodynamics and Planck's relation form not only a convenient but also an essentially complete foundation for the theory of black body radiation [10], [11]. Since they are so fundamental, it is perhaps appropriate to restate and discuss them.

Manuscript received September 5, 1963; revised February 15, 1965.

The author is with the Hewlett-Packard Company, Palo Alto, Calif.

1) *In any closed system the total energy is constant.* This, the first law, is perhaps the most familiar. Since a "closed" system is one in which neither energy nor mass enter or leave, this law is simply a restatement of the principle of the conservation of energy.

2) *In any closed system the entropy eventually maximizes.* Since entropy is a measure of randomness, this, the second law, says that any closed system tends toward a state in which its total energy is randomly distributed among the various degrees of freedom present.

In general the total energy of a system can be written as the sum of a number of terms each involving the square of a system coordinate or its derivative. For example, the total energy of a system of n free particles is

$$\sum_{i=1}^{n} m_i \left(\frac{\dot{x}_i^2}{2} + \frac{\dot{y}_i^2}{2} + \frac{\dot{z}_i^2}{2} \right), \qquad (1)$$

while the energy of a spring-mass resonator is

$$\tfrac{1}{2}(Sd^2 + m\dot{d}^2),$$

where S is the spring stiffness, m the mass, and d the deflection. The energy represented by any particular quadratic term is not constant, but fluctuates as energy is exchanged with other terms. However, in thermal equilibrium, *all quadratic terms will have the same average energy: $kT/2$*, where $k = 1.38 \times 10^{-23}$ J/deg is Boltzmann's constant, and T is the absolute temperature. This is known as the principle of *equipartition of energy*. Any other statistical distribution results in lower entropy. Thus the *average* energy in any mode involving n quadratic terms is $nkT/2$.

A monatomic gas illustrates these principles nicely. Assume an initial state in which all the molecules in an enclosure are at rest save one, so that the entire thermal energy of the gas is in one molecule. This situation is short lived for, like a cue ball in a three dimensional billiard game, the one energetic molecule will stir up all the rest, losing energy to them as it does so. Any molecule having more than average energy will tend to lose energy on collision, while any having less than average energy will tend to gain. In the end, all the molecules will be ricocheting around the enclosure, gaining or losing energy at random on collision, but always tending toward the same average energy $3kT/2$. ($kT/2$ for each quadratic term in (1).)

In the above example we could equally well have said: "Assume an initial state in which all the molecular velocities, $\dot{x}_i$, $\dot{y}_i$, and $\dot{z}_i$, are known. This situation will be very short lived however, for very soon innumerable collisions will occur producing a completely new set of

Reprinted from *Proc. IEEE*, vol. 53, pp. 436–454, May 1965.

velocities. In a very short time our knowledge of the particular velocities will be obliterated and we will only be able to speak of probability distributions." As the entropy increases, so does our ignorance of the state of the system.

Assume that a system is able to assume any of N discrete states and that p_i is the probability of its being in the ith state. Boltzmann, in his famous H theorem was able to show that as a closed system approaches thermodynamic equilibrium, the quantity

$$H = - \sum_{i=1}^{N} p_i \log p_i \qquad (2)$$

never decreases, but tends to increase toward a maximum value, which is attained at thermodynamic equilibrium. In this respect H resembles entropy, and in fact $S = kH$, where S is the physical entropy and k is Boltzmann's constant [10].

If the system state is known, then all the p_i are zero, save the specified one, which has the value unity. In this case H and S are zero. The entropy of a completely specified system is zero. As the state becomes more and more unknown, the probability "flows" from the initial state into neighboring states. In the limit, the p_i become as nearly equal as other physical constants allow, and this maximizes H and S. In a system having a continuous distribution of states an integral expression analogous to (2) holds.

Equation (2) is identical with Shannon's expression [12] for the average information of messages from a discrete source, where N is the number of possible messages and p_i is the probability of the ith message. This is no coincidence. To describe the ith state of the system, using an ideal code, would take $-\log p_i$ natural units of information (bits, if the logarithm is to the base 2) so, on the average, H units are required. Thus the entropy of a physical system is proportional to the information we lack about it, the rate of exchange being k J/deg/natural unit (or $k \log 2$ J/deg/bit).

As a system approaches equilibrium, S, as given by kH, approaches the familiar value of thermodynamics

$$S = \int \frac{\delta Q}{T} \qquad (3)$$

where δQ is an increment of energy (as heat) supplied to the system and T is the system temperature. Unless the system is at equilibrium there is no unique system temperature so (3) cannot be applied to the system as a whole. However, the system may often be divided into a number of isothermal regions and the partial entropies determined by (3) can then be added. When this is done the result agrees with kH, so the latter is considered a more basic definition of entropy than (3).

When attempts were first made to apply the laws of

thermodynamics to systems involving electromagnetic waves, it soon became apparent that something was radically wrong. Consider a cavity enclosed by (imperfectly) conducting walls at a temperature T. The electrons in the walls are in thermal equilibrium and their motions will excite electromagnetic waves in the enclosed space. The expression for the energy in each normal mode of the cavity contains two quadratic terms, one for the electric field and one for the magnetic, and therefore in equilibrium each mode should have an average energy kT. But there are an infinite number of normal modes! Thus, if classical principles applied, the equipartitioning of energy would distribute the finite energy of a closed system among an infinite number of modes; the average energy per quadratic term, and hence the absolute temperature would approach zero. All heat would leak away to infinite frequencies, producing the so-called "ultra-violet catastrophe." The answer to this dilemma was found in the quantization of radiation as expressed in Planck's law. In attempting to derive a theoretical expression for black body radiation, Planck assumed that a simple harmonic oscillator could have only certain discrete energy levels differing by $h\nu$, where ν is the oscillator frequency. His original intent was to let $h \to 0$ in his results, but he found to his surprise that the correct law was obtained for a finite h. This led to the following concept, since amply justified.

3) *Electromagnetic energy is radiated and absorbed in discrete quanta (photons) of energy $h\nu$.* Here ν is the frequency of the radiation and $h = 6.624 \times 10^{-34}$ Joule-seconds is Planck's constant.

The effect of this quantization on the thermal energy in a single resonant mode will be treated quantitatively in the next section. Qualitatively it is apparent that, since any quadratic term rarely has an instantaneous energy of more than a few kT, modes having frequencies so high that $h\nu \gg kT$ are rarely excited thermally.

Thermal Excitation of a Resonant Mode

Consider now any sort of electrical resonator in thermal equilibrium with its environment. Because the resonator can gain or lose energy only in discrete amounts, $h\nu$, where ν is the resonant frequency, if we drain off all the energy available at any instant of time we will find an amount $nh\nu$ to have been present, where n is an integer. By the second law, the probability distribution, $p(n)$, must be such as to maximize the system entropy. We will determine $p(n)$ in two steps. First we will find the shape of the distribution that maximizes the resonator entropy for a given average energy (i.e., for a given expectation, $\bar{n}$, of photons in the resonator). Then using this distribution we will adjust the average energy so as to maximize the system entropy.

The entropy of the distribution $p(n)$ is

$$S_r = kH_r = - k \sum_{n=0}^{\infty} p(n) \log p(n). \qquad (4)$$

We wish to maximize S_r subject to the constraints

$$\sum_{n=0}^{\infty} p(n) = 1 \tag{5}$$

and

$$\sum_{n=0}^{\infty} n p(n) = \bar{n}. \tag{6}$$

The first constraint arises because any probability distribution must add up to unity; i.e., n must be *some* integer. The second constraint says that as we vary the distribution $p(n)$ we will fix its first moment, which is the average number $\bar{n}$ of photons present. Note that $\bar{n}$ need not be an integer.

If the constraints are satisfied, the sums on the left of (5) and (6) will not vary as $p(n)$ is varied, while if S_r has an extreme value, the sum in (4) will not vary for infinitesimal changes in $p(n)$. Thus any linear sum of (4), (5) and (6) will have zero variation for small perturbations of $p(n)$ about the desired solution. Accordingly, we let

$$U = \sum p(n) \log p(n) + \alpha \sum n \, p(n) + \beta \sum p(n)$$

where α and β are constants, and require that

$$\delta U = \sum \left[\log p(n) + 1 + \alpha n + \beta\right]\delta p(n) = 0.$$

This equation will be satisfied for any $\delta p(n)$ if

$$\log p(n) + 1 + \alpha n + \beta = 0$$
$$p(n) = e^{-1-\beta}e^{-\alpha n} = K u^n \tag{7}$$

where $K = e^{-1-\beta}$ and $u = e^{-\alpha}$. The constants K and u may now be evaluated in terms of the constraints. Substitution of (7) into (5) yields $K = 1 - u$, so

$$p(n) = (1 - u)u^n. \tag{8}$$

Substitution of (8) into (6) yields $\bar{n} = u/(1-u)$. Thus adjusting u is equivalent to adjusting $\bar{n}$.

If we substitute (8) into (4), we find for the entropy

$$S_r = - k\left[\log(1 - u) + \frac{u}{1 - u}\log u\right].$$

If we now imagine that the resonator initially contains zero energy ($n = 0$), then the above resonator entropy is produced by draining an energy $\bar{n}h\nu$ from the rest of the system. If the system is at temperature T this produces an entropy change

$$\Delta S_0 = - \frac{\bar{n}h\nu}{T} = - \frac{u}{1 - u}\frac{h\nu}{T}$$

in the rest of the system. By the second law, the "filling" of the resonator will proceed until, in equilibrium

$$\Delta S \equiv S_r + \Delta S_0$$

$$= k\left[-\log(1 - u) - \frac{u}{1 - u}\left(\log u + \frac{h\nu}{kT}\right)\right]$$

is a maximum. Differentiating with respect to u, we find this requires

$$\log u + \frac{h\nu}{kT} = 0$$
$$u = e^{-h\nu/kT}. \tag{9}$$

Using this result, we may rewrite (8)

$$p(n) = (1 - e^{-h\nu/kT})e^{-n(h\nu/kT)}. \tag{10}$$

Since $nh\nu$ is the actual energy W in the resonator, we see from (10) that the probability of a state falls off as $e^{-W/kT}$, as is true in the absence of quantum effects. The effect of quantization is to make only certain discrete energy levels possible. As the frequency ν increases the available levels thin out until at very high frequencies $(\alpha = (h\nu/kT) \gg 1)$ even the lowest excited level, $n = 1$, represents so much energy that $e^{-h\nu/kT}$ is vanishingly small. Figure 1 shows the behavior of $p(n)$ as a function of $\alpha = h\nu/kT$ for $n = 0$ to 5. Figure 2 shows three particular distributions for $\alpha = \frac{1}{4}$, 1, and 4.

When $(h\nu/kT) \ll 1$, the energy levels become very numerous and closely spaced. It is then both permissible and convenient to assume a continuous distribution of energy, $q(W)$, rather than the true discrete distribution, $p(n)$. Since both distributions must give the same probability that the energy $W = nh\nu$ lies in the range $\Delta W = h\nu\Delta n$, we require $q(W)\Delta W = p(n)\Delta n$. Thus from (9)

$$q(W) = \frac{\Delta n}{\Delta W} p\left(\frac{W}{h\nu}\right) = \frac{1}{h\nu}(1 - e^{-h\nu/kT})e^{-W/kT}.$$

Taking the limit as $h\nu \to 0$, we find the Boltzmann distribution

$$q(W) = \frac{1}{kT} e^{-W/kT}. \tag{11}$$

Returning now to (9) we find that the average energy in the resonator is

$$\overline{W} = \bar{n}h\nu = \frac{u}{1 - u}h\nu = \frac{h\nu}{e^{h\nu/kT} - 1}. \tag{12}$$

Figure 3 is a normalized plot of (12) and shows the manner in which the average energy falls with increasing frequency. Later we shall see that this same curve gives the power spectrum for thermal noise.

Mode Densities in One and Three Dimensions

Equations (10) and (12) give the distribution of energy and the average energy in a particular resonant mode. To compute the energy densities and power flows in transmission lines and cavities in thermal equilibrium, we now need to know how the resonant modes in such systems are distributed in frequency.

Let us consider first a lossless transmission line shorted at both ends and capable of supporting a single

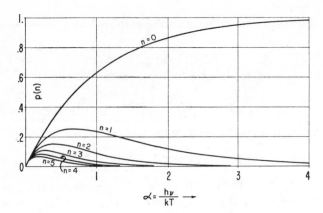

Fig. 1. Probability of resonator containing *n* photons due to thermal excitation.

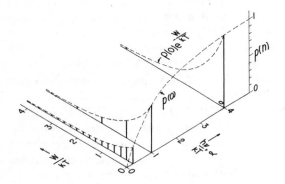

Fig. 2. Probability distributions for three values of α.

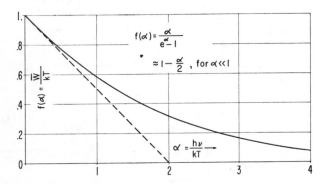

$$f(\alpha) = \frac{\alpha}{e^\alpha - 1}$$

$$\approx 1 - \frac{\alpha}{2}, \text{ for } \alpha \ll 1$$

Fig. 3. Decrease in average thermal energy in a resonator at high frequencies due to quantum effects.

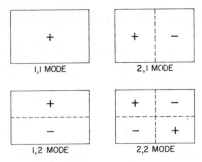

Fig. 4. Modes of a rectangular membrane (indices refer to the number of cells per side).

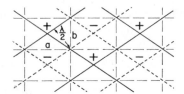

Fig. 5. Waves on a membrane.

density is a constant, independent of frequency.

In three dimensions the situation is more complicated, but is readily analyzed by first considering an intermediate two-dimensional case. Imagine a thin elastic diaphragm stretched over a rectangular frame. Such a diaphragm has an infinite number of normal modes of vibration, each corresponding to a particular standing wave pattern. In the lowest frequency mode, the whole diaphragm vibrates in phase, with the greatest amplitude at the center, the edges remaining motionless. In the next higher mode, a nodal line connects the midpoint of the longer sides, and the two halves of the diaphragm vibrate in opposite phase. Higher normal modes involve one or more nodal lines on each side dividing the entire area into rectangular cells, with adjacent cells vibrating in opposite phase as shown in Fig. 4. Just as standing waves on a transmission line may be resolved into a pair of equal amplitude, oppositely traveling waves, so each of these modes of a diaphragm may be resolved into *two* pairs of oppositely traveling waves as indicated in Fig. 5. The horizontal and vertical dashed lines represent the cell boundaries, while each set of diagonal lines represents the alternate crests (solid) and troughs (dotted) of the standing wave pattern caused by each *pair* of oppositely traveling waves. Note that at the instant shown the crests of all four waves combine to form the maxima in the centers of alternate cells, while the troughs all combine to form the minima in the centers of the other cells. One half period later all crests and troughs will have changed places and, in accomplishing this, each traveling wave will have advanced the distance $\lambda/2$. By similar triangles we see that

$$\frac{\lambda/2}{a} = \frac{b}{\sqrt{a^2 + b^2}}$$

propagation mode. The line will be resonant whenever the length *l* is an integral number, *n*, of half wavelengths. So for any resonant mode, we have $l = n(\lambda/2)$, or $n = (2l/c)\nu$ where *c* is the velocity of propagation. In the second form of the equation it is apparent that *n* is the total number of modes up to the frequency ν. While *n* is limited to integral values, the error in assuming that it varies continuously with ν will be negligible if we take *l* sufficiently large. We conclude, therefore, that there are

$$m_1 = \lim_{l \to \infty} \frac{1}{l} \frac{dn}{d\nu} = \frac{2}{c} \tag{13}$$

modes per unit length per unit bandwidth. Thus in one dimension—in a single propagation mode—the mode when *a* and *b* are the cell dimensions. This may be re-written

$$\left(\frac{2}{\lambda}\right)^2 = \frac{1}{a^2} + \frac{1}{b^2}.$$

Now the electromagnetic or acoustic modes in a rectangular cavity are entirely analogous. Each mode divides the cavity into a number of cells having dimensions which are submultiples of the sides of the cavity as shown in Fig. 6. There are now three pairs of plane waves involved for each mode and the wavelength is given by

$$\left(\frac{2}{\lambda}\right)^2 = \frac{1}{a^2} + \frac{1}{b^2} + \frac{1}{c^2}$$

where *a*, *b*, and *c* are the sides of the cell. Now let $a = A/n_a$, $b = B/n_b$, $c = C/n_c$ where n_a, n_b, and n_c are the numbers of cells along the *A*, *B*, and *C* sides of the cavity. Then

$$\left(\frac{2}{\lambda}\right)^2 = \left(\frac{n_a}{A}\right)^2 + \left(\frac{n_b}{B}\right)^2 + \left(\frac{n_c}{C}\right)^2.$$

This expression will be familiar to those who have worked with microwave cavities.

To find the mode density we now resort to an auxiliary geometric construction. In the first octant of a cartesian coordinate system, we mark off all lattice points whose coordinates are n_a/A, n_b/B, n_c/C where n_a, n_b, and n_c are positive integers, as shown in Fig. 7. The number of modes of wavelength greater than λ is the number of lattice points included by an octant of a sphere whose radius is $2/\lambda$. Neglecting surface and edge effects, which are negligible in the limit as A, B, $C \to \infty$, this number *n* is given by

$$n = \frac{\text{Volume of octant of sphere}}{\text{Volume of unit cell}}$$

$$= \frac{\dfrac{1}{8}\dfrac{4\pi}{3}\left(\dfrac{2}{\lambda}\right)^3}{\dfrac{1}{ABC}} = \frac{4\pi V}{3\lambda^3} = \frac{4\pi\nu^3}{3c^3}V$$

where $V = ABC$ is the volume of the cavity. Thus the mode density is

$$m_3' = \lim_{V\to\infty}\frac{1}{V}\frac{dn}{d\nu} = \frac{4\pi\nu^2}{c^3} \tag{14}$$

modes per unit bandwidth per unit volume. Actually, this result is true for acoustic waves in fluids. For electromagnetic waves there are two independent polarizations so the total number of modes is

$$m_3 = 2m_3' = \frac{8\pi\nu^2}{c^3}. \tag{15}$$

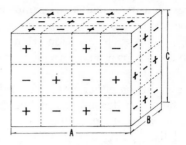

Fig. 6. 4,3,3 mode in a rectangular cavity.

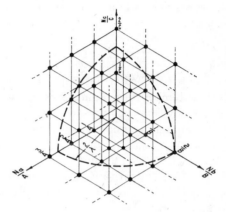

Fig. 7. Lattice points enclosed by octant of sphere.

We note that whereas in one dimension, the mode density is independent of frequency, in three dimensions it increases as the square of frequency. It is this difference that causes the different spectral distribution of black body radiation and thermal noise.

In the above analysis we have assumed lossless lines and cavities. The normal modes are then true spectral lines. Once excited they would continue to resonate at a constant level indefinitely. In any physical system there will be loss and the normal modes will be resonances of finite bandwidth. It is this loss that couples the normal modes to the rest of the system and enables thermal equilibrium to be established. The average energy per mode will be given by (12), the distribution of energy by (10), and the time scale of the fluctuations will depend on the bandwidth of the modes, i.e., upon the degree of coupling.

Thermal Noise

Following the method of Nyquist [7], let us now consider a slightly lossy transmission line shorted at both ends. There will be, as we have seen, $m_1 = 2/c$ normal modes per unit length per unit bandwidth, and each of these modes in thermal equilibrium will, according to (12), have an average energy $\overline{W}$. Hence the thermal energy density on the line will be $\rho_1 = m_1\overline{W}$ J/m/unit bandwidth. As the line is made longer and longer the spectral density of normal modes increases in proportion to length so the energy density along the line is unaffected. Ultimately the line loss prevents resonances, the modes overlap to produce a continuous spectrum, but the energy density remains the same. Half of this energy will be accounted for by waves propagating at the speed of light in each direction. Hence the power flow in one direction past any point is $\psi_1 = c(\rho_1/2) = \overline{W}$ watts/c/s.

If the short circuit on either end is now replaced by a termination, the power density on the line must remain unaffected. For were it to change, the termination would either deliver power to the (lossy) line or absorb energy from it, which cannot happen if everything is at the same temperature. We conclude that the matched load both absorbs and radiates an average power.

$$\psi_1(\nu) = \overline{W} = \frac{h\nu}{e^{h\nu/kT} - 1} \text{ watts/c/s.} \quad (16)$$

If a lossless line is matched at both ends, the two terminations, both at temperature T, are exchanging energy at this rate. We now have an unresonant system and the line length is immaterial. If one termination is cooled to absolute zero so that it can no longer radiate, (16) gives the rate of energy flow from the hot load to the cold one.

If $(h\nu/kT) \ll 1$, we can write with little error $\psi_1 = kT$ watts/c/s, which is the expression normally used in most electronics work.

It is of interest to compute the *total* thermal noise power available from a resistor. This is not infinite as the approximate expression kTB would suggest, but is rather

$$P_1 = \int_0^\infty \psi_1(\nu) d\nu = \frac{(kT)^2}{h} \int_0^\infty \frac{\alpha}{e^\alpha - 1} d\alpha$$

$$= \frac{(kT)^2}{h} \frac{\pi^2}{6} \text{ watts.} \quad (17)$$

At a temperature of 290°K we find $P_1 \approx 4 \times 10^{-8}$ watts. This is the rate at which a resistor would cool if its only heat loss were through electromagnetic radiation over a matched transmission line to a load at $T = 0$. If a transmission line is matched at both ends with terminations at temperature T, the total power flow in both directions is $2P_1$ and the rms voltage on the line is $\sqrt{2P_1R}$. For $T = 290°K$ and $R = 50 \ \Omega$ we find $\sqrt{2P_1R} = 2$ mV.

Expression (16) can be derived in another way that does not make use of the artifice of normal modes. Consider an ideal band-pass filter, having a bandwidth $\Delta\nu$, connected between two matching terminations. As is well known from the sampling theorem, the signal transmitted by such a filter can assume $2\Delta\nu$ statistically independent values per second. The received wave can be thought of as the superposition of a train of elementary pulses having the shape of the filter impulse response. The expression for the energy of the elementary pulse contains one quadratic term, so the pulse energy will have an expectation $\overline{W}/2$. Combining these facts, we see that the filter will deliver from each load to the other a power $\overline{W}\Delta\nu$ watts, so the power spectrum has a density $\overline{W}$ watts/c/s, as given by (16).

Black Body Radiation

In a slightly lossy cavity all the normal modes will be in thermal equilibrium with the conducting walls, which we assume to be at a uniform temperature T. Thus from (12) and (15) we see that the average electromagnetic energy density in the cavity is

$$\rho_3 = m_3\overline{W} = \frac{8\pi h\nu^3}{c^3} \frac{1}{e^{h\nu/kT} - 1} \quad (18)$$

J/m³/c/s. Obviously, this density must be independent of the size and shape of the cavity, for otherwise, when two different cavities were connected by an iris, there would be a net power flow from the one with the higher density to the one with the lower. With both cavities at the same temperature, this is contrary to the second law and we conclude that the mode density m_3, which we derived for a rectangular cavity, is independent of the cavity shape, and indeed, is a property of space itself.

If a lossy patch is applied to one of the walls this cannot change the energy density in the cavity, for were this to happen, there would be a net power flow between the walls and the lossy patch and a temperature difference would develop, contrary to the second law. Thus, radiation and absorption are always balanced at every point on the walls, regardless of the lossiness. The lossiness of a cavity does not affect the energy density inside, but merely (by affecting the Q of the modes) alters the rate at which energy is exchanged between the space and the walls. We conclude that (18) is a universal expression for the energy density in any isothermal enclosure.

Since the radiation field inside the cavity is isotropic (equal power flow per solid angle in all directions) it is a simple matter to compute the power incident on each unit area of wall. Thus

$$\psi_3 = c\rho_3 \frac{\displaystyle\int_{\theta=0}^{\pi/2} \int_{\phi=0}^{2\pi} \cos\theta \sin\theta d\phi d\theta}{\displaystyle\int_{\theta=0}^{\pi} \int_{\phi=0}^{2\pi} \sin\theta d\phi d\theta}$$

$$= \frac{c\rho_3}{4} = \frac{2\pi\nu^2}{c^2} \frac{h\nu}{e^{h\nu/kT} - 1}. \quad (19)$$

If the surface is black all this incident power will be absorbed. However, as we have seen, precisely this same power would be reradiated to maintain thermal equilibrium. Thus (19) is the radiation law for a black body at temperature T.

If a small hole is cut in the wall of a large cavity, (19) gives the rate at which energy will escape through the hole. A small hole in a large cavity always approximates a black body regardless of the "color" of the walls. Naturally, if the walls themselves are nearly black—that is, good absorbers and radiators—at all frequencies in question, the energy exchange between the walls and the enclosed space is rapid, and the presence of a small hole produces little effect on the energy density in the cavity. In other words, the blackest thing in the world is a small hole in a large black box.

Figure 8 shows the black body radiation as given by (19) for two temperatures, 290°K and 5800°K, which are approximately, the surface temperatures of the earth and sun.

Equation (19) can be rewritten

$$\psi_3 = 2\pi \frac{(kT)^3}{h^2c^2} \frac{\alpha^3}{e^\alpha - 1} \qquad (20)$$

where, as before, $\alpha = h\nu/kT$. By differentiating with respect to α, the peak power per cycle is found to occur at $\alpha = 2.82144\cdots$, giving $(\alpha^3/e^\alpha - 1) = 1.4214\cdots$. The peak power density is thus proportional to the cube of the temperature, and the frequency at which this peak occurs is directly proportional to temperature.

Quite frequently the black body radiation law is expressed as a function of wavelength. To do this we note that the new function, $\Phi(\lambda)$ must satisfy the relation

$$\left| \Phi(\lambda)d\lambda \right| = \left| \psi_3(\nu)d\nu \right|$$

where $\nu = c/\lambda$, and $d\nu = -(c/\lambda^2)d\lambda$. Simple substitution gives

$$\Phi(\lambda) = \frac{2\pi hc^2}{\lambda^5} \frac{1}{e^{hc/\lambda kT} - 1}. \qquad (21)$$

This expression peaks at $\alpha = 4.965\cdots$, the shifting of the peak to higher frequencies being a result of the fact that a given increment in wavelength is a greater increment in frequency the higher the frequency.

The total radiated power is found by integrating (19)

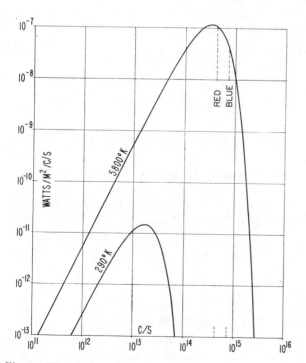

Fig. 8.   Black body radiation at earth and sun temperatures.

or (20) over all frequencies, with the result

$$P_3 = \frac{2\pi^5}{15} \frac{(kT)^4}{h^3c^2}$$

$$= 5.67 \times 10^{-8}T^4 \text{ watts/m}^2 \qquad (22)$$

This is the famous Stefan-Boltzmann radiation law and shows that the total radiation from a black body increases as the fourth power of absolute temperature. Substituting the temperatures 290°K and 5800°K we find from (22) that the radiation from a square meter at room temperature is about 400 watts, while each square meter of the surface of the sun radiates approximately 64 MW!

Returning to (19) we see that at low frequencies, where $(h\nu/kT) \ll 1$,

$$\psi_3(\nu) \approx \frac{2\pi\nu^2}{c^2} kT. \qquad (23)$$

At low frequencies, then, the radiated power per unit bandwidth is proportional to the absolute temperature. Were it not for the high frequency cutoff due to quantum effects, this relation would hold at all frequencies, and among the many catastrophic effects which would ensue would be that of everything on earth, including the retinae of our eyes, being about 290/5800 = 1/20 as bright as the sun. Because of the quantum cutoff, the brightness of a black body at 290°K is less than 10^{-30} that of the sun (or more than 300 dB down) and the total radiation per square meter in the visible range is about 10^{-23} watts, or about three photons per day.

An Antenna Theorem and its Consequences

An antenna is a device that couples a single propagation mode to space. As a transmitting antenna it will, if matched, radiate all the power delivered to it in the proper propagation mode. As a receiving antenna it will capture a certain amount of energy from incident electromagnetic waves and deliver this power in a single propagation mode to the transmission line. The effectiveness of a receiving antenna is conveniently described in terms of its capture cross section, or effective aperture, A. This may be defined as follows:

$$A(\theta, \phi) = \frac{\text{Power received by the antenna}}{\substack{\text{Power per unit area in properly polarized} \\ \text{wave incident from direction } \theta, \phi}}$$

Since antennas are directive, A will in general be a function of the direction of arrival, (θ, ϕ), of the wave, and may be greater or less than the physical cross section of the antenna. The average capture cross section $\overline{A}$ is the value of A averaged over all directions of arrival

$$\overline{A} = \frac{1}{4\pi} \int_{\theta=0}^{\pi} \int_{\phi=0}^{2\pi} A(\theta, \phi) \sin\theta d\phi d\theta.$$

A fundamental antenna theorem states that, for any matched antenna, $\overline{A}$ is a universal constant dependent only on wavelength. We are now in a position to evaluate this constant from purely thermodynamic considerations [13].

Figure 9 shows an antenna matched to a load over a transmission line, the whole assembly enclosed in a cavity at temperature T. Consider now the power passing a plane S-S' on the transmission line. As we have seen earlier, the load at temperature T will radiate a (thermal noise) power

$$P_{\text{rad}} = m_1 c \frac{\overline{W}}{2}. \qquad (24)$$

With a matched antenna, all this power passing S-S' will be radiated; none will be reflected. The thermal power received by the antenna will be

$$P_{\text{rec}} = \tfrac{1}{2} m_3 c W \overline{A}, \qquad (25)$$

where the factor $\tfrac{1}{2}$ arises because only one polarization is effective. With a matched load all this power will pass the plane S-S' and none will be reflected. By the second law, $P_{\text{rad}} = P_{\text{rec}}$, for otherwise the load would grow hotter or cooler than the cavity. Hence, equating (24) and (25) we find

$$\overline{A} = \frac{m_1}{m_3} = \frac{\lambda^2}{4\pi}. \qquad (26)$$

If the antenna is not matched to the transmission line, a similar analysis shows that $\overline{A} = (\lambda^2/4\pi)(1 - |\rho|^2)$ where ρ is the reflection coefficient due to the mismatch. Antenna losses will further reduce $\overline{A}$, so that $\lambda^2/4\pi$ is the maximum possible value.

Just as in a transmitting antenna the power radiated in a given direction can be increased only at the expense of power radiated in other directions, so in a receiving antenna the capture cross section in a particular direction can be increased only at the expense of capture cross section in other directions.

The directive gain of an antenna is the ratio of the power received or radiated in a given direction to that which would be received or radiated if the antenna were nondirectional and is given simply by

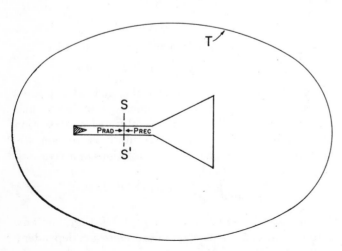

Fig. 9.   An antenna is an isothermal enclosure.

$$g(\theta, \phi) = \frac{4\pi}{\lambda^2} A(\theta, \phi). \qquad (27)$$

This equation may be used to find either A or g if the other quantity is known.

The free space radio transmission law follows simply from the above relations. Assume a transmitting and receiving antenna separated by a distance $d \gg \lambda$, so that the received waves may be considered plane over the receiving antenna cross section. If the transmitting antenna were nondirectional, the transmitted power P_T would be spread uniformly over a sphere, so the power per unit area at the receiver would be $P_T/4\pi d^2$. The actual density will be g_T times this value. Hence the received power will be $P_R = (P_T/4\pi d^2)g_T A_R$ where A_R is the capture cross section of the receiving antenna. Using (27) this may be rewritten

$$\frac{P_R}{P_T} = \left(\frac{\lambda}{4\pi d}\right)^2 g_T g_R, \qquad (28)$$

or in terms of capture cross sections

$$\frac{P_R}{P_T} = \frac{A_T A_R}{\lambda^2 d^2}. \qquad (29)$$

For some antennas, such as large, uniformly illuminated horns or dishes, the capture cross section is approximately equal to the physical area, so (29) is particularly convenient.

When a receiving antenna is in a thermal radiation field, the power received from any direction will depend on the emissivity and temperature of the source and the antenna capture cross section in that direction. If the sources are good absorbers (black) and if we are concerned with frequencies well below the quantum cutoff, then from (23) the power received will be proportional to the source temperature. As a result, to the receiver, a matched, lossless antenna appears as a source with an effective temperature

$$T_{\text{eff}} = \frac{1}{\overline{A}} \int_{\theta=0}^{\pi} \int_{\phi=0}^{2\pi} T(\theta, \phi) A(\theta, \phi) \sin\theta d\theta d\phi$$

$$= \int_{\theta=0}^{\pi} \int_{\phi=0}^{2\pi} T(\theta, \phi) g(\theta, \phi) \sin\theta d\theta d\phi. \qquad (30)$$

where $T(\theta, \phi)$ is the temperature distribution seen by the antenna. If $T(\theta, \phi)$ is a constant T then $T_{\text{eff}} = T$. When the main lobe of an antenna is pointed at a cold source, minor lobes may point at much hotter sources and raise T_{eff} considerably. Thus receiving antennas for radio astronomy or space communication in frequency bands where the sky is cold should have very small minor lobes.

Statistics of Thermal Noise

Noise in the output of a linear amplifier is usually observed as a fluctuating *amplitude*, i.e., a voltage or current or some linearly related quantity. In a receiver, the output fluctuations are the low-pass filtered response of a linear or square law detector to the thermal noise in a band of frequencies. The statistics of the fluctuations in these cases are all different, but can all be derived from the probability distribution of the energy in a thermally excited resonant mode. For simplicity we will consider only frequencies for which $\alpha \ll 1$ and quantum effects may be ignored. This will enable us to deal with continuous distributions. Later we will find that our results apply to the quantum case with only slight modification.

To avoid talking about voltage or current specifically, and to avoid continually worrying about impedance level, let us define an amplitude a as a real quantity which may have either sign and whose square is instantaneous power. Thus $a^2 = e^2/R = i^2 R$ and at any time one may write $e = a\sqrt{R}$ or $i = a/\sqrt{R}$. A sinusoid of power P (peak power $2P$), frequency ν and phase ϕ may then be written as $a = A \cos(\omega t + \phi)$ or as $a = \overline{A}e^{i\omega t}$ real part, where $A = \sqrt{2P}$, $\overline{A} = Ae^{i\phi}$, and $\omega = 2\pi\nu$.

Let us now couple a single thermally excited resonant mode of frequency ν_0 to a transmission line so that, when the energy in the mode is W, a power $P = BW$ is delivered to the line. For the moment we will define B simply as a coupling constant having the dimensions of frequency. The signal on the line will consist of an amplitude and phase modulated sinusoid of frequency ν_0. The instantaneous power on the line will therefore vary cyclically at a frequency $2\nu_0$. The total energy in the resonant mode does not have this cyclic variation, instead it oscillates between electric and magnetic energy at this rate. Since it is the *total* energy in the mode that is distributed according to (11), we must define P as the *average* power in a sinusoid of peak amplitude A, i.e., $P = A^2/2$. If $q(P)$ is the probability distribution of $P = BW$, we then require that $q(P)dP = q(W)dW$, so that from (11)

$$q(P) = \frac{1}{kTB} e^{-P/kTB}. \tag{31}$$

Likewise the peak amplitude (or envelope) distribution, $p(A)$, must satisfy the condition $p(A)dA = q(P)dp$. Since $P = A^2/2$ and $dP = AdA$, we find

$$p(A) = \frac{A}{kTB} e^{-A^2/2kTB}, \qquad A \geq 0. \tag{32}$$

This distribution is of the form $xe^{-x^2/2}$, known as a Rayleigh distribution.

Because, in the thermal process, there is nothing in the nature of a clock to synchronize the excitation of the mode, we conclude that ϕ will vary randomly and is equally likely to have any value from 0 to 2π. This is also the maximum entropy distribution. If we plot the vector $\overline{A} = Ae^{i\phi}$ the tip will be equally likely to lie anywhere on a circle of radius A as shown in Fig. 10. $p(A)dA$ is thus the probability that the tip of $\overline{A}$ lies anywhere in the annulus of radius A and width dA.

Since the area of this annulus is $2\pi A dA$, the probability that the tip lies in an elementary area, dS, is

$$q(A)dS = \frac{dS}{2\pi A dA} p(A)dA = \frac{p(A)}{2\pi A} dS.$$

We conclude that the probability density distribution over the plane is

$$q(A) = \frac{1}{2\pi kTB} e^{-A^2/2kTB}. \tag{33}$$

This is a two dimensional Gaussian distribution. Now let us take as the elemental area $dS = da_x da_y$ located at a_x and a_y. We then find, since $A^2 = a_x^2 + a_y^2$, that

$$q(a_x a_y)dS = \frac{1}{2\pi kTB} e^{-(a_x^2 + a_y^2)/2kTB} da_x da_y$$

$$= p(a_x)da_x p(a_y)da_y$$

where

$$p(a_x) = \frac{1}{\sqrt{2\pi kTB}} e^{-a_x^2/2kTB} \tag{34a}$$

and:

$$p(a_y) = \frac{1}{\sqrt{2\pi kTB}} e^{-a_y^2/2kTB}. \tag{34b}$$

These are both Gaussian distributions of the form $(1/\sigma\sqrt{2\pi})\, e^{-(x^2/2\sigma^2)}$, with $\sigma = \sqrt{kTB}$. Since $q(A)$ is their simple product, a_x and a_y are statistically independent, a consequence of the complete randomness of ϕ.

If instead of plotting the vector $\overline{A}$, we plot $\overline{A}e^{i\omega t}$ we obtain the diagram shown in Fig. 11, which may be thought of as produced by Fig. 10 rotating counterclockwise at ν_0 r/s. The projection A_R of $Ae^{i(\omega_0 t + \phi)}$ onto the real axis is the instantaneous amplitude, a. Since the probability distribution $q(A)$ is a function of the radius A only, the distribution obtained by rotating it at any speed will be the same function of radius. It follows that the distribution of the instantaneous amplitude a is the same as that of a_x or a_y, that is

$$p(a) = \frac{1}{\sqrt{2\pi kTB}} e^{-a^2/2kTB}; \tag{34c}$$

again a Gaussian distribution. Thus the instantaneous amplitude produced by the excitation of a single mode has a Gaussian distribution.

Now consider the signals from two normal modes. Each will have a Gaussian distribution and the two will be statistically independent. If a_1 and a_2 are the two amplitudes, and $p_1(a_1)$ and $p_2(a_2)$ are their distributions, then the distribution of their sum a is

$$p(a) = \int_{-\infty}^{\infty} p_2(a - a_1)p_1(a_1)da_1. \tag{35}$$

$$p(a) = \frac{1}{\sqrt{2\pi kT(B_1 + B_2)}}\, e^{-a^2/(2kT(B_1+B_2))}. \quad (36)$$

Since the above process may be repeated any number of times to include the signals from any number of normal modes, and since the shape of the resulting distribution is independent of the weighting factors or delays that may have been applied to the signals from any of the modes, we conclude that *any linearly filtered or amplified thermal noise wave has a Gaussian distribution of instantaneous amplitude.*

A Gaussian distribution has the greatest entropy for a given average power. The proof is exactly parallel to that given on page 437. In (4) and (5) $p(n)$ is replaced by $p(a)$. The sums are replaced by integrals, and (6) is replaced by a constraint on the second moment rather than the first, i.e., we require

$$P = \int_{-\infty}^{\infty} a^2 p(a)\, da = \text{constant}.$$

Usually the Gaussian amplitude distribution of thermal noise is developed on the basis of a model source containing a very large number of independent generators each of which produces an infinitesmal contribution to the resultant amplitude. For example, in a resistor, each conduction band electron as it is buffeted about produces a random current wave. The total current is then shown to have a Gaussian distribution by the Central Limit Theorem. This theorem says that the distribition resulting from the convolution of a large number of elementary distributions approaches a Gaussian distribution even though the elementary distributions themselves may be non-Gaussian and all different. While a rigorous proof of this theorem (or even a rigorous statement) is lengthy, the central idea is that since probability distribituions are always positive functions, their transforms have their greatest value when the argument is zero, and fall off on either side as the square of the argument. The product of a large number of such transforms tends toward a Gaussian shape and therefore so does its inverse transform which is the resultant distribution. Figure 12 shows an example of repeated convolution. Two rectangular distributions convolve to form a triangular distribution. A second convolution of this with a rectangular distribution produces a distribution composed of three parabolic segments. The rapid trend toward a Gaussian shape is already evident.

Our derivation of the Gaussian distribution arising from coupling to a single resonant mode would seem at first to indicate that one need not assume a complicated model source. However, one must remember that the distribution of a thermally excited mode is the result of a large number of independent statistical forces acting on it at random. The resonant mode model was chosen not to obscure this complexity, but merely because quantum effects are easily included in the analysis.

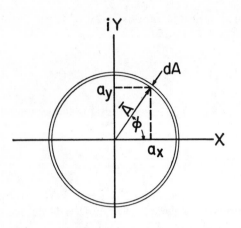

Fig. 10. How the Rayleigh distribution of A produces Gaussian distributions for a_x and a_y.

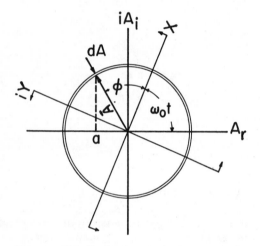

Fig. 11. The distribution of a is the same as the distribution of a_x and a_y.

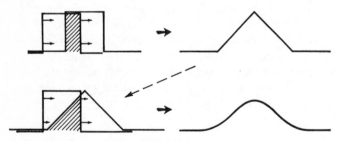

Fig. 12. Approach to a Gaussian distribution produced by successive convolution.

This is because each value of a_1 that produces a requires that $a_2 = a - a_1$. The total probability of a is thus the product of $p_1(a_1)$ and $p_2(a_2)$ integrated over all a_1. Equation (35) will be recognized as a convolution integral. The Fourier transform of p is therefore the product of the transforms of p_1 and p_2. Since p_1 and p_2 are both Gaussian, their transforms are also Gaussian and so is the product. The inverse transform of the product, (i.e., p itself) is therefore also Gaussian. Carrying out these steps, we find

Assuming the amplitude a includes the equally weighted signals from n normal modes, (36) would become

$$p(a) = \frac{1}{\sqrt{2\pi kT \sum_n B_i}} e^{-a^2/2kT\Sigma_n B}. \tag{36a}$$

The average power $\overline{a^2}$ is the second moment of (36a), that is

$$\overline{a^2} = \int_{-\infty}^{\infty} a^2 p(a) da.$$

Evaluating this we find

$$\overline{a^2} = kT \sum_{i=1}^{n} B_i.$$

Since the total thermal noise power in a bandwidth B is kTB, we conclude that $\sum B_i = B$ is the total bandwidth of the ideal filter through which the noise is viewed. Thus B in (34) and the preceding equations may be identified as bandwidth, and a as the total noise amplitude in this bandwidth. Working back through these equations we see now that *the envelope, $A(t)$, of any thermal noise wave has a Rayleigh distribution and the power, $P = A^2/2$ represented by this envelope is exponentially distributed.*

Consider now a noise wave that has been passed through an ideal filter of bandwidth B centered at ν_0 as shown in Fig. 13(a). If $B = 2\nu_0$ we have a low-pass filter, if $B < 2\nu_0$ we have a band-pass filter. In any case the wave may be written as

$$\begin{aligned} a &= A \cos(\omega_0 t + \phi) \\ &= A \cos\phi \cos\omega_0 t - A \sin\phi \sin\omega_0 t \\ &= a_x \cos\omega_0 t - a_y \sin\omega_0 t, \end{aligned} \tag{37}$$

where again

$$\omega_0 = 2\pi\nu_0.$$

The power spectrum of $a(t)$ is flat from $\nu_0 - (B/2)$ to $\nu_0 + (B/2)$, so the power spectra of $a_x(t)$ and $a_y(t)$ (which amplitude modulate $\cos\omega_0 t$ and $\sin\omega_0 t$) must be flat from 0 to $B/2$. If we multiply $a(t)$ by $2\cos\omega_0 t$ and $2\sin\omega_0 t$, as in homodyne detection, we obtain

$$a_c = 2a\cos\omega_0 t = a_x(1 + \cos 2\omega_0 t) - a_y \sin 2\omega_0 t \tag{38a}$$

$$a_s = 2a\sin\omega_0 t = a_x \sin 2\omega_0 t - a_y(1 - \cos 2\omega_0 t). \tag{38b}$$

The spectra of a_c and a_s consist of two portions, one extending from 0 to $B/2$ and the other from $2\nu_0 - (B/2)$ to $2\nu_0 + (B/2)$ as shown in Fig. 13(b). If $(B/2) < 2\nu_0 - (B/2)$, i.e., if $B < 2\nu_0$, these two portions may be separated by filters. The upper portion of (38a) is a replica of (37) with ν_0 replaced by $2\nu_0$. The same is true of (38b) except that ϕ is decreased by $\pi/2$. The lower portion consists of the independent, Gaussian distributed, low-pass filtered noise waves $a_x(t)$ and $a_y(t)$.

Using nonlinear devices one could, from these waves,

generate new amplitudes proportional to $(a_x^2 + a_y^2)/2 = P$ and $\sqrt{a_x^2 + a_y^2} = A$. As we have seen, P has an exponential distribution and A a Rayleigh distribution. However $a_x(t)$ and $a_y(t)$ might as well be *any two independent noise waves* having the same statistics. We conclude that the sum of the squares of any two independent, similarly band-limited, low-pass thermal noise waves has an exponential distribution, while their quadratic sum has a Rayleigh distribution.

If a thermal noise wave $a(t)$ has a power spectrum, $w_a(\nu)$, then it can be shown [8], [9] that the power spectrum of a new amplitude $b(t) = a^2(t)/a_0$, where a_0 is a constant of proportionality, is given by

$$w_b(\nu) = \frac{1}{a_0^2} \int_{-\infty}^{\infty} w_a(x) w_a(\nu - x) dx, \tag{39}$$

with the understanding that $\nu \neq 0$ in the integral, and that $w(\nu) = w(-\nu) = \psi(\nu)/2$. Thus the continuous part of the power spectrum of $b(t)$ is the convolution of the power spectrum of $a(t)$ with itself, when both are expressed as symmetrical functions about $\nu = 0$. At $\nu = 0$, there is a dc component

$$\overline{b(t)} = \frac{\overline{a^2(t)}}{a_0} = \frac{\int \psi(\nu) d\nu}{a_0} = \frac{kTB}{a_0}.$$

From the above we conclude that if the wave $b(t)$ is generated from the $a(t)$ of (37) the spectrum will consist of a dc component kTB/a_0, and a continuous component

$$\begin{aligned} \psi_b(\nu) &= \frac{k^2 T^2 B}{4a_0^2}\left(1 - \frac{\nu}{B}\right), & 0 < \nu < B \\ &= \frac{k^2 T^2 B}{8a_0^2}\left(1 - \frac{|\nu - 2\nu_0|}{B}\right), & 2\nu_0 - B < \nu < 2\nu_0 + B, \end{aligned}$$

as shown in Fig. 14. If $B < 2\nu_0 - B$, i.e., if $B < \nu_0$, the lower portion of the spectrum may be separated from the upper by filtering. The wave thus obtained (including the dc component) will be proportional to P and hence will be exponentially distributed. This is therefore the distribution of amplitude in the output of a square law detector, provided all low frequency components (and no frequencies around $2\nu_0$) are passed by the output filter.

The simple square of a thermal noise wave does *not* have an exponential distribution. If $q(b)$ is the distribution of $b(t) = a^2(t)/a_0$, we require that $q(b)db = [q(a) + q(-a)]da$, since both a and $-a$ produce b. Now since $da = (\sqrt{a_0/b})(db/2)$ we find

$$q(b) = \begin{cases} \dfrac{1}{\sqrt{2\pi kTB}} \sqrt{\dfrac{a_0}{b}}\, e^{-a_0 b/2kTB}, & b > 0 \\[2mm] 0, & b < 0. \end{cases} \tag{40}$$

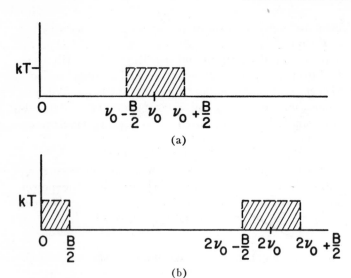

Fig. 13. (a) Power spectrum of a band of thermal noise. (b) Effect on power spectrum of multiplication by $2 \sin \omega_0 t$ or $2 \cos \omega_0 t$.

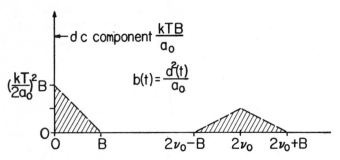

Fig. 14. Power spectrum of the square $b(t)$ of a band-passed noise wave $a(t)$.

Since $a(t)$ crosses the axis frequently, $a^2(t)$ tends to dwell there and $q(b) \to \infty$ as b decreases to zero. When two independent noises are squared and added, the probability of both being nearly zero at the same time is small enough to remove the singularity at $b=0$. If a great many noise waves are squared and added the sum becomes nearly constant with only a small Gaussian fluctuation about the mean.

The spectrum of the envelope A in (37) is not definitely frequency limited as is the spectrum of P. This can be seen qualitatively from Fig. 10. Even though a_x and a_y are smooth band-limited functions, each time the tip of the vector $\overline{A}$ passes through (or near) the origin, a sharp cusp (or near cusp) will be produced for the minimum of A. Thus $A(t)$ contains frequency components greater than B. On the other hand, if a linear detector is supplied with a thermal noise wave (37) for which $\nu_0 \gg B$, and if the low-pass filter following detection also has a bandwidth much greater than B but less than $\nu_0 - (B/2)$, the output will be a good replica of A and hence be Rayleigh distributed. A "half wave" linear detector characterized by

$$b(t) = a(t), \qquad a > 0$$
$$\qquad\qquad = 0, \qquad a < 0$$

will produce, after filtering, the wave

$$b_1(t) \approx \frac{1}{\pi} A(t). \tag{41}$$

Similarly a full wave detector characterized by $b(t) = |a(t)|$ will produce

$$b_2(t) \approx \frac{2}{\pi} A(t). \tag{42}$$

In the latter case the output filter may pass frequencies up to $2\nu_0 - B$.

When a sinusoid of frequency ν_0 is also present in the input to a linear or square law detector the output noise is considerably altered. The distribution of Fig. 10 is then displaced to the tip of the carrier vector A_c. If the carrier power is much greater than the thermal noise power in the bandwidth B, i.e., if $A_c^2 \gg \overline{A^2}$, then the resultant envelope will be very nearly A_c plus the projection of A on A_c. A wave of the form $a_x(t)$ or $a_y(t)$ will thus be added to A_c. Both square law and linear detectors will produce a constant output with an added noise wave having a nearly Gaussian distribution. Modulation of A_c will produce proportional modulation of this noise wave in the case of the square law detector, while very little cross modulation will result with a linear detector so long as the minimum value of A_c is much greater than $\sqrt{\overline{A^2}}$. The case for which A_c and $\sqrt{\overline{A^2}}$ are comparable is more complicated and will not be considered here [14].

There are of course many other statistical properties of thermal noise that we have not discussed. To do so would lead us too far from our central theme. The interested reader will find these treated in many definitive books and papers on the subject [8], [9]. The main purpose of this section has been to relate the maximum entropy statistics produced by thermal equilibrium to the noise waves observed in communications circuits.

Part II: Quantum Noise

In Part I we found that thermal noise falls off exponentially with increasing frequency when $(h\nu/kT) \gg 1$. If thermal noise were the only noise present, then the received signal power necessary for a given communication rate would also fall exponentially with an increase in channel frequency. By choosing ν high enough one could receive the entire contents of the *Encyclopedia Britannica*, say, with a total received energy less that of one photon of frequency ν. Clearly there must be another source of noise to prohibit such impossible economies. This additional noise arises from the quantization itself.

Radiation sensitive devices fall into two fairly distinct categories. One class of devices responds to the amplitude of the wave function associated with the radiation and produces an output statistically related to the amplitude in both magnitude and phase. At high enough input levels, the output will be a good replica of the input wave. These devices we will call (linear) *amplifiers*. Examples are ordinary triode or transistor amplifiers, masers, and lasers. The other class of devices responds to the square of the magnitude of the input wave amplitude (or the product of the amplitude and its complex conjugate) and produces an output statistically related to this quantity, i.e., to the power. These devices we will call *power*, or *energy detectors*. Examples are the eye, photocells, and photon counters. Power detectors can be used as mixers to provide linear amplification, the output being proportional to the input amplitude, or phase, or both. We shall discuss this case at length later since it illustrates how the noise of a detector is converted to the noise of an amplifier.

In power detectors the noise may be considered to be shot noise arising from the arrival of discrete photons. In amplifiers the input signal (i.e., the wave function) may be considered noiseless, but the amplifying mechanism itself introduces a form of shot noise. Since shot noise is involved in either case, let us first study its more important characteristics.

Shot Noise

Whenever discrete particles arrive at random times there will be fluctuations in the rate of arrival. It is these fluctuations that constitute shot noise. The hailstorm of electrons arriving at the plate of an emission limited diode is perhaps the most familiar example [5], [8].

Let us assume that a particle is equally likely to arrive at any time, and that the average rate of arrival is r. By this we mean that if we measure the number n that arrive in a time t the quantity $n(t)/t$ approaches a limit, r, as $t \to \infty$. This implies that the process is statistically stationary, at least during the time of observation. Under these conditions the numbers of arrivals in a given length of time are distributed according to the well-known Poisson distribution.

Under the assumption given, the probability of receiving a particle in the time $\Delta t \ll (1/r)$ is then $r\Delta t$. The probability of *not* receiving a particle during this time is $1 - r\Delta t$. Thus the probability, $p_0(t)$, of no arrivals during the time t is

$$p_0(t) = \lim_{\Delta t \to 0} (1 - r\Delta t)^{t/\Delta t}$$

since there are $t/\Delta t$ independent intervals during which no arrival must occur. Taking the limit we find

$$p_0(t) = e^{-rt}.$$

The probability $p_1(t)$ of exactly one arrival in the time t, is the probability of receiving a particle in the interval τ to $\tau + d\tau$ and none before or after, summed over all τ. That is

$$p_1(t) = \int_0^t p_0(\tau)[r d\tau] p_0(t - \tau)$$

$$= rt e^{-rt}.$$

The probability $p_2(t)$ of two arrivals is the probability

of getting one up to time τ, one during τ to $\tau + d\tau$, and none thereafter, summed over all τ. Thus:

$$p_2(t) = \int_0^t p_1(\tau)[r d\tau] p_0(t - \tau)$$

$$= \frac{(rt)^2}{2!} e^{-rt}.$$

Continuing in this fashion we find that the probability of exactly n arrivals in the time t is

$$p_n(t) = \frac{(rt)^n}{n!} e^{-rt}. \tag{43}$$

Figure 15 shows the behavior of $p_n(t)$ for values of n from 0 to 4. (It is interesting to note that $p_n(t)$ has the same form as the impulse response of a low-pass filter having $n+1$ simple poles, e.g., a multistage amplifier with simple RC cutoffs.)

The *expectation*, i.e., the average number $\bar{n}$ of arrivals during a time t that we would expect to find on repeated trials, is the first moment of the distribution (43). Evaluating this, we find

$$\bar{n} = \sum_{n=0}^{\infty} n p_n(t) = rt, \tag{44}$$

as it should be from the definition of r. Using (44) we may rewrite (43) in the form

$$p(n) = \frac{\bar{n}^n}{n!} e^{-\bar{n}} = \frac{\bar{n}}{n} p(n - 1). \tag{45}$$

Figure 16 shows $p(n)$ for $\bar{n} = 1, 3,$ and 10.

Now let $\delta n = n - \bar{n}$ be the fluctuation about the mean $\bar{n}$. The mean square of the fluctuation is then

$$\overline{(\delta n)^2} = \overline{(n - \bar{n})^2} = \overline{n^2 - 2\bar{n}n + (\bar{n})^2}.$$

Since $\overline{2\bar{n}n} = 2\bar{n}\bar{n}$ we have

$$\overline{(\delta n)^2} = \overline{n^2} - (\bar{n})^2. \tag{46}$$

This is analogous to the familiar statement that the ac power is the total power minus the dc power. Of course $\overline{n^2}$ (the expectation of n^2) is the second moment of (45)

$$\overline{n^2} = \sum_{n=0}^{\infty} n^2 p(n) = \bar{n} + (\bar{n})^2. \tag{47}$$

Substituting in (46) we find the remarkably simple result

$$\overline{(\delta n)^2} = \bar{n}. \tag{48}$$

The rms dispersion of a Poisson distribution is the square root of the mean.

When $n \gg 1$ the factorial in (45) may be replaced by Stirling's approximation

$$n! = \sqrt{2\pi n}\, n^n e^{-n}.$$

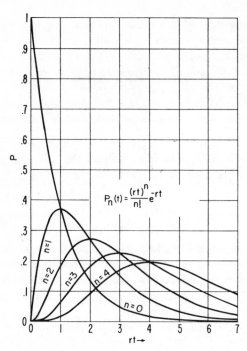

Fig. 15.   Probability of receiving n arrivals in an interval rt.

$$P_n(t) = \frac{(rt)^n}{n!} e^{-rt}$$

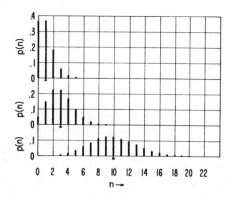

Fig. 16.   Poisson distributions for $\bar{n}=1, 3,$ and 10.

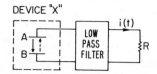

Fig. 17.   Model circuit for shot noise.

Upon substituting $n = \bar{n} + \delta n$ and observing that

$$\lim_{\bar{n} \to \infty} \left(1 + \frac{\delta n}{\bar{n}}\right)^{\bar{n}+\delta n+(1/2)} = e^{\delta n+(\delta n)^2/2\bar{n}},$$

as may be seen by taking the logarithm of both sides and noting that

$$\log(1+x)^m = m\log(1+x) = m\left(x - \frac{x^2}{2} + \cdots\right),$$

we find

$$p(\delta n) = \frac{1}{\sqrt{2\pi\bar{n}}} e^{-(\delta n)^2/2\bar{n}}. \tag{49}$$

Thus when the average number of arrivals during the observing time is large, the fluctuations approach a Gaussian distribution about the mean with $\sigma = \sqrt{\bar{n}}$. This trend is apparent in Fig. 16.

Let us now use the above results to calculate the shot noise in a device X shown in Fig. 17. For the moment we need only say that in device X, electrons pass from A to B, or holes pass from B to A, or both, at random times and with negligible transit time. Each carrier that arrives produces an impulse of current. If we were to observe the current with a device having an impulse response short compared to the mean time between impulses, we would see the individual arrivals. However, we will choose to observe the current after passage through a low pass filter, so that the current at any time is a weighted sum of many past arrivals.

For simplicity we will choose a $\sin x/x$ filter; one whose impulse response is a unit area rectangle of height $1/T$ and duration T. The output current at any instant is then qn/T where q is the charge per carrier and n is the number of arrivals in the last T seconds. We thus have

$$i = \frac{q}{T}n, \tag{50a}$$

$$I = \frac{q}{T}\bar{n}, \tag{50b}$$

$$\overline{(\delta i)^2} = \frac{q^2}{T^2}\overline{(\delta n^2)} = \frac{q^2}{T^2}\bar{n} = \frac{q}{T}I. \tag{50c}$$

Now our filter has a frequency response

$$F(s) = \frac{1}{sT}(1 - e^{-sT})$$

$$F(\nu) = \frac{\sin \pi\nu T}{\pi\nu T} e^{-i\pi\nu T}.$$

Since we have assumed the transit time negligible, the individual current impulses will have a flat spectrum, and so will the noise they produce. At each frequency the filter attenuates the noise *power* by the factor $|F(\nu)|^2$. Hence the effective bandwidth is

$$B = \int_0^\infty |F(\nu)|^2 d\nu = \frac{1}{2T}. \tag{51}$$

Substituting $1/2B$ for T in (50c), we find

$$\overline{(\delta i)^2} = 2qIB. \tag{52}$$

A more sophisticated analysis [4], [8] shows that this result is independent of the filter shape, provided B is defined by the first equality in (51). In particular, it holds for an ideal filter of bandwidth B so the spectral power density is $2qI$. If the number of arrivals in the time $1/2B$ is large, i.e., if $I \gg 2qB$ the fluctuations will have a Gaussian distribution, while if $I \ll 2qB$ individual arrivals will be resolved.

It is important to note that I in (52) is the absolute sum of all currents produced by randomly arriving carriers. The dc components of two such currents may subtract in an external circuit but their noise powers add. As an example let us suppose that device X is a p-n junction diode. The current in such a diode obeys the law

$$i = I_s(e^{qV/kT} - 1), \tag{53}$$

where V is the voltage applied to the junction, and I_s is the saturation current under reverse bias. The current i is the difference of two currents: a forward current $I_s e^{qV/kT}$ due to energetic majority carriers that make their way across the junction against the retarding field, and a reverse current I_s due to minority carriers that get swept back across the junction by the field, as shown in Fig. 18. These two currents are statistically independent and both consist of randomly arriving carriers. Hence the effective shot-noise current is

$$I = I_s(e^{qV/kT} + 1) = i + 2I_s, \tag{54}$$

and

$$\overline{(\delta i)^2} = 2qB(i + 2I_s). \tag{55}$$

By differentiating (53) we find the (incremental) junction conductance

$$G = \frac{q}{kT} I_s e^{-qV/kT} = \frac{q}{kT}(i + I_s). \tag{56}$$

The greatest noise power will be received from the diode if the load resistance $R = 1/G$, and will be given by $P = \frac{1}{4}\overline{(\delta i)^2}/G$. Hence dividing (55) by four times (56) we find

$$P = \frac{kTB}{2} \frac{i + 2I_s}{i + I_s}. \tag{57}$$

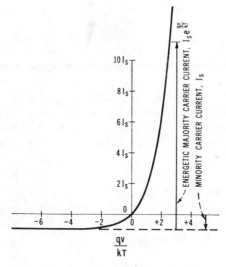

Fig. 18.   Junction diode characteristic.

If $i = 0$, we see that $P = kTB$, which is the *thermal* noise power available from any resistor. We should be very distressed to find any other result since, with no applied voltage, the diode is in thermal equilibrium. We note from (57) that under heavy forward bias ($i \gg I_s$) the diode is half as noisy as a resistor. This may seem a bit surprising, and in many diodes may in fact not be true because of surface effects. However, the result is theoretically possible. There are other examples of *active* devices producing impedances quieter than kTB. The input resistance presented by an amplifier with shunt feedback may have a noise much less than kTB. Under reverse bias ($i \rightarrow -I_s$) the available noise power increases without limit. This is because the incremental conductance, G, approaches zero as the external current approaches $-I_s$. The diode then behaves as a current generator delivering a current I_s into an arbitrarily large load resistor. A similar analysis can be made for the shot noise of thermionic emission.

Quantum Noise in Detectors

A photocell is a good example of a radiation power detector in that the current out is directly proportional to the received power. So let us now assume that device X is a photocell having a quantum efficiency η. This means that every received photon has a probability η of releasing an electron (or forming a hole-electron pair). Before proceeding further, we must ask whether or not the electrons will be regularly or randomly emitted.

If $\eta = 1$ then each photon produces an electron and the series of electrons produced will bear a one to one correspondence with the series of photons captured. In this case, one might think that under constant illumination a regularly spaced stream of electrons would result. No such orderly situation exists, however. According to a well established principle of quantum mechanics, amply verified by experiment, the intensity of a radiation field, i.e., the product of the amplitude vector by its complex conjugate, merely specifies the *probability* of intercepting a photon. Thus even if the illumination is a coherent monochromatic wave, of constant power P, photons will be received at random, at an average rate $\bar{n}/T = P/h\nu$, and the probability that one will be received in the time dt is $(P/h\nu)\, dt$. If $\eta < 1$, not all photons will produce electrons but the misses produce random deletions in an already random series of events.

From the above we conclude that, for a photocell, (50a) and (50b) may be rewritten

$$i = \frac{q}{T}\eta n, \tag{58a}$$

$$I = \frac{q}{T}\eta\bar{n} = \frac{\eta q}{h\nu}P, \tag{58b}$$

where n and $\bar{n}$ now refer to the number of photons producing ηn and $\eta\bar{n}$ electrons, respectively. Similarly (52) becomes

$$\overline{(\delta i)^2} = 2qB\frac{\eta q}{h\nu}P. \qquad (59)$$

We see that both the average current I and the mean square fluctuation $\overline{(\delta i)^2}$, are proportional to the received power. The signal-to-noise ratio is

$$\mathrm{SNR} \equiv \frac{I^2}{\overline{(\delta i)^2}} = \eta\,\frac{P}{2h\nu B} \qquad (60)$$

and increases linearly with the received power. Thus, in a sense, the photocell behaves as if there were an input spectral noise power density of $2h\nu/\eta$ watts/c/s.

In order to facilitate later comparison of the performance of a detector with that of a linear amplifier, let us assume that the filtered photocell current is passed through a nonlinear device, which forms an amplitude, a, proportional to the square root of i. That is, we assume $a(t) = \sqrt{E_0 i(t)}$, where E_0 is a constant of proportionality. The combination is the equivalent of an ordinary "linear" detector. We then have

$$\delta a = \frac{1}{2}\sqrt{\frac{E_0}{i}}\,\delta i \qquad (61\mathrm{a})$$

and

$$\bar{a} = \sqrt{E_0 I} = \sqrt{E_0\,\frac{\eta q}{h\nu}\,P}. \qquad (61\mathrm{b})$$

From which it follows that

$$\overline{(\delta a)^2} = \frac{1}{4}\,\frac{E_0}{I}\,\overline{(\delta i)^2} = \frac{qE_0 B}{2}, \qquad (62)$$

and that

$$\mathrm{SNR} = \frac{\bar{a}^2}{\overline{(\delta a)^2}} = \eta\,\frac{2P}{h\nu B}. \qquad (63)$$

From (62) we see that *the noise power is now independent of received power level* (though the statistics of the noise will change as $P \to 0$) so (63) is more nearly comparable with linear amplifiers. The equivalent noise power spectral density referred to the input is now $h\nu/2\eta$ watts/c/s, or one-fourth that found from (60). While the present device is an absolute-value detector, we shall see later that the same result can be achieved for a truly linear amplitude (or phase) detector, i.e., one that preserves sign.

The Uncertainty Principle

The limiting performance of linear amplifiers, mixers, and homodyne detectors is set by the uncertainty principle, which, for photons at least, can be regarded as a purely mathematical consequence of Planck's law, $E = h\nu$.

Fourier analysis shows that the shorter a pulse is, the broader will be its spectrum, and conversely, the narrower the spectrum, the longer the pulse must be. The question then arises: What pulse has the least duration for a given bandwidth? Before an answer can be given we must define precisely what we mean by the duration or bandwidth of continuous functions.

A convenient and meaningful definition of the duration, Δt, of a function $f(t)$ is the radius of gyration of $f^2(t)$. This is the square root of the second moment of $f^2(t)$ divided by the square root of the integral of $f^2(t)$:

$$\Delta t = \left[\frac{\displaystyle\int_{-\infty}^{\infty} t^2 f^2(t)\,dt}{\displaystyle\int_{-\infty}^{\infty} f^2(t)\,dt}\right]^{1/2}. \qquad (64\mathrm{a})$$

Similarly the bandwidth, $\Delta\nu$, may be defined as

$$\Delta\nu = \left[\frac{\displaystyle\int_{-\infty}^{\infty} \nu^2\,|F(\nu)|^2\,d\nu}{\displaystyle\int_{-\infty}^{\infty} |F(\nu)|^2\,d\nu}\right]^{1/2}, \qquad (64\mathrm{b})$$

where $F(\nu)$ is the Fourier transform of $f(t)$.

With these definitions the pulse shape that minimizes $\Delta\nu\Delta t$ turns out to be Gaussian, that is $f(t) = Ae^{-(t^2/2t_0^2)}$. The proof is a straightforward problem in the Calculus of Variations. For this optimum pulse, $\Delta\nu\Delta t = 1/4\pi$. This means that for any pulse whatever

$$\Delta\nu\Delta t \geq \frac{1}{4\pi}. \qquad (65)$$

In these expressions, $\Delta\nu$ and Δt may be thought of as the rms dispersions of the pulse energy in frequency and in time. The more we localize the energy in frequency the less we are able to localize it in time, and we have only the mathematics to blame for this.

Now if we are talking about a photon, for which $E = h\nu$, localization in frequency means determination of its energy. Since $\Delta E = h\Delta\nu$, the physical counterpart of (65) is

$$\Delta E\Delta t = \frac{h}{4\pi}. \qquad (66)$$

The more accurately we know the energy of a photon the more nearly monochromatic its wave function becomes, and the less accurately can we know its (probable) time of arrival. Planck's law $+$ (65) $=$ (66).

Let us assume photons are being produced by a coherent source of known frequency so that there is no doubt as to their individual energy. If, after considerable inverse square attenuation, say, we receive a small fraction of the emitted photons, there will be random fluctuations in the energy received in a given time, corresponding to the fluctuations in the number of photons received. Substituting $\Delta E = h\nu\Delta n$ in (66) and letting $2\pi\nu\Delta t = \Delta\phi$ we find

$$\Delta n\Delta\phi \geq \tfrac{1}{2}. \qquad (67)$$

This says that if we wish to know the phase accurately we must tolerate a correspondingly large uncertainty in the number of photons. Since

$$\Delta n \equiv \sqrt{\overline{(\delta n)^2}} = \sqrt{\bar n}$$

this means we must receive a large number of photons to know the phase of the wave accurately. If we forego any phase knowledge, we can know n exactly. This is the situation with a particle counter, which gives no phase information. The exact count of *received* photons is of course only a statistical measure of the number sent, so there is still noise present, in the communication sense. Equation (60) is an example of a general quantum mechanical relation known as the *principle of complementarity.*

Noise in Linear Amplifiers

The principle of complementarity shows that a noiseless amplifier is impossible [15]. A noiseless amplifier would have the property that each received photon would produce exactly G output photons, where G is the power gain. Thus n_1 input photons would produce $n_2 = G n_1$ output photons. Further the output phase ϕ_2 would be equal to the input phase ϕ_1 plus some constant phase shift θ. An ideal detector attached to the output would permit us to measure n_2 and ϕ_2 with an uncertainty $\Delta n_2 \Delta \phi_2 = \frac{1}{2}$. But the uncertainty Δn_2 corresponds to an input uncertainty $\Delta n_1 = \Delta n_2 / G$, while the output uncertainty $\Delta \phi_2$ corresponds to an input uncertainty $\Delta \phi_1 = \Delta \phi_2$. Using the combination of a noiseless amplifier and an ideal detector, we would be measuring n_1 and ϕ_1 with an uncertainty $\Delta n_1 \Delta \phi_1 = 1/2G$ in violation of (60). Thus a noiseless amplifier is impossible.[1]

Carrying the analysis further, we can determine exactly how much noise an amplifier must add to satisfy (67). The equality in (67) obtains only if δn and $\delta \phi$ have Gaussian distributions. Hence an ideal amplifier, one that minimizes $\Delta n \Delta \phi$, will have an added Gaussian noise. With this the case, the minimum noise that must be present in the output is $(G-1)h\nu B$, which corresponds to a power $[1 - (1/G)]h\nu B$ watts referred to the input. If further devices are to contribute negligible noise we must let $G \to \infty$. We conclude that the best of all possible linear amplifiers will add a Gaussian noise corresponding to an input spectral power density of

$$\psi_q(\nu) = h\nu \text{ watts/c/s.} \qquad (68)$$

Many analyses have shown that masers and lasers are capable of achieving this limiting performance when complete population inversion of the energy states involved is achieved [16]. The noise in masers and lasers is incoherent radiation produced by spon-

taneous emission, i.e., spontaneous transitions from the upper to the lower state.

Photocells can also achieve the limiting performance given by (68) by using them as mixers. Consider the arrangement shown in Fig. 19. Signal power $P_s = A_s^2/2$ and local oscillator power $P_o = A_o^2/2$ are combined in a half-reflecting, half-transmitting mirror, $M/2$. The combined beams will then have the powers

$$P_1 = \tfrac{1}{2}(A_o \cos \omega_o t + A_s \cos \omega_s t)^2 \qquad (69a)$$

$$P_2 = \tfrac{1}{2}(A_o \sin \omega_o t - A_s \sin \omega_s t)^2 \qquad (69b)$$

where ω_o and ω_s are the local oscillator and signal frequencies. These beams are detected by two photocells of quantum efficiency η to produce currents $i_1 = \eta q(P_1/h\nu)$ and $i_2 = \eta q(P_2/h\nu)$. Ignoring any frequency components higher than $\delta = \omega_s - \omega_o$ we find that

$$i_1 + i_2 = \frac{\eta q}{2h\nu}(A_o^2 + A_s^2) \qquad (70)$$

while the signal current is

$$i_s = i_1 - i_2 = \frac{\eta q}{h\nu} A_o A_s \cos \delta t, \qquad (71)$$

which is linear in A_s and proportional to A_o. The mean square value is

$$\overline{i_s^2} = \left(\frac{\eta q}{h\nu}\right)^2 \frac{A_o^2 A_s^2}{2}. \qquad (72)$$

If the bandwidth of the circuits following the mixer is B, the shot noise affecting the output will be

$$\overline{i_n^2} = 2qB(i_1 + i_2) = \eta \frac{q^2 B}{h\nu}(A_o^2 + A_s^2). \qquad (73)$$

Taking the ratio of (72) to (73) we find for the SNR

$$\text{SNR} = \frac{\eta}{2h\nu B} \frac{A_o^2 A_s^2}{A_o^2 + A_s^2}. \qquad (74)$$

As A_o is increased, both the signal current and the SNR

---

[1] Parametric up-converters satisfy complementarity but need not add noise. They convert the input photons to higher frequency photons on a one for one basis. They thus preserve the quantum noise while raising the signal power (by the frequency ratio) with respect to thermal noise.

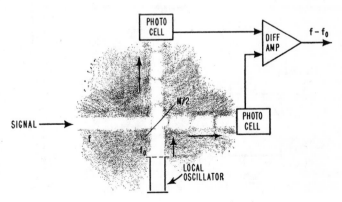

Fig. 19. A balanced photo-electric mixer.

increase, the former without limit, the latter approaching the limit

$$\text{SNR}_{\text{max}} = \frac{\eta}{h\nu B} \frac{A_s^2}{2} = \eta \frac{P}{h\nu B} . \tag{75}$$

Thus the noise power spectral density referred to the input is $h\nu/\eta$ watts/c/s. If $\eta \to 1$ (as is the case for solid state photo-diodes) the performance approaches the ideal limit given by (68).

Homodyne Detection

If we let $\omega_o = \omega_s$ in the previous example we will have a homodyne (or synchronous) detector. It is then necessary to maintain a prescribed phase relation between the local oscillator and the signal. Figure 20 shows a receiver capable of simultaneous in-phase and quadrature amplitude detection. Before entering the balanced homodyne mixer on the right, a fraction, ϵ, of the received signal power, P_s, is diverted by a partially reflecting mirror, ϵM, into an auxiliary balanced mixer. The relative optical path lengths are adjusted by rotating the compensating plate, C, so that, when the phase angle between the signal and local oscillator is zero at the mixer on the right it is $\pm(\pi/2)$ at the auxiliary mixer. With these phase relations the mixer on the right responds to in-phase (amplitude) variations of the signal, while the auxiliary mixer responds to quadrature (phase) variations. For the moment assume P_s is a CW signal and that differential amplifier 2 is connected directly to low-pass filter (LPF2). If the bandwidth of this filter is much less than ϵB, where B is the bandwidth of the low-pass signal filters (LPF1), a relatively noise free control signal will be produced, which can vary the local oscillator frequency by any suitable means. We thus have a closed-loop phase lock system whose response time is inversely proportional to ϵB. Assuming sufficient frequency stability in the signal and local oscillator, we can let $\epsilon \to 0$, thus producing negligible degradation of the SNR at the in-phase detector. We are taking advantage of the fact that the information rate necessary to maintain phase-lock can be made

negligible compared with the signaling rate, so a negligible fraction of the power need be diverted for this function.

If P_s is a carrier suppressed AM signal (for example a binary channel using positive and negative pulses), A_s will assume positive and negative values and we must reverse the polarity in the feedback loop as A_s reverses in order to maintain negative feedback. This is the function of the $\pm$ modulator. The system will now lock in one of two possible states so there is a polarity ambiguity in the output, which, if important, can be corrected once and for all at the start of each transmission.

The signal current in the in-phase mixer will now be given by (71) with $\delta = 0$ and A_s reduced to $A_s\sqrt{1-\epsilon}$. Since $\cos \delta t$ is now replaced by unity, the factor 2 disappears from the denominator of (72). The SNR is now

$$\text{SNR}_{\text{max}} = \frac{\overline{i_s^2}}{\overline{i_n^2}} = \frac{2\eta(1-\epsilon)P_s}{h\nu B} . \tag{76}$$

As $\epsilon \to 0$ the noise power spectral density referred to the input becomes $h\nu/2\eta$ watts/c/s, which is the result (63) obtained for a simple photocell. However, we now have a strictly linear system capable of distinguishing between positive and negative amplitudes A_s and hence suitable for more efficient modulation methods. As $\eta \to 1$, (76) shows that we can approach half the noise power spectral density associated with a linear amplifier.

To see why this is so let us examine the SNR in both outputs of the receiver, assuming $\eta = 1$. If $W_s = P_s/2B$ is the signal energy per elementary pulse (or per Nyquist interval), then from (76) we find

$$\frac{\overline{i_n^2}}{\overline{i_s^2}} = \frac{h\nu}{4(1-\epsilon)W_s} = \frac{1}{4(1-\epsilon)\bar{n}} , \tag{77}$$

where $\bar{n} = W_s/h\nu$ is the input photon expectation per pulse. Now

$$\frac{\overline{i_n^2}}{\overline{i_s^2}} = \frac{\overline{(\delta n^2)}}{(\bar{n})^2} , \tag{78}$$

where $\overline{(\delta n)^2} = (\Delta n)^2$ is the *apparent* mean square fluctuation in input photon count. Combining (77) and (78) we get

$$(\Delta n)^2 = \frac{\bar{n}}{4(1-\epsilon)} . \tag{79}$$

In the quadrature channel we find similarly that the ratio of mean square noise current to the square of the signal current, i_s', corresponding to unit phase deviation, is

$$\frac{\overline{i_n^2}}{(i_s')^2} = \frac{1}{4\epsilon\bar{n}} . \tag{80}$$

This ratio is the apparent mean square phase fluctuation $(\Delta\phi)^2$ of the input. Thus from (79) and (80) we obtain

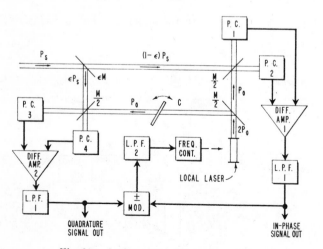

Fig. 20. A phase locked homodyne mixer.

$$\Delta n \Delta \phi = \frac{1}{4\sqrt{\epsilon(1-\epsilon)}} \cdot \qquad (81)$$

The minimum value of this expression occurs when $\epsilon = \frac{1}{2}$ corresponding to equal power into both channels. We then have $\Delta n \Delta \phi = \frac{1}{2}$, or the limiting performance allowed by the principle of complementarity. With $\epsilon = \frac{1}{2}$ both signals will now contain a noise of $h\nu$ watts/c/s referred to the input. The two signals jointly completely describe the input wave modulation, and could be used to reconstruct it aside from the added noise. Thus this system with $\epsilon = \frac{1}{2}$ is directly comparable with a linear amplifier or heterodyne mixer and gives the same performance.

It is evident that by letting $\epsilon \to 1$ we could detect quadrature modulation with a noise of $h\nu/2$ watts/c/s referred to the input. Thus we can detect *either* amplitude modulation *or* phase modulation with a noise power spectral density half as great as that of an ideal amplifier. If we wish to do *both* equally well we are reduced to the performance of the linear amplifier.

Total Noise

Combining the results of Part I and Part II we see that the total noise of an ideal amplifier is given by

$$\psi_a(\nu) = \frac{h\nu}{e^{h\nu/kT} - 1} + h\nu, \qquad (82a)$$

where T is now the temperature that the input faces in the propagation mode involved. Similarly the total noise of an ideal linear amplitude or phase detector is

$$\psi_d(\nu) = \frac{h\nu}{e^{h\nu/kT} - 1} + \frac{h\nu}{2} \cdot \qquad (82b)$$

Figure 21 shows a normalized plot of (82a) and (82b) while Fig. 22 is a plot on log scales at three particular source temperatures. We see that the total noise power spectral density increases monotonically with frequency. There are two distinct regions of the noise spectrum: a "thermal" region where $(h\nu/kT) < 1$ and $\psi(\nu) \approx kT$, and a "quantum" region where $(h\nu/kT) > 1$, $\psi_a(\nu) \approx h\nu$, and $\psi_d(\nu) \approx (h\nu/2)$. At all frequencies the noise is Gaussian. In the thermal region the noise is "white," i.e., has equal power at all frequencies. In the quantum region the noise is "blue," i.e., has a power proportional to frequency.

In spite of the frequency proportionality of quantum noise, the total noise power passed by an amplifier in the quantum region is proportional to bandwidth, since

$$\int_{\nu_0-(B/2)}^{\nu_0+(B/2)} h\nu d\nu = \frac{h\nu^2}{2} \int_{\nu_0-(B/2)}^{\nu_0+(B/2)} = h\nu_0 B.$$

Further, the noise in the output of a homodyne detector will be white since each modulation frequency is produced by frequencies symmetrically disposed about ν_0 and the total noise power from these two sidebands is independent of modulation frequency.

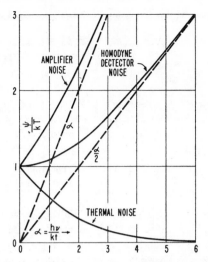

Fig. 21.   Comparison of total noise spectra.

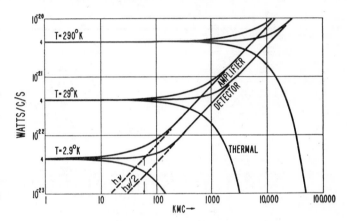

Fig. 22.   Thermal, detector, and amplifier noise at various temperatures.

REFERENCES

[1] Einstein, A., Zum gegenwartigen Stand des Strahlungsproblems, *Physik Zeitschrift*, vol 10, 1909, pp 185–193.
[2] Einstein, A., and L. Hopf, Uber einen Satz der Wahrscheinlichkeitsrechnung und seine Anwendung in der Strahlungstheorie, *Ann. Phys.*, vol 33, Aug–Dec 1910.
[3] Einstein, A., *Investigations on the Theory of the Brownian Movement.* New York: Dover, 1956.
[4] Campbell, N. R., Discontinous phenomena, *Proc. Cambridge Phil. Soc.*, vol 15, 1909, p 117; 1910, p 310.
[5] Schottky, W., Uber Spotane Stromschwankungen in Verschiedenen Elektrizitatsleitern, *Ann. Phys.*, vol 57, 1918, pp 541–546.
[6] Johnson, J. B., Thermal agitation of electricity in conductors, *Phys. Rev.*, vol 32, Jul 1928, p. 97.
[7] Nyquist, H., Thermal agitation of electric charge in conductors, *Phys. Rev.*, vol 32, Jul 1928, p 110.
[8] Rice, S. O., Mathematical analysis of random noise, *Bell Sys. Tech. J.*, vol 22, 1944, p 282; vol 25, 1945, p 25.
[9] Bennett, W. R., *Electrical Noise.* New York: McGraw-Hill, 1960.
[10] Tolman, R. C., *The Principles of Statistical Mechanics.* New York: Oxford, 1962.
[11] Richtmeyer and Kennard, *Introduction to Modern Physics.* New York: McGraw-Hill, ch 5.
[12] Shannon, C. E., A mathematical theory of communication, *Bell Syst. Tech. J.*, vol 27, Jul, Oct 1948.
[13] Slater, J. C., *Microwave Transmission.* New York: McGraw-Hill, pp 252–256.
[14] Bennett, W. R., Responses of a linear rectifier to signal and noise, *Bell Sys. Tech. J.*, vol 23, Jan 1944, pp 97–113.
[15] Heffner, H., The fundamental noise limit of linear amplifiers, *Proc. IRE*, vol 50, Jul 1962, pp 1604–1608.
[16] Strandburg, M. W. P., Inherent noise of quantum-mechanical amplifiers, *Phys. Rev.*, vol 10, May 15, 1957, pp 617–620.

Noise in Semiconductors and Photoconductors*

K. M. VAN VLIET†

Summary—A survey is given of theory and experiments on noise in bulk semiconductors and photoconductors. This paper is divided into four parts, including generation-recombination (gr) noise in semiconductors, gr noise in photoconductors, 1/f noise in single crystals, and modulation noise in granular materials. In the first part an account is given of the appropriate analyses and the results are applied to extrinsic as well as intrinsic fluctuations, generated either in the bulk or at the surface. In the part about photoconductors the limiting sensitivity caused by photon noise is calculated and present infrared detectors are discussed. Next, a survey is given about present understanding of 1/f noise, and of its relation to the field effect as proposed by McWhorter and others. Finally, some remarks are made about 1/f noise in granular material and the proposed theories are briefly reviewed.

I. INTRODUCTION

IN 1932 Williams and Thatcher [107] observed that current carrying carbon resistors generated a large amount of noise. When the current was absent, on the contrary, the noise satisfied Nyquist's formula, according to which the noise of any resistive device in thermal equilibrium equals $4kTR$ in unit bandwidth.

The first extensive investigation of noise in current-carrying nonmetallic resistors was made by Bernamont in 1934 [5]. He found that the spectral density of the noise was approximately inversely proportional to the frequency in the audio-frequency range. In that respect the spectrum looked similar to the flicker effect in vacuum tubes discovered nine years before by Johnson [44]. Surdin [88] extended the frequency range of Bernamont's data and observed this noise up to a few mc beyond which thermal noise usually predominated. Only rarely a change from a 1/f spectrum into a 1/f² spectrum was observed.

The first theory for noise in carbon and evaporated metal layer resistors was also developed by Bernamont. He predicted a spectrum of the form constant/$(1+\omega^2\tau^2)$ where τ is the time constant of the fluctuations. In order to fit this result to the experimental data, Bernamont [5] and Surdin [87] suggested that there should be a distribution of time constants, ranging from ~1 sec up to ~1 μsec. However, as is known today, Bernamont's theory does not apply to carbon resistors but to other noise processes (see Section II-A).

The cause of the noise in carbon and similar resistors

* Original manuscript received by the IRE, March 3, 1958.
† Dept. Elec. Eng., University of Minnesota, Minneapolis, Minn.

Reprinted from *Proc. IRE*, vol. 46, pp. 1004–1018, June 1958.

became clearer from the experiments of Christenson and Pearson [27] in 1936. They could locate the source of the noise in the contacts between the grains in these materials and introduced the name contact noise. Their observations were verified by many other research workers, for a variety of materials. Shortly after World War II, when single crystals of semiconducting materials became available, the hope was raised that this noise would be absent. It was found, however, that again noise of a 1/f nature was present; a detailed investigation on germanium single crystals was reported by Montgomery [60], [81]. Montgomery could show that the noise was associated with the surface of the crystals. This noise apparently was no contact noise and was generally named excess noise.

Deviations from the 1/f dependence in germanium were first reported by Herzog and van der Ziel [39]. The spectrum showed a characteristic time constant of $\sim 1\ \mu\text{sec}$ which later on was identified with the minority carrier lifetime. The noise was thus attributed to the random excitation and capture of free carriers, which during their stay in the conduction band (or valence band) give rise to a current pulse in the output circuit. Because of the resemblance to the random emission of electrons in vacuum tubes this noise was called shot noise. The effect had been predicted before by Gisolf [36] in 1949. His theory, being in error, was modified by van der Ziel [93], [95], who also showed that Bernamont's original theory was equivalent to the modified Gisolf theory [92], [95]. There are two reasons for such a late discovery of this noise. First, as mentioned above, 1/f noise depends on the surface conditions of the crystals. Therefore, 1/f noise masked all other effects until considerable progress had been made in various laboratories in improving the surface properties. Secondly, it is known that 1/f noise is roughly independent of temperature in contrast to the spontaneous fluctuations in excitation and capture of carriers which give appreciable noise in certain temperature regions only.

In photoconductors similar noise effects have been reported. Here the facts are more complicated because of the fluctuations in the incident radiation field (photon noise). Completely contradictory opinions have been stated about the effect of photon noise on the performance of photoconductors. Buttler [24], [25] assumes that the effect is negligible. Shulman [84], on the contrary, states that under ideal conditions all the noise results from photon fluctuations. The author holds the opinion that photon fluctuations can account for up to 50 per cent of the observed noise in an ideal trap free photoconductor (see Section III-B).

In the above survey, several noises have been introduced in a more or less historical order. Before proceeding, it seems appropriate to introduce a terminology which better agrees with present understanding of the various processes. Noise above thermal noise will generally be called *current noise*. This will be divided into two main areas.

A. Generation-Recombination Noise

This is henceforth denoted as gr noise. This noise is caused by spontaneous fluctuations in the generation rates, recombination rates, trapping rates, etc., thus producing fluctuations in the free carrier densities. The term gr noise seems more adequate than shot noise since these carrier fluctuations exist even in equilibrium when there is no applied field. There is some difference in treatment of gr noise in semiconductors and in photoconductors. In semiconductors the noise is of thermal origin and can therefore be calculated with the generalized Nyquist formula (Section II-C). In photoconductors only statistical arguments can be applied.

B. Modulation Noise

This term introduced by Petritz [70] will refer to noise which is not *directly* caused by fluctuations of the carrier transition rates, but instead is due to carrier density fluctuations or current fluctuations caused by some modulating effect. Let us give some examples. According to some investigators 1/f noise in germanium filaments is primarily caused by fluctuations in the occupancy of the "slow surface states" (Section IV-B). When a carrier is trapped in these states it produces a change in the number of carriers and as such is gr noise. However, this is not the full story. Owing to the change in surface charge the surface recombination velocity will change. The bulk conductivity in turn is affected more severely by this effect than just by the gain or loss of one carrier. Thus, the "slow surface states" cause conductivity modulation of the bulk. As another example we turn to some mechanisms proposed for contact noise (Section V-A). Suppose that the current between two grains is affected by molecules diffusing over the contact area. This will produce current modulation. Most of these modulation effects found in practice are of the $1/f^{\alpha}$ type (with $\alpha \approx 1$) and most of the theories fail to give a 1/f dependence over a long frequency range. Some modulation noises have been observed which do not have a 1/f spectrum (*e.g.*, heat conductivity induced fluctuations) [99], [4]. Although more effects of such a nature may be found in the future, at the present time the name modulation noise may be considered a synonym with "1/f noise."

As illustrated by the two previous examples it may well be that 1/f noise in single crystals and in granular materials is of quite different origin. Nothing definite is known about this today, however.

In accordance with the above classification and comments this paper will be divided into four parts. In Section II we discuss theory and experiments of gr noise in semiconductors. In Section III we consider gr noise in photoconductors and the necessarily related topic of photon noise. In Section IV, 1/f noise in single crystals is discussed and in Section V we mention very briefly some modulation effects in granular material. The order of the topics indicates, apart from the author's preference, the order of decreasing understanding and increasing need for future research.

II. Generation-Recombination Noise in Semiconductors

A. Older Theories

Gisolf [36] considered the current pulses caused by the individual carriers during their lifetime. For simplicity it was assumed that only one type of carrier participated in the current. The current pulse caused by a carrier with lifetime τ_p is

$$F_p(t) = e\mu EL^{-1} \qquad t_0 \leq t \leq t_0 + \tau_p \Big\}$$
$$F_p(t) = 0 \qquad \text{elsewhere,} \qquad \Big\} \qquad (1)$$

where E is the electric field, L the electrode distance, and μ the mobility. The total current can be written as

$$i(t) = \Sigma_p \Sigma_r F_p(t - t_r).$$

Assuming that the elementary pulses are independent and occur at random instants t_r with a rate $\overline{N}_p$, we can apply Campbell's theorem [34] for the fluctuations $\Delta i(t) = i(t) - \langle(i)\rangle$

$$\langle \Delta i(t)^2 \rangle = \Sigma_p \overline{N}_p \int_{-\infty}^{\infty} |F_p(\xi)^2| \, d\xi. \qquad (2)$$

The next step is to make a Fourier integral analysis of $F_p(\xi)$ and apply Parseval's theorem. This leads to the generalized Carson's theorem:

$$S_i(f) = 2\Sigma_p \overline{N}_p |A_p(f)^2| \qquad (3)$$

where $A_p(f)$ is the Fourier coefficient of $F_p(\xi)$ and $S_i(f)$ is the spectral intensity of the fluctuations defined by

$$\langle \Delta i(t)^2 \rangle = \int_0^{\infty} S_i(f) df. \qquad (4)$$

Assuming a statistical weight factor for τ_p [95], [10] and carrying out the above procedure one is led to the simple result

$$S_i(f) = 4I^2\tau/n_0(1 + \omega^2\tau^2) \qquad (5)$$

which was first found by Bernamont [5]. Here n_0 is the total steady-state number of carriers in the sample.

The above theory, interesting for historical reasons and for its simplicity, is incorrect since the elementary pulses are not independent. The possible transitions are limited by the Pauli exclusion principle. A better result was obtained by van der Ziel [93] and by Machlup [52], who suggested that the current carriers were subject to binomial statistics. Let λ be the probability that a carrier is free and $1-\lambda$ the probability that a carrier is bound. Then the noise is found to be

$$S_i(f) = 4(1 - \lambda)I^2\tau/n_0(1 + \omega^2\tau^2). \qquad (6)$$

Obviously, for $\lambda \approx 1$ the noise is negligible. This is the case for extrinsic germanium and silicon at room temperature when all the donors are ionized or all the acceptors filled. Although (6) gives a good approximation, it was first shown by Burgess [18], [21] that an improvement could be obtained.

B. The Langevin Equations [100]

The relation between the instantaneous current and the instantaneous number of carriers is

$$i(t) = e\mu_n n(t)EL^{-1} + e\mu_p p(t)EL^{-1}. \qquad (7)$$

Consequently, denoting by S_{nn} the spectrum of $\langle \Delta n^2 \rangle$, by S_{pp} the spectrum of $\langle \Delta p^2 \rangle$ and by S_{np} the spectrum of $\langle \Delta n \Delta p \rangle$, which quantities are defined analogous to (4), we have

$$S_i = \lfloor I/(bn_0 + p_0) \rfloor^2 (b^2 S_{nn} + 2b S_{np} + S_{pp}). \qquad (8)$$

Note that n_0 and p_0 are not densities but mean *total numbers*; I is the dc current and $b = \mu_n/\mu_p$. In the particular case that $p_0 = 0$ we have

$$S_i = (I/n_0)^2 S_{nn} \qquad (8a)$$

and in another practical case where $\Delta n(t) = \Delta p(t)$ we have

$$S_i = I^2[(b + 1)/(bn_0 + p_0)]^2 S_{nn}. \qquad (8b)$$

In accordance with what has been said in the introduction, it will suffice to find S_{nn}, S_{np}, and S_{pp}. The current noise spectrum S_i follows from the above relations.

First we will consider "single step processes," *i.e.*, transitions occur only between the conduction band (containing n electrons) and localized levels, or between the conduction band and the valence band. Noise in a p-type semiconductor involving transitions between the valence band and localized levels can be found *mutatis mutandis*. Let $g(n)$ be the generation rate of electrons and $r(n)$ the recombination rate with either holes in the localized levels or with holes in the valence band. It will be assumed that the mass action law applies to the rates $g(n)$ and $r(n)$. The effect of different spins will be neglected since the variances and covariances are not affected by this refinement in statistics. Hence, the kinetic equation is

$$dn/dt = g(n) - r(n) + f(t) \qquad (9)$$

where $f(t)$ is a stochastic source function for the fluctuations. Expanding $g(n)$ and $r(n)$ up to first-order terms in $n - n_0 = \Delta n$, we obtain the Langevin equation

$$d(\Delta n)/dt = -\Delta n/\tau + f(t) \qquad (10)$$

where τ is the lifetime of added current carriers

$$\tau = \left\{ \left(\frac{dr}{dn}\right)_0 - \left(\frac{dg}{dn}\right)_0 \right\}^{-1}; \qquad (11)$$

the subscript zero refers to the values of the derivatives for $n = n_0 = \langle n \rangle$. The solution of (10) yields:

$$\langle \Delta n(t + s)\Delta n(t) \rangle = \langle \Delta n(t)^2 \rangle e^{-s/\tau}. \qquad (12)$$

With the Wiener-Khintchine theorem we find

$$S_{nn}(f) = 4 \int_0^{\infty} \langle \Delta n(t + s)\Delta n(t) \rangle \cos \omega s \, ds$$
$$= 4\langle (\Delta n)^2 \rangle \tau/(1 + \omega^2\tau^2). \qquad (13)$$

The method does not give a complete answer unless $\langle(\Delta n)^2\rangle$ is specified. Eq. (9) suggests the following relation between the statistical properties of $g(n)$ and $r(n)$

$$W(n)r(n) - W(n - 1)g(n - 1) = 0, \qquad (14)$$

where $W(n)$ is the probability distribution of n. Since (14) is a recurrent relation, $W(n)$ and $\langle(\Delta n)^2\rangle$ can be found as was shown by Burgess [18], [21]; the result is the gr theorem:

$$\langle(\Delta n)^2\rangle = g_0 \left\{ \left(\frac{dr}{dn}\right)_0 - \left(\frac{dg}{dn}\right)_0 \right\}^{-1} = g_0\tau. \qquad (15)$$

Hence, the spectrum (13) can be written as

$$S_{nn}(f) = 4g_0\tau^2/(1 + \omega^2\tau^2). \qquad (16)$$

The Langevin method can easily be extended to multistep processes [100]. Let there be s energy levels (including the conduction band and the valence band) between which transitions occur. The electron concentrations in these levels will be denoted by $n_1 \cdots n_s$. Since $\Sigma_j n_j$ is constant one variable can be eliminated. Let the independent variables then be $n_1 \cdots n_{s-1}$. If p_{ij} is the transition rate per second from i to j then

$$dn_i/dt = \sum_{j=1}^{s}{}' p_{ji} - \sum_{j=1}^{s}{}' p_{ij} + f_i(t). \qquad (17)$$

The Langevin equations follow from expansion of p_{ji} up to first-order terms in the fluctuations. They can be written as

$$d(\Delta n_i)/dt = \sum_{j=1}^{s-1} a_{ij}\Delta n_j + f_i(t) \qquad (18)$$

where

$$a_{ij} = \sum_{k=1}^{s}{}' \left\{ \left(\frac{\partial p_{ki}}{\partial n_j}\right)_0 - \left(\frac{\partial p_{ik}}{\partial n_j}\right) \right\}. \qquad (19)$$

Eq. (18) can be formally solved, given the values of $\Delta n_i(t)$ at some time t_0. It is then found that the correlation matrix has the following form:

$$\langle\Delta n_i(t + s)\Delta n_j(t)\rangle = \sum_{k=1}^{s-1} c_k{}^{ij}e^{-t/\tau_k} \qquad (20)$$

where the quantities $-1/\tau_k$ are the eigenvalues of the matrix $[a_{ij}]$. The spectrum follows from the Wiener-Khintchine theorem:

$$S_{ij}(f) = 4 \sum_{k=1}^{s-1} c_k{}^{ij}\tau_k/(1 + \omega^2\tau_k^2). \qquad (21)$$

The general form is depicted in Fig. 1. The quantities $c_k{}^{ij}$ depend on the variances $\langle\Delta n_i^2\rangle$ and the covariances $\langle\Delta n_i\Delta n_j\rangle$ which have again to be found by other means. Van Vliet and Blok [100] calculated the variances and the covariances from the Fokker Planck equation and generalized the gr theorem (15). The spectrum (21) is therefore completely known once the transition rates p_{ij} between the various energy levels can be specified ac-

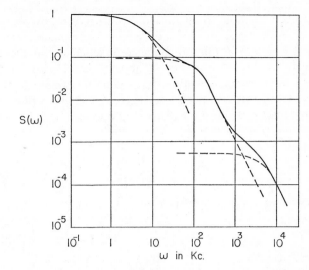

Fig. 1—Illustration of possible spectra in the case of multistep processes or several simultaneous single-step processes. The following values are chosen: $\tau_1 = 2\times10^{-4}$, $\tau_2 = 10^{-5}$; $\tau_3 = 2\times10^{-7}$ sec and $c_1 = 5\times10^3$, $c_2 = 10^4$, $c_3 = 3\times10^3$; these numbers are relative values.

cording to the mass action law. A detailed example employing this method was given for noise in cadmium sulphide [101]. The calculations are cumbersome, however, and comparison with experiment is not readily achieved. In semiconductors the situation is slightly better than in photoconductors since an alternate method based on thermodynamics can be applied (see below).

C. Application of the Generalized Nyquist Formula

It has been emphasized already that it is sufficient to calculate the spectral densities S_{nn} under the conditions that there is no electric field. For moderate electric fields S_{nn}, etc., will not change and (8) can be applied to find the current noise. Consequently for the calculation of S_{nn}, etc., we may assume that the crystal is in thermal equilibrium. Obviously, we restrict ourselves here to semiconductors.

As is well known, the steady-state electron distribution in semiconductors can be obtained thermodynamically (minimization of the electronic free energy) and from mass action law considerations. In accordance with this, two methods which are entirely different in principle can be employed to find the noise. This was shown by Burgess [19] for the variances $\langle\Delta n^2\rangle$, etc. Recently, the author [104] proved that the noise spectra S_{nn}, etc., can also be found from thermodynamic arguments; the results are in complete agreement with the previous section in which the mass action law was used. The method is based on the generalized Nyquist formula, established first by Callen and Welton [26] in 1951. According to this relation the spectral density of a single fluctuating extensive thermodynamic variable can be expressed as

$$S(f) = 4kT\omega^{-2} \, \text{Re}\,(Y). \qquad (22)$$

The admittance Y relates a sinusoidal variation of the

external variable to a sinusoidal perturbation by a generalized force acting on the system. In semiconductors such a perturbation can be accomplished if the Fermi level is changed into two or more quasi-Fermi levels for the various carrier groups. The response of the transition rates to such a perturbation defines the admittance Y. If only two carrier groups are considered, *e.g.*, electrons and bound or free holes, the admittance is found to be [104]

$$Y = \frac{g_0}{kT} \cdot \frac{j\omega\tau}{1 + j\omega\tau} .$$ (23)

This equation and (22) immediately yield the result (16).

The method has also been applied to more variable cases.

D. Extrinsic Semiconductors

With the framework developed in the previous sections the noise in extrinsic semiconductors is readily found. We will assume that the semiconductor is *n*-type and has N_d donors all at a single level E_d below the conduction band. It will first be assumed that no intrinsic transitions occur. The generation rate and recombination rate can then be expressed as follows:

$$g(n) = \gamma(N_d - n); \qquad r(n) = \delta n^2$$ (24)

and in equilibrium we have

$$g_0 = \gamma(N_d - n_0) = \delta n_0^2 = r_0.$$ (25)

The quantities γ and δ are transition constants which include cross sections, etc. The constant δ depends only weakly on the temperature; γ, on the contrary, depends on T according to a Boltzmann factor $\gamma = \gamma_0 \exp(-E_d/kT)$. Substituting (24) and (25) into (11) and (15) yields

$$\langle \Delta n^2 \rangle = \frac{n_0(N_d - n_0)}{2N_d - n_0} = N_d \frac{\lambda(1 - \lambda)}{2 - \lambda}$$ (26)

$$\tau = \frac{1}{\delta n_0} \cdot \frac{N_d - n_0}{2N_d - n_0} = \frac{1}{\delta N_d} \cdot \frac{1 - \lambda}{\lambda(2 - \lambda)}$$ (27)

where $\lambda = n_0/N_d$ is the fractional ionization of the impurities and has the same meaning as in Section II-A. The noise spectrum is found from (8a), (16), (26), and (27) which give after some arrangement:

$$S_i(f) = 4\left(\frac{I^2}{n_0}\right)\left(\frac{1 - \lambda}{2 - \lambda}\right)\left(\frac{\tau}{1 + \omega^2\tau^2}\right).$$ (28)

The conclusions drawn in Section II-A that the noise occurs only if the fraction λ of ionized donors is low is still valid. This explains why hardly any noise measurements on strongly extrinsic material have been reported. A characteristic feature of this noise is the strong temperature dependence of τ which is the majority carrier lifetime. According to (27) τ decreases rapidly with increasing temperature since λ approaches 1. Noise of this nature was reported by Gebbie [35].

However, Gebbie attributed the noise to trapping effects.

In the above analysis it was assumed that the donor levels are filled at absolute zero. If the donor levels contain already m_0 holes at $T=0$, the properties of the semiconductor are markedly different. The Fermi level, instead of lying approximately midway between the conduction band and the donor levels, is now clamped at the position of the donor levels for sufficiently low temperatures such that $n_0 \ll m_0$ [66]. The noise properties are also different, since the transition rates, instead of (24), are now given as

$$\left.\begin{array}{l} g(n) = \gamma(N_d - m_0 - n_0) \approx \gamma(N_d - m_0) \\ r(n) = \delta n_0(m_0 + n_0) \approx \delta n_0 m_0 \end{array}\right\} .$$ (29)

The reaction mechanism is now approximately monomolecular. One finds from (29)

$$\langle \Delta n^2 \rangle \approx n_0; \qquad \tau \approx 1/\delta m_0.$$ (30)

Instead of (28) the noise is now given by Bernamont's formula (5) as is easily shown from (30) and (16). Moreover, the lifetime τ is approximately constant. For *p*-type material similar arguments apply. Measurements were recently performed on manganese-doped germanium in the range 77°K–125°K by Fassett [31a] at the University of Minnesota. It was estimated that 50 per cent of the acceptor levels were filled at absolute zero. The spectra were flat from 1 kc–10 mc for temperatures ~120°K. For temperatures below 100°K the lifetime increased, in contrast to (30). The possibility that hole trapping is present or that excited states of the impurity centers play a role is being investigated.

Noise in cases where more donor levels participate in the transitions can also be calculated from the general analysis in the preceding sections. Also, an analysis can be given for the case that intrinsic transitions and impurity transitions occur simultaneously. Since most likely the donors are completely ionized before intrinsic transitions start to be present abundantly, the noise arising from the transitions to and from the donors or acceptors can usually be neglected in these cases. Hence, only the intrinsic transitions cause noise. Consequently, without treating the complete case, it is evident that the noise output of a semiconductor is a complicated function of the temperature. Appreciable noise is found only in the strongly extrinsic and in the near-intrinsic or intrinsic region.

E. Near Intrinsic and Intrinsic Crystals

In these crystals the transition rates are, neglecting first recombination centers:

$$g(n) = \alpha; \qquad r(n) = \beta n^2.$$ (31)

The noise can be found in the same way as before. The relation between the density fluctuations and the current fluctuations appropriate to this case is (8b), since the electron and hole densities fluctuate coherently; *i.e.*,

$\Delta n(t) = \Delta p(t)$. The following result is easily obtained in the same manner as in the previous section

$$S_i(f) = 4I^2 \frac{(b+1)^2 n_0 p_0}{(b n_0 + p_0)^2 (n_0 + p_0)} \cdot \frac{\tau}{1 + \omega^2 \tau^2}. \quad (32)$$

For intrinsic semiconductors ($n_0 = p_0$) this reduces to

$$S_i(f) = 2I^2 \tau / n_0 (1 + \omega^2 \tau^2). \quad (33)$$

In contrast to the above assumptions usually recombination in germanium and silicon occurs via recombination centers. This changes the problem into a two-variable problem, in which two of the three variables n, p, n_t (where n_t is the number of electrons in the recombination centers) can be taken as the independent ones. The variances $\langle \Delta n^2 \rangle$, $\langle \Delta p \rangle^2$ and the covariance $\langle \Delta n \Delta p \rangle$ were calculated by Burgess [19] and by van Vliet [103]. If, however, the number of electrons in the recombination centers is small compared to the number of majority carriers, we have $\Delta n \approx \Delta p$. Then (32) and (33) apply where τ is the Shockley-Read lifetime [83], [104].

Noise in intrinsic or nearly intrinsic material has been observed by several investigators [39], [55], [94], [79], [43], [40]. In the last two references the noise is compared with (32) and (33) and good agreement is found in most cases. In some cases, the spectrum did not fall off as rapidly as $1/\omega^2$ at high frequencies, indicating that the recombination centers in the bulk or at the surface (see below) are slightly distributed in energy.

F. Surface GR Noise

In many germanium crystals generation and recombination of electrons and holes involve surface centers. This complicates the theory considerably. However, some simplifying argument will be given. First of all, it is noted that the noise is not determined by the local number of carriers. Secondly, as for bulk recombination centers, we will assume that the population of the surface states is not too large. Then the decay of fluctuations can again be described with (10) where τ is a surface lifetime. Although $r(n)$ and $g(n)$ are not specified, the variances $\langle \Delta n^2 \rangle$, $\langle \Delta p^2 \rangle$, and $\langle \Delta n \Delta p \rangle$ will be the same since these expressions can be found from thermodynamic arguments. Consequently, (31) and (32) will be approximately valid. Hyde [43] has pointed out that the spectra may reveal some smaller lifetimes since the solution of the continuity equations for injected carriers give rise to higher order modes [82]. It is also possible to solve the stochastic partial differential equations governing the generation, recombination, and diffusion of carriers exactly, but this would be beyond the scope of this paper.

G. Influence of the Drift of the Carriers on the Noise

In the previous section it was assumed that the probability distribution $W(n)$ for the carriers was not distorted by the presence of the field. This is the case up to very high fields, apart from heat dissipation effects. At much lower field strengths, however, the spectral distribution of the noise may change since carriers are swept to the electrodes of the sample in a time smaller than τ. Let us first consider a semiconductor containing only one type of carriers, say electrons. What will happen if $\tau \gtrsim \tau_d$ where $\tau_d = L/\mu_n E$? Davydov and Gurevich [29] first treated the problem. Their work implies the correlation function

$$\langle \Delta n(t) \Delta n(t + s) \rangle = \langle \Delta n^2 \rangle e^{-s/\tau}(1 - s/\tau_d) \quad (34)$$

for $s \leq \tau_d$ and zero for $s > \tau_d$. The term $(1 - s/\tau_d)$ takes into account that carriers are swept out of the sample in a time τ_d. It has been pointed out before [100], [84], [11], [77] however, that (34) is erroneous if the dielectric relaxation time of the sample is much shorter than τ_d. In that case carriers are not simply swept out at the contacts and their life terminated. As soon as some extra electrons Δn are generated at a spot x and begin to move under the influence of the applied field, electrons start to move everywhere in the crystal; at the negative electrode, electrons flow into the crystal and space charge neutrality is preserved. Hence, a carrier lives virtually its full lifetime τ and the noise is the same as for $\tau \ll \tau_d$. If both electrons and holes are present, things may change, however [62]. Consider, e.g., an n-type sample in which some extra hole-electron-pairs are created somewhere in the middle between the two electrodes. The electrons, starting to move to the positive electrode cannot produce a space charge so that other electrons come in at the negative contact whereas electrons are carried off at the positive electrode. If the holes start to move to the negative electrode they cannot produce a space charge either; their charge will be neutralized mainly by electrons flowing in the opposite direction. Therefore, virtually a hole and a neutralizing electron are flowing to the negative electrode with the so-called ambipolar mobility [98], given by

$$\mu_a = \frac{|n_0 - p_0| \mu_n \mu_p}{\mu_n n_0 + \mu_p p_0}; \quad \tau_a = \frac{L}{\mu_a E}. \quad (35)$$

The ambipolar mobility is equal to the minority carrier mobility in highly extrinsic material in accordance with the above picture. The correlation function is now given by (34) if we replace τ_d by the ambipolar drift time τ_a. The spectrum can be calculated with the Wiener-Khintchine theorem; the integration involved is found in a paper by Burgess [20]. For $\tau \gg \tau_a$ the spectrum becomes oscillatory

$$S_i(f) = \text{constant} \times \sin^2 \left(\tfrac{1}{2}\omega\tau_a\right) / \left(\tfrac{1}{2}\omega\tau_a\right)^2. \quad (36)$$

The effect has been experimentally verified by Hill. His measurements and an exact theory will be published elsewhere [105].

H. Degenerate Semiconductors

The analysis given in Section II-B is not immediately applicable to degenerate semiconductors like InSb. A more careful averaging process over the electrons in a degenerate band has to be carried out than is implied by the simple terms $g(n)$ and $r(n)$ in previous sections. The variances $\langle\Delta n^2\rangle$ and $\langle\Delta p^2\rangle$ for hole electron pair transitions were calculated by Oliver [68] and by Burgess [22]. The spectrum was recently derived with the approach of irreversible thermodynamics (Section II-C) by the author [104]. It was shown that (16) is still valid. However, g_0 and τ depend on the carrier concentrations in a more complicated way than before. The noise is found to be given by [instead of (32)]:

$$S_i(f) = 4I^2 \frac{(b+1)^2 \xi_n \xi_p n_0 p_0}{(bn_0 + p_0)^2(\xi_n n_0 + \xi_p p_0)} \cdot \frac{\tau}{1 + \omega^2\tau^2} ; \quad (37)$$

here ξ_n and ξ_p depend on the position of the Fermi level of the degenerate electron and hole distributions. For complete degeneracy of an n-type sample $\xi_n = 3kT/2\epsilon_F$ and $\xi_p = 1$, where ϵ_F is the position of the Fermi level with respect to the bottom of the conduction band. The result (37) has not as yet been verified experimentally since other noise sources dominate in InSb [69], [86].

III. GR Noise in Photoconductors

A. Circuitry and Figures of Merit

A photoconductor is usually incorporated in a simple circuit, consisting of the photoconductive film or crystal, a load resistor, and a battery. The signal can be taken either from the crystal terminals or from the load resistor. If the signal and the rms noise both result from conductivity changes, both these effects are proportional to the current and the signal-to-noise ratio does not depend on the load resistor, providing the thermal noise is negligible. There is, however, another reason to make the load resistor large. Often, $1/f$ noise of the current carrying contacts is a dominating noise effect. This can be suppressed by measuring the signal between probes on the crystal. If the load resistor is so large that the supply acts as a constant current generator, then resistance fluctuations at the current contacts will not show up as noise between the probes, unless concentration disturbances are swept into the probe region. Therefore, if T_p is the lifetime and $E\mu_p$ the drift velocity of the minority carriers, and d the distance between probes and the adjacent end contacts, we should have $\mu_p E \tau_p \ll d$ (compare Fig. 2).

Since, assuming that no contact noise is present, the signal-to-noise ratio does not depend on the circuitry, we will find it convenient to represent both signal and noise by short-circuited current generators.

We will assume that the light signal is being chopped with an angular frequency $\omega = 2\pi f$ and that the signal passes through an amplifier, with bandwidth $(f - \frac{1}{2}\Delta f, f + \frac{1}{2}\Delta f)$. The signal will be denoted by $i(\omega)$ and the rms noise in this bandwidth by $\sqrt{S_i(\omega)\Delta f}$. The signal-to-noise ratio is then

$$\frac{\text{signal}}{\text{noise}} = \frac{i(\omega)}{\sqrt{S_i(\omega)\Delta f}}. \quad (38)$$

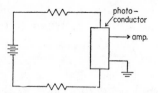

Fig. 2—Probe method for photoconductive cells.

With Jones [46] we will define the responsivity as the signal per unit radiation input. The dimension of the signal has already been fixed (amps). Since photoconductors are quantum detectors we will express the input radiation intensity on the whole area of the cell[1] by J_s. Hence the responsivity is expressed in amps/quanta sec^{-1}:

$$i(\omega) = J_s(\omega)\mathcal{R}(\omega). \quad (39)$$

It should be noted that $\mathcal{R}$ also depends on the optical frequency of the light signal which will be denoted by ν. The noise equivalent radiation intensity P_{eq} is defined as

$$P_{eq} = [S_i(\omega)\Delta f]^{1/2}/\mathcal{R}(\omega). \quad (40)$$

As expected, the signal-to-noise ratio (38) is unity when $J_s = P_{eq}$. The noise equivalent power in watt is simply $h\nu P_{eq}$.

$$P_{eq}' = h\nu P_{eq}. \quad (41)$$

It will turn out (Section III-C) that P_{eq} is proportional to $A^{1/2}(\Delta f)^{1/2}$. Therefore, to compare cells it is meaningful to evaluate P_{eq} for $A = 1$ cm² and $\Delta f = 1$ cycle per second. The inverse of P_{eq} under these conditions is called the specific sensitivity [46], [74] and can be seen as a figure of merit. In simple cases the signal and noise depend on ω in the same way and P_{eq} does not depend on the response time τ. In other cases, however, P_{eq} does depend on ω and thus on τ. (This is always the case if $1/f$ noise predominates.) Figures of merit for special reference conditions, including the dependence on τ have been given by Jones in his survey paper on the performance of radiation detectors [46]. If the signal-to-noise ratio is smaller than unity, τ determines the information rate. Petritz [73] and MacQuiston [53] have discussed the performance using an information theory approach.

The value of P_{eq} determined from the observed noise and the measured response according to (40) is not necessarily the limit that can be reached with a particular photoconductive material at a particular temperature. First of all, the detector should not be limited by $1/f$ noise but by gr noise. This is the case for several

---

[1] Since we are dealing with fluctuations we get into trouble when we work with radiation per unit area or with carrier densities. E.g., the simple relation $\langle\Delta n^2\rangle \simeq \langle n\rangle$ gives dimension inconsistencies when applied to densities. Although unorthodox with respect to other fields, our quantities J_s and J_r will refer to intensities on a given cell area A, and likewise our quantities n, p refer to total numbers of electrons and holes in a given cell volume $V = AD'$ in which the absorbed radiation is supposed to be approximately uniform. The conversion of the the results to the usual units is straightforward and left to the reader.

materials today (see Section III-C). Furthermore, as has been especially emphasized by Petritz [73], in most of the present materials fluctuations in the carrier densities are caused by interaction with lattice vibrations (phonons) and not by interaction with the background black body radiation field (photons). The limiting sensitivity is reached when all carrier excitations involve photon absorption. The noise is then closely related to photon fluctuations of the background radiation. Photon noise will be defined here as the fluctuations in background radiation intensity J_r *incident* on the detector and is indicated by $S_{J_r}(f)$. Since the effect is caused by the arrival of the individual photons, we have for the fluctuations in any part ϵ of the stream:

$$S_{\epsilon J_r} = \epsilon S_{J_r}. \tag{42}$$

In a photocell of area A the fraction of the incident intensity effective in producing free carriers, will be denoted by ηJ. The quantum efficiency η is related to the absorption coefficient α and the reflection coefficient R. However, each absorbed quantum need not be effective since part of the energy may be exchanged with phonons. (See also next section.) As an estimate, we may put, however

$$\eta(\nu) \sim [1 - R(\nu)][1 - \exp(-\alpha(\nu)d)] \tag{43}$$

where d is the thickness of the layer. Usually the simplifying assumption is made[2] that the absorption is uniform over an equivalent layer $d' \simeq 1/\alpha$.

The noise in that part of the photon stream that is effective in excitation is found from (42), setting $\epsilon = \eta$. If η_s is the quantum efficiency at the signal frequency then the number of excitations per second due to J_s is $\eta_s J_s$; when this number is equal to twice the rms fluctuations induced by the photon noise we will denote it by Q_{eq}. The factor 2 will be explained in the next section. According to (42) and the above reasoning we have

$$Q_{eq} = \left[2 \int \eta(\nu) S_{J_r}(\nu) d\nu \Delta f \right]^{1/2} \Big/ \eta_s. \tag{44}$$

The integral is to be extended over all frequencies of the background radiation that is permitted to fall on the detector. A performance figure of a particular cell is

$$F = P_{eq}/Q_{eq}. \tag{45}$$

Jones calls this factor the noise figure. It is worthwhile to compare this figure with the noise figure in communication receivers. In that case the limiting input noise is the thermal noise of the transformed antenna resistance [96], which depends on the particular antenna and on its temperature. Likewise, the photon-induced noise with which the actual noise is compared

according to (44) depends on the operation of the particular detector (its quantum efficiency, cutoff wavelength, and temperature). The situation is more complicated here, however. A value of F close to unity may indicate that the optimum performance under the particular operating conditions is attained but these conditions themselves may be rather unfavorable. As an example, let us assume that a lead sulphide cell cooled to liquid nitrogen temperature is seeing radiation noise of its room temperature surroundings. Noise figures close to unity have been reported experimentally under these conditions [32], [65]. A better performance could be obtained, however, by introducing a radiation shield with a transmission filter that only passes the signal wavelength, both of which are kept at low temperatures. The cell may then no longer be limited by photon noise, but the noise equivalent power has been reduced below the radiation limit under previous conditions. This will be worked out more quantitatively in Section III-C.

B. Interaction Between Photons and a Solid

Let the photoconductor exchange energy with a black-body radiation field of temperature T_r. We assume thermal equilibrium and also that all transitions involve photon emission or absorption. The density of photons $q(\nu)$ is given by Planck's law

$$q(\nu) = 8\pi(\nu^2/c^3)/(e^{h\nu/kT_r} - 1). \tag{46}$$

Let J_r be again the radiation intensity incident on the detector area A; then

$$J_r(\nu) = cAq(\nu)/4 = A2\pi(\nu^2/c^2)/(e^{h\nu/kT_r} - 1). \tag{47}$$

The spectrum of the photon noise is found with Bose-Einstein statistics [91], [50], [45], [101].

$$S_{J_r}(f) = 2J_r(\nu)[e^{h\nu/kT_r}/(e^{h\nu/kT_r} - 1)]. \tag{48}$$

For Q_{eq} we find from (48) and (44)

$$Q_{eq} = \left[4 \int \frac{\eta(\nu)J_r(\nu)(\Delta f)\exp(h\nu/kT_r)d\nu}{\exp(h\nu/kT_r) - 1} \right]^{1/2} \frac{1}{\eta_s}. \tag{49}$$

The spectrum of the photon noise as expressed by (48) is white as for shot noise. Apparently, this stream is noisier than one would expect from a Poisson distribution since the Bose factor [ ] in (48) is larger than unity. It is remarkable that the noise of the effective photon stream $\eta_r J_r$ reflects these Bose-Einstein fluctuations [compare (42)] whereas the electrons follow Fermi statistics. Several authors have neglected these effects and assumed for simplicity that photons as well as electrons followed Poisson statistics. The consistency is then obvious [73], [77]. However, there is no discrepancy, as we will now show. To that purpose we will adopt Einstein's notation [31], [3], [22]. Let the radiation induce transitions between two levels $\mathcal{E}_1$ and $\mathcal{E}_2$ with

---

[2] If the signal and the background radiation are composed of different wavelength regions, the inhomogeniety cannot be neglected. This can be one of the reasons why the response time and the time τ from noise measurements are seldom in complete agreement.

carrier occupancy n' and n, respectively; then we may write[3]

$$\frac{dn}{dt} = n'J_r B_{12} - n[J_r B_{21} + A_{21}]. \qquad (50)$$

The first term gives the photon induced excitations, the first term in the bracket is the "stimulated emission," and the last term is the spontaneous recombination rate. The quantum efficiency η is, expressed in this notation, the factor that relates the net excitation rate to J_r, hence

$$\eta = (n'B_{12} - nB_{21}). \qquad (51)$$

We will identify $n'J_r B_{12}$ with $g^*(n)$ and $n(J_r B_{21}+A_{21})$ with $r^*(n)$ to conform with the notation of Section II-B; the asterisk is added to indicate that these quantities refer to photon induced excitation and to radiative recombination. From (50) we have:

$$\frac{d(\Delta n)}{dt} = \left(\frac{\partial g^*}{\partial n} - \frac{\partial r^*}{\partial n}\right)\Delta n + \left(\frac{\partial g^*}{\partial J_r} - \frac{\partial r^*}{\partial J_r}\right)\Delta \dot{J}_r - \Delta F_{sp}$$

$$= -\Delta n/\tau^* + (n_0'B_{12} - n_0 B_{21})\Delta J_r - \Delta F_{sp}. \qquad (52)$$

The interpretation is obvious: the first term gives the regression of any disturbance to equilibrium, τ^* being the radiative lifetime; the second term gives the fluctuations of the induced net excitation rate F_q; and the last term accounts for the random fluctuations in the spontaneous recombination rate F_{sp}. The subscript zero refers again to average values. Note that this equation is of the general form (10), the origin of the random function $f(t)$ being specified in this case. For the noise in the net excitation rate we found already [(48), (42), and (51)]:

$$S_{\eta J_r} = 2(n_0'B_{12} - n_0 B_{21})J_r(\nu)e^{h\nu/kT_r}/(e^{h\nu/kT_r} - 1). \qquad (53)$$

Following Einstein's treatment the following relations are known

$$n_0/n_0' = K \exp(-h\nu/kT_r); \quad B_{12} = KB_{21}. \qquad (54)$$

Hence one shows easily

$$S_{\eta J_r} = 2B_{12}n_0'J_0 \equiv 2g_0^*. \qquad (55)$$

Likewise we find for the fluctuations in the spontaneous recombination rate:

$$S_{F_{sp}} = 2n_0 A_{21}e^{h\nu/kT_r}/(e^{h\nu/kT_r} - 1)$$

$$= 2n_0(J_r B_{21} + A_{21}) \equiv 2r_0^*. \qquad (56)$$

The contributions of (55) and (56) are equal since $g_0^* = r_0^*$. Making a Fourier analysis of the various fluctuating quantities in (52) one finds for S_{nn}:

$$S_{nn}(\omega) = 4g_0^*\tau^{*2}/(1 + \omega^2\tau^{*2}) \qquad (57)$$

---

[3] In Einstein's paper the photon field is represented by the radiation density $q(\nu)$ instead of $J_r(\nu)$. Both quantities are proportional (47) but differ in dimension by [sec]. It should also be noted that the Einstein A's and B's applied to this case are not constants but depend on the occupancy of the levels to which transitions are made and hence on n since $n+n'=$constant.

in complete agreement with (16). We thus proved: 1) the photon induced transitions and the spontaneous transitions give equal contributions to the noise; 2) the noise of the transition rates behaves as if two uncorrelated and completely random electron currents "flow" between the energy states $\mathcal{E}_1$ and $\mathcal{E}_2$ [multiplying with e^2 in (55) gives the shot noise formula $S_i = 2e(eg_0^*) = 2eI$]; 3) there is complete reconciliation between the Bose statistics of the photons and the Fermi statistics of the electrons. The factor 2 in (44) is now explained: *the total noise can never be less than twice the photon induced noise.* In the above treatment we assumed that the photoconductor interacted with background radiation of a single frequency only. Obviously, the result remains the same when more frequencies are present providing the excitation mechanism is the same for all photon energies. Finally, we have to go into the case that the photoconductor is cooled and sees black-body radiation of room temperature. In these cases there is no thermal equilibrium. The absorption process is the same but the emission is usually radiationless. Since the emitted phonons have the same statistics as the photons (see also [34a]) the same result is found. If the absorption act also involves phonons, the detector is obviously not limited by photon noise and the noise equivalent power $P_{eq} \ll Q_{eq}$.

It is instructive to also express Q_{eq} in photoconductor attributes. Let us specify to this purpose the detector as an intrinsic or near-intrinsic semiconductor. Then (Section II-B) the quantity g_0^* can be expressed as $g_0^* = \langle\Delta n^2\rangle/\tau^* = n_0 p_0/\tau^*(n_0+p_0)$. Since a signal J_s gives an increase in excitation rate of $\eta_s J_s$, we find for Q_{eq} from (55) and (56).

$$Q_{eq} = (4g_0^*\Delta f)^{1/2}/\eta_s = [4n_0 p_0 \Delta f/\tau^*(n_0 + p_0)\eta_s^{-2}]^{1/2}. \qquad (58)$$

C. Infrared Photoconductors

Since semiconductors have a bandgap of ~ 1 ev or smaller, these materials are suitable to detect infrared radiation. A review of the optical properties of germanium and silicon is found in a paper by Burstein, Picus, and Sclar [23]. The long wavelength limits of Ge and Si are 1.8 μ and 1.2 μ, respectively. It is known that the quantum efficiency for *absorbed* quanta is close to unity [37]. At room temperature these detectors see no radiation noise, since the lifetimes usually measured are orders of magnitude less than the radiative lifetimes which are of the order of one second [98a]. Obviously, the photon noise limit Q_{eq} will not be reached at room temperature. Since the absorption coefficient is high ($\approx 10^4$ cm^{-1}) the crystals should be very thin (about 1 micron) in order to reach the radiative limit at all. An alternative is offered by the use of junction diodes which do not have such a large dark current.

PbS, PbSe, and PbTe are intrinsic photoconductors which have been widely used [65a]. GR noise in lead sulphide was measured by Lummis and Petritz [51]. This is somewhat surprising since these materials are applied in the form of microcrystalline films. It is assumed that in these materials the electrons (minority carriers) are mainly trapped. At room temperature the films are limited by dark current gr noise caused by lattice vibrations [51]. At low temperatures the photon limit has been reached for films exposed to radiation of room temperature or lower. Noise equivalent powers as low as 2×10^{-14} watt have been reported [32], [106]. Lead selenide and telluride are less sensitive at comparable temperatures. In general, Q_{eq} increases with decreasing bandgap as was discussed by Petritz [73] [compare also (62b) below].

Impurity semiconductors with deep donors and acceptors have been prepared in great variety in recent years. Germanium doped with nickel, iron, cobalt, manganese, gold, etc., has been prepared by Tyler, *et al.* [90]. These impurities generally give rise to double acceptor levels; n or p-type materials can be obtained by counterdoping. The impurity centers are of the order of 0.2–0.4 ev from the conduction and valence band. Silicon has been doped similarly [28] although less is known than for Ge. To observe photoconductivity the temperature must be so low that most impurities are unionized. For the above materials 77°K is sufficient. Since the absorption constant is of the order of 1 cm^{-1}, crystals can have the usual thickness (~1 mm). Noise in these crystals has not been reported as far as the author knows. In manganese doped germanium Fassett found that the noise was gr noise [31a] (Section II-D). This noise was measured with the crystal at low temperature and shielded from background radiation. It is known that the resistance decreases when the crystal at 77°K is exposed to room temperature radiation. Whether these crystals are still limited by photon noise when shielded by a cooled transmission filter (Section III-A) is not known yet. The noise in the germanium specimen doped with other impurities is being investigated. An estimate of the sensitivity limit will be made below.

We will calculate the quantities mentioned in Section III-A to predict the behavior of these detectors. For the signal response an equation analogous to (52) can be stated

$$d(\Delta n)/dt = -\Delta n/\tau + \eta_s J_{s0} e^{j\omega t} \tag{59}$$

or, putting $\Delta i = i_0 e^{j\omega t} = (I/n_0)\Delta n$ for a semiconductor with a simple type of carrier, we find:

$$i_0(j\omega + 1/\tau) = (I/n_0)\eta_s J_{s0} \tag{60}$$

from which we find for the responsivity (39)

$$\mathcal{R}(\omega) = I\eta_s \tau / n_0 (1 + \omega^2 \tau^2)^{1/2}. \tag{61}$$

With the noise given by (28) we obtain for P_{eq}:

$$P_{eq} = \frac{2}{\eta_s} \left[\frac{n_0}{\tau} \left(\frac{1-\lambda}{2-\lambda} \right) \Delta f \right]^{1/2}. \tag{62a}$$

In a similar way one obtains for near-intrinsic semiconductors

$$P_{eq} = \frac{2}{\eta_s} \left[\frac{n_0 p_0 \Delta f}{\tau(n_0 + p_0)} \right]^{1/2}. \tag{62b}$$

Since p_0 and n_0 are total numbers P_{eq} is actually proportional to $A^{1/2}(\Delta f)^{1/2}$.

In PbS the result is slightly different since electron trapping and "barrier amplification" may occur [74]. For an intrinsic photoconductor, the noise figure is, according to (58) and (62b)

$$F = (\tau^*/\tau)^{1/2} \tag{63}$$

and this simple formula also holds for other two level cases. As emphasized by Petritz [73] it is not necessary to measure the absolute responsivity to find P_{eq}. If the shape of the spectrum shows that (62a) or (62b) applies, then the quantity P_{eq} may be calculated from n_0 and τ (which follow from the noise spectrum) and from an estimate of η_s. In case the spectrum is not of the above simple form, this indicates that transitions between more than two energy levels occur. The noise and the responsivity can still be calculated if the proper electron mechanism is known but this is usually complicated. Moreover, the noise and responsivity may differ in frequency dependence (next section). Lower sensitivity limits for intrinsic photoconductors have been listed in Petritz [73]. We will now make an estimate of P_{eq} for p-type Ge-Mn, using (62a). [The factor $(1-\lambda)/(e-\lambda)$ has to be omitted in this case; compare Section II-D]. At 77°K we observed: $p_0 \approx 10^9$ holes; $A \approx 0.3$ cm^2 $\tau \approx 10^{-5}$ sec. Putting $\Delta f = 1$ sec^{-1}, $\eta_s \approx 0.2$, $h\nu = 0.16$ ev, we find for P_{eq}:

$$P_{eq} \approx 2.5 \times 10^{-12} \text{ watt.}$$

This detector should be useful up to about 7 microns.

D. Photoconducting Insulators

The best known photoconductive insulator is CdS. Noise and response measurements have been reported by Shulman [84], Böer [9], and by van Vliet and Blok [102]. The latter investigators found that the noise and response were quite different for wavelengths smaller or larger than the absorption limit. The results were explained on account of the two-center model for CdS. Response times varying from 0.1 sec ($\lambda > 5100$ Å) to 10^{-4} sec ($\lambda < 5100$ Å) were observed. The response and the noise were parallel up to about 10 kc. Above that frequency the response dropped faster than the noise. This could be explained on account of electron trapping processes. In accordance with this the signal-to-noise ratio is constant up to 10 kc, even though the response time could be of the order of 1 second. Since CdS has a negligible dark resistance (depending on doping) all excitations are photon induced. Thus, this detector is definitely limited by photon noise. The introduction of concepts like P_{eq} and Q_{eq} is not unambiguous, however. The problem here is not to detect a light signal so small that quasi-thermal equilibrium can still be assumed. The dark current fluctuations, moreover, would have a complicated behavior since these currents are space charge limited [85]. Usually CdS is illuminated strongly till reasonable conductivities (*e.g.*, 10 KΩ for a cell) are obtained. One can, therefore, define a noise equivalent power depending on a certain amount of "bias light" that plays the same role as background radiation in semiconductors. Since the response as a function of the light intensity is not linear, this value $P_{eq}(J)$ will depend on J in a complicated way. From absolute response measurements we obtained values of P'_{eq} of 10^{-11} watt for the highest light levels which may decrease by several orders of magnitude for the lowest practical light levels.

IV. 1/f Noise in Semiconductor Filaments

A. Experimental Data

When the noise in a semiconductor filament is measured between probes with a constant current generator applied to the crystal end contacts, any noise resulting from the current carrying contacts is eliminated or at least heavily reduced (similar to Fig. 2). It is thus supposed that the remaining low-frequency noise is characteristic of the filament. Whether the noise associated with the contacts is of the same nature as the noise that is attributed to the filament itself has not been systematically investigated. There is no reason to believe, however, that this necessarily should be the case. Here we turn our attention to the crystal 1/f noise.

It seems quite clear at the present time that 1/f noise in germanium single crystals is, at least for the greater part, caused by the surface conditions. Probably the most direct proof was given by Maple, Bess, and Gebbie [54] who reported a 10 to 20 db increase in 1/f noise by switching the filament from a dry nitrogen ambient to one of carbon tetrachloride. The effect of proper etching of a crystal surface on 1/f noise is also well known. Montgomery [60] could produce changes in 1/f noise by concentrating the carriers on the surface by means of a magnetic field.

The frequency range over which 1/f noise extends is quite remarkable. It has been observed down to 2.10^{-4} by Rollin and Templeton [76] for germanium filaments. For point contact diodes it has been found even at 6×10^{-6} cycles per second [1], [33]. As to the upper limit more doubt exists. In carbon resistors and other materials 1/f noise was found up to 1 mc [99], [59] above which thermal noise drowned out the effect. However, this does not mean that 1/f noise in germanium filaments extends that far, since this noise may be of other origin as remarked above. Montgomery [60] measured 1/f noise to quite high frequencies in germanium crystals. In those days germanium crystals were much less perfected than nowadays, however. Hyde [42] observed 1/f noise in two-terminal germanium crystals up to 4 mc. Beyond this frequency the noise changed into a $1/f^2$ spectrum. Since the noise was not measured between probes it is uncertain whether

this noise should be considered to be characteristic for the filament. In many recent observations the 1/f noise does not extend beyond 100 to 10,000 cycles per second where it is masked by gr noise. Bess [8] reports an upper turnover frequency of ~1000 cycles per second for crystals at low temperatures. In the experiments of Maple, *et al.* [54], the l/f noise for a near intrinsic crystal emerged in carbon tetrachloride changed from $1/f^{1.22}$ dependence into a $1/f^3$ dependence at the frequency where the gr noise also started to decrease. This is in accordance with various proposed theories in which a lower limit for the upper turnover frequency is set by $\omega_0 = 1/\tau_{gr}$ where τ_{gr} is the lifetime of the carriers involved. We return to this in the next section.

Another remarkable effect is the slight temperature dependence of 1/f noise. In germanium little change has been found between liquid nitrogen temperature and room temperature. Templeton and MacDonald [89] measured noise in carbon resistors between 20 cycles per second and 10 kc and found little variation in magnitude. Similar results were reported by Russel [78] for ZnO crystals.

The current dependence of 1/f noise in germanium differs somewhat from case to case. Often, the noise is proportional to the square of the dc current as one would expect for true conductivity fluctuations. Sometimes, however, a I^4 dependence is reported. Brophy [14] has shown that such a higher power dependence is quite often found in plastically deformed crystals. Bess [8] attributes this behavior to edge dislocations. His theory will be discussed in Section IV-C. Since dislocations exist throughout the crystal, the noise is in his theory a combined bulk and surface effect. It was noticed by Brophy [16] that 1/f noise, if due to conductance modulation, should not necessarily depend on the presence of an electric field. He could demonstrate that 1/f noise also occurs when conductivity changes are detected by placing the crystal in a temperature gradient rather than in an electric field, which is an important clue as to the mechanism of 1/f noise.

Significant experiments have been performed in order to identify the carrier that is responsible for 1/f noise. The first experiments performed by Montgomery seemed to indicate that the noise is associated with the minority carriers. In his setup the noise was measured between each pair of three closely spaced terminals, A, B, and C on *n*-type germanium samples. The noise intensities S_{AB}, S_{BC}, and S_{AC} indicated considerable correlation. From the drift length of the carriers the correlation was also calculated and found to be in good agreement with the results, if the carrier lifetime was that of the minority carriers. In more recent measurements, on the contrary, this correspondence could no longer be established [61]. Another approach to this problem was made by Brophy [13], Rostoker [15], and Bess [7] who measured the Hall effect noise. They concluded that the noise in their samples was mainly due to majority carriers. In nearly intrinsic material fluctuations in the minority carrier density also contributed. The experimental data could best be fitted by assuming correlation between hole and electron fluctuations, such that

$$\frac{\Delta n}{n_0} = - \frac{\Delta p}{p_0}.\qquad(64)$$

Bess [8] has pointed out that such a relation would be expected for slow fluctuations which modulate the Fermi level in a quasi-equilibrium-like fashion. In that case $pn = n_i^2 = $ constant from which (64) immediately follows. It is interesting to note that Bess [7] also measured the Hall effect noise at frequencies where gr noise predominated; in this case the fact that $\Delta n = \Delta p$, as stated in Section II-E, was corroborated.

Some general conclusions may be drawn from these experimental data. First of all, the effect cannot be caused by random events with a single time constant, since this results in a spectrum of the form $c\tau/(1+\omega^2\tau^2)$. Moreover, the mechanism should explain the fact that the spectrum is usually not exactly $1/f$, but of the form $1/f^\alpha$, where α varies somewhere between 0.7 and 1.5 for different materials and specimen. It has been known for a long time [87], [92], [30] that a superposition of $\tau/(1+\omega^2\tau^2)$ spectra can result in a $1/f^\alpha$ law. Formally, one may write

$$S(f) = \int_{\tau_1}^{\tau_2} g(\tau) \frac{\tau}{1+\omega^2\tau^2} d\tau.\qquad(65)$$

For $g(\tau) = A/\tau$ this results in a $1/f$ spectrum for $1/\tau_2 < f < 1/\tau_1$. This transfers the problem into the finding of a mechanism for $g(\tau)$. This is not simple either. Electronic transitions between traps and the conduction and valence band as suggested by Baumgartner and Thoma [2] may give long trapping times but this is not observed in the noise since the average *free* time of the carriers is much smaller. The time constant which determines the noise spectrum is always the smaller one of the two time constants involved. This is the main reason that $1/f$ noise cannot be caused by a superposition of gr noise terms, involving deep traps. Presuming with Brophy that the noise is not inherent in the passage of current but can be attributed to conductivity fluctuations, the two alternatives left are: either the carrier densities themselves are modulated (*e.g.*, by the random creation and disappearance of donor centers [8]), or the *rates* of the carrier transitions are modulated in some way. Bess [8] opposes this idea, arguing that because of detailed balance such fluctuations are smoothed out within a few carrier lifetimes. This may be true for the behavior of the bulk where indeed, for times large in comparison with the carrier lifetime the occupancy of all states and consequently the rate of the transitions only depends on the Fermi level. At the surface, on the contrary, the relative position of the Fermi level depends on the surface charges and can fluctuate. This will be considered in more detail in the next section.

B. McWhorter's and North's Analyses

McWhorter [58], [57] has attributed the noise to the trapping and untrapping of so-called "slow surface states" [47], [49]. Before discussing his theory we first briefly review what is known about the surface of semiconductors such as germanium. It has been known for some time that the energy bands in a semiconductor are curved at the surface due to charges trapped in surface states. By a suitable choice of the ambient it is possible to make the surface either n-type or p-type irrespective of the bulk conductivity. The space charge region can be described by the parameters ϕ_B and ϕ_S (see Fig. 3).

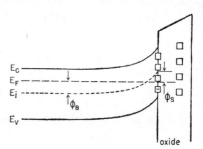

Fig. 3—Band picture for a germanium surface (after Kingston, [47]). The oxide layer is 20–40 Å.

In addition there is an oxide layer at the surface indicated by a surface barrier. The nature of the surface states was particularly investigated with the aid of the field effect [48], [63]. In this effect the conductivity of the germanium sample is modulated by changing the surface charges with the aid of a pulsed or sinusoidally modulated transverse electric field (perpendicular to the surface). The response is usually quite complex. If a pulsed field is applied there is first a relatively rapid response, reaching a value corresponding to some quasi-equilibrium state of the carriers and the surface recombination centers. Then the conductance decays slowly to its original value with a half life for the decay ranging from milliseconds to several seconds, depending on the surface treatment and the gaseous ambient. The effect has been generally analyzed, assuming that there are two groups of surface states, the "fast states" which are responsible for the recombination velocity of the carriers, and the "slow states" which give the tail in the response curve. It is further assumed that the "fast states" are at the germanium-oxide interface, and that the "slow states" are in the oxide layer, or at the outside. McWhorter assumes that free carriers communicate with the slow states by tunneling through the barrier. The attractive feature of this assumption is the temperature independence of this process. McWhorter measured the response to a sinusoidally varying field. In many cases the response could be approximated by

$$\Delta\sigma(\omega) = a \log b\omega \qquad (66)$$

for frequencies $f < f_{max}$. If the slow surface states would have a single capture time constant τ, then McWhorter shows that the response should be of the form $j\omega\tau/(1+j\omega\tau)$. Note that this is the same as found in (23). The form (66) can only be explained if we introduce a distribution of τ's:

$$\Delta\sigma(\omega) = a' \int_{\tau_1}^{\tau_2} \frac{g(\tau)j\omega\tau}{1 + j\omega\tau} d\tau. \qquad (67)$$

For $g(\tau) \sim 1/\tau$ the result (66) is approximately found. Apparently, this is just the distribution of time constants needed to obtain $1/f$ noise. It is accordingly very promising to assume that $1/f$ noise is caused by spontaneous fluctuations in the capture and release of carriers by the slow surface states. If τ would be due to tunneling, then, according to quantum mechanics

$$\tau = \tau_0 \exp (2mV/\hbar^2)^{1/2}w \qquad (68)$$

where w is the barrier width, V the barrier height, and $\tau_0 \approx 10^{-12}$ sec. If w varies between 20 and 40 Angstrom, τ varies between 10^{-4} sec and 10^6 sec. Unfortunately, the field experiments seem to indicate an upper value of f_{max} lower than usually found for $1/f$ noise. However, the analysis of the field effect for frequencies close to the carrier lifetimes is not unambiguous. Since also the transition of $1/f$ noise into gr noise is not well known, no discrepancy may exist at all. McWhorter has also given a quantitative calculation of the expected noise. It is felt that this noise can be found in an easier way from the general procedures outlined in Sections II-B and II-C. The slow fluctuations in electrons and holes in the surface region are then easily found. The next step is to solve for the bulk conductivity fluctuations with the aid of Poisson's equation as is also done by McWhorter.

Closely related to this procedure is a theory developed by North [67]. North assumes that the fluctuations in the surface potential ϕ_S are thermal. Hence, the fluctuations in surface recombination velocity s follow from

$$\langle\Delta\phi_s^2\rangle = 4kTR_{eq}\Delta f \quad \langle\Delta s^2\rangle = \left(\frac{\partial s}{\partial \phi_s}\right)^2\langle\Delta\phi_s^2\rangle. \qquad (69)$$

The quantity R_{eq} is the real part of an equivalent impedance into which ϕ_s looks. To calculate R_{eq} an equivalent network is developed by North in which the transition rates serve as conductances and the barrier capacitance and the time constants determine the capacitances. The theory is closely related to that of Section II-C. His basic idea was applied with success by Fongers to noise in transistors and junction diodes [97].

C. Other Theories

Bess [6], [8] has proposed an entirely different interpretation of $1/f$ noise. In accordance with the observation that the amount of $1/f$ noise can be changed by plastic deformation Bess assumed that the noise was associated with edge dislocations. Impurities should be diffusing along the edge dislocation line to and from the surface where they undergo some type of Brownian motion. With a highly specialized mathematical model this results in $1/f$ noise. Although the application of Bess' mathematical model is doubtful, his basic idea to associate the noise with dislocations is very attractive. As pointed out by Morrison [64a] the energy band structure in the neighborhood of a dislocation is similar to that of the surface (Fig. 4) a fluctuation of the trapped charge will thus modulate ϕ_D and the recombination velocity as in previous theories. This effect has

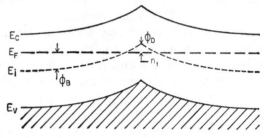

Fig. 4—Band picture at a dislocation (after Morrison, [64a]).

also been worked out by Morrison himself in a different way [64]. He assumes that the transition rates follow a relation of the Elovich type:

$$d(\Delta n_t)/dt = B(e^{b\Delta n_t} - 1) \qquad (70)$$

where n_t is the trapped charge and B and b are constants. From this the correlation function and spectrum can be found. This results in a $1/f$ law over several decades. In contrast to North's theory one must assume very large deviations from thermal equilibrium in order to explain the nonlinear behavior.

Schönfeld [80] has found the interesting result that random events of a $1/\sqrt{i}$ character result in a $1/f$ spectrum according to Carson's theorem (Section II-A). However, no elementary events of such a form are known.

V. Modulation Effects in Granular Material

A. Proposed Theories

So far we have not mentioned several of the older theories for $1/f$ noise which were largely based on diffusion mechanisms. If these theories are applicable at all, then they might have some value for granular or microcrystalline material. MacFarlane [56] and Richardson [75] considered the diffusion of atoms or ions over the contact area of the grains. The spectra, however, are not $1/f$ like over many decades as pointed out by Burgess [17]. Moreover, the region in which a reasonable $1/f$ approximation is to be found is strongly temperature dependent. Petritz [72] has given a similar theory involving heat diffusion. McWhorter [56], [57] proposes that the tunnel processes also play a role in the passage of current between contacting grains. Experiments performed on a single mercury-aluminum contact gave support to this idea. Various other ideas have been suggested but nothing definite about the nature of the noise is known yet. The current dependence seems to be somewhat characteristic for the material. Carbon resistors invariably give a I^2 dependence. PbS films often show a stronger current dependence, thermistors gave noise proportional to $I^{1.25}$. Several features are discussed in a wartime report by Harris, Abson, and Roberts [38].

VI. BIBLIOGRAPHY

[1] Baker, D. "Flicker Noise in Germanium Rectifiers at Very Low Frequencies," *Journal of Applied Physics*, Vol. 25 (July, 1954), pp. 922–924.

[2] Baumgartner, W., and Thoma, H. U. "Zum Stromrauschen von Halbleitern," *Zeitschrift für angew. Physik*, Vol. 6 (1955), p. 66.

[3] Becker, R. *Theorie der Elektrizität*. Leipzig and Berlin: B. G. Teubner, sec. 70, 1933. Lithoprinted in the U.S.A. by Edwards Brothers, Inc., Michigan, 1945.

[4] Becking, A. G. T. "Fluctuatieverschijnselen by Bolometers," thesis, University of Utrecht, The Netherlands, 1953.

[5] Bernamont, J. "Fluctuations du Potential aux Bornes d'un Conducteur Métallique de Faible Volume, Parcouru par un Courant," *Annales de Physique*, Vol. 7 (1937), pp. 71–140.

[6] Bess, L. "A Possible Mechanism of 1/f Noise Generation in Semiconductor Filaments," *Physical Review*, Vol. 91 (September, 15, 1953), p. 1569.

[7] ——. "Relative Influence of Majority and Minority Carriers on Excess Noise in Semiconductor Filaments," *Journal of Applied Physics*, Vol. 26 (November, 1955), pp. 1377–1381.

[8] ——. "Study of 1/f Noise in Semiconductor Filaments," *Physical Review*, Vol. 103 (July 1, 1956), pp. 72–82.

[9] Böer, K. W., and Junge, K. "Zur Frequenzabhängigkeit von Elektronenschwankungserscheinungen in Halbleitern," *Zeitschrift für Naturforschung*, Vol. 8A (November, 1953), pp. 753–755.

[10] ——. "Einige Bemerkungen zur Gisolfschen Theorie der Elektronenschwankungserscheinungen von Halbleitern," *Annalen der Physik*, Vol. 14 (1954), pp. 87–96; also Vol. 15 (1954), pp. 55–56.

[11] ——, Kummel, U., and Molgedey, G. "Elektronenrauschen von CdS Einkristallen bei Hohen Feldstärken," *Annalen der Physik*, Vol. 17 (1956), pp. 344–356.

[12] Brophy, J. J. "Current Noise in Thermistor Bolometer Flakes," *Journal of Applied Physics*, Vol. 25 (February, 1954), pp. 221–224.

[13] ——, and Rostoker, N. "Hall Effect Noise," *Physical Review*, Vol. 100 (October 15, 1955), pp. 754–756.

[14] ——. "Excess Noise in Deformed Germanium," *Journal of Applied Physics*, Vol. 27 (November, 1956), pp. 1383–1384.

[15] ——. "Excess Noise in n-Type Germanium," *Physical Review*, Vol. 106 (May 15, 1957), pp. 675–678.

[16] ——. "Experimental Investigation of Excess Noise in Semiconductors." Presented before the American Physical Society, Boulder, Colo., September 5–7, 1957.

[17] Burgess, R. E. "Contact Noise in Semiconductors," *Proceedings of the Physical Society*, Vol. B66 (April, 1953), pp. 334–335.

[18] ——. "Fluctuations in the Number of Charge Carriers in a Semiconductor," *Physica*, Vol. 20 (November, 1954), pp. 1007–1010.

[19] ——. "Fluctuations of the Number of Electrons and Holes in a Semiconductor," *Proceedings of the Physical Society*, Vol. B68 (September, 1955), pp. 661–671.

[20] ——. "Electronic Fluctuations in Semiconductors," *British Journal of Applied Physics*, Vol. 6 (June, 1955), pp. 185–190.

[21] ——. "The Statistics of Charge Carrier Fluctuations in Semiconductors," *Proceedings of the Physical Society*, Vol. B69 (October, 1956), pp. 1020–1027.

[22] ——. "Statistical Fluctuations in Semiconductors." Presented before the American Physical Society, Boulder, Colo., September 5–7, 1957.

[23] Burstein, E., Picus, G., and Sclar, N. "Optical and Photoconductive Properties of Silicon and Germanium," in *Photoconductivity Conference*, R. G. Breckenridge *et al.*, eds. New York: John Wiley and Sons, Inc., 1956.

[24] Buttler, W. M. "Über das Randschichtrauschen in Halblei-

[25] tern," *Annalen der Physik*, Vol. 11 (1952), pp. 362–367·

[25] ——, and Muscheid, W. "Die Bedeutung des elektrischen Kontaktes bei Untersuchungen an Kadmium Sulfid-Einkristallen," *Annallen der Physik*, Vol. 14 (1954), pp. 215–219; also Vol. 15 (1954), pp. 82–111.

[26] Callen, H. B. and Welton, T. A. "Irreversibility and Generalized Noise," *Physical Review*, Vol. 83 (July 1, 1951), pp. 34–39.

[27] Christenson, C. J., and Pearson, G. L. "Spontaneous Fluctuations in Carbon Microphones and Other Granular Resistors," *Bell System Technical Journal*, Vol. 15 (April, 1936), pp. 197–223.

[28] Collin, C. B., and Carlson, R. O. *Bulletin of the American Physical Society II*, Vol. 1 (March 15, 1955), p. 127.

[29] Davydov, B., and Gurevich, B. "Voltage Fluctuations in Semiconductors," *Journal of Physics of the USSR*, Vol. 7 (1943), pp. 138–140.

[30] duPré, F. K. "A Suggestion Regarding the Spectral Density of Flicker Noise," *Physical Review*, Vol. 78 (June 1, 1950), p. 615.

[31] Einstein, A. *Physikalisch Zeitschrift*, Vol. 18 (1917), p. 121.

[31a] Fassett, J. R., M.Sc. thesis, University of Minnesota, Minneapolis, Minn., 1958, unpublished.

[32] Fellgett, P. B. "On the Ultimate Sensitivity and Practical Performance of Radiation Detectors," *Journal of the Optical Society of America*, Vol. 39 (November, 1949), p. 970.

[33] F rle, T., and Winston, H. "Noise Measurements in Semiconductors at Very Low Frequencies," *Journal of Applied Physics*, Vol. 26 (June, 1955), p. 716.

[34] Fowler, R. "Statistical Mechanics," Cambridge: University Press, sec. 20.71, 1936.

[34a] Fröhlich, H. "Elektronentheorie der Metalle," Berlin: Springer Verlag, 1936. See especially sec. 13.

[35] Gebbie, H. A. "Excess Noise and Trapping in Germanium," *Physical Review*, Vol. 98 (June 1, 1955), p. 1567.

[36] Gisolf, J. H. "On the Spontaneous Current Fluctuations in Semiconductors," *Physica*, Vol. 15 (September, 1949), pp. 825–832.

[37] Gouscher, F. S. "The Photon Yield of Electron-Hole Pairs in Germanium," *Physical Review*, Vol. 78 (June 15, 1950), p. 816.

[38] Harris, E. J., Abson, W., and Roberts, W. J. Report, Telecommunications Research Establishment, 1946 (unpublished).

[39] Herzog, G. B., and van der Ziel, A. "Shot Noise in Germanium Single Crystals," *Physical Review*, Vol. 84 (December 15, 1951), p. 1249.

[40] Hill, J. E., and van Vliet, K. M. "Generation-Recombination Noise in Intrinsic and Near-Intrinsic Germanium Crystals," *Journal of Applied Physics*, Vol. 29 (February, 1958), pp. 177–182.

[41] Hyde, F. J. "Measurements of Noise Spectra of a Point Contact Germanium Rectifier," *Proceedings of the Physical Society*, Vol. B66 (December, 1953), pp. 1017–1024.

[42] ——. "Excess Noise Spectra in Germanium," *Proceedings of the Physical Society*, Vol. B69 (February, 1956), pp. 242–245.

[43] "Shot Noise in a Germanium Filament," *Report of the Conference on Semiconductors of the Physical Society*, Rugby, England (1956), pp. 57–64.

[44] Johnson, J. B. "The Schottky Effect in Low Frequency Circuits," *Physical Review*, Vol. 26 (July, 1925), pp. 71–85.

[45] Jones, R. C. "The Ultimate Sensitivity of Radiation Detectors," *Journal of the Optical Society of America*, Vol. 37 (November, 1947), pp. 879–890.

[46] ——. "Performance of Radiation Detectors" in *Advances of Electronics*. New York: Academic Press, Vol. 5, 1953.

[47] Kingston, R. H. "Review of Germanium Surface Phenomena," *Journal of Applied Physics*, Vol. 27 (February, 1956), pp. 101–114.

[48] ——, and McWhorter, A. L. "Relaxation Time of Surface States on Germanium," *Physical Review*, Vol. 103 (August 1, 1956), pp. 534–540.

[49] ——, *et al. Surface Physics*. Philadelphia: University of Pennsylvania Press, 1957.

[50] Lewis, W. B. "Fluctuations in Streams of Thermal Radiation," *Proceedings of the Physical Society*, Vol. 59 (January, 1947), pp. 34–40.

[51] Lummis, F. L., and Petritz, R. L. "Noise, Time Constant and Hall Studies on Lead Sulfide Photoconductive Films," *Physical Review*, Vol. 105 (January 15, 1957), pp. 502–508.

[52] Machlup, S. "Noise in Semiconductors; Spectrum of a Two Parameter Random Signal," *Journal of Applied Physics*, Vol. 25 (March, 1954), pp. 341–343.

[53] MacQuiston, R. B. Private communication.

[54] Maple, T. G., Bess, L., and Gebbie, H. A. "Variation of Noise with Ambient in Germanium Filaments," *Journal of Applied Physics*, Vol. 26 (April, 1955), p. 490.

[55] Mattson, R. H., and van der Ziel, A. "Shot Noise in Germanium Filaments," *Journal of Applied Physics*, Vol. 24 (February, 1953), p. 222.

[56] McFarlane, G. G. "A Theory of Contact Noise in Semiconductors," *Proceedings of the Physical Society*, Vol. B63 (October, 1950), pp. 807–814.

[57] McWhorter, A. L. "1/f Noise and Related Surface Effects in Germanium," Lincoln Laboratory, Massachusetts Institute of Technology, Lexington, Mass., Report No. 80 (May, 1955), unpublished.

[58] ———. "1/f Noise and Germanium Surface Properties," in *Semiconductor Surface Physics*, R. H. Kingston *et al.*, eds. Philadelphia: University of Pennsylvania Press, 1957.

[59] Miller, P. H., Jr. "Noise Spectrum of Crystal Rectifiers," PROCEEDINGS OF THE IRE, Vol. 35 (March, 1947), pp. 252–256.

[60] Montgomery, H. C. "Electrical Noise in Semiconductors," *Bell System Technical Journal*, Vol. 31 (September, 1952), pp. 950–975.

[61] ———. Private communication to A. L. McWhorter.

[62] ———. Private communication.

[63] ———, and Brown, W. L. "Field-Induced Conductivity Changes in Germanium," *Physical Review*, Vol. 103 (August 15, 1956), pp. 865–870.

[64] Morrison, S. R. "Generation of 1/f Noise by Levels in a Linear or Planar Array," *Physical Review*, Vol. 99 (September 15, 1955), p. 1904.

[64a] ———. "Recombination of Electrons and Holes at Dislocations," *Physical Review*, Vol. 104 (November 1, 1956), pp. 619–623.

[65] Moss, T. S. "The Ultimate Limits of Sensitivity of PbS and PbTe Photoconductive Detectors," *Journal of the Optical Society of America*, Vol. 40 (September, 1950), pp. 603–607.

[65a] ———. "Lead Salt Photoconductors," PROCEEDINGS OF THE IRE, Vol. 43 (December, 1955), pp. 1869–1881.

[66] Mott, N. F., and Gurney, R. W. "Electronic Processes in Ionic Crystals." New York: Oxford University Press, second edition, 1948.

[67] North, D. O. "Theory of Noise Processes in Diodes and Transistors." Presented before the meeting of the American Physical Society, Boulder, Colo., September 5–7, 1957.

[68] Oliver, D. J. "Fluctuations in the Number of Electrons and Holes in a Semiconductor," *Proceedings of the Physical Society*, Vol. B70 (February, 1957), pp. 244–247.

[69] ———. "Current Noise in Indiumantimonide," *Proceedings of the Physical Society*, Vol. B70 (March, 1957), pp. 331–332.

[70] Petritz, R. L. "On the Theory of Noise in P-N Junctions and Related Devices," PROCEEDINGS OF THE IRE, Vol. 40 (November, 1952), pp. 1440–1456.

[71] ———. "On the Diffusion Theory of Noise in Rectifiers and Transistors," *Physical Review*, Vol. 87 (July 1, 1952), p. 189.

[72] ———. "Theory of Contact Noise," *Physical Review*, Vol. 87 (August 1, 1952), p. 535.

[73] ———. "The Relation Between Lifetime, Limit of Sensitivity and Information Rate in Photoconductors," in *Photoconductivity Conference*, R. G. Breckenridge *et. al.*, eds. New York: John Wiley and Sons, Inc., 1956.

[74] ———. "Theory of Photoconductivity in Semiconductor Films," *Physical Review*, Vol. 104 (December 15, 1956), pp. 1508–1516.

[75] Richardson, J. M. "The Linear Theory of Fluctuations Arising from Diffusion Mechanisms; An Attempt at a Theory of Contact Noise," *Bell System Technical Journal*, Vol. 29 (January, 1950), pp. 117–141.

[76] Rollin, R. V., and Templeton, I. M. "Noise in Semiconductors at Very Low Frequencies," *Proceedings of the Physical Society*, Vol. B66 (March, 1953), pp. 259–261.

[77] Rose, A. "Performance of Photoconductors," in *Photoconductivity Conference*, R. G. Breckenridge *et al.*, eds. New York: John Wiley and Sons, Inc., 1956.

[78] Russell, R. R. Tenth Annual Conference on Physical Electronics, M.I.T., Cambridge, Mass., March 30–April 1, 1950.

[79] Sautter, D., and Seiler, K. "Über das Rauschen von Germanium Einkristallen," *Zeitschrift für Naturforschung*, Vol. 12A (June, 1957), p. 490.

[80] Schönfeld, H. "Beitrag zum 1/f Gesetz beim Rauschen von Halbleitern," *Zeitschrift für Naturforschung*, Vol. 10A (April, 1955), pp. 291–300.

[81] Shockley, W. *Electrons and Holes in Semiconductors.* New York: D. van Nostrand Co., Inc., pp. 342–346, 1950.

[82] *Ibid.*, pp. 319–325.

[83] ———, and Read, W. T. "Statistics of the Recombination of Holes and Electrons," *Physical Review*, Vol. 87 (September 1, 1952), pp. 835–842.

[84] Shulman, G. I. "Shot Noise in CdS Crystals," *Physical Review*, Vol. 98 (April 15, 1955), pp. 384–386.

[85] Smith, R. W., and Rose, A. "Space-Charge-Limited Currents in Single Crystals of Cadmium Sulfide," *Physical Review*, Vol. 97 (March 15, 1955), pp. 1531–1537.

[86] Smits, G. H., *et al.* "Excess Noise in Indium-Antimonide," *Journal of Applied Physics*, Vol. 27 (1956), p. 1385.

[87] Surdin, M. M. "Fluctuations de Courant Thermionique et le Flicker Effect," *Journal de Physique et le Radium*, Vol. 10 (April, 1939), pp. 188–189.

[88] ———. *Revue Générale d'Electricité*, Vol. 47 (1940), p. 97.

[89] Templeton, I. M., and McDonald, D. K. C. "The Electrical Conductivity and Current Noise of Carbon Resistors," *Proceedings of the Physical Society*, Vol. B66 (August, 1953), pp. 680–684.

[90] *E.g.*, Tyler, W. W., and Woodbury, H. H. "Properties of Germanium Doped with Iron; I Electrical Conductivity," *Physical Review*, Vol. 96 (November 15, 1954), pp. 874–886.

Newman, R., and Tyler, W. W. "Properties of Germanium Doped with Iron; II Photoconductivity," *Physical Review*, Vol. 96 (November 15, 1954), pp. 882–886. The same authors have published similar papers on many other impurity elements in germanium. They have appeared in recent volumes of the *Physical Review*.

[91] Tolman, R. C. *The Principles of Statistical Mechanics*. New York: Oxford University Press, 1938.

[92] van der Ziel, A. "On the Noise Spectra of Semiconductor Noise and of Flicker Effect," *Physica*, Vol. 16 (April, 1950), pp. 359–372.

[93] ———. "Shot Noise in Semiconductors," *Journal of Applied Physics*, Vol. 24 (February, 1953), pp. 222–223.

[94] ———. "Simpler Explanation of the Observed Shot Effect in Germanium Filaments," *Journal of Applied Physics*, Vol. 24 (August, 1953), p. 1063.

[95] ———. "Note on the Shot Effect in Semiconductors and Flicker Effect in Oxide Cathodes," *Physica*, Vol. 19 (August, 1953), pp. 742–744.

[96] ———. *Noise*. Englewood Cliffs: Prentice Hall, Inc., 1954.

[97] ———. "Noise in Junction Transistors," this issue, p. 1019.

[98] van Roosbroeck, W. "The Transport of Added Carriers in a Homogeneous Semiconductor," *Physical Review*, Vol. 91 (July 15, 1953), pp. 282–288.

[98a] ———, and Shockley, W. "Photon Radiative Recombination of Electrons and Holes in Germanium," *Physical Review*, Vol. 94 (June 15, 1954), pp. 1558–1561.

[99] van Vliet, K. M., van Leeuwen, C. J., Blok, J., and Ris, C. "Measurements on Current Noise in Carbon Resistors and in Thermistors," *Physica*, Vol. 20 (August, 1954), pp. 481–496.

[100] ———, and Blok, J. "Electronic Noise in Semiconductors," *Physica*, Vol. 22 (March, 1956), pp. 231–242.

[101] ———, ———. "Electronic Noise in Photoconducting Insulators," *Physica*, Vol. 22 (June, 1956), pp. 525–540.

[102] ———, ———, Ris, C., and Steketee, J. "Measurements of Noise and Response to Modulated Light of Cadmium Sulphide Single Crystals," *Physica*, Vol. 22 (August, 1956), pp. 723–740.

[103] ———. "On the Equivalence of the Fokker-Planck Method and the Free Energy Method for the Calculation of Carrier Density Fluctuations in Semiconductors," *Physica*, Vol. 23 (March, 1956), pp. 248–252.

[104] ———. "Irreversible Thermodynamics and Carrier Density Fluctuations in Semiconductors," *Physical Review*, Vol. 110 (April 1, 1950), pp. 50–60.

[105] ———, and Hill, J. E. "Ambipolar Transport of Carrier Density Fluctuations in Germanium," to be published.

[106] Watts, B. N. "Increased Sensitivity of Infrared Photoconductive Receivers," *Proceedings of the Physical Society*, Vol. A62 (July, 1949), pp. 456–457.

[107] Williams, N. H., and Thatcher, E. W. "On Thermal Electronic Agitation in Conductors," *Physical Review*, Vol. 40 (April 1, 1932), p. 121.

DISCUSSION OF RECENT EXPERIMENTS ON $1/f$ NOISE

F. N. HOOGE[+]

*Philips Research Laboratories, N. V. Philips Gloeilampenfabrieken,
Eindhoven, Nederland*

Received 29 September 1971

Synopsis

The experimental evidence for the existence of bulk $1/f$ noise is discussed. An empirical relation between the magnitude of the noise and the number of free charge carriers is given. The $1/f$ noise of point contacts can be treated quantitatively using this relation. A comparison of the $1/f$ noise in the conductance and in the voltage of thermo cells and concentration cells leads to the conclusion that the $1/f$ noise is due to fluctuations in the mobility of the free carriers, and not to fluctuations in their number as has been assumed up to now.

1. *Introduction.* During the last five years experiments on $1/f$ noise have been done which throw a new light upon the ill-understood phenomenon of $1/f$ noise. This noise is named after its power spectrum, which is inversely proportional to the frequency. The $1/f$ noise of a quantity X (X may represent current, voltage or conductivity) will be expressed here as relative noise defined by the relation

$$\left\langle \left(\frac{\Delta X}{X}\right)^2 \right\rangle = C_x \frac{\Delta f}{f}, \tag{1}$$

where f is frequency and Δf is bandwidth.

Our experiments show that $1/f$ noise is always present when an electric current is flowing. A simple empirical relation for the magnitude of the noise has been found, which, however, does not suggest a theoretical model. No generally applicable model has been proposed until now.

The purpose of the present paper is to discuss systematically experiments the results of which are scattered over different publications. Arguments used in earlier papers can sometimes be replaced by stronger arguments based on more recent measurements. In this way a consistent picture can be given of our present knowledge of $1/f$ noise. We shall not discuss $1/f$ noise in semiconductor devices. The main part of this paper is devoted to a discussion of the noise in concentration cells and in thermo e.m.f. for which only experimental results have been published.

[+] Present address: Eindhoven University of Technology, Eindhoven, Netherlands.

It is not our intention to discuss all models proposed, but we shall devote our attention to McWhorter's surface-noise model[1]), since that is the generally accepted model at the moment. The greater part of the work on $1/f$ noise has been done on MOSTs (metal–oxide–silicon transistors) in which the $1/f$ noise is indeed probably McWhorter's surface noise. A MOST is a capacitor with one metal plate and one semiconductor plate. The voltage across the insulating oxide between the two plates (gate voltage) induces a mobile charge in the semiconductor. A current can then flow in the semiconductor immediately beneath the oxide when a voltage is applied across the semiconductor parallel to the oxide (source-drain voltage). The many studies on MOSTs give the mistaken impression that there is no $1/f$ noise other than the surface noise. Our experiments have shown not only that bulk $1/f$ noise exists but also that it is present whenever a current is carried by a small number of charge carriers. The number of the carriers determines the magnitude of the $1/f$ noise. Although McWhorter's model may be right in some specific cases (however interesting these cases may be for transistor technology), it is not the answer to the more fundamental problem of bulk $1/f$ noise.

2. *Some problems of $1/f$ noise and McWhorter's model.* One of the main problems of $1/f$ noise is its frequency dependence $\propto 1/f$ over a large number of decades. In some cases $1/f$ noise has been measured from the 10 kHz region down to 5×10^{-5} Hz[2]). The correlation function leading to an $1/f$ spectrum cannot be found by directly applying the Wiener–Khinchine theorem. The only obvious way to obtain $1/f$ noise is to consider it as the superposition of spectra of the type

$$\langle (\Delta X)^2 \rangle = A \{\tau/[1 + (2\pi f\tau)^2]\} \Delta f, \tag{2}$$

which corresponds to a simple exponential correlation function

$$\langle \Delta X(t) \rangle = \Delta X(0) \, e^{-t/\tau}. \tag{3}$$

For $1/f$ noise we need a continuous set of τ values with a statistical weight

$$g(\tau) \propto \frac{1}{\tau} \, d\tau. \tag{4}$$

The correlation function for $1/f$ noise is then

$$\langle \Delta X(t) \rangle = \Delta X(0) \left(1/\ln \frac{\tau_b}{\tau_a} \right) \int_{\tau_a}^{\tau_b} \frac{1}{\tau} \, e^{-t/\tau} \, d\tau, \tag{5}$$

where τ_a and τ_b are the limits of the range of τ values. This leads to a spectrum which is white up to $f_b = 1/2\pi\tau_b$ and proportional to $1/f$ between f_b and $f_a = 1/2\pi\tau_a$, while from f_a onwards it is proportional to $1/f^2$.

The problem now is to find a physical model which satisfies (4). The only model that gives the wide range of relaxation times with the correct weights naturally is McWhorter's surface model[1]. Its simplest version is as follows. It is assumed first that in the oxide layer on the semiconductor there are electron traps with a constant concentration through the whole layer. The second assumption is that the probability of penetration into the oxide layer decreases exponentially with distance. These two assumptions give the required weight for the relaxation times. The traps close to the interface have small τ's and the farthest traps have large τs, leading to the low-frequency part of the spectrum.

$$\text{Probability of penetration} \propto 1/\tau \propto e^{-x/x_0}, \tag{6}$$

$$g(\tau) = \frac{dN}{d\tau} = \frac{dN}{dx}\frac{1}{d\tau/dx} = C\frac{1}{\tau/x_0} \propto \frac{1}{\tau}. \tag{7}$$

A condition for the model to give $1/f$ noise is that the traps must be independent. Electrons in each element of the interface must interact with one specific trap. If an electron could interact with several traps having different τ's, the result would be a simple generation–recombination spectrum (2) with an effective τ which would be close to the smallest τ since

$$1/\tau_{\text{eff}} = \sum_i 1/\tau_i. \tag{8}$$

Many refinements of this model have been proposed[3-6]. One feature common to them all is that the slow states required are assumed to be at the surface. Slow surface states have been found experimentally in silicon and germanium, whereas it is impossible to conceive such slow states in the bulk. The location of the noise sources at the surface follows from the long relaxation times required rather than from direct experimental evidence.

In some cases (all involving MOSTs) a close relation between $1/f$ noise and measured surface state concentration has been established experimentally[7-9] but the work of Feigt and Jäntsch[10] shows a large discrepancy between the smallest relaxation time and the frequency at which the $1/f$ noise changes to $1/f^2$ noise. Berz and Prior[11] found that the $1/f$ noise can on the whole be described quantitatively by the McWhorter model, but that a second mechanism may dominate at higher voltages.

3. *1/f noise in homogeneous resistors.* From the very start[12] of the study of $1/f$ noise there has always been a school of thought which considered $1/f$ noise to be a bulk effect[13,14]. Bell even tried to find a relation between $1/f$ noise and

the number of charge carriers[15]). He did not succeed, however, since the data available to him at that time (1955) were few in number and mostly unreliable.

There are several recent publications which point to the bulk as the place where 1/f noise originates[16-18]).

A few years ago we collected all the published data on 1/f noise of well defined homogeneous samples[19]). The survey suggested that for those samples the noise could be expressed by the empirical relation:

$$\left\langle \left(\frac{\Delta G}{G}\right)^2 \right\rangle = \frac{\alpha}{N} \frac{\Delta f}{f}, \tag{9}$$

where G is the conductance, N the total number of mobile charge carriers in the sample and α a dimensionless constant with the value 2×10^{-3}.

The survey did not provide absolute evidence that relation (9) is correct. The noise of some samples differed considerably from relation (9), mostly on the high side. Now, knowing the noise properties of point contacts, we would suggest that those samples may have displayed considerable contact noise either because of a grainy structure or because only two measuring contacts were used. On the other hand, though no unfavourable data were excluded from the survey, the general picture was nevertheless promising. It formed the starting point for a systematic study of the correctness of the empirical relation.

Published silicon and germanium data already agreed reasonably well with (9). Still better agreement was obtained on our own epitaxial specimen[20]). The data on compound semiconductors are more uncertain. Their α values might possibly differ from 2×10^{-3}. InSb in particular is always too high[19,21]). Main and Owen[22]) applied (9) to vitreous semiconductors and estimated the number of free electrons. Noise measurements on GaAs Gunn devices by Meade[23]) agree rather well with the predicted noise if (as is correct), the number of electrons is used instead of the concentration.

If (9) also holds true for metals, it is possible to calculate how small a metal sample must be in order to show a detectable 1/f noise (above 100 Hz, with a current which does not heat the sample excessively). The answer is that 10^{13} electrons at most may be present, which means volumes of not more than 10^{-10} cm^3. This explains why previously measured metal samples did not show any 1/f noise[24]). We made small continuous samples of this kind by evaporating gold and using photo-etching techniques[25]). The samples, designed for four-probe measurements, had a central part measuring $800\,\mu \times 10\,\mu$ and a thickness of several hundred Å. The experiments showed that the 1/f noise of gold is quantitatively described by relation (9). A surprising result was that cooling had no effect on α. Even at 4 K, with the sample immersed in liquid helium, we found about the same value for α. The accuracy is such that a weak temperature dependence might be present but at most this could only be proportional to $T^{\frac{1}{2}}$.

It is difficult to make reliable continuous samples of other metals with such small dimensions. We therefore did not investigate other metals as homogeneous resistors but studied them with the aid of a point-contact arrangement (see next section).

4. 1/f contact noise. A convenient way to obtain small samples for 1/*f* noise studies is to use point contacts. Two relatively large bodies, spheres or crossed cylindrical bars, made of the same material are then pressed together.

Assuming that the contact area is circular with a radius a, the contact can be treated as a set of hemispherical shells[26,27]. The contact resistance follows from the integration of the resistance of the shells from a to ∞.

$$R = \varrho/\pi a. \tag{10}$$

(10) can be applied in each shell. $\langle (\Delta R)^2 \rangle$ for the complete contact is found by integration of $\langle (\Delta R)^2 \rangle$ for the shells. The final result for a contact with two similar contact members is

$$\left\langle \left(\frac{\Delta R}{R} \right)^2 \right\rangle = \left\langle \left(\frac{\Delta G}{G} \right)^2 \right\rangle = \frac{\alpha \pi^2}{20 \pi \varrho^3} R^3 \frac{\Delta f}{f}, \tag{11}$$

where n is the concentration of the electrons and ϱ the resistivity.

The factor $R^3/n\varrho^3$ originally appears as $1/na^3$. Since a cannot easily be measured, it is eliminated by applying (10). The factor na^3 shows that the noise is inversely proportional to the number of electrons in the contact region. Relation (11) is essentially the same as the general relation (9). This is an example of an inhomogeneous sample, showing that for such samples the noise is determined not by an average over N, but by an average over $1/N$.

The results obtained with point contacts of 10 different metals are shown in fig. 1. The R^3 dependence is clearly demonstrated. The absolute values of the noise agree with the expected values given by the straight lines calculated from (11) with $\alpha = 2 \times 10^{-3}$ for all the metals. It is surprising that the experimental results are so well described by relation (11), which is based on the simple assumption that in the contact region under high pressure the values for n and ϱ (and hence for μ) are the same as in the bulk.

Such good agreement was not found with semiconductors. A surface influence exists, as shown by the fact that etching can lower the noise[27]. For "clean" semiconductors the noise is close to the noise expected on the basis of (11), but the R^3 dependence is not found[27,28]. Relation (11) is too simple for this case. The influence of the pressure on the local concentration and mobility is obviously of more importance for semiconductors than for metals. One may wonder whether our simple model is relevant when the radius of the contact area is of the order of the free path of the electrons.

In the case of metals, especially, the contact-noise arrangement is highly suited for demonstrating the correctness of (11) and therefore of (9). It shows that contact noise is nothing special, but that it is normal $1/f$ bulk noise. This general conclusion is also correct for semiconductors, but here further refinements of (11) will have to be made before quantitative agreement can be obtained.

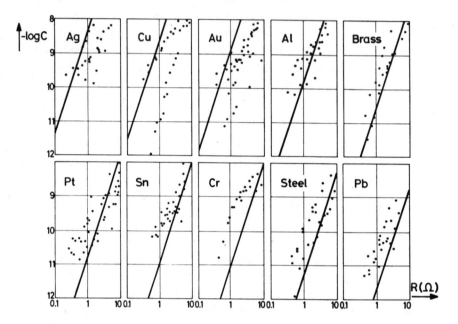

Fig. 1. $1/f$ noise of point constants of 10 metals. The solid lines represent values expected on the basis of eq. (11). C is defined by $\langle(\Delta G/G)^2\rangle = C\Delta f/f$.

Another area where the point-contact arrangement was succesfully used was in the study of ionic solutions in water[29-31]. Measurements on solutions were effected in a cell with two compartments separated by a thin membrane with a hole of about $10\,\mu$ diameter. We found $1/f$ noise in the conductance of solutions of HCl, $AgNO_3$, $CuSO_4$, $Cu(NO_3)_2$, $NiSO_4$ and $Ni(NO_3)_2$. If the noise in the conductance is expressed in the usual way as $C_G = \alpha/N$, α is found to be the same for all ions investigated; moreover, it proves to be proportional to the concentration. For 1-molar solutions $\alpha = 10$. The consequence of the concentration dependence is that the relative noise is constant if N is varied by using different concentrations and the same hole diameter. The usual expression α/N is relevant only if N is varied by varying the contact area without varying the concentration.

The concentration dependence of α is difficult to interpret, but the mere existence of $1/f$ noise in the ionic conductance is an important experimental fact. A model for $1/f$ noise can now no longer be based on some sort of refinement of the treatment of electrons in periodic lattices. The 1/f noise in the ionic conductance strongly supports the idea that 1/f noise is always present whenever a current is carried by a limited number of charge carriers.

5. *1/f noise in the cell voltage of a concentration cell and in thermo e.m.f.* Since measurement of the voltage across the membranes of Ranvier's nodes in nerve cells has revealed 1/f noise[32], it was to be expected that the cell voltage of concentration cells would also have 1/f noise. This proves to be true[30,31]. For the study of the noise in concentration cells we can use the same cell as in ionic conductance studies, the solutions in the two compartments now having different concentrations. The experimental result is

$$\left\langle \left(\frac{\Delta V_{cell}}{V_{cell}} \right)^2 \right\rangle_{c_1 > c_2} = \left\langle \left(\frac{\Delta G}{G} \right)^2 \right\rangle_{c_2}. \tag{12}$$

In this comparison the concentration used for measurement of the conductivity fluctuations is the same as the lower concentration of the concentration cell. This seems fair, since the noise and the resistance of the concentration cell are mainly determined by the lower concentration.

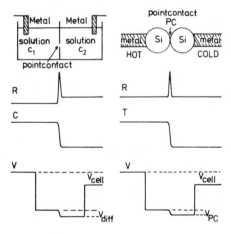

Fig. 2. Point-contact arrangement for noise studies in concentration cells and in thermo e.m.f.

For the noise study of the thermo e.m.f. we used a point-contact arrangement, made by pressing together a "cold" and a "hot" cylindrical bar, both of the same semiconductor material[28] (see fig. 2). The temperature difference was about 7°C. We did not find 1/f noise in the voltage of an open thermo cell, which is in contrast to Brophy's earlier results[33]. We further studied the noise of a thermo cell (with internal resistance R_i) when it was short-circuited by a parallel resistor R_p. V_p, the voltage drop across R_p, and ΔV_p were measured. A typical result is given in fig. 3, again showing that for $R_p \to \infty$ (open cell) the $1/f$ noise approaches zero. These short-circuiting experiments answer the question whether the $1/f$ noise is originally in the current, the voltage or the conductance.

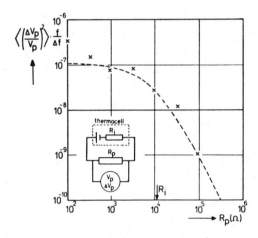

Fig. 3. A typical result of $1/f$ noise measurements in thermo e.m.f.

In this arrangement V_p is given by

$$V_p = \frac{R_p}{R_i + R_p} V_{cell}. \tag{13}$$

The experimental dependence on R_p (fig. 3) can only be explained when the noise is in R_i because then

$$\left\langle \left(\frac{\Delta V_p}{V_p}\right)^2 \right\rangle = \left(\frac{R_i}{R_i + R_p}\right)^2 \left\langle \left(\frac{\Delta R_i}{R_i}\right)^2 \right\rangle = \left(\frac{R_i}{R_i + R_p}\right)^2 \left\langle \left(\frac{\Delta G}{G}\right)^2 \right\rangle. \tag{14}$$

Fluctuations in i or in V do not yield the experimentally found dependence on R_p. The situations with $R_p \ll R_i$ show that for short-circuited thermo cells ($R_p \to 0$).

$$\left\langle \left(\frac{\Delta i}{i}\right)^2 \right\rangle = \left\langle \left(\frac{\Delta G}{G}\right)^2 \right\rangle. \tag{15}$$

$\langle (\Delta G/G)^2 \rangle$ was measured on the same contact without temperature difference, using an externally applied voltage.

$1/f$ noise measurements on short-circuited concentration cells could not be made because of the occurrence of a pronounced $1/f^2$ noise.

The experimental results for thermo cells and concentration cells are summarized in table I.

TABLE I

	Concentration cell	Thermo cell
Open cell	$\left\langle \left(\frac{\Delta V}{V}\right)^2 \right\rangle_{c_1 > c_2} = \left\langle \left(\frac{\Delta G}{G}\right)^2 \right\rangle_{c_2}$	$\left\langle \left(\frac{\Delta V}{V}\right)^2 \right\rangle = 0$
Short-circuited cell $R_p \to 0$	no $1/f$ noise measured because of heavy $1/f^2$ noise	$\left\langle \left(\frac{\Delta i}{i}\right)^2 \right\rangle = \left\langle \left(\frac{\Delta G}{G}\right)^2 \right\rangle$

The important difference between an open concentration cell and an open thermo cell is that in the stationary state in a concentration cell there is a continuous current of positive ions compensated on the average by a continuous current of negative ions, whereas there is no current flowing in the thermo cell. Current flow is an essential condition for $1/f$ noise. A voltage drop or electric field is not necessary, as is proved by the $1/f$ noise in a concentration cell with KCl[30]). V_{diff}, the voltage drop over the contact region, for KCl is practically zero, hence the field in the contact region is zero. Nevertheless there is $1/f$ noise in V_{diff}, the magnitude being the same as with other salts where V_{diff} is about 10% of V_{cell}.

The experimental results of Schrenk and Waidelich[34]) who studied $1/f$ noise in the photovoltaic effect are in line with our view. We expect no $1/f$ noise to be present in the photo voltage when no current is flowing through the sample. The experiments showed a weak $1/f$ noise, much weaker than the $1/f$ noise measured in the conductance when a current is passed through the sample. The magnitude of the observed $1/f$ noise depended on the location of the illuminated spot on the sample. The noise was therefore attributed to internal currents due to inhomogeneities in the sample. It is obviously also the opinion of Schrenk and Waidelich that there is no $1/f$ noise without a current.

6. *Fluctuations in number or in mobility?* Having experimentally established that the $1/f$ fluctuations are in the conductance, as has generally been assumed, we now come to the next question, "Are the fluctuations in the number of the free carriers or in their mobility?" In order to answer this question we shall further analyse the concentration cell. The cell voltage is made up of three components (see fig. 2)

$$V_{\text{cell}} = V_{\text{metal}-c_1} + V_{\text{diff}} + V_{c_2-\text{metal}}. \tag{16}$$

V_{diff} is the voltage drop across the contact region where a concentration gradient exists from c_1 to c_2. The only place where $1/f$ noise can be generated is in the contact region, because the number of ions there is much lower than at the metal-solution interface. The magnitude of the noise has been experimentally found to depend on the diameter of the hole. Our main interest, therefore, is in V_{diff} and its fluctuations. This will be treated here as a one-dimensional problem.

The following relations hold for the average currents, concentrations and mobilities

$$j_+ = pq\mu_+ E - qD_+ \, dp/dx, \tag{17}$$

$$j_- = nq\mu_- E + qD_- \, dn/dx, \tag{18}$$

$$j = qE(p\mu_+ + n\mu_-) - q(D_+ \, dp/dx - D_- \, dn/dx). \tag{19}$$

For an open cell $j = 0$. Application of the Einstein relation

$$D/\mu = kT/q, \tag{20}$$

results in

$$E = \frac{kT}{q} \frac{\mu_+ \, dp/dx - \mu_- \, dn/dx}{\mu_+ p + \mu_- n}. \tag{21}$$

Assuming that $n(x) = p(x) = c(x)$ everywhere and hence that $dn/dx = dp/dx = dc/dx$ we obtain

$$E = \frac{kT}{q} \frac{1}{c} \frac{dc}{dx} \left(\frac{\mu_+ - \mu_-}{\mu_+ + \mu_-} \right), \tag{22}$$

$$V_{\text{diff}} = \frac{kT}{q} \left(\frac{\mu_+ - \mu_-}{\mu_+ + \mu_-} \right) \ln \frac{c_1}{c_2}. \tag{23}$$

The factor $(\mu_+ - \mu_-)/(\mu_+ + \mu_-)$ is usually about 0.1. For KCl it is negligible compared to 1. The contribution of the metal–liquid interface to the cell voltage is $-kT/q \ln c_1/c_2$. Hence the cell voltage is

$$V_{\text{cell}} = \left(-1 + \frac{\mu_+ - \mu_-}{\mu_+ + \mu_-} \right) \frac{kT}{q} \ln \frac{c_1}{c_2}. \tag{24}$$

Relation (21) has been derived using $j = 0$ and the Einstein relation. These two relations also hold when there are $1/f$ fluctuations. For $j = 0$ this is obvious because the cell is open. We believe that the Einstein relation is still valid because the $1/f$ fluctuations are extremely slow. Relation (21) is therefore true not only for the average values but also for quantities with $1/f$ fluctuations.

We shall now analyse the consequences of fluctuations in the mobility and in the number of free carriers. The two results will be compared with the experimental findings, which will show that mobility fluctuations can explain the experimental results whereas number fluctuations cannot.

6.1. Fluctuations in number. We are interested in a comparison of $\langle(\Delta G/G)^2\rangle$ and $\langle(\Delta V/V_{\text{cell}})^2\rangle$.

If the fluctuations are in the number of the charge carriers, then the conductance fluctuations are given by

$$\left\langle\left(\frac{\Delta G}{G}\right)^2\right\rangle = \left\langle\left(\frac{\Delta c}{c}\right)^2\right\rangle, \tag{25}$$

the average of the concentration fluctuations being taken over the contact region.

The fluctuations in the cell voltage follow from relation (21).

$$E + \Delta E = \frac{kT}{q}\left[\frac{\mu_+\,(dp/dx) + \mu_+\,(d\Delta p/dx) - \mu_-\,(dn/dx) - \mu_-\,(d\Delta n/dx)}{\mu_+\,(p + \Delta p) + \mu_-\,(n + \Delta n)}\right]. \tag{26}$$

The 1/f fluctuations are slow, so we may assume charge neutrality in any space element at any moment.

$$\Delta p = \Delta n, \qquad d\,(\Delta p)/dx = d\,(\Delta n)/dx = d\,(\Delta c)/dx, \tag{27}$$

$$E + \Delta E = \frac{kT}{q}\frac{(\mu_+ - \mu_-)\,[dc/dx + d\,(\Delta c)/dx]}{(\mu_+ + \mu_-)\,(c + \Delta c)}, \tag{28}$$

$$V + \Delta V = \frac{kT}{q}\left(\frac{\mu_+ - \mu_-}{\mu_+ + \mu_-}\right)\ln\frac{c_1 + \Delta c_1}{c_2 + \Delta c_2}$$

$$= \frac{kT}{q}\left(\frac{\mu_+ - \mu_-}{\mu_+ + \mu_-}\right)\left(\ln\frac{c_1}{c_2} + \frac{\Delta c_1}{c_1} - \frac{\Delta c_2}{c_2}\right), \tag{29}$$

$$\Delta V = \frac{kT}{q}\left(\frac{\mu_+ - \mu_-}{\mu_+ + \mu_-}\right)\left(\frac{\Delta c_1}{c_1} - \frac{\Delta c_2}{c_2}\right), \tag{30}$$

$$\langle(\Delta V)^2\rangle = \left(\frac{kT}{q}\right)^2\left(\frac{\mu_+ - \mu_-}{\mu_+ + \mu_-}\right)^2\left[\left\langle\left(\frac{\Delta c_1}{c_1}\right)^2\right\rangle + \left\langle\left(\frac{\Delta c_2}{c_2}\right)^2\right\rangle\right], \tag{31}$$

$$\left\langle\left(\frac{\Delta V}{V_{\text{cell}}}\right)^2\right\rangle = \left(\frac{\mu_+ - \mu_-}{\mu_+ + \mu_-}\right)^2\frac{1}{[\ln(c_1/c_2)]^2}\left[\left\langle\left(\frac{\Delta c_1}{c_1}\right)^2\right\rangle + \left\langle\left(\frac{\Delta c_2}{c_2}\right)^2\right\rangle\right]. \tag{32}$$

Comparison of (32) with (25) yields

$$\left\langle\left(\frac{\Delta V}{V_{\text{cell}}}\right)^2\right\rangle \ll \left\langle\left(\frac{\Delta G}{G}\right)^2\right\rangle,$$

for the following reasons:
1) the factor $[(\mu_+ - \mu_-)/(\mu_+ + \mu_-)]^2$ is of the order of 10^{-2}; 2) the factor $[\ln(c_1/c_2)]^{-2}$ is of the order of 10^{-1} or 10^{-2}; 3) $\langle(\Delta c_1/c_1)^2\rangle$ and $\langle(\Delta c_2/c_2)^2\rangle$

are averaged over a large volume at the edge of the contact region. They will therefore be much smaller than $\langle (\Delta c/c)^2 \rangle$ averaged over only the small contact region in calculating $\langle (\Delta G/G)^2 \rangle$.

A dip in the concentration inside the contact region has no influence on the voltage measured outside the contact region. The factor $\ln (c_1/c_2)$ will appear again, originating from $\ln (c_1/c_{dip}) + \ln (c_{dip}/c_2)$. Fluctuations in the concentration lead to $\langle (\Delta V/V_{cell})^2 \rangle$, which is almost zero compared to $\langle (\Delta G/G)^2 \rangle$.

6.2. Fluctuations in mobility. Also in this case we shall compare $\langle (\Delta G/G)^2 \rangle$ with $\langle (\Delta V/V_{cell})^2 \rangle$.

If we assume that the fluctuations are in the mobility and that Einstein's relation still stands, then

$$\left\langle \left(\frac{\Delta G}{G} \right)^2 \right\rangle = \left\langle \left(\frac{\Delta \mu_+}{\mu_+ + \mu_-} \right)^2 \right\rangle + \left\langle \left(\frac{\Delta \mu_-}{\mu_+ + \mu_-} \right)^2 \right\rangle. \tag{33}$$

The voltage fluctuations follow again from relation (21)

$$E + \Delta E = \frac{kT}{q} \frac{d \ln c}{dx} \left(\frac{\mu_+ + \Delta \mu_+ - \mu_- - \Delta \mu_-}{\mu_+ + \Delta \mu_+ + \mu_- + \Delta \mu_-} \right), \tag{34}$$

$$E + \Delta E = \frac{kT}{q} \frac{d \ln c}{dx} \left(\frac{\mu_+ - \mu_-}{\mu_+ + \mu_-} \right) \left(1 + \frac{\Delta \mu_+ - \Delta \mu_-}{\mu_+ - \mu_-} \right), \tag{35}$$

$$\Delta V = \frac{kT}{q} \ln \frac{c_1}{c_2} \left(\frac{\Delta \mu_+}{\mu_+ + \mu_-} - \frac{\Delta \mu_-}{\mu_+ + \mu_-} \right), \tag{36}$$

$$\left\langle \left(\frac{\Delta V}{V_{cell}} \right)^2 \right\rangle = \left\langle \left(\frac{\Delta \mu_+}{\mu_+ + \mu_-} \right)^2 \right\rangle + \left\langle \left(\frac{\Delta \mu_-}{\mu_+ + \mu_-} \right)^2 \right\rangle = \left\langle \left(\frac{\Delta G}{G} \right)^2 \right\rangle. \tag{37}$$

It is only for mobility fluctuations that the experimental findings summarized in table I can be explained.

The experimental results of the noise measurements on the thermo e.m.f. are also in agreement with fluctuations in the mobility. The relation for the current in a thermocell is

$$j = \sigma \left[E + \frac{kT}{q} \frac{d \ln n}{dx} + \frac{S^*}{q} \frac{dT}{dx} \right], \tag{38}$$

where S^* is the entropy of transport. [Eq. (24.5) in ref. 35, eq. (2.42) in ref. 36.] a. For an open cell we have $j = 0$. This makes the expression between square brackets zero, enabling E to be calculated. Fluctuations in μ only influence σ

outside the square brackets. They do not effect E. This explains the experimental result

$$\left\langle \left(\frac{\Delta V}{V_{\text{cell}}}\right)^2 \right\rangle = 0. \tag{39}$$

b. For a short-circuited cell where

$$V_{\text{PC}} = -(V_{\text{metal}-\text{Si}(T_1)} + V_{\text{metal}-\text{Si}(T_2)}), \tag{40}$$

the terms within the bracket are free of $1/f$ noise. This leads to the simple relation

$$\left\langle \left(\frac{\Delta i}{i}\right)^2 \right\rangle = \left\langle \left(\frac{\Delta \mu}{\mu}\right)^2 \right\rangle = \left\langle \left(\frac{\Delta G}{G}\right)^2 \right\rangle, \tag{41}$$

in agreement with experiment.

7. *Present status of the $1/f$ noise problem.* As experiments have shown that bulk $1/f$ noise is due to fluctuations in the mobility of free carriers we are now interested in bulk models in which the current fluctuations are not the result of fluctuations in the number of free carriers. A model of this class was recently proposed by Handel[37-39]. In this model a "magnetic-barrier instability" causes a turbulence in the plasma of the current carriers. The sources of the turbulence have no $1/f$ spectrum, but they create hydromagnetic turbulences with a $1/f$ spectrum in the plasma of the carriers of the whole circuit connected to the sample. This provides the large k values corresponding to the low frequencies. Arguments against the general validity of this model in its present state are: 1) the instabilities will occur only for high currents[16]; 2) carriers of both signs must be present, but experiments show that the relative noise of metals is the same as the relative noise of nearly intrinsic semiconductors; 3) the number of carriers in the sample completely determines the magnitude of the noise, whereas the model requires imperfections for the instabilities to develop.

Much work of a more mathematical nature has been done on $1/f$ noise[40-46]. These papers do not propose specific physical mechanisms for the noise and are thus not restricted to number fluctuations. However, we do not see how any of these mathematical treatments could be immediately translated into a realistic mobility model.

Finally if we must accept McWhorter for the greater part of the $1/f$ noise in MOS transistors, we still have the unsolved problem of how much the bulk $1/f$ noise contributes to the total noise of a MOST.

8. *Summary*. Comparison of $1/f$ noise in the conductance and in the voltage of thermocells and concentration cells leads to the conclusion that the mobility of a free charge carrier fluctuates with a $1/f$ spectrum.

$$\left\langle \left(\frac{\Delta \mu}{\mu} \right)^2 \right\rangle = \alpha \frac{\Delta f}{f}. \tag{42}$$

For electrons and holes in semiconductors and metals $\alpha = 2 \times 10^{-3}$. It is not certain that all materials have the same α, but the order of magnitude, at least, is the same. For ions in water α is roughly proportional to the concentration; $\alpha = 10$ for 1-molar solutions.

For homogeneous samples with N free charge carriers (42) becomes

$$\left\langle \left(\frac{\Delta G}{G} \right)^2 \right\rangle = \frac{\alpha}{N} \frac{\Delta f}{f}. \tag{9}$$

The fact that the experimental results of bulk $1/f$ noise studies on well-known samples can generally be expressed by the simple relation (9) shows that $1/f$ noise is not some spurious effect due to accidental impurities or imperfections, but that it is a systematic effect inherent in electrical conduction.

REFERENCES

1) McWhorter, A.L., Semiconductor Surface Physics, University of Pennsylvania Press (1957) p. 207.
2) Mansour, I.R.M., Hawkins, R.J. and Bloodworth, G.G., Radio Electronic Engr. (1968) 212.
3) Berz, F., Solid State Electron. **13** (1970) 631.
4) Christensson, S., Lundström, I. and Svensson, C., Solid-State Electronics **11** (1968) 797.
5) Hsu, S.T., Solid-State Electronics **13** (1970) 1451.
6) Mansour, I.R.M., Hawkins, R.J. and Bloodworth, G.G., Brit. J. appl. Phys. **2** (1969) 1063.
7) Hsu, S.T., Fitzgerald, D.J. and Grove, A.S., Appl. Phys. Letters **12** (1968) 287.
8) Abowitz, G., Arnold, E. and Leventhal, E.A., IEEE Trans. Electron Devices ED **14** (1967) 775.
9) Klaassen, F.M., IEEE Trans. Electron Devices ED **18** (1971) 887.
10) Feigt, I. and Jäntsch, O., Solid-State Electronics **14** (1971) 391.
11) Berz, F. and Prior, C.G., Electronics Letters **6** (1970) 595.
12) Bernamont, J., CR Acad. Sci. **198** (1934) 1755.
13) Bernamont, J., Ann. Physique **7** (1937) 71.
14) Bell, D.A., Electrical Noise, D. van Nostrand Co. (Toronto, London, New York, 1960).
15) Bell, D.A., Brit. J. appl. Phys. **6** (1955) 284.
16) Ganefel'd, R.V. and Repa, I.I., Soviet Phys. Semiconductors **3** (1969) 22.
17) Conti, M., Solid-State Electronics **13** (1970) 1461.
18) Jaeger, R.C. and Brodersen, A.J., IEEE Trans. Electron Devices ED **17** (1970) 128.
19) Hooge, F.N., Phys. Letters **29A** (1969) 139.

20) Hooge, F.N.,Van Dijk, H.J.A. and Hoppenbrouwers, A.M.H., Philips Res. Rep. **25** (1970) 81.

21) Heyke, K., Lautz, G. and Schumny, H., Phys. Status solidi (a) **1** (1970) 459.

22) Main, C. and Owen, A.E., Phys. Status solidi (a) **1** (1970) 297.

23) Meade, M.L., Radio Electronic Engr. **41** (1971) 126.

24) Bittel, H. and Scheidhauer, H., Z. angew. Phys. **8** (1956) 417.

25) Hooge, F.N. and Hoppenbrouwers, A.M.H., Physica **45** (1969) 386.

26) Hooge, F.N. and Hoppenbrouwers, A.M.H., Phys. Letters **29A** (1969) 642.

27) Hoppenbrouwers, A.M.H. and Hooge, F.N., Philips Res. Rep. **25** (1970) 69.

28) Hooge, F.N. and Gaal, J.L.M., Philips Res. Rep. **26** (1971) 345.

29) Hooge, F.N., Phys. Letters **33** (1970) 165.

30) Hooge, F.N. and Gaal, J.L.M., Philips Res. Rep. **26** (1971) 77.

31) DeFelice, L.J. and Firth, D.R., IEEE Trans. bio-med. Engng. **18** (1971) 339.

32) Verveen, A.A. and Derksen, H.E., Proc. IEEE **56** (1968) 906.

33) Brophy, J.J., Phys. Rev. **111** (1958) 1050.

34) Schrenk, H. and Waidelich, W., Phys. Status solidi **32** (1969) K99.

35) Tauc, J., Photo and Thermoelectric Effects in Semiconductors, Pergamon Press (Oxford, London, New York, Paris, 1962).

36) Heikes, R.R. and Ure, R.W., Thermoelectricity, Science and Engineering, Interscience Publ. (New York, London, 1961).

37) Handel, P.H., Z. Naturforsch. **21a** (1966) 561.

38) Handel, P.H., Phys. Status solidi **29** (1968) 299.

39) Handel, P.H., Phys. Rev. **A 3** (1971) 2066.

40) Barnes, J.A. and Allen, D.W., Proc. IEEE **54** (1966) 176.

41) Mandelbrot, B., IEEE Trans. Inform. Theory IT **13** (1967) 289.

42) Halford, D., Proc. IEEE **56** (1968) 251.

43) Heiden, C., Phys. Rev. **188** (1969) 319.

44) Offner, F.F., J. appl. Phys. **41** (1970) 5033.

45) Teitler, S. and Osborne, M.E.M., J. appl. Phys. **41** (1970) 3274.

46) Offner, F.F., Biophys. J. **11** (1971) 123.

1/|f| NOISE IN PHYSICAL MEASUREMENTS *

Veljko Radeka
Brookhaven National Laboratory
Upton, N. Y.

Summary

The purpose of this paper is to provide an insight into low frequency divergent noises with spectral density $|f|^{\alpha}$, where $\alpha \leq -1$, and into their effect on physical measurements, with special reference to 1/|f| noise. This class of noise is widespread in nature, and it presents unique limitations to the measurement accuracy. In an attempt to present a picture of this class of noise with regard to the measurements of observable physical quantities, the questions about generation of noise, its divergence, correlation properties and measurements of variance are discussed.

A statistical model for generation of low frequency divergent noises is used to consider the divergence problem in both the frequency and time domain. It is shown that 1/|f| noise is "weakly divergent," and that power limitation presents no reason to impose a low frequency limit within time intervals observable in nature. Correlation properties are discussed in terms of the time-dependent correlation function, using an ideal impulse response which generates low frequency noise from white noise. Two general models for generation of 1/|f| noise are summarized and discussed. Generation of 1/|f| noise from white noise over a limited frequency range by distributed and lumped-parameter filters is described.

It is shown that the variance (i.e. mean square noise) is determined by the frequency limits of the observation method. The variance is independent of the low frequency limit of noise, if such a limit exists, and if the frequency limit of noise is lower than the low frequency limit of the measurement process. If the ratio of the high frequency limit and the low frequency limit of the measurement process is constant, the variance is a function of one parameter, τ, which is proportional to the "measurement time." For power-law noises, the variance $\sigma \propto \tau^{-1-\alpha}$. Variance in the case of a general power spectral density function can be represented by the power series, where each term $\tau^{-1-\alpha}$ may be associated with a power-law noise component. Thus, the measurement of variance as a function of measurement time represents a method for identification of power-law noises, and their effect under actual measurement conditions.

1. Introduction

The purpose of this paper is to provide an insight into low frequency divergent noises with spectral density $|f|^{\alpha}$, where $\alpha \leq -1$, and into their effect on physical measurements, with special reference to 1/|f| noise. The class of noise $|f|^{\alpha}$ with α close to minus one is widespread in nature and it presents unique limitations to the measurement accuracy. Even in cases where the signal energy can be increased arbitrarily by extending the measurement time, the accuracy of the measurement of the signal magnitude cannot be improved by increasing the measurement time.

The basic distinction between white noise and 1/|f| noise is that the span of interdependence between samples is very large for 1/|f| noise, while sufficiently spaced samples for white noise (and for bandlimited white noise) are independent. This can be expressed in terms of correlation functions, that, while the correlation function for white noise is the delta function (in the limit of infinite bandwidth), the correlation function for

*This work was performed under the auspices of the U.S. Atomic Energy Commission.

Reprinted from *IEEE Trans. Nucl. Sci.*, vol. NS-16, pp. 17–35, Oct. 1969.

178

low frequency noises (to the extent that it can be defined) decreases only very slowly with the interval between samples. The basic feature of any noise-generating mechanism for low frequency divergent noises is an "infinitely long memory" for individual independent perturbations.

Integration of white noise generates noises with spectral density $|f|^\alpha$, where $\alpha = -2i$, and i is the order of integration. One "integer-order" integration converts white noise into $1/f^2$ noise, or "random walk." One "half-order integration" is required to obtain $1/|f|$ noise from white noise. Noises which are related to white noise by an "integration" of a fractional-order are sometimes referred to as "fractional noises." Noises with spectral density $|f|^\alpha$, where $-\alpha$ is close to unity, observed over a certain frequency region, have been referred to as "flicker effect," "excess noise," "low frequency noise," "contact noise," and "pink noise."

$1/|f|$ noise has been observed as electrical noise or as fluctuations of some other physical quantity, or, more generally, of a "process variable" in a number of different devices and physical and other systems. It was discovered in electron tubes as "flicker effect."[1,2] It was also observed in all semiconductor devices which could have any application for amplification and detection of small signals.[3,4,5] Some semiconductor devices have particularly large $1/|f|$ noise, for example, MOS transistors as compared to junction field-effect transistors. Noise in MOS transistors has recently been a subject of extensive studies.[6,7,8,9] Low frequency fluctuations appear to be the principal limiting factor on the accuracy of frequency and time measurements and on the stability of precision signal generators.[10,11,12,13,14] $1/|f|$ fluctuations have been found also in a biological system.[15] Fluctuations in the frequency of rotation of the earth[16,17,14] appear to have a frequency spectrum with $\alpha = -2$ over a certain region. The concept of fractional noises has also been invoked

to study the fluctuations of variables in economics.[18] It is also known that nuclear reactors with their feedback control exhibit power fluctuations which have $1/|f|$ spectral density over a range of frequencies. This noise presents a limitation in some nuclear physics experiments. Noises with spectral density represented approximately by the sum of white, $1/|f|$ and $1/f^2$ densities impose a limit on the measurement resolution of charge-sensitive amplifiers for nuclear radiation detectors. One source of $1/|f|$ noise in this case seems to be frequency-dependent thermal noise of solid dielectrics.[20]

The mechanisms of generation of $1/|f|$ noise in particular physical situations have been little understood. In the vast amount of literature on observed low frequency noises a satisfactory explanation of mechanisms was provided only in a few cases. Recently more insight has been gained as to how $1/|f|$ noise could be generated, and several models and interpretations have been proposed.[13,14,19] The most general model is the mechanical model proposed by Halford,[14] in which broad classes of perturbations are shown to be able to generate a given spectrum. This model leads also to analog and digital schemes for generation of low frequency noise for simulation purposes.

$1/|f|$ noise has received very little attention from the point of view of signal processing in the literature on statistical communication theory. Some mathematical studies of fractional noises have been published (other references are given in Ref. 18), where a number of problems of mathematical nature have been raised.

Measurements of physical quantities in the presence of low frequency fluctuations have been considered in greater detail in some special areas. The measurements of frequency and time have been analyzed extensively,[10-14,16,17] and will not be discussed here. In pulse amplitude measurements a general low

frequency spectrum represented by a negative power series has been considered by
Gatti and Svelto.[21] Dependence of the
noise power (variance) on the measurement
time has been used for some time in this
field to identify $|f|^{\alpha}$ noise components
with different exponents. Consideration
of dependence of noise on other variables
has led to identification of physical
sources of noise in charge amplifiers.[20]
The effect of $1/|f|$ noise on nuclear
magnetic resonance (NMR) measurements was
discussed by Klein and Barton.[22]

In this paper an interpretation of low
frequency noises and of the divergence
problem is presented. A note is made on
the difficulties with the definition of
the correlation function for divergent
noises, and on the characterization of
their correlation properties. Models and
mechanisms for generation of $1/|f|$ noise
are reviewed, and an example of a circuit
is given for generation of $1/|f|$ noise.
Some effects of $1/|f|$ noise on the pulse
amplitude measurements and on repetitive
measurements in NMR are discussed. The
method for identification of power-law
noises based on time domain measurements
is described. The interpretation and the
models discussed here are based on linear
superposition of perturbations generated
by a stationary process, which should
result in a divergent noise process with
stationary increments. Problems arising
from more complex processes with nonstationary behaviour are indicated.

2. An Interpretation of Low Frequency Divergent Noises

Representation of Noise and Divergence Tests

Power-law noises are represented as

$$w_{\alpha}(f) = |f|^{\alpha} \tag{1}$$

$w_{\alpha}(f) = W_{\alpha}(f)/W_{\alpha}(0)$ is the normalized one-
sided spectral density as a function of
cycle frequency. ($W_{\alpha}(0)$ is the mean
square value of the fluctuating variable

at 1 Hz per $Hz^{\alpha+1}$; for voltage $1/|f|$
noise $w_{-1}(f) = \overline{v^2}(f)/A_f$, where $A_f = W_{-1}(0)$
is expressed in $volts^2$.) Of particular
interest here are the cases of white noise
($\alpha=0$), $1/|f|$ noise ($\alpha = -1$), and "random
walk" ($\alpha = -2$), since they appear most
frequently and since white noise and
random walk represent interesting limit
cases for comparison with $1/|f|$ noise.
Power-law noise can be considered as
being generated by a Poisson process acting upon an appropriate filter.[13] In the
frequency domain, the transfer function
of the filter which converts white noise
into power-law noise is

$$H(\omega) = (j\omega)^{\frac{\alpha}{2}} \tag{2}$$

In the time domain, we think of the power-
law noise as being generated by the random
sequence of impulses, each impulse generating at the output of the filter an
impulse response, $h(t)$, which is the
Fourier transform of the frequency domain
transfer function $H(\omega)$,

$$h(t) = \frac{2^{\frac{-\alpha-1}{2}} (\pi)^{-\frac{\alpha}{2}}}{\Gamma(\frac{\alpha}{2})} t^{-\frac{\alpha}{2} - 1} \tag{3}$$

(Campbell and Foster,[23] pair 516.) For
$\alpha = -2$, $\Gamma(\frac{\alpha}{2}) = 1$, for $\alpha = -1$, $\Gamma(\frac{\alpha}{2}) = \pi^{1/2}$.
(The coefficient $2^{(-\alpha-1)/2} \cdot \pi^{-\alpha/2}$ is due
to the normalization of all the variances,
calculated in the following, to the one-
sided spectral density $W_{\alpha}(0)$ at 1 Hz.)

If the input white process is $x(t)$, the
process at the output of the filter is
convolution $y(t) = x(t) * h(t)$,

$$y(t) = \frac{2^{\frac{-\alpha-1}{2}} (\pi)^{-\frac{\alpha}{2}}}{\Gamma(\frac{\alpha}{2})} \int_0^t (t-u)^{-\frac{\alpha}{2}-1} x(u) du \tag{4}$$

Thus power-law noise can be obtained by
fractional-order integration of white
noise.[13] (For $\alpha = -2$, random walk is
obtained by 1st order integration.) For
$\alpha = -1$, $1/|f|$ noise is obtained by half-
order integration,

$$y(t) = \int_0^t \frac{1}{(t-u)^{1/2}} \, x(u) \, du \qquad (5)$$

The impulse response of the filter for conversion of white noise into $1/|f|$ noise is

$$h(t) = \frac{1}{t^{1/2}} \qquad \text{for } t > 0$$

$$\qquad \qquad \qquad \qquad \qquad \qquad (6)$$

$$\quad\;\; = 0 \qquad\quad \text{for } t \le 0$$

The concept of the generation of various basic noises by a Poisson process is illustrated in Fig. 1. Fig. 1·(b) shows the physical white noise - with high-frequency cutoff [impulse response $\frac{1}{\tau_h} \exp(-t/\tau_h)$ is shown resulting in the spectrum $1/(1 + \omega^2 \tau_h^2)$ and autocorrelation function $r(\tau) = \exp(-|\tau|/\tau_h)$.] The main distinction of low-frequency noises is that each impulse of the random process produces an effect of infinite duration (constant for random walk, and a slowly decaying one for $1/|f|$ noise).

As a divergence test in the frequency domain, total power (variance) in the frequency band limited by frequencies f_ℓ and f_h can be calculated

$$\sigma^2(f_h, f_\ell) = \int_{f_\ell}^{f_h} |f|^\alpha \, df$$

$$\qquad \qquad \qquad \qquad \qquad (7)$$

$$= \begin{cases} \dfrac{1}{1+\alpha} [f_h^{1+\alpha} - f_\ell^{1+\alpha}] & \text{for } \alpha \ne -1 \\[2mm] \ell n \dfrac{f_h}{f_\ell} & \text{for } \alpha = -1 \end{cases}$$

Extending the frequency band $f_\ell \to 0$ and $f_h \to \infty$, the total power tends to infinity, at the low frequency limit for $\alpha < -1$, at the high frequency limit for $\alpha > -1$, and

at both limits for $\alpha = -1$. Thus, the unique place of $1/|f|$ noise among power-law noises is that it is divergent at both frequency limits. Low frequency divergent noises are characterized by $\alpha \le -1$. The high frequency divergence presents no actual problem, since in any physical system there is a high frequency limit, and infinite power at high frequencies cannot exist. For analytical purposes it can be handled by introducing an appropriate high frequency cutoff (which is a realistic solution, since any method for observation of noise introduces a high frequency limit).

To get closer to the substance of low frequency divergence, a time domain divergence test can be applied. In this test we imagine that the Poisson process is switched-on at the input of the filter $h(t)$, which converts it into power-law noise, and then we observe the output mean square noise power (variance) as a function of time, as illustrated in Fig. 2. There are a number of different approaches to calculate this. The most plausible one in this case is to apply Campbell's theorem,[24] and to determine the effect of input impulses occurring in the interval $0,t$ on the output at time t, as the sum of mean square contributions by independent impulses. The output due to a single impulse $q \cdot \delta(\lambda)$ will be $q \cdot h(\lambda)$ at a time (λ) after the impulse occurred. For the random sequence with mean rate $\bar{n}$, the mean square output due to $\bar{n} d\lambda$ impulses in the interval $d\lambda$ will be $\bar{n} q^2 h^2(\lambda) d\lambda$, and the output for the whole interval $(0,t)$ is obtained by the integral[26]

$$\sigma_w^2(t,0) = \bar{n} q^2 \int_0^t h^2(\lambda) \, d\lambda \qquad (8)$$

Noting that $\bar{n} q^2$ is the (two-sided) power spectral density of the input process, the normalized output variance is

$$\sigma^2(t,0) = \int_0^t h^2(\lambda) \, d\lambda \qquad (9)$$

For power-law noise, using Eq. (3), it follows,

$$\sigma^2(t,0) = \frac{2^{-\alpha-1}\pi^{-\alpha}}{[\Gamma(\frac{\alpha}{2})]^2} \int_0^t \lambda^{-\alpha-2} \, d\lambda$$

(10)

$$= \frac{2^{-\alpha-1}\pi^{-\alpha}}{[\Gamma(\frac{\alpha}{2})]^2} \frac{1}{-\alpha-1} t^{-\alpha-1} \quad \text{for } \alpha < -1$$

For $\alpha = -1$, integral (10) gives $\ln \lambda \rfloor_0^t$, and the lower limit presents a problem due to infinite power of $1/|f|$ noise at high frequencies. This can be solved by introducing a high frequency cutoff. One way is to introduce averaging over a short time interval δ. The "smoothed" impulse response for $\alpha = -1$ is

$$h(t,\delta) = \frac{1}{\delta} \int_t^{t+\delta} h(u) \, du$$

$$= \frac{2}{\delta} [(t+\delta)^{1/2} - t^{1/2}]$$

(11)

Impulse response $(t/\delta)^{-1/2}$ and the "smoothed" response $h(t,\delta)$ are shown in Fig. 3, curves a and c. While Eq. (11) gives the effect of sampling with a finite integration time δ, a function similar to this (curve b) can be used, as it results in somewhat simpler calculations. (Both curves affect only the high frequency response.)

Variance as a function of time for $1/|f|$ noise is then

$$\sigma^2(t,0) = \frac{1}{\delta} \int_0^t d\lambda + \int_\delta^t \frac{1}{\lambda} d\lambda$$

$$= \begin{cases} t/\delta & \text{for } t \leq \delta \\ \\ 1 + \ln(t/\delta & \text{for } t \geq \delta \end{cases}$$

(12)

Thus, for various low frequency divergent noises, we have:

$w_\alpha(f) =$	$\alpha =$	$\sigma^2(t,0) \propto$
$1/|f|$	-1	$\ln(t/\delta)$
$1/f^2$	-2	t
$1/|f|^3$	-3	t^2
$1/f^4$	-4	t^3

(13)

For low frequency divergent noise, the variance increases in time without limit. This brings up the question of the existence of the low frequency limit. Must it exist? The divergence of the $1/|f|$ noise is so weak that the variance increases very little over a large range of time. According to the argument of Flinn,[25] if the low frequency limit corresponds to current estimates of the age of the universe ($\sim 10^{-17}$ Hz), and the highest frequency to the time taken for light to traverse the classical radius of the electron ($\sim 10^{23}$ Hz), this would represent 40 decades, and in that time the rms noise would increase only $\sqrt{40}$ times the value for one decade. In many measurements the averaging time δ is increased with the observation interval t (the ratio of cut-off frequencies remains the same). In that case one measures a constant power for $1/|f|$ noise.

Noises with $\alpha = -2$ and $\alpha = -3$ are strongly divergent, and they cannot exist without a low frequency limit for any variable which has an upper bound, or a dynamic range limit in the sense of electronic systems. However, they can exist if the variable has no such bound (variables accumulated in time, phase of an oscillator, for example).

Among these noises there is a distinct difference in the ratio of power contained at high and at low frequencies. This is apparent also from sample waveforms shown in Fig. 4. An interesting effect is that the appearance of $1/|f|$ changes little with time scale, Fig. 5. $1/|f|$ noise in this case was generated by an MOS transistor. The divergence of some noises is shown in Fig. 6, which presents a demonstration of the experiment in Fig. 2. According to relations (13), rms noise increases as $t^{1/2}$, t, $t^{3/2}$ respectively for the three types of noise presented. The low frequency limit of the circuits used to generate these noises was about 2 orders of magnitude lower than the inverse of the (gating) time interval shown on Fig. 6, and the high frequency limit was about 3 orders of magnitude higher.

A Note on the Autocorrelation Function of Divergent Noises

According to the Wiener-Khintchine theorem, the autocorrelation function of a random process is the Fourier transform of its power spectrum. For the spectral density obtained by passing a uniform spectral density (white noise) through a power-law filter according to Eq. (2)

$$R(\tau) = 2 \frac{1}{(2\pi)} \int_0^\infty H^2(\omega) \, \cos \, \omega\tau \, d\omega \qquad (14)$$

for $\tau = 0$, $R(0) = 2 \frac{1}{(2\pi)} \int_0^\infty H^2(\omega) \, d\omega$ is the

total noise power, which for low frequency divergent noises tends to infinity in the sense discussed in the preceding section. An alternative way of defining the autocorrelation function is based on the concept of the filter autocorrelation function, which is defined as,[27,28]

$$R(\tau) = \int_0^\infty h(u) \cdot h(u+\tau) \, du \qquad (15)$$

where $h(t)$ is the impulse response of the filter. If R_x is the autocorrelation function of the process at the input of the filter, the autocorrelation function of the output process is given by the convolution of the input correlation function and the filter correlation function,

$$R_y(\tau) = \int_0^\infty R_x(\tau-\lambda) \cdot R_h(\lambda) \, d\lambda \qquad (16)$$

For white noise $R_x(\tau) = \delta(\tau)$, and the autocorrelation function of the output process equals the autocorrelation function of the filter.

Relation (15) is of no help, however, since it is equivalent to (14). The problem is only a little better illustrated if one tries to calculate $R(\tau)$ for $1/f^2$ and for $1/|f|$ noise using Eq. (15), as shown in Fig. 7. It is obvious that $R(\tau)$ is undefined for any impulse response which results in $\sigma^2(t,0) \to \infty$.

A somewhat better insight into the correlation properties of low frequency divergent noises can be gained by using the concept of the "time-dependent autocorrelation function." The time-dependent autocorrelation function was introduced by Lampard.[28] The time-dependent autocorrelation function is based on the assumption of a finite observation interval, and can be defined as follows,

$$R(t,\tau) = \int_0^{t-\tau} h(u) \cdot h(u+\tau) \, du \qquad (17)$$

For the case of a random walk, $\alpha = -2$, the impulse response of the filter which converts white noise into $1/f^2$ noise is the unit step function $\sqrt{2\pi} \, U(u)$, Eq. (3), and the function $R(t,\tau)$, according to Fig. 7(a), is

$$R(t,\tau) = (2\pi^2)(t-|\tau|) \qquad (18)$$

For $\tau = 0$ $\quad R(t,0) = (2\pi^2)t , \qquad (19)$

which is the variance as a function of time, as derived from Eq. (10). However, the quantity

$$R(t,0) - R(t,\tau) = (2\pi^2)|\tau| \qquad (20)$$

is independent of t.

For $1/|f|$ noise, $\alpha = -1$, the impulse response $h(u) = u^{-1/2}$ can be used to determine $R(t,\tau)$.

The high frequency limit can be handled in one of the ways discussed in the previous section. Using the function b in Fig. 3, the calculation of $R(t,\tau)$ can be performed according to Fig. 7(b) and Eq. (16). An unimportant term results from the part of the function $0 < u < \delta$, so that this part can be ignored, and the integration carried on from δ to $t-\tau$. (δ corresponds to the inverse of the high frequency limit and can be made arbitrarily low.) Then,

$$R(t,\tau) = \int_{\delta}^{t-\tau} \frac{1}{u^{1/2}(u+\tau)^{1/2}}\, du \qquad (21)$$

The result of this integration (Peirce,[29] integral 160) is

$$R(t,\tau) = \ell n\, \frac{(t^2-t\tau)^{1/2} + t - \frac{\tau}{2}}{(\delta+\delta\tau)^{1/2} + \delta + \frac{\tau}{2}} \qquad (22)$$

It is reasonable to assume that $t \gg \tau$, so that

$$R(t,\tau) = \ell n\, \frac{2t}{(\delta+\delta\tau)^{1/2}+\delta+\frac{\tau}{2}} \qquad (23)$$

Several results are of interest. For $\tau = 0$,

$$R(t,0) = \ell n(\tfrac{t}{\delta}) \qquad (24)$$

which equals the noise power (variance) as a function of time, Eq. (12).

$$R(t,0) - R(t,\tau) = \ell n\, \frac{(\delta+\delta\tau)^{1/2}+\delta+\frac{\tau}{2}}{2\delta} \qquad (25)$$

For $\delta \ll \tau$, the expressions assume a simple form,

$$R(t,\tau) = \ell n4 + \ell n(t/\delta) - \ell n(|\tau|/\delta) \qquad (26)$$

$$R(t,0) - R(t,\tau) = \ell n(|\tau|/\delta) - \ell n4 \qquad (27)$$

Both in the case of $1/f^2$ noise and of $1/|f|$ noise, the quantity characteristic of correlation, $R(t,0) - R(t,\tau)$ is independent of t, and therefore of the "age" of the process. This quantity is significant for the calculation of the differences of instantaneous values of noise spaced by τ, since

$$\overline{[x(t+\tau) - x(t)]^2} = 2[R(t,0) - R(t,\tau)] \qquad (28)$$

One assumes here (without proof!) that the translation of the values of x in time is justified since $R(t,0) - R(t,\tau)$ is independent of t. Finite differences as statistically well-behaved quantities for low frequency divergent noise have been used by Barnes[12] and Allan.[11] Barnes[12] was the first to introduce and make use of the function $R(t,0) - R(t,\tau)$. It has been shown,[31] that in these cases the generalized Fourier transform[30] of the generalized spectral density function gives formally $R(t,0) - R(t,\tau)$. This does not imply that the term $R(t,0)$ is not significant. The significance of $R(t,0)$ as considered above is that it represents the variance of the process at time t after the inception of the process. While the knowledge of $R(t,0)$ is not necessary for analysis of differences, it may be required for solution of some time domain problems.

3. Generation of $1/|f|$ Noise

Distributed RC-line Model (Diffusion Model)

The filter which converts a Poisson process into noise with $1/|f|$ spectral density has transfer function $(j\omega)^{-1/2}$, and impulse response $t^{-1/2}$, according to Eqs. (2) and (3). This also represents the required relation between the voltage and current (impedance) for a two-pole (single port) network. It can be shown[32,31] that an "infinitely long" distributed RC line has such an impedance. Referring to Fig. 8(a) and (b), one can write for an infinitesimal section of the line, which is followed by the remainder of the line with line impedance Z,

$$Z = r\delta x + \frac{1}{j\omega c\delta x} \;\|\; Z, \qquad (29)$$

where r and c are resistance and capacitance per unit length. The equation for Z is then,

$$Z^2 = Zr\delta x + \frac{r}{j\omega c} \qquad (30)$$

For $\delta \to 0$, it follows,

$$Z = \sqrt{\frac{r}{j\omega c}} \qquad (31)$$

If a noise current is applied to the driving point (input), the mean square noise voltage is, Fig. 8(c),

$$\overline{v_n^2} = \overline{i_n^2} \, |Z|^2 = \overline{i_n^2} \cdot \frac{r}{c} \cdot \frac{1}{|\omega|} \qquad (32)$$

Thus, for a uniform spectral density of the current, $1/|f|$, spectral density is obtained for the voltage.

We can consider now the noise of the RC-line itself. The noise current of the line, which is in thermal equilibrium, is determined by the real component of its admittance,

$$Y = G + jB = \frac{1}{Z} = \left(\frac{\omega c}{r}\right)^{1/2} \cdot j^{1/2}$$

$$= \left(\frac{\omega c}{r}\right)^{1/2} \frac{\sqrt{2}}{2} (1+j) \qquad (33)$$

The mean square noise current density is then

$$\overline{i_n^2} = 4kT\, G(\omega) = \frac{\sqrt{2}}{2}\, 4kT\left(\frac{c}{r}\right)^{1/2} \cdot |\omega|^{1/2} \qquad (34)$$

The mean square noise voltage at the driving point, Fig. 8(d) is then,

$$\overline{v_n^2} = \frac{\sqrt{2}}{2} \cdot 4kT\left(\frac{r}{c}\right)^{1/2} \frac{1}{|\omega|^{1/2}} \qquad (35)$$

If this noise current flows into a large capacitance C_o, Fig. 8(e), such that $\frac{1}{\omega C_o} \ll |Z|$, then the voltage on that capacitance is

$$\overline{v_n^2} = \frac{\sqrt{2}}{2}\, 4kT\left(\frac{c}{r}\right)^{1/2} \frac{1}{C_o^2} \cdot \frac{1}{|\omega|^{3/2}} \qquad (36)$$

Thus, a number of different fractional noises may result from the RC-line noise, depending on the impedance of the input termination.

An RC-line is an electrical equivalent to the diffusion process, and to the heat conduction process. They are all described by the same differential equation, which can be written in the normalized form as

$$\frac{\partial^2 v(x,\theta)}{\partial x^2} - \frac{\partial v(x,\theta)}{\partial \theta} = 0 \qquad (37)$$

where v is the variable equivalent to potential (temperature, concentration of ions), x is distance, and θ is normalized time.

It is of interest to determine the validity of the above expressions and conclusions for a line of finite length (diffusion over a finite distance). Detailed behavior as a function of end termination is quite complicated and has been treated in the literature from various aspects (temperature,[33] impedance,[34] and, recently, impulse response[35,36]). Fortunately, a simple orientative rule can be developed. An RC-line of finite length, Fig. 8(f), is characterized by the time constant $R_\ell C_\ell$, where $R_\ell = r \cdot \ell$ and $C_L = c \cdot \ell$. It can be shown[34] that the impedance of the finite-length line follows the relation for the infinite line, Eq. (31), independently of the end termination, above a lower limit frequency, which is given by

$$\frac{1}{\omega_\ell} \approx \frac{R_\ell C_\ell}{10} \qquad (38)$$

(It may be interesting to note that $R_\ell C_\ell / \pi^2$ appears as a dominant time constant in the expressions for the propagation time of such a line.)

This parameter is significant since it determines the low frequency limit of

1/|f| noise generated by such a network or physical process.

The diffusion model for generation of 1/|f| noise is very attractive since it covers a wide range of physical systems, and since various interactions among electrical, thermal and chemical quantities could be considered in some systems in the analysis of the noise generating process. However, detailed analysis and justification of this mechanism in particular cases is quite difficult.

Halford's Mechanical Model

The model which assumes impulse response $t^{-1/2}$ is too restricted in the sense that each perturbation constituting a white process produces the same effect, which has a particular shape in time.

In some studies of noise in solid-state devices (Refs. 3, 4, 5, and others) it has been recognized that 1/|f| can be generated by perturbations occurring randomly in time which have certain specific distributions of lifetimes. Halford[14] has proposed a general model in which broad classes of perturbations are found to satisfy the criteria for generation of 1/|f| noise. This model is summarized in the following.

According to this model, any class of "reasonable perturbations" occurring at random, under certain constraints, generates random noise having a spectral density $|f|^\alpha$ over an arbitrarily large range of frequency only for $-2 \le \alpha \le 0$. A class is the set of all perturbations which are equivalent under some individual independent scaling of amplitude, scaling of time and translation of time. A subclass of perturbations is characterized by $P(\tau)$ and $A^2(\tau)$, where $P(\tau)$ is the probability density of perturbations with lifetime τ, and $A^2(\tau)$ is a mean-square amplitude of perturbations having lifetime τ. For a given class, $|f|^{\alpha_\infty}$ and $|f|^{\alpha_0}$ are the frequency-smoothed laws in the limits of infinite and zero frequencies which specify the cutoff properties. Any

reasonable perturbation must have $\alpha_\infty \le -2$ and $\alpha_0 \ge 0$, which means that there should be no divergence at either the low or high frequency limit. To generate random noise having an $|f|^\alpha$ law over an arbitrarily large range of f from a subclass chosen from any class characterized by α_∞ and α_0, it is necessary that $\alpha_\infty \le \alpha \le \alpha_0$. For $P(\tau)$ and $A^2(\tau)$, it is then necessary and sufficient to satisfy the condition,

$$P(\tau) \cdot A^2(\tau) \approx B\tau^{-\alpha-3} \tag{39}$$

where B is a constant. This condition should be satisfied over a range of τ, which determines the range of f over which the $|f|^\alpha$ law is obeyed. Outside of this range of τ, the condition

$$P(\tau) A^2(\tau) \le B \cdot \tau^{-\alpha-3} \tag{40}$$

should be satisfied. A reasonable perturbation is any perturbation which satisfies the requirements that it is everywhere finite, and that it has a finite integral, finite energy and finite nonzero lifetime.

For 1/|f| noise $\alpha = -1$, and the condition (39) can be satisfied over an arbitrarily large range. This can be illustrated by the example of a very common perturbation $A(\tau) U(t) \exp(-t/\tau)$, which represents a step change with exponential decay. The condition (39) for 1/|f| noise is

$$P(\tau) A^2(\tau) \approx B \frac{1}{\tau^2} \tag{41}$$

Assuming a restricted case where all lifetimes are equally probable, $P(\tau) = 1$, it follows

$$A(\tau) \propto \frac{1}{\tau} \tag{42}$$

The spectral density of the perturbations $\frac{1}{\tau} \exp(-t/\tau)$ for the values of lifetime between τ and $\tau+d\tau$ is $d\tau/(1+\omega^2\tau^2)$. Then the spectral density for the perturbations over the range of τ from 0 to ∞ will be

$$\int_0^\infty \frac{1}{1+\omega^2\tau^2} \, d\tau = \frac{1}{|\omega|} \tan^{-1}(\omega\tau) \Big]_0^\infty = \frac{\pi}{2} \frac{1}{|\omega|}$$

(The same result is obtained in this example for $A(\tau) = $ const. and $P(\tau) \propto 1/\tau^2$). It should be noted here that in any class of perturbations the low frequency limit of the spectral density is determined by the maximum value of lifetime τ.

The mechanism of $1/|f|$ noise in MOS transistors has been recently described[6,7] in terms which fall within the frame of this model. The particular mechanism is explained in terms of tunneling of carriers at the silicon-silicon oxide interface to traps located inside the oxide. The dispersion of the time constant was found to be 10^{17}, and the low frequency limit corresponds to $\tau \approx 10^9$ sec ($\approx$ 30 years).

An interesting case of this model is for constant amplitude perturbations, $A(\tau) = 1$. Then, for $1/|f|$ noise, the condition to be satisfied is,

$$P(\tau) \propto \frac{1}{\tau^2} \qquad (43)$$

This case is suitable for simulation of $1/|f|$ noise on the computer, although an amplitude distribution may also be used.

The models described in this section are elegant and analytically neat. The actual physical mechanisms in particular cases might be considerably more involved. Some complex mechanisms are discussed in Ref. 15.

Lumped-Parameter Approximations of Impulse Response $t^{-1/2}$

The realization of impedances and filters with transfer functions with $(j\omega)^{-1/2}$ frequency dependence ($t^{1/2}$ impulse response) by using distributed and lumped parameter networks to approximate this law over a limited range of frequencies for various purposes has received much attention in the literature (further references are given in Ref. 34 of this paper). These networks are variously referred to as "constant-argument impedance," "constant-angle impedance," "fractional capacitance," "fractional integral and derivative operators," or "power-law magnitude impedance." Each of these terms is meaningful in some way of representing $1/|f|$ noise.

For generation of $1/|f|$ noise in the low frequency range, digital circuits, or digital computers, are more suitable because perturbations (impulse responses) with arbitrarily long lifetimes can be realized, aside from other processing advantages. Analog circuits are generally simpler, but unsuitable for the low frequency range. The choice between them is somewhat arbitrary in the frequency range where both can be realized. Analog circuits are the only solution in the frequency range where the speed of digital circuits is not sufficient. (The high frequency limit of the noise generated by digital operations is about an order of magnitude lower than the rate of digital operations, if gaussian distribution of amplitudes is to be achieved.)

An example of a lumped-parameter network is shown in Fig. 9. The ratio of time constants is the same between successive RC networks. A large value was selected for this ratio to emphasize the effects of the lumped-parameter approximation, as shown in Fig.10. In spite of this, the departure of the spectral density from the $1/|f|$ law is no larger than 0.7 dB over 4 decades (Fig. 10). An advantage of this circuit is that independent adjustment of the attenuation coefficients and the time constants is possible. Switches (realized by junction field-effect transistors) are included for time-domain and time-variant filter studies. A point sometimes neglected in realizations of such circuits is that the resistors which cover a wide range of values should be real (a ladder network attenuator may be necessary, as shown in Fig. 9, to avoid the effect of stray capacitances, which can be significant in the range above 10^5 Hz).

4. Conclusions on Measurements

From the discussion about power-law noises $|f|^\alpha$, the conclusion is that $1/|f|$

noise occupies a central place between the noises which are high frequency divergent ($\alpha \geq -1$) and the noises which are low frequency divergent ($\alpha \leq -1$). The unique property of $1/|f|$ noise is that it is divergent at both limits. The relations derived for the variance in the frequency domain, Eq. (7), and in the time domain, Eqs. (10) and (12), and the relations describing correlation properties, are important in the measurements of physical quantities in the presence of noise and in the measurements of noise. In the case of narrow-band measurements, where $\Delta f = f_h - f_\ell << \frac{1}{2}(f_h + f_\ell)$, an obvious result follows for the variance

$$\sigma^2(f, \Delta f) \approx |f|^\alpha \Delta f \qquad (44)$$

For a given bandwidth, the variance is determined by the spectral density.

In the case of wide-band measurements, the variance is determined by both frequency limits. In the case of wide-band measurements where $f_\ell << f_h$, the variance is determined by the low frequency limit for low frequency divergent noises $\alpha < -1$, and by the high frequency limit for high frequency divergent noises $\alpha > -1$. Both limits should always be considered when α is close to -1.

In the time domain measurements, the length of the measurement interval T, or, in other words, the observation time [0,t in Eqs. (10) and (12)] corresponds to the low frequency limit. In the time domain measurements, instantaneous values are observed by some sampling method. Each sample represents some averaging function of all the instantaneous values in a short time interval δ, which then corresponds to the high frequency limit. The variance is then determined by T and δ.

The class of measurements which is of particular interest is the one where the ratio of the high frequency limit and the low frequency limit is maintained constant ($f_h/f_\ell = $ const., or $T/\delta = $ const.) as the limits are varied. The frequency limits are characterized by one parameter, τ, which is related to $1/f_\ell$ and $1/f_h$ (T and δ) by a constant. Using the

relation (7), the relation of the variance to the parameter τ ("measurement time") is determined for power-law noises,

$$\sigma^2(\tau) = K_\alpha \tau^{-1-\alpha} \qquad (45)$$

K_α is a constant determined by α from Eq. (7) and by the relation of τ to the frequency limits. More generally, K_α is a constant characteristic of the impulse response of the particular bandpass filter, or of the weighting function in pulse measurements and for time-variant filters.

The importance of the relation (45) is that the observed dependence of variance as a function of measurement time τ can be used to identify a particular noise component, which would otherwise be impossible to measure separately. Any physical power spectral density function can be represented by the power series,

$$W(f) = --- + W_1|f|^1 + W_0 f^0 + W_{-1}|f|^{-1} + W_{-2}f^{-2} + --- \qquad (46)$$

The corresponding power series for variance as a function of τ is

$$\sigma^2(\tau) = --- + K_1\tau^{-2} + K_0\tau^{-1} + K_{-1}\tau^0 + K_{-2}\tau^1 + --- \qquad (47)$$

The meaning of this power series is not the mere approximation of a continuous function. Each term is meant to correspond to a particular physical source of noise or to a group of sources. To fulfill this, in some cases, a power series with fractional exponents may be more appropriate. As an example, the noise of the feedback resistor in charge amplifiers can be considered. A resistor in the range of $10^9 - 10^{10}$ ohms represents a distributed RC-line due to its stray capacitance. Its noise current is integrated on the amplifier input capacitance, and, according to Fig. 8(e) and Eq. (36), it produces $|f|^{-3/2}$ noise in the frequency range of interest in that case.[39] The term $K_{-3/2} \cdot \tau^{1/2}$ added to the series (47) would in such a case result in a better agreement with the measured function $\sigma^2(\tau)$.

This method, based on three terms of a power series ($\alpha = 0$, -1, -2), was first used in measurements with charge amplifiers for nuclear radiation detectors.[37,38] It has been used extensively in frequency and time data analysis and measurements.[10,11,12,17] If this method is extended to measure $\sigma^2(\tau)$ as a function of some other system variables (temperature of input amplifier components), as has been done with charge amplifiers,[20] it can be a powerful tool in identifying some physical sources of noise. The parameter τ is equivalent to various commonly used terms in different areas of measurements. It is referred to as "filter time-constant" in pulse amplifiers for nuclear detectors. It corresponds to the "sweep time," or to the "filter integration time-constant" in nuclear magnetic resonance and in electron paramagnetic resonance measurements.

Relation (45) was derived from Eq. (7), and it holds for any α. However, it must be noted that the bandpass frequency characteristic implied by the integral (7) is a "window" limited by frequencies f_ℓ and f_h. If the lowest order bandpass filter is used with one pole and one zero (one RC integration and one RC differentiation), the variance for that filter would not converge for $\alpha \geq 1$ at the high frequency limit, and for $\alpha < -2$ at the low frequency limit. The relation (45) holds for this range of α, and such a filter is in most cases satisfactory. If noises beyond this range of α are expected, filters with a sharper cutoff (multiple poles and zeros) should be used. In time domain measurements, sampling with simple integration of instantaneous values in each sample represents a first order filter, which is not sufficient for noises with $\alpha > 1$. If f^2 noise is expected, instantaneous values from the interval δ should be passed through a second order low-pass filter, which can be realized in a number of ways. These remarks about the sampling apply also to the derivation of Eq. (45) from the time domain expression (10), where higher-order smoothing should be applied to $h(u)$ for high frequency divergent noises with $\alpha > 1$.

Having the noise defined, the usual procedure is to find an optimum filter for a given signal (filter matched to signal and noise). One of the most common signals is a step function (or a dc quantity resulting from rectification of an ac signal), the amplitude of which is to be measured in the presence of noise. An intuitive step usually undertaken is to integrate this signal and noise. This is a correct procedure for white noise since integration represents a matched filter for this case. If $1/|f|$ noise is present, the variance decreases as τ^{-1} until the variance due to $1/|f|$ noise is reached, which is independent of τ. Little advantage can be gained by changing the ratio f_h/f_ℓ since variance is a logarithmic function of this ratio. Thus, in the case of $1/|f|$ noise, the most important aspect is to find its physical source, with the aim of reducing it at the source if possible.

According to relation (7) the variance is a function of the frequency limits of the measurement for a given type of noise. In the measurements of differences, Eq. (28), the variance is determined by the observation interval τ and by the averaging interval δ (for noises where averaging or smoothing is important), Eqs. (20) and (27); that is, by the time domain equivalents of the frequency limits. This also applies to the general relation (45) for power-law noises in all the frequency domain and time domain measurements in which the ratio of the high frequency limit and the low frequency limit is constant (or where one of the limits is unimportant). For low frequency noise, this means that the part of the spectrum below the low frequency limit of the measurement process is unimportant (paying attention to the exact nature of the low frequency cutoff in relation to the power-law exponent α). For an imagined truly divergent low frequency noise (where the spectrum follows the power-law to zero frequency), the variance in such measurements would be independent of the "age" of the noise process. Thus, for purposes of the measurements, that is, for physical observations, it is unimportant whether the low frequency noise has a low frequency limit or not (whether it is convergent or divergent), provided that it obeys its power-law within the observable frequency range. Consequently, all physical evidence relating to low frequency noises is limited by the low frequency limit of the measurement process; that is, by the maximum observation time.

Some Further Questions about $1/|f|$ Noise

In this paper the qualification of $1/|f|$ noise as being stationary or non-stationary has not been explicitly considered so far. We can first specify that in the representation in Section 2 the imagined Poisson process which generates $1/|f|$ noise is stationary in time (in the strict sense) and that the parameters of the "transforming filter" are constant (in the case of Halford's model, for example, we assume that the distribution functions remain constant in time). The resulting low frequency divergent process cannot be called stationary in the strict sense since its variance increases with time. Some of its features are stationary, however. From the discussion of the correlation function it follows that finite differences of $1/|f|$ noise are stationary. From this, and from the fact that the underlying mechanism which generates $1/|f|$ noise by a linear transformation is stationary, one can consider this case as a "divergent process growing in a stationary way" or as a divergent process with "stationary increments."[31]

A stationary characteristic of such a process can be expressed in the following way. The true variance (mean value of noise power) for a given frequency range is given by Eq. (7), or more generally, including time domain measurements by Eq. (45). An estimate of this variance is obtained by measurements in which the noise power (variance) is averaged over a measurement time T_m. An underlying assumption in this procedure is that the spectral density function of the fluctuations of the variance is white, which should be the case for a stationary process. This should apply to finite-bandwidth measurements on low frequency noises, since these measurements involve only the noise increments which are stationary. In such a case the variance of these fluctuations ("variance of variance") should decrease with the measurement time as $\sim 1/T_m$. [In the case where variance (mean square noise) is determined from a number of samples, N, the "variance of variance" should decrease as $\sim 1/N$] $1/|f|$ noise satisfying this requirement would correspond to what is sometimes vaguely referred to as "well-behaved" $1/|f|$ noise.

There is some evidence that there are noises whose spectral density function is generally categorized as $1/|f|$ noise (under unspecified measurement conditions in many cases), and which are not "stationary" or "well-behaved" in the sense discussed above. Brophy[40,41,42] has reported more detailed measurements on $1/|f|$ noise in carbon resistors, in which he finds the fluctuations of the variance to be larger than for white noise under equivalent measurement conditions. He also finds the dependence of variance of these fluctuations on the measurement interval to be different from $1/T_m$.

For such noises, one is led to assume that there might be some low frequency fluctuations in the parameters of the process generating $1/|f|$ noise. Alternatively, a nonlinear mechanism may be involved, where low frequency components would cause variations in the amplitude of the high frequency components.

There is too little statistically meaningful experimental evidence on such behavior so far. To specify the noise more precisely, the measurements of "variance of variance" should be performed as a function of measurement time, after the essential noise components have been identified by spectrum measurements or by the method discussed in the preceding section. Such measurements are easier for the high frequency portion of the spectrum, since the large amount of data required can be collected in a shorter time. In these measurements it becomes increasingly difficult to separate "causal" changes in process parameters (due to temperature) from random changes. Nevertheless, the significance of such information is that it would lead closer to the sources of the fluctuations, whether they are random or not, and thus increase the possibility of their reduction.

In some particular cases of measurements of physical quantities, the knowledge of the spectrum of fluctuations of the variance may result in better optimization of measurement and data processing procedures than by just assuming that the spectrum of fluctuations of the variance is white.

Acknowledgements

It is a pleasure to acknowledge many discussions with R. L. Chase and M. J. Rosenblum. A useful discussion with J. A. Barnes and D. Halford of NBS, Boulder, Colorado is gratefully acknowledged.

DISCUSSION

S.S. Klein : - I understand from Mr. Radeka's talk that there is an optimal time for the minimal variance of a meaurement. However, it is necessary to indicate the accuracy of the measurement. As I understand it now, you are not sure the value obtained from any measurement will be repeated to a given accuracy in ten years or in thirty years or in a time comparable to the age of the universe, so leading to a state of despair about ever measuring anything with any accuracy at all.

Radeka : - Well, I understand this more as a comment than as a question.

S.S.Klein :-Give me hope, please.

Radeka : - Well, the things may be not so bad as I presented them. We are still making some measurements within the time which is available to us and the effects of noise with the different power laws is a matter for a longer discussion. The effects, of any variance in the noise power as a function of measurement parameters would depend on a specific case. In some cases the $1/f^n$ noise may be quite stationary and in some other cases it may not be. If they are stationary then, any measurement that we make with the measurement time of one second would always give the same probability distribution of results now and after ten years. If however we have a process which is not stationary in this sense, that is a process which generates low frequency noise, then, it may not be so.

References

1. J. B. Johnson, "The Schottky Effect in Low Frequency Circuits," Phys. Rev. 26, 71 (1925)

2. W. Schottky, "Small-Shot Effect and Flicker Effect," Phys. Rev. 28, 75 (1926)

3. A. van der Ziel, Fluctuation Phenomena in Semiconductors, Academic Press, New York 1959

4. A. L. McWhorter, "1/f Noise and Related Surface Effects in Germanium," M.I.T. Lincoln Lab., Lexington, Mass., Tech. Rep. 80, May 1955

5. A. L. McWhorter, "1/f Noise and Germanium Surface Properties," Semiconductor Surface Physics, R. H. Kingston, Ed., Philadelphia, Pa., Univ. of Pennsylvania Press, 1957, pp. 207-228

6. C. Christensson, I. Lundström and C. Svensson, "Low Frequency Noise in MOS Transistors-I," Solid-State Electronics 11, 797 (1968)

7. S. Christensson and I. Lundström, "Low Frequency Noise in MOS Transistors-II," Solid-State Electronics 11, 813 (1968)

8. I. R. M. Mansour, R. J. Hawkins and G. G. Bloodworth, "Digital Analysis of

Current Noise at Very Low Frequencies," The Radio & Electronic Engineer, 35, 201 (1968)

9. I. R. M. Mansour, R. J. Hawkins and G. G. Bloodworth, "Measurement of Current Noise in M.O.S. Transistors from 5×10^{-5} to 1 Hz," The Radio and Electronic Engineer, 35, 212 (1968)

10. L. S. Cutler and C. L. Searle, "Some Aspects of the Theory and Measurement of Frequency Fluctuations in Frequency Standards," Proc. IEEE 54, 136, Feb. 1966

11. D. W. Allan, "Statistics of Atomic Frequency Standards," Proc. IEEE 54, 221 (1966)

12. J. A. Barnes, "Atomic Timekeeping and the Statistics of Precision Signal Generators," Proc. IEEE 54, 207 (1966)

13. J. A. Barnes and D. W. Allan, "A Statistical Model of Flicker Noise," Proc. IEEE 54, 176 (1966)

14. D. Halford, "A General Mechanical Model for $|f|^{\alpha}$ Spectral Density Random Noise with Special Reference to Flicker Noise $1/|f|$," Proc. IEEE 56, 251 (1968)

15. H. E. Derksen, "Axon Membrane Voltage Fluctuations," Acta Physiol. Pharmacol. Neerl. 13, 373 (1965)

16. D. Brouwer, "A Study of the Changes in the Rate of Totation of the Earth," Astron. J. 57, 125 (1952)

17. J. A. Barnes and D. W. Allan, "An Approach to the Prediction of Coordinated Universal Time," Frequency, pp 3-8, Nov./ Dec. 1967

18. B. B. Mandelbrot and J. W. van Ness, "Fractional Brownian Motions, Fractional Noises and Applications," Siam Rev. 10, No. 4, 422 (Oct. 1968)

19. B. B. Mandelbrot, "Noises with an 1/f Spectrum, a Bridge Between Direct Current and White Noise," IEEE Trans. Information Theory, IT-13, 289 (1967)

20. V. Radeka, "State of the Art of Low Noise Amplifiers for Semiconductor Radia-tion Detectors, Proc. Internat. Symposium on Nuclear Electronics, Versailles, 1968, p. 46-1

21. E. Gatti and V. Svelto, "Resolution as a Function of Noise Spectrum in Amplifiers for Particle Detection," Energia Nucleare, 8, 505 (1961)

22. M. P. Klein and G. W. Barton, Jr., "Enhancement of Signal-to-Noise Ratio by Continuous Averaging: Application to Magnetic Resonance, Rev. Sci. Instr. 34, 754 (1963)

23. G. A. Campbell and R. M. Foster, "Fourier Integrals for Practical Applications," D. Van Nostrand Co., Inc., New York, 1961

24. N. Campbell, Proc. Cambridge Phil. Soc. 15, 117 (1909)

25. I. Flinn, "Extent of the 1/f Noise Spectrum" Nature, 219, 1356, Sept.28,1968

26. J. S. Bendat, "Principles and Applications of Random Noise Theory, John Wiley & Sons, Inc., New York 1958

27. J. R. Schwarz and B. Friedland, Linear Systems, McGraw-Hill, New York, 1965

28. D. G. Lampard, "The Response of Linear Networks to Suddenly Applied Stationary Random Noise," IRE Trans. on Circuit Theory, CT-2, 43 (1955)

29. B. O. Peirce, A Short Table of Integrals, Ginn & Co., New York 1929

30. M. Lighthill, Introduction to Fourier Analysis and Generalized Functions, New York: Cambridge 1962

31. J. A. Barnes, Private Communication

32. O. Heavyside, Electromagnetic Theory, Dover Publications, New York 1950

33. H. S. Carslaw and J. C. Jaeger, Conduction of Heat in Solids, Clarendon Press, Oxford, 1947, p. 76

34. S. C. Dutta Roy and B. A. Shendi, "Distributed and Lumped RC Realization

of a Constant Argument Impedance," Jour. of the Franklin Institute, 282, No. 5, 318 (1966)

35. W. W. Happ and S. C. Gupta, "Time-Domain Analysis and Measurement Techniques for Distributed RC Structures. I. Analysis in the Reciprocal Time Domain," Jour. Applied Phys. 40, 109 (1969)

36. R. C. Perison and E. C. Bertnulli, "Time-Domain Analysis and Measurement Techniques for Distributed RC Structures. II. Impulse Measurement Techniques," Jour. Applied Phys. 40, 118 (1969)

37. A. B. Gillespie, Signal, Noise and Resolution in Nuclear Counter Amplifiers, Pergamon Press, London, 1953

38. E. Baldinger and W. Franzen, Advances in Electronics and Electron Physics, Vol. 8, 256 (1956)

39. V. Radeka, to be published

40. J. J. Brophy, "Statistics of 1/f Noise," Phys. Rev. 166, 827, (1968)

41. J. J. Brophy, "Zero-Crossing Statistics of 1/f Noise," J. Appl. Phys. 40, 567 (1969)

42. L. J. Greenstein and J. J. Brophy, "Influence of Lower Cutoff Frequency on the Measured Variance of 1/f Noise," J. Appl. Phys. 40, 682 (1969)

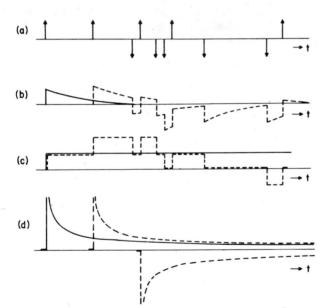

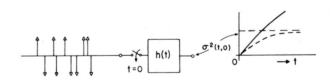

Fig. 2 - Time domain divergence test.

Fig. 1 - Generation of $1/|f|$ noise and $1/f^2$ noise from a Poisson process by a filter with appropriate impulse response. a) Poisson process, random sequence of impulses ; b) Band-limited white noise ; c) $1/f^2$ noise (random walk) ; d) $1/|f|$ noise.

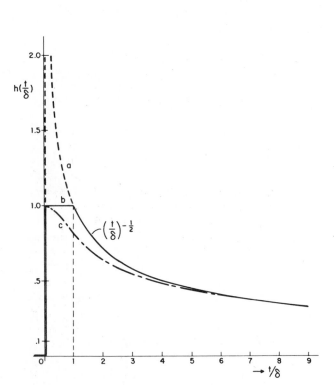

Fig. 3 - Impulse response for generation of $1/|f|$ noise (curve a), and with high frequency cutoff (curves b and c);

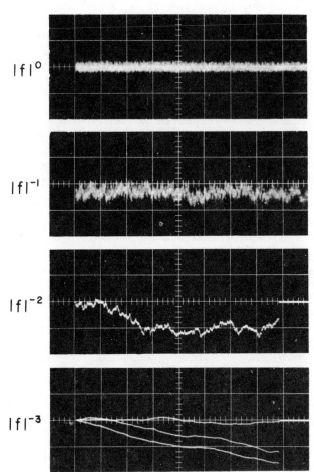

Fig. 4 - Samples of noise waveforms for $\alpha = 0$, -1, -2, -3 (white noise, $1/|f|$ noise, random walk, $1/|f|^3$ noise)

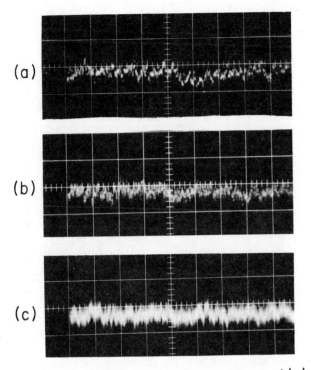

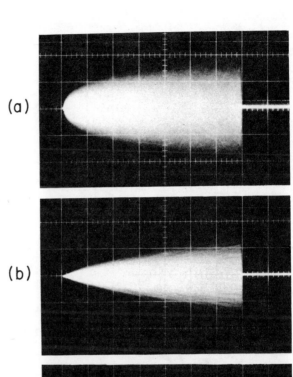

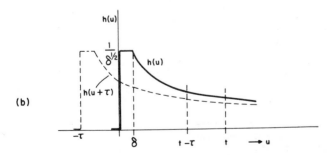

Fig. 5 - Samples of noise waveforms for $1/|f|$ noise at different time scales :
a) 10 μsec/div. , b) 50 μsec/div. ,
c) 500 μsec/div. .

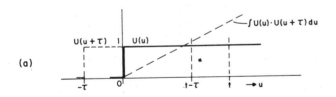

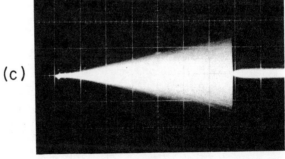

Fig. 6 - Standard deviation (rms noise) as a function of time for some low frequency divergent noises. (envelope of bright area ≈ σ(t,0)). a) $1/f^2$ noise ; white noise switched into an integrator, $h(t) = U(t)$; b) $1/f^3$ noise switched on $h(t) = t^{1/2}$; c) $1/f^4$ noise ; white noise switched on $h(t) = t$.

Fig. 7 - Calculation of "time-dependent correlation function", a) $h(u) = \sqrt{2\pi}\, U(u)$ for $1/f^2$ noise ; b) $h(u) = u^{-1/2}$ for $1/|f|$ noise.

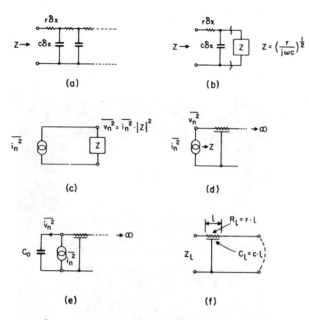

(a) (b)

(c) (d)

(e) (f)

Fig. 8 - Distributed RC-line model for generation of $1/|f|$ noise , a) Distributed RC-line ; b) Impedance of distributed RC-line ; c) Noise voltage-current relation for impedance Z ; d) Thermal noise of RC-line acting upon the line itself ; e) Thermal noise of RC-line acting upon a capacitance ; f) RC-line of finite length.

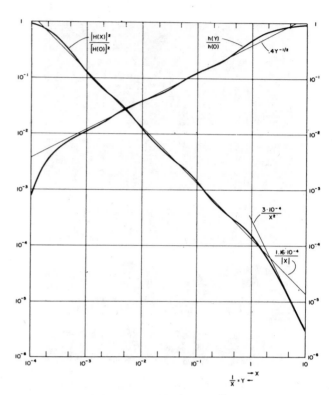

Fig. 10 - Spectral density $|H(X)|^2$ and impulse response $h(Y)$ for the filter in Fig.9. Coefficients $B_1 = B_4 = 1$, $B_2 = B_3 = .7$; ratio of time constants $N = 15$; $X = \omega\tau$; $Y = t/\tau$.

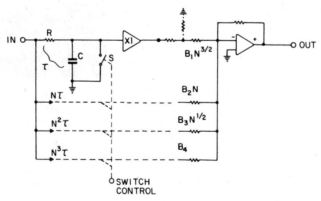

Fig. 9 - A lumped parameter circuit for generation of $1/|f|$ noise from white noise. N - ratio of time constants.

196

Bibliography for Part III

```
******************************************************************************
     3-A.  THERMAL NOISE
******************************************************************************
```
ALKEMADE, C. T. J.
     ON THE PROBLEM OF BROWNIAN MOTION OF NONLINEAR SYSTEMS
          PHYSICA, VOL. 24, NO. 12, PP. 1029-1034, DECEMBER 1958.
BAKKER, C. J. AND G. HELLER
     ON THE BROWIAN MOTION IN ELECTRIC RESISTANCES
          PHYSICA, VOL. 6, NO. 3, PP. 262-274, 1939.
BALAZS, N. L.
     THERMAL FLUCTUATIONS IN CONDUCTORS
          PHYSICAL REVIEW, VOL. 105, NO. 3, PP. 896-899, FEBRUARY 1, 1957.
BELL, D. A.
     THERMAL FLUCTUATIONS IN SPACE CHARGE CONTROLLED DIODES
          JOUR. ELECTRONICS, VOL. 2, NO. 5, PP. 477-488, MARCH 1957.
BERNARD, W. AND H. B. CALLEN
     IRREVERSIBLE THERMODYNAMICS OF NONLINEAR PROCESSES AND NOISE IN
     DRIVEN SYSTEMS
          REV. MODERN PHYS., VOL. 31, NO. 4, PP. 1017-1044, OCTOBER 1959.
BILGER, H. R.
     THERMAL NOISE OF A SILICON DOUBLE-INJECTION PLASMA DIODE
          SOLID-STATE ELECTRONICS, VOL. 16, NO. 12, PP. 1363-1365,
          DECEMBER 1973.
BUNEMAN, O.
     RESISTANCE AS DISSIPATION INTO MANY REACTIVE CIRCUITS   LANDAU
     DAMPING AND NYQUIST'S NOISE THEOREM
          JOUR. APPL. PHYS., VOL. 32, NO. 9, P. 1783, SEPTEMBER 1961.
BURCKHART, C. B. AND M. J. O. STRUTT
     NOISE IN NONRECIPROCAL TWO PORTS BASED ON HALL EFFECT
          IEEE TRANS. ELECTRON DEVICES, VOL. ED-11, NO. 2, PP. 47-50,
          FEBRUARY 1964.
BURGESS, R. E.
     ELECTRICAL CONDUCTIVITY AND THERMAL FLUCTUATIONS
          PHYSICA, VOL. 23, NO. 8, PP. 705-706, AUGUST 1957.
CALLEN, H. B. AND T. A. WELTON
     IRREVERSIBILITY AND GENERALIZED NOISE
          PHYSICAL REVIEW, VOL. 83, NO. 1, PP. 34-40, JULY 1, 1951.
CASE, K. M.
     ON FLUCTUATION-DISSIPATION THEOREMS
          TRANSPORT THEORY AND STATISTICAL PHYSICS, VOL. 2, NO. 2, PP. 129-
          176, 1972.
COLE, E. A. B.
     THE QUANTUM MECHANICAL FLUCTUATION-DISSIPATION THEOREM
          PHYSICA, VOL. 38, NO. 3, PP. 349 - 356, MAY 1, 1968.
CONVERT, G.
     ON THE EXTENTION OF NYQUIST THEOREM TO SYSTEMS WITH NEGATIVE
     TEMPERATURE (IN FRENCH)
          COMPTES RENDUS ACAD. SCIENCE (PARIS), VOL. 261, NO. 1, PP. 58-61,
          5 JULY 1965.
DEKER, U. AND F. HAAKE
     FLUCTUATION - DISSIPATION THEOREMS FOR CLASSICAL PROCESSES
          PHYSICAL REV. A, VOL. 11, NO. 6, PP. 2043-2056, JUNE 1975.
DEMASSA, T. A. AND S. R. IYER
     THERMAL NOISE IN THE LINEAR GRADED CHANNEL JUNCTION FET
          SOLID-STATE ELECTRONICS, VOL. 18, NO. 11, PP. 933-935, NOVEMBER
          1975.
GOLDER, J. AND M-A. NICOLET
     NOISE OF SPACE-CHARGE-LIMITED CURRENT IN SOLIDS IS THERMAL
          SOLID-STATE ELECTRONICS, VOL. 16, NO. 10, PP. 1151-1157, OCTOBER
          1973.
GOLDER, J., M-A. NICOLET, AND A. SHUMKA
     THERMAL NOISE MEASUREMENTS ON SPACE-CHARGE-LIMITED HOLE CURRENT IN
     SILICON
          SOLID-STATE ELECTRONICS, VOL. 16, NO. 5, PP. 581-585, MAY 1973.
GRIMM, H. H.
     NOISE TEMPERATURE IN PASSIVE CIRCUITS
          MICROWAVE JOUR., VOL. 3, NO. 2, PP. 52-54, FEBRUARY 1960.
GUTTLER, P.
     THE NYQUIST FORMULA - POSSIBILITIES OF ITS DERIVATION (IN GERMAN)
          NACHRICHTENTECHNIK, VOL. 18, NO. 11, PP. 425-432, NOVEMBER 1968.
GUTTLER, P.
     THE NYQUIST FORMULA - VALIDITY AND UPPER FREQUENCY LIMIT (IN GERMAN)
          NACHRICHTENTECHNIK, VOL. 19, NO. 3, PP. 103-106, MARCH 1969.
HAMMAD, P.
     THE THEOREM OF USEFUL ENERGY AND BROWNIAN MOTION,
          C.R. ACAD. SCI. B(FRANCE), VOL. 266, NO 15, PP 981-983, 8 APRIL
          1968.
HARWIT, M.
     MEASUREMENT OF THERMAL FLUCTUATIONS IN RADIATION
          PHYSICAL REV., VOL. 120, NO. 5, PP. 1551-1556, DECEMBER 1, 1960.
HASHITSUME, N.
     A STATISTICAL THEORY OF LINEAR DISSIPATIVE SYSTEMS

PROGRESS THEORETICAL PHYSICS, VOL. 8, PP. 461-478, OCTOBER 1952.
HAUS, H. A.
    THERMAL NOISE IN DISSIPATIVE MEDIA
        JOUR. APPL. PHYS., VOL. 32, NO. 3, PP. 493-500, MARCH 1961.
HENRY, R. J.
    THE GENERALIZED LANGEVIN EQUATION AND THE FLUCTUATION-DISSIPATION
    THEOREMS
        JOUR. PHYSICS A, VOL. 4, NO. 5, PP. 685-694, SEPTEMBER 1971.
HUANG, C. AND A. VAN DER ZIEL,
    THERMAL NOISE IN ION-IMPLANTED MOSFETS,
        SOLID STATE ELECTRONICS, VOL. 18, NO. 6, PP. 509-510, JUNE 1975.
HUANG, C. H., A. VAN DER ZIEL, AND K. M. VAN VLIET
    DIFFUSION NOISE IN DOUBLE INJECTION DIODES IN THE OHMIC RELAXATION
    REGIME
        ELECTRONICS LETTERS, VOL. 7, NO. 11, PP. 291-292, 3JUNE 1971.
JOHNSON, J. B.
    THERMAL AGITATION OF ELECTRICITY IN CONDUCTORS
        PHYS. REV., VOL. 32, NO. 1, PP. 97-107, JULY 1928.
KLAASSEN, F. M.
    THERMAL NOISE IN SPACE-CHARGE-LIMITED SOLID STATE DEVICES
        SOLID STATE ELECTRONICS, VOL. 11, NO. 3, PP. 377-378, MARCH 1968.
KRASOVSKII, A. A.
    PASSIVE LINEAR SYSTEMS UNDER THE ACTION OF INTERNAL THERMAL NOISE
        AUTOMATION AND REMOTE CONTROL, VOL. 26, NO. 6, PP. 1017-1027,
        JUNE 1965.
KUBO, R.
    THE FLUCTUATION-DISSIPATION THEOREM AND BROWNIAN MOTION
        TOKYO SUMMER LECTURES IN THEORETICAL PHYSICS, PT. 1, PP. 1-16,
        W. A. BENJAMIN, INC., NEW YORK, 1966.
LANDSBERG, P. T. AND E. A. B. COLE
    THE STEADY STATE FLUCTUATION-DISSIPATION THEOREM
        PHYSICA, VOL. 37, NO. 2, PP. 309-319, 1967.
LENK, R.
    A SIMPLE PROOF OF THE CLASSICAL FLUCTUATION DISSIPATION THEOREM
        PHYS. LETTERS, A, VOL. 25, NO. 3, PP. 198-199, AUGUST 14, 1967.
LEVIN, M. L. AND S. M. RYTOV
    'KIRCHOFF FORM' OF THE FLUCTUATION - DISSIPATION THEOREM FOR
    DISTRIBUTED SYSTEMS
        SOVIET PHYSICS - JETP, VOL. 38, NO. 4, PP. 688-692, APRIL 1974.
LUKES, T.
    FLUCTUATIONS AND THE ELECTRICAL CONDUCTIVITY OF METALS
        PHYSICA, VOL. 27, NO. 3, PP. 319-328, MARCH 1961.
MAC DONALD, D. K. C.
    THE BROWNIAN MOVEMENT AND SPONTANEOUS FLUCTUATIONS OF ELECTRICITY
        RESEARCH, VOL. 1, NO. 5, PP. 194-203, FEBRUARY 1948.
MAC DONALD, D. K. C.
    SPONTANEOUS FLUCTUATIONS
        REPTS. PROGRESS OF PHYSICS, VOL. 12, PP. 56-81, 1948-1949.
MAC DONALD, D. K. C.
    A SURVEY OF RECENT PROGRESS IN THE STUDY OF FLUCTUATION NOISE
        JOUR. IEE (GB), VOL. 95, PT. III, NO. 37, PP. 330-331, SEPTEMBER
        1948.
MATHEWS, J. AND M-A. NICOLET
    CURRENT CORRELATION FUNCTION J(X,T)J(X',T') DERIVED FROM A MODEL
    BASED ON BROWNIAN MOTION
        AMERICAN JOUR. PHYSICS, VOL. 44, NO. 5, PP. 448-454, MAY 1976.
MEIXNER, J.
    LINEAR PASSIVE SYSTEMS
        IN STATISTICAL MECHANICS OF EQUILIBRIUM AND NON-EQUILIBRIUM,
        J. MEIXNER, ED., NORTH-HOLLAND PUBLISHING CO., AMSTERDAM, 1965,
        PP. 52-68.
MORRIS, E. AND R. FURTH
    SPATIAL CORRELATIONS IN THE THERMAL FLUCTUATIONS OF ELECTRICAL
    QUANTITIES IN THE PRESENCE OF CONDUCTING SURFACES
        PHYSICA, VOL. 26, NO. 9, PP. 687-697, SEPTEMBER 1960.
NAG, B. R. AND P. N. ROBSON
    CALCULATION OF HOT-ELECTRON NOISE IN SEMICONDUCTORS
        PHYSICS LETTERS A, VOL. 43A, NO. 6, PP. 507 - 508, 23 APRIL 1973.
NEWMAN, N., AND G. T. CONRAD, JR.
    DISCUSSION OF ERRORS OF A RECOMMENDED STANDARD RESISTOR - NOISE TEST
    SYSTEM
        IRE TRANS. COMPONENTS AND PARTS, VOL. CP - 9, NO. 4, PP. 180 - 192,
        DECEMBER 1962.
NICOLET, M.-A.AND J. GOLDER
    THERMAL NOISE CALCULATION OF SINGLE-CARRIER SPACE-CHARGE LIMITED
    CURRENT LN A NON-INSULATING SOLID
        PHYSICA STATUS SOLIDI, VOL. 15, NO. 2, PP. 565-572, FEBRUARY 16,
        1973.
NOUGIER, J. P. AND M. ROLLAND
    MOBILITY, NOISE TEMPERATURE, AND DIFFUSIVITY OF HOT HOLES
    IN GERMANIUM
        PHYSICAL REVIEW B, VOL. 8, NO. 12, PP. 5728-5737, 15 DECEMBER
        1973.
NYQUIST, H.
    THERMAL AGITATION OF ELECTRIC CHARGE IN CONDUCTORS
        PHYS. REV. VOL. 32, NO. 1, PP. 110-113, JULY 1928.

PLATONOV, A. A. AND R. L. STRATONOVICH
    COLLIDING PARTICLE MODEL IN THE THEORY OF NONLINEAR CONDUCTIVITY
    AND ELECTRICAL FLUCTUATIONS
        RADIO ENGINEERING AND ELECTRONIC PHYSICS, VOL. 18, NO. 2,
        PP. 238-243, FEBRUARY 1973.
POZHELA, YU. K., V. A. BAREIKIS AND I. B. MATULENENE
    ELECTRICAL FLUCTUATIONS OF HOT ELECTRONS IN SILICON
        SOVIET PHYS. - SEMICONDUCTORS, VOL. 2, NO. 4, PP. 503-504,
        OCTOBER 1968.
PRICE, P. J.
    NOISE THEORY FOR HOT ELECTRONS
        IBM JOUR. RES. DEVEL., VOL. 3, NO. 2, PP. 191-193, APRIL 1959.
SENITZKY, B.
    NOISE IN THE PRESENCE OF SMALL SIGNALS
        IEEE TRANS. MICROWAVE THEORY AND TECHNIQUES, VOL. MTT-16, NO. 9,
        PP. 728-732, SEPTEMBER 1968.
SHARMA, Y.K. AND V.K. SHARMA,
    THERMAL NOISE IN THE RADIAL FLOW OF CURRENT IN THE INSULATOR,
        ACTA PHYSICA, VOL 37, NO 1-2, PP 181-182, 1974.
SHARMA, Y. K. AND B. B. SRIVASTAVA
    THERMAL NOISE IN THE TRAP-FILLED DIODE
        JOUR. APPLIED PHYSICS, VOL. 45, NO. 9, P. 4123, SEPTEMBER 1974.
SHARMA, Y. K., B. B. SRIVASTAVA, AND V. K. SHARMA
    THERMAL NOISE IN AN INSULATED DIODE WITH MOBILITY AS A FUNCTION OF
    CARRIER DENSITY
        CANADIAN JOUR. PHYSICS, VOL. 52, NO. 18, P. 1844-1845,
        15 SEPTEMBER 1974.
SIEGMAN, A. E.
    THERMAL NOISE IN MICROWAVE SYSTEMS- PARTS I,II AND III.
        MICROWAVE JOURNAL, VOL. 4, NO. 3, PP. 81-90, MARCH 1961.
        MICROWAVE JOURNAL, VOL. 4, NO. 4, PP. 66-73, APRIL 1961.
        MICROWAVE JOURNAL, VOL. 4, NO. 5, PP. 93-104, MAY 1961.
SOLOMON, S. S.
    THERMAL AND SHOT FLUCTUATIONS IN ELECTRICAL CONDUCTORS AND VACUUM
    TUBES
        JOUR. APPL. PHYS., VOL. 23, NO. 1, PP. 109-112, JANUARY 1952.
STRATONOVICH, R. L. AND A. A. PLATONOV
    THERMAL FLUCTUATIONS IN NONLINEAR MACROSCOPIC ELECTRODYNAMICS
        RADIO ENGINEERING AND ELECTRONIC PHYSICS, VOL. 18, NO. 2,
        PP. 231-238, FEBRUARY 1973.
TANDON, J. L., H. R. BILGER, AND M.-A. NICOLET
    THERMAL EQUILIBRIUM NOISE OF SPACE-CHARGE LIMITED CURRENT IN SILICON
    FOR HOLES WITH FIELD-DEPENDENT MOBILITY
        SOLID-STATE ELECTRONICS, VOL. 18, NO. 2, PP. 113-118, FEBRUARY 1975.
TUNALEY, J. K. E.
    CLASSICAL MICROSCOPIC MODELS OF JOHNSON NOISE
        AMERICAN JOUR. PHYSICS, VOL. 43, NO. 5, PP. 446-448, MAY 1975.
TWISS, R. Q.
    NYQUIST'S AND THEVENIN'S THEOREMS GENERALIZED FOR NONRECIPROCAL
    LINEAR NETWORKS
        JOUR. APPLIED PHYSICS, VOL. 26, NO. 5, PP. 599-602, MAY 1955.
ULLERSMA, P.
    AN EXACTLY SOLVABLE MODEL FOR BROWNIAN MOTION, PTS. I TO IV.
        PHYSICA, VOL. 32, NO. 1, PP. 27-96, JANUARY 1966.
VAN DER ZIEL, A.
    THERMAL NOISE AT HIGH FREQUENCIES
        JOUR. APPL. PHYS., VOL. 21, NO. 5, PP. 399-401, MAY 1950.
VAN DER ZIEL, A.
    THERMAL NOISE IN FIELD EFFECT TRANSISTORS
        PROC. IRE, VOL. 50, NO. 8, PP. 1808-1812, AUGUST 1962.
VAN DER ZIEL, A.
    THERMAL NOISE IN SPACE CHARGE LIMITED DIODES
        SOLID STATE ELECTRONICS, VOL. 9, NO. 9, PP. 899-900, SEPTEMBER 1966.
VAN DER ZIEL, A.
    H.F. THERMAL NOISE IN SPACE-CHARGE-LIMITED SOLID STATE DIODES
        SOLID STATE ELECTRONICS, VOL. 9, NO. 11-12, PP. 1139-1140,
        NOVEMBER-DECEMBER 1966.
VAN DER ZIEL, A.
    MORE RIGOROUS PROOF OF THE THERMAL NOISE FORMULA FOR DOUBLE-INJECTION
    DIODES FOR WT 1
        ELECTRONICS LETTERS, VOL. 6, NO. 8, PP. 231-232, APRIL 16, 1970.
VAN DER ZIEL, A.
    NYQUIST THEOREM FOR NON-LINEAR RESISTORS
        SOLID-STATE ELECTRONICS, VOL. 16, NO. 6, PP. 751-752, JUNE 1973.
VAN DER ZIEL, A. AND K. M. VAN VLIET
    H.F. THERMAL NOISE IN SPACE CHARGE LIMITED SOLID STATE DIODES - II
        SOLID-STATE ELECTRONICS, VOL. 11, NO. 4, PP. 508-509, APRIL 1968.
VAN NIE, A. G.
    REPRESENTATION OF LINEAR PASSIVE NOISY 1-PORT BY TWO CORRELATED
    NOISE SOURCES
        PROC. IEEE, VOL. 60, NO. 6, PP. 751-753, JUNE 1972.
VAN VLIET, K. M.
    DERIVATION OF THE FLUCTUATION DISSIPATION THEOREM
        PHYS. REV., VOL. 109, NO. 4, PP. 1021-1022, FEBRUARY 15, 1958.

VAN VLIET, K. M.
   IRREVERSIBLE THERMODYNAMICS AND CARRIER DENSITY FLUCTUATIONS IN
   SEMICONDUCTORS
      PHYSICAL REVIEW, VOL. 110, NO. 1, PP. 50-61, APRIL 1, 1958.
VAN VLIET, K. M.
   MARKOVIAN APPROACH TO DENSITY FLUCTUATIONS DUE TO TRANSPORT AND
   SCATTERING. I. MATHEMATICAL FORMALISM
               II. APPLICATIONS
      JOUR. MATHEMATICAL PHYSICS, VOL. 12, NO. 9, PP. 1981-1998 AND
      PP. 1998-2012, SEPTEMBER 1971.
VAN VLIET, K. M. AND E. R. CHENNETTE
   NOISE SPECTRA RESULTING FROM DIFFUSION PROCESSES IN A CYLINDRICAL
   GEOMETRY
      PHYSICA, VOL. 31, NO. 7, PP. 985-1001, JULY 1965.
VAN VLIET, K. M. AND A. VAN DER ZIEL
   ON THE NOISE GENERATED BY DIFFUSION MECHANISM;
      PHYSICA, VOL. 24, NO. 6, PP. 415-421, JUNE 1958.
      CORRECTION - PHYSICA, VOL. 24, NO. 7, P. 556, JULY 1958.
VANWORMHOUD, M. AND H. A. HAUS
   THERMAL NOISE IN LINEAR, LOSSY, ELECTROMAGNETIC MEDIA
      JOUR. APPLIED PHYSICS, VOL. 33, NO. 8, PP. 2572-2577, AUGUST 1962.
VYSIN, V. AND V. JANKU
   NOTE ON THE FLUCTUATIONS IN SYSTEMS WITH NEGATIVE ABSOLUTE TEMPERATURES,
      PHYS. LETTERS, VOL 9, NO 1, PP 19-20, 15 MARCH 1964.
WANG, M. C. AND G. E. UHLENBECK
   ON THE THEORY OF BROWNIAN MOTION
      PHYSICAL REVIEW, VOL. 36, NO. 5, PP. 823-841, SEPTEMBER 1, 1930.
WANG, M. C. AND G. E. UHLENBECK
   ON THE THEORY OF BROWNIAN MOTION. II.
      REVIEWS OF MODERN PHYSICS, VOL. 17, NO. 2/3, PP. 323-342, 1945.
WEBER, J.
   FLUCTUATION THEOREM
      PHYSICAL REVIEWS, VOL. 101, NO. 6, PP. 1620-1626, MARCH 15, 1956.
YAMAMOTO, S., S. T. LIU AND A. VAN DER ZIEL
   THERMAL NOISE IN GE P-V-P SCL DIODES
      APPL. PHYS. LETTERS, VOL. 11, NO. 4, PP. 140-141, AUGUST 15, 1967.

****
   3-B. 1/F NOISE
****

ABDUVAKHIDOV, KH. M., A. S. VOLKOV AND V. V. GALAVAVANOV
   INVESTIGATION OF THE PHOTOCONDUCTIVITY KINETICS AND OF THE NOISE
   SPECTRUM OF N-TYPE INSB
      SOVIET PHYSICS - SEMICONDUCTORS, VOL. 2, NO. 1, PP. 103-105,
      JULY 1968.
ABOWITZ, G., E. ARNOLD, AND E. A. LEVENTHAL
   SURFACE STATES AND 1/F NOISE IN MOS TRANSISTORS
      IEEE TRANS. ELECTRON DEVICES, VOL. ED-14, NO. 11, PP. 775-777,
      NOVEMBER 1967.
AFANAS'EV, V. F.
   INFLUENCE OF DISLOCATIONS ON EXCESS NOISE OF SURFACE-BARRIER
   GOLD-SILICON JUNCTIONS
      SOVIET PHYSICS - SEMICONDUCTORS, VOL. 4, NO. 1, PP. 99-101, JULY
      1970.
AGU, M.
   A SIMPLE MODEL FOR THE 1/F-TYPE POWER SPECTRUM
      PHYSICS LETTERS A, VOL. 51A, NO. 2, PP. 77-78, FEBRUARY 10, 1975.
ALBERDING, N.
   RANDOM WALK MODELS OF SEMICONDUCTOR NOISE
      JOUR. APPLIED PHYSICS, VOL. 44, NO. 4, PP. 1911-1912, APRIL 1973.
ALBERDING, N., J. K. E. TUNALEY.
   COMPARISONS OF THEORY WITH SIMULATIONS OF 1/F NOISE.
      JOUR. APPLIED PHYSICS, VOL. 44, NO. 8, PP. 3788-3789, AUGUST 1973.
ALLAN, D. W.
   STATISTICS OF ATOMIC FREQEUNCY STANDARDS
      PROC. IEEE, VOL. 54, NO. 2, PP. 221-230, FEBRUARY 1966.
AMAKASU, K. AND M. ASANO
   TEMPERATURE DEPENDENCE OF FLICKER NOISE OF P-N-P JUNCTION TRANSISTORS
      JOUR. APPL. PHYS., VOL. 27, NO. 10, P. 1249, OCTOBER 1956.
AMIRYAN, R. A.
   THE FLICKER NOISE OF A THERMIONIC OXIDE CATHODE
      RADIO ENGINEERING AND ELECTRONIC PHYSICS, VOL. 14, NO. 11,
      PP. 1758-1762, NOVEMBER 1969.
AMIRYAN, R. A. AND M. D. VOROB'EV
   ANOMALOUS FLICKER NOISE IN OXIDE CATHODE TUBES
      RADIO ENGINEERING AND ELECTRONIC PHYSICS, VOL. 14, NO. 4,
      PP. 735-737, APRIL 1969.
ANDERSON, L. H. AND K. M. VAN VLIET
   NOISE IN ZNO   A SURFACE PHENOMENON
      PHYSICA, VOL. 34, NO. 3, PP. 445-455, 1967.

ATTKINSON, W. R., L. FEY, AND J. NEWMAN
    SPECTRUM ANALYSIS OF EXTREMELY LOW FREQUENCY VARIATIONS OF QUARTZ
    OSCILLATORS
        PROC. IEEE, VOL. 51, NO. 2, P. 379, FEBRUARY 1963.
BADLINGER, E. AND E. NUESCH
    RAUSCHEN VON TRANSISTOREN BEI SEHR TIEFEN FREQUENZEN
        HELV. PHYS. ACTA, VOL. 41, PP. 313-322, 1968.
BAKER, D.
    FLICKER NOISE IN GERMANIUM RECTIFIERS AT VERY LOW AND AUDIO FREQUENCIES,
        JOUR. APPLIED PHYSICS, VOL. 25, NO. 7, PP. 922-924, JULY 1954.
BARBER, D.
    MEASUREMENTS OF CURRENT NOISE IN LEAD SULPHIDE AT AUDIO FREQUENCIES
        PROC. PHYSICAL SOC. (LONDON), VOL. B68, PT. 11, PP. 898-907,
        NOVEMBER 1955.

BARNES, J. A. AND D. W. ALLEN
    A STATISTICAL MODEL OF FLICKER NOISE
        PROC. IEEE, VOL. 54, NO. 2, PP. 176-178, FEBRUARY 1966.
BARNES, J. A.
    ATOMIC TIMEKEEPING AND THE STATISTICS OF PRECISION SIGNAL GENERATORS
        PROC. IEEE, VOL. 54, NO. 2, PP. 207-220, FEBRUARY 1966.
BARNES, J. A. AND S. JARVIS, JR.
    EFFICIENT NUMERICAL AND ANALOG MODELING OF FLICKER NOISE PROCESSES
        NBS TECHNICAL NOTE 604, NATIONAL BUREAU OF STANDARDS, U. S. DEPT.
        OF COMMERCE, BOULDER, COLO., U.S.A., JUNE 1971.
BELL, D. A.
    DISTRIBUTION FUNCTION OF SEMICONDUCTOR NOISE
        PROC. PHYSICAL SOC. (LONDON), VOL. B68, PT. 9, PP. 690-691,
        SEPTEMBER 1955.
BELL, D. A.
    SEMICONDUCTOR NOISE AS A QUEUING PROBLEM
        PROC. PHYS. SOC., VOL. 72, PT. 1, PP. 27-32, JULY 1958.
BELL, D. A.
    THE NOISINESS OF MATERIALS FOR FIXED RESISTORS
        PROC. IEE (LONDON), VOL. 109B, SUPPL. 21, PP. 9-13, 1962.
BELL, D. A.
    COMMENTS ON 1/F NOISE GENERATED BY HIGH-FREQUENCY CURRENTS
        ELECTRONICS LETTERS, VOL. 12, NO. 9, P. 235, APRIL 29, 1976.

BELL, D. A. AND K. Y. CHONG
    CURRENT NOISE IN COMPOSITION RESISTORS
        WIRELESS ENGINEER, VOL. 31, NO. 6, PP. 142-144, JUNE 1954.
BELL, D. A. AND S. P. B. DISSANAYAKE
    VARIANCE FLUCTUATIONS OF 1/F NOISE
        ELECTRONICS LETTERS, VOL. 11, NO. 13, P. 274, JUNE 26, 1975.
BELL, J. H.
    NOISE IN SEMICONDUCTOR MATERIALS AND DEVICES
        PROC. NATIONAL ELECTRONICS CONFERENCE, VOL. 13, 1957, PP. 218-234.
BERNAMONT, J.
    FLUCTUATIONS DE RESISTANCE DANS UN CONDUCTEUR METALLIQUE DE FAIBLE
    VOLUMNE
        COMPTES RENDUS ACAD. SCI., VOL. 198, PP. 1755-1758, 1934.
BERNAMONT, J.
    FLUCTUATIONS IN THE RESISTANCE OF THIN FILMS
        PROC. PHYSICAL SOC. (LONDON), VOL. 49 SUPPL., PP. 138-139, JULY 1937.
BERZ, F.
    THEORY OF LOW FREQUENCY NOISE IN SI MOST'S
        SOLID-STATE ELECTRONICS, VOL. 13, NO. 5, PP. 631-647, MAY 1970.
BERZ, F.
    FLUCTUATIONS IN NOISE LEVEL
        ELECTRONICS LETTERS, VOL. 8, NO. 21, PP. 515-517, 19 OCTOBER 1972.

BERZ, F. AND C. G. PRIOR
    TEST OF MC WHORTER'S MODEL OF LOW-FREQUENCY NOISE IN SI MOST'S
        ELECTRONICS LETTERS, VOL. 6, NO. 19, PP. 595-597, SEPTEMBER 17,
        1970.
BESS, L.
    STUDY OF 1/F NOISE IN SEMICONDUCTOR FILAMENTS
        PHYSICAL REV., VOL. 103, NO. 1, PP. 72-82, JULY 1, 1956.
BESS, L.
    A POSSIBLE MECHANISM FOR 1/F NOISE GENERATION IN SEMICONDUCTOR
    FILAMENTS
        PHYS. REV., VOL. 91, NO. 6, P. 1569, SEPTEMBER 15, 1953.
BESS, L.
    RELATIVE INFLUENCE OF MAJORITY AND MINORITY CARRIERS ON EXCESS NOISE
    IN SEMICONDUCTOR FILAMENTS
        JOUR. APPLIED PHYSICS, VOL. 26, NO. 11, PP. 1377-1381, NOVEMBER
        1955.
BESS, L. AND L. S. KISNER
    INVESTIGATION OF 1/F NOISE SPECTRA
        JOUR. APPL. PHYS., VOL. 37, NO. 9, PP. 3451-3462, AUGUST 1966.

BHAT, T. A. R.
    CURRENT NOISE IN SI-MOSFETS
        ARCH. ELEKTRONIK UND UBERTRANGUNGSTECHNIK, VOL. 26, NO. 7/8,
        PP. 343-348, JULY/AUGUST 1972.

BIRD, J. F.
    NOISE SPECTRUM ANALYSIS OF A MARKOV PROCESS VS RANDOM WALK COMPUTER
    SOLUTIONS SIMULATING 1/F NOISE SPECTRA
        JOUR. APPLIED PHYSICS, VOL. 45, NO. 1, PP. 499-500, JANUARY 1974.
BITTEL, H. AND N. MULLER
    SO-CALLED 1/DELTA F NOISE
        APPLIED PHYSICS, VOL. 5, NO. 3, PP. 283-284, DECEMBER 1974.
BITTEL, H. AND L. STORM,
    INVESTIGATIONS OF CURRENT NOISE IN RESISTORS (IN GERMAN)
        ZEIT. ANGEW. PHYS., VOL. 7, NO. 1, PP. 27-32, JANUARY 1955.
BLANC-LAPIERRE, A. AND A. FIFONTOFF
    MECANISME DE CONDUCTIBILITE DANS LES LAMES MINCES GRANULAIRES ET
    RELATION AVEC L'EFFECT DE SCINTILLATION
        JOUR. PHYSIQUE RADIUM, VOL. 17, NO. 3, PP. 230-236, MARCH 1956.
BLOODWORTH, G. G. AND R. J. HAWKINS
    THE PHYSICAL BASIS OF CURRENT NOISE
        RADIO AND ELECTRONIC ENGINEER, VOL. 38, NO. 1, PP. 17-22, JULY
        1969.
BLOODWORTH, G. G. AND R. J. HAWKINS
    DRIFT AND LOW-FREQUENCY NOISE
        RADIO AND ELECTRONIC ENGINEER, VOL. 41, NO. 2, PP. 61-64,
        FEBRUARY 1971.
BOREL, J. P.
    EFFET DE SCINTILLATION ET STRUCTURE DE COUCHES MINCES METALLIQUES
        JOUR. PHYSIQUE RADIUM, VOL. 17, NO. 3, PP. 224-229, MARCH 1956.
BOZIC, S. M.
    NOISE IN METAL-OXIDE-SEMICONDUCTOR TRANSISTOR
        ELECTRONIC ENGINEERING, VOL. 38, NO. 455, PP. 40-41, JANUARY 1966.
BROPHY, J. J.
    CURRENT NOISE IN THERMISTOR BOLOMETER FLAKES
        JOUR. APPLIED PHYSICS, VOL. 25, NO. 2, PP. 222-224, FEBRUARY 1954.
BROPHY, J. J.
    EXCESS NOISE IN DEFORMED GERMANIUM
        JOUR. APPLIED PHYSICS, VOL. 27, NO. 11, PP. 1383-1384, NOVEMBER
        1956.
BROPHY, J. J.
    EXCESS NOISE IN N-TYPE GERMANIUM
        PHYSICAL REV., VOL. 106, NO. 4, PP. 675-678, MAY 15, 1957.
BROPHY, J. J.
    SEEBECK EFFECT FLUCTUATIONS IN GERMANIUM
        PHYSICAL REV., VOL. 111, NO. 4, PP. 1050-1052, AUGUST 15, 1958.
BROPHY, J. J.
    EXPERIMENT SHOWING THE INFLUENCE OF SURFACES ON 1/F NOISE IN
    GERMANIUM
        JOUR. APPL. PHYS., VOL. 29, NO. 9, PP. 1377-1378, SEPTEMBER 1958.
BROPHY, J. J.
    CRYSTALLINE IMPERFECTIONS AND 1/F NOISE
        PHYSICAL REV., VOL. 115, NO. 5, PP. 1122-1125, SEPTEMBER 1, 1959.
BROPHY, J. J.
    TRAPPING AND DIFFUSION IN THE SURFACE REGION OF CADMIUM SULFIDE
        PHYSICAL REV., VOL. 119, NO. 2, PP. 591-596, JULY 15, 1960.
BROPHY, J. J.
    CURRENT NOISE AND DISTRIBUTED TRAPS IN CADMIUM SULFIDE
        PHYSICAL REV., VOL. 122, NO. 1, PP. 26-30, APRIL 1, 1961.
BROPHY, J.J.,
    CURRENT NOISE IN PYROLYSED POLYACRYLONITRILE,
        IN ORGANIC SEMICONDUCTORS (NEW YORK MACMILLAN,1962),PP 169-179.
BROPHY, J. J.
    STATISTICS OF 1/F NOISE
        PHYSICAL REV., VOL. 166, NO. 3, PP. 827-831, FEBRUARY 15, 1968.
BROPHY, J. J.
    ZERO CROSSING STATISTICS OF 1/F NOISE
        JOUR. APPLIED PHYS., VOL. 40, NO. 2, PP. 567-569, FEBRUARY 1969.
BROPHY, J. J.
    VARIANCE FLUCTUATIONS IN FLICKER NOISE AND CURRENT NOISE
        JOUR. APPL. PHYS., VOL. 40, NO. 9, PP. 3551-3553, AUGUST 1969.
BROPHY, J. J.
    LOW-FREQUENCY VARIANCE NOISE
        JOUR. APPLIED PHYSICS, VOL. 41, NO. 7, PP. 2913-2916, JUNE 1970.
BROPHY, J. J.
    STATISTICAL RANDOMNESS AND 1/F NOISE
        NAVAL RESEARCH REV., VOL. 23, NO.   , PP. 8- , 1970.
BROPHY, J. J. AND R. J. ROBINSON
    FREQUENCY FACTOR AND ENERGY DISTRIBUTION OF SHALLOW TRAPS IN CADMIUM
    SULFIDE
        PHYSICAL REV., VOL. 118, NO. 4, PP. 959-966, MAY 15, 1960.
BROPHY, J. J. AND N. ROSTOKER
    HALL EFFECT NOISE
        PHYSICAL REV., VOL. 100, NO. 2, PP. 754-756, OCTOBER 15, 1955.
BROUX, G., R. VAN OVERSTRAETEN AND G. DECLERCK
    EXPERIMENTAL RESULTS ON FAST SURFACE STATES AND 1/F NOISE IN MOS-
    TRANSISTORS
        ELECTRONICS LETTERS, VOL. 11, NO. 5, PP. 97-98, MARCH 6, 1975.
BRUNCKE, W. C., E. R. CHENETTE, AND A. VAN DER ZIEL
    TRANSISTOR NOISE AT LOW TEMPERATURES

IEEE TRANS. ELECTRON DEVICES, VOL. ED-11, NO. 2, PP. 50-53,
FEBRUARY 1964.
BULL, C. S.
    SPACE CHARGE AS A SOURCE OF FLICKER EFFECT
        PROC. IEE, VOL. 105, PT. B, PP. 190-194, MARCH 1958.
BULL, C. S. AND S. M. BOZIC
    EXCESS NOISE IN SEMICONDUCTING DEVICES DUE TO FLUCTUATIONS IN THEIR
    CHARACTERISTICS WHEN SIGNALS ARE APPLIED
        BRIT. JOUR. APPLIED PHYSICS, VOL. 18, NO. 7, PP. 883-895, JULY 1967
BURCKHART, C. B. AND M. J. O. STRUTT
    NOISE IN NONRECIPROCAL TWO PORTS BASED ON HALL EFFECT
        IEEE TRANS. ELECTRON DEVICES, VOL. ED-11, NO. 2, PP. 47-50,
        FEBRUARY 1964.
BURGESS, R. E.
    CONTACT NOISE IN SEMICONDUCTORS
        PROC. PHYSICAL SOC. (LONDON), VOL. B66, PT. 4, PP. 334-335, APRIL
        1953.
BURGESS, R. E.
    ELECTRONIC FLUCTUATIONS IN SEMICONDUCTORS
        BRIT. JOUR. APPL. PHYS., VOL. 6, NO. 6, PP. 185-190, JUNE 1955.
BURGESS, R. E.
    FLUCTUATIONS IN NUMBER OF CHARGE CARRIERS IN A SEMICONDUCTOR
        PHYSICA, VOL. 20, PP. 1007-1010, 1955.
BURGESS, R. E.
    FLUCTUATIONS IN NUMBER OF ELECTRONS AND HOLES IN A SEMICONDUCTOR
        PROC. PHYSICAL SOC. (LONDON), VOL. B68, PT. 9, PP. 661-671,
        SEPTEMBER 1955.
BURGESS, R. E.
    THE STATISTICS OF CHARGE CARRIER FLUCTUATIONS IN SEMICONDUCTORS
        PROC. PHYSICAL SOC. (LONDON), VOL. B69, PT. 10, PP. 1020-1027,
        OCTOBER 1956.
BUTTERWECK, H. J.
    NOISE VOLTAGES OF BULK RESISTORS DUE TO RANDOM FLUCTUATIONS
    OF CONDUCTIVITY
        PHILIPS RESEARCH REPTS., VOL. 30, SPECIAL ISSUE, PP. 316-321, 1975.
BUTZ, A. R.
    A THEORY OF 1/F NOISE.
        JOUR. STATISTICAL PHYSICS, VOL. 4, NO. 2-3, PP. 199-216,
        MARCH-APRIL 1973.
BYRNE, J.
    NOISE SPECTRUM OF A STATIONARY CURRENT WITH FLUCTUATING MEAN
        JOUR. APPL. PHYS., VOL. 38, NO. 5, PP. 2247-2251, APRIL 1967.
BYRNE, J.
    A PROPOSAL CONCERNING THE ORIGINS OF 1/F CURRENT NOISE IN
    SEMICONDUCTORS
        PHYSICS LETTERS, VOL. 30A, NO. 8, PP. 457-458, DECEMBER 15, 1969.
CALOYANNIDES, M. A.
    MICROCYCLE SPECTRAL ESTIMATES OF 1/F NOISE IN SEMICONDUCTORS
        JOUR. APPLIED PHYSICS, VOL. 45, NO. 1, PP. 307-316, JANUARY 1974.
        PROC. IRE, VOL. 37, NO. 8, PP. 938-942, AUGUST 1949.
CAMPBELL, R. H. AND R. A. CHIPMAN
    NOISE FROM CURRENT CARRYING RESISTORS, 20 TO 300 KC
CARINGELLA, P.C. AND W.L. EISENMAN,
    SYSTEM FOR LOW-FREQUENCY NOISE MEASUREMENTS.
        REVIEWS OF SCIENTIFIC INSTRUMENTS, VOL. 33, NO. 6, PP. 654-655,
        JUNE 1962.
CARRUTHERS, T.
    BIAS-DEPENDENT STRUCTURE IN EXCESS NOISE IN GAAS SCHOTTKY TUNNEL
    JUNCTIONS
        APPLIED PHYSICS LETTERS, VOL. 18, NO. 1, PP. 35-37, JANUARY 1, 1971.

CARUTHERS, J. W. AND J. S. HAM
    ELECTRICAL FLUCTUATIONS IN IODINE
        JOUR. APPLIED PHYSICS, VOL. 41, NO. 5, PP. 1984-1989, APRIL 1970.

CELASCO, M. AND F. FIORILLO
    CURRENT NOISE MEASUREMENTS IN CONTINUOUS METAL THIN FILMS
        APPLIED PHYSICS LETTERS, VOL. 26, NO. 4, PP. 211-212, FEBRUARY 15,
        1975.
CHENNETE,E.R.,
    MEASUREMENT OF THE CORRELATION BETWEEN FLICKER NOISE SOURCES IN
    TRANSISTORS
        PROC. IRE, VOL. 46, NO. 6, P. 1304, JUNE 1958.
CHOUDHURY, N. K. AND A. DAW
    LOW-FREQUENCY NOISE OF TRANSISTORS AT LOW CURRENT
        INDIAN JOUR. PURE & APPLIED PHYSICS, VOL. 8, NO. 10, PP. 656-658,
        OCTOBER 1970.
CHRISTENSEN, C. J. AND G. L. PEARSON
    SPONTANEOUS RESISTANCE FLUCTUATIONS IN CARBON MICROPHONES AND OTHER
    GRANULAR RESISTANCES
        BELL SYSTEM TECH. JOUR., VOL. 15, NO. 2, PP. 197-223, APRIL 1936.
CHRISTENSSON, S. AND I. LUNDSTROM
    LOW FREQUENCY NOISE IN MOS TRANSISTORS - II. EXPERIMENTS
        SOLID-STATE ELECTRONICS, VOL. 11, NO. 9, PP. 813-820, SEPTEMBER
        1968.
CHRISTENSSON, S., I. LUNDSTROM, AND C. SVENSSON
    LOW FREQUENCY NOISE IN MOS TRANSISTORS - I. THEORY

        SOLID-STATE ELECTRONICS, VOL. 11, NO. 9, PP. 797-812, SEPTEMBER
        1968.
CLARKE, J. AND G. HAWKINS
    LOW FREQUENCY NOISE IN JOSEPHSON JUNCTIONS
        IEEE TRANS. MAGNETICS, VOL. MAG-11, NO. 2, PP. 841-844, MARCH 1975.
CLARKE, J. AND T. Y. HSIANG
    LOW-FREQUENCY NOISE IN TIN FILMS AT THE SUPERCONDUCTING TRANSITION
        PHYSICAL REV. LETTERS, VOL. 34, NO. 19, PP. 1217-1220, 12 MAY 1975.
CLARKE, J. AND R. F. VOSS
    1/F NOISE FROM THERMAL FLUCTUATIONS IN METAL FILMS
        PHYSICAL REV. LETTERS, VOL. 33, NO. 1, PP. 24-27, JULY 1, 1974.
CLAY, J. R. AND M. F. SHLESING
    ANALYTICAL THEORY OF 1/F NOISE IN NEURAL MEMBRANE
        FEDERATION PROC., VOL. 33, NO. 5, PP. 1339-   , 1974.
CLAY, J. R. AND M. F. SHLESINGER
    THEORETICAL MODEL OF IONIC MECHANISM OF 1/F NOISE IN NERVE MEMBRANE -
        BIOPHYSICS JOUR., VOL. 16, NO. 2, PP. 121-136, 1976.
CONRAD, G. T., JR.
    A PROPOSED CURRENT-NOISE INDEX FOR COMPOSITION RESISTORS
        IRE TRANS. COMPONENTS AND PARTS, VOL. CP-3, NO. 1, PP. 14-20,
        MARCH 1956.
CONTI, M.
    SURFACE AND BULK EFFECTS IN LOW FREQUENCY NOISE IN NPN PLANAR
    TRANSISTORS
        SOLID-STATE ELECTRONICS, VOL. 13, NO. 11, PP. 1461-1469,
        NOVEMBER 1970.
CUTLER, L. S. AND C. L. SEARLE
    SOME ASPECTS OF THE THEORY AND MEASUREMENT OF FREQUENCY FLUCTUATIONS
    IN FREQUENCY STANDARDS
        PROC. IEEE, VOL. 54, NO. 2, PP. 136-154, FEBRUARY 1966.
DAS, D. M.
    FET NOISE SOURCES AND THEIR EFFECT ON AMPLIFIER PERFORMANCE AT LOW
    FREQUENCIES
        IEEE TRANS. ELECTRON DEVICES, VOL. ED-19, NO. 3, PP. 338-348,
        MARCH 1972.
DECKER, M., L. GONSOV, AND D. RIGAUD
    NOISE MEASUREMENTS IN METAL-OXIDE-SEMICONDUCTOR TRANSISTORS BELOW
    SATURATION
        ELECTRONICS LETTERS, VOL. 3, NO. 12, PP. 565-566, DECEMBER 1967.
DE FELICE, L. J.
    1/F RESISTOR NOISE
        JOUR. APPLIED PHYSICS, VOL. 47, NO. 1, PP. 350-352, JANUARY 1976.
DE FELICE, L. J. AND D. R. FIRTH
    SPONTANEOUS VOLTAGE FLUCTUATIONS IN GLASS MICROELECTRODES
        IEEE TRANS. BIOMEDICAL ENGG., VOL. BME-18, NO. 5, PP. 339-351,
        SEPTEMBER 1971.
DE LA FUENTE
    QUASISTATIONARY NOISE IN RESISTORS
        ELECTRONICS LETTERS, VOL. 5, NO. 12, PP. 263-265, JUNE 12, 1969.
DELL, R. A., M. EPSTEIN, AND C. R. KANNEWURF
    EXPERIMENTAL STUDY OF 1/F NOISE STATIONARITY BY DIGITAL TECHNIQUE
        JOUR. APPL. PHYS., VOL. 44, NO. 1, PP. 472 - 476, JANUARY 1973.
DERKSEN, H. E.
    AXON MEMBRANE VOLTAGE FLUCTUATIONS
        ACTA PHYSIOL. PHARMACOL. NEERL., VOL. 13, PP. 373-466, 1965.
DERKSEN, H. E. AND A. A. VERVEEN
    FLUCTUATIONS OF RESTING NEURAL MEMBRANE POTENTIAL
        SCIENCE, VOL. 151, NO. 3716, PP. 1388-1399, MARCH 18, 1966.
DILL, H. G.
    NOISE CONTRIBUTION OF THE OFFSET-GATE I.G.F.E.T. WITH AN ADDITIONAL
    GATE ELECTRODE
        ELECTRONICS LETTERS, VOL. 3, NO. 7, PP. 341-342, JULY 1967.
DORSET, D. L. AND H. M. FISHMAN
    EXCESS ELECTRICAL NOISE DURING CURRENT FLOW THROUGH POROUS MEMBRANES
    SEPARATING IONIC SOLUTIONS
        JOUR. MEMBRANE BIOLOGY, VOL. 21, NO. 3-4, PP. 291-309, 1975.
DU PRE, F. K.
    A SUGGESTION REGARDING THE SPECTRAL DENSITY OF FLICKER NOISE
        PHYSICAL REV., VOL. 78, NO. 5, P. 615, JUNE 1, 1950.
ELLIOTT, D. A.
    LOW TEMPERATURE NOISE IN MOST AMPLIFIERS
        SOLID-STATE ELECTRONICS, VOL. 14, NO. 10, PP. 1042-1047, OCTOBER
        1971.
ENG, S. T.
    A NEW LOW 1/F NOISE MIXER DIODE   EXPERIMENTS, THEORY AND PERFORMANCE
        SOLID-STATE ELECTRONICS, VOL. 8, NO. 1, PP. 59-77, JANUARY 1965.
EPSTEIN, M.
    CURRENT NOISE IN EVAPORATED FILMS OF INSB AND INAS
        JOUR. APPLIED PHYSICS, VOL. 36, NO. 8, PP. 2590-2591, AUGUST 1965.
FARRIS, J.J. AND J.M. RICHARDSON,
    EXCESS NOISE IN MICROWAVE DETECTOR DIODES,
        IRE TRANS. MICROWAVE THEORY AND TECHNIQUES, VOL. MTT-9, NO. 4,
        PP. 312-314, JULY 1961.
FAULKNER, E. A. AND D. W. HARDING
    FLICKER NOISE IN SILICON PLANAR TRANSISTORS

ELECTRONICS LETTERS, VOL. 3, NO. 2, PP. 71-72, FEBRUARY 1967.
CORRECTION IN VOL. 3, NO. 4, P. 176, APRIL 1967.

FAULKNER, E. A. AND M. L. MEADE
FLICKER NOISE IN GUNN DIODES
ELECTRONICS LETTERS, VOL. 4, NO. 11, PP. 226-227, MAY 31, 1968.

FEIGT, I. AND O. JANTSCH
THE UPPER FREQUENCY LIMIT OF THE 1/F NOISE AND THE SURFACE RELAXATION
TIME
SOLID-STATE ELECTRONICS, VOL. 14, NO. 5, PP. 391-396, MAY 1971.

FETINA, V. N.
ANOMALIES IN THE LOW-FREQUENCY FLUCTUATIONS OBSERVED IN NARROW
GERMANIUM P-N JUNCTIONS
RADIO ENGINEERING AND ELECTRONIC PHYSICS, VOL. 11, NO. 9, PP.
1444-1448, SEPTEMBER 1966.

FIRLE, T. E. AND H. WINSTON
NOISE MEASUREMENTS IN SEMICONDUCTORS AT VERY LOW FREQUENCIES
JOUR. APPLIED PHYSICS, VOL. 26, NO. 6, PP. 716-717, JUNE 1955.

FLINN, I.
EXTENT OF THE 1/F NOISE SPECTRUM
NATURE, VOL. 219, PP. 1356-1357, SEPTEMBER 28, 1968.

FLINN, I., G. BEW, AND F. BERZ
LOW FREQUENCY NOISE IN M.O.S. FIELD EFFECT TRANSISTORS
SOLID-STATE ELECTRONICS, VOL. 10, NO. 8, PP. 833-845, AUGUST 1967.

FONGER, W.H.,
A DETERMINATION OF 1/F NOISE SOURCES IN SEMICONDUCTOR DIODES AND
TRIODES
IN   TRANSISTORS , VOL. 1, (RCA LABS, PRINCETON, N.J., 1956),
PP. 239-295

FRENKEL, I. YA
APPARATUS FOR MEASURING SPECTRAL DISTRIBUTION OF 1/F**(ALPHA) NOISE
OF P-N-P AND N-P-N TRANSISTORS IN 20 HZ - 30 KHZ FREQUENCY BAND
MEASUREMENT TECHNIQUES, VOL. 18, NO. 6, PP. 854-857, 1975.

FRY, P. W.
LOW-FREQUENCY NOISE MEASUREMENTS ON THE P-CHANNEL M.O.S.T.
ELECTRONIC ENGINEERING, VOL. 38, NO. 464, PP. 650-653, OCTOBER 1966

FU, H-S. AND C. T. SAH
THEORY AND EXPERIMENTS ON SURFACE 1/F NOISE
IEEE TRANS. ELECTRON DEVICES, VOL. ED-19, NO. 2, PP. 273-285,
FEBRUARY 1972.

GANEFEL'D, R. V. AND I. I. REPA
EXCESS NOISE IN SEMICONDUCTORS
SOVIET PHYSICS - SEMICONDUCTORS, VOL. 3, NO. 1, PP. 22-24, JULY
1969.

GIBBONS, J. F.
LOW-FREQUENCY NOISE FIGURE AND ITS APPLICATION TO THE MEASUREMENT OF
CERTAIN TRANSISTOR PARAMETERS
IRE TRANS. ELECTRON DEVICES, VOL. ED-9, NO. 3, PP. 308-315,
MAY 1962.

GREENSTEIN, L. J. AND J. J. BROPHY
INFLUENCE OF LOWER CUTOFF FREQUENCY ON THE MEASURED VARIANCE OF 1/F
NOISE
JOUR. APPLIED PHYSICS, VOL. 40, NO. 2, PP. 682-685, FEBRUARY 1969.

GUDKOV, I. D.
INVESTIGATION OF THE 1/F LOW FREQUENCY NOISE OF BACK BIASED
GERMANIUM P-N JUNCTIONS
RADIO ENGINEERING AND ELECTRONIC PHYSICS, VOL. 12, NO. 5,
PP. 880-882, MAY 1967.

HALFORD, D.
A GENERAL MECHANICAL MODEL FOR F **A SPECTRAL DENSITY RANDOM NOISE
WITH SPECIAL REFERENCE TO FLICKER NOISE 1/ F
PROC. IEEE, VOL. 56, NO. 3, PP. 251-258, MARCH 1968.

HALLADAY, E. H. AND W. C. BRUNCKE
EXCESS NOISE IN FIELD-EFFECT TRANSISTORS
PROC. IEEE, VOL. 51, NO. 11, P. 1671, NOVEMBER 1963.

HALLADAY, H. E. AND A. VAN DER ZIEL
TEST OF THERMAL-NOISE HYPOTHESIS IN M.O.S.F.E.T.S
ELECTRONICS LETTERS, VOL. 4, NO. 17, PP. 366-367, AUGUST 23, 1968.

HANDEL, P. H.
INSTABILITIES, TURBULENCE AND FLICKER-NOISE IN SEMICONDUCTORS
PHYSICS LETTERS, VOL. 18, NO. 3, PP. 224-225, SEPTEMBER 1, 1965.

HANDEL, P. H.
INSTABILITATEN, TURBULENZ UND FUNKELRAUSCHEN IN HALBLEITERN III.
TURBULENZ IM HALBLEITERPLASMA UND FUNKELRAUSCHEN
ZEIT. NATURFORSCH., VOL. 21A, NO. 5, PP. 579-593, MAY 1966.

HANDEL, P. H.
INSTABILITIES AND TURBULENCE IN SEMICONDUCTORS
PHYSICA STATUS SOLIDI, VOL. 29, NO. 1, PP. 299-306, SEPTEMBER 1,
1968.

HANDEL, P. H.
TURBULENCE THEORY FOR THE CURRENT CARRIERS IN SOLIDS AND A THEORY
OF 1/F NOISE
PHYSICAL REV. A, VOL. 3, NO. 6, PP. 2066-2073, JUNE 1971.

HANDEL, P. H.
1/F NOISE - AN INFRARED PHENOMENON

PHYSICAL REV. LETTERS, VOL. 34, NO. 24, PP. 1492-1495, 16 JUNE
1975
HANDEL, P. H.
NATURE OF 1/F PHASE NOISE
PHYSICAL REV. LETTERS, VOL. 34, NO. 24, PP. 1495-1498, 16 JUNE
1975
HANDEL, P. H.
QUANTUM THEORY OF 1/F NOISE
PHYSICS LETTERS, VOL. 53A, NO. 6, PP. 438-440, JULY 28, 1975.
HANDEL, P. H.
1/F MICROSCOPIC QUANTUM FLUCTUATIONS OF ELECTRIC CURRENTS DUE TO
BREMSSTRAHLUNG WITH INFRARED RADIATIVE CORRECTIONS
ZEITSCH. NATURFORSCH. A, VOL. 30A, NO. 9, PP. 1200-1201, SEPTEMBER
1975.
HANNAM, H.J. AND A. VAN DER ZIEL,
ON THE FLICKER NOISE GENERATED IN AN INTERFACE LAYER,
JOUR. APPLIED PHYSICS, VOL. 25, NO. 10, PP. 1336-1340, OCTOBER
1954.
HANNAM, H. J. AND A. VAN DER ZIEL
ANODE EFFECT   THE INFLUENCE OF ELECTRON BOMBARDMENT OF THE ANODE ON
FLICKER NOISE
IRE TRANS. ELECTRON DEVICES, VOL. ED-8, NO. 3, PP. 230-233, MAY
1961.
HANSON, D. G. AND A. VAN DER ZIEL
LOW-FREQUENCY NOISE IN GAS DISCHARGES
JOUR. APPLIED PHYSICS, VOL. 39, NO. 3, PP. 1683-1688, 15 FEBRUARY
1968.
HARRIS, E. J.
CIRCUIT AND CURRENT NOISE
ELECTRONIC ENGINEERING, VOL. 20, NO. 243, PP. 145-148, MAY 1948.
HARRIS, E. J. AND P. O. BISHOP
LOW-FREQUENCY NOISE FROM THERMIONIC VALVES WORKING UNDER AMPLIFYING
CONDITIONS
NATURE, VOL. 161, P. 971, JUNE 19, 1948.
HASLETT, J. W. AND E. J. M. KENDALL
TEMPERATURE DEPENDENCE OF LOW-FREQUENCY EXCESS NOISE IN JUNCTION-GATE
FETS
IEEE TRANS. ELECTRON DEVICES, VOL. ED-19, NO. , PP. 943- , 1972.
HAWKINS, R. J.
MODIFIED RANDOM-WALK MODEL FOR 1/F NOISE
JOUR. APPLIED PHYSICS, VOL. 43, NO. 3, PP. 1276-1277, MARCH 1972.
HAWKINS, R. J. AND G. G. BLOODWORTH
TWO COMPONENTS OF 1/F NOISE IN MOS TRANSISTORS
SOLID-STATE ELECTRONICS, VOL. 14, NO. 10, PP. 929-932, OCTOBER 1971.
HAWKINS, R. J. AND G. G. BLOODWORTH
MEASUREMENT OF LOW-FREQUENCY NOISE IN THICK-FILM RESISTORS
THIN SOLID FILMS, VOL. 8, NO. 3, PP. 193-197, SEPTEMBER 1971.
HAWKINS, R. J., I. R. M. MANSOUR, AND G. G. BLOODWORTH
THE SPECTRUM OF CURRENT NOISE IN M.O.S. TRANSISTORS AT VARY LOW
FREQUENCIES
JOUR. PHYSICS D (APPL. PHYS.), SER. 2, VOL. 2, NO. 8, PP.
1059-1062, AUGUST 1969.
HAYASHI, A. AND A. VAN DER ZIEL
CORRELATION COEFFICIENT OF GATE AND DRAIN FLICKER NOISE
SOLID-STATE ELECTRONICS, VOL. 17, NO. 6, PP. 637-639, JUNE 1974.
HERZOG, G. B. AND A. VAN DER ZIEL
SHOT NOISE IN GERMANIUM SINGLE CRYSTALS
PHYSICAL REV., VOL. 84, NO. 6, PP. 1249-1250, DECEMBER 15, 1951.
HOFFAIT, A. H. AND R. D. THORNTON
LIMITATIONS OF TRANSISTOR DC AMPLIFIERS
PROC. IEEE, VOL. 52, NO. 2, PP. 179-184, FEBRUARY 1964.
HOLDEN, A. V.
FLICKER NOISE AND STRUCTURAL-CHANGES IN NERVE MEMBRANE
JOUR. THEORETICAL BIOLOGY, VOL. 57, NO. 1, PP. 243-246, 1976.
HOOGE, F. N.
1/F NOISE IS NO SURFACE EEEECT
PHYSICS LETTERS, VOL. 29A, NO. 3, PP. 139-140, APRIL 21, 1969.
HOOGE, F. N., H. J. A. VAN DIJK, AND A. M. H. HOPPENBOUWERS
1/F NOISE IN EPITAXIAL SILICON
PHILIPS RESEARCH REPTS., VOL. 25, NO. 2, PP. 81-86, APRIL 1970.
HOOGE, F. N.
1/F NOISE IN THE CONDUCTANCE OF IONS IN AQUEOUS SOLUTIONS
PHYSICS LETTERS A, VOL. 33A, NO. 3, PP. 169-170, OCTOBER 19, 1970.
HOOGE, F. N.
DISCUSSION OF RECENT EXPERIMENTS ON 1/F NOISE
PHYSICA, VOL. 60, PP. 130-144, 1972.
HOOGE, F. N. AND J. L. M. GALL
FLUCTUATIONS WITH A 1/F SPECTRUM IN THE CONDUCTANCE OF IONIC
SOLUTIONS AND IN THE VOLTAGE OF CONCENTRATION CELLS
PHILIPS RESEARCH REPTS., VOL. 26, NO. 2, PP. 77-90, APRIL 1971.
HOOGE, F. N. AND J. L. M. GALL
EXPERIMENTAL STUDY OF 1/F NOISE IN THERMO E. M. F.
PHILIPS RESEARCH REPTS., VOL. 26, NO. 5, PP. 345-358, OCTOBER 1971.
HOOGE, F. N. AND A. M. H. HOPPENBROUWERS
CONTACT NOISE

PHYSICS LETTERS A, VOL. 29, PP. 642-643, 1969.

HOOGE, F. N. AND A. M. H. HOPPENBROUWERS
AMPLITUDE DISTRIBUTION OF 1/F NOISE
PHYSICA, VOL. 42, NO. 3, PP. 331-339, MAY 15, 1969.

HOOGE, F. N. AND A. M. H. HOPPENBROUWERS
1/F NOISE IN CONTINUOUS THIN GOLD FILMS
PHYSICA, VOL. 45, PP. 386-392, 1969.

HOOGE, F. N. AND T. G. M. KLEINPENNING
COMMENTS ON 'CURRENT NOISE MEASUREMENTS IN CONTINUOUS METAL THIN FILMS
APPLIED PHYSICS LETTERS, VOL. 27, NO. 3, P. 160, AUGUST 1, 1975.

HOPPENBROUWERS, A. M. H. AND F. N. HOOGE
1/F NOISE OF SPREADING RESISTANCES
PHILIPS RESEARCH REPTS., VOL. 25, NO. 2, PP. 69-80, APRIL 1970.

HSIEH, K. C., E. R. CHENETTE, AND A. VAN DER ZIEL
CURRENT NOISE IN SURFACE LAYER INTEGRATED RESISTORS
SOLID-STATE ELECTRONICS, VOL. 19, NO. 6, PP. 451-453, JUNE 1976.

HSU, S. T.
LOW FREQUENCY EXCESS NOISE IN METAL-SILICON SCHOTTKY BARRIER DIODES
IEEE TRANS. ELECTRON DEVICES, VOL. ED-17, NO. 7, PP. 496-506,
JULY 1970.

HSU, S. T.
SURFACE STATE RELATED 1/F NOISE IN P-N JUNCTIONS
SOLID-STATE ELECTRONICS, VOL. 13, NO. 6, PP. 843-855, JUNE 1970.

HSU, S. T.
SURFACE STATE RELATED 1/F NOISE IN MOS TRANSISTORS
SOLID-STATE ELECTRONICS, VOL. 13, NO. 11, PP. 1451-1459, NOVEMBER
1970.

HSU, S. T.
FLICKER NOISE IN METAL SEMICONDUCTOR SCHOTTKY BARRIER DIODES DUE TO
MULTISTEP TUNNELING PROCESSES
IEEE TRANS. ELECTRON DEVICES, VOL. ED-18, NO. 10, PP. 882-887,
OCTOBER 1971.

HSU, S. T., D. J. FITZGERALD, AND A. S. GROVE
SURFACE STATE RELATED 1/F NOISE IN P-N JUNCTIONS AND MOS TRANSISTORS
APPLIED PHYSICS LETTERS, VOL. 12, NO. 9, PP. 287-289, MAY 1, 1968.

HSU, S. T., R. J. WHITTIER, AND C. A. MEAD
PHYSICAL MODEL FOR BURST NOISE IN SEMICONDUCTOR DEVICES
SOLID-STATE ELECTRONICS, VOL. 13, NO. 7, PP. 1055-1071, JULY 1970.

HUANG, C. & A. VAN DER ZIEL
A NEW TYPE OF FLICKER NOISE IN MICROWAVE MOSFETS,
SOLID-STATE ELECTRONICS, VOL. 18, NO. 10, PP. 885-886, OCTOBER 1975.

HUEBENER, R. P. AND D. E. GALLUS
CURRENT-INDUCED INTERMEDIATE STEPS IN THIN-FILM TYPE-I
SUPERCONDUCTORS   ELECTRICAL RESISTANCE AND NOISE
PHYSICAL REV. B, VOL. 7, NO. 9, PP. 4089-4099, MAY 1, 1973.

HYDE, F. J.
MEASUREMENTS OF NOISE SPECTRA OF A POINT CONTACT GERMANIUM RECTIFIER
PROC. PHYSICAL SOC. (LONDON), VOL. B66, PT. 12, PP. 1017-1024,
DECEMBER 1953.

HYDE, F. J.
MEASUREMENT OF NOISE SPECTRA OF A GERMANIUM P-N JUNCTION DIODE
PROC. PHYSICAL SOC. (LONDON), VOL. B69, PP. 231-241, FEBRUARY 1956.

HYDE, F. J.
EXCESS NOISE SPECTRA IN GERMANIUM
PROC. PHYSICAL SOC. (LONDON), VOL. B69, PP. 242-245, FEBRUARY 1956.

INUISHI, Y. AND T. C. YANG
THE INFLUENCE OF IMPURITY ATOMS ON FLICKER NOISE,
JOUR. PHYSICAL SOC. JAPAN, VOL. 8, PP. 565-567, JULY-AUGUST 1953.

JAEGER, R. C. AND A. J. BRODERSEN
LOW-FREQUENCY NOISE SOURCES IN BIPOLAR JUNCTION TRANSISTORS
IEEE TRANS. ELECTRON DEVICES, VOL. ED-17, NO. 2, PP. 128-134,
FEBRUARY 1970.

JANTSCH, O.
A THEORY OF 1/F NOISE AT SEMICONDUCTOR SURFACES
SOLID-STATE ELECTRONICS, VOL. 11, NO. 2, PP. 267-272, FEBRUARY 1968

JANTSCH, O. AND I. FEIGT
1/F NOISE IN SILICON DIODE
SOLID-STATE ELECTRONICS, VOL. 16, NO. 12, PP. 1517 - 1520, DECEMBER
1973.

JANTSCH, O. AND I. FEIGT
1/F NOISE OBSERVED ON SEMICONDUCTOR SURFACES
PHYSICAL REV. LETTERS, VOL. 23, NO. 16, PP. 912-913, OCTOBER 20,
1969.

JOHNSON, J. B.
THE SCHOTTKY EFFECT IN LOW FREQUENCY CIRCUITS
PHYSICAL REVIEWS, VOL. 26, NO. 1, PP. 71-85, JULY 1925.

JONES, B. K.
1/F AND 1/DELTA F NOISE PRODUCED BY A RADIO-FREQUENCY CURRENT IN A
CARBON RESISTOR
ELECTRONICS LETTERS, VOL. 12, NO. 4, PP. 110-111, FEBRUARY 19, 1976.

JONES, B. K. AND J. D. FRANCIS
DIRECT CORRELATION BETWEEN 1/F NOISE AND OTHER NOISE SOURCES
JOUR. PHYSICS D, VOL. 8, NO. 10, PP. 1172-1176, JULY 11, 1975.

JONES, B. K. AND J. D. FRANCIS
HARMONIC GENERATION IN 1/F NOISE
JOUR. PHYSICS D, VOL. 8, NO. 16, PP. 1937-1940, NOVEMBER 11, 1975.

JONES, B. L. AND R. E. HURLSTON
    VHF NOISE DUE TO SURFACE STATES IN MOS DEVICES
        PROC. IEEE, VOL. 58, NO. 1, PP. 152-153, JANUARY 1970.
JORDAN, A. G. AND N. A. JORDAN
    THEORY OF NOISE IN METAL OXIDE SEMICONDUCTOR DEVICES
        IEEE TRANS. ELECTRON DEVICES, VOL. ED-12, NO. 3, PP. 148-156,
        MARCH 1965.
JORDAN, N. A.
    CURRENT NOISE IN EVAPORATED INAS POLYCRYSTALLINE FILMS
        SOLID-STATE ELECTRONICS, VOL. 10, NO. 5, PP. 503-504, MAY 1967.
KATZ, R. M. AND K. ROSE
    COMPARATIVE STUDIES OF NOISE LIMITATIONS IN SUPERCONDUCTING THIN-FILM
    RADIATION DETECTORS
        PROC. IEEE, VOL. 61, NO. 1, PP. 55-58, JANUARY 1973.
KEONJIAN, E. AND J. S. SCHAFFNER
    AN EXPERIMENTAL INVESTIGATION OF TRANSISTOR NOISE
        PROC. IRE, VOL. 40, NO. 11, PP. 1456-1460, NOVEMBER 1952.
KHAJEZADEH, H. AND T. T. MC CAFFREY
    MATERIAL AND PROCESS CONSIDERATIONS FOR MONOLITHIC LOW-1/F-NOISE
    TRANSISTORS
        PROC. IEEE, VOL. 57, NO. 9, PP. 1518-1522, SEPTEMBER 1969.
KISER, K.M.,
    1/F NOISE IN THIN FILMS OF SEMICONDUCTORS.
        JOUR. ELECTROCHEMICAL SOCIETY, VOL. 111, NO. 5, PP. 556-560,
        MAY 1964.
KLAASSEN, F. M.
    ON GEOMETRICAL DEPENDENCE OF 1/F NOISE IN MOS TRANSISTORS
        PHILIPS RESEARCH REPTS., VOL. 25, NO. 3, PP. 171-174, JUNE 1970.
KLAASSEN, F. M.
    CHARACTERIZATION OF LOW 1/F NOISE IN MOS TRANSISTORS
        IEEE TRANS. ELECTRON DEVICES, VOL. ED-18, NO. 10, PP. 887-891,
        OCTOBER 1971.
KLAASEN, F. M. AND J. BLOK

    CURRENT FLUCTUATIONS IN PBS CELLS
        PHYSICA, VOL. 24, NO. 12, PP. 975-984, DECEMBER 1958.
KLAASSEN, F. M., J. BLOK, AND F. J. DE HOOG
    GENERATION-RECOMBINATION NOISE IN P-TYPE INSB
        PHYSICA, VOL. 27, PP. 185-196, 1961.
KLEINPENNING, T. G.
    CURRENT NOISE IN SOME TRANSITION-METAL COMPOUNDS
        PHYSICA, VOL. 59, PP. 370-378, 1972.
KLEINPENNING, TH. G. M.
    1/F NOISE OF THERMO EMF OF INTRINSIC AND EXTRINSIC SEMICONDUCTORS
        PHYSICA, VOL. 77, NO. 1, PP. 78-98, OCTOBER 1, 1974.
KLEINT, C.,
    THE TEMPERATURE DEPENDENCE OF FLICKER NOISE IN FIELD EMMISION (IN
    GERMAN)
        CZECH. J. PHYS. B, VOL. 14, NO. 4, PP. 256-266, 1964.
KLEINT, CH., R. MECLEWSKI, AND R. BLASZCZYSZYN
    COMPARISON OF EXPERIMENTAL NOISE SPECTRAL DENSITIES OF K-COVERED
    TUNGSTEN EMITTERS WITH FIELD-EMISSION FLICKER-NOISE THEORIES
        PHYSICA, VOL. 68, NO. 2, PP. 382-391, SEPTEMBER 1, 1973.
KNOTT, K. F.
    1/F VOLTAGE NOISE IN SILICON PLANAR BIPOLAR TRANSISTORS
        ELECTRONICS LETTERS, VOL. 4, NO. 4, PP. 555-556, DECEMBER 13, 1968.
KNOTT, K. F.
    LEAKAGE CURRENTS, SURFACE CURRENT AND 1/F NOISE IN PLANAR BIPOLAR
    TRANSISTORS
        ELECTRONICS LETTERS, VOL. 6, NO. 25, PP. 825-826, DECEMBER 10, 1970
KNOTT, K. F.
    EXPERIMENTAL LOCATION OF THE SURFACE AND BULK LOCATION OF 1/F NOISE
    CURRENTS IN LOW-NOISE, HIGH-GAIN NPN PLANAR TRANSISTORS
        SOLID-STATE ELECTRONICS, VOL. 16, NO. 12, PP. 1429 - 1434, DECEMBER
        1973.
KOBAYASHI, I., M. NAKAHARA AND M. ATSUMI
    AN INVESTIGATION OF THE RELATION BETWEEN THE INTERFACE STATE AND
    1/F NOISE BEHAVIOURS IN SILICON MOS TRANSISTOR WITH THE NEGATIVE
    BIAS-HEAT TREATMENT.
        PROC. IEEE, VOL. 61, NO. 8, PP. 1145-1146, AUGUST 1973.
KROUPA, J.
    MEASURING 1/F NOISE IN ZENER DIODES
        SLABOPROUDY OBZOR, VOL. 31, NO. 12, PP. 519 - 525, 1970.
KUDABA, V. E.
    LOW-FREQUENCY CURRENT NOISE IN CDSE SINGLE-CRYSTAL AND THIN-LAYERS
    IN A VARYING ELECTRIC FIELD
        RADIO ENGINEERING AND ELECTRONIC PHYSICS, VOL. 16, NO. 8,
        PP. 1412-1413, AUGUST 1971.
LAX, M. AND P. MENGERT
    INFLUENCE OF TRAPPING, DIFFUSION AND RECOMBINATION ON CARRIER
    CONCENTRATION FLUCTUATIONS
        JOUR. PHYSICS AND CHEMISTRY OF SOLIDS, VOL. 14, PP. 248-267,
        JULY 1960.
LAURITZEN, P. O. AND J. F. GIBBONS
    THE 1/F NOISE ON SURFACE DOPED GERMANIUM FILAMENTS
        JOUR. APPLIED PHYSICS, VOL. 33, NO. 2, PP. 758-759, FEBRUARY 1962.

LEBERWURST, K.,
    CURRENT NOISE IN OXIDE SEMICONDUCTORS AT LOW FREQUENCIES (IN GERMAN)
        NACHRICHTENTECHNIK, VOL. 8, NO. 12, PP. 568-580, DECEMBER 1958.
LEE, S. J. AND A. VAN DER ZIEL
    FLICKER NOISE COMPENSATION IN HIGH-IMPEDANCE MOSFET CIRCUITS
        SOLID-STATE ELECTRONICS, VOL. 16, NO. 11, PP. 1301-1302,
        NOVEMBER 1973.
LEUENBERGER, F.
    1/F NOISE IN GATE-CONTROLLED PLANAR SILICON DIODES
        ELECTRONICS LETTERS, VOL. 4, NO. 13, P. 280, JUNE 28, 1968.
LEUENBERGER, F.
    1/F-RAUSCHEN VON MOS-TRANSISTOREN
        HELV. PHYS. ACTA, VOL. 41, PP. 448-450, 1968.
LEUENBERG, F.
    CHARGE PUMPING AND LOW-FREQUENCY NOISE IN MOS STRUCTURES
        PHYSICA STATUS SOLIDI A, VOL. 8, NO. 2, PP. 545-550, DECEMBER 16,
        1971.
LEUENBERG, F.
    ABSENCE OF 1/F NOISE IN MOS TRANSISTORS OPERATED IN SATURATION
        ELECTRONICS LETTERS, VOL. 7, NO. 18, P. 561, SEPTEMBER 9, 1971.
LEVENTHAL, E. A.
    DERIVATION OF 1/F NOISE IN SILICON INVERSION LAYERS FROM CARRIER
    MOTION IN A SURFACE BAND
        SOLID-STATE ELECTRONICS, VOL. 11, NO. 6, PP. 621-627, JUNE 1968.
LINDEMANN, W.W. AND VAN DER ZIEL, A.
    ON THE FLICKER NOISE CAUSED BY AN INTERFACE LAYER,
        JOUR. APPLIED PHYSICS, VOL. 23, NO. 12, PP. 1410-1411, DECEMBER
        1952.
LORTEIJE, L. H. J. AND A. M. H. HOPPENBROUWERS
    AMPLITUDE MODULATION BY 1/F NOISE IN RESISTORS RESULTS IN 1/DELTA F
    NOISE
        PHILIPS RESEARCH REPTS., VOL. 26, NO. 1, PP. 29-36, FEBRUARY 1971.

LOSEV, V. V. AND A. OKSMAN
    UNIVERSAL MECHANISM FOR GENERATION OF EXCESS NOISE IN INHOMOGENEOUS
    SEMICONDUCTORS
        SOVIET PHYSICS - SEMICONDUCTORS, VOL. 6, NO. , PP. 442-443, 1972.
LUK'YANCHIKOVA, N. B., E. B. KAGANOVICH, N. P. GARBAR, M. K. SHEYNKMAN,
AND S. V. SVECHNIKOV
    GENERATION-RECOMBINATION NOISE AND THRESHOLD SENSITIVITY OF CADMIUM
    SULFIDE LAYERS
        RADIO ENGINEERING AND ELECTRONIC PHYSICS, VOL. 14, NO. 6,
        PP. 927-   , JUNE 1969.
LUKYANCHIKOVA, N. B., A. A. KONOVAL, AND M. K. SHEINKMAN
    HIGH-FREQUENCY 1/F NOISE OF PHOTOCURRENT AND RESIDUAL CONDUCTIVITY
    IN CDS
        SOLID-STATE ELECTRONICS, VOL. 18, NO. 1, PP. 65-70, JANUARY 1975.
LUKYANCHIKOVA, N. B., M. K. SHEINKMAN, N. P. GABAR, AND S. V. SVECHNIKOV
    EXCESS CURRENT AND LUMINESCENCE NOISE OF P-N JUNCTIONS IN GAP
        PHYSICA, VOL. 58, PP. 219-224, 1972.
LUK'YANCHIKOVA, N. B., M. K. SHEYNKMAN, A. M. PAVELETS, A. A. KONOVAL,
AND G. A. FEDORUS
    THE NOISE AND THRESHOLD SENSITIVITY OF SINTERED CDS LAYERS
        RADIO ENGINEERING AND ELECTRONIC PHYSICS, VOL. 18, NO. 7,
        PP. 1062-1065, JULY 1973.
LUK'YANCHIKOVA, N. B., B. D. SOLGANIK, M. K. SHEINKMAN, I. I. PROTASOV,
AND V. G. TROFIM
    EXCESS NOISE GENERATED IN P-AL(X)GA(1-X)AS--N-GAAS HETEROJUNCTION
    PHOTODIODES.
        SOVIET PHYSICS - SEMICONDUCTORS, VOL. 6, NO. 10, PP. 1599-1602,
        APRIL 1973.
LUNDSTROM, I. AND D. MC QUEEN
    A PROPOSED 1/F NOISE MECHANISM IN NERVE-CELL MEMBRANES
        JOUR. THEORETICAL BIOLOGY, VOL. 45, NO. 2, PP. 405-409, JUNE 1974.
LUNDSTROM, I., D. MC QUEEN, AND C. KLASON
    ON LOW FREQUENCY AND 1/F NOISE FROM DIFFUSION LIKE PROCESSES
        SOLID STATE COMMUNICATIONS, VOL. 13, NO. 12, PP. 1941-1944,
        DECEMBER 15, 1973.
MAC FARLANE, G. G.
    A THEORY OF FLICKER NOISE IN VALVES AND IMPURITY SEMI-CONDUCTORS
        PROC. PHYSICAL SOC. (LONDON), VOL. 59, PT. 3, PP. 366-375, MAY 1947
MAC FARLANE, G. G.
    A THEORY OF CONTACT NOISE IN SEMICONDUCTORS
        PROC. PHYSICAL SOC. (LONDON), VOL. 63B, PP. 807-814, OCTOBER 1950.
MACHLUP, S.
    NOISE IN SEMICONDUCTORS   SPECTRUM OF A TWO-PARAMETER RANDOM SIGNAL
        JOUR. APPL. PHYS., VOL. 25, NO. 3, PP. 341-343, MARCH 1954.
MAC RAE, A. U.
    1/F NOISE FROM VACUUM-CLEANED SILICON
        JOUR. APPLIED PHYSICS, VOL. 33, NO. 8, PP. 2570-2572, AUGUST 1962.
MAC RAE, A. U. AND H. LEVINSTEIN
    SURFACE DEPENDENT 1/F NOISE IN GERMANIUM
        PHYSICAL REV., VOL. 119, NO. 1, PP. 62-69, JULY 1, 1960.
MANDELBROT, B.
    SOME NOISES WITH 1/F SPECTRUM. A BRIDGE BETWEEN DIRECT CURRENT AND
    WHITE NOISE

        IEEE TRANS. INFORMATION THEORY, VOL. IT-13, NO. 2, PP. 289-298,
        APRIL 1967.
MANDELBROT, B. B. AND J. W. VAN NESS
    FRACTIONAL BROWNIAN MOTIONS, FRACTIONAL NOISES, AND APPLICATIONS
        SIAM REVIEWS, VOL. 10, NO. 4, PP. 422-437, OCTOBER 1968.
MANIUS, C., J.P. BOREL AND R. MERCIER,
    A SPECTRUM ANALYSER FOR THE STUDY OF FLICKER NOISE (IN FRENCH)
        HELV. PHYS. ACTA, VOL. 27, NO. 6, PP. 497-502, 1954.
MANSOUR, I. R. M., R. J. HAWKINS, AND G. G. BLOODWORTH
    PHYSICAL MODEL FOR THE CURRENT NOISE SPECTRUM OF M.O.S. TRANSISTORS
        JOUR. PHYSICS D (APPL. PHYS.), SER. 2, VOL. 2, NO. 8,
        PP. 1063-1082, AUGUST 1969.
MANSOUR, I. R. M., R. J. HAWKINS, AND G. G. BLOODWORTH
    DIGITAL ANALYSIS OF CURRENT NOISE AT VERY LOW FREQUENCIES
        RADIO & ELECTRONIC ENGINEER, VOL. 35, NO. 4, PP. 201-211, APRIL
        1968.
MANSOUR, I. R. M., R. J. HAWKINS, AND G. G. BLOODWORTH
    MEASUREMENT OF CURRENT NOISE IN M.O.S. TRANSISTORS FROM 5X10**(-5) TO
    1 HZ
        RADIO & ELECTRONIC ENGINEER, VOL. 35, NO. 4, PP. 212-216, APRIL
        1968.
MANTENA, N. R. AND R. C. LUCAS
    EXPERIMENTAL STUDY OF FLICKER NOISE IN M.I.S. FIELD-EFFECT TRANSISTORS
        ELECTRONICS LETTERS, VOL. 5, NO. 24, PP. 607-608, NOVEMBER 27, 1969
MAPLE, T. G., L. BESS, AND H. A. GEBBIE
    VARIATION OF NOISE WITH AMBIENT IN GERMANIUM FILAMENTS
        JOUR. APPLIED PHYSICS, VOL. 26, NO. 4, PP. 490-491, APRIL 1955.
MARTIN, J. C., D. ESTEVE, G. BLASQUEZ, AND J. M. RIBEYROL
    THEORY OF 1/F NOISE IN BIPOLAR SILICON PLANAR TRANSISTORS (IN FRENCH)
        ELECTRONICS LETTERS, VOL. 6, NO. 5, PP. 128-130, MARCH 5, 1970.
MARTIN, J. C., F. X. MATEU-PEREZ, AND F. SERRA-MESTIES
    VERY LOW-FREQUENCY MEASUREMENTS OF FLICKER NOISE IN PLANAR
    TRANSISTORS (IN FRENCH)
        ELECTRONICS LETTERS, VOL. 2, NO. 9, PP. 343-345, SEPTEMBER 1966.
MATOBERTI, F., F. MONTECCHI, AND V. SVELTO
    FLICKER NOISE IN THICK FILM RESISTORS
        ALTA FREQUENZA, VOL. 44, NO. 11, PP. 391-   , NOVEMBER 1975.
MATSUMO, K.
    LOW FREQUENCY CURRENT FLUCTUATIONS IN A GAAS GUNN DIODE
        APPLIED PHYS. LETTERS, VOL. 12, NO. 12, PP. 403-404, JUNE 15, 1968.
MAY, E. J. P. AND H. G. MORGAN
    1/F NOISE IN MULTICARRIER SYSTEMS
        ELECTRONICS LETTERS, VOL. 12, NO. 1, PP. 8-9, JANUARY 8, 1976.
MAY, E. J. P. AND W. D. SELLARS
    1/F NOISE PRODUCED BY RADIO-FREQUENCY CURRENT IN RESISTORS
        ELECTRONICS LETTERS, VOL. 11, NO. 22, PP. 544-545, OCTOBER 30, 1975
MC DONALD, B. A.
    AVALANCHE-INDUCED 1/F NOISE IN BIPOLAR TRANSISTORS
        IEEE TRANS. ELECTRON DEVICES, VOL. ED-17, NO. 2, PP. 134-136,
        FEBRUARY 1970.
MC WHORTER, A. L.
    1/F NOISE AND RELATED SURFACE EFFECTS IN GERMANIUM
        SC. D. DISSERTATION, M.I.T., CAMBRIDGE, MASS.,   ALSO AS
        M.I.T. LINCOLN LAB., LEXINGTON, MASS., TECH. REPT. 80, MAY 1955.
MC WHORTER, A. L.
    1/F NOISE AND GERMANIUM SURFACE PROPERTIES
        IN SEMICONDUCTOR SURFACE PHYSICS , R. H. KINGSTON, ED., PP. 207-
        228, UNIVERSITY OF PENNSYLVANIA PRESS, PHILADELPHIA, PENN., 1957.
MECLEWSKI, R., CH. KLEINT, AND R. BLASZCYZSZYN
    CORRELATION OF FIELD-EMISSION FLICKER NOISE AND WORK FUNCTION FOR
    TUNGSTEN SINGLE-CRYSTAL PLANES WITH ADSORBED POTASSIUM
        SURFACE SCIENCE, VOL. 52, NO. 2, PP. 365-376, OCTOBER 1975.
MEYER, R. G., L. NAGEL AND S. K. LIU
    COMPUTER SIMULATION OF 1/F NOISE PERFORMANCE OF ELECTRONIC CIRCUITS
        IEEE JOUR. SOLID-STATE CIRCUITS, VOL. SC-8, NO. 3, PP. 237-240,
        JUNE 1973.
MILLER, P. H.
    NOISE SPECTRUM OF CRYSTAL RECTIFIERS
        PROC. IRE, VOL. 35, NO. 3, PP. 257-265, MARCH 1947.
MIRCEA, A., A. ROUSSEL, AND A. MITONNEAU
    1/F NOISE - STILL A SURFACE EFFECT
        PHYSICS LETTERS A, VOL. 41A, NO. 4, PP. 343-346, OCTOBER 9, 1972.
MONTAGNON, N.B.,
    L.F. NOISE IN RESISTORS.
        WIRELESS ENGINEER, VOL. 31, NO. 10, PP. 253-263, OCTOBER 1954.
        WIRELESS ENGINEER, VOL. 31, NO. 11, PP. 301-305, NOVEMBER 1954.
MONTGOMERY, H. C.
    ELECTRICAL NOISE IN SEMICONDUCTORS
        BELL SYSTEM TECH. JOUR., VOL. 31, NO. 5, PP. 950-975, SEPTEMBER
        1952.
MOORE, W. J.
    STATISTICAL STUDIES OF 1/F NOISE FROM CARBON RESISTORS
        JOUR. APPLIED PHYSICS, VOL. 45, NO. 4, PP. 1896-1901, APRIL 1974.
MORCOM, W. R. AND T. M. CHEN
    MEASUREMENT OF EXCESS NOISE IN THIN GERMANIUM FILMS
        SOLID-STATE ELECTRONICS, VOL. 14, NO. 4, PP. 337-340, APRIL 1971.

MORRISON, S. R.
    GENERATION OF 1/F NOISE BY LEVELS IN A LINEAR OR PLANAR ARRAY
        PHYSICAL REV., VOL. 99, NO. 6, P. 1904, SEPTEMBER 15, 1955.
MUELLER, O.
    RECOMBINATION, THERMAL FEEDBACK, AND FLICKER NOISE IN TRANSISTORS
        ARCH. ELEKTRONIK UND UBERTRAGUNGSTECHNIK, VOL. 24, NO. 4,
        PP. 169-178, APRIL 1970.
MUELLER, O.
    TEMPERATURE FLUCTUATIONS AND FLICKER NOISE IN P-N JUNCTION DIODES
        IEEE TRANS. ELECTRON DEVICES, VOL. ED-21, NO. 8, PP. 539-540,
        AUGUST 1974.
MUELLER, O.
    A FORMULA FOR 1/F-FLICKER NOISE IN P-N JUNCTIONS
        ARCHIV FUR ELKTRONIK UBERTRAGUNGSTECHNIK, VOL. 28, NO. 10, PP.
        429 - 432, OCTOBER 1974.
MUELLER, O.
    CALCULATING FLICKER NOISE OF P-N JUNCTION DIODES
        ARCH. ELEKTRON. UND UBERTRAGUNGSTECH., VOL. 28, NO. 11,
        PP. 450-454, NOVEMBER 1974.
MUELLER, O.
    COMMENTS ON 1/F NOISE
        ELECTRONICS LETTERS, VOL. 12, NO. 2, PP. 48-49, 22 JANUARY 1976.
MUSHA, T.
    THEORY OF 1/F FLUCTUATION OF PARTICLE CONCENTRATION IN THERMAL
    EQUILIBRIUM
        PHYSICA A, VOL. 80A, NO. 4, PP. 387-397, 15 MAY 1975.
MUSHA, T. AND H. HIGUCHI
    THE 1/F FLUCTUATION OF A TRAFFIC CURRENT ON AN EXPRESSWAY
        JAPANESE JOUR. APPLIED PHYSICS, VOL. 15, NO. 7, PP. 1271-1275,
        JULY 1976.
MYTTON, R. J. AND R. K. BENTON
    HIGH 1/F NOISE ANOMALY IN SEMICONDUCTING BARIUM STRONTIUM TITANATE
        PHYSICS LETTERS A, VOL. 39A, NO. 4, PP. 329-330, MAY 22, 1972.
NAKAHARA, M.
    ANOMALOUS LOW-FREQUENCY NOISE IN MOS TRANSISTORS AT LOW TEMPERATURES.
        ELECTRONICS AND COMMUNICATIONS IN JAPAN, VOL. 55, NO. 6, PP. 99-
        105, JUNE 1972.
NFUMCKE, B.
    1/F MEMBRANE NOISE GENERATED BY DIFFUSION PROCESSES IN UNSTIRRED
    SOLUTION LAYERS
        BIOPHYSICS OF STRUCTURE AND MECHANISM, VOL. 1, NO. 4, PP. 295-309,
        1975.
NICOLLIAN, E. H. AND H. MELCHIOR
    A QUANTITATIVE THEORY OF 1/F TYPE NOISE DUE TO INTERFACE STATES IN
    THERMALLY OXIDIZED SILICON
        BELL SYSTEM TECH. JOUR., VOL. 46, NO. 9, PP. 2019-2033, NOV. 1967.
NOBLE, V. E. AND J. E. THOMAS, JR.
    EFFECT OF GASEOUS AMBIENT UPON 1/F NOISE IN GERMANIUM FILAMENTS
        JOUR. APPLIED PHYSICS, VOL. 32, NO. 9, PP. 1709-1714, SEPTEMBER
        1961.
OFFNER, F. F.
    1/F NOISE IN SEMICONDUCTORS
        JOUR. APPLIED PHYSICS, VOL. 41, NO. 12, PP. 5033-5034, NOVEMBER
        1970.
OFFNER, F. F.
    1/F FLUCTUATIONS IN MEMBRANE-POTENTIAL AS RELATED TO MEMBRANE THEORY
        BIOPHYSICS JOUR., VOL. 11, PP. 123-124, 1971.
OFFNER, F. F.
    QUANTITATIVE MEASUREMENT OF 1/F NOISE AND MEMBRANE THEORY
        BIOPHYSICS JOUR., VOL. 11, PP. 969-971, 1971.
OFFNER, F. F.
    COMMENTS ON MODIFIED RANDOM-WALK MODEL OF 1/² NOISE
        JOUR. APPLIED PHYSICS, VOL. 43, NO. 3, PP. 1277-1278, MARCH 1972.
OLIVER, D. J.
    CURRENT NOISE IN INDIUM ANTIMONIDE
        PROC. PHYSICAL SOC. (LONDON), VOL. B70, PT. 3, PP. 331-332, MARCH
        1957.
ORTMANS, L. H. F. AND L. K. VANDAMME
    CHARACTERIZATION OF IMPULSE-FRITTING PROCEDURES OF CONTACTS BY
    MEASURING 1/F NOISE
        APPLIED PHYSICS, VOL. 9, NO. 2, PP. 147-151, FEBRUARY 1976.
PEARSON, G. L., H. C. MONTGOMERY AND W. L. FELDMAN
    NOISE IN SILICON P-N JUNCTION PHOTOCELLS
        JOUR. APPLIED PHYSICS, VOL. 27, NO. 1, PP. 91-92, JANUARY 1956.
PETRITZ, R. L.
    A THEORY OF CONTACT NOISE
        PHYSICAL REV., VOL. 87, NO. 3, PP. 535-536, AUGUST 1, 1952.
PLUMB, J.L. AND E.R. CHENETTE,
    FLICKER NOISE IN TRANSISTORS,
        IEEE TRANS. ELECTRON DEVICES, VOL. ED-10, NO. 5, PP. 304-308,
        SEPTEMBER 1963.
PURCELL, W. E.
    VARIANCE NOISE SPECTRA OF 1/F NOISE
        JOUR. APPLIED PHYSICS, VOL. 43, NO. 6, PP. 2890-2895, JUNE 1972.
RADEKA, V.
    1/F NOISE IN PHYSICAL MEASUREMENTS

IEEE TRANS. NUCLEAR SCIENCE, VOL. NS-16, NO. 5, PP. 17-35,
OCTOBER 1969.

RICHARDSON, J. M.
THE LINEAR THEORY OF FLUCTUATIONS ARISING FROM DIFFUSIONAL MECHANISMS
- AN ATTEMPT AT A THEORY OF CONTACT NOISE
BELL SYSTEM TECH. JOUR., VOL. 29, NO. 1, PP. 117-141, JANUARY 1950.

RICHARDSON, J. M. AMD J. J. FARIS
EXCESS NOISE IN MICROWAVE CRYSTAL DIODES USED AS RECTIFIERS AND
HARMONIC GENERATORS
IRE TRANS. MICROWAVE THEORY AND TECHNIQUES, VOL. MTT-5, NO. 3,
PP. 208-212, JULY 1957.

RINGO, J. A. AND P. O. LAURITZEN.
1/F NOISE IN UNIFORM AVALANCHE DIODES
SOLID-STATE ELECTRONICS, VOL. 16, NO. 3, PP. 327-328, MARCH 1973.

ROLLIN, B.V. AND J.P. RUSSELL.
THE CURRENT NOISE SPECTRUM OF COPPER DOPED GERMANIUM AT 20 DEGREE K,
PROC. PHYS. SOC.(GB), VOL. 81, PT. 3, PP. 578-581, MARCH 1963.

ROLLIN, B. V. AND I. M. TEMPLETON
NOISE IN SEMICONDUCTORS AT VERY LOW FREQUENCIES
PROC. PHYSICAL SOC. (LONDON), VOL. B66, PT. 3, PP. 259-261, MARCH
1953.

ROLLIN, B. V. AND I. M. TEMPLETON
NOISE IN GERMANIUM FILAMENTS AT VERY LOW FREQUENCIES
PROC. PHYSICAL SOC. (LONDON), NO. B67, PT. 3, P. 271, MARCH 1954.

ROGERS, C. G.
LOW FREQUENCY NOISE IN MOST'S AT CRYOGENIC TEMPERATURES
SOLID-STATE ELECTRONICS, VOL. 11, NO. 12, PP. 1099-1104, DECEMBER
1968.

RONEN, R. S.
LOW-FREQUENCY 1/F NOISE IN MOSFETS
RCA REVIEW, VOL. 34, NO. 2, PP. 280-307, JUNE 1973.

ROTHE, H., W. DAHLKE AND J. SCHUBERT,
MEASUREMENT OF THE FLICKER-EFFECT CONSTANTS (IN GERMAN)
TELEFUNKEN ZTG, VOL. 26, PP. 77-84, MARCH 1953.

SAH, C. T.
THEORY OF LOW-FREQUENCY GENERATION NOISE IN JUNCTION-GATE
FIELD-EFFECT TRANSISTORS
PROC. IEEE, VOL. 52, NO. 7, PP. 795-814, JULY 1964.

SAH, C. T. AND F. H. HIELSCHER
EVIDENCE OF THE SURFACE ORIGIN OF 1/F NOISE
PHYSICAL REV. LETTERS, VOL. 17, NO. 18, PP. 956-958, OCTOBER 31,
1966.

SCHICK, K. L. AND A. A. VERVEEN
1/F NOISE WITH A LOW-FREQUENCY WHITE NOISE LIMIT
NATURE, VOL. 251, NO. 5476, PP. 599-600, 1974.

SCHONFELD, H.
BEITRAG ZUM 1/F-GESETZ BEIM RAUSCHEN VON HALBLEITERN
ZEIT. FUR NATURFORSCH., VOL. 10A, PP. 291-300, APRIL 1955.

SCHOTTKY, W.
SMALL-SHOT EFFECT AND FLICKER EFFECT
PHYSICAL REVIEWS, VOL. 28, NO. , PP. 75-103, JULY 1925.

SCHWANTES, R. C., H. J. HANNAM AND A. VAN DER ZIEL
FLICKER NOISE IN SECONDARY EMISSION TUBES AND MULTIPLIER PHOTOTUBES
JOUR. APPL. PHYS., VOL. 27, NO. 6, PP. 573-577, JUNE 1956.

SCHWANTES, R. C. AND A. VAN DER ZIEL
SECONDARY EMISSION FLICKER NOISE
PHYSICA, VOL. 26, NO. 12, PP. 1162-1166, DECEMBER 1960.

SCHWANTES, R. C. AND A. VAN DER ZIEL
FLICKER NOISE IN PENTODES FLICKER PARTITION NOISE
PHYSICA, VOL. 26, NO. 12, PP. 1157-1161, DECEMBER 1960.

SCHWANTES, R. C. AND A. VAN DER ZIEL
FLICKER NOISE IN TRIODES WITH POSITIVE GRID
PHYSICA, VOL. 26, NO. 12, PP. 1143-1156, DECEMBER 1960.

SHULMAN, C. I.
MEASUREMENT OF SHOT NOISE IN CDS CRYSTALS
PHYSICAL REV., VOL. 98, NO. 2, PP. 384-386, APRIL 15, 1955.

SHURMER, H. V.
BACKWARD DIODES AS MICROWAVE DETECTORS.
PROC. IEE (GB), VOL. 111, NO. 9, PP. 1511-1516, SEPTEMBER 1964.

SHURMER, H.V.,
A CRYSTAL MIXER FOR CW RADARS.
PROC. IEE (GB), VOL. 110, NO. 1, PP. 117-122, JANUARY 1963.

SIKULA, J., B. KOKTAVY, L. KRATENA, AND J. MISEK
STOCHASTIC PHENOMENA IN EPITAXIAL P-N JUNCTIONS IN GAP
PHYSICA STATUS SOLIDI A, VOL. 29, NO. 1, PP. 41-46, MAY 16, 1975.

SLOCUM, A. AND J. N. SHIVE
SHOT DEPENDENCE OF P-N JUNCTION PHOTOTRANSISTOR NOISE
JOUR. APPLIED PHYSICS, VOL. 25, NO. 3, P. 406, MARCH 1954.

STEPANESCU, A.
1/F NOISE AS A TWO-PARAMETER STOCHASTIC PROCESS
NUOVO CIMENTO B, VOL. 23B, SER. 2, NO. 2, PP. 356-364, OCTOBER 11,
1974.

STEPANESCU, A.
SURFACE TRAPPING PHENOMENA IN THERMIONIC EMISSION GENERATING 1/F NOISE
REVUE ROUMAINE PHYSIQUE, VOL. 20, NO. 4, PP. 387-391, 1975.

STEPHANY, J. F.
   ORIGIN OF 1/F NOISE
      JOUR. APPLIED PHYSICS, VOL. 46, NO. 2, PP. 665-667, FEBRUARY 1975.
STEPHANY, J. F.
   ORIGIN OF 1/F NOISE - PART III   SOURCES OF MAGNETICALLY GENERATED
   1/F NOISE
      JOUR. APPLIED PHYSICS, VOL. 46, NO. 11, PP. 5010-5011, NOVEMBER
      1975.
STOISIEK, M. AND D. WOLF
   RECENT INVESTIGATION ON THE STATIONARITY OF 1/F NOISE
      JOUR. APPLIED PHYSICS, VOL. 47, NO. 1, PP. 362-364, JANUARY 1976.
STRASILLA, U. J. AND M. J. O. STRUTT
   NARROW-BAND VARIANCE NOISE
      JOUR. APPLIED PHYSICS, VOL. 45, NO. 3, PP. 1423-1428, MARCH 1974.
STRASILLA, U. J. AND M. J. O. STRUTT
   MEASUREMENT OF WHITE AND 1/F NOISE WITHIN BURST NOISE
      PROC. IEEE, VOL. 62, NO. 12, PP. 1711-1713, DECEMBER 1974.
SUHL, H.
   THEORY OF MAGNETIC EFFECTS ON THE NOISE IN A GERMANIUM FILAMENT
      BELL SYSTEM TECH. JOUR., VOL. 32, NO. 3, PP. 647-664, MAY 1953.
SUITS, G. H., W. D. SCHMITZ, AND R. W. TERHUNE
   EXCESS NOISE IN INSB
      JOUR. APPLIED PHYSICS, VOL. 27, NO. 11, P. 1385, NOVEMBER 1956.
SURDIN, M.
   UNE THEORIE DES FLUCTUATIONS ELECTRIQUES DANS LES SEMI-CONDUCTEURS
      JOUR. PHYSIQUE RADIUM, VOL. 12, NO. 8, PP. 773-783, OCTOBER 1951.
SUTCLIFFE, H.
   MEASUREMENT OF FLICKER NOISE SPECTRUM OF TRANSISTORS
      SYMPOSIUM ON TEST METHODS AND MEASUREMENTS OF SEMICONDUCTOR
      DEVICES, BUDAPEST, HUNGARY, 1967, PAPER NO. 104.
SUTCLIFFE, H.
   CURRENT-INDUCED RESISTOR NOISE NOT ATTRIBUTABLE ENTIRELY TO
   FLUCTUATIONS OF CONDUCTIVITY
      ELECTRONICS LETTERS, VOL. 7, NO. 7, PP. 160-161, APRIL 8, 1971.
TANAKA, T., K. NAGANO, AND N. NAMEKI
   THE 1/F NOISE MOS TRANSISTORS
      JAPANESE JOUR. APPLIED PHYSICS, VOL. 8, NO. 8, PP. 1020-1026,
      AUGUST 1969.
TANDON, J. L. AND H. R. BILGER
   1/F NOISE AS A NONSTATIONARY PROCESS   EXPERIMENTAL EVIDENCE AND SOME
   ANALYTICAL CONDITIONS
      JOUR. APPLIED PHYSICS, VOL. 47, NO. 4, PP. 1697-1701, APRIL 1976.
TANDON, J. L., H. R. BILGER & M. A. NICOLET.
   EXCESS NOISE SPECTRAL ANALYSIS BORON-IMPLANTED LAYERS IN SILICON.
      1973 SWIEEECO RECORD OF TECHNICAL PAPERS, HOUSTON, TEX., APRIL 4-6
      1973 (IEEE, NEW YORK, 1973), PP. 561-568.
TARATUTA, A. S. AND G. E. CHAIKA
   SURFACE CURRENT NOISE
      SOVIET PHYSICS - SEMICONDUCTORS, VOL. 5, NO. 3, PP. 333-338,
      SEPTEMBER 1971.
TARATUTA, A. S. AND G. E. CHAIKA
   INFLUENCE OF SURFACE RECOMBINATION ON LOW-FREQUENCY SURFACE CURRENT
   NOISE
      SOVIET PHYSICS - SEMICONDUCTORS, VOL. 7, NO. 3, P. 411, SEPTEMBER
      1973.
TEITLER, S. AND M. F. M. OSBORNE
   PHENOMENOLOGICAL APPROACH TO LOW-FREQUENCY ELECTRICAL NOISE
      JOUR. APPLIED PHYSICS, VOL. 41, NO. 8, PP. 3274-3276, JULY 1970.
TEITLER, S. AND M. F. M. OSBORN
   SIMILARITY ARGUMENTS AND AN INVERSE FREQUENCY NOISE SPECTRUM FOR
   ELECTRICAL CONDUCTORS
      PHYSICAL REV. LETTERS, VOL. 27, NO. 14, PP. 912-915, OCTOBER 4,
      1971.
TEMPLETON, I. M. AND D. K. C. MAC DONALD
   THE ELECTRICAL CONDUCTIVITY AND CURRENT NOISE OF CARBON RESISTORS
      PROC. PHYSICAL SOC. (LONDON), VOL. B66, PT. 8, PP. 680-687,
      AUGUST 1953.
THEOBOLD, G.
   APPLICATION OF THE BUTZ MODEL TO THE STUDY OF THE INFLUENCE OF
   VELOCITY DISTRIBUTION ON CURRENT NOISE (IN FRENCH)
      C. R. HEBD. SEANCES ACAD. SCI. B, VOL. 279, NO. 25, PP. 577-580,
      FEBRUARY 16, 1974.
TOMLINSON, T.B.,
   PARTITION COMPONENTS OF FLICKER NOISE,
      JOUR. BRIT. IRE, VOL. 14, NO. 11, PP. 515-526, NOVEMBER 1954.
TOMLINSON, T. B. AND W. L. PRICE
   THEORY OF THE FLICKER EFFECT
      JOUR. APPLIED PHYSICS, VOL. 24, NO. 8, PP. 1063-1065, AUGUST 1953.
TUNALEY, J. K. E.
   A PHYSICAL PROCESS FOR 1/F NOISE IN THIN METALLIC FILMS
      JOUR. APPLIED PHYSICS, VOL. 43, NO. 9, PP. 3851-3855, SEPTEMBER
      1972.
TUNALEY, J. K. E.
   SOME STOCHASTIC PROCESS YIELDING A F**(-NU) TYPE OF SPECTRAL DENSITY
      JOUR. APPLIED PHYSICS, VOL. 43, NO. 11, PP. 4777-4783, NOVEMBER
      1972.

TUNALEY, J. K. E.
    NYQUIST THEOREM AND 1/F NOISE
        JOUR. APPL. PHYS., VOL. 45, NO. 1, PP. 482-483, JANUARY 1974.
TUNALEY, J. K. E.
    A THEORY OF 1/F CURRENT NOISE BASED ON A RANDOM WALK MODEL
        JOUR. STATISTICAL PHYSICS, VOL. 15, NO. 2, PP. 149-156, AUGUST
        1976.
UCHIDA, I. AND T. ISHIGURO
    EXCITED CURRENT NOISE IN NON-OHMIC REGION OF CDS
        JOUR. PHYSICAL SOC. JAPAN, VOL. 24, NO. 3, PP. 661-662, MARCH 1968.
VANDAMME, L. K. J.
    1/F NOISE OF POINT CONTACTS AFFECTED BY UNIFORM FILMS
        JOUR. APPLIED PHYSICS, VOL. 45, NO. 10, PP. 4563-4565, OCTOBER 1974
VANDAMME, L. K. J.
    1/F NOISE IN HOMOGENEOUS SINGLE CRYSTALS OF III-V COMPOUNDS
        PHYSICS LETTERS A, VOL. 49A, NO. 3, PP. 233-234, 9 SEPT. 1974
VANDAMME, L. K. J.
    1/F NOISE AND CONSTRICTION RESISTANCE OF ELONGATED CONTACT SPOTS
        ELECTRONICS LETTERS, VOL. 12, NO. 4, PP. 109-110, FEBRUARY 19, 1976.
VAN DER ZIEL, A.
    ON THE NOISE SPECTRA OF SEMICONDUCTOR NOISE AND OF FLICKER EFFECT
        PHYSICA, VOL. 16, NO. 4, PP. 359-372, APRIL 1950.
VAN DER ZIEL, A.
    A SIMPLER EXPLANATION FOR THE OBSERVED SHOT EFFECT IN GERMANIUM
    FILAMENTS
        JOUR. APPLIED PHYSICS, VOL. 24, NO. 8, P. 1063, AUGUST 1953.
VAN DER ZIEL, A.,
    NOISE MECHANISMS IN OXIDE-COATED CATHODES,
        PHYSICA, VOL. 20, PP. 327-336, JUNE 1954.
VAN DER ZIEL, A.
    FLICKER NOISE IN THIN SUPERCONDUCTING FOILS
        PHYS. LETTERS (NETHERLANDS), VOL. 25A, NO. 9, PP. 672-673,
        NOVEMBER 6, 1967.
VAN DER ZIEL, A.
    SURFACE RECOMBINATION MODEL OF P-N DIODE FLICKER NOISE
        PHYSICA, VOL. 48, NO. 2, PP. 242-246, AUGUST 25, 1970.
VAN DER ZIEL, A.
    PROOF OF BASIC SEMICONDUCTOR FLICKER NOISE FORMULAE
        SOLID-STATE ELECTRONICS, VOL. 17, NO. 1, PP. 110-111, JANUARY 1974.
VAN DER ZIEL, A.
    LIMITING FLICKER NOISE IN MOSFETS
        SOLID-STATE ELECTRONICS, VOL. 18, NO. 11, P. 1031, NOV. 1975.
VAN VLIET, K. M.
    IRREVERSIBLE THERMODYNAMICS AND CARRIER DENSITY FLUCTUATIONS IN
    SEMICONDUCTORS
        PHYSICAL REV., VOL. 110, NO. 1, PP. 50-61, APRIL 1, 1958.
VAN VLIET, K. M., J. BLOK, C. RIS, AND J. STEKETEE
    MEASUREMENT OF NOISE AND RESPONSE TO MODULATED LIGHT OF
    CADMIUMSULPHIDE SINGLE CRYSTALS
        PHYSICA, VOL. 22, NO. 9, PP. 723-740, SEPTEMBER 1956.
VAN VLIET, K. M. AND J. R. FASSETT
    FLUCTUATIONS DUE TO ELECTRONIC TRANSITIONS AND TRANSPORT IN SOLIDS
        IN FLUCTUATION PHENOMENA IN SOLIDS , R. E. BURGESS, ED., ACADEMIC
        PRESS, NEW YORK, 1965, PP. 267-354.
VAN VLIET, K.M. AND R.R. JOHNSON,
    FLICKER NOISE IN OXIDE CATHODES ARISING FROM DIFFUSION AND DRIFT OF
    IONIZED DONORS
        JOUR. APPLIED PHYSICS, VOL. 35, NO. 7, PP. 2039-2051, JULY 1964.
VAN VLIET, K. M. AND A. VAN DER ZIEL
    ON THE NOISE GENERATED BY DIFFUSION MECHANISMS
        PHYSICA, VOL. 24, PP. 415-421, 1958.
VAN VLIET, K. M., C. J. VAN LEEUWEN, J. BLOK, AND C. RIS
    MEASUREMENTS ON CURRENT NOISE IN CARBON RESISTORS AND IN THERMISTORS
        PHYSICA, VOL. 20, NO. 8, PP. 481-496, AUGUST 1954.
VAN WIJNGAARDEN J.G. AND K.M. VAN VLIET,
    LOW FREQUENCY NOISE IN ELECTRON TUBES. A. SPACE CHARGE REDUCTION OF FLICKER
    NOISE EFFECT,
        PHYSICA, VOL. 18, PP. 683-688, OCTOBER 1952.
VERSTER, T. C.
    ANOMALIES IN TRANSISTOR LOW-FREQUENCY NOISE
        PROC. IEEE, VOL. 55, NO. 7, PP. 1204-1205, JULY 1967.
VERVEEN, A. A. AND H. E. DERKSEN
    FLUCTUATION PHENOMENA IN NERVE MEMBRANE,
        PROC. IEEE, VOL 56, NO 6, PP 906-916, JUNE 1968.
VOSS, R. F.
    COMMENT ON A SIMPLE MODEL FOR THE 1/F TYPE POWER SPECTRUM BY M. AGU
        PHYSICS LETTERS, VOL. 53A, NO. 4, P. 277, 30 JUNE 1975.
VOSS, R. F. AND J. CLARKE
    1/F NOISE FROM SYSTEMS IN THERMAL EQUILIBRIUM
        PHYSICAL REV. LETTERS, VOL. 36, NO. 1, PP. 42-45, JANUARY 5, 1976.
VOSS, R. F. AND J. CLARKE
    FLICKER (1/F) NOISE EQUILIBRIUM TEMPERATURE AND RESISTANCE
    FLUCTUATIONS
        PHYSICAL REV. B, VOL. 13, NO. 2, PP. 556-573, 15 JUNE 1976.
WADE, J. M. A.
    FLUX-FLOW NOISE IN MAGNETICALLY COUPLED SUPERCONDUCTORS. OBSERVATION

OF A 1/F SPECTRUM
      PHILOSOPHICAL MAG., VOL. 23, NO. 185, PP. 1029-1040, MAY 1971.
WALL, E. L.
   EDGE INJECTION CURRENTS AND THEIR EFFECT ON 1/F NOISE IN PLANAR
   SCHOTTKY DIODES
      SOLID-STATE ELECTRONICS, VOL. 19, NO. 5, PP. 389-396, MAY 1976.
WATKINS, T. B.
   1/F NOISE IN GERMANIUM DEVICES
      PROC. PHYS. SOC., VOL. 73, PT. 1, NO. 469, PP. 59-68, JANAURY 1,
      1959.
WEISSMAN, M. B.
   SIMPLE MODEL FOR 1/F NOISE
      PHYSICAL REV. LETTERS, VOL. 35, NO. 11, PP. 689-692, SEPTEMBER 15,
      1975.
WHITFIELD, S.J.M.,
   L.F. NOISE IN TRANSISTOR DIRECT COUPLED AMPLIFIERS.
      ELECTRONIC ENGINEERING, VOL. 36, PP. 558-559, AUGUST 1964.
WIGGINS, M. J.
   AN EXPERIMENTAL STUDY OF 1/F NOISE IN TRANSISTORS
      IEEE INTERNATIONAL CONVENTION RECORD, VOL. 12, PT. 2, PP. 102-110,
      1964.
WIGGINS, M. J.
   AN EXPERIMENTAL STUDY OF 1/F NOISE IN TRANSISTORS
      IEEE TRANS. BROADCAST AND TELEVISION RECEIVERS, VOL. BTR-10, NO. 1,
      PP. 84-92, MAY 1964.
WIGGINS, M. J.
   1/F NOISE IN TRANSISTORS
      ELECTRO-TECHNOLOGY, VOL. 74, NO. 6, PP. 34-38, DECEMBER 1964.
WILLIAMS, J. L. AND R. K. BURDETT
   CURRENT NOISE IN CERMET RESISTIVE FILMS
      BRITISH JOUR. APPLIED PHYSICS, VOL. 17, NO. 7, PP. 977-978,
      JULY, 1966.
WILLIAMS, J. L. & R. K. BURDETT
   CURRENT NOISE IN THIN GOLD FILMS
      JOUR. PHYSICS C, VOL. 2, NO. 2, PP. 298-307, FEBRUARY 1969.
WILLIAMS, J. L. AND I. L. STONE
   CURRENT NOISE IN THIN DISCONTINUOUS FILMS
      JOUR. PHYSICS C, VOL. 5, NO. 16, PP. 2105-2116, AUGUST 21, 1972.
WILLIAMS, T. R. AND J. B. THOMAS
   A COMPARISON OF THE NOISE AND VOLTAGE COEFFICIENTS OF PRECISION METAL
   FILM AND CARBON FILM RESISTORS
      IRE TRANS. COMPONENT PARTS, VOL. CP-6, NO. 2, PP. 58-62, JUNE 1959.
WOLL, H.J. AND F.L. PUTZRATH,
   A NOTE ON NOISE IN AUDIO AMPLIFIERS,
      IRE TRANS. AUDIO, VOL. AU-2, NO. 2, PP. 39-42, MARCH-APRIL 1954.
WU, S-Y.
   THEORY OF GENERATION-RECOMBINATION NOISE IN MOS TRANSISTORS
      SOLID-STATE ELECTRONICS, VOL. 11, NO. 1, PP. 25-32, JANUARY 1968.
YAJIMA, T. AND L. ESAKI
   EXCESS NOISE IN NARROW GERMANIUM P-N JUNCTIONS
      JOUR. PHYS. SOC. JAPAN, VOL. 13, NO. 11, PP. 1281-1287, NOVEMBER
      1958.
YAKOVCHUK, N. S.
   ESTIMATION OF FLICKER NOISE IN SOME AMPLIFIER TUBES
      TELECOMMUNICATIONS AND RADIO ENGINEERING, PT. 2 - RADIO ENGG.,
      VOL. 19, NO. 3, PP. 112-113, MARCH 1964.
YAU, L. D. AND C. T. SAH
   ON THE 'EXCESS WHITE NOISE' IN MOS TRANSISTORS
      SOLID-STATE ELECTRONICS, VOL. 12, NO. 12, PP. 927-936, DECEMBER
      1969.
YOKOTA, M.
   GENERAL THEORY OF 1/F NOISE
      PROGRESS THEORETICAL PHYSICS, VOL. 54, NO. 4, PP. 1237-1238, 1975.
YU, K. K., A. G. JORDAN, AND R. L. LONGINI
   RELATIONS BETWEEN ELECTRICAL NOISE AND DISLOCATIONS IN SILICON
      JOUR. APPLIED PHYSICS, VOL. 38, NO. 2, PP. 572-583, FEBRUARY 1967.
ZAKLIKIEWICZ, A. M.
   CONTRIBUTION TO 1/F NOISE THEORY
      ELCTRON TECHNOLOGY, VOL. 6, NO. 1/2, PP. 91-95, 1973.
ZILSTRA, R. J. J. AND A. VAN DER ZIEL
   NOISE IN THIN N-TYPE CDS LAYERS ON AN INSULATING CDS SUBSTRATE
      PHYSICA, VOL. 29, NO. 8, PP. 851-856, AUGUST 1963.

****
      3-C. BURST OR POPCORN NOISE
****

BRODERSEN, A. J., E. R. CHENETTE, AND R. C. JAEGER
   NOISE IN INTEGRATED OPERATIONAL AMPLIFIERS
      PROC. IEEE REGION 6 CONF., PP. 224-226, PHOENIX, ARIZ., APRIL 1969.
BRODERSEN, A. J., E. R. CHENETTE, AND R. C. JAEGER
   NOISE IN INTEGRATED-CIRCUIT TRANSISTORS
      IEEE JOUR. SOLID-STATE CIRCUITS, VOL. SC-5, NO. 2, PP. 63-66,
      APRIL 1970.
CARD, W. H. AND P. K. CHAUDHARI
   CHARACTERISTICS OF BURST NOISE
      PROC. IEEE, VOL. 53, NO. 6, PP. 652-653, JUNE 1965.

GIRALT, G., J. C. MARTIN, AND F. X. MATEU-PEREZ
    SUR UN PHENOMENE DE BRUIT DANS LES TRANSISTORS, CARACTERISE PAR DES
    CRENEAUX DE COURANT DAMPLITUDE CONSTANTE
        COMPTES RENDUS ACAD. SCI. (PARIS), VOL. 261, PP. 5350-5353, 1965.
GIRALT, G., J. C. MARTIN, AND F. X. MATEU-PEREZ
    BURST NOISE OF SILICON PLANAR TRANSISTORS
        ELECTRONICS LETTERS, VOL. 2, NO. 6, PP. 228-230, JUNE 1966.
HSU, S. T. AND R. J. WHITTIER
    CHARACTERIZATION OF BURST NOISE IN SILICON DEVICES
        SOLID-STATE ELECTRONICS, VOL. 12, NO. 11, PP. 867-878,
        NOVEMBER 1969.
HSU, S. T., R. J. WHITTIER, AND C. A. MEAD
    PHYSICAL MODEL FOR BURST NOISE IN SEMICONDUCTOR DEVICES
        SOLID-STATE ELECTRONICS, VOL. 13, NO. 7, PP. 1055-1071, JULY 1970.
JAEGER, R. C. AND A. J. BRODERSEN
    LOW-FREQUENCY NOISE SOURCES IN BIPOLAR JUNCTION TRANSISTORS
        IEEE TRANS. ELECTRON DEVICES, VOL. ED-17, NO. 2, PP. 128-134,
        FEBRUARY 1970.
KNOTT, K. F.
    BURST NOISE AND MICROPLASMA NOISE IN SILICON PLANAR TRANSISTORS
        PROC. IEEE, VOL. 58, NO. 9, PP. 1368-1369, SEPTEMBER 1970.
KOJI, T.
    EFFECT OF EMITTER CURRENT DENSITY ON POPCORN NOISE IN TRANSISTORS
        IEEE TRANS. ELECTRON DEVICES, VOL. ED-22, NO. 1, PP. 24-25,
        JANUARY 1975.
KOJI, T.
    POPCORN NOISE AND GENERATION-RECOMBINATION NOISE OBSERVED IN ION-
    IMPLANTED SILICON RESISTORS
        ELECTRONICS LETTERS, VOL. 11, NO. 9, PP. 185-186, 1 MAY 1975.
LEONARD, P.L. AND S. V. JASKOLSKI
    AN INVESTIGATION INTO THE ORIGIN AND NATURE OF POPCORN NOISE
        PROC. IEEE, VOL. 57, NO. 10, PP. 1786-1788, OCTOBER 1969.
LUQUE, A., J. MULET, T. RODRIGUEZ, AND R. SEGOVIA
    PROPOSED DISLOCATION THEORY OF BURST NOISE IN PLANAR TRANSISTORS
        ELECTRONICS LETTERS, VOL. 6, NO. 6, PP. 176-178, 19 MARCH 1970.
OREN, R.
    DISCUSSION OF VARIOUS VIEWS ON POPCORN NOISE
        IEEE TRANS. ELECTRON DEVICES, VOL. ED-18, NO. 12, PP. 1194-1195,
        DECEMBER 1971.
RODRIGUEZ, T. AND A. LUQUE
    BEHAVIOR OF BURST NOISE UNDER UV AND VISIBLE RADIATION
        SOLID-STATE ELECTRONICS, VOL. 19, NO. 7, PP. 573-575, JULY 1976.
WOLF, D. AND E. HOLLER
    BISTABLE CURRENT FLUCTUATIONS IN REVERSE BIASED P-N JUNCTIONS OF
    GERMANIUM
        JOUR. APPLIED PHYSICS, VOL. 38, NO. 1, PP. 189-192, JANUARY 1967.
WOLF, D. AND E. HOLLER
    BISTABLE CURRENT FLUCTUATIONS IN REVERSE BIASED P-N JUNCTIONS OF
    GERMANIUM
        JOUR. APPL. PHYS., VOL. 38, NO. 1, PP. 189-192, JANUARY 1967.

The Subject

The literature of the last half century on the subject of electron devices shows that the inception of almost every new electron device has been followed by a study of the various noise mechanisms in it (the only exceptions being some power devices). Furthermore, in almost every case, the ultimate goal of the device noise studies is to find ways of reducing noise (unless the device finds an application as a noise generator). But within this general pattern, the individual investigations of device noise have a varied set of aims and methods, from theoretical modeling to experimental measurement, and from studies of the fundamental physical mechanism of noise to the limitation imposed by noise in a particular device application.

The very large number of electron devices that have been studied (or even the smaller number of devices that are of practical interest) makes it unfeasible to provide a comprehensive survey of the entire device noise literature in this volume. At best, the survey can encompass only two or three major families of devices. However, this predicament is not so unfortunate in view of the objectives of this volume. In a broad study of electrical noise, the devices can be viewed as means for understanding rather than the main objective of the study. Then, the study of noise in a subset of devices is sufficient if it brings up most of the significant issues and principles.

The selection of device families for inclusion in this part is dictated by the above criterion as well as by tne available literature on the devices and the limitations of space. The selected families are: 1) electron-beam devices (microwave tubes), 2) semiconductor junction devices (diodes and transistors), and 3) radiation-sensitive devices (optical detectors). The electron beam devices illustrate the space-charge effects and noise smoothing, the semiconductor junction devices illustrate the combining of noise contributions from several individual processes, and the radiation-sensitive devices exemplify the case where one must study the individual carrier pulses rather than their aggregate effect. Several important classes of devices have been left out, including the superconducting and Josephson junction devices, the magnetic devices, the quantum mechanical devices, bulk semiconductor devices of the hot-electron type, and gaseous plasmas.

The subject of noise in electron tubes is well established and mature as measured by the rate of work in the field. This is so in part due to the trend towards replacing electron tubes by solid-state devices, although some electron tubes continue to be heavily used, including photomultiplier tubes, television camera tubes, cathode-ray tubes, and electron-beam tubes which are used as microwave sources and amplifiers. Perhaps a more important reason is that the noise performance of many

types of electron tubes is close to the theoretical optimum. Stated conversely, this implies a good understanding of the noise mechanisms in electron tubes. In particular, the noise in the various types of electron-beam tubes has been studied in detail, and is well explained in terms of shot noise subjected to transformation by the potential minimum in the tube. While an exact analysis of the noise in crossed-field tubes would be very complex, the experimental results are in fair agreement with less sophisticated analyses employing such simplifying approximations as the single-valued velocity assumption.

Semiconductor junction devices form the largest class of electron devices, and the study of noise in semiconductor devices is correspondingly extensive. The noise in semiconductor diodes and transistors can be generally understood in terms of the several basic noise mechanisms that were taken up in the previous part.

The term "radiation detector" also denotes a very wide class of devices. For a microwave engineer, detectors are either linear or square law devices like Schottky-barrier diodes which are small compared to the radiation wavelength. For an optical engineer, radiation detector devices are exemplified by photodiodes where the device dimensions are not negligible compared to radiation wavelength, and the issue of spatial coherence becomes significant. For a nuclear engineer, radiation detectors count (rather than "detect") individual events or particles such as the quanta of gamma "radiation." The characterization, measurement, and analysis of noise in detector devices is correspondingly different for the different classes of detectors. The study of noise in optical detectors serves well as an introduction to the consequences of energy quantization and wave coherence, and the term detector refers to optical detectors here. The generation of noise in photoconductors was the subject of a paper in the last part. The two other significant detectors are avalanche photodiodes and photomultiplier tubes. Avalanche photodiodes are included with semiconductor junction devices and the photomultiplier tubes are separately taken up here.

The Reprinted Papers

The three papers reprinted in this part are devoted to noise in the above three families of devices. Whinnery's paper on microwave tube noise is a survey of the important results obtained before 1959. It gives a historical account of the developments leading to the successive improvements in the noise figures of available tubes, and it emphasizes the physical concepts behind the results.

The paper on noise in semiconductor devices, by van der Ziel, is more recent as well as more extensive in the number and types of devices discussed. In addition to the classical semiconductor junction devices, it covers Gunn diodes, photodiodes, LED's, and lasers, as well as the calculation of noise for devices embedded in circuits. It has a high reference value due to its summaries of a large number of recent results and its bibliography. It also points out some areas requiring further investigation, although the more recent literature should also be consulted in this connection; much of this later development is surveyed in another very recent review article by van der Ziel [1]. An earlier review of noise in transistors is available for a more tutorial introduction [2].

The review of noise in photomultiplier tubes by Kovaleva *et al.* is an exhaustive survey of the early literature on the subject. It covers the noise sources in the tubes and their dependence upon the operating conditions. The pulse height distribution of noise pulses in the tube output is particularly emphasized.

The Bibliography

The first bibliography included in this part are devoted to the noise in various electron-beam microwave tubes. Papers pertaining to the noise in electron beams are included even if they do not refer to a particular microwave tube. But papers concerning nonbeam-type microwave tubes (e.g., microwave triodes) are mostly excluded. No attempt is made to subclassify the bibliography according to the type of tube (klystrons, O-type, and M-type) because of the common noise related ideas. A large fraction of the papers in this bibliography are more than a decade old, correctly reflecting the shift in research interests. However, the subject is still useful from a pedagogical point of view because of the issues it brings out.

No separate bibliography devoted to semiconductor devices is provided in this part. The sizable bibliographies appended to van Vliet's paper in the previous part and van der Ziel's paper in this part should suffice for most purposes. The bibliography on avalanche diode noise generators, incorporated in the next part, is also relevant to noise in semiconductor devices. A significantly more exhaustive bibliography would occupy an intolerably large space. The list of references cited by Kovaleva *et al.* in their paper on photomultiplier tubes is also very comprehensive, although now somewhat dated.

A large number of electron devices have been left out from this part. Understandably, not all electron devices have attracted equal attention in noise studies; for example, the noise in "low-level" or "front-end" electron devices has been investigated in greater detail where it may set the ultimate limits to the system noise performance. Fortunately, the more actively pursued areas of device noise studies are also more likely to have been surveyed. Therefore, a short list of such resource materials is included in this part with the hope of partially compensating for the more significant omissions. This list contains references to tutorial/review articles and bibliographies, devoted either entirely or in part, to noise in devices. Some research papers, usable in this way, are also included in this list, although the list remains inevitably short. To this list may be added some very recent review papers on magnetic and dielectric materials [3], two-terminal active microwave devices [4], photoconductors [5], and other solid-state devices [1].

References

[1] A. van der Ziel, "The state of solid state device noise research," *Physica B & C*, vol. 83, pp. 41–51, May 1976.
[2] A. van der Ziel, "Noise in junction transistors," *Proc. IRE*, vol. 46, pp. 1019–1038, June 1958.
[3] H. Bittel, "Noise in magnetic and dielectric materials," *Physica B & C*, vol. 83, pp. 6–13, May 1976.
[4] E. Constant, "Noise in microwave injection, transit-time, and transferred-electron devices," *Physica B & C*, vol. 83, pp. 24–40, May 1976.
[5] K. M. van Vliet, "Photon fluctuations and their interaction with solids," *Physica B & C*, vol. 83, 52–69, May 1976.

History and problems of microwave tube noise

by *J. R. Whinnery*

Invited lecture given at the Institute for Advanced Electrical Engineering at the Swiss Federal Institute
of Technology, Zurich, on May 4, 1959

Der Artikel gibt eine Übersicht über die neuern Fortschritte für die Klärung der Rauschvorgänge in
Mikrowellenröhren, welche zur Reduktion des Rauschfaktors von etwa 30 db für die ersten Wander-
wellenröhren bis auf 3,5 db für die neuesten Entwicklungen im cm-Wellengebiet führten. Das Nieder-
frequenzverhalten des Rauschens von Elektronenströmen wird besonders im Hinblick auf das Phänomen
der Rauschverminderung im Potentialminimum behandelt. Je nach den Parametern für das Minimum
können sich dabei eine Rauschverminderung oder Instabilitäten einstellen. Entsprechende Theorien
führten auf die Konstruktion von Elektronenkanonen mit einem einzigen Geschwindigkeitssprung oder von
solchen mit progressiver Geschwindigkeit. Für den minimalen Rauschfaktor wurden inzwischen verschie-
dene Ausdrücke gefunden, welche die Wichtigkeit der Verminderung der Stromschwankungen oder der
Korrelation zwischen Geschwindigkeit und Strom zeigen. Die Korrelation zwischen Geschwindigkeit und
Strom ist ein wichtiges Phänomen im Driftgebiet kleiner Geschwindigkeiten; dies scheint einer der Haupt-
gründe für das niedrige Rauschen der Elektronenkanonen vom Currie-Typ zu sein. Schliesslich führt der
Autor einige noch ungelöste Aufgaben in der Theorie des Hochfrequenzrauschens von Elektronenröhren an.

Cet article traite des récents progrès dans l'explication des processus de bruit dans les tubes hyperfréquences·
Ces études ont conduit à une réduction du facteur de bruit de quelque 30 db pour les premiers tubes à ondes
progressives jusqu'à 3,5 db pour les développements modernes dans la région des tubes centimétriques. Les
propriétés de bruit de faisceaux électroniques sont considérées spécialement pour le phénomène de la réduction du bruit au minimum de potentiel. Suivant les paramètres du minimum, on peut obtenir ou une réduction du bruit ou des instabilités. Les expressions trouvées pour le facteur de bruit minimum montrent l'importance d'une atténuation des fluctuations du courant ou d'une corrélation entre le courant et la vitesse. Cette corrélation est un phénomène de première importance dans l'espace de glissement à vitesse basse; ceci semble
être la principale raison pour le faible niveau de bruit des canons du type Currie. L'article se termine par
l'énumération de quelques problèmes non résolus dans l'étude des phénomènes du bruit à haute fréquence.

This paper deals with the recent progress in the understanding of the noise processes in microwave
tubes, which has led to improvements in noise figures from values near 30 db for first traveling-wave
tubes to present record noise figures of the order of 3,5 db in the centimeter-wave region. The low-
frequency behaviour of noise in electron streams is then reviewed, especially with respect to the noise
reduction phenomena at the potential minimum and the studies which led to the single-valued velocity
theory of noise. Verifications of this theory and advancements through its use are described, considering
especially the noise reduction by either velocity-jump or tapered-velocity guns. Further are discussed
expressions for the minimum noise figure, showing the importance of current-smoothing or correlation
between current and velocity. In studying the noise compensation phenomenon at the minimum, it
appears that either noise reduction or instabilities may be possible depending upon the parameters of the
minimum. A most important phenomenon in the low-velocity drift region is the correlation between
current and velocity in such regions; this is believed to be an important reason for low-noise performance
in the Currie guns. Finally, some remaining problems in the understanding of high-frequency noise
are presented.

1. Introduction

The importance of the internal noise problem in the first amplifier of receivers for use with
radar, radio astronomy or communications over links in which external static is at a minimum,
is well known. In the last two cases an improvement in receiver noise figure by 3 db is equivalent
to a doubling of transmitter power; in the first it determines the limits of smallness or distance
of sources from which signals may be received. At this writing, the record for low noise figure
in the microwave range is hold by masers [References 1, 2], equivalent noise temperatures of
only a few degrees absolute having been achieved in these [3]. Present masers are relatively
complicated devices because of the cryogenic temperatures and the high magnetic fields required
but, in spite of this, are finding immediate important applications to radio astronomy. Improvements in packaging on one hand, and promise of future research in materials which might relax
the temperature and field requirements on the other, would make other systems applications
very likely also were it not for the competition of the excellent and presently much simpler
parametric amplifier, which makes the future of masers less clear.

The old idea of amplification by parametric variation of an energy storage element [4] has in the last two years been recognized as a very important one, in spite of its use of r.f. rather than d.c. power as the source of amplified energy, because it too is a low noise method of amplification [2]. Presently obtainable noise temperatures with these are considerably higher than with masers, but appreciably less than room temperature. This is good enough for many radar and communication purposes where external noise sources set the limit of sensitivity, and the strong case for such devices is in their simplicity, particularly for those using semiconducting diodes as the variable element [2, 5]. Thus, these latter are finding immediate systems applications, and it seems certain that this use will grow rapidly in the next few years.

Semiconducting devices of the transistor class have not yet found practical application at microwaves, but the extension to higher frequencies continues each year, and also the understanding of the noise processes in such devices [6, 7, 8], so these also must be considered from the long-term point of view.

In the face of the above new and dramatic developments, it might not seem that conventional microwave tube types, using energy interchange between electromagnetic fields and electrons in vacuum, would be much longer of interest for low-noise amplifiers. The possibility must be considered that the very simple semiconducting parametric amplifier with its excellent noise figure will in fact displace tubes in most preamplifier applications in the forseeable future. However there have also been important developments in the achievement and understanding of results for low-noise tubes in the last several years. The presently attained minimum noise figures are in the same range as for the parametric amplifiers cited [9, 10], and present theory indicates that still lower limits are possible in principle. Recognizing the great flexibility in design for various applications that has been the characteristic of vacuum-electronic devices, it thus seems too early to rule these out of consideration and most certainly from a research point of view the limits of their performance must be understood.

It is the purpose of this survey to describe some of the important steps of the last decade or so in the development of the presently attainable low-noise tubes, and to describe some of the remaining problems that need attention. In giving this survey it is recognized that the very recent book edited by *Smullin* and *Haus* [11] presents an excellent and very complete summary of the status at the time it went to press, but a few important developments have taken place in the brief period since then, and for some, there may also be an advantage in a briefer statement by way of introduction.

2. Basic background of low-frequency noise theory

For comparison of later formulas, we would like to record the well-known formulas for shot and thermal noise. The statistical fluctuation in a temperature-limited current, for which the emission of each electron may be considered independent of the emission of all others, was shown by *Schottky* [12] to give rise to a time-varying fluctuation current having mean-square value

(1)
$$\overline{i^2} = 2\,e\,I_0\,B$$

where e = electronic charge magnitude, I_0 = d.c. current,
and B = equivalent bandwidth of the measuring device.

The formula for mean-square short-circuit fluctuation current from an admittance with real part G at absolute temperature T_e, usually attributed to *Nyquist*[1]), is

(2)
$$\overline{i^2} = 4\,k\,T_e\,G\,B$$
$$k = Boltzmann\text{'s constant} = 1{,}38 \cdot 10^{-23}\ \text{Joule}/°\text{K}$$

---

[1]) Prof. *Strutt* has pointed out to me much earlier work of *Lorentz* [13] and *Slingelandt* [14] giving essentially the result derived by *Nyquist* [15].

In comparing the shot noise predicted by (1) with measured values on actual tubes, it was found that agreement was good when the current was temperature-limited and when other extraneous effects such as interception of current, secondaries, etc. were minimized. When the current was space-charge limited however, very important reductions in the value of noise occurred, the noise in such cases amounting to only a few percent of the temperature-limited value. It is not surprising in a general way to find a modification, since the passages of electrons across the diode are no longer independent events, the emission of one charge group affecting potential and consequently space-charge limited current, as will be described in more detail in the following section.

The analysis of the space-charge reduced shot noise for low frequencies was given by *Schottky* and *Spenke* [16], *Thompson* and *North* [17], and *Rack* [18] and amounted to the solution for the new equilibrium current when an excess charge in any velocity class was introduced at the cathode of the planar diode with *Maxwell*ian distribution of velocities in the unperturbed state. The results could be written either as comparisons with the shot noise or thermal noise formulas:

(3a) $$\overline{i^2} = \Gamma^2\, 2\, e\, I_0\, B$$
(3b) $$= \Theta\, 4\, k\, T_c\, g$$

where
g = small-signal conductance of the diode or transconductance of triode or other multi-element tube
T_c = cathode temperature

When plotted as a function of tube operating conditions as shown in Fig. 1, it is found that for essentially the entire practical operating range of practical tubes asymptotic values of Γ^2 or Θ are valid, shown dotted on the figure. Most interesting and also useful for calculation purposes is the fact that this asymptotic value for Θ is a constant:

(4) $$\Theta \to 3\,(1 - \pi/4) \approx 0.644$$

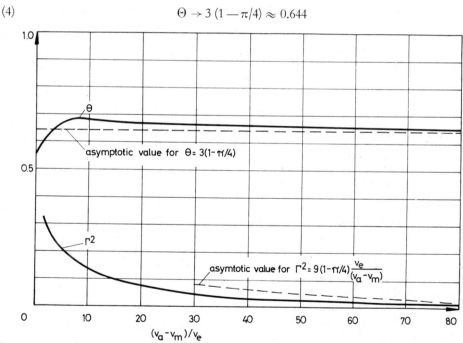

Fig. 1. Space-charge reduction factors for shot noise compared with asymtotic values. Data from [17] for anode current small compared with emission current. V_a = anode potential, V_m = potential minimum value, V_e = voltage equivalent of thermal energy = kT_c/e

221

There had actually been much early speculation from a thermodynamic point of view [19] leading to the form (3b) with remarkably correct values of Θ (around $\frac{2}{3}$), but as the thermionic diode was not a system in equilibrium, the arguments for this had remained in some doubt until verified by the analysis of the physical processes taking place.

For purposes of the later discussions of high frequency phenomena we wish to stress two things: the physical picture of the noise compensation phenomena inherent in all of the above analyses but stated most specifically by *Thompson* and *North* [17], and the very important single-valued velocity hypothesis drawn by *Rack* [18] from the nature of the asymptotic solution.

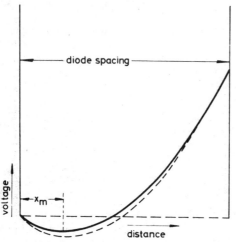

Fig. 2. Illustration of curve of potential versus distance under space-charge limited conditions and perturbed potential by an excess negative charge (dashed curve)

Physical Picture of Noise Compensation

The physical picture of the noise smoothing process by the potential minimum is easily visualized by the sketch of Fig. 2. A noise event consisting of an instantaneous excess group of electrons emitted in any velocity class will, as it proceeds towards the anode, lower the potential distribution everywhere within the diode as shown by the dotted curve of potential versus distance. In particular the decrease in this potential at the plane of the minimum will cause more electrons to be turned back, since the continuous distribution of electron velocities (given to a high degree of approximation by the *Maxwell-Boltzmann* distribution) means that the number which surmount the potential barrier of the minimum depends upon the value of potential there. Thus this self-regulating phenomenon causes a compensation phenomenon for the noise disturbance. It seems clear that the compensation does not take place in exact time synchronysm with the original perturbing noise event, but the time shift is a small part of a period at the lower frequencies so that the perturbance and its compensation may there be considered to add in phase. The modification of this at the higher frequencies will be the subject of later discussion but it is clear in a general way that one might expect the compensation currents to add in the wrong phase and thus destroy the compensation process or even produce an increase in noise in such cases.

The single-valued velocity hypothesis

As noted above, all the low-frequency analyses of space-charge smoothed shot noise, which solved the *Poisson* equation perturbed by the excess charge of the noise event [16, 17, 18] showed the same character of results and in particular the relatively simple asymptotic behaviour described. It is obviously desirable to think about the reasons for such asymptotic behaviour, and a number of physical pictures with approximate analyses which lead just to this asymptotic solution have been discussed in the literature[1]. Of these, most important has been that given by *Rack* who noticed that the asymptotic result could be obtained by considering the noise as an equivalent signal with single-valued velocity and current modulation impressed on the diode at the potential minimum, and solving in conventional manner for the a.c. response to such a signal. He then extended this concept, hypothesizing that the same procedure might hold at the

[1] For a good summary giving several of these pictures see C. F. *Quate*'s discussion in Chapter I of [11]

222

higher transit-time frequencies, and in the last part of his paper [18] applied it to calculate the effects of transit-time on space-charge limited shot noise. Current and velocity modulation equivalent to the r.m.s. values of the noise phenomena, but otherwise treated as signals, are applied at the input plane corresponding to the position of the potential minimum. The a. c. behaviour is found for such a case from the wellknown transit-time equations for a planar diode given in convenient form by *Llewellyn* and *Peterson* [20]. Of the two input modulations, it turned out that the noise velocity modulation was the important one, the current modulation producing negligible effects for large anode voltages. Unfortunately it was not easy to check *Rack*'s results on the high frequency diodes or triodes of the time because of complicating factors such as secondaries in diodes, non-uniformities due to grids of triodes, and total emission effects in all the close-spaced tubes of interest at the high frequencies. Thus his bold and very important assumption remained unverified for several years. It has since had experimental and theoretical justification, and within wide limits remains a most important principle. Certainly it was responsible for nearly the entire progress in noise reduction for the decade 1946...1956, which developments will be described at more length.

3. Single-valued velocity theory of noise

First application to traveling-wave tubes

Although the *Rack* hypothesis had not been completely verified at the time *Pierce*'s first important and extensive analyses of traveling-wave tube behaviour were formulated [21], he felt it the most promising method for attacking noise in such tubes, or at any rate worth trying to see if predicted results could be verified. In its application to such a tube, the mean square velocity fluctuation is impressed as an equivalent signal at the potential minimum (substantially the cathode), and handled up to the accelerating anode by the *Llewellyn-Peterson* equations [20] as in the diode. In the drift region which follows it is handled by the *Hahn-Ramo* space-charge wave equations [22, 23]. And upon entrance into the slow-wave circuit, it is analyzed by the *Pierce* small-signal equations [21] for such interaction devices. Formulation in this way makes it at once clear that the position of the cathode in relation to that of the circuit, and the nature of the accelerating and drifting regions in between, will affect very greatly the noise induced on the circuit in comparison with the useful signal which is injected directly on the slow-wave circuit (Fig. 3). This seems an obvious observation now, but it was not so obvious at the time; for to many the beam noisiness was of such a disorganized nature that it seemed unlikely that position of the circuit along the beam would matter. However the prediction was immediately verified and by careful optimizations using the *Pierce* equations, *Pierce* [21], *Field* and *Quate* [24] among others were able to bring the noise figures of centimeter-wave traveling-tubes from previous typical values of 25...30 db down to values in the range of 15 db.

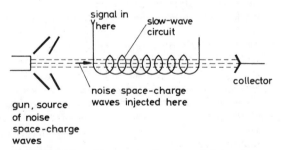

Fig. 3. Illustration of importance of spacing and design of gun to noise induced on r. f. circuit of a traveling-wave tube

The Cutler-Quate experiment and theoretical justifications

A most important experiment was made which showed very clearly that the assumptions inherent in the above picture were valid to a high degree. In this a cavity coupled to the beam by a narrow gap was moved along the beam and the noise output measured (Fig. 4). If the noise does behave to any extent like an equivalent signal, space-charge waves of noise would be predicted in the drifting region with very high ratios of maxima to minima. In the experiment of *Cutler* and *Quate* [25], such standing waves of noise were indeed observed, with the correct spacing between maxima as given by the *Hahn-Ramo* theory, and ratios of maxima to minima of about 20 db.

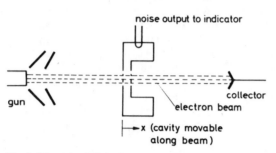

Fig. 4. Schematic of *Cutler-Quate* experiment [25] for measurement of noise standing-waves along a beam

If the beam were an ideal single-velocity, infinite and plane medium and the noise behaved exactly as a signal with only velocity modulation impressed at the initial plane, there would be perfect zeros of noise at the minima, so that standing-wave ratio would be infinite. Possible contributing factors to the finite minimum measured are:

1. Finite gap in the probing cavity,
2. Distribution of d.c. velocities because of thermal effects and space-charge depression at the beam center,
3. Excitation of noise waves by current modulation as well as velocity modulation at the initial plane,
4. Effects from finite beam diameter and space-charge waves with higher-order radial and circumferential variations, each of which waves has a slightly different wavelength,
5. Miscellaneous complicating factors such as secondaries, ions, lens effects and the like.

The first three of the above are subject to fairly good estimate and do not explain standing wave ratios as low as the values measured. The last is difficult to estimate but attempts were of course made to reduce such complications to a minimum. Work of *Rowe* [26] and *Haffter* [21] indicate that the higher order waves (point 4 above) do yield results of the correct order and so is the most likely explanation of the measurements. This is of considerable importance since most noise analyses have been based upon the one-dimensional model with a single space-charge mode of propagation and it is easy to forget the complications of the higher-order modes.

To complete the support of the single-valued velocity theory of noise, we mention here the theoretical justification that has grown up from study of the *Boltzmann* equation by *Knipp* [28], *Gray* [29], and *Siegman* [30], among others. From this it is clear that the consideration of noise as an equivalent signal on a beam is a good approximation for the part of the beam where average velocity is much greater than the velocity spread, which includes the interaction region, the usual drift regions and most of a typical gun except for the region in the immediate vicinity of the potential minimum. The importance of the departures from such conditions in the vicinity of the minimum as well as in artificially produced regions of low velocity will be the subject of later discussion. Much progress was however made by operations on the beam in the region where the single-valued velocity theory was clearly applicable.

Low-noise guns of velocity-jump type

Following the work of *Pierce* et al on application of the single-valued velocity theory to optimization of noise figure by choice of spacing and potential of the drift space between gun and active circuit, and its justification by the *Cutler-Quate* experiment, *Watkins* made the most important single contribution to the problem of noise reduction [31, 32]. He recognized that if noise behaves substantially as a signal on the beam, and that the noise wave excited by velocity modulation is the most important component in normal guns, that portion could be deamplified by proper operations on the beam. The space-charge wave deamplification was achieved in his first guns by allowing the beam to drift a quarter of a plasma wavelength at one d.c. voltage, then suddenly jump to a lower voltage, drift again for a quarter of a plasma wavelength for that voltage, jump up again in voltage, and repeat again if desired.

The basis for the noise reduction in the above type of gun can be seen by referring to the equations for a.c. current and velocity in space-charge waves along a beam, which we will give for simplicity in the form for infinite plane beams [33]:

$$i_1(z) = i_1(0) \cos \beta_p z + j \frac{u_0 v_1(0)}{Z_0 \eta} \sin \beta_p z$$

(5)

$$v_1(z) = v_1(0) \cos \beta_p z + j \frac{\eta Z_0 i_1(0)}{u_0} \sin \beta_p z$$

where

$$Z_0 = \frac{2 V_0 \omega_p}{|i_0| \omega} = \frac{\eta^{1/4}(2 V_0)^{3/4}}{\omega |\varepsilon_0 i_0|^{1/2}}; \qquad \beta_p = \frac{\omega_p}{u_0}, \qquad \eta = \left| \frac{e}{m} \right|$$

u_0 = d. c. velocity, i_0 = d. c. current density, V_0 = d. c. voltage, ε_0 = permittivity of space. Also, at a sudden change in d. c. velocity, energy and continuity conditions show the following:

(6)
$$i_1(z + \delta) = i_1(z - \delta)$$

$$u_0(z + \delta) \cdot v_1(z + \delta) = u_0(z - \delta) \cdot v_1(z - \delta) \qquad \text{with } \delta \to 0.$$

Thus referring to Fig. 5 which shows one section of the gun, we can use the above relations to write:

(7)
$$\left| \frac{v_e}{v_a} \right| = \left| \frac{v_e}{v_d} \right| \cdot \left| \frac{v_d}{v_c} \right| \cdot \left| \frac{i_c}{i_b} \right| \cdot \left| \frac{i_b}{v_a} \right| = \left(\frac{u_{02}}{u_{01}} \right) \left(\frac{Z_{02} \eta}{u_{02}} \right) (1) \left(\frac{u_{01}}{Z_{01} \eta} \right) = \frac{Z_{02}}{Z_{01}} = \left(\frac{V_{02}}{V_{01}} \right)^{3/4}$$

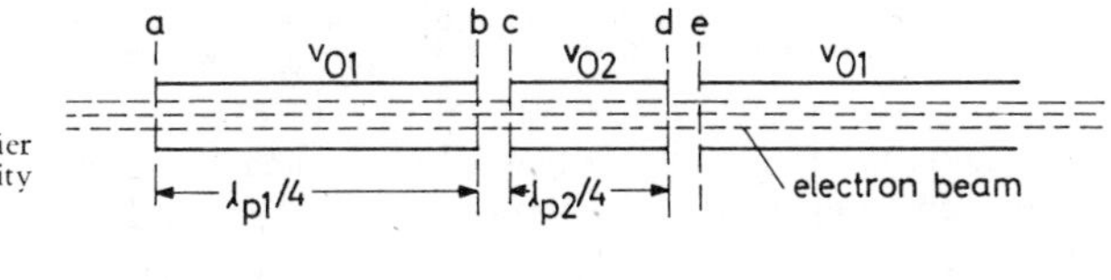

Fig. 5. Schematic of velocity-jump deamplifier for space-charge waves entering with velocity modulation

From this we see that the wave component arising from velocity modulation at the input can be deamplified as desired if $V_{02} < V_{01}$. However at the same time the wave component due to current modulation at the input is amplified since

(8)
$$\left| \frac{i_e}{i_a} \right| = \left| \frac{i_e}{i_d} \right| \cdot \left| \frac{i_d}{v_c} \right| \cdot \left| \frac{v_c}{v_b} \right| \cdot \left| \frac{v_b}{i_a} \right| = (1) \left(\frac{u_{02}}{Z_{02} \eta} \right) \left(\frac{u_{01}}{u_{02}} \right) \left(\frac{\eta Z_{01}}{u_{01}} \right) = \frac{Z_{01}}{Z_{02}} = \left(\frac{V_{01}}{V_{02}} \right)^{3/4}$$

Although this latter component was of small importance in normal guns, it will be increased by the same process that decreases the first contribution and a limit is finally set when the two components become equal. *Watkins* [31] calculated this limit for typical traveling-wave tubes

225

and klystrons, obtaining a value of about 6 db minimum noise figure under the assumption that velocity and current modulations at the potential minimum are uncorrelated, that current modulation corresponds to that of pure shot noise, and that the behaviour beyond the potential minimum in the gun region is given by the *Llewellyn-Peterson* equation. For practical reasons his first guns did not achieve noise figures as low as this, but did yield noise figures in the range of 10 db for the centimeter wave range.

Low-noise guns of tapered-velocity type

Although we have referred to the above low-noise gun as a noise-wave deamplifier, we see that while it is deamplifying one component of the wave it is amplifying the other, so in reality it is changing the ratio of current to velocity modulation in the noise on the beam to achieve the optimum for excitation of minimum noise on the slow-wave circuit. The problem is then similar to one in transmission line matching where one wishes to obtain the correct ratio of voltage to current (impedance) for a given purpose. In fact the development of the transmission line analogue for space charge waves in very complete form, even to the analogy of non-uniform line theory for accelerated beams [34], made this relationship very clear. The low-noise gun from this point of view is then merely a «transducer» for the space-charge waves, and as in transmission line matching sections, might be achieved in many different forms.

A number of workers, and especially *Peter* and co-workers, recognized the advantages of sections with continuous variation of d. c. potential, analogous to a tapered or non-uniform transmission line matching section, as compared with the velocity-jump type which is analogous to a sequence of quarter-wave matching sections. A gun for achievement of the continuous taper might be as pictured in Fig. 6, where almost any desired potential profile along the beam can be obtained with proper d. c. potentials applied to the several discs. However as the exponential taper for a transmission line matching section is known to be simple and close to the optimum in providing wide bandwidth for matching between two impedance levels with a given length of matching section, so the exponential taper on the beam impedance also is one of the best in the sense of providing matching over a wide range of currents and voltages for a given transducer length [35]. This requires a d. c. voltage variation for the infinite planar beam of

(9) $$V = c \, (z - z_0)^{4/3}$$

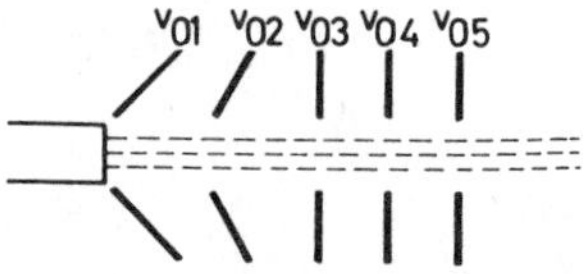

Fig. 6. Schematic of low-noise gun with several electrodes which permits obtaining of a smooth variation of d.c. velocity along the beam

Optimum results were obtained for voltage profiles very near the above, although one of the important practical features of this type of gun is that the several electrodes provide flexibility enough so that almost any potential profile can be obtained along the beam, and the best one adjusted for minimum noise by trial and error. The noise figures achieved by such guns, after attention to other complicating factors which will be discussed in Part 5, were in the range of 5 or 6 db, near enough the ultimate limit calculated by *Watkins* to give some support to the rather drastic assumptions made in that calculation, and for a time it appeared that this would indeed represent the lower practical limit on noise figure for beam type amplifiers at microwaves.

Theories of minimum noise figure

At this point a number of theories on the theoretical minimum noise figure appeared, making the same assumptions as *Watkins* and leading to about the same general conclusion of a minimum noise figure around 6 db under these assumptions [36, 37]. The contribution over the work of

Watkins was in considering general transducers and amplifying means so that the minimum value did not depend upon the specific configurations chosen by *Watkins* for the calculation. Others considered the importance of modifying the basic assumptions especially with regard to correlation [38, 39, 40]. A most general formulation has been given by *Haus* and *Robinson* [40] which made it possible to see clearly what might be done to reduce noise figure beyond the above «minimum» if other input conditions than those first assumed could be obtained.

According to the *Haus-Robinson* formulation, the theoretical minimum noise figure could be written in the form:

$$(10) \qquad F_{\min} = 1 + \left(1 - \frac{1}{G}\right) \frac{2\pi}{k T_c} (S - \Pi) \cdot$$

where

$$G = \text{amplifier power gain}$$
$$T_c = \text{cathode temperature}$$
$$S(\omega) = \sqrt{Z_0 (\Psi_{\max} \cdot \Psi_{\min})}$$
$$\Pi(\omega) = \text{Re } \Theta(\omega)$$
$$\Psi = \text{self-power density spectrum of current fluctuations}$$
$$\Theta = \text{cross-power density spectrum of current and velocity}$$
$$Z_0 = \text{beam characteristic impedance defined in eq. (5)}$$

That is, Ψ and Θ are respectively the *Fourier* transforms of the auto-correlation function of current and the cross-correlation function between current and velocity:

$$(11) \qquad \Psi(\omega) = \frac{1}{2\pi} \int_{-\infty}^{+\infty} f_{ii}(\tau) e^{-j\omega\tau} d\tau$$

$$\Theta(\omega) \quad \frac{1}{2\pi} \int_{-\infty}^{+\infty} f_{iv}(\tau) e^{-j\omega\tau} d\tau$$

$$(12) \qquad f_{ii}(\tau) = \lim_{T\to\infty} \frac{1}{2T} \int_{-T}^{+T} i(t)\, i(t-\tau)\, dt$$

$$f_{iv}(\tau) = \lim_{T\to\infty} \frac{1}{2T} \int_{-T}^{+T} i(t)\, v(t-\tau)\, dt$$

The factors S and Π appearing in (10) are invariants of the beam in the region of drift velocities large compared with velocity spread [30, 40], so nothing can be done to decrease this minimum by operations in that region, although of course the transducer operations to attain that minimum are performed there. Operations in the low-velocity region where S and Π are not invariants are attractive, however. From (10) one could imagine several possibilities for decreasing the minimum noise figure. One could try to maintain the current and velocity uncorrelated ($\Pi = 0$) and decrease the measure of the current fluctuations through some sort of smoothing action to decrease S. The other possibility is to introduce a mechanism which will produce correlation between current and velocity, the ideal of course being a cancellation of the factor S by Π yielding a theoretical minimum noise figure of unity (zero db). In principle it seems possible to achieve both some smoothing action and correlation between current and velocity by operations in the low-velocity region, and the recent important advances have been made by attention to such regions.

Since the «theoretical» minimum noise figure of 6 db is still quoted on occasion, we would like to stress again before leaving this part that the more complete expression (10) gives a theoretical minimum of 0 db once one allows the possibility of correlation. There will of course be important practical problems in achieving the very low noise figures, but this result does encourage continued work on the problem.

4. Importance of the low-velocity region of the gun

In considering improvements in noise figure which might come from phenomena in the low-velocity region of the gun, we wish to examine the high-frequency compensation effects in the vicinity of the potential minimum, the correlation between velocity and current fluctuations introduced in low-velocity drift regions, and possible sorting schemes to change the equivalent velocity distribution.

High-frequency compensation effects of the potential minimum

The picture of the compensation phenomenon for shot noise at low frequencies was given in Part 2. Let us now pay somewhat more attention to the dynamic aspect of this as it is of importance to the high-frequency problem. An excess group of perturbing electrons in a given velocity class emitted at the cathode will begin to move toward the anode, and will either reach it or be returned to the cathode depending upon whether the velocity is greater or less than the value needed to surmount the potential barrier at the minimum. At any place between anode and cathode it will produce electric fields which act to accelerate or decelerate all other electrons within the diode. Of most importance is the deceleration of the electrons behind the perturbing group in time and in a velocity class such that in the unperturbed state they had just sufficient energy to surmount the potential barrier at the minimum. This latter group will be referred to in the following as the group of «critical» electrons. The deceleration of this group will cause some part of it to be turned back toward the cathode. This was the origin of the compensation effect already described, but to follow its behaviour in time, we recognize two important things:

1. The electrons returning to the cathode leave a hole in the distribution passing on to the anode, which acts as positive perturbing charge (Fig. 7), and the «compensation pair» of returning charge and hole in turn influence other critical electrons, leading to a kind of feedback phenomenon.

2. The total compensation current at the potential minimum at any instant is a result of the combined

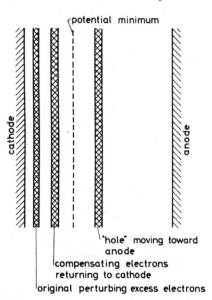

Fig. 7. Diode with perturbation charge and a typical elemental pair of compensation charge and corresponding hole which left the minimum at a certain earlier time

action of the original perturbing excess charge and the totality of all earlier compensation pairs, and could scarcely have a time function that would cancel the effect of the original disturbance for all times (i. e. for all frequency ranges).

Looking more specifically at Fig. 7, we see that critical electrons arriving at the plane of the minimum at time t will have had an integrated effect on velocity both from fields of the original disturbance and all earlier compensation pairs. An approximate integral equation for the compensation current at the plane of the minimum may then be derived [41, 45] in the form

$$(13) \qquad j(\tau) = \frac{1}{2\sqrt{\pi}}\left[h_0(\tau) - \int_0^{\tau} j(\tau')\, h_1(\tau - \tau')\, d\tau'\right]$$

where $j(\tau)$ is the normalized compensation current as a function of normalized time τ

$h_0(\tau)$ is a function of diode parameters giving effect of original disturbance on j

$h_1(\tau)$ is a function of diode parameters giving effect of one compensation pair on j

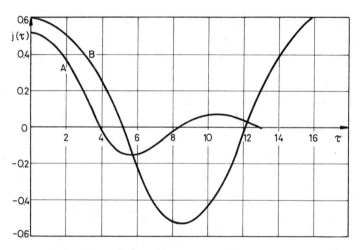

Fig. 8. Curves from approximate analysis of [45] for compensation current as a function of time for two different conditions of the minimum:

curve A: (Saturation current/anode current) = 5, initial velocity of compensation charges = $\sqrt{2\,\tau_i\,V_e}$
curve B: (Saturation current/anode current) = 11, initial velocity of compensation charges = $\sqrt{2\,\tau_i\cdot 0{,}215\,V_e}$

A plot of compensation current from (13) for typical parameters of the minimum and for an infinite-spaced diode short-circuited to a. c. shows a curve somewhat of the form of a damped sinusoid (curve A of Fig. 8), showing that there is at first an overcompensation phenomenon with a required reversal in current direction, repeating until an equilibrium is reached. However for certain parameters of the minimum, the curve may not be damped as the feedback phenomenon of compensation currents upon themselves may cause a continued oscillation. Such a case is shown in curve B of Fig. 8. The frequency spectra of the two curves of Fig. 8 are shown in Fig. 9, along with corresponding curves from a similar approximate method by *Siegman* and *Watkins* [43] done for the open-circuit diode, and the results of *Tien* and *Moshman* [42] from a very elegant Monte Carlo method for a short-circuited diode of specific spacing. Results of the last-named analysis have been verified at least in part by a very careful distribution function analysis of *Löcherer* [44]. A study of the approximate picture of the high-frequency compensation phenomenon in its relation to the various analyses cited has led us to these conclusions:

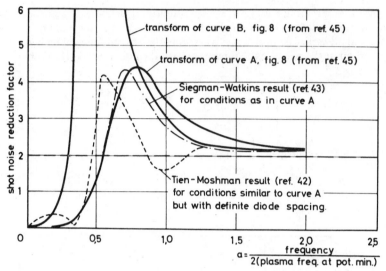

Fig. 9. Shot noise reduction factor as a function of frequency compared for several analyses

1. Analysis of the diode open-circuited to a. c. gives the same results as for one short-circuited to a. c. when diode spacing is infinite, as might be expected.

2. The results for short- and open-circuited analyses are different on diodes of finite spacing, the total transit angle across the diode becoming an important parameter. It is thus believed that the dips in noise at certain frequencies predicted by the analysis of *Tien* and *Moshman* (occurring at total diode transit angles of approximately π and 3π in their example) are a consequence of the finite diode spacing.

3. In more general terms it appears that depending upon diode parameters either noise compensation or noise build-up phenomena can be obtained by the interaction of the original charge and the compensation pairs. The build-up phenomenon is believed to be responsible for the high noise observed in certain magnetron-type cathodes when a potential minimum is formed. The compensation phenomenon may help to reduce noise figures at the higher frequencies, at least in certain frequency ranges.

Correlation between velocity and current in low-velocity drift regions

The expression (10) shows that correlation between the velocity and the current fluctuations can lead to reduced noise figures. Arguments have been given that correlation between velocity and current fluctuations should be zero in temparature-limited emission [46]. Because of the invariance of the quantities S and Π in regions where velocity is large compared with velocity spread, no correlation is introduced there. However, it seems likely that correlation will be introduced in the low-velocity region where average velocity and velocity spread are of the same order.

The first reason for expecting correlation between velocity and current in the low velocity regions might be the processes occurring at the potential minimum itself, described in the last section. This is because a fluctuation in number of electrons in a given velocity class will affect the instantaneous average velocity and also the current passing the minimum at any instant. However *Watkins* has pointed out [47] that if the velocity sorting action at the minimum simply works on the *Maxwell*ian distribution, there is no change in average instantaneous velocity since the tail of a *Maxwell*ian distribution is also *Maxwell*ian and will have the same instantaneous average, only with smaller current. The approximate analysis of [41, 45] also indicates that the compensation process is not a very strong function of the velocity group in which the excess charge finds itself, so this also supports the view of small correlation introduced by the action of the potential minimum plane. Finally it is noted that *Tien* and *Moshman* [42] calculated correlation between velocity and current at a plane just to the right of the minimum and found a negligible amount. (The distance selected by them was $1,5 \cdot 10^{-3}$ mm beyond the minimum, and by the argument below, not much correlation would be expected by drifting in such a distance.)

For the low velocity region just to the right of the minimum, or an artificially lengthened low-velocity drifting region, correlation might be expected by conversion of velocity variations into current variations, by drifting and bunching actions as in klystrons, but here in very short distances because of the low velocities. (The equations (5) would show conversion of a. c. velocity to current in 0.012 mm for current density of 0.3 A/cm² and average velocity corresponding to 0.1 V.) *Siegman, Watkins* and *Hsieh* [48] made a calculation by numerical methods for such correlation effects in the region to the right of a minimum, using the *Fry-Langmuir* potential variation of potential, and found correlation coefficients of the order of 0.3. A somewhat different model studied by *Pierce* [49], consisting of a beam suddenly accelerated to the small average velocity, then drifting at this low velocity, also showed appreciable correlation effect in the low-velocity region. A measurement was made by *Saito* [50] on a fairly typical gun which might be assumed to have a potential variation in the low-velocity part similar to the *Fry-Langmuir* solution. This measurement was made by comparing outputs in magnitude and phase from two cavities, one placed to be responsive to convection current arising from current fluctuations at the input plane and the other related to the velocity fluctuations there. The amount of measured correlation was close to that calculated in the *Siegman-Watkins-Hsieh* model, and would predict a minimum attainable noise figure from such guns (eq. 10) in the range of 4 db.

The above, and particularly the calculations of *Pierce* [49], indicate that more correlation might be obtained and still lower noise figures achieved, if the low-velocity drift region could be artificially lengthened. At this time it is believed that this is probably the main effect in the *Currie* guns [51] which have given the 3.5 db noise figures, since the high focusing electrode potential, followed by a low first anode potential, gives a potential profile somewhat as shown in Fig. 10, with the typical lengthened low-velocity region. A gun with many electrodes to permit the obtaining of nearly any potential profile along the beam has been tried by *Shaw, Siegman* and *Watkins* [52] and has also confirmed the importance of low-velocity regions.

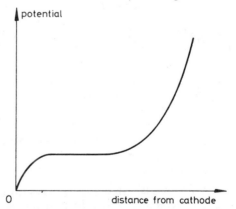

Fig. 10. Potential profiles obtainable in *Currie* gun [51] showing low-potential drift region

Velocity sorting and other schemes

In *Currie*'s first work on his gun, one attempt was to produce a velocity sorting scheme by the action of the strong transverse electric gradients and the focusing magnetic field near the cathode, in an attempt to produce a smaller noisiness by a narrower velocity spread [51]. It is possible that this is in fact one of the processes in his gun, or if not that such velocity sorting schemes might be made to be successful. A second extension would be to study the effect when velocity in later sections of the gun is reduced to such a low value that other virtual cathodes are obtained. Whatever phenomena take place in the original potential minimum, smoothing or instability, might be expected in exaggerated form in the case of one or more additional virtual cathodes.

5. Miscellaneous complicating effects in low-noise tubes[1]

Partition Noise

The classical analysis of partition noise is made for an infinitely fine grid in which every electron is assumed to have the same probability of collection as all others [53]. This leads to the formula for shot noise in such cases:

$$(14) \qquad \overline{i_n^2} = \left[1 - \frac{I_n}{I_a}(1 - \Gamma^2)\right] 2 e I_n B$$

where

Γ^2 = space-charge reduction factor as defined in eq. (3a)

I_n = current in beam *after* passing through grid

I_a = current in beam *before* passing trough grid

A considerable increase in noisiness occurs if $\Gamma^2 \ll 1$ and $\left(1 - \dfrac{I_n}{I_a}\right)$ becomes appreciable.

In microwave beam-type tubes the above model is not applicable as no grids are used as such, a magnetic constraining field is applied, and the outer electrons near the beam edge have the greatest probability of collection. *Beam* [54] has analyzed a model more realistic for such cases with the result for current and velocity fluctuations after interception:

$$\overline{i_2^2} = \Omega_1^2 \, \overline{i_1^2} + \Omega_2 \, 2 e I_0 B$$

$$(16) \qquad \overline{v_2^2} = \overline{v_1^2} + \frac{\Omega_2}{\Omega_1^2} \frac{2 e B}{I_0} \left(\frac{k T_c}{m u_0}\right)^2$$

where

$$\Omega_1 = \frac{1}{I_0} \int_A k_0 \, i_0 \, dA$$

$$(17) \qquad \Omega_2 = \frac{1}{I_0} \int_A k_0 (1 - k_0) \, i_0 \, dA$$

and I_0 is total current, i_0 the current density at any position of the beam cross-section and k_0 the average fraction of the current intercepted at that position.

For low-noise operation it is still important to keep current collection on the gun electrodes and initial part of the circuit to a small value. This is usually done by application of a confining magnetic field and raising this to a sufficiently high value to obtain the minimum collection.

---

[1] An excellent summary of all of the effects to be discussed in this part is given by R. W. *Peter* in Chapter 5 of [11]

Obviously gun electrodes and circuit diameter must not be too close to the cathode diameter to make such adjustments impossible, although for good interaction it is desirable to have them as close as possible to the beam edge. Careful alignment of the tube is clearly important in keeping collection to a minimum. It is believed that partition noise represents a negligible contribution in low-noise tubes designed with these precautions.

Secondary electrons

Assuming current collection on gun electrodes and r. f. circuit kept to a minimum, the only secondaries of importance are those from the collector itself, but if these are allowed to return through the interaction circuit they can have serious effects on noise [55]. Observed degradations in noise figure of several decibels from such returned secondaries are common in such cases. It is not hard to keep the secondaries of low velocity from returning by applying only a small difference of potential between collector and r. f. circuit. However the group of secondaries which have velocities of the same order as the primary velocity (frequently called reflected primaries) are not helped much by this process. The additional acceleration of the striking electrons is canceled by an equal deceleration of the returning group, and the yield of this group is insensitive to primary voltage. Selection of anode material can help in the reduction of this group of secondaries, but it has frequently been found necessary to design the magnetic fringing field in the vicinity of the collector to give more definite assurance that such electrons would not be returned [55].

Non-uniformity of cathode emission

It has been found by many developers of low-noise tubes that smoothness of the cathode surface and uniformity of cathode emission are very important to low-noise performance. Special attention to these factors has frequently given improvements in noise figure of a decibel or more after other parameters were carefully adjusted. Conversely, poor noise performance of a tube otherwise electrically similar to good ones has frequently been traced to poor uniformity in the emission. Beam [56] has analyzed some of the degradation that might be expected from excitation of higher order space-charge waves by variation of emission across the cathode surface, and found important effects. There are probably still other important effects associated with the behavior of the potential minimum in cases of such non-uniform emission, particularly if current smoothing by the minimum or correlation effects of the type described in Part 4 are utilized.

Lens effects

Inhomogeneous electric fields, as from an electrostatic lens, would be expected to have effects upon the beam somewhat like partition noise: electron rays in different parts of the beam would be delayed by different amounts. Such effects have occasionally been called «partition noise of the second kind», since there is some similarity to the inhomogeneous effects across the beam cross-section produced by the more drastic collection of electrons described above. Most important is the coupling of the transverse fluctuations to the longitudinal fluctuations, the latter of these having been reduced by the beam transducers and any space-charge smoothing.

Knechtli [57] has analyzed the importance of such lens effects and found theoretically and experimentally increases in beam noisiness around 2.5 db in passing through an Einzel lens. It is thus desirable to avoid strong lens effects, as from sudden velocity changes in the beams of low-noise tubes, and it is one of the incidental reasons why the tapered velocity schemes described in Part 3 are to be preferred to the velocity-jump schemes for noise reduction.

Growth of noise waves along Brillouin-flow beams

Smullin and co-workers in attempting the equivalent of the *Cutler-Quate* experiment on *Brillouin*-flow beams, found conditions for which the noise suddenly started to grow to saturation levels at certain critical positions along the beam [58]. This phenomenon was especially puzzling since a single-frequency signal, lying in the same band as used for the noise measurements, did not exhibit such growth phenomena. *Rigrod* [59] and *Louis* [60], in studying this problem, have attributed the basic growth phenomenon to rippled-stream amplification [61] in the beam of rippling diameter, this growth taking place at a much higher frequency than that of measurement and hence its lack of observance with the chosen signal. *Rigrod* postulated the non-linear mixing phenomena as the mechanism for transfer of the energy to the band of measurement, whereas *Louis* has made a calculation indicating that collision phenomena may actually be important in this transfer process. Because of these effects, most low-noise work has been done on beams with confining-flow focusing.

Ion noise

Noise from the statistical «collision» of electrons with residual ions is also somewhat similar to partition noise and may cause a measurable effect in tubes with an imperfect vacuum. This can be eliminated by proper attention to the vacuum, *Peter* [11] stating that it is negligible for pressures lower than 10^{-6} mm Hg.

6. Conclusions

This brief survey of the state of the art in low-noise microwave tube research was intended to give some feeling for the rapid rate of change in this field over the last decade especially, and which is still taking place. As we have indicated, there is now believed to be no basic lower limit to the theoretically obtainable noise figures of such tubes, smoothing and correlation phenomena allowing a noise figure of unity (zero db) in principle. There are obviously many difficult practical matters to be faced in making any significant improvements over present record low noise figures of the order of 3.5 db, but the flexibility of design and use which are characteristic of vacuum electronic devices makes such effort—in at least understanding the remaining problems—very much worth while.

Some of the problems for additional study which seem important to this writer are:

1. Completion of understanding of the processes in the potential minimum, not only for the planar model but also with nonplanar effects, and with a transverse magnetic field present. Many ingenious measurements as well as theoretical analysis will be necessary for this understanding.

2. Continuing study of the limits of correlation possible by drifting in low velocity regions, and the best ways of designing such regions.

3. Study of the effects of virtual cathodes following the first potential minimum.

4. Study of the significance of all of the above advances in noise behavior to transverse interaction devices and electron beam type parametric amplifiers.

References

[1] *Gordon J. P.; Zeiger H. J.* and *Townes C. H.*: «The Maser—a new type of microwave amplifier, frequency standard, and spectrometer.» Phys. Rev. *99* (1955), p. 1264...1274.

[2] *Heffner H.*: «Solid-state microwave amplifiers.» Trans. IRE, MTT-7 (1959), p. 83...91.

[3] *McWhorter A. L.* and *Arams F. R.*: «System noise measurement of a solid state maser.» Proc. IRE *46* (1958), p. 913...914.

[4] *Hartley R. V. L.*: «Oscillations in systems with nonlinear reactance.» Bell Syst. Tech. Jour. *15* (1936), p. 424...440.

[5] *Knechtli R. C.* and *Weglein W. G.*: «Low noise parametric amplifier.» Proc. IRE *47* (1959), p. 584...585.

[6] *Schneider B.* and *Strutt M. J. O.*: «Theory and experiments on shot noise in silicon p-n junction diodes and transistors.» Proc. IRE *47* (1959), p. 546...554.

[7] *Guggenbühl W.* and *Strutt M. J. O.*: «Theory and experiments on shot noise in semiconductor diodes and transistors.» Proc. IRE *45* (1957), p. 839...854.

[8] *van der Ziel A.*: «Shot noise in junction diodes and junction transistors.» Scientia Electrica IV (1958), p. 11...21.

[9] *Currie M. R.*: «A new type of low-noise electron gun for microwave tubes.» Proc. IRE (1958), p. 911.

[10] *Caulton M.* and *John G. E. St.*: «S-band traveling-wave tube with noise figure below 4 db.» Proc. IRE *46* (1958), p. 911.

[11] *Smullin L. D.* and *Haus H. A.* (eds.): «Noise in electron devices.» Technology Press MIT and Wiley 1959.

[12] *Schottky W.*: «Über spontane Stromschwankungen in verschiedenen Elektrizitätsleitern.» Ann. Phys. *44* (1918), p. 541...567.

[13] *Lorentz H. A.*: «Les théories statistiques en thermodynamique.» Teubner, Leipzig 1916.

[14] *van Slingelandt J.*: «Fluctuations in electrical and optical processes.» (in Dutch) Dissertation Leiden 1919.

[15] *Nyquist H.*: «Thermal agitation of electric charge in conductors.» Phys. Rev. *32* (1928), p. 110...113.

[16] *Schottky W.* and *Spenke E.*: «Die Raumladungsschwächung des Schroteffektes.» (I und II) Wiss. Veröff. Siemens-Werke *16*, II, p. 1...41.

[17] *Thompson B. J.* and *North D. O.*: «Fluctuations in space-charge limited currents at moderately high frequencies.» (Parts I and II) RCA Rev. *4* (1940), p. 269...285 and 441...472.

[18] *Rack A. J.*: «Effect of space charge and transit time on the shot noise in diodes.» Bell Syst. Tech. Jour. *17* (1938), p. 592...619.

[19] *Llewellyn F. B.*: «A study of noise in vacuum tubes and attached circuits.» Proc. IRE *18* (1930), p. 243...265.

[20] *Llewellyn F. B.* and *Peterson L. C.*: «Vacuum tube networks.» Proc. IRE *32* (1944), p. 144...166.

[21] *Pierce J. R.*: «Traveling-wave tubes.» Van Nostrand 1950.

[22] *Hahn W. C.*: «Small signal theory of velocity-modulated electron beams.» Gen. Elec. Rev. *42* (1939), p. 258...270.

[23] *Ramo S.*: «Space charge and field waves in an electron beam.» Phys. Rev. *56* (1939), p. 276...283.

[24] *Field M. L.* and *Quate C. F.*: Quarterly progress reports, Stanford University Electronics Research Laboratory 1948...1949.

[25] *Cutler C. C.* and *Quate C. F.*: «Experimental verification of space-charge and transit-time reduction of noise in electron beams.» Phys. Rev. *80* (1950), p. 875...878.

[26] *Rowe H. E.*: «Noise analysis of a single-velocity electron gun of finite cross section in an infinite magnetic field.» IRE Trans. ED-2, Nr. 1 (1953), p. 36...46.

[27] *Haffter T. F.*: «Über das Rauschen von Elektronenstrahlen.» Diss. ETH Zürich 1956.

[28] *Knipp J. K.* Chapters 3 and 5 of *Hamilton D. R.*, *Knipp K.* and *Kuper J. B. H.*: «Klystrons and microwave triodes.» Mc Graw-Hill 1948.

[29] *Gray F.*: «Electron streams in a diode.» Bell Syst. Tech. Jour. *30* (1951), p. 830...854.

[30] *Siegman A. E.*: «Analysis of multivelocity electron streams by the density function method.» Jour. Appl. Phys. *28* (1957), p. 1132...1138.

[31] *Watkins D. A.*: Dissertation Stanford University 1951.

[32] *Watkins D. A.*: «Noise reduction in beam type amplifiers.» Proc. IRE *40* (1952), p. 65...70.

[33] *Beck A. H. W.*: «Space-charge waves.» Pergamon Press 1958.

[34] *Bloom S.* and *Peter R. W.:* «Transmission-line analog of a modulated electron beam.» RCA Rev. *15* (1954), p. 95...112.

[35] *Eichenbaum A.* and *Peter R. W.:* «The exponential gun — a low noise gun for traveling-wave amplifiers.» RCA Rev. *20* (1959), p. 18...56.

[36] *Pierce J. R.:* «A theorem concerning noise in electron streams.» Jour. Appl. Phys. *25* (1954), p. 931...933.

[37] *Bloom S.* and *Peter R. W.:* «A minimum noise figure for the traveling-wave tube.» RCA Rev. *15* (1954), p. 252...267.

[38] *Bloom S.:* «The effect of initial noise current and velocity correlation on the noise figure of traveling-wave tubes.» RCA Rev. *16* (1955), p. 179...196.

[39] *König H. W.:* «Über die Beeinflussbarkeit der Korrelationsverhältnisse in Raumladungswellen.» Arch. Elektr. Übertr. *10* (1956), p. 339...342.

[40] *Haus H. A.* and *Robinson F. N. H.:* «The minimum noise figure of microwave beam amplifiers.» Proc. IRE *43* (1955), p. 981...991.

[41] *Whinnery J. R.:* «Noise phenomena in the region of the potential minimum.» Trans. IRE, ED-1 (No. 4) (1954), p. 221...237.

[42] *Tien P. K.* and *Moshman J.:* «Monte Carlo calculation of noise near the potential minimum in a high-frequency diode.» Jour. Appl. Phys. *27* (1956), p. 1067...1078.

[43] *Siegman A. E.* and *Watkins D. A.:* «Potential-minimum noise in the microwave diode.» Trans. IRE, ED-4 (1957), p. 82...86.

[44] *Löcherer K. H.:* «Rauschen von Raumladungsdioden im Laufzeitgebiet bei Berücksichtigung der Maxwellschen Geschwindigkeitsverteilung der Elektronen.» Arch. Elektr. Übertr. *12* (1958), p. 225...236 and 265...270.

[45] *Whinnery J. R.:* «High-frequency effects of the potential minimum on noise.» Submitted for publication to the Transactions on Electron Devices of the IRE.

[46] *Yadavalli S. V.:* «Cross-correlation between velocity and current fluctuations in tube noise.» Jour. Appl. Phys. *26* (1955), p. 605...608.

[47] *Watkins D. A.:* «Noise at the potential minimum in the high-frequency diode.» Jour. Appl. Phys. *26* (1955), p. 622...624.

[48] *Siegman A. E., Watkins D. A.* and *Hsieh H. C.:* «Density function calculations of noise propagation on an accelerated electron beam.» Jour. Appl. Phys. *28* (1957), p. 1138...1148.

[49] *Pierce J. R.:* Paper presented at International Congress on Microwave Tubes. London, May 1958.

[50] *Saito S.:* «New method of measuring the noise parameters of an electron beam.» Trans. IRE, ED-5 (1958), p. 264...275.

[51] *Currie M. R.* and *Forster D. C.:* «New mechanism of noise reduction in electron beams.» Jour. Appl. Phys. *30* (1959), p. 94...103.

[52] *Shaw A. W., Siegman A. E.* and *Watkins D. A.:* «Reduction of electron-beam noisiness by means of a low-potential drift region.» Proc. IRE *41* (1959), p. 334...335.

[53] *North D. Ò.:* «Fluctuations in space-charge-limited currents at moderately high frequencies. Part III: Multi-collectors.» RCA Rev. *5* (1940), p. 244...260.

[54] *Beam W. R.:* «Interception noise in electron beams at microwave frequencies.» RCA Rev. *16* (1955), p. 551...579.

[55] *Peter R. W.* and *Ruetz J. A.:* «Influence of secondary electrons on noise factor and stability of traveling-wave tubes.» RCA Rev. *14* (1953), p. 441...452.

[56] *Beam W. R.:* «Noise wave excitation at the cathode of a microwave beam amplifier.» Trans. IRE, ED-4 (1957), p. 226...234.

[57] *Knechtli R. C.:* «Effect of electron lenses on beam noise.» Trans. IRE, ED-5 (1958), p. 84...88.

[58] *Smullin L. D.* and *Fried C.:* «Microwave-noise measurements on electron beams.» Trans. IRE, ED-1, No. 4 (1954), p. 168...183.

[59] *Rigrod W. W.:* «Noise spectrum of electrical beam in longitudinal magnetic field.» Bell. Syst. Tech. Jour. *36* (1957), p. 831...878.

[60] *Louis H.:* «Untersuchungen über das anomale Rauschen in magnetisch fokussierten Elektronenstrahlen.» Tech. Mitt. Schweiz. PTT *36* (1958), p. 333...348.

[61] *Birdsall C. K.:* «Rippled wall and rippled stream amplifiers.» Proc. IRE *42* (1954), p. 1628 ff.

Address of the Author:

J. R. Whinnery, Prof. of El. Engrg., Univ. of California, Berkeley, Calif. USA.

Noise in Solid-State Devices and Lasers

ALDERT VAN DER ZIEL, FELLOW, IEEE

Invited Paper

Abstract—A survey is given of the most important noise problems in solid-state devices. Section II discusses shot noise in metal-semiconductor diodes, p-n junctions, and transistors at low injection; noise due to recombination and generation in the junction space-charge region; high-level injection effects; noise in photodiodes, avalanche diodes, and diode particle detectors, and shot noise in the leakage currents in field-effect transistors (FETs). Section III discusses thermal noise and induced gate noise in FETs; generation–recombination noise in FETs and transistors at low temperatures; noise due to recombination centers in the space-charge region(s) of FETs, and noise in space-charge-limited solid-state diodes. Section IV attempts to give a unified account of 1/f noise in solid-state devices in terms of the fluctuating occupancy of traps in the surface oxide; discusses the kinetics of these traps; applies this to flicker noise in junction diodes, transistors, and FETs; and briefly discusses flicker noise in Gunn diodes and burst noise in junction diodes and transistors. Section V discusses shot noise in the light emission of luminescent diodes and lasers, and noise in optical heterodyning. Section VI discusses circuit applications. It deals with the noise figure of negative conductance amplifiers (tunnel diodes and parametric amplifiers), and of FET, transistor, and mixer circuits. In the latter discussion capacitive up-converters, and diode, FET, and transistor mixers are dealt with.

TABLE OF CONTENTS

Manuscript received January 12, 1970; revised April 27, 1970. *This invited paper is one of a series planned on topics of general interest—The Editor.*

The author is with the Department of Electrical Engineering, University of Minnesota, Minneapolis, Minn. 55455.

I. INTRODUCTION

THERE are five main sources of noise that play a part in solid-state devices. The first source is *shot noise*, caused by the random emission of electrons or photons, or the random passage of carriers across potential barriers. The second source is *thermal noise*, usually caused by the random collision of carriers with the lattice, but generally found in conditions of thermal equilibrium. The third source is *partition noise*, caused when a carrier current is split into two parts that flow to different electrodes, or a photon current that is split into two parts, e.g., a part that emits zero electrons per photon and a part that emits one electron per photon in photoemission processes. The fourth source is *generation–recombination noise* (GR noise) that is caused by the random generation and recombination of hole–electron pairs, or the random generation of carriers from traps, or the random recombination of carriers with empty traps. Finally there is *flicker noise or 1/f noise*, characterized by a $1/f^{\alpha}$ spectrum with $\alpha \simeq 1$.

Reprinted from *Proc. IEEE*, vol. 58, pp. 1178–1206, Aug. 1970.

It is the aim of this paper to review the most important effects of these noise sources in lasers and solid-state circuits and devices. It should be clear to the reader that even a rather lengthy review paper cannot cover *all* the detailed aspects of noise in electron devices. For this the reader is referred to the original literature.

II. Shot Noise in Diodes and Transistors [155], [170]

In discussing noise in p-n junction diodes and transistors one must distinguish between low injection and high injection. Low injection in an abrupt p-n junction means that $p \ll N_d$ everywhere in the n-region and $n \ll N_a$ everywhere in the p-region; here p and n are the hole and electron concentrations and N_d and N_a are the donor and acceptor concentrations, respectively. At high-injection levels $p > N_d$ in part of the n-region of a p^+n diode and $n > N_a$ in part of the p-region of a pn^+ diode.

At low injection the passage of carriers across barriers can be considered as a series of independent random events [147], [152]. As a consequence full shot noise should be associated with each current flowing in the device. At high frequencies the noise may be modified by transit time or diffusion time effects. This approach is called the "corpuscular approach"; it breaks down at high injection, since the carriers are no longer independent in that case. For many cases, however, it should be a good first approximation.

Another approach considers the noise as being caused by the random processes of diffusion and/or recombination [114], [120], [121], [151]. The noise sources to be introduced here are diffusion and generation–recombination noise sources. This is called the "collective approach."

It has been shown that the two approaches are equal at low-injection levels [157]. At high-injection levels only the collective approach can be used and the diffusion and recombination noise sources must be properly modified (Section II-E) [171].

A. Shot Noise in Metal-Semiconductor (Schottky Barrier) Diodes

The current in a metal-semiconductor diode[1]

$$I = I_0(V)\left[\exp\left(\frac{qV}{mkT}\right) - 1\right] \quad (1)$$

consists of two parts, a current $I_0(V) \exp(qV/mkT) = (I + I_0)$ due to carriers flowing from the semiconductor into the metal, and a current $-I_0(V)$ due to carriers flowing from the metal into the semiconductor. Both currents should fluctuate independently and each should show full shot noise. Hence one would expect

$$S_i(f) = 2q[(I + I_0) + I_0] = 2q(I + 2I_0). \quad (2)$$

This agrees essentially with Weisskopf's early formula [175]. Since the low-frequency (LF) conductance g_0 may be written

$$g_0 = \frac{dI}{dV} \simeq \frac{q}{mkT} I_0 \exp\left(\frac{qV}{mkT}\right) = \frac{q}{mkT}(I + I_0), \quad (3)$$

(2) may be written

$$S_i(f) = 2mkTg_0\left(\frac{I + 2I_0}{I + I_0}\right) \quad (4)$$

corresponding to full thermal noise at zero current ($I = 0$) and to half thermal noise at large currents ($I \gg I_0$) if $m = 1$. In addition the series resistance r of the diode should show full thermal noise:

$$S_e(f) = 4kTr. \quad (5)$$

In metal-semiconductor mixer diodes one should make the series resistance r small in comparison with the LF diode resistance $R_0 = 1/g_0$. This can be done by proper diode design. For large forward bias (R_0 small) the effect of the series resistance r can be quite significant. The spectral intensity $S_{sc}(f)$ of the short-circuit noise current is then, if $R_0 = 1/g_0$,

$$S_{sc}(f) = \frac{2mkTR_0 + 4kTr}{R^2} \quad (6)$$

where $R = R_0 + r$. This increases from the value $2mkT/R_0$, for $R_0 \gg r$, to the value $4kT/r$, for $R_0 \ll r$.

In point contact diodes the current I flows through a very small cross section at the contact and hence considerable heating occurs. As a consequence T can be larger than the temperature T_0 of the environment. Therefore, if (6) is written

$$S_{sc}(f) = n_1 \cdot 4kT_0/R, \quad (6a)$$

it may sometimes happen that n_1 is somewhat larger than unity for forward bias. For a well designed Schottky barrier diode, n_1 can be as low as $1/2$ if $m \simeq 1$. This is one of the reasons why Schottky barrier diodes are gradually replacing point contact diodes as microwave mixers. One of the other reasons is low $1/f$ noise.

At high frequencies, transit-time effects across the barrier should be taken into account. This problem, which should be significant in the high gigahertz range, has not yet been solved. Equations (2) through (6a) have been well verified by experiment [87].

B. Shot Noise in p-n Junction Diodes at Low Injection

The theory of the previous section can also be applied to p-n junction diodes at low injection. For the sake of simplification we assume that the diode is a p^+n diode so that practically all current is carried by holes. The current

$$I = I_0\left[\exp\left(\frac{qV}{kT}\right) - 1\right] \quad (7)$$

then consists of two parts, a part $(I + I_0)$ due to holes injected into the n-region and recombining there, and a part $-I_0$ due to holes generated in the n-region and being collected by the p-region. Both currents should show full shot noise and

---

[1] In this equation the factor m, unity at zero current, increases slowly with current. Also, $I_0(V)$ is slowly voltage dependent.

hence (2) through (4) should be valid with $m=1$. Again the series resistance r of the diode should show thermal noise.

At higher frequencies an additional source of noise occurs. It is due to holes injected into the n-region and returning to the p-region by back diffusion before having recombined. This changes the junction admittance from the low-frequency value $Y_0 = g_0$ to the high-frequency (HF) value $Y = g + jb$, where g increases with increasing frequency. The increase $(g - g_0)$ in g comes from these back-diffused holes. Since diffusion is a thermal process, one would expect the conductance $g - g_0$ to have thermal noise. Consequently, at higher frequencies (2) should be written

$$S_i(f) = 2q(I + 2I_0) + 4kT(g - g_0). \qquad (8)$$

Van der Ziel and Becking proved this in greater detail [152] from the corpuscular point of view.

In long p^+n junctions, that is, junctions where the width of the n-region is larger than a few times the diffusion length $L_p = (D_p \tau_p)^{1/2}$ of holes, the junction admittance Y is

$$Y = g_0(1 + j\omega\tau_p)^{1/2} \qquad (9)$$

where τ_p is the lifetime of the holes in the n-region, so that

$$g = g_0[\tfrac{1}{2}(1 + \omega^2\tau_p^2)^{1/2} + \tfrac{1}{2}]^{1/2}. \qquad (9a)$$

These equations have been verified by experiment [4], [57], [58], [88].

For p-i-n diodes a somewhat different behavior is found [119]. Here the characteristic is of the form

$$I = I_0\left[\exp\left(\frac{qV}{mkT}\right) - 1\right] \qquad (10)$$

where m is unity for low currents and increases slowly with increasing current [43a]. In that case (8) must be replaced by

$$S_i(f) = 2q(I + 2I_0) + 4mkT(g - g_0). \qquad (11)$$

This has been discovered by experiment for $I \gg I_0$. The occurrence of the factor m in (11) has been understood from a slight extension of the Becking–van der Ziel theory [119].

The same shot noise approach also applies to tunnel diodes. Here there are two opposing currents flowing that depend on bias in a different way and that are equal at zero bias. For forward bias [2], [125], [145]

$$S_i(f) = 2qI \coth(qV/2kT) \qquad (12)$$

which reduces to full shot noise of the current I for $qV/kT \gg 1$:

$$S_i(f) = 2qI. \qquad (12a)$$

This is the case in the negative conductance region. Equation (12) has been verified by experiment [2].

In the valley region the noise is sometimes larger than shot noise [2]. This behavior is not yet fully understood; it seems that trapping states in the space-charge region are involved. It would be worthwhile to investigate whether this condition still exists for modern tunnel diodes.

C. Shot Noise in Transistors at Low Injection

For the sake of simplicity we assume the transistor to be a p-n-p transistor in which practically all the current is carried by holes. The emitter current I_E then consists of a part $I_{ES} \exp(qV_{EB}/kT)$ due to holes injected into the base, where V_{EB} is the emitter-base voltage, and a part $-I_{BE}$ due to holes generated in the base region and collected by the emitter. The emitter current I_C consists of a part $\alpha_F I_{ES} \exp(qV_{EB}/kT)$ due to the collection of holes injected by the emitter, where α_F is the forward-current amplification factor, and a part I_{BC} due to holes generated in the base region and collected by the collector. Therefore,

$$I_E = I_{ES} \exp\left(\frac{qV_{EB}}{kT}\right) - I_{BE} \qquad (13)$$

$$I_C = \alpha_F I_{ES} \exp\left(\frac{qV_{EB}}{kT}\right) + I_{BC} = \alpha_F I_E + I_{CO} \qquad (14)$$

where

$$I_{CO} = \alpha_F I_{BE} + I_{BC} \qquad (14a)$$

is the collector saturated current. The LF emitter conductance g_{eb0} is

$$g_{eb0} = \frac{\partial I_E}{\partial V_{EB}} = \frac{q}{kT} I_{ES} \exp\left(\frac{qV_{EB}}{kT}\right) = \frac{q}{kT}(I_E + I_{EB}), \quad (15)$$

the LF transconductance g_{m0} is

$$g_{m0} = g_{ec0} = \frac{\partial I_c}{\partial V_{EB}} = \frac{q}{kT} \alpha_F I_{ES} \exp\left(\frac{qV_{EB}}{kT}\right) = \alpha_F g_{eb0}, \quad (16)$$

and the LF current amplification factor is

$$\alpha_0 = \frac{\partial I_C}{\partial I_E} = \frac{\partial I_C / \partial V_{EB}}{\partial I_E / \partial V_{EB}} = \alpha_F. \qquad (17)$$

All these currents should show full shot noise. Consequently, at low frequencies

$$S_{I_E}(f) = 2qI_{ES} \exp\left(\frac{qV_{EB}}{kT}\right) + 2qI_{BE} = 2q(I_E + 2I_{BE}) \quad (18)$$

$$S_{I_C}(f) = 2q\alpha_F I_{ES} \exp\left(\frac{qV_{EB}}{kT}\right) + 2qI_{BC} = 2qI_C. \qquad (19)$$

Since the current I_E and I_C have the component $\alpha_F I_{ES} \exp(qV_{EB}/kT)$ in common, and this current should show full shot noise, the cross-correlation current should be

$$S_{I_C, I_E}(f) = 2q\alpha_F I_{ES} \exp\left(\frac{qV_{EB}}{kT}\right) = 2kT g_{ec0}. \qquad (20)$$

At high frequencies the emitter admittance Y_{eb} becomes complex and its real part g_{eb} increases with increasing frequency. This effect is due to holes injected by the emitter and returning by back-diffusion before having been collected. The noise associated with these holes corresponds to thermal noise of the increment $(g_{eb} - g_{eb0})$ in the emitter

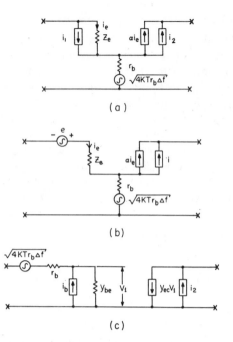

Fig. 1. Equivalent transistor circuits. (a) General equivalent circuit. (b) Common base equivalent circuit. (c) Common emitter equivalent circuit.

conductance since diffusion is a thermal process. Hence

$$S_{I_E}(f) = 2q(I_E + 2I_{BE}) + 4kT(g_{eb} - g_{eb0}). \quad (18a)$$

In addition the HF transfer admittance changes from the LF value g_{ec0} to the complex HF value Y_{ec}, because diffusion gives rise to a random delay time between emission and collection so that the coherence of the ac collector current deteriorates. Since one would expect the noise to behave in the same way as the signal, (20) should be written

$$S_{I_C, I_E}(f) = \frac{Y_{ec}}{g_{ec0}} \cdot 2kT g_{ec0} = 2kT Y_{ec} = 2kT\alpha Y_{eb}. \quad (20a)$$

Finally, since the passage of carriers across the collector junction still consists of a series of independent random events, (19) should remain valid. Van der Ziel and Becking have proven (18a) and (20a) in greater detail [152]. The HF current amplification factor α is here defined as

$$\alpha = \frac{Y_{ec}}{Y_{eb}} \simeq \frac{\alpha_0}{1 + jf/f_\alpha} \quad (21)$$

where f_α is the alpha cutoff frequency of the transistor. Especially in cases where drift (e.g., due to built-in fields) and diffusion in the base region are comparable, it is found that (21) must be replaced by

$$\alpha = \frac{\alpha_0 \exp(-j\omega\tau)}{1 + jf/f_\alpha} \quad (21a)$$

where τ is a time constant associated with the drift effect. This does not alter most noise effects, however, since $\overline{i^2}$ depends on $|\alpha|^2$ only. (See (23) for the definition of i.)

The noise can now be represented by two current generators i_1 and i_2 in parallel with the emitter and the collector

junctions (Fig. 1(a)), respectively, where

$$\overline{i_1^2} = S_{I_E}(f)\Delta f; \quad \overline{i_2^2} = S_{I_C}(f)\Delta f; \quad \overline{i_1^* i_2} = S_{I_C, I_E}(f)\Delta f. \quad (22)$$

In Fig. 1(a) the thermal noise of the base resistance r_b has been included.[2]

The noise can also be represented by an electromotive force (EMF) e in series with the emitter junction and a current generator i in parallel with the collector junction (Fig. 1(b)), where

$$e = i_1 Z_e; \quad i = (i_2 - \alpha i_1), \quad (23)$$

where $Z_e = 1/Y_e$ is the emitter impedance. The advantage of this representation is that e and i are only weakly correlated. At low frequencies

$$\overline{e^2} = \overline{i_1^2}|Z_e|^2 = 2kT g_{ebo}\left[\frac{I_E + 2I_{BE}}{I_E + I_{BE}}\right]\Delta f;$$

$$\overline{i^2} = 2q[I_E\alpha_F(1 - \alpha_F) + I_{CO}]\Delta f; \quad \overline{ei^*} \simeq 0 \quad (24)$$

since $\alpha_0 = \alpha_F$ in this approximation. This representation corresponds to the LF equivalent circuit developed by Montgomery and Clark [111] and van der Ziel [150]. It has been well verified by experiment, even in cases where α_0 differs from α_F (see Section II-D). It is obvious that the equivalent circuit of Fig. 1(b) is best adapted to the common-base circuit.

At high frequencies (24) must be modified [30]. A detailed calculation shows that $\overline{e^2}$ decreases with increasing frequency, whereas $\overline{ei^*}$ can have a significant reactive component [29], [34]. Finally, $\overline{i^2}$ depends on $|\alpha|^2$ and may be written

$$\overline{i^2} = 2q\Delta f\left[\alpha_F I_E \frac{(1 - \alpha_F + f^2/f_\alpha^2)}{1 + f^2/f_\alpha^2} + I_{CO}\right] \quad (24a)$$

so that it reaches the limiting value $2qI_C\Delta f$, for $f \gg f_\alpha$.

To measure $\overline{i^2}$, a large resistance R is inserted in series with the emitter and the output noise of the transistor is measured. The resistance R eliminates the effect of the noise EMF e. The only remaining noise source is then the current source i as long as $4kT\Delta f/R \ll \overline{i^2}$. The measurements have verified (24a) very accurately even at relatively high injection [146], and even when recombination in the emitter space-charge region is significant [22].

To measure $\overline{e^2}$ and the cross correlation $\overline{ei^*}$ one must use the transistor as an HF amplifier. This problem will be discussed in Section VI-C.

In common emitter connection another representation, suggested by Giacoletto, is often used [50]. It consists of a current generator $i_b = i_1 - i_2$ in parallel with the base—

---

[2] In a more accurate description of the equivalent circuit one sometimes splits the base resistance into two parts. This should have little effect on the noise properties It should be noted that the values of r_b deduced from the noise data and those obtained by other methods often do not agree too well [107].

emitter admittance $Y_{be} = Y_{eb} - Y_{ce}$ and the current generator i_2 between emitter and collector (Fig. 1(c)). It is easily evaluated that

$$\overline{i_b^2} = 2q(I_B + 2I_{BE} + 2I_{BC})\Delta f + 4kT[g_{eco} - \mathrm{Re}(Y_{ec})]\Delta f \\ + 4kT(g_{eb} - g_{ebo})\Delta f \tag{25}$$

$$\overline{i_b^* i_c} = 2kT(Y_{ec} - g_{eco})\Delta f - 2qI_{BC}\Delta f. \tag{26}$$

At low frequencies this can be approximated as

$$\overline{i_b^2} = 2qI_B\Delta f; \quad \overline{i_b^* i_c} = 0. \tag{27}$$

At high frequencies this circuit leads to very complicated calculations and hence a transformation back to the common-base circuit is advisable (Section VI-C).

D) Noise Due to Recombination in the Space-Charge Region of a Junction Diode

This problem was discussed by van der Ziel [156], by Scott and Strutt [136], and by Lauritzen [98]. The effect is especially significant in silicon devices. Since Lauritzen has given the most detailed discussion, we state essentially his results.

The generation and recombination of carriers in the space-charge region comes about because of the presence of recombination centers, the so-called Shockley–Hall–Read (SHR) centers. For back-biased diodes these centers alternately generate an electron and a hole with the relaxation time τ_t of the centers relatively large. For forward-biased diodes these centers alternately capture an electron and a hole with the relaxation time τ_t of the centers relatively small.

To calculate the noise, the space-charge region is split into small sections Δx, the contribution of each section Δx to the noise is evaluated, and then the result is integrated over the width of the space-charge region. It must be taken into account in the calculation that an emission or capture event displaces a total charge that depends on the position of the center and that is only a fraction of the electronic charge q.

For back-biased diodes and low frequencies ($\omega\tau_t \ll 1$), Lauritzen finds

$$S_i(f) = 2qI \frac{e_p^2 + e_n^2}{(e_p + e_n)^2} \tag{28}$$

where e_p and e_n are the hole and electron emission rates of the center. For $e_p \gg e_n$ or $e_p \ll e_n$ this corresponds to full shot noise, for $e_p = e_n$ it corresponds to half shot noise. For high frequencies ($\omega\tau_t \gg 1$),

$$S_i(f) = \tfrac{2}{3} \cdot 2qI. \tag{29}$$

This expression has been verified by Scott and Strutt [136].

It is interesting to note that at low frequencies ($\omega\tau_t \ll 1$), the alternate emission of a hole and an electron appears as a single event and any possible reduction comes about because the SHR center tends to "smooth" the emission of carriers if $e_p \simeq e_n$. At high frequencies ($\omega\tau_t \gg 1$), the carriers appear

to be generated independently; the shot noise reduction in this case comes about because an emission event displaces a charge that is only a fraction of the electronic charge q.

For forward-biased diodes the current I is partly due to carriers injected into the space-charge region; the part I_R is due to recombination in the space-charge region. The first part gives full shot noise

$$S(f) = 2q(I - I_R). \tag{30}$$

According to Lauritzen [98] the spectral intensity of the recombination noise for $\omega\tau_t \ll 1$ is

$$S_{I_R}(f) = 2qI_R \tag{31}$$

for relatively small forward bias, and

$$S_{I_R}(f) = \tfrac{3}{4} \cdot 2qI_R \tag{32}$$

for larger forward bias. Since the relaxation time τ_t in these cases is much smaller than in the back-biased condition, the frequency range $\omega\tau_t \gg 1$ is not very important. For $I \gg I_R$ little error is made by assuming full shot noise of the current I. It is doubtful that (31) and (32) can be verified experimentally.

In p-n-p transistors recombination can occur in the emitter space-charge region, giving rise to a component I_R in the emitter current I_E. Since the transistor has a low current-amplification factor α_F if I_R is comparable to I_E, modern transistors are so designed that in the useful range of operation I_R is small in comparison with the emitter current I_E.

Nevertheless it has a significant effect. For since I_R increases slower with increasing emitter–base voltage V_{EB} than I_E, $\alpha_F = I_C/I_E$ increases with increasing V_{EB}, or α_F increases with increasing I_E. Consequently, the LF current amplification factor α_0 is

$$\alpha_0 = \frac{\partial I_C}{\partial I_E} = \alpha_F + \frac{\partial \alpha_F}{\partial I_E} I_E \tag{33}$$

so that $\alpha_0 > \alpha_F$ and $\beta_0 = \alpha_0/(1-\alpha_0) > \beta_F = \alpha_F/(1-\alpha_F)$. The effect of the recombination current on β_F and β_0 can be appreciable.

As far as the noise is concerned, (18) through (20a) remain essentially correct. But instead of (15), one now has

$$g_{ebo} \simeq \frac{\alpha_F}{\alpha_0} \frac{q(I_E + I_{BE})}{kT}. \tag{34}$$

Finally, if one calculates $\overline{i^2}$, one finds that (24a) is essentially correct as long as $(\alpha_0 - \alpha_F)^2 \ll (1-\alpha_F)$; usually this is the case [22], [31].

E. High-Level Injection Effects

At high currents so much hole charge can be stored in the n-region of a p^+n diode that $p > N_d$ in the part of the n-region adjacent to the space-charge region. This effect is readily observable in Ge diodes. It has an interesting effect on the input admittance. According to Schneider and Strutt [134], [135], the equivalent circuit of the diode for that case

241

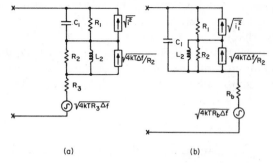

(a)   (b)

Fig. 2. (a) Equivalent junction diode circuit at high injection. (b) Partial equivalent transistor circuit at high injection.

is shown in Fig. 2(a). Their impedance measurements seem to indicate that the device can be well represented by the equivalent circuit; their noise measurements indicate that thermal noise is associated with the impedances $Z_2 = R_2 + j\omega L_2$ and the series resistance R_3.

Similar effects were found in p-n-p and n-p-n transistors. Here there is an HF impedance associated with the emitter junction; the equivalent circuit is shown in Fig. 2(b). This effect gives rise to a current dependence of the HF current-amplification factor α. Shot noise is associated with the diode proper and thermal noise is associated with R_2 and the base resistance R_b. This circuit has also been well verified by Schneider and Strutt [134], [135].[3]

Nevertheless, there is something peculiar about these circuits, for the storage capacitance C_1 is in parallel with R_1 in Fig. 2(a) and in parallel with the series connection of R_1 and R_2, and L_2 in Fig. 2(b). It is difficult to account for this difference. For that reason a careful investigation of this problem would seem appropriate.

The effect does not seem to be present in modern transistors. In modern transistors other high-level injection effects have been found. Agouridis and van der Ziel [3] found a saturation effect in the collection region of microwave p-n-p transistors. It comes about as follows. Most microwave transistors have a narrow p^- collector region adjacent to the base and a wider p^+ collector region adjacent to the contact. At high current the p^- collector region becomes space-charge limited; as a consequence the effective path length of carriers in the base region becomes larger, resulting in a considerable drop in the alpha cutoff frequency. This effect has been studied with the help of noise figure measurements.

Another effect was found by Tong and van der Ziel [146]. They found an appreciable correlation between the emitter noise EMF e and the collector noise current generator i and they demonstrated that this was a high-level injection effect. The magnitude of the effect is somewhat disturbing and, therefore, it would be worthwhile to study it in greater detail.

---

[3] Johnson and van der Ziel [91] have reported that noise measurements in 2N332 transistors indicated that a slight modification in Fig. 2(b) was necessary. Later measurements on this transistor type did not verify the need for such a modification.

The theory of high-level injection is presently being developed. Van Vliet [171] has recently evaluated the noise sources. Since electrons and holes are present in about equal numbers, the ambipolar diffusion equation must be used. One then obtains for hole injection into n-type material

$$\frac{\partial p}{\partial t} = g - r - \mu_a \mathbf{E} \cdot \nabla p + \gamma(\mathbf{r}, t) + \nabla \cdot \boldsymbol{\eta}_a(\mathbf{r}, t) + D_a \nabla^2 p \quad (35)$$

where g and r are the generation and recombination rates, $\mathbf{E}$ is the field strength, μ_a the ambipolar mobility, D_a the ambipolar diffusion constant, $\gamma(\mathbf{r}, t)$ the generation–recombination noise source, and $\nabla \cdot \boldsymbol{\eta}_a(\mathbf{r}, t)$ the diffusion noise source. Van Vliet has shown that these sources have the following spatial cross-correlation spectra:

$$S_\gamma(\mathbf{r}, \mathbf{r}', f) = \frac{2p}{\tau_H} \frac{\kappa^2 + \delta^2}{(\kappa + \delta)^2} \delta(\mathbf{r} - \mathbf{r}'), \quad (36)$$

for recombination by SRH centers [compare (28)], and

$$s_{\Delta \cdot \eta_a}(\mathbf{r}, \mathbf{r}', f) = 4\nabla\nabla' \left[D_a \frac{np}{n + p} \delta(\mathbf{r} - \mathbf{r}') \right] \colon \boldsymbol{i}. \quad (37)$$

Here τ_H is the high-injection lifetime, δ and κ are transition constants, ∇ means differentiation with respect to $\mathbf{r}$, and ∇' differentiation with respect to $\mathbf{r}'$, whereas $\delta(\mathbf{r} - \mathbf{r}')$ is the Dirac delta function and $\boldsymbol{i}$ is the unit tensor.[4] Usually $\kappa \neq \delta$ and $(\kappa^2 + \delta^2)/(\kappa + \delta)^2 \simeq 1$.

Tarng [141] evaluated the noise spectrum for a p^+n diode in the high-injection limit and obtained:

1) at low frequencies ($\omega \tau_H \ll 1$)

$$S_I(f) = \frac{2}{3}\left(1 + \frac{1}{b}\right) \cdot 2eI \cdot \quad (38)$$

2) at high frequencies ($\omega \tau_H \gg 1$)

$$S_I(f) = \frac{1}{2}\left(1 + \frac{1}{b}\right) \cdot 2eI(2\omega \tau_H)^{1/2}, \quad (39)$$

whereas the low-injection values are, if τ is the low-injection lifetime,

$$S_I(f) = 2eI \quad \text{(for} \quad \omega \tau \ll 1) \quad (38a)$$

$$S_I(f) = 2eI(2\omega \tau)^{1/2} \quad \text{(for} \quad \omega \tau \gg 1). \quad (39a)$$

Here $b = \mu_n/\mu_p$ is the mobility ratio. For n^+p diodes the term $1/b$ must be replaced by b.

This theory has not yet been verified experimentally, and for that reason a further study seems warranted.

In addition the effect of the stored charge on the noise must be evaluated. Van der Ziel [153], [155] calls this effect a *modulation effect* and has suggested that the diode noise current generator i should extend not only over the junction proper but also over the impedance due to the stored charge. This problem requires a much more detailed analysis and a

---

[4] For the one-dimensional case the tensor notation can be dropped, $\nabla = \partial/\partial x$ and $\nabla' = \partial/\partial x'$.

careful experimental study that should verify and/or extend Schneider and Strutt's results [134], [135].

F. Noise in Photodiodes and Avalanche Diodes

Junction diodes are often used as photodetectors. If the photocurrent generated in the photodiode by amplitude-modulated light is

$$I_s(t) = I_{s0} + I_{s1} \cos \omega_p t \qquad (40)$$

where ω_p is the modulation frequency and I_0 is the dark current of the diode, then the noise is [116]

$$S_I(f) = 2q(I_{s0} + I_0). \qquad (41)$$

Another way of using the diode as a photodetector is the photovoltaic mode of operation. In that case the photo-diode is used open circuited and the dc current is zero, or

$$I = -I_{s0} + I_0 \left[\exp\left(\frac{eV}{kT}\right) - 1 \right] = 0. \qquad (42)$$

The noise is now shot noise of the independent currents I_{s0}, I_0, and $I_0 \exp(eV/kT)$ and hence one obtains [51], instead of (41),

$$S_I(f) = 4q(I_{s0} + I_0). \qquad (43)$$

It often happens that the signal coming out of a photo-diode drowns in the noise background of the associated receivers and amplifiers. In that case it is convenient to use a multiplication process before the signal is fed into the receiver. One way of doing so is to use avalanche multiplication.

Early avalanche diodes (or Zener diodes) showed breakdown spots, so-called "microplasmas," in the space-charge region. These microplasmas switched on and off at random and thus produced large amounts of noise. This noise process was first studied in detail by Champlin [26] and has since been used for noise generators [67]–[69]. It has been possible, however, to design avalanche diodes that show uniform breakdown. Such diodes can be used for avalanche multiplication [13]–[15] in the prebreakdown region and as oscillators and negative resistance amplifiers in the breakdown region [83].

To calculate the multiplied signal and the noise, we assume first for the sake of simplicity that the electron and the hole have equal ionizing power. If $I_s(t)$ is again the photo-current generated in the diode and I_0 the dark current of the diode, and if p hole–electron pairs are generated when an individual hole–electron pair traverses the space-charge region once, then the multiplied ac current is

$$I_{sc} \cos \omega_p t = I_{s1} \cos \omega_p t [1 + p + p^2 + \cdots] \qquad (44)$$
$$= M I_s \cos \omega_p t$$

so that

$$M = \frac{1}{1 - p} \qquad (44a)$$

is the multiplication factor; it can be quite large if p is close to unity.

For the noise we start with the shot noise of $I_{s0} + I_0$, and this is multiplied. In addition, since each generated hole–electron pair is generated at random and traverses the space-charge region, all the generated currents should show full shot noise before being multiplied. One would thus expect

$$S_I(f) = 2[(I_{s0} + I_0) + p(I_{s0} + I_0) + p^2(I_{s0} + I_0) \cdots] M^2 \qquad (45)$$
$$= 2q(I_{s0} + I_0)M^3.$$

The case in which the electrons and holes have different ionizing power has been evaluated by McIntyre [105] and by Gummel and Blue [65]. McIntyre finds that the noise can be considerably better or considerably worse than given by (45), depending on the situation. If α and β are the ionization coefficients for electrons and holes, respectively, and $\beta = k\alpha$, then

$$S_I(f) = 2eI_{in}M^3 \left[1 + \left(\frac{1 - k}{k}\right)\left(\frac{M - 1}{M}\right)^2 \right] \qquad (45a)$$

if the injected current I_{in} consists of holes and

$$S_I(f) = 2eI_{in}M^3 \left[1 - (1 - k)\left(\frac{M - 1}{M}\right)^2 \right] \qquad (45b)$$

if the injected current I_{in} consists of electrons. The first expression is quite large when $k \ll 1$ and much smaller if $k \gg 1$, so that the best results are obtained if the holes have the largest ionizing power. The second expression is quite large when $k \gg 1$ and much smaller if $k \ll 1$, so that the best results are obtained if the electrons have the largest ionizing power. These predictions have been well verified by experiment [13]–[15].

If the devices are used in the breakdown region as microwave negative-conductance amplifiers, one obtains large noise figures. It seems that silicon avalanche amplifiers have the highest noise figure, germanium avalanche amplifiers are better, and gallium arsenide avalanche amplifiers are still better [96]. This seems to be associated with (45a) and (45b). If the avalanche devices are used as oscillators, the microwave signals thus generated have amplitude noise modulation and frequency noise modulation. Again, silicon avalanche oscillators are the noisiest, germanium avalanche oscillators are quieter, and gallium arsenide oscillators are much quieter.

G. p-n Diodes as Particle Detectors [46]

When p-n junctions are bombarded with high-energy particles a considerable number of hole–electron pairs will be generated in the space-charge region. Most of these will be collected, and as a consequence a large number of random current pulses will pass through the device. This generates a considerable amount of noise. In analogy with the noise generated in secondary electron multiplication, the spectral intensity of the noise is

$$S_I(f) = 2e^2 \overline{N_B} \, \overline{m^2} \qquad (46)$$

where $\overline{N_B}$ is the rate of incidence of bombarding particles and m the number of collected hole–electron pairs per primary particle. This has been verified experimentally.

H. Shot Noise in FETs [167]

In junction FETs thermal generation of carriers gives rise to a leakage current flowing from gate to channel, which contributes to the gate noise and to some extent to the drain noise. In metal-oxide-semiconductor (MOS) FETs thermal generation of carriers gives rise to a leakage current flowing from substrate to channel, resulting in an additional term in the drain noise. In modern silicon FETs both effects are insignificant at room temperature but may become important at elevated temperatures.

According to van der Ziel [167] a gate current δi_g flowing into the gate between x_0 and $x_0+\Delta x_0$ gives rise to a drain current δi_d such that

$$\delta i_d = \frac{x_0}{L}\,\delta i_g. \qquad (47)$$

If the gate is slightly back biased, δi_g will show full shot noise. One can thus evaluate the self-spectral intensities $\Delta S_{gg}(f)$, $\Delta S_{dd}(f)$, and the cross-spectral intensity $\Delta S_{gd}(f)$ of δi_g and δi_d. Integrating over the length of the channel, assuming that the gate current density $J_g(x_0)$ is independent of x_0, yields the spectral intensities

$$S_{gg}(f)= 2qI_g; \qquad S_{dd} = \tfrac{1}{3}\cdot 2qI_g; \qquad S_{gd} = \tfrac{1}{2}\cdot 2qI_g \quad (48)$$

so that the correlation coefficient is

$$C = \frac{S_{gd}}{\sqrt{S_{gg}\cdot S_{dd}}} = \tfrac{1}{2}\sqrt{3}. \qquad (48a)$$

For junction FETs the device is generally used at a high input impedance level. Therefore even a relatively small gate leakage current I_g can be quite harmful, so that the gate leakage sets a serious limit to the high temperature operation of the device.

For MOS FETs there is no gate noise, but the drain noise has a component due to the current I_s flowing to the substrate. It consists of three parts:

1) a leakage current I_{ss} from source to substrate, that does not contribute to $S_{dd}(f)$;
2) a leakage current I_{cs} from channel to substrate, that contributes $\tfrac{1}{3}\cdot 2qI_{cs}$ to $S_{dd}(f)$;
3) a leakage current I_{ds} from drain to substrate, that contributes $2qI_{ds}$ to $S_{dd}(f)$.

The total contribution of the substrate leakage current to $S_{dd}(f)$ is, therefore,

$$2q(\tfrac{1}{3}I_{cs} + I_{ds}) = 2q\beta I_s \qquad (49)$$

where β is a factor somewhat smaller than unity. This must be compared with the term due to the thermal noise of the channel, which may be written as

$$\tfrac{2}{3}\cdot 4kT g_{max}, \qquad (50)$$

which is usually much larger. It may thus be concluded that the substrate leakage current gives only a small contribu-

tion to $S_{dd}(f)$ and therefore hardly sets a high temperature limitation to the operation of the device.

I. Summary

We have seen how a simple shot noise picture can give a reasonable description of most noise phenomena in diodes and transistors, and of some noise phenomena in FETs. It was shown that noise in avalanche diodes could also be described in terms of an "amplified shot noise" picture.

Only in the case of recombination–generation in the space-charge region of a junction is it necessary to develop a more detailed picture. While recombination–generation in the emitter space-charge region of a transistor has a profound effect on the current amplification factor, the noise description in terms of shot noise mechanisms remains practically valid.

It is not fully clear how this picture must be modified at high injection, since the theory for this case has not yet been developed. Also the effect of the stored charge on the noise has not been explored sufficiently. Experiments seem to indicate that high-level injection effects affect the noise in long junction diodes much more than in transistors with a narrow base region. More work is needed before a unified picture can be developed.

III. Thermal and Generation–Recombination Noise in FETs and Other Devices [170]

Junction FETs operate on the principle that the width of a conducting channel is modulated in the rhythm of the applied gate voltage. Since the conducting channel should show thermal noise, one would expect the limiting noise of junction FETs to be thermal noise, modified perhaps by the modulation effect.

MOS FETs operate on the principle that the mobile charge present under the gate is modulated in the rhythm of the applied gate voltage. Since the conducting channel is only present when the gate voltage is properly chosen, it is not a priori certain that the device should have thermal noise. Rather, one would expect the noise to be diffusion noise, caused by the collisions of carriers with the lattice. It has been shown, however, that this diffusion noise corresponds to thermal noise as long as the Einstein relation $qD_n=kT\mu_n$ holds [164]. Since this is usually the case, one may assume that the limiting noise in the channel is thermal noise. Deviations may occur, of course, near saturation, where the electric field strength near the drain may be so high that the mobility becomes field dependent. In that case it is not certain that the Einstein relation holds, and even if it does, the electron temperature will not be equal to the device temperature.

A similar situation occurs in single- and double-injection space-charge-limited solid-state diodes. Here the carriers are injected and the steady-state situation is again a nonequilibrium situation. The limiting noise should again be diffusion noise, but as long as the Einstein relation holds this corresponds to thermal noise. Again, in some instances, especially in single-injection diodes at high currents, the electric field strength in the device can be so high that the mobility becomes field dependent. It has been shown experimentally that this results in a higher "effective noise temperature" of the device.

At room temperature practically all the impurity atoms in the channel of a junction FET are ionized, but at lower temperatures part of the atoms are neutral. As a consequence the carrier density will fluctuate by processes of the type

neutral donor + energy $\rightleftarrows$ ionized donor + electron,

etc. This corresponds to a local change in the conductance of the channel, which shows up as noise in the external circuit if dc current is passed through the device. The effect becomes more and more pronounced at lower temperatures. It should not occur in MOS FETs, but may be significant in the base region of transistors at low temperatures.

In the space-charge region of the junction FET and in the space-charge region between channel and substrate of the MOS FET recombination centers will be present. Due to the fluctuating occupancy of these centers, the channel conductance will be modulated, resulting in a noise current in the external circuit.

A. Operation of the FET

In order to understand the noise behavior of the FET, one has to know something about the operation of the device. We first introduce some concepts associated with the FET.

Since the device operates on the principle of modulation of a conducting channel, we introduce the conductance per unit length $g(x)$, at a point at a distance x from the source end of the channel. If $V_0(x)$ is the channel potential at the point, taken with respect to the source, we may write $g(V_0)$ instead of $g(x)$. For an n-type channel, we may then write for the current I_d

$$I_d = g(V_0)\frac{dV_0}{dx}; \quad \text{or} \quad I_d dx = g(V_0)dV_0 \quad (51)$$

so that, since I_d is independent of x,

$$I_d L = \int_0^{V_d} g(V_0)dV_0; \quad \text{or} \quad I_d = \frac{1}{L}\int_0^{V_d} g(V_0)dV_0 \quad (52)$$

where V_d is the drain voltage. Here I_d flows from drain to source.

For a junction FET with an n-type channel

$$g(V_0) = g_{00}\left\{1 - \left[\frac{-V_g + V_{\text{dif}} + V_0}{V_{00}}\right]^{1/2}\right\} \quad (53)$$

where g_{00} is the conductance of the fully open channel, V_g the gate potential, V_{dif} the diffusion potential of the junction, and V_{00} the potential difference between gate and channel for zero channel conductance. The channel is thus cut off at the drain (that is, it has zero conductance at the drain) if

$$V_d + V_{\text{dif}} - V_g = V_{00}, \quad \text{or} \quad V_d = V_{d0} = V_{00} - V_{\text{dif}} + V_g. \quad (53a)$$

For a MOS FET with n-type channel and a low-conductivity substrate

$$g(V_0) = \mu w C_{\text{ox}}(V_g + V_{g0} - V_0) \quad (54)$$

where V_g is the gate potential, V_{g0} the offset potential between gate and channel, C_{ox} the capacitance per unit area of the gate-oxide-channel system, w the width of the channel, and μ the carrier mobility. In this case the channel is cut off at the drain if

$$V_d = V_{d0} = V_g + V_{g0}. \quad (54a)$$

For the case of arbitrary substrate conductivity the expression for $g(V_0)$ is more complicated.

It is thus possible to calculate I_d as a function of V_g and V_d in each case, and to introduce the transconductance g_m and the drain conductance g_d by the definitions

$$g_m = \frac{\partial I_d}{\partial V_g}; \quad g_d = \frac{\partial I_d}{\partial V_d} = \frac{g(V_d)}{L}. \quad (55)$$

A calculation shows that g_d has its maximum value

$$g_{d0} = \frac{g_0}{L} \quad (55a)$$

at zero drain bias and zero value if the channel is cut off at the drain, whereas g_m has zero value at zero drain bias and its maximum value g_{max} if the channel is cut off at the drain. The device is said to be *saturated* if the channel is cut off at the drain [$g(V_d)=0$]. Beyond saturation, I_d and g_m are practically independent of V_d. The device is usually operated under that condition.

To understand the high-frequency behavior of the FET, the device must be treated as an active distributed line. One can derive the wave equation for this line and find solutions in terms of special transcendental functions [49]. For the junction FET, the wave equation for the small signal voltage distribution $\Delta V(x)$ along the channel is

$$\frac{d^2}{dx^2}\left[g(W_0)\Delta V(x)\right] \doteq \frac{j\omega\rho_0 wa\Delta V(x)}{V_{00}^{1/2}W_{00}^{1/2}(x)} \quad (56)$$

where ρ_0 is the charge density in the space-charge region, w the width of the channel, $2a$ the height of the open channel, $W_0(x) = -V_g + V_{\text{dif}} + V_0$ is the bias between channel and gate, ω is the frequency, and $j=\sqrt{-1}$. For the MOS FET one has instead

$$\frac{d^2}{dx^2}\left[V_1(x)\Delta V(x)\right] = \frac{j\omega}{\mu}\Delta V(x) \quad (57)$$

where μ is the mobility of the carriers, and $V_1(x) = V_g + V_{\text{dif}} - V_0(x)$.

When these equations are solved [49], it is found that the transconductance becomes complex and may be approximated as

$$Y_m = \frac{g_{m0}}{1 + jf/f_0} \quad (58)$$

where g_{m0} is the transconductance at low frequencies and f_0 a kind of cutoff frequency which is a few times larger than $g_m/(2\pi C_{gs})$. Furthermore one finds for the gate-source admittance Y_{gs},

$$Y_{gs} = j\omega C_{gs} + g_{gs} \quad (59)$$

where C_{gs} is independent of frequency and g_{gs} varies as ω^2, both over a wide frequency range.[5]

B. Thermal Noise in FETs [158]

We follow here the treatment given by Klaassen and Prins [95]. To that end we extend (51) as follows:

$$I_d + \Delta I_d(t) = g(V_0 + \Delta V)d\left(\frac{V_0 + \Delta V}{dx}\right) + h(x, t). \quad (60)$$

Here $h(x, t)$ is the thermal noise source operating on the channel, $\Delta I_d(t)$ the noise current flowing in the external circuit and $\Delta V(x, t)$ the noise voltage along the channel, both caused by the noise source $h(x, t)$. Making a Taylor expansion of $g(V_0 + \Delta V)$, ignoring higher order terms in ΔV, and bearing in mind that

$$I_d = g(V_0)\frac{dV_0}{dx}, \quad (60a)$$

yields

$$\Delta I_d(t) = \frac{d}{dx}\left[g(V_0)\Delta V\right] + h(x, t). \quad (61)$$

Here $\Delta I_d(t)$ flows from drain to source.

We now integrate this equation with respect to x between the limits 0 and L for the case that the drain is HF short-circuited to the source, so that $\Delta V = 0$ at $x = 0$ and $x = L$. This yields

$$\Delta I_d(t)L = \int_0^L h(x, t)dx \quad (62)$$

or

$$\overline{\Delta I_d(t)\Delta I_d(t + s)} = \frac{1}{L^2}\int_0^L \int_0^L S_h(x, x', f)dxdx'. \quad (63)$$

But since $h(x, t)$ is a thermal noise source one would expect for the spatial cross-correlation spectrum of $h(x, t)$

$$S_h(x, x', f) = 4kTg(x')\delta(x' - x) \quad (64)$$

where $\delta(x' - x)$ is the Dirac delta function. Equation (64) expresses the fact that the thermal noise source at the points x and x' at a given instant t are uncorrelated.

Carrying out the calculation yields

$$S_i(f) = \frac{4kTg_0}{L}\frac{\int_0^{V_d}[g(V_0)/g_0]^2dV_0}{\int_0^{V_d}[g(V_0)/g_0]dV_0} = \gamma 4kTg_{d0}. \quad (65)$$

It is easily seen that $\gamma = 1$ for $V_d = 0$, since in that case

[5] Van der Ziel and Ero [162] have developed an approximation method in which one writes

$$\Delta V(x) = \Delta V_0(x) + j\omega\Delta V_1(x) + (j\omega)^2\Delta V_2(x) + \cdots.$$

By collecting equal terms in $j\omega$ one then obtains differential equations in $\Delta V_0(x), \Delta V_1(x), \Delta V_2(x) \cdots$, that can be solved successively. This method is especially useful in the case of substrates of arbitrary conductivity [125a].

$g(V_0) = g_0$ everywhere. Also $\gamma < 1$ for $V_d > 0$, since in that condition

$$g(V_0) \leqq g_0 \quad \text{or} \quad \left[\frac{g(V_0)}{g_0}\right]^2 \leqq \frac{g(V_0)}{g_0}$$

so that the denominator in γ is larger than the numerator. Carrying out the calculation one finds that γ decreases monotonically with increasing V_d and reaches a minimum value γ_{sat} at saturation. For a junction FET one finds

$$\gamma_{sat} = \frac{1}{2}\frac{(1 + 3z^{1/2})}{1 + 2z^{1/2}}$$

where

$$z = \frac{-V_g + V_{dif}}{V_{00}} \quad (66)$$

so that $\gamma_{sat} = \frac{1}{2}$ for $z = 0$ (channel fully open at the source) and $\gamma_{sat} = \frac{2}{3}$ for $z = 1$ (channel fully cut off at the source) [158]. For an MOS FET with a low-conductivity substrate one finds $\gamma = \frac{2}{3}$ for all values of V_g [92]. For a MOS FET with a channel of higher substrate conductivity γ_{sat} lies slightly lower.

If one plots γ versus V_d one should thus obtain a curve that decreases monotonically with increasing V_d. Experimentally one finds that this is not true for a junction FET; the reason is not that the preceding theory is incorrect, but rather that it is incomplete, due to the presence of series resistances at the end of the channel. When this series resistance effect was taken into account, Bruncke found good agreement between theory and experiment [21].

For MOS FETs there are also deviations in some instances. Often γ decreases first with increasing V_d and then rises rapidly near saturation [73]. Halladay and van der Ziel [74] attributed this to mobility fluctuations, and Takagi and van der Ziel [140] showed that such fluctuations, if present, should have the strongest effect for substrates of larger conductivity. Recent experiments [179] have indicated, however, that MOS FETs can be built on substrates of higher conductivity that show only thermal noise. While this eliminates the idea of noise caused by mobility fluctuations, it leaves open the question of the source of the excess noise found in earlier devices. While the tail end of the flicker noise could perhaps explain some of the data, it cannot explain all. A further investigation in this problem would thus be worthwhile.

Another way of representing the noise is to write

$$S_i(f) = \alpha_{sat}4kTg_{max} \quad (67)$$

where g_{max} is the maximum transconductance at saturation. A comparision with (65) indicates that

$$\alpha_{sat} = \gamma_{sat}\frac{g_{d0}}{g_{max}}. \quad (67a)$$

MOS FETs with low-conductivity substrates have $g_{d0} = g_{max}$ so that $\alpha_{sat} = \gamma_{sat}$. For substrates of higher conductivity [95] $g_{max} < g_{d0}$ so that $\alpha_{sat} > \gamma_{sat}$.

Introducing the noise resistance R_n of the device by equating $S_i(f) = 4kTR_ng_{max}^2$ yields

$$R_n = \alpha_{sat}/g_{max},$$

or

$$\alpha_{sat} = R_n g_{max} \qquad (68)$$

so that α_{sat} can be determined from noise resistance data. Since the noise resistance R_n is a useful quantity for the design engineers, the quantity α_{sat} is of greater practical significance than γ_{sat}. The latter parameter, however, gives more physical insight.

C. Induced Gate Noise in FETs

At high frequencies the noise voltage distribution along the channel gives, by capacitive coupling, rise to a noise current i_g flowing to the gate that is partially correlated with the drain noise i_d. One must thus calculate the mean square values of these current generators and their cross correlation, i.e., one needs the quantities $\overline{i_g^2}$, $\overline{i_d^2}$, and $\overline{i_g i_d^*}$.

The rigorous way of solving this problem is to introduce localized noise EMFs ΔV_{x0} between x_0 and $x_0 + \Delta x_0$, calculate the contributions Δi_g and Δi_d by solving the wave equation, evaluate $\overline{\Delta i_g^2}$, $\overline{\Delta i_d^2}$, and $\overline{\Delta i_g \Delta i_d^*}$, and then sum over all the sections Δx_0, which corresponds to integration over the length of the channel. This procedure is quite tedious and often not very rewarding. (For the rigorous solution of the junction FET problem, see Klaassen [94a].)

If one wants Δi_g and Δi_d up to first-order terms in $j\omega$, one can apply a procedure developed by van der Ziel [125a], [159]. The starting point is now (61), slightly modified by omitting the distributed noise source and by changing the direction of positive current flow. We thus have

$$\frac{d}{dx}\left[g(V_0)\Delta V(x)\right] = -\Delta I_d(t) \qquad (69)$$

where $I_d(t)$ now flows *out of* the drain. The initial condition is that $\Delta V(x)$ shows a jump ΔV_{x0} between x_0 and $x_0 + \Delta x_0$. Since the drain is HF connected to the source, if we want the short-circuit noise current, we must require $\Delta V(0) = \Delta V(L) = 0$. Integration yields

$$\begin{aligned} g(V_0)\Delta V &= -\Delta I_d x, & \text{for } 0 < x < x_0, \\ g(V_0)\Delta V &= -\Delta I_d(x - L), & \text{for } x_0 + \Delta x_0 < x < L. \end{aligned} \qquad (70)$$

Bearing in mind the condition at $x = x_0$, yields immediately

$$\Delta I_d(t) = \frac{g(V_0)}{L}\,\Delta V_{x0}. \qquad (71)$$

Moreover, if $C(x)$ is the capacitance per unit area between channel and gate and w is the channel width, then the potential distribution $\Delta V(x)$ induces a charge $d\Delta Q_g = -wC(x)dx\Delta V(x)$ in the gate. The total charge ΔQ_g is obtained by integrating over the channel length,

$$\Delta Q_g = -w\int_0^L C(x)\Delta V(x)dx. \qquad (72)$$

The current flowing into the gate is therefore $\Delta I_g(t) = d\Delta Q_g/dt$.

Making a Fourier analysis of $\Delta V_{x0}(t)$, $\Delta I_d(t)$, and $\Delta I_g(t)$, introducing the Fourier components Δv_{x0}, Δi_d, and Δi_g, and bearing in mind that

$$\overline{\Delta v_{x0}^2} = \frac{4kT\Delta x\Delta f}{g(V_0)}, \qquad (73)$$

one can calculate $\overline{\Delta i_g^2}$, $\overline{\Delta i_d^2}$ and $\overline{\Delta i_g \Delta i_d^*}$. By integrating over the channel length L one obtains $\overline{i_g^2}$, $\overline{i_d^2}$, and $\overline{i_g i_d^*}$. The value of $\overline{i_d^2}$ corresponds to (65) but the other results are new. Since d/dt becomes $j\omega$ in the Fourier analysis it is obvious that $\overline{i_g^2}$ varies as ω^2 and $\overline{i_g i_d^*}$ as $j\omega$.

One may now introduce the correlation coefficient

$$c = \frac{\overline{i_g i_d^*}}{\sqrt{\overline{i_g^2}\cdot\overline{i_d^2}}}. \qquad (74)$$

In the approximation used here c is imaginary. Evaluation of $|c|$ indicates that for a junction FET the factor $|c|$ decreases monotonically with increasing $z = (-V_g + V_{dif})/V_{00}$, starting from the value $|c| = 0.445$ for $z = 0$ to the value [159] $|c| = 0.395$ for $z = 1$. For a MOS FET with a low-conductivity substrate [74] $|c| = 0.395$, for a MOS FET with substrates of higher conductivity $|c|$ is about the same.

The reason why $|c|$ is so much smaller than unity is that the sign of $\Delta Q_g(x_0, t)$ changes with increasing x_0, whereas the sign of $\Delta I_d(x_0, t)$ does not depend on x_0. Consequently the contributions of different sections Δx_0 to $\overline{i_g i_d^*}$ partly cancel each other. The value of c can be determined from noise figure measurements of the device (see Section VI), but not very accurately since $|c|$ is so small.

Since the input conductance g_{gs} and $\overline{i_g^2}$ both vary as ω^2 over a wide frequency range, it is convenient to introduce the equivalent noise temperature T_{ng} of g_{gs}, by putting

$$(\overline{i_g^2})_{sat} = 4kT_{ng}(g_{gs})_{sat}\Delta f. \qquad (75)$$

A long calculation shows that for a junction FET[6] at temperature T

$$T_{ng} = T\frac{1 + 9z^{1/2} + 15z + 7z^{3/2}}{1 + 8z^{1/2} + 10\frac{5}{7}z + 4\frac{2}{7}z^{3/2}} \qquad (76)$$

where $z = (-V_g + V_{dif})/V_{00}$ as before, so that $T_{ng} = T$ for $z = 0$ (channel fully open at the source) and $T_{ng} = \frac{4}{3}T$ for $z = 1$ (channel fully cut off at the source). This agrees roughly with the measurements [23].

For a MOS FET with a low-conductivity substrate the

---

[6] This result can be evaluated from van der Ziel and Ero's paper if a misprint in that paper is corrected. The expression $(\frac{1}{6}z^{1/2} - \frac{1}{6}z^2 + \frac{1}{10}z^{5/2})$ in (37a) of that paper should have a minus sign.

value of T_{ng}/T is evaluated [74], [138] at 4/3. Experimentally one sometimes finds a much higher value [74]. The reason for this is not quite clear; it is unlikely that this is a substrate effect, however [179].

D. Generation–Recombination (GR) Noise in FETs and Transistors [161]

If we consider a section Δx of the channel of a junction FET then the carrier density in that section will show fluctuations due to processes of the type:

free electron + ionized donor $\rightleftarrows$ neutral donor + energy.

Hence the conductance

$$g(W_0) = q\mu \frac{\Delta N}{\Delta x} \tag{77}$$

where ΔN is the number of carriers in the section Δx, will show fluctuations around the equilibrium value $g(W_0)$ because ΔN shows fluctuations $\delta\Delta N$. As a consequence the resistance $\Delta R = \Delta x/g(W)$ will show fluctuations $\delta\Delta R$,

$$\delta\Delta R = -\frac{\Delta x}{[g(W_0)]^2}\delta g = -\frac{\Delta x}{g(W_0)}\frac{\delta\Delta N}{\Delta N}, \tag{78}$$

which in turn gives rise to a fluctuating EMF

$$\Delta V_{x0} = -I_d\delta\Delta R = \frac{I_d\Delta x}{g(W_0)}\frac{\delta\Delta N}{\Delta N}. \tag{79}$$

According to (71) this gives rise to a drain current fluctuation

$$\delta\Delta I_d(t) = \frac{I_d\Delta x}{L\Delta N}\delta\Delta N(t) \tag{80}$$

so that

$$\overline{\delta\Delta I_d(t)\delta\Delta I_d(t+s)} = \frac{I_d^2\Delta x^2}{L^2\Delta N^2}\overline{\delta\Delta N(t)\delta\Delta N(t+s)}. \tag{80a}$$

But if $\overline{\delta\Delta N^2} = \alpha\Delta N$, and if $\delta N(t)$ decays exponentially,

$$\overline{\delta\Delta N(t)\delta\Delta N(t+s)} = \alpha\Delta N\exp\left(-\frac{s}{\tau}\right). \tag{81}$$

Substituting into (80a) and applying the Wiener–Khintchine theorem, yields for the spectrum $S_i(f)$ of $\delta\Delta I_d(t)$, since $I_d\Delta x = g(W_0)\Delta W_0$ and $g(W_0) = q\mu\Delta N/\Delta x$,

$$\Delta S_i(f) = \frac{4q\mu I_d\alpha}{L^2}\frac{\tau}{1+\omega^2\tau^2}\Delta W_0. \tag{82}$$

By integrating over the length of the channel, one obtains for the spectrum of the drain noise

$$S_i(f) = \frac{4q\mu I_d(W_d - W_s)}{L^2}\frac{\tau}{1+\omega^2\tau^2} \tag{83}$$

where $W_s = -V_g + V_{dif}$ and $W_d = V_d - V_g + V_{dif}$, if α and τ are independent of x. In saturation W_d must be replaced by V_{00} and I_d by its saturated value.

If we apply this to the case of deep lying donors in the channel, where the generation rate $g(N)$ and the recombination rate $r(N)$ are given by

$$g(N) = \gamma(N_d - N), \qquad r(N) = \rho N^2, \tag{84}$$

we find after some calculations [24], [170], if N_0 is the equilibrium carrier concentration,

$$\tau = \frac{1}{dr/dN - dg/dN}\bigg|_{N=N_0} = \frac{N_d - N_0}{\rho N_0(2N_d - N_0)}$$

and $\tag{85}$

$$\alpha = \frac{N_d - N_0}{2N_d - N_0}.$$

In silicon junction FETs the effect begins to set in around 100°K and it increases rapidly with decreasing temperature. At the onset of the effect $N_d - N_0$ is still small, and hence τ will be so small that the noise is practically white. This effect has been studied by Shoji [137]. Since germanium has much shallower donor levels than silicon, one will have to go to very low temperatures to see the effect in germanium junction FETs. This has been verified experimentally.

Another effect studied by Shoji [137] is due to deep lying traps in the channel. It should not occur in well-designed units. The effect also should not occur in MOS FETs. To understand this, consider an MOS FET with an n-type channel on a p-type substrate. The electrons in the channel now come from the source and drain, which are usually made of n^+ material, so that the donors lie very close to the bottom of the conduction band. The only effect that could occur is that part of the acceptors in the channel and in the space-charge region between channel and substrate may not be ionized. It is presently unknown what this would do to the noise.

GR noise of this type may also occur in the base of transistors at low temperature. The generation–recombination processes of the type

free electron + ionized donor $\rightleftarrows$ neutral donor + energy

gives rise to a fluctuation δr_b in the base resistance r_b, which is converted into a noise EMF $I_B\delta r_b$ due to the flow of base current. This results, therefore, in a noise EMF in series with r_b that should have a spectral intensity proportional to I_B^2. This effect is presently under study at the University of Florida, Gainesville, Fla.

E. Noise Due to Recombination Centers in the Space-Charge Region of an FET [126]

We follow here essentially Sah's treatment of the problem [126].

Consider an n-type channel of length L, width W, and height $2a$. Let the x axis be in the direction of the channel and the y axis perpendicular to it. Let n_T be the density of centers, then a small volume element $\Delta V = \Delta x\Delta yw$ around the point (x_1, y_1) in the space-charge region contains $\Delta N_T = n_T\Delta V$ centers. Let ΔN_t of these centers be occupied, and let $\delta \Delta N_t$ be the fluctuation in N_t, then

$$\overline{\delta \Delta N_t^2} = \Delta N_T f_t (1 - f_t) = n_T \Delta V f_t (1 - f_t). \qquad (86)$$

The reader will recognize this as a partition noise term. Hence the spectral intensity of $\delta \Delta N_t$ will be

$$S_{\delta \Delta_t(f)} = 4 n_T f_t (1 - f_t) \Delta V \frac{\tau_t}{1 + \omega^2 \tau_t^2}. \qquad (87)$$

The fluctuating occupancy of the centers modulates the width $2b$ of the conducting channel by an amount $2\delta b$, where[7]

$$\delta b = - \frac{\delta \Delta N_t}{N_d \Delta x_1 W} \frac{(a - y_1)}{(a - b)}. \qquad (88)$$

The fluctuation δb gives rise to a fluctuation $\delta \Delta R$ in the resistance $\Delta R = \Delta x_1 / [2 q \mu_n N_d b w]$ of the section Δx_1, where μ_n is the carrier mobility. This gives, for $\delta \Delta R$,

$$\delta \Delta R = - \Delta R \frac{\delta b}{b} = \frac{\delta \Delta N_t}{2 q \mu_n N_d^2 w^2} \frac{a - y_1}{(a - b) b^2}. \qquad (89)$$

The flow of the dc current I_d gives a fluctuating EMF $\delta \Delta V_{x0} = I_d \delta \Delta R$ in the section Δx, and, by virtue of (71), to a fluctuating current $\delta \Delta I_d = g(W_0) \delta \Delta V_{x0}/L$ in the external circuit, so that, substituting for $g(W_0)$

$$\delta \Delta I_d = \frac{I_d}{N_d w L} \frac{a - y_1}{(a - b) b} \delta \Delta N_t. \qquad (90)$$

Making a Fourier analysis one obtains for the spectrum

$$\Delta S_{I_l}(f) = \frac{I_d^2}{N_d^2 w^2 L^2} 4 n_T f_t (1 - f_t)$$
$$\cdot \frac{\tau_t}{1 + \omega^2 \tau_t^2} \frac{(a - y_1)^2}{(a - b)^2 b^2} \Delta x_1 \Delta y_1. \qquad (91)$$

Assuming that $f_t(1 - f_t)$ and τ_t are independent of y_1 and x_1 and integrating over the full space-charge region yields

$$S_{I_d}(f) = \frac{8 q \mu_n V_{00} I_d}{3 L^2 N_d} n_T f_t (1 - f_t) \frac{\tau_t}{1 + \omega^2 \tau_t^2}$$
$$\left[-(y - z) - 2 (y^{1/2} - z^{1/2}) + 2 \ln\left(\frac{1 - y^{1/2}}{1 - z^{1/2}} \right) \right] \qquad (92)$$

---

[7] δb is evaluated by solving Poisson's equation

$$\nabla^2 \psi = - \frac{q}{\varepsilon \varepsilon_0} \left[N_d - \frac{\delta \Delta N_t}{\Delta x_1 \Delta y_1} f(x - x_1) f(y - y_1) \right]$$

where N_d is the donor concentration, and $f(x - x_1)$ and $f(y - y_1)$ are unity in the volume element ΔV and zero elsewhere. This two-dimensional problem can be reduced approximately to a one-dimensional problem by assuming that $|d\psi/dx| \ll |d\psi/dy|$ and $|d^2\psi/dx^2| \ll |d^2\psi/dy^2|$ and that $\delta b(x) = \delta b$ inside the interval $x_1 < x < x_1 + \Delta x_1$ and zero outside that interval. Poisson's equation between x_1 and $x_1 + \Delta x_1$ may be written

$$\frac{d^2 \psi}{dy^2} = - \frac{q}{\varepsilon \varepsilon_0} \left[N_d - \frac{\delta \Delta N_t}{\Delta x_1 \Delta y_1 w} f(y - y_1) \right]$$

with $\psi = 0$ and $d\psi/dy = 0$ at $y = b + \delta b$ and $\psi = V_g - V_{dif}$ at $y = a$. The solution of this equation corresponds to (88).

where $y = (-V_g + V_{dif} + V_d)/V_{00}$ and $z = (-V_g + V_{dif})/V_{00}$ and V_{00} is the cutoff voltage.

This shows a logarithmic divergence for $y \to 1$ (saturation). Actually $S_{I_d}(f)$ converges to a finite value because the mobility becomes field dependent near the drain and hence the equations for ΔR and $\delta \Delta R$ must be modified accordingly [72]. Beyond saturation $S_{I_d}(f)$ is practically independent of V_d.

We now discuss a few additional corrections [170] of this result.

1) The fluctuation δb of (88) is calculated for one side of the channel. Since both sides of the channel fluctuate independently with fluctuations δb_1 and δb_2 and the total width is $2b_1$, we have

$$2\delta b = \delta b_1 + \delta b_2, \quad \text{or} \quad \overline{\delta b^2} = \tfrac{1}{4}(\overline{\delta b_1^2} + \overline{\delta b_2^2}). \qquad (93)$$

Therefore, (92) must be divided by a factor 4 if $\overline{\delta b_1^2} \ll \overline{\delta b_2^2}$ and by a factor 2 if $\overline{\delta b_1^2} = \overline{\delta b_2^2}$.

2) $f_t(1 - f_t)$ is not independent of y_1, as assumed, but is very small except near the point where the quasi-Fermi level of the traps crosses the trapping level, at which point $f_t(1 - f_t)$ has its maximum value $\tfrac{1}{4}$. It is, in principle, not difficult to take this effect into account, and it would be worthwhile doing it.

3) Often there is a distribution in time constants τ_t. One then obtains a somewhat "smeared-out" $1/(1 + \omega^2 \tau_t^2)$ spectrum.

A similar effect also occurs in the space-charge region between channel and substrate of a MOS FET. This problem has been studied by Yau and Sah [178].

F. Noise in Space-Charge-Limited Solid-State Diodes

In space-charge-limited diodes the limiting noise is diffusion noise. In single-injection diodes only one type of carrier is present, in double-injection diodes both electrons and holes are present in about equal numbers.

In the case where the carriers are electrons, the noise in a section Δx can be represented by a current generator $\overline{i^2}$ in parallel with Δx. As van der Ziel and van Vliet [163], [164] indicated,

$$\overline{i^2} = 4 \frac{q^2 D_n n(x) A}{\Delta x} \Delta f = 4kT \left[\frac{q \mu_n n(x) A}{\Delta x} \right] \Delta f = \frac{4kT \Delta f}{\Delta R} \qquad (94)$$

where ΔR is the resistance of the section Δx, $n(x)$ the carrier density, A is the cross-sectional area of the device, and D_n the diffusion constant for electrons. We have here made use of the Einstein relation $q D_n = kT \mu_n$.

In the case of double injection the electrons and holes fluctuate independently and hence [169]

$$\overline{i^2} = 4 \frac{q^2 D_n n(x) A}{\Delta x} \Delta f + \frac{4 q^2 D_p p(x) A}{\Delta x} \Delta f$$
$$= 4kT \left[\frac{q \mu_n n(x) + q \mu_p p(x)}{\Delta x} \right] A \Delta f = \frac{4kT \Delta f}{\Delta R}. \qquad (95)$$

Here D_p is the hole diffusion constant, $p(x)$ the hole concentration, and ΔR the resistance of the section Δx. We have here again made use of the Einstein relation.

The open-circuit EMF in series with the section Δx is thus, in either case,

$$\overline{\Delta v^2} = \overline{i^2}\Delta R^2 = 4kT\Delta R\Delta f = 4kT\frac{\Delta V}{I_a}\Delta f \qquad (96)$$

where I_a is the device current and ΔV the dc voltage developed across the section Δx. Summing over all sections Δx, bearing in mind that the fluctuations in individual sections are independent, yields for the mean square open-circuit voltage

$$\overline{v_a^2} = 4kT\frac{V_a}{I_a}\Delta f. \qquad (97)$$

This should hold for both single and double injection.

For single injection, at relatively low frequencies where transit-time effects do not play a part, $g=2I_a/V_a$ is the ac conductance so that the mean square short-circuit current is

$$\overline{i_a^2} = g^2\ \overline{v_a^2} = 8kTg\Delta f. \qquad (98)$$

This equation remains valid when transit-time effects become important. This has been verified experimentally [85], [103].

For double-injection diodes, at frequencies for which $\omega\tau \gg 1$, where τ is the carrier lifetime, $g=I_a/V_a$ and, hence [169],

$$\overline{i_a^2} = \overline{v_a^2}g^2 = 4kTg\Delta f. \qquad (99)$$

This equation has also been verified experimentally [18], [37], [104]. Note that the last result is half the previous one.

G. Summary

We have seen that a simple thermal noise picture can give a reasonable description of most noise phenomena in FETs and in space-charge-limited solid-state diodes.

Moreover, a detailed application of the theory of GR noise can explain the low-temperature noise behavior of junction FETs. The same theory, when applied to recombination centers in the space-charge regions of junction and MOS FETs, can describe some low-frequency noise phenomena in these devices. In both cases one has to know the statistics of recombination centers and traps, and one must evaluate how the fluctuating occupancy of recombination centers or traps modulates the local channel conductance. The theory is well understood in principle, but some details need further clarification.

It is not clear at present how a field-dependent mobility affects the noise. A very likely possibility is that the noise remains thermal noise but at a temperature corresponding to the effective temperature of the "hot" electrons. This seems to be a fruitful field of further study.

IV. FLICKER NOISE IN SOLID-STATE DEVICES

Besides shot noise and/or thermal noise all solid-state devices show a noise component with a $1/f^\alpha$ spectrum, where $\alpha \simeq 1$. This type of noise is known as "flicker noise"

or "$1/f$-noise." It has been demonstrated in many instances that this $1/f$ noise spectrum holds down to extremely low frequencies [108], [109].

General discussions about $1/f$ noise have been given in the past and several formal theories have been presented. Recent progress has come from a more detailed study of the mechanisms operating in devices showing $1/f$ noise.

It was first suggested that a wide distribution in time constants might be responsible for the $1/f$ spectrum. This idea was specified by McWhorter who demonstrated that a tunneling mechanism in the surface oxide of the material was a very likely cause for such a distribution in time constants [94], [106]. This was reinforced by field-effect experiments which indicated for the first time that a wide distribution of time constants was actually present at the surface of many semiconductors.

The idea behind this $1/f$ noise source is that the carriers in the material communicate with trapping levels at some depth in the surface oxide by tunneling. The time constants in the process are of the form

$$\tau = \tau_0 \exp(\alpha x) \qquad (100)$$

where x is the distance between the trap and the semiconductor-oxide interface, α is of the order of 10^8 cm^{-1}, and τ_0 is the time constant for a trap at the surface. If x varies between 0 and 40 Å, τ will vary over many orders of magnitude, from very small time constants (less than 10^{-6} seconds) to very large time constants (more than 10^6 seconds). The traps that are most effective in the process lie near the Fermi level in the oxide, for those more than a few kT above it are permanently empty and those more than a few kT below it are permanently filled.

This is directly applicable to FETs and to semiconductor filaments. But in junction diodes and transistors the current flow is caused by minority carriers, and holes and electrons are generated in pairs and recombine in pairs, mostly at the surface. Since the work of McWhorter, Fonger [45], and Watkins [174] it is held that the traps in the oxide, just mentioned, modulate the generation and recombination processes, and that this gives rise to flicker noise in these devices. This has recently been challenged by new experiments, but we shall see that these experiments can be incorporated with the theory if it is further developed.

A. Kinetics of the Oxide Traps [155], [170] and Explanation of the 1/f Noise

Let $\Delta N_T = n_T\Delta S\Delta x$ be the number of traps in a small surface element ΔS at a distance between x and $x + \Delta x$ from the semiconductor–oxide interface and let ΔN_t be the number of electrons trapped in that volume element. If E_t is the depth of the trap below the bottom of the conduction band in the semiconductor, and $g(\Delta N_t)$ and $r(\Delta N_t)$ are the generation and release rates of the trapped electrons, respectively, then

$$g(\Delta N_t) = C_1 n(\Delta N_T - \Delta N_t)\exp(-\alpha x);$$
$$r(\Delta N_t) = C_2\Delta N_t\exp\left(-\frac{E_t}{kT}\right)\exp(-\alpha x) \qquad (101)$$

where C_1 and C_2 are constants and n is the electron density in the semiconductor. Putting $g(\Delta N_t) = r(\Delta N_t)$ yields the average number of trapped electrons as

$$\overline{\Delta N_t} = \lambda \Delta N_T; \quad \lambda = \frac{C_1 n}{C_1 n + C_2 \exp(-E_t/kT)}. \quad (102)$$

ΔN_t fluctuates by an amount $\delta \Delta N_t$, and this amount and the time constant τ of trapping processes are given by [24]

$$\overline{\delta \Delta N_t^2} = \Delta N_T \lambda (1 - \lambda) \quad (103)$$

$$\tau = \frac{1}{\dfrac{dr(\Delta N_t)}{dt} - \dfrac{dg(\Delta N_t)}{dt}}$$

$$= \frac{\exp(\alpha x)}{C_2 \exp(-E_t/kT) + C_1 n} \quad (104)$$

$$= \frac{\lambda \exp(\alpha x)}{C_1 n}.$$

Only those traps are effective which lie near the Fermi level in the oxide; for those, $\lambda \simeq 1/2$ and, hence,

$$\overline{\Delta N_t^2} = \tfrac{1}{4}\Delta N_T; \quad \tau = \frac{\exp(\alpha x)}{2C_1 n}. \quad (105)$$

We thus see that (100) is valid and that $\tau_0 = 1/(2C_1 n)$.

The spectral intensity $S_{\Delta N_t}(f)$ of the fluctuation $\delta \Delta N_t$ is [155]

$$S_{\Delta N_t}(f) = 4\overline{\delta N_t^2} \frac{\tau}{1 + \omega^2 \tau^2} = n_T \Delta S \Delta x \frac{\tau}{1 + \omega^2 \tau^2}. \quad (106)$$

Since the corresponding fluctuation δN in N equals $-\delta \Delta N_t$, the spectral intensity of δN equals (106). Here $N = nV$ is the average number of carriers in the sample and V is the sample volume. The fluctuation in N thus has a spectral intensity, obtained by integrating over the surface S,

$$S_N(f) = n_T S x_1 \int_0^{x_1} \frac{\tau}{1 + \omega^2 \tau^2} \frac{dx}{x_1} \quad (106a)$$

if n_T is constant for $0 < x < x_1$ and zero otherwise. But in view of (100), if $\tau_1 = \tau_0 \exp(\alpha x_1)$,

$$\frac{dx}{x_1} = \frac{d\tau/\tau}{\ln \tau_1/\tau_0}. \quad (106b)$$

Carrying out the integration yields

$$S_N(f) = \frac{n_T S x_1}{\omega \ln(\tau_1/\tau_0)} \left[\tan^{-1}(\omega \tau_1) - \tan^{-1}(\omega \tau_0) \right] \quad (107)$$

which has a $1/f$ spectrum for $1/\tau_1 \ll \omega \ll 1/\tau_0$. The model thus explains the occurrence of $1/f$ noise in a very natural manner.

B. Flicker Noise in Junction Diodes

As mentioned in the introduction of this section, flicker noise occurs here due to fluctuations in the surface recombination processes, since these processes are modulated by the fluctuating occupancy of the traps in the oxide. Since

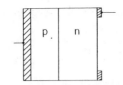

Fig. 3.   Cross section of a p-n diode with surface recombination at the exposed face of the n-region.

the surface recombination is measured by the surface recombination velocity s, the flicker noise occurs because of fluctuations δs in s.

Fonger [45] and Watkins [174] found that the flicker noise in diodes could be zero at a particular current; earlier indications of such an effect are already found in measurements by Anderson and van der Ziel [4]. Fonger attributed it to a modulation effect, but Watkins demonstrated that it was caused by the fact that $dV/ds = 0$ at that point; here V is the total diode voltage.

We shall now expand [168] Watkins' theory, using the geometry of Fig. 3; this geometry is chosen because it reduces the problem to a one-dimensional one. It is further assumed that the diode is a p^+n diode and that the surface recombination occurs at the exposed side of the n-region. The boundaries of the n-region are located at $x=0$ and $x=w$.

The current density equations are

$$J_p = q\mu_p pF - qD_p \frac{dp}{dx} \quad (108)$$

$$J_n = q\mu_n nF + qD_n \frac{dn}{dx} = 0 \quad (109)$$

so that the electric field strength F is

$$F = -\frac{D_p}{\mu_p} \frac{1}{(N_d + p)} \frac{dp}{dx}. \quad (110)$$

We have here made use of the Einstein relation and have assumed that space-charge neutrality prevails, so that $n = N_d + p$, where N_d is the donor concentration.

Substituting (110) into (108) yields

$$J_p = qD_p \left(1 + \frac{p}{N_d + p}\right) = qsp(w). \quad (111)$$

Integration gives

$$2p(0) - 2p(x) - N_d \ln \left[\frac{N_d + p(0)}{N_d + p(x)}\right] = \frac{s}{D_p} p(w)x \quad (112)$$

so that at $x = w$

$$2p(0) - 2p(w) - N_d \ln \left[\frac{N_d + p(0)}{N_d + p(w)}\right] = \frac{sw}{D_p} p(w). \quad (112a)$$

If p_n is the equilibrium hole concentration in the n-region, then the junction voltage

$$V_g = \frac{kT}{q} \ln \left[\frac{p(0)}{p_n}\right], \quad (113)$$

and the voltage V_b in the bulk n-region is

$$V_b = -\int_w^0 F\,dx = \frac{kT}{q}\ln\left[\frac{N_d + p(0)}{N_d + p(w)}\right], \quad (114)$$

so that the total diode voltage V is

$$V = V_g + V_b = \frac{kT}{q}\left\{\ln\frac{p(0)}{p_n} + \ln\left[\frac{N_d + p(0)}{N_d + p(w)}\right]\right\}. \quad (115)$$

Differentiating (112a) with respect to s for constant J_p, that is for a constant product $sp(w)$, yields

$$\left[\frac{2p(0) + N_d}{p(0) + N_d}\right]\frac{dp(0)}{ds} - \left[\frac{2p(w) + N_d}{p(w) + N_d}\right]\frac{dp(w)}{ds} = 0. \quad (116)$$

Evaluating dV/ds and eliminating $dp(0)/ds$ with the help of (116) yields

$$\frac{dV}{ds} = \frac{kT}{q}\left[\frac{2p(w) + N_d - p(0)}{p(0)}\right]\frac{dp(w)/ds}{N_d + p(w)} \quad (117)$$

which is zero for

$$2p(w) + N_d - p(0) = 0. \quad (118)$$

A fluctuation in s will not give any fluctuation in V in that case.

To see whether this leads to a realizable value of $p(w)$, we substitute into (112a). This yields

$$2p(w) + N_d(2 - \ln 2) = \frac{sw}{D_p}p(w) \quad (119)$$

which gives a positive solution for $p(w)$ only if $sw/D_p > 2$. Consequently the minimum in the flicker noise occurs only if the surface recombination velocity is sufficiently large.

Some time ago Guttkov [66] published flicker noise measurements on germanium p-n junctions. He found that the spectrum of the current fluctuation could be represented as

$$S_I(f) \simeq \frac{I^2}{s^2}B(f) \quad (120)$$

where $B(f)$ depends on frequency as $1/f^\alpha$ with $\alpha \simeq 1$; he claimed that this was incompatible with Fonger's surface recombination model.

We shall see that this claim is invalid if the surface recombination model is properly extended [168]. For small injection, $p(0) \ll N_d$ and $p(w) \ll N_d$, and hence (117) may be written

$$\frac{dV}{ds} = \frac{kT}{q}\frac{dp(w)/ds}{p(0)} = -\frac{kT}{q}\frac{p(w)}{p(0)}\frac{1}{s} \quad (121)$$

since, according to (111),

$$\frac{dp(w)}{ds} = -\frac{J_p/q}{s^2} = -\frac{p(w)}{s}. \quad (121a)$$

Making a Taylor expansion of (112a) for $p(0) \ll N_d$ and $p(w) \ll N_d$ yields

$$p(0) = \left(1 + \frac{sw}{D_p}\right)p(w) \quad (122)$$

so that

$$\frac{dV}{ds} = -\frac{kT}{q}\frac{1}{s(1 + sw/D_p)}. \quad (123)$$

Therefore a fluctuation δs in s gives a fluctuation δV in V

$$\delta V = -\frac{kT}{q}\frac{\delta s}{s(1 + sw/D_p)} \quad (123a)$$

so that the spectral intensities $S_v(f)$ and $S_s(f)$ are related as

$$S_v(f) = \left(\frac{kT}{q}\right)^2\frac{S_s(f)}{s^2(1 + sw/D_p)^2}. \quad (124)$$

Now $S_s(f)$ should be independent of current at low injection and, according to Section IV-A, it should be proportional to the trap density in the oxide which, in turn, would be proportional to s. Putting $S_s(f) = sB(f)$ yields

$$S_v(f) = \left(\frac{kT}{q}\right)^2\frac{B(f)}{s(1 + sw/D_p)^2} \quad (124a)$$

and, hence,

$$S_I(f) = \left(\frac{qI}{kT}\right)^2 S_v(f) = \frac{I^2 B(f)}{s(1 + sw/D_p)^2}. \quad (125)$$

This varies as $1/s$ for $sw/D_p \ll 1$ and as $1/s^3$ for $sw/D_p \gg 1$. Intermediately there will be a region where $S_I(f)$ varies approximately as $1/s^2$, as found by Guttkov.

Guttkov mentioned that in his diodes s was fairly large. If this geometry was not one-dimensional, which most likely will have been the case, $S_I(f)$ must be averaged over a wide range of values of w. This has the tendency of extending the $1/s^2$-region. We thus conclude that Guttkov's results can be brought into agreement with the surface recombination model.

Hsu and his coworkers [86] found that the noise spectrum $S_I(f)$ in silicon diodes at constant current was proportional to the surface recombination velocity s. This seems to be in disagreement with Guttkov's results and with the surface recombination model. But we must now bear in mind that in silicon diodes a significant carrier recombination occurs in the space-charge region, especially at the surface of that region. If the flicker noise is due to fluctuations in the recombination current produced in that region, as likely is the case, the discrepancy disappears [168].

The recombination occurs in a well defined part of the space-charge region. Let this region be characterized by the coordinate x_1 and let it have an effective area A_{eff}.

Let for an applied potential V the change in potential at x_1 be V_1, and let the hole concentration $p(x)$ at x_1 be denoted by $p(x_1)$. It varies with V_1 as

$$p(x_1) = p_1(x_1)\exp\left(\frac{qV_1}{kT}\right) \quad (126)$$

where $p_1(x_1)$ is the value of $p(x_1)$ for $V_1 = 0$. The recombination current I_R is then

$$I_R = qsp(x_1)A_{eff} = qp_1A_{eff}s\exp\left(\frac{eV_1}{kT}\right). \qquad (127)$$

If now s fluctuates by an amount δs, the fluctuation δI_R in I_R is

$$\delta I_R = qp_1A_{eff}\exp\left(\frac{qV_1}{kT}\right)\delta s \qquad (128)$$

so that

$$S_{I_R}(f) = [qp_1A_{eff}]^2\exp\left(\frac{2qV_1}{kT}\right)S_s(f). \qquad (129)$$

If all flicker noise is associated with δI_R, the spectrum $S_I(f)$ of the diode current I is equal to (129). Putting

$$I = I_0\exp\left(\frac{qV}{mkT}\right) \qquad (130)$$

with $m \geq 1$ and substituting $S_s(f) = sB(f)$ yields for $S_I(f)$

$$S_I(f) = \text{const } I^{2mV_1/V}B(f)s, \qquad (130a)$$

so that $S_I(f)$ at constant current is indeed proportional to s, as required by the experiments.

For a symmetrical junction $V_1 \simeq \frac{1}{2}V$ and, hence, for $m \simeq 1$, $S_I(f)$ in a given diode is proportional to I. In an asymmetric diode, such as a p^+n diode, V_1 may be different and, hence, $S_I(f)$ may have a somewhat different current dependence. This is significant for understanding the current dependence of flicker noise in transistors.

We thus conclude that the surface recombination model of flicker noise, if properly applied, can explain the experiments on flicker noise in germanium and silicon diodes.

C. Flicker Noise in Transistors

In analogy with the shot noise case one would like to represent flicker noise in transistors by two current generators i_{f_1} and i_{f_2} in parallel with the emitter and collector junction, respectively. Chenette [28] demonstrated that i_{f_1} and i_{f_2} were almost fully correlated, but his experiments are equally compatible with the idea that most of the $1/f$ noise comes from the current generator i_{f_1} only. In addition he showed that the base resistance r_b could be determined from flicker noise measurements. This was further substantiated by the work of Gibbons [52].

The clearest demonstration is found in a paper by Plumb and Chennette [122] who showed that i_{f_2}/i_{f_1} was quite small. In their experiments the collector was HF connected to ground, a large resistance R_E was inserted in the emitter lead, and a variable resistor R_B was inserted between base and ground. The noise was measured between emitter and ground and R_B was so adjusted that the measured noise was a minimum. As seen from Fig. 4, the noise voltage appearing at the emitter terminal is

$$v = i_{f_1}r_{e0} - (i'_{f_2} + \alpha_0 i_{f_1})(R_B + r_b) - i''_{f_2}(R_B + r_b) \qquad (131)$$

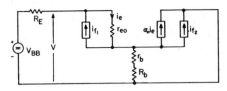

Fig. 4. Circuit for locating flicker noise sources in transistors.

where $r_{e0} = kT/qI_E$ and i_{f_2} has been split into two parts i'_{f_2} and i''_{f_2}. Here i'_{f_2} is fully correlated with i_{f_1}, and i''_{f_2} is uncorrelated. Therefore, $\overline{v^2}$ will go through a deep minimum if

$$r_{e0} - \left(\alpha_0 + \frac{i'_{f_2}}{i_{f_1}}\right)(R_b + r_b) = 0$$

or $\qquad\qquad\qquad\qquad\qquad\qquad\qquad\qquad (131a)$

$$R_B = (R_B)_{min} = \frac{1}{1 + i'_{f_2}/(\alpha_0 i_{f_1})}\cdot\frac{r}{\alpha_0} - r_b.$$

Hence, by plotting $(R_B)_{min}$ versus $r_{e0}/\alpha_0 = kT/qI_C$, one should obtain a straight line intercepting the vertical axes at $-r_b$, whereas the slope of the line gives a measure for $[1 + i'_{f_2}/(\alpha_0 i_{f_1})]$. This agreed very well with the experimental data and the values found for i'_{f_2} were only a few percent of i_{f_1}, so that one can neglect i'_{f_2} for all practical purposes. Measurements of $\overline{i_{f_1}^2}$ were obtained for small values of I_E by omitting R_B altogether; they indicated that $\overline{i_{f_1}^2}$ varied as I_E^β with β somewhat smaller than unity. This is compatible with (130a). Other units give $\beta > 1$.

Measurements by Viner [172] on silicon transistors operated at elevated temperatures indicated another interesting feature. At elevated temperatures the base current I_B may be written as $-I_{CBO} + I'_B$ where $-I_{CBO}$ is the base current for zero emitter current and I'_B the injected base current. Viner showed that the currents I_{CBO} and I'_B fluctuated independently and that each current showed flicker noise.

Flicker noise in silicon transistors must be interpreted in terms of a fluctuation in the recombination current I_R in the emitter-junction space-charge region, presumably mostly at the surface. Therefore the surface model of Section IV-B should be applicable.

The effect can be reduced by reducing I_R; this has been achieved by surface pacification techniques. It is not quite clear where the remaining noise comes from. Is it generated at the surface or in the interior of the space-charge region? A careful comparison of r_b data obtained from flicker noise and HF noise measurements may give the answer. Also a further comparison between p-n-p and n-p-n transistors would be useful; it seems that the latter show more $1/f$ noise.

D. Flicker Noise in FETs [32], [112]

The theory of flicker noise in silicon junction FETs should be similar to the theory for transistors. The flicker noise will be generated in the space-charge region between gate and channel, presumably mostly at the surface. It can be reduced by surface pacification techniques and is presently not very large in well-designed units.

In MOS FETs the $1/f$ noise is often much larger than in junction FETs. The reason for this is that the MOS FET is a surface device; the fluctuating occupancy of traps in the oxide can modulate the conducting surface channel all along the channel.

The fact that the $1/f$ noise in MOS FETs was a surface effect was first demonstrated by Sah and Hielscher [129], who showed that quantitative correlation existed between the $1/f$ noise power spectrum and the lossy part of the gate impedance due to carrier recombination in the surface states at the oxide–silicon interface. The $1/f$ noise is proportional to the surface-state density [1]; this is important, for it means that the $1/f$ noise can be reduced by proper choice of the crystal orientation of the substrate. Since (100) surfaces have a much smaller surface-state density than the more commonly used (111) surfaces, the $1/f$ noise can be lowered by making the MOS FETs on (100) surfaces. Other effects also seem to play a part, however.

The theory of Section III-E, if properly modified, can be applied to this case [32]. Let $\Delta N_T = n_T w \Delta x \Delta y$ be the number of traps in a volume element $w\Delta x\Delta y$ of the oxide, at a point $P(x, y)$ in the oxide; here n_T is the trap density, w is the channel width, and the x axis is directed along the channel. Let ΔN_t electrons be trapped and let $\delta\Delta N_t$ be the fluctuation in ΔN_t due to the random processes of trapping and detrapping. Then, if f_t is the fractional occupancy of the traps,

$$\overline{(\delta\Delta N_t)^2} = n_T f_t(1 - f_t)w\Delta x\Delta y;$$

$$\overline{\delta\Delta N_t(t)\delta\Delta N_t(t + s)} = \overline{(\delta\Delta N_t)^2} \exp\left(-\frac{s}{\tau}\right) \quad (132)$$

where τ is the time constant of the trapping process. A fluctuation $\delta\Delta N_t$ in the number of trapped electrons gives rise to a fluctuation $\delta\Delta N = -\delta\Delta N_t$ in the section Δx of the channel.

We now apply (80a), replace W_0 by V_0, the dc potential of the section Δx of the channel, and write

$$\frac{I_d\Delta x}{L\Delta N} = \frac{q\mu I_d}{Lg(V_0)}$$

where $g(V_0)$ is the conductance per unit length of the channel. Then (80a) may be written

$$\overline{\delta\Delta I_d(t)\delta\Delta I_d(t + s)} = \left[\frac{q\mu I_d}{Lg(V_0)}\right]^2 n_T f_t(1 - f_t)w\Delta x\Delta y$$
$$\cdot \exp\left(-\frac{s}{\tau}\right). \quad (133)$$

Applying the Wiener–Khintchine theorem yields

$$\Delta S_{I_d}(f) = 4\left[\frac{q\mu I_d}{Lg(V_0)}\right]^2 n_T f_t(1 - f_t)w\Delta x\Delta y\frac{\tau}{1 + \omega^2\tau^2}. \quad (134)$$

By integrating over the oxide one obtains the total spectrum $S_{I_d}(f)$,

$$S_{I_d}(f) = 4\left(\frac{qI_d}{L}\right)^2 n_T w \int_0^L dx \int_0^d \frac{f_t(1 - f_t)}{[g(V_0)]^2} \frac{\tau}{1 + \omega^2\tau^2} dy \quad (134a)$$

where d is the thickness of the oxide. Note that τ depends on y, since the time constant is governed by a tunneling process; integration with respect to y then gives the $1/f$ spectrum, as before. Integration with respect to x gives the dependence of the noise in the operating conditions.

It is here not allowed to put $f_t = 1/2$, but it is necessary to evaluate the dependence of f_t on the operating conditions explicitly. It is thus necessary to introduce a distribution in the energies of the trapping level and to integrate over that distribution. Only if one does that, does one obtain a finite spectrum. If one merely puts $f_t = 1/2$ and integrates, one obtains a logarithmic divergence at saturation. See Christenson et al. [32] for details.

An alternate theory of flicker noise in MOS FETs by Leventhal [100] assumes that the carriers move in a surface band. He then obtains a $1/f$ frequency dependence and a linear dependence on the surface-state density. Moreover, for a given sample the noise should be inversely proportional to the absolute temperature. It would seem, however, that a tunnel mechanism is a more likely cause of $1/f$ noise.

E. Flicker Noise in Gunn Diodes [43], [110]

Flicker noise has also been observed in Gunn diodes. The flicker noise, measured between the terminals of the device is almost independent of whether or not the device is oscillating. This $1/f$ noise can contribute heavily to the FM noise of the oscillator.

F. Burst Noise [25], [53], [90], [131], [176]

Burst noise has been observed in planar silicon and germanium diodes and transistors. The phenomenon consists of a random turning on and off of a current pulse; this can be described by the random telegraph signal approach. This is not properly flicker noise, since it has a $\text{const}/(1 + \omega^2\tau^2)$ spectrum. It is believed that a current pulse is caused by a single trapping center in the space-charge region.

G. Summary

We have seen how the model of distributed traps in the surface oxide layer of a semiconductor can give a natural explanation of $1/f$ noise. The fluctuating occupancy of the traps corresponds to a fluctuating carrier density in the material, which is detected as noise when dc current is passed through the material. The communication of carriers inside the semiconductor with the traps comes about by tunnel effect; this gives the wide distribution in time constants needed for the $1/f$ spectrum.

This picture is directly applicable to noise in FETs, but for noise in junction diodes and transistors one must bear in mind that the current flow is by *minority* carriers instead of by *majority* carriers. One must now take into account that the fluctuating occupancy of the traps in the oxide modulates the surface recombination velocity. We have seen how this model can explain a variety of seemingly contradictory experimental data. The model, when applied to transistors, indicates how the flicker noise source is located in the equivalent circuit.

Further work to clarify these flicker noise models would be useful.

V. Light Noise in Luminescent Solid-State Diodes and Solid-State Lasers

The general noise problem of lasers is too complicated to be treated in this survey. We shall restrict ourselves chiefly to light diodes and laser diodes. It will be seen that the photons always show shot noise, but in addition, especially near threshold, there is also spontaneous emission noise. Finally we discuss optical heterodyning as a means of receiving very weak optical signals.

A. Luminescent Diodes

Luminescent diodes should show shot noise. The reason for that is a very simple one: the carriers crossing the barrier and recombining after crossing show shot noise.

To discuss this problem we need a property of shot noise that has not directly concerned us up to now. Let a certain series of events occur at the average rate $\bar{n}$ and let n be the fluctuating number of events occurring during a given second. We can then make up

$$\text{var } n = \overline{n^2} - (\bar{n})^2 \tag{135}$$

for this fluctuating process. If the events are independent and occur at random (Poisson process or shot noise process), one finds

$$\text{var } n = \bar{n}. \tag{136}$$

A process satisfying condition (136) gives *shot noise*. We can now apply this to photon emission and see when the photon emission process is a shot noise process.

Let us for the sake of simplicity assume that the injected carriers are electrons. According to Section II-B, those electrons give shot noise; i.e.,

$$\text{var } n = \bar{n}. \tag{137}$$

Let it further be assumed that all injected electrons recombine under the generation of a photon and that all generated photons are actually emitted. Then the rate n_p of emission of photons equals the rate n of injection of electrons so that

$$n_p = n, \qquad \text{var } n_p = \text{var } n = \overline{n_p}. \tag{138}$$

In other words the emitted photons show shot noise in this case.

Let it now be assumed that each injected electron has a probability λ of giving rise to an emitted photon. Then the partition noise theorem applies according to which

$$\overline{n_p} = \lambda \bar{n}; \qquad \overline{\Delta n_p^2} = \text{var } n_p = \lambda^2 \text{ var } n + \bar{n}\lambda(1 - \lambda). \tag{139}$$

But since var $n = \bar{n}$ this may be written

$$\overline{\Delta n_p^2} = \text{var } n_p = \lambda \bar{n} = \overline{n_p} \tag{139a}$$

so that again the photons show shot noise.

This is not altered if optical feedback is applied so that the luminescent diode becomes a laser, since the emitted photons still obey Poisson statistics (see the following section). Near threshold there is perhaps some additional noise

due to spontaneous emission, but far away from threshold, that is, at large emitted powers of the laser diode, the diode should show shot noise (see Section V-B).

This is quite different from a black-body radiator. If n quanta of frequency between v and $v + \Delta v$ are received per second from a blackbody radiator at the temperature T, then according to statistical mechanics

$$\text{var } n = \frac{\bar{n}}{1 - \exp(-hv/kT)}. \tag{140}$$

This reduces to var $n = \bar{n}$ for large hv/kT but var $n \gg \bar{n}$ for small hv/kT. In the latter case the noise of the received photons is thus much larger than would be expected from shot noise considerations. Equation (140) is one of the oldest formulas of quantum theory.

The fluctuations in the diode current (in n) and in the emitted radiation (in n_p) are correlated. This is easily seen as follows. Let during a given second n electrons be injected. We then write $n = \bar{n} + \Delta n$. The Δn electrons will give rise to $\lambda \Delta n$ emitted photons, so that

$$\Delta n_p = \lambda \Delta n + \Delta n_p' \tag{141}$$

is the fluctuation in n_p. Here $\Delta n_p'$ is the noise caused by the randomness in the recombination of the carriers and/or the emission of the photons. It is of course independent of Δn; consequently, $\overline{\Delta n \Delta n_p'} = 0$ and, hence,

$$\overline{\Delta n \Delta n_p} = \lambda \overline{\Delta n^2} + \overline{\Delta n \Delta n_p'} = \lambda \bar{n} \tag{142}$$

so that the correlation coefficient

$$c = \frac{\overline{\Delta n \Delta n_p}}{\sqrt{\overline{\Delta n^2} \, \overline{\Delta n_p^2}}} = \frac{\lambda \bar{n}}{\sqrt{\bar{n} \, \lambda \bar{n}}} = \sqrt{\lambda}. \tag{142a}$$

According to Guekos and Strutt the pattern of light noise fluctuations and voltage fluctuations across GaAs laser diodes closely parallel each other [55], which suggests a strong correlation between the two noise phenomena. Actual correlation measurements indicate that the correlation coefficient fluctuates strongly with varying current [56].

B. Shot Noise and Spontaneous Emission Noise in Lasers

A laser far above threshold can be considered as a generator of a stable amplitude. In that case the emitted radiation obeys Poissonian statistics, that is,

$$\text{var } n = \bar{n}$$

and the device shows shot noise of the emitted photons.

We can give two reasons why a laser should show shot noise. First consider an ideal laser with 100 percent efficiency and zero optical loss. Then the rate n of emission of photons is equal to the pumping rate W. Since the pumping can be represented as a series of independent random events, we have var $W = \overline{W}$ and, hence, var $n = \bar{n}$. If there are optical losses, so that the part λ of the produced quanta is actually emitted, then the partition noise theorem gives

$$\bar{n} = \overline{W}\lambda; \text{ var } n = \lambda^2 \text{ var } W + \overline{W}\lambda(1 - \lambda) = \overline{W}\lambda = \bar{n}$$

since var $W = \overline{W}$, as stated before. Hence, n shows full shot noise.

One can also use the following reasoning. An arbitrary spectral line always gives at least shot noise of the quanta contained in the line. In addition beats occur between the frequencies within the line (photon bunching); this is called *wave interaction noise* [3a] and gives an additional contribution to var n. A single-mode laser sufficiently far above threshold is a single-frequency oscillator that produces no beats and, hence, no wave interaction noise. The spontaneous emission noise observed near threshold can be interpreted as wave interaction noise. Another interpretation, to be given in the following, provides additional insight.

Closer to threshold there is excess noise caused by spontaneous emission. The laser can then be considered a van der Pol oscillator driven by spontaneous emission noise [80]. The noise was first observed by Prescott and van der Ziel [124] in a gas laser and by Smith and Armstrong in a GaAs laser [139]. According to Freed and Haus [47] the bandwidth B of the spontaneous emission noise spectrum is inversely proportional to the average laser power $\overline{P}$ below threshold and proportional to $\overline{P}$ above threshold, having a minimum value at threshold. The low-frequency ratio $S_e(0)/S_s(0)$, where $S_e(f)$ is the spontaneous emission noise power and $S_s(f)$ the shot noise power of the laser, is proportional to $(\overline{P})^2$ below threshold and inversely proportional to $(\overline{P})^2$ above threshold, with a maximum value near threshold. The spectrum $S_e(f)$ is of the form

$$S_e(f) = \frac{S_e(0)}{1 + (f/B)^2} \tag{143}$$

where B is the bandwidth of the spectrum.[8] For gas lasers the minimum bandwidth B is relatively small and, hence, $S_e(0)/S_s(0)$ near threshold is reasonably large. For GaAs lasers the bandwidth B is very much larger and, hence, the observed value of $S_e(0)/S_s(0)$ near threshold observed by Armstrong and Smith was only 0.017. This small value was determined with the help of a counting technique. Elaborate photon counting experiments on gas lasers were reported by Arecchi *et al.* [5].

Because of the dependence of $S_e(0)/S_s(0)$ on average laser power, $S_e(0)$ becomes negligible sufficiently far above threshold. If the laser radiation is then detected by a photodetector, the noise observed in the detector current is full shot noise. The same is true for noise measurements for low laser powers at sufficiently high frequencies ($f \gg B$) [124].

Additional noise has been observed in a multimode gas laser when the modes are not locked [84]. The effect disappears when the modes are locked. This effect has not been observed in GaAs lasers, but Smith and Armstrong [139] have found two additional interaction effects in these devices:

1) low-frequency nonstationary noise which occurs when a weak mode is lasing in competition with a strong mode,
2) broad-band stationary noise which occurs when two modes are about equal in intensity.

The authors believe that the first type of noise arises from heat transfer processes in the diodes and dewar, while the second is thought to be partition noise which must occur when a photon can be stimulated into one of a number of lasing modes. The total noise for all modes is very small, being comparable to that of a single mode with the same total power.

According to Haug the junction current noise is larger than shot noise [77], [78]. He finds two additional noise terms:

1) a term due to the fluctuations in emission and absorption rates,
2) a term due to light-field fluctuations.

It is interesting to note that Guekos and Strutt [54] found that the current noise levels of GaAs laser diodes was much higher than expected from shot noise considerations. The same is true for the voltage noise observed in the lasing mode. This may have some connection with Haug's prediction.

In their lasing diodes the ratio of excess noise over shot noise in the light output and in the diode voltage was quite large and fluctuated wildly with current. The authors suggest that this may be associated with the switching of lasing filaments in the junction.

In conclusion it would seem that a further study of these noise phenomena would be worthwhile.

C. Optical Heterodyning

When photomultipliers are used in the reception of weak laser signals, enough built-in gain is present in the photomultiplier to make the output noise of the multiplier large in comparison with the noise of the associated amplifier used for processing the detected signal. If direct photodetectors, either photodiodes or photoconductors, are used, the detected signal may drown in the receiver noise. If an optical heterodyning method is used, whereby one beats the incoming signal with a local oscillator laser (pump) signal, one obtains a large power gain so that the noise of the amplifier can be overcome. It has the disadvantage, however, of requiring accurate alignment of the two beams.

Teich *et al.* [142] did an optical heterodyning experiment at 10.6 micron in photoconductive copper-doped germanium. They measured the signal-to-noise power ratio as a function of the incident power and obtained an experimental curve that was about a factor two below the theoretical curve, well within the experimental error, however.[9]

---

[8] Since $S_e(0)/S_s(0)$ varies as $1/B^2$ below threshold, $S_e(0)/S_s(0)$ is very small if the minimum bandwidth is large, as is the case in GaAs lasers.

[9] Even this discrepancy disappears when one corrects the theoretical (signal-to-noise) power ratio used by the authors. The correct expression is $\eta P_s/(2h\nu\Delta f)$ for a photodiode and $\eta P_s/(4h\nu\Delta f)$ for a photoconductive detector [see (145)].

Lee and van der Ziel [99] measured the noise at the output of a photodiode detector carrying a current I_p due to the local oscillator signal and an additional current I_1 due to the incoming radiation. The power gain, defined as the ratio[10]

$$\frac{\text{output signal power of optical heterodyne detector}}{\text{output signal power of direct detector}}$$

turns out to be $2I_p/I_1$, as long as I_1 is not too small, whereas the spectral intensity $S_m(f)$ of the optical heterodyne detector is

$$S_m(f) = 2eI_{eq} = 2eI_p + 4eI_p\eta. \tag{144}$$

Here the first term is the shot noise due to the pump signal and the second term is the amplified shot noise of the incoming radiation. By measuring I_{eq} versus I_p they could demonstrate the presence of the second term. They also showed that I_{eq} was independent of I_1.

The noise measurements indicate that the optical heterodyne detector is operating closely to the limit set by the noise of the incoming radiation. The signal-to-noise ratio of the optical heterodyne detector is found to be

$$\left(\frac{S}{N}\right)_m = \left(\frac{2}{1+2\eta}\right)\frac{I_1}{2eB} = \left(\frac{2}{1+2\eta}\right)\frac{\eta P_s}{2h\nu B}, \tag{145}$$

where η is the quantum efficiency of the detector, $I_1 = \eta eP_s/h\nu$ the detected current for direct detection, and B the bandwidth. The reader will recognize $P_s/(2h\nu B)$ as the signal-to-noise power ratio of the incoming radiation for the bandwidth B. (See footnote nine.)

D. Summary

We have seen that GaAs light diodes and laser diodes should always show at least shot noise of the emitted photons. In other lasers there may also be spontaneous emission noise near threshold, but in GaAs diodes this type of noise is very small. There should be additional terms in the current noise of the GaAs laser. These, together with the correlation between current noise and laser light noise, should be studied further.

In a good optical mixer one can see the contribution of the noise of the incoming radiation if the quantum efficiency of the detector is sufficiently large. Such a mixer therefore operates closely to the theoretical limit.

VI. SOLID-STATE DEVICES IN CIRCUITS

It is here assumed that the reader is familiar with the concepts of noise figure F and noise measure M. He should also be familiar with Friiss' formula for the noise figure of a combination of stages [79], [149], [170]. Using these concepts we now discuss the noise figure of various solid-state amplifiers.

---

[10] The reason for the large power gain is that the direct detector is quadratic, whereas the optical mixer is linear. Therefore, at low input power the mixer is much better than the direct detector.

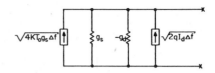

Fig. 5. Equivalent circuit of tunnel diode amplifiers.

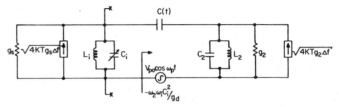

Fig. 6. Equivalent circuit of parametric amplifier.

A. Noise Figure of Negative Conductance Amplifiers

As a first example we consider a tunnel diode amplifier [27], [118], [145], [170]. The equivalent circuit is shown in Fig. 5, the output circuit is considered to belong to the next stage. We assume that the circuit has been so designed that $g_s > g_d$, where $-g_d$ is the negative conductance of the tunnel diode. It is then seen by inspection that the available gain is

$$G_{av} = \frac{g_s}{g_s - g_d} \tag{146}$$

which approaches infinity if g_s approaches g_d. It is also seen by inspection that

$$F = \frac{4kT_0g_s\Delta f + 2qI_d\Delta f}{4kT_0g_s\Delta f} = 1 + \frac{q}{2kT_0}\frac{I_d}{g_s}. \tag{147}$$

It would thus seem at first sight that a low noise figure would be obtained for large values of g_s. However, since G_{av} approaches unity if g_s approaches infinity, one must be careful. One finds the noise measure of the stage

$$M = \frac{F-1}{1-1/G_{av}} = \frac{q}{2kT_0}\frac{I_d}{g_d} \tag{148}$$

independent of g_s, so that it is cheapest to use a single stage and choose g_s relatively close to g_d. In that case

$$F \simeq F_\infty = 1 + \frac{q}{2kT_0}\frac{I_d}{g_d}. \tag{148a}$$

One should choose the operating point of the diode in such a way that I_d/g_d attains its minimum value. Noise figures of the order of 2 (3 dB) are achievable in this manner.

As a second example we consider a parametric amplifier [81], [82], [154] made by connecting a varactor diode driven by a pump signal of frequency ω_p and with a tuned output (idler) circuit of tuned conductance g_2 and tuning frequency $\omega_2 = (\omega_p - \omega_i)$ in series with the pump, in parallel with a signal source and a tuned circuit tuned at the frequency ω_i (Fig. 6). Let the time dependent capacitance $C(t)$ of the varactor diode have a Fourier expansion

$$C(t) = C_0 + 2C_1\cos\omega_p t + 2C_2\cos 2\omega_p t \cdots. \tag{149}$$

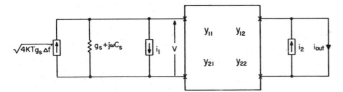

Fig. 7. Equivalent circuit of FET amplifier for determining noise figure.

Then the tuned idler circuit with tuned circuit conductance g_2 produces a negative conductance

$$-g_d = -\frac{\omega_2 \omega_i C_1^2}{g_2} \qquad (150)$$

in parallel with the source and the input circuit. The theory of the tunnel-diode amplifier can thus be applied, provided that the noise $\sqrt{\overline{i_d^2}}$ associated with $-g_d$ is known. Since this represents the converted thermal noise EMF $4kT\Delta f/g_2$ of the tuned idler circuit, we have

$$\overline{i_d^2} = \frac{4kT\Delta f}{g_2} \omega_i^2 C_1^2 = 4kT\Delta f g_d \frac{\omega_i}{\omega_2} \qquad (151)$$

and, hence,

$$F = \frac{4kTg_s\Delta f + \overline{i_d^2}}{4kTg_s\Delta f} \simeq 1 + \frac{\omega_i}{\omega_2} \qquad (152)$$

if g_s approaches g_d. Very low noise figures can thus be obtained by choosing $\omega_2 \gg \omega_i$.

It is beyond the scope of this review to include the many practical realizations of this principle or the evaluation of the effect of the shot noise and of the series resistance of the varactor diode on the performance of the circuit [118a].

B. Field-Effect Transistor Circuits [102], [159], [170]

The equivalent circuit of the field-effect transistor in common source connection is shown in Fig. 7. This equivalent circuit is purposely chosen different from the circuit recommended by the IRE Standards Committee on Noise [180] because the noise parameters hereby introduced relate closely to the physics of the device.[11] Fig. 7 shows the equivalent circuit of the FET in common source connection. Let the source-free two-port in this circuit be represented by the admittance matrix

$$\begin{pmatrix} Y_{11} & Y_{12} \\ Y_{21} & Y_{22} \end{pmatrix}, \qquad (153)$$

and let i_1 be split into a part i_1', fully correlated with i_2, and a part i_1'' correlated with i_2. Let a source, consisting of a current generator $\sqrt{4kT_0 g_s \Delta f}$ in parallel with the source

admittance $Y_s = g_s + j\omega C_s$, be connected to the input and let the output be short-circuited. Collecting the various contributions to the short-circuited current i_{out}, one easily obtains for the noise figure

$$F = 1 + \frac{\overline{i_1''^2}}{4kT_0 g_s \Delta f}$$

$$+ \frac{\overline{i_2^2}}{4kT_0 g_s |Y_{21}|^2} \left| Y_s + Y_{11} + \frac{i_1'}{i_d} Y_{21} \right|^2. \qquad (154)$$

Introducing the noise parameters g_n, R_n, and Y_{cor} by the definitions

$$\overline{i_1''^2} = 4kT_0 g_n \Delta f;$$

$$\overline{i_2^2} = 4kT_0 R_n \Delta f |Y_{21}|^2;$$

$$Y_{\text{cor}} = Y_{21} \left| \frac{i_1'}{i_2} \right| = Y_{21} \frac{\overline{i_1 i_2^*}}{\overline{i_2^2}}, \qquad (155)$$

one obtains

$$F = 1 + \frac{g_n}{g_s} + \frac{R_n}{g_s} |Y_s + Y_{11} + Y_{\text{cor}}|^2 \qquad (156)$$

which is easily optimized as a function of b_s and g_s. Note that in the representation g_n, R_n, and Y_{cor} are directly related to the physics of the device; R_n measures the output noise, g_n the uncorrelated part of the input noise, and the correlation admittance Y_{cor} measures the correlated part of the input noise.

Substituting $Y_{11} = g_{11} + j\omega C_{11}$; $Y_{\text{cor}} = g_{\text{cor}} + j\omega C_{\text{cor}}$, one obtains that F as a function of C_s has a minimum value

$$F_t = 1 + \frac{g_n}{g_s} + \frac{R_n}{g_s} (g_s + g_{11} + g_{\text{cor}})^2 \qquad (157)$$

for

$$C_s + C_{11} + C_{\text{cor}} = 0. \qquad (157a)$$

It turns out that C_{cor} is not very large if Y_{12} is neutralized, so that tuning for maximum signal then increases the noise figure only slightly. By measuring F_t as a function of g_s one can determine the other noise parameters. F_t as a function of g_s has a minimum value

$$F_{\min} = 1 + 2R_n(g_{11} + g_{\text{cor}}) + 2\sqrt{g_n R_n + (g_{11} + g_{\text{cor}})^2 R_n^2} \qquad (158)$$

for

$$g_s = \sqrt{(g_{11} + g_{\text{cor}})^2 + \frac{g_n}{R_n}}. \qquad (158a)$$

For a junction FET $g_{\text{cor}} \ll g_{11}$ and $g_n \simeq g_{11}$, so that $F_{\min}$ can be simplified to

$$F_{\min} = 1 + 2R_n g_{11} + 2\sqrt{g_{11}R_n + (g_{11}R_n)^2}. \qquad (159)$$

In this case only a simple noise parameter $g_{11}R_n$ is needed

---

[11] In my opionion the settling down to one particular equivalent circuit representation was a mistake. The IRE Standards Committee should have made several options available, so that the device and circuit engineer could choose which option should be used in a particular case. We demonstrate this freedom of choice in our treatment of the FET and the transistor problem.

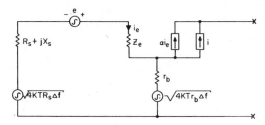

Fig. 8. Equivalent circuit of common base transistor for determining noise figure.

to describe the minimum noise figure over a wide range of frequencies. This agrees very well with experiment. In MOS FETs, however, it is not always true that $g_n \simeq g_{11}$. It is often necessary to neutralize the feedback admittance Y_{12} for greater stability.

We see that $F_{min} \simeq 6$, for $g_{11}R_n = 1$. One can consider this condition to be the cutoff condition for low-noise operation. To achieve low-noise operation over a wide frequency range one should therefore make $g_{11}R_n$ as small as possible.

If one investigates how the product $g_{11}R_n$ depends on the device parameters, one finds that it decreases sharply with decreasing channel length. To achieve microwave operation of FETs one should make the channel as short as feasible. Channel lengths of two microns bring the operation well into the microwave range [95a].

Since g_{11} varies as f^2 and R_n is practically frequency independent for the frequency range where the device is useful, it is seen from (159) that F_{min} increases rapidly for $g_{11}R_n > 1$.

It can be demonstrated easily that the noise figure of the common gate circuit can be transformed back to that of the common source circuit, if it is assumed that the admittance Y_{12} is neutralized in each case [170]. It is therefore not worthwhile to give that circuit a separate treatment.

Another popular circuit is the *cascode* circuit. It consists of a common source FET as the first stage and a common gate FET as the second stage with the drain of the first stage directly connected to the source of the second stage. Units in which the two devices are manufactured on a single chip are known as *tetrode FETs*.

The noise figure of this circuit has been calculated by van der Ziel and Takagi [166], who have shown that improvement in noise figure can be obtained by neutralizing the first (equal to common source) stage and by tuning the interstage network. This agrees with experiment.

C. Transistor Circuits [155]

We first consider the common base circuit of Fig. 8 which incorporates the transistor equivalent circuit of Fig. 1(b).[12] Open-circuiting the output, and neglecting the noise developed across r_b in comparison with the noise voltage across the collector junction, one easily obtains for the noise figure, if $Z_s = R_s + jX_s$,

[12] Again, we deviate here from the equivalent circuit recommended by the IRE Standards Committee on Noise [180].

$$F = 1 + \frac{r_b}{R_s} + \frac{\overline{e''^2}}{4kT_0 R_s \Delta f}$$
$$+ \frac{\overline{i^2}/|\alpha|^2}{4kT_0 R_s \Delta f} \left| Z_s + r_b + Z_e + \frac{e'\alpha}{i} \right|^2 \quad (160)$$

where the EMF e has been split into a part e' fully correlated with the current generator i and a part e'' uncorrelated with i. We now introduce the parameters R_n, g_n, and Z_{cor} by the definitions

$$\overline{e''^2} = 4kT_0 R_n \Delta f; \quad \frac{\overline{i^2}}{|\alpha|^2} = 4kT_0 g_n \Delta f;$$
$$Z_{cor} = \alpha \frac{e'}{i} = \alpha \frac{\overline{ei^*}}{\overline{i^2}}. \quad (161)$$

These noise parameters again relate directly to the physics of the device. Equation (160) may now be written

$$F = 1 + \frac{r_b + R_n}{R_s} + \frac{g_n}{R_s} |Z_s + r_b + Z_{cor}|^2. \quad (162)$$

If we put $Z_e = R_e + jX_e$, the correlation impedance $Z_{cor} = R_{cor} + jX_{cor}$, then (162), when considered as a function of X_s, has a minimum value

$$F = F_t = 1 + \frac{r_b + R_n}{R_s} + \frac{g_n}{R_s} (R_s + r_b + R_e + R_{cor})^2 \quad (163)$$

if

$$(X_s + X_e + X_{cor}) = 0. \quad (163a)$$

Usually X_{cor} is quite small so that its effect is small, little deterioration in noise figure will occur if the input is tuned for maximum signal transfer. By measuring F_t as a function of R_s, one can determine the other noise parameters. Considered as a function of R_s, F_t has a minimum value

$$F_{min} = 1 + 2g_n(r_b + R_e + R_{cor})$$
$$+ 2\sqrt{g_n(r_b + R_n) + g_n^2(r_b + R_e + R_{cor})^2} \quad (164)$$

if

$$R_s = (R_s)_{opt} = \sqrt{(r_b + R_e + R_{cor})^2 + \frac{(r_b + R_n)}{g_n}}. \quad (164a)$$

Usually the correlation resistance R_{cor} is small so that it can be neglected [113]. Moreover, it is often assumed [113] that $R_e \simeq \frac{1}{2}R_{e0}$, whereas g_n follows from (24a) as

$$g_n = \frac{\alpha_F}{R_{e0}\alpha_0^2} \left(1 - \alpha_F + \frac{f^2}{f_\alpha^2} \right). \quad (165)$$

Another approximation [48], [107] which may be useful at higher frequencies, is to neglect $R_e + R_{cor}$ and R_n with respect to r_b, so that

$$F_{min} = 1 + 2g_n r_b + 2\sqrt{g_n r_b + g_n^2 r_b^2}. \quad (166)$$

259

Then only one parameter $g_n r_b$ is needed to characterize the noise over a wide frequency range. Both approximations give reasonable results up to near the cutoff frequency of the transistor.

If one substitutes $g_n r_b = 1$, $F_{\min} = 6$ is obtained. We can thus define this condition as the cutoff condition for low-noise operation. To achieve low-noise operation one should therefore make $g_n r_b$ as small as possible. This can be achieved by making the alpha cutoff frequency f_α of the transistor as large as possible. Essentially this amounts to making the base region very thin. Operation up to 10 GHz is feasible with present-day techniques.

Since g_n varies as $(f/f_\alpha)^2$ over the frequency range of interest, and r_b is practically frequency independent, a raising of the alpha cutoff frequency has the additional benefit that the frequency range of very low noise figure (say $F_{\min} < 1.5$) is materially extended.

Next we consider the common emitter transistor circuit of Fig. 9. We now revert back to the current generators i_1 and i_2 by substituting $i_b = i_1 - i_2$. After some manipulating whereby one takes the feedback through the capacitance C_{cb} into account and puts

$$e = i_1 Z_e; \quad i' = i_2 - \alpha' i_1; \quad \alpha' = \alpha - j\omega C_{cb} Z_e, \quad (167)$$

one obtains a result similar to (160) with the only difference being α must be replaced by α' and i by i'. Noise parameters can now be introduced by putting

$$\overline{e''^2} = 4kT_0 R_n \Delta f; \quad \frac{\overline{i'^2}}{|\alpha|^2} = 4kT_0 g_n' \Delta f;$$

$$Z'_{cor} = \frac{\alpha' e'}{i'} = \alpha' \frac{\overline{ei'^*}}{\overline{i'^2}}. \quad (168)$$

One then obtains (162) except that g_n and Z_{cor} are replaced by g_n' and Z'_{cor}. Here there is quite a difference between the tuning for minimum noise figure and the tuning for maximum signal transfer.

When comparing theory and experiments one finds good agreement for $F_{\min}$ up to near the cutoff frequency, even at current levels where one expects high-level injection effects in the base. This is especially true for the parameter g_n. In some units a collector saturation effect occurs at the collector [3], as mentioned in Section II. The only high-level injection effect that has not been explained is that in some units R_{cor} is much larger than anticipated [146].

At very high frequencies one gets into trouble because of the header parasitics. At most frequencies these parasitics act as lossless transformers that affect $(R_s)_{opt}$ but do not affect $F_{\min}$. At the highest frequencies, however, the parasitics act as attenuators; that increases the noise figure to a value larger than expected from (164). Malaviya [107] found good agreement between theory and experiment up to 3 GHz for a 4-GHz transistor, but at 4 GHz, where the transistor was near the end of its operating range, the experimental value of $F_{\min}$ was larger than the theoretical one, as expected from the preceding discussion.

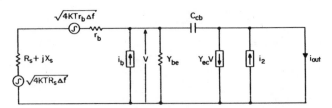

Fig. 9. Equivalent circuit for common emitter transistor for determining noise figure.

D. Mixer Circuits [149], [170]

In a mixer a large pump (or local oscillator) signal of frequency ω_p is applied to a nonlinear device, an input signal of frequency ω_i is applied to the input and an output signal of frequency, ω_o is taken from the output where either $\omega_p = |\omega_i - \omega_o|$, or $\omega_p = \omega_i + \omega_o$. In the latter case the phase angle ϕ_i of the input signal gives rise to a phase angle $-\phi_i$ in the converted output signal, whereas there is no phase reversal in the first case.

We first turn to the diode mixer [149]. If the instantaneous conductance $g(t)$ of the diode has a Fourier representation

$$g(t) = g_0 + 2g_1 \cos \omega_p t + 2g_2 \cos 2\omega_p t + \cdots, \quad (169)$$

then the diode mixer can be represented by a two-port active network with an admittance matrix

$$\begin{pmatrix} g_0 & -g_1 \\ -g_1 & g_0 \end{pmatrix}. \quad (169a)$$

If a source current generator i_s in parallel with a conductance g_s is applied to the input of this network, the output is connected to a load conductance g_L, and the maximum power gain, obtained by matching the source to the input and the load to output, is evaluated, one obtains

$$G_{\max} = \frac{\beta}{[1 + (1 - \beta)^{1/2}]^2}, \quad \text{for } g_s = g_L = g_0(1 - \beta)^{1/2} \quad (170)$$

where $\beta = (g_1/g_0)^2$. We note that $G_{\max}$ approaches unity if β approaches unity. We shall see that a large value of $G_{\max}$ is beneficial for the noise figure.

In order to evaluate the noise figure one must bear in mind that the noise has a component of frequency ω_o and a component of frequency ω_i. The latter passes through the input circuit, gives rise to a noise voltage of frequency ω_i at the input, and by mixing gives rise to a converted noise signal of frequency ω_o that is correlated with the original component of frequency ω_o. If one now puts for the diode noise[13]

$$\overline{i^2} = n \cdot 4kTg_0 \Delta f \quad (171)$$

where $n \simeq 1$ for a point contact diode and $n \simeq \frac{1}{2}$ for a Schottky

---

[13] The n-value here represents an averaging over the periodic local oscillator waveform. It can be as low as $\frac{1}{2}$ in devices with a low series resistance and a low back current, about unity in devices with an appreciable series resistance and a low back current, and much higher than unity in devices with a large back current.

barrier diode, one obtains for the minimum noise figure of the mixer

$$F_{\min} = 1 + n\left(1 - \frac{1}{G_{\max}}\right), \quad \text{for } g_s = g_0(1 - \beta)^{1/2} \quad (172)$$

so that the noise figure approaches unity if $G_{\max}$ approaches unity.

If the mixer is followed by an intermediate frequency (IF) amplifier of noise figure F_2, then according to Friiss' formula the noise figure of the combination is

$$F = F_{\min} + \frac{F_2 - 1}{G_{\max}} = 1 + \frac{n(1 - G_{\max}) + F_2 - 1}{G_{\max}} \quad (173)$$

which approaches F_2 if $G_{\max}$ approaches unity. Recent decreases in the noise figure of diode mixers have come about by lowering n (by going to Schottky barrier diodes), by raising $G_{\max}$ (by going to diodes with low series resistance), and by lowering F_2 (by going to low-noise transistor IF amplifiers) [16].

In Doppler radar receivers one usually starts with a mixer, chooses $\omega_o = 0$, and obtains beat frequencies. One then wants to avoid that the beat frequencies drown in the noise background of the receiver. To that end one must require that the mixer diodes have low $1/f$ noise. Well constructed Schottky barrier diodes and backward diodes seem to meet this requirement best [38]–[40], [44].

The impedance of the input circuit of a mixer diode is usually so low that it not only responds to the wanted frequency $\omega_i = \omega_p - \omega_o$ but also to the frequency $\omega_i' = \omega_p + \omega_o$, which is called the *image* frequency. A lower value of the noise figure $F_{\min}$ is now obtainable by proper choice of the admittance of the input circuit for the frequency ω_i'. The best results are obtained if the input circuit has an infinite impedance (i.e., is open-circuited) for the frequency ω_i' [170].

Since a point contact or Schottky barrier diode also has a voltage-dependent capacitance, the net result is that the diode mixer is a combination of a variable conductance and a variable capacitance mixer. This may alter the above considerations to some extent in some cases.

Next we consider the variable capacitance mixer [148], where the diode is replaced by a voltage-dependent capacitor. If a large pump voltage of frequency ω_p is applied, the instantaneous capacitance of the diode has a Fourier representation

$$C(t) = C_0 + 2C_1 \cos \omega_p t + 2C_2 \cos 2\omega_p t + \cdots. \quad (174)$$

Then for $\omega_p = |\omega_i - \omega_o|$ the mixer can be represented by a two-port active network with an admittance matrix

$$\begin{pmatrix} j\omega_i C_0 & -j\omega_i C_1 \\ -j\omega_o C_1 & j\omega_o C_0 \end{pmatrix}. \quad (174a)$$

If a source current generator i_s in parallel with a conductance g_s is applied to the input of this network, the output is connected to the load conductance g_L, and the maximum power gain, obtained by matching the source to the input and the

load to the output, is evaluated, one obtains

$$G_{\max} = \frac{\omega_o}{\omega_i} \quad (175)$$

so that the up-converter ($\omega_o > \omega_i$) has power gain.

For the case $\omega_p = \omega_i + \omega_o$, the load conductance g_L gives rise to a negative conductance $-\omega_o\omega_i C_1^2/g_L$ seen at the input side of the circuit. This is the basis of the parametric amplifier (Section VI-A).

An ideal voltage-dependent capacitor has no noise sources associated with it, and as a consequence an ideal variable capacitance mixer should have a noise figure of unity [148]. Because of the unavoidable losses in the tuned circuits and in the nonlinear device, the noise figure of actual circuits will be somewhat larger than unity [82], [118a], [154]; in practice, however, very low noise figures have been obtained.

We now turn to the FET and the transistor mixer [170]. The best results are obtained for the common source FET and for the common emitter transistor circuits. The common gate FET and the common base transistor mixer have too much current flowing through the input of the device; the noise due to these currents seriously deteriorates the noise figure of the device unless the current flows in very short pulses.

The general approach to noise problems in FET and transistor mixers in the two recommended circuits is the same as before. One constructs an equivalent circuit with an input noise current generator i in parallel with the input admittance for the frequency ω_i, and an output noise current generator i_2 in parallel with the output admittance for the frequency ω_o. The correlation between the two noise sources is usually so small that it can be neglected. The theory of Section VI-B is then applicable, with $Y_{cor} = 0$, for the FET mixer [170]. The transistor mixer is slightly more complicated because of the base resistance r_b [170], [173].

The noise figure of these devices when used as a mixer is somewhat larger than when used as amplifiers. This is seen as follows. If the instantaneous transconductance $g_m(t)$ of the device is represented by its Fourier series

$$g_m(t) = g_{m0} + 2g_{m1} \cos \omega_p t + 2g_{m2} \cos 2\omega_p t + \cdots, \quad (176)$$

then the output noise of the mixer can be represented as

$$\overline{i_2^2} = n_2 \cdot 4kT_0 g_{m0}\Delta f = 4kT_0 R_{nm}\Delta f g_{m1}^2 \quad (177)$$

where n_2 is of the order of unity, so that the mixer noise resistance R_{nm} is

$$R_{nm} = n_2 \frac{g_{m0}}{g_{m1}^2}, \quad (177a)$$

whereas the noise resistance R_{na} of the amplifier is

$$R_{na} = \frac{n_2}{g_{m0}}. \quad (177b)$$

Since $g_{m1} < g_{m0}$, $R_{nm} > R_{na}$ and, hence, the noise figure of the mixer is somewhat higher than the noise figure of the corresponding amplifier.

The effect can be eliminated by HF feedback from output to input. If so much feedback is applied that the circuit is at its limit of stability, the noise figure of the mixer becomes equal to that of the HF amplifier [115], [165].

E. Summary

The noise figure of various amplifier and mixer circuits is discussed. The tunnel diode amplifier and the parametric amplifier have quite acceptable noise figures at microwave frequencies.

In FETs the noise figure is determined by the parameter $g_{11}R_n$, where g_{11} is the HF input conductance and R_n is the noise resistance of the device. For low-noise operation $g_{11}R_n \leqq 1$. The cutoff condition $g_{11}R_n = 1$ leads to the requirement that the channel length should be made very small. Low-noise microwave FETs are feasible.

In transistors the noise figure is essentially determined by the parameter $g_n r_b$, where g_n is defined in the text and r_b is the base resistance. Again, for low-noise operation $g_n r_b \leqq 1$. The cutoff condition $g_n r_b = 1$ leads to devices with a very high alpha cutoff frequency. Microwave operation is quite feasible and the frequency range of low-noise operation is increasing steadily.

Diode mixers and variable capacitance mixers are also discussed. Lossless capacitance mixers should have a noise figure of unity. In diode mixers the noise of the IF amplifier gives a contribution to the noise figure of the system.

In common source FET mixers and in common emitter transistor mixers, the noise figure is somewhat larger than in the corresponding HF amplifier.

VII. REFERENCES

[1] G. Abowitz, E. Arnold, and E. A. Leventhal, "Surface states and $1/f$ noise in MOS transistors," *IEEE Trans. Electron Devices*, vol. ED-14, pp. 775–777, November 1967.
[2] D. C. Agouridis and K. M. van Vliet, "Noise measurements on tunnel diodes," *Proc. IRE* (Correspondence), vol. 50, p. 2121, October, 1962.
[3] D. C. Agouridis and A. van der Ziel, "Noise figure of UHF transistors as a function of frequency and bias conditions," *IEEE Trans. Electron Devices*, vol. ED-14, pp. 808–816, December 1967.
[3a] C. T. J. Alkemade, "On the excess photon noise in single-beam measurements with photoemissive and photoconductive cells," *Physica*, vol. 25, pp. 1145–1158, 1959.
[4] R. L. Anderson and A. van der Ziel, "On the shot effect of p-n junctions," *IRE Trans. Electron Devices*, vol. ED-1, pp. 20–24, November 1952.
[5] F. T. Arecchi, A. Berné, A. Sona, and P. Burlamacchi, "Photocount distributions and field statistics," *IEEE J. Quantum Electron.*, vol. QE-2, pp. 341–350, September 1966.
[6] W. Bächtold and M. J. O. Strutt, "Analog model for the signal-and-noise equivalent circuit of VHF transistors," *Electron. Lett.*, vol. 2, pp. 335–336, September 1966.
[7] ——, "Noise parameter measurements of microwave transistors up to 2.4 GHz," *Electron. Lett.*, vol. 3, pp. 323–324, July 1967.
[8] ——, "Simplified equivalent circuit for the noise figure calculation of microwave transistors," *Electron. Lett.*, vol. 4, pp. 209–210, May 17, 1968.
[9] ——, "Optimum source admittance for minimum noise figure of microwave transistors," *Electron. Lett.*, vol. 4, pp. 346–348, August 1968.
[10] ——, "Noise in microwave transistors," *IEEE Trans. Microwave Theory Tech.*, vol. MTT-16, pp. 578–585, September 1968.
[11] A. Baelde, "The influence of non-uniform base width on the noise of transistors," *Philips Res. Rept.*, vol. 16, pp. 225–236, June 1961.
[12] ——, Theory and experiments of noise in transistors," Ph.D. dissertation, *Philips Res. Rept.* (Suppl.), no. 4, p. 130, 1965.
[13] R. D. Baertsch, "Low-frequency noise measurements in silicon avalanche photodiodes," *IEEE Trans. Electron Devices* (Correspondence), vol. ED-13, pp. 383–384, March 1966.
[14] ——, "Noise and ionization rate measurements in silicon photodiodes," *IEEE Trans. Electron Devices* (Correspondence), vol. ED-13, p. 987, December 1966.
[15] ——, "Noise and multiplication measurements in InSb avalanche photodiodes," *J. Appl. Phys.*, vol. 38, pp. 4267–4273, October 1967.
[16] M. R. Barber, "Noise figure and conversion loss of the Schottky barrier mixer diode," *IEEE Trans. Microwave Theory Tech.*, vol. MTT-15, pp. 629–635, November 1967.
[17] R. N. Beatie, "A lumped model analysis of noise in semiconductor devices," *IRE Trans. Electron Devices*, vol. ED-6, pp. 133–140, April 1959.
[18] H. R. Bilger, D. H. Lee, M. A. Nicolet, and E. R. McCarter, "Noise and equivalent circuit of double injection," *J. Appl. Phys.*, vol. 39, pp. 5913–5918, December 1968.
[19] J. Borel, "Alloy transistor equivalent base resistance and noise input," *Solid-State Electron.*, vol. 8, pp. 31–40, January 1965.
[20] ——, "Consideration of the equivalent circuit and noise of a field effect transistor," *Onde Elec.*, vol. 46, pp. 1190–1201, November 1966.
[21] W. C. Bruncke, "Noise measurements in field-effect transistors," *Proc. IEEE* (Correspondence), vol. 51, pp. 378–379, February 1963.
[22] W. C. Bruncke, E. R. Chenette, and A. van der Ziel, "Transistor noise at low temperatures," *IEEE Trans. Electron Devices*, vol. ED-11, pp. 50–53, February 1964.
[23] W. C. Bruncke and A. van der Ziel, "Thermal noise in junction-gate field-effect transistors," *IEEE Trans. Electron Devices*, vol. ED-13, pp. 323–329, March 1966.
[24] R. E. Burgess, "The statistics of charge carrier fluctuations in semiconductors," *Proc. Phys. Soc. London*, vol. B-69, pp. 1020–1027, October 1956.
[25] W. H. Card and P. K. Chaudhari, "Characteristics of burst noise," *Proc. IEEE* (Correspondence), vol. 53, pp. 652–653, June 1965.
[26] K. S. Champlin, "Microplasma fluctuations in silicon," *J. Appl. Phys.*, vol. 30, pp. 1039–1050, July 1959.
[27] K. K. N. Chang, "Low-noise tunnel-diode amplifier," *Proc. IRE* (Correspondence), vol. 47, pp. 1268–1269, July 1959.
[28] E. R. Chenette, "Measurement of the correlation between flicker noise sources in transistors," *Proc. IRE* (Correspondence), vol. 46, p. 1304, June 1958.
[29] ——, "The influence of inductive source reactance on the noise figure of a junction transistor," *Proc. IRE* (Correspondence), vol. 47, pp. 448–449, March 1959.
[30] ——, "Frequency dependence of the noise and the current amplification factor of silicon transistors," *Proc. IRE* (Correspondence), vol. 48, pp. 111–112, January 1960.
[31] E. R. Chenette and A. van der Ziel, "Accurate noise measurements on transistors," *IRE Trans. Electron Devices*, vol. ED-9, pp. 123–128, March 1962.
[32] S. Christenson, I. Lundström, and C. Svensson, "Low frequency noise in MOS transistors, I: Theory," *Solid-State Electron.*, vol. 11, pp. 797–812, September 1968.
[33] S. Christenson and I. Lundström, "Low frequency noise in MOS transistors, II: Experiments," *Solid-State Electron.*, vol. 11, pp. 813–820, September 1968.
[34] W. N. Coffey, "Behavior of noise figure in junction transistors," *Proc. IRE* (Correspondence), vol. 46, pp. 495–496, February 1958.
[35] I. Constant, B. Kramer, and I. Raczy, "Avalanche noise from silicon junctions in the radio range and beyond," *C.R. Acad. Sci.* (France), vol. 265, pp. 385–388, August 16, 1967.
[36] H. G. Dill, "Noise contribution of the offset-gate IGFET with an additional gate electrode," *Electron. Lett.*, vol. 3, pp. 341–342, July 1967.
[36a] C. Dragone, "Analysis of thermal and shot noise in pumped resistive diodes," *Bell Syst. Tech. J.*, vol. 47, pp. 1883–1902, November 1968.

[37] F. Driedonks, R. J. J. Zijlstra, and C. T. J. Alkemade, "Double injection and high frequency noise in germanium diodes," *Appl. Phys. Lett.*, vol. 11, pp. 318–319, November 15, 1967.

[38] S. T. Eng, "Low-noise properties of microwave backward diodes," *IRE Trans. Microwave Theory Tech.*, vol. MTT-9, pp. 419–425, September 1961.

[39] ——, "A new low $1/f$ noise mixer diode: experiments, theory and performance," *Solid-State Electron.*, vol. 8, pp. 59–77, January 1965.

[40] ——, "Recent results on low $1/f$ noise mixer diodes," *Proc. IEEE* (Letters), vol. 54, pp. 1968–1970, December 1966.

[41] J. W. Englund, "Noise considerations for p-n-p junction transistors," in *Transistors I.* Princeton, N. J.: RCA Laboratories, 1956, pp. 309–321.

[42] E. A. Faulkner and D. W. Harding, "Flicker noise in silicon planar transistors," *Electron. Lett.*, vol. 3, pp. 71–72, February 1967.

[43] E. A. Faulkner and M. L. Meade, "Flicker noise in Gunn diodes," *Electron. Lett.*, vol. 4, pp. 226–227, May 31, 1968.

[43a] N. H. Fletcher, "The high current limit for semiconductor junction devices," *Proc. IRE*, vol. 45, pp. 862–872, June 1957.

[44] W. C. Follmer, "Low-frequency noise in backward diodes," *Proc. IRE* (Correspondence), vol. 49, pp. 1939–1940, December 1961.

[45] W. H. Fonger, "A determination of $1/f$ noise sources in semiconductor diodes and transistors," in *Transistors I.* Princeton, N. J.: RCA Laboratories, 1956, pp. 239–297.

[46] W. H. Fonger, J. J. Loferski, and P. Rappaport, "Radiation induced noise in p-n junctions," *J. Appl. Phys.*, vol. 29, pp. 588–591, March 1958.

[47] C. Freed and H. A. Haus, "Photocurrent spectrum and photoelectron counts produced by a gas laser," *Phys. Rev.*, vol. 141, pp. 287–298, January 1966.

[48] H. Fukui, "The noise performance of microwave transistors," *IEEE Trans. Electron Devices*, vol. ED-13, pp. 329–341, March 1966.

[49] J. A. Geurst, "Calculation of high-frequency characteristics of thin film transistors," *Solid-State Electron.*, vol. 8, pp. 88–90, January 1965.

—, "Calculation of high-frequency characteristics of field-effect transistors," *Solid-State Electron.*, vol. 8, pp. 563–566, June 1965.

[50] L. J. Giacoletto, "The noise factor of junction transistors," in *Transistors I.* Princeton, N. J.: RCA Laboratories, 1956, pp. 296–308.

[51] U. F. Gianola, "Photovoltaic noise in silicon broad area p-n junctions," *J. Appl. Phys.*, vol. 27, pp. 51–54, January 1956.

[52] J. F. Gibbons, "Low-frequency noise figure and its application to the measurement of certain transistor parameters," *IRE Trans. Electron Devices*, vol. ED-9, pp. 308–315, May 1962.

[53] G. Giralt, J. C. Martin, and F. X. Mateu-Perez, "Burst noise of silicon planar transistors," *Electron. Lett.*, vol. 2, pp. 228–230, June 1966.

[54] G. Guekos and M. J. O. Strutt, "Current noise spectra of GaAs laser diodes in the luminescence mode," *IEEE J. Quantum Electron.* (Correspondence), vol. QE-4, pp. 502–503, August 1968.

[55] ——, "Laser emission noise and voltage noise of GaAs C.W. laser diodes," *Electron. Lett.*, vol. 4, pp. 408–409, September 1968.

[56] ——, "Correlation between laser emission noise and voltage noise in GaAs CW laser diodes," *IEEE J. Quantum Electron.* (Correspondence), vol. QE-5, pp. 129–130, February 1969.

[57] W. Guggenbuehl and M. J. O. Strutt, "Messungen der spontanen Schwankungen bei Strömen mit verschiedenen Trägern in Halbleitersperrschichten," *Helv. Phys. Acta*, vol. 28, pp. 694–704, 1955.

[58] ——, "Experimentelle Bestätigung der Schottkyschen Rauschformeln an neueren Halbleiterflächendioden im Gebiet des weissen Rauschspektrums," *Arch. Elek. Übertragung*, vol. 9, pp. 103–108, March 1955.

[59] ——, "Experimentelle Untersuchung und Trennung der Rauschursachen in Flächentransistoren," *Arch. Elek. Übertragung*, vol. 9, pp. 259–269, June 1955.

[60] ——, "Theorie des Hochfrequenz rauschens von Transistoren bei kleinen Stromdichten," *Nachrichtentech. Z.*, vol. 5, pp. 30–33, 1956.

[61] W. Guggenbuehl, B. Schneider, and M. J. O. Strutt, "Messungen über das Hochfrequenzrauschen von Transistoren," *Nachrichtentech. Z.*, vol. 5, pp. 34–36, 1956.

[62] W. Guggenbuehl, "Theoretische Überlegungen zur physikalischen Begründung des Ersatzschaltbildes von Halbleiterflächendioden

bei hohen Stromdichten," *Arch. Elek. Übertragung*, vol. 10, pp. 433–435, November 1956.

[63] W. Guggenbuehl and M. J. O. Strutt, "Theory and experiments of shot noise in semiconductor junction diodes and transistors," *Proc. IRE*, vol. 45, pp. 839–854, June 1957.

[64] ——, "Transistors in high frequency amplifiers," *Electron. Radio Eng.*, vol. 34, pp. 258–267, July 1957.

[65] H. K. Gummel and J. L. Blue, "A small-signal theory of avalanche noise in IMPATT diodes," *IEEE Trans. Electron Devices*, vol. ED-14, pp. 569–580, September 1967.

[66] I. D. Guttkov, "Investigation of the $1/f$ low frequency noise of back biased germanium p-n junctions," *Radiotekh., Elektron.*, vol. 12, pp. 946–948, May 1967; *Radio Eng. Electron. Phys.* (English Transl.), vol. 12, pp. 880–882, May 1967.

[67] R. H. Haitz, "Mechanisms contributing to the noise pulse ratio of avalanche diodes," *J. Appl. Phys.*, vol. 36, pp. 3123–3131, October 1965.

[68] ——, "Controlled noise generation with avalanche diodes. I. Low pulse rate design," *IEEE Trans. Electron Devices*, vol. ED-12, pp. 198–207, April 1965.

[69] ——, "Controlled noise generation with avalanche diodes II. High pulse rate design," *IEEE Trans. Electron Devices*, vol. ED-13, pp. 342–346, March 1966.

[70] R. H. Haitz and F. W. Voltmer, "Noise of self-sustaining avalanche discharge in silicon: studies at microwave frequencies," *J. Appl. Phys.*, vol. 39, pp. 3379–3384, June 1968.

[71] H. E. Halladay and W. C. Bruncke, "Excess noise in field-effect transistors," *Proc. IEEE* (Correspondence), vol. 51, p. 1671, November 1963.

[72] H. E. Halladay and A. van der Ziel, "Field-dependent mobility effects in the excess noise of junction-gate field-effect transistors," *IEEE Trans. Electron Devices* (Correspondence), vol. ED-14, pp. 110–111, February 1967.

[73] ——, "Test of the thermal noise hypothesis in MOS FET's," *Electron. Lett.*, vol. 4, pp. 366–367, August 1968.

[74] ——, "On the high-frequency excess noise and equivalent circuit representation in MOS-FET's with n-type channel," *Solid-State Electron.*, vol. 12, pp. 161–176, March 1969.

[75] G. H. Hanson, "Shot noise in p-n-p transistors," *J. Appl. Phys.*, vol. 26, pp. 1338–1339, November 1955.

[76] G. H. Hanson and A. van der Ziel, "Shot noise in transistors," *Proc. IRE*, vol. 45, pp. 1538–1542, November 1957.

[77] H. Haug, "Population and current noise in semiconductor laser junctions," *Z. Phys.*, vol. 206, pp. 163–176, September 21, 1967.

[78] ——, "Noise in semiconductor lasers," *IEEE J. Quantum Electron.*, vol. QE-4, p. 168, April 1968.

[79] H. A. Haus and R. B. Adler, *Circuit Theory of Linear Noisy Networks.* New York: Wiley, 1959.

[80] H. A. Haus, "Amplitude noise in laser oscillators," *IEEE J. Quantum Electron.*, vol. QE-1, pp. 179–180, July 1965.

[81] H. Heffner and G. Wade, "Minimum noise figure of a parametric amplifier," *J. Appl. Phys.*, vol. 29, p. 1262, August 1958.

[82] ——, "Gain, bandwidth and noise characteristics of variable parameter amplifiers," *J. Appl. Phys.*, vol. 29, pp. 1321–1331, September 1958.

[83] M. E. Hines, "Noise theory for the Read type avalanche diode," *IEEE Trans. Electron Devices*, vol. ED-13, pp. 158–163, January 1966.

[84] H. Hodara and N. George, "Excess photon noise in multimode lasers," *IEEE J. Quantum Electron.*, vol. QE-2, pp. 337–340, September 1966.

[85] S. T. Hsu, A. van der Ziel, and E. R. Chenette, "Noise in space-charge-limited solid-state devices," *Solid-State Electron.*, vol. 10, pp. 129–135, February 1967.

[86] S. T. Hsu, D. J. Fitzgerald, and A. S. Grove, "Surface state related $1/f$ noise in p-n junctions and MOS transistors," *Appl. Phys. Lett.*, vol. 12, pp. 287–289, May 1, 1968.

[87] F. J. Hyde, "Measurements of noise spectra of a point contact germanium rectifier," *Proc. Phys. Soc. London* (Gen.), vol. 66, pp. 1017–1024, December 1953.

[88] ——, "Measurement of noise spectra of a germanium p-n junction diode," *Proc. Phys. Soc. London* (Gen.), pt. 2, vol. 69, pp. 231–241, February 1956.

[89] F. J. Hyde, H. J. Roberts, and B. E. Buckingham, "Excess high frequency noise in junction transistors," *Proc. Phys. Soc. London*, pt. 5, vol. 78, pp. 1076–1077, November 15, 1961.

[90] R. C. Jaeger, A. J. Brodersen, and E. R. Chenette, "Low frequency noise in integrated transistors," *1968 Region III Conv. Rec.*, pp. S8–1pl, November 1968.

[91] K. H. Johnson, A. van der Ziel, and E. R. Chenette, "Transistor noise at high injection levels," *IEEE Trans. Electron Devices* (Correspondence), vol. ED-12, pp. 387–388, June 1965.

[92] A. G. Jordan and N. A. Jordan, "Theory of noise in metal oxide semiconductor devices," *IEEE Trans. Electron Devices*, vol. ED-12, pp. 148–156, March 1965.

[93] R. H. Kingston, "Review of germanium surface phenomena," *J. Appl. Phys.*, vol. 27, pp. 101–114, February 1956.

[94] R. H. Kingston and A. L. McWhorter, "Relaxation time of surface states on Ge," *Phys. Rev.*, vol. 103, pp. 534–540, August 1956.

[94a] F. M. Klaassen, "High-frequency noise of the junction field-effect transistor," *IEEE Trans. Electron Devices*, vol. ED-14, pp. 368–373, July 1967.

[95] F. M. Klaassen and J. Prins, "Thermal noise of MOS transistors," *Philips Res. Rep.*, vol. 22, pp. 505–514, October 1967.

[95a] ——, "Noise in VHF and UHF MOS tetrodes," *Philips Res. Rep.*, vol. 23, pp. 478–484, December 1968.

[96] H. F. Kumo, J. R. Collard, and A. R. Golat, "Low-noise epitaxial GaAs avalanche diode amplifiers," *1968 IEEE Internatl. Electron Devices Meeting* (Washington, D. C.), October 23–25, p. 104.

[97] P. O. Lauritzen, "Low frequency generation noise in junction field effect transistors," *Solid-State Electron.*, vol. 8, pp. 41–58, January 1965.

[98] ——, "Noise due to generation and recombination of carriers in p-n junction transition regions," *IEEE Trans. Electron Devices*, vol. ED-15, pp. 770–776, October 1968.

[99] S. J. Lee and A. van der Ziel, "Noise in optical mixing," *Physica*, vol. 45, pp. 379–386, 1969.

[100] E. A. Leventhal, "Derivation of 1/f noise in silicon inversion layers from carrier motion in a surface band," *Solid-State Electron.*, vol. 11, pp. 621–627, June 1968.

[101] F. Leuenberger, "1/f noise in gate-controlled planar silicon diodes," *Electron. Lett.*, vol. 4, p. 280, June 28, 1968.

[102] A. Leupp and M. J. O. Strutt, "Noise behavior of the MOS FET at V.H.F. and U.H.F.," *Electron. Lett.*, vol. 4, pp. 313–314, July 1968.

[103] S. T. Liu, A. van der Ziel, and G. U. Jatnieks, "On the limiting noise of space-charge-limited solid-state diodes," *Physica*, vol. 38, pp. 279–284, April 1968.

[104] S. T. Liu, S. Yamamoto, and A. van der Ziel, "Noise in double injection space-charge limited diodes," *Appl. Phys. Lett.*, vol. 10, pp. 308–309, June 1, 1967.

[105] R. J. McIntyre, "Multiplication noise in uniform avalanche diodes," *IEEE Trans. Electron Devices*, vol. ED-13, pp. 164–168, January 1966.

[106] A. L. McWhorter, "1/f noise and related surface effects in germanium," MIT Lincoln Lab., Lexington, Mass., Rep. 80, May 1955.

[107] S. D. Malaviya, "Study of noise in transistors at microwave frequencies up to 4 GHz," Ph.D. dissertation, University of Minnesota, Minneapolis, 1969; *Solid-State Electron.*, 1970, in press.

[108] I. R. M. Mansour, R. J. Hawkins, and G. J. Bloodworth, "Measurement of current noise in MOS transistors," *Radio Electron. Eng.*, vol. 35, pp. 212–216, April 1968.

[109] J. C. Martin, F. X. Mateu-Perez, and F. Serra-Mesties, "Very low-frequency measurements of flicker noise in planar transistors," *Electron. Lett.*, vol. 2, pp. 343–345, September 1966.

[110] K. Matsumo, "Low frequency current fluctuations in a GaAs Gunn diode," *Appl. Phys. Lett.*, vol. 12, pp. 403–404, June 15, 1968.

[111] H. C. Montgomery and M. A. Clark, "Shot noise in junction transistors," *J. Appl. Phys.*, vol. 24, pp. 1337–1338, October 1953.

[112] E. M. Nicollian and H. Melchior, "A quantitative theory of 1/f type noise due to interface states in thermally oxidized silicon," *Bell Syst. Tech. J.*, vol. 46, pp. 2019–2033, November 1967.

[113] E. G. Nielsen, "Behavior of noise figure in junction transistors," *Proc. IRE*, vol. 45, pp. 957–963, July 1957.

[114] D. O. North, "A physical theory of noise in transistors," presented at the 1955 IRE-AIEE Conf. on Semiconductor Device Research, University of Pennsylvania, Philadelphia, Pa.

[115] M. Okamoto and A. van der Ziel, "Noise figure of FET mixers with HF feedback from output to input," *IEEE J. Solid-State Circuits*, vol. SC-3, pp. 300–302, September 1968.

[116] G. L. Pearson, H. C. Montgomery, and W. L. Feldmann, "Noise in silicon p-n junction photocells," *J. Appl. Phys.*, vol. 27, pp. 91–92, January 1956.

[117] D. G. Pederson, "Noise performance of transistors," *IRE Trans. Electron Devices*, vol. ED-9, pp. 296–303, May 1962.

[118] P. Penfield, "Noise performance of tunnel-diode amplifiers," *Proc. IRE* (Correspondence), vol. 48, pp. 1478–1479, August 1960.

[118a] P. Penfield and P. Rafuse, *Varactor Applications*. Cambridge, Mass.: M.I.T. Press, 1962.

[119] R. A. Perala and A. van der Ziel, "Noise in p-i-n junction diodes," *IEEE Trans. Electron Devices* (Correspondence), vol. ED-14, pp. 172–173, March 1967.

[120] R. L. Petritz, "On the theory of noise in p-n junctions and related devices," *Proc. IRE*, vol. 40, pp. 1440–1456, November 1952.

[121] ——, "On noise in p-n junction rectifiers and transistors, I. Theory," (*abstract*) *Phys. Rev.*, vol. 91, p. 204, July 1953. See especially the correction shown on p. 204.

[122] J. L. Plumb and E. R. Chenette, "Flicker noise in transistors," *IEEE Trans. Electron Devices*, vol. ED-10, pp. 304–308, September 1963.

[123] D. Polder and A. Baelde, "Theory of noise in transistor-like devices," *Solid-State Electron.*, vol. 6, pp. 103–110, March–April 1963.

[124] L. J. Prescott and A. van der Ziel, "Detection of spontaneous emission noise in He-Ne lasers," *Phys. Lett.*, vol. 12, pp. 317–319, October 1964.

[125] R. A. Pucel, "The equivalent noise current of Esaki diodes," *Proc. IRE* (Correspondence), vol. 49, pp. 1080–1081, June 1961.

[125a] P. S. Rao, "The effect of the substrate upon the gate and drain noise of MOS FET's," *Solid-State Electron.*, vol. 12, pp. 549–555, July 1969.

[126] C. T. Sah, "Theory of low-frequency generation noise in junction-gate field-effect transistors," *Proc. IEEE*, vol. 52, pp. 795–814, July 1964.

[127] C. T. Sah and H. C. Pao, "The effects of fixed bulk charge on the characteristics of metal-oxide-semiconductor transistors," *IEEE Trans. Electron Devices*, vol. ED-13, pp. 393–409, April 1966.

[128] C. T. Sah, S. Y. Wu, and F. H. Hielscher, "The effects of fixed bulk charge on the thermal noise in metal-oxide-semiconductor transistors," *IEEE Trans. Electron Devices*, vol. ED-13, pp. 410–414, April 1966.

[129] C. T. Sah and F. H. Hielscher, "Evidence of the surface origin of the 1/f noise," *Phys. Rev. Lett.*, vol. 17, pp. 956–958, October 31, 1966.

[130] M. Savelli and M. Teboul, "Study of the background noise in germanium p-n junctions with current generation," *Ann. Telecommun.*, vol. 18, pp. 163–172, July–August 1963.

[131] J. F. Schenck, "Burst noise and walkout in degraded silicon devices," *Proc. 1967 IEEE Sixth Ann. Reliability Physics Symp.* (Los Angeles, Calif.), pp. 31–39, 1968.

[132] B. Schneider and M. J. O. Strutt, "The characteristics and noise of silicon p-n diodes and silicon transistors," *Arch. Elek. Übertragung*, vol. 12, pp. 429–440, October 1958.

[133] ——, "Theory and experiments on shot noise in silicon p-n junction diodes and transistors," *Proc. IRE*, vol. 47, pp. 546–554, April 1959.

[134] ——, "Noise in germanium and silicon transistors in the high current range," *Arch. Elek. Übertragung*, vol. 13, pp. 495–502, December 1959.

[135] ——, "Shot and thermal noise in germanium and silicon transistors at high-level current injections," *Proc. IRE*, vol. 48, pp. 1731–1739, October 1960.

[136] L. Scott and M. J. O. Strutt, "Spontaneous fluctuations in the leakage current due to charge generation and recombination in semiconductor diodes," *Solid-State Electron.*, vol. 9, pp. 1067–1073, November–December 1966.

[137] M. Shoji, "On the limiting noise of field effect transistors," Ph.D. dissertation, University of Minnesota, Minneapolis, 1965, unpublished.

[138] ——, "Analysis of high-frequency thermal noise of enhancement mode MOS field-effect transistors," *IEEE Trans. Electron Devices*, vol. ED-13, pp. 520–524, June 1966.

[139] A. W. Smith and J. A. Armstrong, "Intensity fluctuations and correlations in a GaAs laser," *Phys. Lett.*, vol. 16, pp. 5–6, May 1, 1965; ——, "Intensity noise in multimode GaAs laser emission," *IBM J. Res. Develop.*, vol. 10, pp. 225–232, May 1966.

[140] K. Takagi and A. van der Ziel, "Non-thermal noise in MOS FET's and MOS tetrodes," *Solid-State Electron.*, vol. 12, pp. 907–913, November 1969.

[141] M. Tarng, "Fluctuations in p-n junctions at high injection levels," Ph.D. dissertation, University of Minnesota, Minneapolis, 1969, to be published.

[142] M. C. Teich, R. J. Keyes, and R. H. Kingston, "Optimum heterodyne detection at 10.6 μm in photoconductive Ge:Cu," *Appl. Phys. Lett.*, vol. 9, pp. 357–360, November 15, 1966.

[143] W. Thommen and M. J. O. Strutt, "Small signal and noise equivalent circuits of germanium U.H.F. transistors at small current densities," *Arch. Elek. Übertragung*, vol. 19., pp. 169–177, April 1965.

[144] ——, "Noise figure of UHF transistors," *IEEE Trans. Electron Devices* (Correspondence), vol. ED-12, pp. 499–500, September 1965.

[145] J. J. Tiemann, "Shot noise in tunnel diode amplifiers," *Proc. IRE*, vol. 48, pp. 1418–1423, August 1960.

[146] A. H. Tong and A. van der Ziel, "Transistor noise at high injection levels," *IEEE Trans. Electron Devices*, vol. ED-15, pp. 307–313, May 1968.

[147] A. Uhlir, "High-frequency shot noise in p-n junctions," *Proc. IRE* (Correspondence), vol. 44, pp. 557–558, April 1956; Erratum, vol. 44, p. 1541, November 1956.

[148] A. van der Ziel, "On the mixing properties of non-linear condensers," *J. Appl. Phys.*, vol. 19, pp. 999–1006, November 1948.

[149] ——, *Noise.* Englewood Cliffs, N. J.: Prentice-Hall, 1954.

[150] ——, "Note on shot and partition noise in junction transistors," *J. Appl. Phys.*, vol. 25, pp. 815–816, June 1954.

[151] ——, "Theory of shot noise in junction diodes and junction transistors," *Proc. IRE*, vol. 43, pp. 1639–1646, 1955, and correspondence item in vol. 45, p. 1011, July 1957.

[152] A. van der Ziel and A. G. T. Becking, "Theory of junction diode and junction transistor noise," *Proc. IRE*, vol. 46, pp. 589–594, March 1958.

[153] A. van der Ziel, "Noise in junction transistors," *Proc. IRE*, vol. 46, pp. 1019–1038, June 1958.

[154] ——, "Noise figure of reactance converters and parametric amplifiers," *J. Appl. Phys.*, vol. 30, p. 1449, September 1959.

[155] ——, *Fluctuation Phenomena in Semiconductors.* London; Butterworth, 1959.

[156] ——, "Shot noise in transistors," *Proc. IRE* (Correspondence), vol. 48, pp. 114–115, January 1960.

[157] ——, "Equivalence of the collective and corpuscular theories of noise in junction diodes," *IRE Trans. Electron Devices*, vol. ED-8, pp. 525–528, November 1961.

[158] ——, "Thermal noise in field-effect transistors," *Proc. IRE*, vol. 50, pp. 1808–1812, August 1962.

[159] ——, "Gate noise in field-effect transistors at moderately high frequencies," *Proc. IEEE*, vol. 51, pp. 461–467, March 1963.

[160] ——, "The system noise temperature of quantum amplifiers," *Proc. IEEE* (Correspondence), vol. 51, pp. 952–953, June 1963.

[161] ——, "Carrier density fluctuation noise in field-effect transistors," *Proc. IEEE* (Correspondence), vol. 51, pp. 1670–1671, November 1963.

[162] A. van der Ziel and J. W. Ero, "Small-signal, high-frequency theory of field-effect transistors," *IEEE Trans. Electron Devices*, vol. ED-11, pp. 128–135, April 1964.

[163] A. van der Ziel, "Thermal noise in space-charge-limited diodes," *Solid-State Electron.*, vol. 9, pp. 899–900, September 1966.

[164] A. van der Ziel and K. M. Van Vliet, "H.F. thermal noise in space-charge-limited solid-state diodes, II," *Solid-State Electron.*, vol. 11, pp. 508–509, April 1968.

[165] A. van der Ziel and M. Okamoto, "Noise in common-base transistor mixers with HF feedback from the output," *IEEE J. Solid-State Circuits* (Correspondence), vol. SC-3, pp. 303–304, September 1968.

[166] A. van der Ziel and K. Takagi, "Improvement in the tetrode FET noise figure by neutralization and tuning," *IEEE J. Solid-State Circuits* (Correspondence), vol. SC-4, pp. 170–172, June 1969.

[167] A. van der Ziel, "Noise in junction and MOS-FET's at high temperatures," *Solid-State Electron.*, vol. 12, pp. 861–866, November 1969.

[168] ——, "The surface recombination model of p-n diode flicker noise," *Physica*, 1970, to be published.

[169] ——, "Thermal noise in double-injection space-charge-limited solid-state diodes," *Electron. Lett.*, vol. 6, pp. 45–46, January 1970.

[170] ——, *Noise; Sources, Characterization, Measurement.* Englewood Cliffs, N. J.: Prentice-Hall, 1970.

[171] K. M. Van Vliet, "Noise sources in transport equations associated with ambipolar diffusion and Shockley-Read recombination," *Solid-State Electron.*, vol. 13, pp. 649–657, May 1970.

[172] J. Viner, "Noise in transistors at elevated temperature," M.Sc. Thesis, University of Minnesota, Minneapolis, 1969, unpublished work.

[173] J. S. Vogel and M. J. O. Strutt, "Noise in transistor mixers," *Arch. Elek. Übertragung*, vol.16, pp. 215–222, May 1962; *Proc. IEEE*, vol. 51, pp. 340–349, February 1963.

[174] T. B. Watkins, "1/f noise in germanium devices," *Proc. Phys. Soc. London*, pt. 1, vol. 73, pp. 59–68, January 1, 1959.

[175] V. F. Weisskopf, "On the theory of noise in conductors, semiconductors, and crystal rectifiers," NDRC no. 14-133, May 15, 1943, unpublished. For a good summary see H. C. Torrey and C. H. Whitmer, *Crystal Rectifiers.* New York: McGraw-Hill, 1948.

[176] D. Wolf and E. Holler, "Bistable current fluctuations in reverse-biased p-n junctions of germanium," *J. Appl. Phys.*, vol. 38, pp. 189–192, January 1967.

[177] S. Y. Wu, "Theory of generation-recombination noise in MOS transistors," *Solid-State Electron.*, vol. 11, pp. 25–32, January 1968.

[178] L. D. Yau and C. T. Sah, "Theory and experiments of low-frequency generation-recombination noise in MOS transistors," *IEEE Trans. Electron Devices*, vol. ED-16, pp. 170–177, February 1969.

[179] ——, "On the excess white noise in MOS transistors," *Solid-State Electron.*, vol. 12, pp. 927–936, December 1969.

[180] IRE Sub-Committee 7.9 on Noise, "Representation of noise in linear twoports," *Proc. IRE*, vol. 48, pp. 69–74, January 1960.

NOISE IN PHOTOMULTIPLIERS (REVIEW)

(UDC 621.383.5)

T. A. Kovaleva, A. E. Melamid, A. N. Pertsev,
and A. N. Pisarevskii

Belorussian State University, Minsk
Translated from Pribory i Tekhnika Éksperimenta, No. 5,
pp. 5-19, September-October, 1966
Original article submitted February 1, 1966

The survey covers the noise sources, the pulse-height distribution of the noise, and the effects of external conditions on the noise, with particular attention to the one-electron component.

The inherent noise level of a photomultiplier (PM) governs its limit of detection and so is one of the most important parameters determining the range of application. This parameter is of especial importance when the PM tube is used at the limit of its performance, as in astronomy, bioluminescence, and tritium counting. It has been usual in discussing PM noise to center the attention on aspects of noise-level determination and signal detection in the presence of noise; but it is necessary to study the nature and origin of the noise pulses in order to provide a rational approach to the problem of reducing inherent noise levels. Here we survey the main trends in the study of PM noise and a method of reducing the latter.

1. Research on PM Noise

Improvements in PM tubes [1, 2] led to their use in scintillation counters for ionizing radiations [3-10]. Many uses of such counters were opened up by very detailed study of the processes in the PM tube itself, and there are now many papers on the working principles of these counters and on the limitations set by the PM [11-33]. From the start it was clear that the fluctuations in the dark current set the fundamental limit of detection for a scintillation counter operating with weak scintillations. Very weak light signals (even individual quanta) have to be recorded in various branches of optics, astronomy, biology, and nuclear physics; this is possible only as a result of careful studies of the causes of the dark current and of the conditions providing minimal dark current.

Dark-Current Sources and Secondary Effects. One of the earliest detailed papers on PM dark currents [34] states that this current has at least four sources: a) leakage, b) currents caused by ions of residual gas in the tube, c) field emission from the electrodes, d) thermionic emission from the photocathode and dynodes. This paper even contains comments on the possible role of secondary processes. For example, a primary photocurrent i_0 produces ions in the residual gas as it passes along the tube, and the ions produce feedback, which increases the initial photocurrent to

$$i = i_0 + \lambda I,$$

in which $\lambda = \alpha\beta\delta\gamma$, $I = Gi$, G is the mean gain of the PM tube, α is the coefficient for the ionization produced by a current I, β is the fraction of the ions that reach the photocathode, δ is the secondary-electron emission factor of that electrode for those ions, and γ is the fraction of those electrons amplified by the dynodes. Then the anode receives a current

$$I = \frac{G}{1-\lambda G} i_0 = G' i_0$$

and a self-sustaining process sets in when $\lambda G = 1$.

An additional cause of feedback is optical coupling, while cosmic rays add to the dark current [35]. Cosmic rays and thermionic emission are phenomena that cannot be suppressed completely, whereas all the other factors are

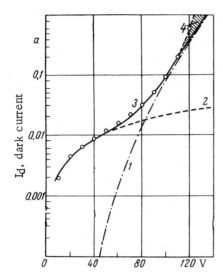

Fig. 1. Relation of dark-current components to dynode voltage: 1) amplified thermionic emission of photocathode and dynodes; 2) leakage current; 3) total dark current. Region 4 is that of unstable operation.

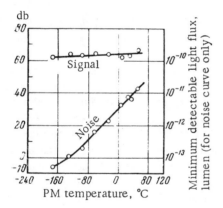

Fig. 2. Relation of signal and noise at PM output to PM temperature for the RCA type IP21.

secondary, being dependent on the details of the design and manufacture of the tube.

Some papers [35-47] deal with the noise arising in the load impedance and first amplifier stage, the signal-to-noise ratio then incorporating these factors. However, the contribution from these can be neglected for existing multipliers operated at reasonably high gain with properly designed recording circuits [37, 43, 48-54].

Voltage-Current Characteristics of the Dark Current. One of the first experimental studies [55] on dark currents in RCA tubes types 931A, IP21, IP22, and IP28 dealt with these as a function of interstage voltage. Three main regions were distinguished (Fig. 1), each dominated by a different component of the dark current: a) up to 50-60 $\dot{V}$ per stage, where the ohmic leakage predominates, this being due to residual Cs and other contaminants on the inner surface of the tube, the current being directly proportional to the potential difference; b) 60-100 V, where thermionic emission predominates, and the output current from this source characterizes the gain of the PM tube; c) above 110 V, in which the regenerative ionization process predominates and there are large variations in the output current. The sums of the extrapolated leakage and thermionic-emission curves agree with the observed curve and, hence, support this interpretation. It will be clear from this that the tube is best operated in the second region, where the fluctuations in the output current are represented approximately [37, 43, 52, 53, 56] by

$$[(I^2)_{av} \Delta f]^{1/2} = G (2 e i_t \Delta f)^{1/2},$$

in which G is as above, e is the electronic charge, i_t is the thermionic emission of the photocathode, and Δf is the bandwidth of the recording system. If the photocurrent at the cathode is i_0, the output current due to the light is $I = G i_0$; then

$$[(I^2)_{av} \Delta f]^{1/2} = G [2 e \Delta f (i_t + i_0)]^{1/2}$$

and the signal-to-noise ratio S is given by

$$S^2 = i_0^2 / 2 e \Delta f (i_t + i_0);$$

and for

$$i_t \to 0 \quad S^2 \to i_0 / 2 e \Delta f.$$

Noise Reduction by Cooling. Cooling the tube to $-175°$ reduces i_t [57-60], which reduces the inherent noise level considerably while leaving the signal virtually unaltered (Fig. 2). The recording of one-electron pulses has been examined with a tube cooled to $-190°$C; Fig. 3 shows the pulse-height distribution in complete darkness (upper curve) and with a light flux of 10^{-13} lm (lower curve). The rise in the dark-current curve at small amplitudes is due to thermionic emission from the dynodes. There is a fall in S for pulses of height less than 3 units, which implies that not all the one-electron pulses are recorded. This is very important to the choice of working point and determination of performance in many applications.

Role of Cathode Emission in Noise Pulse-Height Distribution. It has been assumed [12] that thermionic emission from the photocathode is the principal source of the dark current; if it were the sole source, the output current

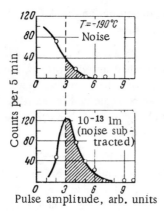

Fig. 3. Amplitude distribution of the signal and noise output pulses for a cooled PM (RCA IP21).

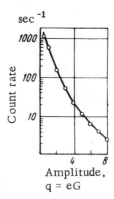

Fig. 4. Typical amplitude distribution for the noise pulses from a good 931A tube at room temperature.

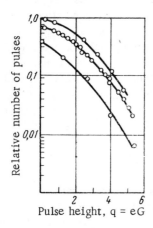

Fig. 5. Amplitude distribution of the one-electron pulses as affected by the potential difference between the photocathode and the first dynode.

could be represented as the superposition of pulses of mean height eG, in which G is the mean gain. The individual dynode secondary-emission factors $\bar{k}_i$ fluctuate from pulse to pulse, so the output pulses produced by single electrons from the photocathode have a certain amplitude distribution. Of course, the actual pulse-height distribution is more complicated than this; Fig. 4 shows a typical distribution, which represents a count rate of some 600 sec^{-1} pulses with an amplitude of eG or more. The current corresponding to these pulses in Fig. 4 is less than 1% of the total dark current, most of which arises from small pulses below the noise level of the amplifier system used. The dark-current count rate is also very much dependent on the temperature; for instance, the rate may increase by a factor of 10-40 in the range 27-50°C, but there is little change in the general shape of the distribution, as when the tube is cooled in solid carbon dioxide, there being less change in the large-amplitude pulses than in the small-amplitude ones. An electrostatic field along the glass causes the number of pulses at the output to be minimal when the screen has a potential close to that of the photocathode. The cause is ascribed to the effect of the potential of the inner surface of the glass near the photocathode. The effect is very much reduced if the glass is of high resistivity, as in the RCA type IP28.

A study has been made [12] of the amplitude distribution in the one-electron pulses produced by weak illumination of the photocathode at dry-ice temperature; Fig. 5 shows this distribution as affected by the potential difference between the photocathode and the first dynode (the effect is small). The illumination produced a pulse distribution whose mean charge per pulse was less than that for the noise pulses. It was assumed for the purposes of comparison with theory that the secondary emission of each dynode varies in accordance with a Poisson distribution:

$$p_k = \bar{k}^k e^{-\bar{k}} / \bar{k}!,$$

in which p_k is the probability of occurrence of $\bar{k}$ secondary electrons when the mean number per primary is $\bar{k}$. Then the relative variance of Z, the number of electrons reaching the anode, is

$$\overline{\Delta Z^2}/Z^2 = \bar{k}/(\bar{k} - 1).$$

The actual distribution had a variance twice that predicted in this way, which has been variously explained: that the secondary emission does not fit a Poisson distribution or that some of the secondaries are lost between the dynodes. It was subsequently [15] assumed that the discrepancy between the actual and theoretical curves for the one-electron pulses arises from differences in the focusing loss for the secondaries. On this basis $\bar{k}$ is replaced by $\varepsilon\bar{k}$, in which ε is a correction factor taking account of the loss. Then

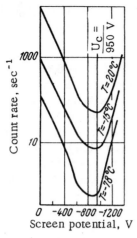

Fig. 6. Relation of number of noise pulses to external potential for several temperatures of the PM tube.

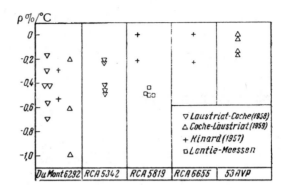

Fig. 7. Observed temperature coefficients for the dynodes of various PM tubes.

$$\overline{\frac{\Delta Z^2}{Z^2}} = \frac{\overline{k}}{\overline{k}-1}\left(1 + \frac{\varepsilon}{\overline{k}-1}\right)$$

and ε may be found by comparing the two curves. The result for a 5819 tube was ε = 1.54; a typical dark-current pulse-height distribution for the 5819 was also reported in this paper, the shape remaining substantially constant for potential differences of 70 to 130 V per stage.

Noise Reduction without Photocathode Cooling. There are many papers on the multiplication process [61-82], but the actual probability distribution of k (secondaries per primary) was not known until the pulse-height distribution of the one-electron pulses at the output was examined. Papers [12, 13, 15, 55, 88] still left open the form of this distribution, and the center of attention turned to methods of reducing the dark current without cooling the photocathode [88-99]. One of these [96] was to use magnetic focusing to insure that the electrons for subsequent multiplication were taken only from the illuminated part of the photocathode. This is not a method of general application, because it is usable only when the light may be focused onto a small area or when magnetic scanning of the electronic image at the photocathode is required.

A second approach was based on a detailed study of the effects of an external electrostatic field on the dark-current count rate N. This number becomes minimal if the external potential is close to that of the photocathode (Fig. 6); this relation of N to U remains applicable in the range −78 to +20°C, so we may put that

$$N = f(U) f(T).$$

This, in turn, indicates that the dependence of N on U is not due to some effect of the residual gas. It has been supposed [93] that thermal electrons are injected from the inner surface of the bulb by a negative potential and themselves eject secondaries from the cathode, while a positive potential causes electrons emitted from the electrodes to produce scintillations in the glass, which then eject electrons from the cathode and, thus, increase the number of dark-current pulses. The number of pulses arising in this way may exceed the number of the others by more than a factor of ten.

Temperature Dependence of Photoelectric and Secondary Emission Processes. It is troublesome to cool the photocathode in many applications; moreover, this tends to affect the response of the photocathode and the gain of the dynodes, which is dependent on the resistivity of the latter. The theory of secondary emission [84-87] gives no direct indication that there should be a temperature dependence of the secondary-emission coefficient, but results for antimony-cesium surfaces [83, 100, 101] show such a dependence for photoelectric and secondary emission processes. Measurements have been reported [101-118] of the temperature coefficients for cathode and dynode systems, as well as of the wavelength dependence of the temperature coefficient (see [119] for survey). The temperature coefficient is defined as $\rho = T^{-1}\, di/dT$, %/°C, in which ρ is referred to the current i at 20°C. The dynodes of these

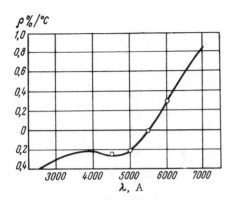

Fig. 8. Temperature coefficient as a function of wavelength for Sb + Cs photocathodes.

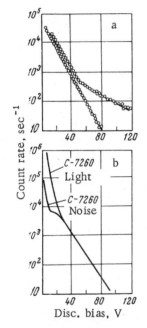

Fig. 9. Pulse-height distribution at the output: (a) noise pulses from two 7264'S; (b) noise and weak light for a C-7260'S.

PM tubes (Fig. 7) had negative coefficients of about 0.4% per °C, which was explained as due to reduction of the surface charge.

Figure 8 shows the temperature coefficient of a Cs + Sb photocathode as a function of wavelength (2500 to 7000 A), with ρ changing sign at 5500 A and being about −0.25% per °C for λ ≤ 5500 A, with a rapid rise for λ > 5500 A. In these cooling experiments it was found that the thermal emission of an alloy does not obey Richardson's law [101, 120]; similar statements have been made for Cs + Sb and Cs + O₂ cathodes below 0°C [101]. Moreover, it is clear [121] that the behavior of standard PM tubes is such as not to be explicable solely in terms of temperature dependence in the response of the photocathode.

Quasiexponential One-Electron Pulse-Height Distribution. It has been shown from theory [122] that the output pulse height has a Poisson distribution if the photocathode emission is a train of separate electrons and if the fluctuations in the secondary emission of the first dynode also have a Poisson distribution. On the other hand, many-electron dark-current pulses from the photocathode have been observed [123, 124]. A paper of 1960 [125] dealt with the sources of dark current in RCA standard and experimental tubes types 7264'S, 7265'S, 6810A'S, and C-7260'S.

The thermionic emission was recorded as a function of voltage and temperature, as were effects associated with the residual gas. Figure 9 shows the noise pulse-height distribution typical of most of the tubes. The linear part at small amplitudes is due to single electrons from the cathode; the region of high amplitudes is due to groups of electrons. All the tubes (over 20) gave a simple exponential distribution for the one-electron pulses, and this was taken to be universal; it was stressed that this cannot occur if secondary emission has a Poisson distribution. The expression used to describe the one-electron distribution was

$$p_1(V) = V_0^{-1} e^{-V/V_0}$$

which by integration gives an output distribution corresponding to the escape of N electrons from the cathode, the maximum occurring at $V_0 (N-1)$:

$$p_N(V) = \frac{1}{(N-1)! V_0^N} V^{N-1} e^{-V/V_0}.$$

Figure 10 gives curves for a 7264'S with the walls cooled by dry ice and the photocathode at various temperatures; similar curves were recorded for improved tubes type 7260'S. All of these curves correspond to one or two simple exponentials, and the voltages or temperatures have no marked effects on the form of the relationship. The reduced count rates on cooling the walls were taken to imply that the residual gas has some effect on the dark current. This was tested by using tubes with pressures of argon and hydrogen in the range 10^{-8} to 10^{-4} torr, but no effect was observed, which was ascribed to the high ionization potentials of Ar and H₂ (13.5 and 15.6 V, as against 3.9 V for Cs).

Poisson Distribution of One-Electron Pulses and Role of Gas Discharges. The results of [125] on the deviation from a Poisson distribution in the secondary emission have been confirmed [126-132], but this does not serve to settle

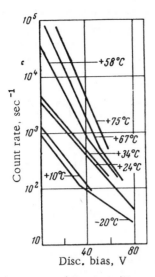

Fig. 10. Integral noise-pulse curves for a 7264'S at various cathode temperatures.

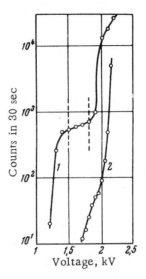

Fig. 11. Noise-selected 13-stage PM tube: 1) count-rate curve; 2) number of gas-discharge pulses.

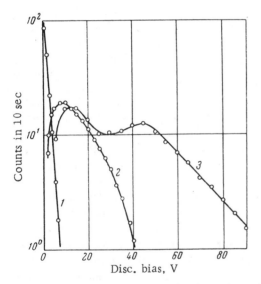

Fig. 12. Pulse-height distributions for the noise pulses from a 13-stage tube operated in the three principal ranges of the count-rate curve.

the question of the secondary-electron distribution, for a paper of 1962 [133] analyzed the results then available for 13-stage tubes (similar to the Russian FÉU-1S) as regards escape of single electrons from the cathode. Over 20 tubes with emission rates below 10^3 sec^{-1} were examined, the supply voltages being chosen to lie in each of the three principal ranges (Fig. 11): a) rapid rise with voltage, due to gain inadequate to record every single electron reaching the first dynode (1); b) region in which every single electron is recorded (2); c) region of rapid rise due to gas discharges (3).

Figure 12 shows the pulse-height distribution for each of these regions, which correspond respectively to an exponential, a peaked curve (close to Poisson form), and a two-peaked curve. Figure 13 illustrates the effects of voltage between photocathode and first dynode; the two curves represent Poisson distributions with $\bar{k}$ of 6 and 3, respectively. It was also shown [133] that the exponential curves of [125] are readily obtained in any PM tube in which gas discharges set in before the gain has become sufficient to record all the secondary electrons from the first dynode. The following assumptions have been made [134] in a detailed discussion of the gas-discharge mechanism and optical feedback: a) positive ions produce ionization and excitation of molecules [139], while photoionization [140] and ion-electron combination [141] are reckoned improbable; b) the gas discharge occurs mainly between the last dynode and the collector [142]; c) the field of the positive-ion space charge has no effect.

A primary electron from the photocathode produces G secondary electrons at the last dynode; if the gas discharge arises in the last gap, the resulting photons produce on average A_0 photoelectrons, which after a delay τ (the transit time of the system) themselves produce A_1 photoelectrons, and so on. The process is regenerative if $A_0 \gg 1$, Let the probability of formation of the i-th avalanche be

$$P(i) = A_i \left/ \sum_{i=0}^{\infty} A_i, \right.$$

271

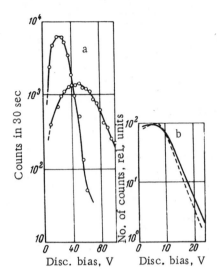

Fig. 13. Pulse-height distribution of the single-electron pulses from a 13-stage multiplier: a) for two different potential differences between photocathode and first dynode; b) for thermionic emission (full line) and photoelectrons (broken line).

in which A_i is the mean number of electrons due to photons in avalanche i; for $A < 1$,

$$\sum_{i=0}^{\infty} A_i = A_0 + A_1 + A_2 + \ldots + A_i + \ldots \simeq 1/(1 - A),$$

so

$$P(i) \simeq A_i(1 - A).$$

This relationship is confirmed by the results. The number of pulses produced by discharge phenomena is about 1% of the total number when the tube is operated in the plateau region but rises rapidly towards 100% above this. A relatively minor proportion of gas discharges will greatly distort the dark-current pulse-height distribution. It is found that A ranges from 0.05 for the best tubes to 0.5 for poor ones. The discharge pulses are delayed relative to the main ones by about 5×10^{-6} sec, so they can be prevented from registering if the paralysis time of the scaling circuit is raised appropriately.

A study of random processes in scintillation counters [135] has shown that $Cu + Al + Mg$ dynodes (FÉU-49) and AMGK ones (FÉU-S, FÉU-1B) give an accurate fit to a Poisson distribution for primary-electron energies in the range 150-350 V. It has several times been reported [136-138, 143-152] that a weak light flux produces an output-pulse distribution with a peak; careful measurements [137] by coincidence methods for the one-electron pulses in RCA tubes also indicate that the secondary emission has a Poisson distribution, in contrast to the assertions of [126-132].

Details of PM noise and one-electron pulse distributions are also given in [153-155]. Single electrons of light and thermal origins give identical distributions, so a component $S_0(I)$ is distinguished in the exponential distribution found for most PM tubes as being the component due to thermal emission from the photocathode, which has a Poisson distribution; the other component $S_0(II)$ has an exponential distribution and arises from various sources. The form appears purely exponential if $S_0(II) \gg S_0(I)$, but the spectrum differs from exponential if $S_0(II)$ and $S_0(I)$ are comparable. The performance (grade) of a PM tube is judged from $S_0(II)$, which is governed by the history of the tube (deposition of volatiles from the light-sensitive layer outside the working area, electrical conductivity of the glass, presence of sharp corners, insulation defects). PM tubes with $S_0(II) \gg S_0(I)$ are unsuitable for detecting weak signals. In the above papers, details are given of the differential noise pulse-height spectra as functions of the temperatures of photocathode and envelope [for tubes with $S_0(II) \gg S_0(I)$], which imply that thermal emission makes only a small contribution under normal conditions for tubes of this type. These conclusions conflict with those of [13, 156-159] but agree with the results of [98, 123, 144, 160].

A study has been made [161] of the one-electron pulse-height distribution for 13-stage (denoted by 1A) PM tubes type FÉU-1S and FÉU-1B as affected by photocathode and bulb temperatures, as well as by external electric fields and by the potential difference between the photocathode and first dynode or between the first two dynodes. It was concluded that this distribution and the secondary-emission process follow a Poisson distribution in the simple form

$$p_k = \bar{k}^k e^{-\bar{k}}/k!,$$

in which $\bar{k}$ is a linear function of the secondary-emission coefficient of the first dynode and is thus dependent on the potential difference between it and the photocathode. In this case the one-electron component made only a small contribution to the total dark current; coincidence measurements [162] confirm these results (Figs. 14 and 15).

Some aspects of the manufacture and use of PM tubes have very marked effects on the level and pulse-height distribution of the noise. For instance, the presence of Cs vapor and the distribution of the Cs within the bulb have marked effects on the general stability and on the dependence of the noise on temperature and supply voltages [163-170], on account of migration of Cs within the tube. It is difficult to obtain comparable results from different tubes,

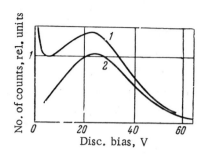

Fig. 14. Pulse-height distributions for an FÉU-1A: 1) noise; 2) one-electron pulses.

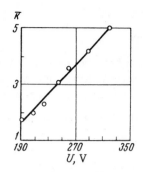

Fig. 15. Relation of $\bar{k}$ of output distribution to potential differ- ence between photocathode and first dynode for an FÉU-1A.

on account of the effects of uncontrolled variations during manufacture, previous running conditions, and previous illumination; moreover, a PM tube is a very labile system at the inherent noise level. All the same, the data reveal certain general trends.

2. Stability of Noise Level

This feature is no less important than the noise level itself or than the pulse-height distribution. Measurements at the limit of detection of the PM tube become very difficult if the threshold is time-varying. The thermal emission is a function of the reciprocal of the temperature alone, but this is not the only noise source, and many of the others are largely independent of temperature. These other effects are due to processes in the dynode system, so it is clear that the noise-level stability (NLS) at the output is closely related to stability in the gain, a subject which has [31] received insufficient attention.

The gain stability is equivalent to the NLS for a PM tube used near its threshold, whereas stable operation far from threshold is no proof of stable operation at the threshold. At present we should discuss not the causes of instability in noise level but rather the practical question of how to select PM tubes for stable noise level. The NLS may be evaluated qualitatively from the count-rate curve; the presence of a plateau is evidence for stability [133], and the NLS is generally the greater the longer this plateau and the less its slope. The absence of a plateau implies that the other effects predominate. The best Russian PM tubes in this respect are the FÉU-1S, FÉU-1B, and others with curved dynodes, which usually show a plateau. Recent work [172, 173] provides better-defined stability criteria. It has been found that there is often a disagreement between the noise levels measured by static and dynamic methods, namely

$$I_d \gg N_t eG \text{ or } \chi = I_d/N_t eG \gg 1,$$

in which I_d is the residual dark current (total less leakage and currents due to dynodes), N_t is the number of thermal electrons, e is the electronic charge, and G is the dynamic gain (deduced from the mean height of the one-electron pulses). Values of $\chi \gg 1$ imply that a major contribution comes from sources other than cathode thermionic emission, perhaps emission effects [174] in the dynodes. The point of importance to us is that χ is directly related to the NLS, the instability increasing the more $\chi \gg 1$. This instability is seen in pulse and dc measurements. The noise level is always stable for $\chi \approx 1$, and a count-rate plateau is always observed; but a plateau is sometimes seen for $\chi \gg 1$, although NLS is then lacking, so the χ criterion is more searching and unambiguous. The reason for this is that χ not only takes account of the number of pulses due to all effects but also of the charge that each pulse carries [175]. To select PM tubes for NLS we may use the criterion

$$10 > \chi \geqslant 1.$$

The NLS is also much dependent on the previous history, being much worse when the photocathode has been exposed to strong light fluxes, although there is extensive recovery if the tube is kept in darkness for 10-15 days. There is generally an irreversible deterioration in NLS if the tube is exposed to strong fluxes of ionizing radiation [176]. Radioactive contamination of the tube components is trivial as regards NLS and the noise level itself, as is the background radiation level in the housing of the tube, if this is low. The best PM tubes show an NLS of 1-2% at gains not exceeding $5 \cdot 10^6$; $\chi \approx 1$ cannot be obtained at higher gains.

3. Conclusions

It will be clear from the above that the noise level of a PM tube may be examined via the mean dark current and via the pulse-height distribution of the noise pulses. The second approach provides much more information about the effects of operating conditions and about the nature of the noise components.

The sources of the dark current may be divided into external (cosmic rays and radiation from nearly radioactive materials) and internal; the latter sources themselves are subdivided into primary (thermal emission from cathode and dynodes, field emission from the dynodes and other parts) and secondary (feedback of optical or ionic origin, gas-discharge effects due to ionization or excitation of residual gas). The importance assigned to each source is dependent on the method used to assess the noise. Pulse-height analysis (measurement of number of pulses rather than amount of charge) indicates that cathode thermal emission and gas-discharge effects are the principal sources. The discharge effects may be minimized by choice of the working voltage (point on the plateau), so the attention in all recent work has centered on the pulse-height distribution for one-electron pulses, on the assumption that the cathode emits electrons independently and at random, and also on the assumption that each electron is multiplied independently by the dynode system.

The response of the dynode system to single electrons indicates that the secondary emission has a Poisson distribution (over a certain range of working voltages) for many tubes differing in pulse-height distribution. The exponential distributions reported [125-132] for one-electron pulses were due either to the use of low gain or to a large probability of zero secondary emission from the first dynode; Stirling's approximation allows us to put the Poisson distribution in the form

$$p_k = \frac{\bar{k}^k e^{-\bar{k}}}{\sqrt{2\pi} \sqrt{k} \; e^{k(\ln k - 1)}} = \frac{f(\bar{k}, k)}{e^{k(\ln k - 1)}} \cdot$$

The result is an exponential law, because $\bar{k}$ is small, as is clear from the data of [137], where $\bar{k} = 1.5$. On the other hand, we cannot rule out the possibility that a distribution at first appearing to fit a Poisson one might on more detailed examination be found to deviate from it. Moreover, studies performed with equal care have given Poisson and quasiexponential distributions for the one-electron amplitudes, so we may expect that the general form of those amplitudes includes both components. Further work is needed to determine this point.

All known PM tubes may [145] be classified in two large groups: the first has the noise pulse-height distribution governed by the one-electron component, while the second has it governed by other components. The one-electron peak can be seen directly for group I, whereas coincidence methods are needed to reveal it for group II. It has been considered [155] that this component is due to thermal emission from the photocathode, but the results of [161] indicate otherwise: the intensity is governed by processes involving the cathode, but the thermal emission plays only a small part. For example, FÉU-1A tubes have this component accounting for only about 10% of the one-electron component, which itself is therefore of multiple origin.

Coincidence methods give a one-electron component with a Poisson distribution, but group I tubes have an exponential branch in the noise-pulse spectrum at low amplitudes. The noise arising from the dynode system also has an exponential distribution, but this is weak and cannot account for the exponential branch. An interesting point is that this branch increases together with the one-electron peak when the cathode is illuminated. It would seem that the passage of a one-electron avalanche is followed by effects in the dynode system that do not involve the cathode. The noise spectrum of a group I tube may [177] be put as

$$N_k = I_g \frac{1}{k_1} e^{-k \, k_1} + (I_t + I_e) \frac{(a\bar{k}_1)^k}{k!} e^{-a\bar{k}_1},$$

in which I_g, I_t, and I_e are respectively the relative contributions from the exponential branch, the thermal emission, and the additional effects contributing to the one-electron component: $I_g + I_t + I_e = 1$; $\bar{k}_1$ is the mean secondary-emission coefficient of the first dynode and $a \leq 1$. For group II we should add an exponential term for which $I \gg (I_t + I_e)$.

The minimum number of one-electron pulses of thermal origin given by an Sb+Cs cathode appears [161, 171] to be about 4 sec^{-1} cm^{-2} at room temperature, which is far less than the figures normally quoted. The high NLS of such tubes indicates that this result is a consequence of more careful manufacture, and also of the elimination of additional effects by proper conditions of use.

However, present manufacturing processes give inadequate control of the details of a PM tube, so the parameters show a large spread; only a few tubes have the dark current dominated by thermal emission. However, the occurrence of specimens with $\chi \approx 1$ gives grounds for hope.

It is already clear [175, 178-180] that there are great advantages in the operation of selected tubes in the one-electron mode; here the useful signals usually do not differ in amplitude from the noise ones, both arising from one-electron processes. The useful signal is found by subtracting the noise manually or automatically; in that case it is essential to make allowance for the change in the noise output of the cathode in response to illumination and any possible instability in the one-electron noise level during use. The latter is possible if the background is recorded almost simultaneously.

Cooling of the photocathode is not nearly as useful as it might be, in view of the numerous effects other than thermal emission still present in current PM tubes.

Although there has been much progress in the study of PM tube noise, the detailed composition of the latter is still far from understood, though advances in pulse-height and time-distribution analysis give us hope that this problem will be solved.

LITERATURE CITED

1. L. A. Kubetskii, Patent (1930).
2. P. T. Farnsworth, J. Franklin Inst., 218, 411 (1934).
3. M. Blau and B. Breyfus, Rev. Scient. Instrum., 16, 245 (1945).
4. S. C. Curran and W. R. Baker, Rev. Scient. Instrum., 19, 116 (1948).
5. I. W. Coltman and F. H. Marshall, Phys. Rev., 72, 528 (1947).
6. H. Kallman, Natur und Technik (July, 1947).
7. M. Deutsch, Phys. Rev., 73, 1240 (1948).
8. P. R. Bell, Phys. Rev., 73, 1405 (1948).
9. R. Hofstadter, Phys. Rev., 74, 100 (1948).
10. G. T. Reynolds, F. B. Harrison, and G. Salvani, Phys. Rev., 78, 488 (1950).
11. P. M. Woodward, Proc. Cambridge Philos. Soc., 44, 404 (1948).
12. G. A. Morton and I. A. Mitchell, RCA Rev., 9, 632 (1948).
13. G. A. Morton and I. A. Mitchell, Nucleonics, 4, 16 (1949).
14. R. C. Hoyt, Rev. Scient. Instrum., 20, 178 (1949).
15. G. A. Morton, RCA Rev., 10, 525 (1949).
16. F. Sauter, Z. Naturforsch., 4a, 682 (1949).
17. F. Seitz and D. W. Mueller, Phys. Rev., 78, 605 (1950).
18. A. W. Schardt and W. Bernstein, Rev. Scient. Instrum., 22, 1020 (1951).
19. P. Maignan, D. Blanc, and J. F. Detoeuf, J. Phys. et Radium, 13, 661 (1952).
20. R. K. Swank and W. L. Buck, Nucleonics, 10, 51 (1959).
21. G. F. Garlick and G. T. Wright, Proc. Phys. Soc., B65, 415 (1952).
22. G. T. Wright, Thesis, Birmingham (1952).
23. P. W. Roberts, Proc. Phys. Soc., A66, 192 (1955).
24. J. G. Campbell and A. J. F. Boyle, Austral. J. Phys., 6, 171 (1953).
25. T. F. Godlove and W. G. Wadey, Rev. Scient. Instrum., 25, 1 (1954).
26. W. Hanle, Naturwissenschaften, 38, 364 (1951).
27. S. C. Curran, Luminescence and the Scintillation Counter, London (1953).
28. A. Krebs, Ergebn.-Exakt. Naturwiss., 27, 361 (1953).
29. E. Breitenberger, Progr. Nucl. Phys., Pergamon Press, London, 4, 56 (1955).
30. J. Birks, Scintillation Counters [Russian translation], Izd. inostr. lit. (1955).
31. V. O. Vyazemskii, I. I. Lomonosov, A. N. Pisarevskii, Kh. V. Protopopov, et al., The Scintillation Method in Radiometry [in Russian], Atomizdat (1961).
32. Yu. K. Akimov, Scintillation Methods of Recording High-Energy Particles [in Russian], Izd. MGU (1963).
33. V. V. Matveev and A. D. Sokolov, Photomultipliers for Scintillation Counters [in Russian], Atomizdat (1962).
34. J. Reichman, Arch. Sci. Phys. et Nat. (5th period)(20 Sept., Oct., Nov., Dec., 1938).
35. N. S. Khlebnikov, Uspekhi fiz. nauk, 24, 358 (1940).
36. A. Sommer and W. E. Turk, J. Scient. Instrum., 27, 113 (1950).
37. W. Shockley and J. R. Pierce, Proc. I.R.E., 26, 321 (1938).

38. F. C. Williams, J. I.E.E., 82, 561 (1938).
39. J. S. Allen, Phys. Rev., 55, 966 (1939).
40. F. Eckart, Physik, 11, Hf. 4-7, 181 (1953).
41. C. J. Bakker and G. R. Van der Pol, Radio Sci. Venise, 5, 217 (1938).
42. V. K. Zworykin, G. A. Morton, and L. Malter, Proc. I.R.E., 24, 351 (1936).
43. F. Preisach, Wireless Engr., 16, 169 (1939).
44. P. A. Sinitsyn, Zh. tekhn. fiz., 10, 76 (1940).
45. W. Schottky, Ann. Physik, 104, 248 (1937).
46. B. Krovelmeyer and L. J. Hagner, Phys. Rev., 32, 952 (1937).
47. R. Champeik and P. Marchet, J. Phys. et Radium, 4, 7, 448 (1954).
48. A. M. Glover, Proc. I.R.E., 29, 413 (1941).
49. I A. Raichman and R. L. Snyder, Electronics, 13, 20 (1940).
50. V. K. Zworykin and I. A. Raichman, Proc. I.R.E., 27, 558 (1939).
51. R. B. Janes and A. M. Glover, RCA Rev., 6, 43 (1941).
52. A. Sommer, Electr. Engng., 17, 164 (1944).
53. R. C. Winaus and J. R. Pierce, Rev. Scient. Instrum., 12, 269 (1941).
54. W. Shockley and J. R. Pierce, Proc. I.R.E., 26, 321 (1938).
55. R. W. Engstrom, J. Opt. Soc. of America, 37, 420 (1947).
56. A. M. Glover, R. W. Engstrom, and W. J. Pietenpol, Final Report on the Investigation and Manufacture of Noise Sources at RCA, OSRD Report 1060-2 (March 7, 1945).
57. D. H. Rank and R. V. Wiegend, J. Opt. Soc. America, 36, 325 (1946).
58. G. E. Kron, Astrophys. J., 103, 324 (1946).
59. A. Blanc-Lapierre and D. Charles, J. Phys. et Radium, 5, 239 (1944).
60. E. F. Coleman, Electronics, 19, 120 (1946).
61. R. Kollath, Z. Phys., 38, 202 (1937).
62. H. Bruining and J. H. de Boer, Physica, 6, 823 (1939).
63. A. V. Afanas'eva and P. V. Timofeev, Zh. tekhn. fiz., 6, 831 (1936).
64. N. S. Khlebnikov, Zh. tekhn. fiz., 8, 11 (1938).
65. K. Siktus, Ann. Physik, 3, 1017 (1929).
66. A. I. Pyatnitskii, Zh. tekhn. fiz., 8, 1014 (1938).
67. P. Farnsworth, Phys. Rev., 31, 419 (1928).
68. V. W. Malter, Phys. Rev., 50, 48 (1936).
69. P. V. Timofeev and A. I. Pyatnitskii, Zh. tekhn. fiz., 10, 1 (1940).
70. P. V. Timofeev and V. V. Nalimov, Zh. tekhn. fiz., 6, 47 (1936).
71. P. V. Timofeev and K. A. Yumatov, Zh. tekhn. fiz., 10, 11 (1940).
72. K. G. McKay, in: Marton, Advances in Electronics, Vol. 1, Academic Press, New York (1948).
73. J. Matthes, Z. tech. Phys., 22, 232 (1941).
74. P. L. Copeland, Phys. Rev., 46, 167 (1934).
75. H. Frohlich, Ann. Phys., 13, 229 (1934).
76. D. E. Wolldridge, Phys. Rev., 56, 562 (1939).
77. F. M. Penning and A. A. Kruithof, Physica, 2, 793 (1935).
78. W. Kluge, O. Beyer, and H. Steyskal, Z. tech. Phys., 18, 219 (1937).
79. M. Ziegler, Physica, 3, 1 (1936).
80. I I. Ryabinin, Fizichni zapiski AN UkrSSR, 9, No. 2, 180 (1940).
81. V. N. Dyatlovitskaya, Dokl. AN SSSR, 13, No. 6, 822 (1948).
82. S. M. Fainshtein and L. I. Tatarinova, DAN SSSR, 79, No. 7, 1211 (1951).
83. N. S. Khlebnikov, Uspekhi fiz. nauk, 21, 301 (1939).
84. A. E. Kadyshevich, Zh. éksperim. i teor. fiz., 9, 930 (1939).
85. P. M. Morozov, Zh. éksperim. i teor. fiz., 11, 402 (1941).
86. Yu. M. Kushnir, V. Milyutin, and V. P. Goncharov, Zh. tekhn. fiz., 9, 158 (1938).
87. G. Blankfeld, Ann. phys., 84, 5, (1951).
88. N. Schaetti, Z. angew. Math. und Phys., 11, 123 (1953).
89. S. Rhodda, J. Scient. Instrum., 26, 65 (1949).
90. M. G. Scroggie, Wireless World, 58, 14 (1952).
91. S. Rhodda and F. C. Heath, British Patent Specification 645, 743.
92. E. H. Belcher, Brit. J. Radiol., 26, 455 (1953).

93. S. Rhodda, Photoelectric Multipliers, London (1953).

94. Z. Naray, Acta phys. Hungar., 4, 255 (1955).

95. Z. Naray, Magy. Fiz., 3, 1 (1955); J. Scient. Instrum., 33, 476 (1956).

96. Z. Naray, Acta phys. Hungar., 5, 159 (1955).

97. L. P. de Valence, Brit. J. Appl. Phys., 6, 311 (1955).

98. Z. Naray and P. Varga, Brit. J. Appl. Phys., 8, 377 (1957).

99. V. V. Matveev, A. D. Sokolov, and N. E. Sulimova, Units of a New Apparatus for Nuclear Radiation Studies [in Russian], Atomizdat (1961).

100. N. D. Margulis and B. N. Dyatlovitskaya, Zh. tekhn. fiz., 10, 657 (1940).

101. Z. Naray, Z. Ann. Phys., 20, 386 (1957).

102. M. Meessen, These, Univ. Louvain (1955).

103. A. Meessen, J. Phys. et Radium, 19, 437 (1958).

104. W. P. Ball, R. Booth, and M. H. McGregor, Nucl. Instrum., 1, 71 (1957).

105. L. A. Webb, U.S. N.R.D.L., Tr. 48 (1955).

106. K. Herold, S. Kropp, and R. Stutheit, E. J. du Pont de Nemours D. P., 47 (1956).

107. F. E. Kinard, Nucleonics, 15, 92 (1957).

108. Macq et Deutsch, Ann. Soc. Scient. Bruxelles, 71, 172 (1957).

109. G. C. Kelley and M. Goodrich, Phys. Rev., 77, 138 (1950).

110. H. H. Seliger and C. A. Ziegler, Nucleonics, 49 (April 14, 1956).

111. P. Beyster, W. Henkel, R. Hobles, and A. Kister, Phys. Rev., 98, 1216 (1955).

112. M. Bailliez (Mme. Lontie), These, Univ. Louvain (1958).

113. G. Laustriat and A. Coche, J. Phys. et Radium, 19, 927 (1958).

114. A. Coche and G. Laustriat, J. Phys. et Radium, 20, 719 (1959).

115. N. Schaetti and W. Baumgartner, Helv. phys. acta, 24, 614 (1951).

116. W. Khazanov and S. Yurov, Zh. Tech. Fiz., 22, 744 (1952).

117. H. Miyazawa, J. Phys. Soc. Japan, 8, 169 (1953).

118. W. E. Spicer, Phys. Rev., 112, 114 (1958).

119. M. Lontie-Bailliez and M. Meessen, Ann. Soc. Scient. Bruxelles, 73, 390 (1960).

120. S. V. Vonsovskii, A. V. Sokolov, and A. S. Veksler, Uspekhi fiz. nauk, 4, 216 (1956).

121. P. Gohrlich, A. Kross, X. I. Pol, and G. Wolf, Nachrichtentechnik, 2, 1-12, 1 (1959).

122. L. Jánossy, Zh. éksperim. i teor. fiz., 28, 599 (1955).

123. E. K. Zavoiskii, M. M. But-slov, and G. E. Smolkin, Dokl. AN SSSR, 111, 996 (1956).

124. J. Pelchowitch, Rev. Scient. Instrum., 26, 470 (1955).

125. J. A. Baicker, I.R.E. Trans. Nucl. Sci., NS-7, 2, 74 (1960).

126. F. J. Lombard and F. Martin, Rev. Scient. Instrum., 32, 2, 200 (1960).

127. K. Livesey (Private comm.), Colloque International d'Electronique Nucleaire, Paris (November 25-27, 1963).

128. J. R. Prescott, Nucl. Instrum. and Methods, 22, 256 (1963).

129. C. G. F. Delaney and P. W. Walton, Nucl. Instrum. and Methods, 25, 353 (1964).

130. M. Rome, IEEE Trans. Nucl. Sci., NS-11, No. 3, 93 (1964).

131. J. C. Barton, C. F. Barnaby, and B. M. Jasani, J. Scient. Instrum., 41, 599 (1964).

132. G. C. Baldwin and S. I. Friedman, Rev. Scient. Instrum., 36, 16 (1965).

133. N. S. Khlebnikov, A. E. Melamid, and T. A. Kovaleva, Radiotekhnika i élektronika, No. 3, 518 (1962).

134. N. S. Khlebnikov, A. E. Melamid, and T. A. Kovaleva, Radiotekhnika i élektronika, No. 6, 1020 (1964).

135. D. G. Fleishman, PTÉ, No. 3, 256 (1963).

136. G. Pfeffer, H. Lami, G. Laustriat, and A. Coche, Compt. Rend. Acad. Sci. Colon., 254, 1035 (1962).

137. R. F. Tusting, Q. A. Kerns, and H. K. Knudsen, I.R.E. Trans. Nucl. Sci., 9, 118 (1962).

138. M. Bertolaccini and S. Cova, Energia Nucl., 10, 259 (1963).

139. V. L. Granovskii, Electrical Conduction by Gases [in Russian], Gostekhizdat (1952).

140. F. L. Mohler, C. Boeckner, R. Stair, and W. W. Coolenz, Science, 69, 479 (1929).

141. F. L. Mohler, J. Res. Bur. Standards, 10, 771 (1933).

142. Radiophysical Electronics (edited by N. A. Kaptsov) [in Russian], Fizmatgiz (1960).

143. I. Koechlin, These de Doctorat, Paris (1961).

144. M. Brault and C. Gazier, Compt. Rend. Acad. Sci. Colon., 256, 1241 (1963).

145. L. M. Bollinger and G. E. Thomas, Rev. Scient. Instrum., 32, 1044 (1961).

146. D. Jeanselme, These 3-e Cycle, Strasbourg (1962).

147. L. G. Hyman, R. M. Schwarcz, and R. A. Schluter, Rev. Scient. Instrum., 35, 393 (1964).

148. L. Colli, U. Facchini, and A. Rossi, Nuovo cimento, 11, 255 (1964).
149. E. H. Eberhardt, IEEE Trans. Nucl. Sci., NS-11, No. 3, 48 (1964).
150. R. M. Matheson, IEEE Trans. Nucl. Sci., NS-11, No. 3, 64 (1964).
151. G. Pietri, IEEE Trans. Nucl. Sci., NS-11, No. 3, 76 (1964).
152. R. Evrard and C. Gazier, J. Phys. et Radium, 26, 37 (1965).
153. M. Duquesne and J. Kaplan, J. Phys. et Radium, 21, 708 (1960).
154. M. Duquesne and J. Tatischeff-Kaplan, Rapport interne (July, 1962).
155. M. Duquesne and J. Tatischeff-Kaplan, Colloque International d'Electronique Nucleaire, Paris (November 25-27, 1963).
156. H. J. J. Braddick, Rep. Progr. Phys., 23, 154 (1960).
157. F. Boeschoten, J. M. W. Milatz, and C. Smit, Physica, 20, 139 (1954).
158. G. Pietri, L'Onde electrique, 38, 606 (1954).
159. F. M. Marschall, J. M. Coltman, and A. J. Bennet, Rev. Scient. Instrum., 19, 744 (1944).
160. J. Sharpe and V. A. Stanley, 5503 EMI, Symposium on the Detection and Use of Tritium in the Physical and Biological Sciences (May, 1961).
161. A. N. Pertsev, A. N. Pisarevskii, and L. D. Soshin, PTÉ, No. 5, 173 (1963).
162. A. N. Pertsev, A. N. Pisarevskii, and L. D. Soshin, No. 3, 132 (1964).
163. S. M. Fainshtein, Zh. tekhn. fiz., 18, 39 (1948).
164. N. O. Chechek, Uspekhi fiz. nauk, 37, 74 (1949).
165. M. Volmer and K. Adhicari, Z. Phys., 35, 722 (1926).
166. J. A. Bekker, Trans. Faraday Soc., 28, 1948 (1932).
167. G. B. Haritonov, N. N. Semenov, and A. I. Schalnikov, Trans. Faraday Soc., 28, 169 (1932).
168. N. O. Chechek, S. M. Fainshtein, and T. M. Lifshits, Electron Multipliers [in Russian], Gostekhizdat (1954).
169. H. Paetow, Z. Phys., 111, 770 (1939).
170. E. L. Stolyarova, G. M. Suchkov, and L. S. Nesterova, in Coll.: Radiation Equipment and Analysis Methods [in Russian], Atomizdat (1960).
171. D. Emberson, A. Todhill, and W. Wilcox, Applications of Cascade Electron-Optical Converters [Russian translation], Izd. Mir (1965).
172. I. I. Lomonosov and G. I. Volina, Radiotekhnika i élektronika, 10, 1024 (1965).
173. I. I. Lomonosov, Dissertation, Belorussian State University, Minsk (1966).
174. J. Cantarell and J. Almodovar, Nucl. Instrum. and Methods, 24, 353 (1963).
175. T. A. Kovaleva, A. E. Melamid, A. N. Pisarevskii, I. V. Reznikov, and S. S. Shushkevich, PTÉ, No. 2, 150 (1965).
176. A. N. Pertsev, A. N. Pisarevskii, and L. D. Soshin, PTÉ, No. 2, 146 (1965).
177. A. N. Pertsev, Dissertation, Belorussian State University, Minsk (1965).
178. A. N. Pertsev, A. N. Pisarevskii, and L. D. Soshin, Zh. prikl. spektroskopii, 2, 396 (1965).
179. A. N. Pertsev, A. N. Pisarevskii, I. V. Reznikov, L. D. Soshin, et al., Zh. prikl. spektroskopii, 1, 303 (1964).
180. L. I. Komarov and A. N Pisarevskii, PTÉ, No. 4, 226 (1965).

All abbreviations of periodicals in the above bibliography are letter-by-letter transliterations of the abbreviations as given in the original Russian journal. *Some or all of this periodical literature may well be available in English translation.* A complete list of the cover-to-cover English translations appears at the back of the first issue of this year.

Bibliography for Part IV

```
**********************************************************************
            4-A.  NOISE IN MICROWAVE TUBES
**********************************************************************
```

ADLER, R., G. HRBEK AND G. WADE
    A LOW-NOISE ELECTRON-BEAM PARAMETRIC AMPLIFIER
        PROC. IRE, VOL. 46, NO. 10, PP. 1756-1757, OCTOBER 1958.
ADLER, R., G. HRBEK AND G. WADE
    THE QUADRUPOLE AMPLIFIER, A LOW-NOISE PARAMETRIC DEVICE
        PROC. IRE, VOL. 47, NO. 10, PP. 1713-1723, OCTOBER 1959.
ADLER, R. AND G. WADE
    BEAM REFRIGERATION BY MEANS OF LARGE MAGNETIC FIELDS
        JOUR. APPLIED PHYSICS, VOL. 31, NO. 7, PP. 1201-1203, JULY 1960.
ADLER, R. AND G. WADE
    BEHAVIOR OF THERMAL NOISE AND BEAM NOISE IN A QUADRUPOLE AMPLIFIER
        PROC. IRE , VOL. 49, NO. 4, P. 802, APRIL 1961.
AITCHISON, C. S.
    FREQUENCY SYNCHRONIZATION OF AN X-BAND REFLEX KLYSTRON
        PROC. IEE (GB), VOL. 105B, SUPPL., PP. 944-951, 1958.
ANDERSON, J. R.
    NOISE MEASUREMENT OF AN M-TYPE BACKWARD WAVE AMPLIFIER
        PROC. IRE, VOL. 48, NO. 5, PP. 946-947, MAY 1960.
ANAND, R. P.
    AN ELECTRON GUN FOR REDUCING ION OSCILLATION IN TRAVELING-WAVE TUBES
        IEEE TRANS. ELECTRON DEVICES, VOL. ED-11, NO. 2, PP. 75-76,
        FEBRUARY 1964.
ANISHCHENKO, V. S. AND A. I. SHTYROV
    EFFECT OF FEEDING ENERGY ALONG COUPLED HELICES ON THE NOISE FIGURE OF
    A TWT
        RADIO ENGINEERING AND ELECTRONIC PHYSICS, VOL. 13, NO. 6,
        PP. 991-993, JUNE 1968.
APUSHINSKII, G. P.
    THE SPECTRAL PROPERTIES OF TRAVELLING-WAVE TUBES
        RADIO ENGINEERING AND ELECTRONICS, VOL. 4, NO. 11, PP. 277-280,
        NOVEMBER 1959.
ARMSTRONG, J. G., B. DUNFORD, AND J. WILLARD
    LOW-NOISE TRAVELLING WAVE TUBES
        PROC. IEE (LONDON), VOL. 117, NO. 2, PP. 285-294, FEBRUARY 1970.
ARNAUD, J.
    ANOMALOUS NOISE IN ELECTRON GUNS IN CROSSED MAGNETIC AND ELECTRIC
    FIELDS (IN FRENCH)
        ANNALES DE RADIOELECTRICITE, VOL. 19, NO. 75, PP. 3-20, JANUARY
        1964.
ARNAUD, J. AND O. DOEHLER
    STUDY OF THE NOISE IN CROSSED-FIELD GUNS
        JOUR. APPLIED PHYSICS, VOL. 33, NO. 1, P. 234, JANUARY 1962.
ASHBY, D. E. T. F. AND R. B. DYOTT
    MEASURING MODULATION NOISE FROM A HIGH-POWER C. W. KLYSTRON AMPLIFIER
        PROC. IEE (LONDON), VOL. 106, PT. B, SUPPL, NO. 12, PP. 879-882,
        1959.
ASHKIN, A.
    A LOW-NOISE MICROWAVE QUADRUPOLE AMPLIFIER
        PROC. IRE, VOL. 49, NO. 6, PP. 1016-1020, JUNE 1961.
ASHKIN, A. AND L. D. WHITE
    PARTITION NOISE IN ELECTRON BEAMS AT MICROWAVE FREQUENCIES
        JOUR. APPLIED PHYSICS, VOL. 31, NO. 8, PP. 1351-1357, AUGUST 1960.
BEAM, W. R.
    NOISE WAVE EXCITATION AT THE CATHODE OF A MICROWAVE BEAM AMPLIFIER
        IRE TRANS. ELECTRON DEVICES, VOL. ED-4, NO. 3, PP. 226-234, JULY
        1957.
BEAM, W. R.
    INTERCEPTION NOISE IN ELECTRON BEAMS AT MICROWAVE FREQUENCIES
        RCA REVIEW, VOL. 16, NO. 4, PP. 551-579, DECEMBER 1955.
BEAM, W. R.
    EXTENSION OF THE EFFECTS OF INITIAL NOISE CURRENT AND VELOCITY
    CORRELATION ON THE NOISE FIGURE OF TRAVELING-WAVE TUBES
        RCA REV., VOL. 16, NO. 3, PP. 458-460, SEPTEMBER 1955.
BEAM, W. R.
    PROGRESS IN LOW-NOISE MICROWAVE TUBE DESIGN
        PROC. IEE (GB), VOL. 105, PT. B, SUPPLEMENT NO. 11, PP. 790-795,
        1958.
BELL, R. L.
    KLYSTRON OSCILLATOR NOISE THEORY
        BRITISH JOUR. APPLIED PHYSICS, VOL. 7, NO. 7, PP. 262-266, JULY
        1956.
BERGHAMMER, J.
    SPACE-CHARGE EFFECTS IN ULTRA-LOW-NOISE ELECTRON GUNS
        RCA REVIEW, VOL. 21, NO. 3, PP. 369-376, SEPTEMBER 1960.
BERGHAMMER, J.
    NOISE SMOOTHING BY REACTIVE DAMPING IN FINITE MULTIVELOCITY ELECTRON
    BEAMS
        RCA REVIEW, VOL. 22, NO. 1, PP. 185-194, MARCH 1961.

BERGHAMMER, J. AND S. BLOOM
    ON THE NONCONSERVATION OF NOISE PARAMETERS IN MULTIVELOCITY BEAMS
        JOUR. APPLIED PHYSICS, VOL. 31, NO. 3, PP. 454-458, MARCH 1960.
BIGUENET, C.
    CONTRIBUTION TO THE STUDY OF PARAMETRIC NOISE PHENOMENA IN BACKWARD
    WAVE OSCILLATORS
        ANNALES DE RADIOELECTRICITE, VOL. 17, PP. 100-118, APRIL 1962.
BIRDSALL, C. K. AND W. B. BRIDGES
    SPACE-CHARGE INSTABILITIES IN ELECTRON DIODES AND PLASMA CONVERTERS
        JOUR. APPLIED PHYSICS, VOL. 32, NO. 12, PP. 2611-2618, DECEMBER
        1961.
BLOOM, S.
    EFFECT OF DISTRIBUTED-LOSS NOISE GENERATORS ON TRAVELING-WAVE TUBE
    NOISE FACTOR
        RCA REVIEW, VOL. 22, NO. 2, PP. 347-349, JUNE 1961.
BLOOM, S.
    THE EFFECT OF INITIAL NOISE CURRENT AND VELOCITY CORRELATION ON THE
    NOISE FIGURE OF TRAVELING-WAVE TUBES
        RCA REVIEW, VOL. 16, NO. 2, PP. 179-196, JUNE 1955.
BLOOM, S. AND R. W. PETER
    A MINIMUM NOISE FIGURE FOR THE TRAVELING-WAVE TUBE
        RCA REV., VOL. 15, NO. 2, PP. 252-267, JUNE 1954.
BLOOM, S. AND B. VURAL
    NOISE ON A DRIFTING MAXWELLIAN BEAM
        JOUR. APPLIED PHYSICS, VOL. 34, NO. 2, PP. 356-363, FEBRUARY 1963.
BLOTEKJAER, K.
    TRANSVERSE ELECTRON BEAM NOISE DESCRIBED BY FILAMENTARY BEAM
    PARAMETERS
        JOUR. APPLIED PHYSICS, VOL. 33, NO. 8 ,PP. 2409-2414, AUGUST 1962.
BOSCH, B. G. AND W. A. GAMBLING
    NOISE IN REFLEX KLYSTRONS AND BACKWARD-WAVE OSCILLATORS
        JOUR. BRIT. I.R.E., VOL. 24, NO. 5, PP. 389-403, NOVEMBER 1962.
BOYD, J. A.
    NOISE CHARACTERISTICS OF A VOLTAGE TUNABLE MAGNETRON
        IRE TRANS. ELECTRON DEVICES, VOL. ED-1, NO. 4, PP. 201-205,
        DECEMBER 1954.
BUCHMILLER, L. D., R. W. DEGRASSE, AND G. WADE
    DESIGN AND CALCULATION PROCEDURES FOR LOW-NOISE TRAVELING-WAVE TUBES
        IRE TRANS. ELECTRON DEVICES, VOL. ED-4, NO. 3, PP. 234-245, JULY
        1957.
CAULTON, M. AND G. E. ST. JOHN
    S-BAND TRAVELING-WAVE TUBE WITH NOISE FIGURE BELOW 4 DB
        PROC. IRE, VOL. 46, NO. 5, PT. 1, PP. 911-912, MAY 1958.
CICCHETTI, J. B. AND J. MUNUSHIAN
    NOISE CHARACTERISTICS OF A BACKWARD-WAVE OSCILLATOR
        IRE NATIONAL CONVENTION RECORD, VOL. 6, PT. 3, PP. 84-93, 1958.
CURRIE, M. R.
    A NEW TYPE OF LOW-NOISE ELECTRON GUN FOR MICROWAVE TUBES
        PROC. IRE, VOL. 46, NO. 5, PT. 1, P. 911, MAY 1958.
CURRIE, M. R. AND D. C. FORSTER
    LOW NOISE TUNABLE PREAMPLIFIERS FOR MICROWAVE RECEIVERS
        PROC. IRE, VOL. 46, NO. 3, PP. 570-579, MARCH 1958.
CURRIE, M. R. AND D. C. FORSTER
    NEW MECHANISM OF NOISE REDUCTION IN ELECTRON BEAMS
        JOUR. APPL. PHYS., VOL. 30, NO. 1, PP. 94-103, JANUARY 1959.
CURRIE, M. R. AND D. C. FORSTER
    CONDITIONS FOR MINIMUM NOISE GENERATION IN BACKWARD-WAVE AMPLIFIERS
        IRE TRANS. ELECTRON DEVICES, VOL. ED-5, NO. 2, PP. 88-98, APRIL
        1958.
CURRIE, M. R. AND J. R. WHINNERY
    THE CASCADE BACKWARD-WAVE AMPLIFIER   A HIGH-GAIN VOLTAGE-TUNABLE
    FILTER FOR MICROWAVES
        PROC. IRE, VOL. 43, NO. 11, PP. 1617-1631, NOVEMBER 1955.
CURTICE, W. R.
    MEASUREMENT OF NOISE QUALITY OF MIG BEAM NOISE
        PROC. IEEE, VOL. 57, NO. 4, PP. 739-740, APRIL 1969.
CURTICE, W. R. AND L. A. MAC KENZIE
    R. F. MEASUREMENTS OF M. I. G. BEAM WITH PINHOLE COLLECTOR
        PROC. IEEE, VOL. 53, NO. 7, PP. 728-729, JULY 1965.
CUTLER, C. C.
    SPURIOUS MODULATION OF ELECTRON BEAMS
        PROC. IRE, VOL. 44, NO. 1, PP. 61-64, JANUARY 1956.
CUTLER, C. C. AND C. F. QUATE
    EXPERIMENTAL VERIFICATION OF SPACE CHARGE AND TRANSIT TIME REDUCTION
    OF NOISE IN ELECTRON BEAMS
        PHYS. REV., VOL. 80. NO. 5, PP. 875-878, DECEMBER 1, 1950.
DE GRASSE, R. W. AND G. WADE
    ELECTRON BEAM NOISINESS AND EQUIVALENT THERMAL TEMPERATURE FOR HIGH-
    FIELD EMISSION FROM A LOW-TEMPERATURE CATHODE
        PROC. IRE, VOL. 44, NO. 8, PP. 1048-1049, AUGUST 1956.
DISHAL, M.
    THEORETICAL GAIN AND SIGNAL TO NOISE RATIO OF THE GROUNDED GRID
    AMPLIFIER AT ULTRA-HIGH FREQUENCIES
        PROC. IRE, VOL. 32, NO. 5, PP. 276-284, MAY 1964.
DOEHLER, O. AND G. CONVERT
    THE SIGNAL TO NOISE RATIO IN THE M-CARCINOTRON

IRE TRANS. ELECTRON DEVICES, VOL. ED-1, NO. 4, PP. 184-188,
DECEMBER 1954.
EICHENBAUM, A. L.
ULTRA LOW NOISE BEAMS FOR SYNTHESIZED PLASMA CATHODES
PROC. IEEE, VOL. 53, NO. 6, PP. 631-632, JUNE 1965.
EICHENBAUM, A. L.
ULTRA-LOW-NOISE TRAVELING-WAVE-TUBE GUN WITH BEAM-FORMING ELECTRODES
PROC. IEEE, VOL. 52, NO. 5, PP. 613-614, MAY 1964.
EICHENBAUM, A. AND R. W. PETER
THE EXPONENTIAL GUN - A LOW NOISE GUN FOR TRAVELING-WAVE AMPLIFIERS
RCA REVIEW, VOL. 20, NO. 1, PP. 18-56, MARCH 1959.
ESPERSEN, G. A.
NOISE STUDIES ON TWO-CAVITY C.W. KLYSTRONS
IRE TRANS. MICROWAVE THEORY AND TECHNIQUES, VOL. MTT-8, NO. 5,
PP. 474-477, SEPTEMBER 1960.
EVERHART, T. E.
CONCERNING THE NOISE FIGURE OF A BACKWARD WAVE AMPLIFIER
PROC. IRE, VOL. 43, NO. 4, PP. 444-449, APRIL 1955.
FELDMANN, G. AND K. H. RATHSMANN
COMPARATIVE NOISE MEASUREMENTS ON OSCILLATOR KLYSTRONS
NACHRICHTENTECH. FACHBER. (NTF), VOL. 35, PP. 337-341, 1968.
FISCHER, K.
THE EQUIVALENT NOISE TWO-PORT OF A TRAVELLING-WAVE-TUBE AMPLIFIER
NACHRICHTENTECH. FACHBER. (NTF), VOL. 35, PP. 342-346, 1968.
FORSTER, D. C.
COOLING OF THE SLOW SPACE-CHARGE WAVE WITH APPLICATION TO THE
TRAVELING-WAVE TUBE
IRE TRANS. ELECTRON DEVICES, VOL. ED-9, NO. 6, PP. 449-453,
NOVEMBER 1962.
FORSTER, D. C.
COOLING OF THE SLOW SPACE-CHARGE WAVE WITH APPLICATION TO THE
TRAVELING-WAVE TUBE
IRE WESCON RECORD, VOL. 4, PT. 3, PP. 90-95, 1960.
FOX, A. J.
THE MINIMUM NOISE FIGURE OF THE BACKWARD WAVE AMPLIFIER
JOUR. ELECTRONICS AND CONTROL, VOL. 7, NO. 3, PP. 270-271,
SEPTEMBER 1959.
FOX, A. J., J. R. MANSELL AND J. L. PHILLIPS
MEASUREMENT OF NOISE PARAMETERS IN A LOW NOISE BACKWARD WAVE
AMPLIFIER
PROC. IEEE, VOL. 53, NO. 12, PP. 2113-2114, DECEMBER 1965.
FROHLICH, A.
NOISE IN ADLER TUBES
PROC. IEEE, VOL. 52, NO. 7, P. 868, JULY 1964.
FUJII, Y. AND S. SAITO
MEASUREMENT OF THE SHOT-NOISE REDUCTION FACTOR
IEEE TRANS. ELECTRON DEVICES, VOL. ED-14, NO. 4, PP. 207-214,
APRIL 1967.
FUJIOKA, T.
THREE-DIMENSIONAL EFFECTS OF POTENTIAL MINIMUM ON ELECTRON BEAM NOISE
ELECTRONICS AND COMMUNICATIONS IN JAPAN, VOL. 51-B, NO. 7, PP. 49-
57, JULY 1968.
GAMBLING, W. A., G. R. NUDD, AND J. E. RYLEY
MEASUREMENT OF THE NOISE PARAMETERS OF AN ELECTRON BEAM
PROC. IEE (LONDON), VOL. 112, NO. 9, PP. 1695-1699, SEPTEMBER 1965.
GANDHI, O. P.
A TWO-DIMENSIONAL THEORY OF NOISE EMISSION FROM THERMIONIC CATHODES
INTERNATIONAL JOUR. ELECTRONICS, VOL. 19, NO. 4, PP. 315-321,
OCTOBER 1965.
GERTSENSHTEIN, M. E.
ELECTRON BEAM NOISE
RADIO ENGINEERING AND ELECTRONICS, VOL. 4, NO. 1, PP. 246-248,
JANUARY 1959.
GOLUBENTSEV, A. F.
ON THE CONCEPT OF NOISE IMPEDANCE OF AN ELECTRON BEAM
RADIO ENGINEERING AND ELECTRONIC PHYSICS, VOL. 15, NO. 5,
PP. 896-900, MAY 1970.
GOLUBENTSEV, A. F. AND L. M. MINKIN
MINIMIZATION OF TWT NOISE FIGURE WITH CONSIDERATION OF CURRENT
SETTLING AT INPUT TO DELAY STRUCTURE
RADIO ENGINEERING AND ELECTRONIC PHYSICS, VOL. 11, NO. 5,
PP. 810-812, MAY 1966.
GOLUBENTSEV, A. F. AND L. M. MINKIN
MINIMIZATION OF THE TWT NOISE FIGURE FOR CONTINUOUS CURRENT SETTLING
ALONG THE SLOW-WAVE SYSTEM
RADIO ENGINEERING AND ELECTRONIC PHYSICS, VOL. 12, NO. 8, PP.
1400-1403, AUGUST 1967.
GOLUBENTSEV, A. F. AND L. M. MINKIN
MEASUREMENT OF THE NOISE INVARIANTS OF AN ELECTRON BEAM BY MEANS OF
AN O-TYPE BEAM AMPLIFIER
RADIO ENGINEERING AND ELECTRONIC PHYSICS, VOL. 13, NO. 6,
PP. 996-998, JUNE 1968.
GOLUBENTSEV, A. F. AND L. M. MINKIN
EFFECT OF CURRENT SETTLING IN AN ELECTRON GUN ON THE MINIMUM NOISE
FACTOR OF A TRAVELLING-WAVE TUBE

RADIOELECTRONICS AND COMMUNICATION SYSTEMS, VOL. 13, NO. 8, AUGUST 1970.

GOPINATH, A.
TRANSFORMATION MATRIX FOR PERIODIC ELECTROSTATICALLY FOCUSED ELECTRON BEAMS
INTERNATIONAL JOUR. ELECTRONICS, VOL. 23, NO. 3, PP. 217-220, MARCH 1968.

GOPINATH, A., S. ONO, AND H. L. HARTNAGEL
MEASUREMENT OF NOISE ALONG ELECTROSTATICALLY FOCUSED ELECTRON BEAMS
IEEE TRANS. ELECTRON DEVICES, VOL. ED-15, NO. 11, PP. 936-938, NOVEMBER 1968.

GORDON, E. I.
NOISE IN BEAM-TYPE PARAMETRIC AMPLIFIERS
PROC. IRE, VOL. 49, NO. 7, P. 1208, JULY 1961.

GOTTSCHALK, W. M.
DIRECT DETECTION MEASUREMENT OF NOISE IN CW MAGNETRONS
IRE TRANS. ELECTRON DEVICES, VOL. ED-1, NO. 4, PP. 91-98, DECEMBER 1954.

GOULD, R. W. AND C. C. JOHNSON
COUPLED-MODE THEORY OF ELECTRON-BEAM PARAMETRIC AMPLIFICATION
JOUR. APPLIED PHYSICS, VOL. 32, NO. 2, PP. 248-258, FEBRUARY 1961.

HAALAND, C. M.
REDUCTION OF THERMAL NOISE IN ELECTRON BEAMS
PROC. NATIONAL ELECTRONICS CONF., VOL. 15, PP. 394-403, 1959.

HAGGBLOM, H.
THE SPECTRAL DENSITY OF THE A.M. NOISE IN REFLEX KLYSTRONS
PROC. IEE (GB), VOL. 106, PT. B, PP. 497-500, NOVEMBER 1959.

HAMMER, J. M.
MEASURED VALUES OF NOISE SPECTRA, S AND PI OF ULTRA-LOW-NOISE BEAMS
PROC. IEEE, VOL. 51, NO. 2, PP. 390-391, FEBRUARY 1963.

HAMMER, J. M.
POWER SPECTRA MEASUREMENTS ON ULTRALOW-NOISE BEAMS
JOUR. APPLIED PHYSICS, VOL. 35, NO. 4, PP. 1147-1152, APRIL 1964.

HAMMER, J. W. AND E. E. THOMAS
TRAVELING-WAVE-TUBE NOISE FIGURE OF 1.0 DB AT S-BAND
PROC. IEEE, VOL. 52, NO. 2, P. 207, FEBRUARY 1964.

HAMMER, J. M. AND C. P. WEN
EFFECT OF HIGH MAGNETIC FIELD ON ELECTRON-BEAM NOISE
RCA REVIEW, VOL. 25, NO. 4, PP. 785-789, DECEMBER 1964.

HARRIS, R. D.
MINIMUM NOISE FIGURE FOR MAGNETRON INJECTION GUNS
IEEE TRANS. ELECTRON DEVICES, VOL. ED-14, NO. 2, PP. 102-109, FEBRUARY 1967.

HARRISON, S. W.
ON THE MINIMUM NOISE NOISE FIGURE OF TRAVELING-WAVE TUBES
PROC. IRE, VOL. 43, NO. 2, P. 227, FEBRUARY 1955.

HART, P. A. H.
ON CYCLOTRON WAVE NOISE REDUCTION
PROC. IRE, VOL. 50, NO. 2, P. 216, FEBRUARY 1962.

HARVEY, A. F.
MICROWAVE TUBES   AN INTRODUCTORY REVIEW WITH A BIBLIOGRAPHY
PROC. IEE (GB), VOL. 107, PT. C, NO. 11, PP. 29-59, MARCH 1960.

HAUS, H. A.
NOISE IN ONE-DIMENSIONAL ELECTRON BEAMS
JOUR. APPLIED PHYSICS, VOL. 26, NO. 5, PP. 560-571, MAY 1955.

HAUS, H. A.
LIMITATIONS ON THE NOISE FIGURE OF MICROWAVE AMPLIFIERS OF THE BEAM TYPE
IRE TRANS. ELECTRON DEVICES, VOL. ED-1, NO. 4, PP. 238-257, DECEMBER 1954.

HAUS, H. A. AND F. N. H. ROBINSON
THE MINIMUM NOISE FIGURE OF MICROWAVE BEAM AMPLIFIERS
PROC. IRE, VOL. 43, NO. 8, PP. 981-991, AUGUST 1955.

HO, R. Y. C. AND T. VAN DUZER
THE EFFECT OF SPACE CHARGE ON SHOT NOISE IN CROSSED-FIELD ELECTRON GUNS
IEEE TRANS. ELECTRON DEVICES, VOL. 15, NO. 2, PP. 75-84, FEBRUARY 1968.

HOK, G.
THE NOISE FACTOR OF TRAVELING WAVE TUBES
PROC. IRE, VOL. 44, NO. 8, PP. 1061-1062, AUGUST 1956.

HSIEH, H. S.
HYDRODYNAMIC ANALYSIS OF NOISE IN A FINITE-TEMPERATURE ELECTRON BEAM
JOUR. APPL. PHYS., VOL. 36, NO. 8, PP. 2414-2421, AUGUST 1965.

ISHII, K.
NOISE FIGURES OF REFLEX KLYSTRON AMPLIFIERS
IRE TRANS. MICROWAVE THEORY AND TECHNIQUES, VOL. MTT-8, NO. 3, PP. 291-294, MAY 1960.

ISRAELSEN, B. P. AND R. W. HAEGELE
LOW-NOISE TRAVELING-WAVE AMPLIFIERS
IEEE TRANS. COMMUNICATION TECHNOLOGY, VOL. COM-14, NO. 3, PP. 308-317, JUNE 1966.

ISRAELSEN, I. P.
THE EFFECT OF HELIX LOSS ON NOISE FIGURE IN TRAVELING WAVE TUBES
IEEE TRANS. ELECTRON DEVICES, VOL. ED-9, NO. 2, PP. 217-221, MARCH 1962.

JOHNSON, S. L., B. H. SMITH AND D. A. CALDER
    NOISE SPECTRUM CHARACTERISTICS OF LOW-NOISE MICROWAVE TUBES AND SOLID
    STATE DEVICES
        PROC. IEEE, VOL. 54, NO. 2, PP. 258-265, FEBRUARY 1966.
JOHNSON, S. L., B. H. SMITH, AND D. A. CALDER
    NOISE SPECTRUM CHARACTERISTICS OF LOW-NOISE MICROWAVE TUBES AND
    SOLID-STATE DEVICES
        PROC. IEEE, VOL. 54, NO. 2, PP. 258-265, FEBRUARY 1966.
JUILLERAT, R.
    MICROWAVE OSCILLATOR NOISE SPECTRUM. NOISE MEASUREMENTS ON TWO TYPES
    OF KA BAND KLYSTRONS (IN FRENCH)
        ONDE ELECTRIQUE, VOL. 45, PP. 101-106, JANUARY 1965.
KANAVETS, V. I., G. A. KUZ'MINA, AND V. M. LOPUKHIN
    NOISE IN DOUBLE-BEAM TUBES PRODUCED BY SHOT FLUCTUATIONS IN THE BEAMS
        RADIO ENGINEERING AND ELECTRONICS, VOL. 3, NO. 6, PP. 106-113,
        1958.
KAWAMURA, Y., T. MISUGI, AND T. YAMAGUCHI
    HUM NOISE IN TRAVELLING-WAVE TUBE
        FUJITSU SCIENTIFIC AND TECHNICAL JOUR., VOL. 6, NO. 2, PP. 75-88,
        JUNE 1970.
KHOMASSEN, K. I. AND D. A. DUNN
    EXPERIMENTAL OBSERVATIONS OF NOISE IN A VOLTAGE-TUNABLE MAGNETRON
    OPERATING IN A NOISY MODE
        PROC. IEEE, VOL. 53, NO. 2, PP. 202-203, FEBRUARY 1965.
KLUVER, J. W.
    A LOW-NOISE M-TYPE PARAMETRIC AMPLIFIER
        IEEE TRANS. ELECTRON DEVICES, VOL. ED-11, NO. 5, PP. 205-215,
        MAY 1964.
KNECHTLI, R. C.
    EFFECT OF ELECTRON LENSES ON BEAM NOISE
        IRE TRANS. ELECTRON DEVICES, VOL. ED-5, NC. 2, PP. 84-88, APRIL
        1958.
KNECHTLI, R. C. AND W. R. BEAM
    VALIDITY OF TRAVELING-WAVE-TUBE NOISE THEORY
        RCA REVIEW, VOL. 18, NO. 1, PP. 24-38, MARCH 1957.
KNECHTLI, R. C. AND W. R. BEAM
    PERFORMANCE AND DESIGN OF LOW-NOISE GUNS FOR TRAVELING-WAVE TUBES
        RCA REV., VOL. 17, NO. 3, PP. 410-424, SEPTEMBER 1956.
KORNELSEN, E. V., R. F. C. VESSOT AND G. WOONTON
    CURRENT AND VELOCITY FLUCTUATIONS AT THE ANODE OF AN ELECTRON GUN
        JOUR. APPL. PHYS., VOL. 28, NO. 10, PP. 1213-1214, OCTOBER 1957.
KORNILOV, S. A.
    COMBINATION NOISE IN A TRANSIT KLYSTRON
        RADIO ENGINEERING AND ELECTRONIC PHYSICS, VOL. 6, NO. 12, PP.
        1807-1816, DECEMBER 1961.
KORNILOV, S. A.
    COMPLETE SPECTRUM OF COMBINATION NOISE IN A FLOATING-DRIFT KLYSTRON
        RADIO ENGINEERING AND ELECTRONIC PHYSICS, VOL. 8, NO. 10, PP.
        1623-1631, OCTOBER 1963.
KORNILOV, S. A.
    LOW-FREQUENCY MODULATION NOISE IN A FLOATING-DRIFT KLYSTRON
        RADIO ENGINEERING AND ELECTRON PHYSICS, VOL. 8, NO. 11, PP.
        1759-1768, NOVEMBER 1963.
KORNILOV, S. A.
    AMPLITUDE AND PHASE FLUCTUATIONS IN A DRIFT KLYSTRON UNDER GENERAL
    INITIAL NOISE CONDITIONS
        RADIO ENGINEERING AND ELECTRONIC PHYSICS, VOL. 10, NO. 7, PP.
        1068-1073, JULY 1965.
KORNILOV, S. A.
    AMPLITUDE AND PHASE FLUCTUATIONS IN A DRIFT KLYSTRON MIXER
        RADIO ENGINEERING AND ELECTRONIC PHYSICS, VOL. 11, NO. 7, PP.
        1073-1081, JULY 1966.
KORNILOV, S. A.
    THE DEPENDENCE OF THE AMPLITUDE AND FREQUENCY NOISE OF A REFLEX
    KLYSTRON ON ITS OPERATING CONDITIONS
        RADIO ENGINEERING AND ELECTRONIC PHYSICS, VOL. 11, NO. 11,
        PP. 1833-1836, NOVEMBER 1966.
KOSMAHL, H.
    CORRELATION FACTOR FOR THE NOISE FLUCTUATIONS AT THE POTENTIAL
    MINIMUM OF A DIODE (IN GERMAN)
        ARCH. ELEKT. UBERTRAGUNG, VOL. 10, NO. 3, PP. 353-357, AUGUST 1956.
KRULEE, R. L.
    CARCINOTRON NOISE MEASUREMENTS
        IRE TRANS. ELECTRON DEVICES, VOL. ED-1, NO. 4, PP. 131-133,
        DECEMBER 1954.
KUYPERS, W. AND M. T. VLAARDINGERBROEK
    MEASUREMENT OF ELECTRICAL BEAM NOISE
        PHILIPS RES. REPTS., VOL. 20, NO. 3, PP. 349-356, JUNE 1965.
LEA-WILSON, C. P.
    SOME POSSIBLE CAUSES OF NOISE IN ADLER TUBES
        PROC. IRE, VOL. 48, NO. 2, PP. 255-257, FEBRUARY 1960.

LEA-WILSON, C. P., T. J. BRIDGES, AND V. C. VOKES
    THE APPLICATION OF BEAM COOLING TO QUADRUPOLE AMPLIFIERS
        JOUR. ELECTRONICS AND CONTROL, VOL. 10, NO. 4, PP. 261-272,
        APRIL 1961.

LEHR, C. G. AND A. L. COLLINS
    PHYSICAL MECHANISMS OF NOISE GENERATION IN MAGNETRONS
        IRE TRANS. ELECTRON DEVICES, VOL. ED-1, NO. 4, PP. 260-268,
        DECEMBER 1954.
LELE, S. G.
    NOISE TRANSPORT IN O-TYPE NONUNIFORM BEAMS
        PROC. IEEE, VOL. 60, NO. 2, PP. 225-226, FEBRUARY 1972.
LESOTA, S. K.
    MINIMUM NOISE FIGURE OF TRAVELLING WAVE TUBES
        RADIO ENGINEERING AND ELECTRONICS, VOL. 3, NO. 9, PP. 126-133,
        SEPTEMBER 1958.
LIEBSCHER, R. AND R. MULLER
    FREQUENCY NOISE IN TRAVELLING-WAVE TUBES
        PROC. IEE (GB), VOL. 105B, SUPPL., PP. 796-799, 1958.
LINDSAY, P. A.
    VELOCITY DESTRIBUTION IN ELECTRON STREAMS
        ADVANCES IN ELECTRONICS AND ELECTRON PHYSICS, L. MARTON, EDITOR,
        VOL. 13, ACADEMIC PRESS, NEW YORK, 1960, P³, 181-315.
LINDSAY, P. A.
    THE PROBLEM OF ELECTRON VELOCITY DISTRIBUTION
        PROC. IEEE, VOL. 51, NO. 12, PP. 1710-1722, DECEMBER 1963.
LINDSAY, P. A.
    THREE DIMENSIONAL ANALYSIS OF ELECTRON-VELOCITY DISTRIBUTION IN
    LOW-NOISE GUNS
        ELECTRONICS LETTERS, VOL. 1, NO. 2, PP. 32-33, APRIL 1965.
LINDSAY, P. A.
    THE CHOICE OF BOUNDARY CONDITIONS IN MONTE CARLO CALCULATIONS OF
    NOISE
        IEEE TRANS. ELECTRON DEVICES, VOL. ED-17, NO. 2, PP. 165-166,
        FEBRUARY 1970.
LINDSAY, P. A.
    NEW FORMULATION OF NOISE IN COLLISION-FREE SYSTEMS
        INTERNATIONAL JOUR. ELECTRONICS, VOL. 32, NO. 2, PP. 186-202,
        FEBRUARY 1972.
LINDSAY, P. A. AND V. PETRIDIS
    NEW FORMULATION OF NOISE IN COLLISION-FREE SYSTEMS II
        INTERNATIONAL JOUR. ELECTRONICS, VOL. 38, NO. 3, PP. 289-291,
        MARCH 1975.
LITTLE, R. P., H. M. RUPPEL AND S. T. SMITH
    BEAM NOISE IN CROSSED ELECTRIC AND MAGNETIC FIELDS
        JOUR. APPL. PHYS., VOL. 29, NO. 9, PP. 1376-1377, SEPTEMBER 1958.
LOPUKHIN, V. M. AND V. P. MARTYNOV
    EVALUATION OF NOISE IN A THREE-FREQUENCY ELECTRON-BEAM BACKWARD-WAVE
    AMPLIFIER
        RADIO ENGINEERING AND ELECTRONIC PHYSICS, VOL. 11, NO. 8,
        PP. 1259-1264, AUGUST 1966.
LOPUKHIN, V. M. AND A. S. ROSHAL
    REMOVAL OF NOISE OF SPACE CHARGE FAST WAVE USING RESONATOR
        RADIO ENGINEERING AND ELECTRONIC PHYSICS, VOL. 9, NO. 2,
        PP. 191-200, FEBRUARY 1964.
LOUIS, H. P.
    CONTRIBUTIONS TO THE KNOWLEDGE OF EXCESS NOISE
        IRE TRANS. ELECTRON DEVICES, VOL. ED-7 , NO. 2, PP. 95-99,
        APRIL 1960.
MAC DONALD, D. K. C.
    TRANSIT TIME DETERIORATION OF SPACE-CHARGE REDUCTION OF SHOT NOISE
        PHILOSOPHICAL MAGAZINE, VOL. 40, PP. 561-568, MAY 1949.
MAC DONALD, D. K. C.
    TRANSIT-TIME PHENOMENA IN ELECTRON STREAMS
        PHILISOPHICAL MAGAZINE (LONDON), SER. 7, VOL. 41, NO. 320,
        PP. 863-872, SEPTEMBER 1950.
MAC INTOSH, B. A.
    AN EXPERIMENTAL STUDY OF INTERCEPTION NOISE IN ELECTRON STREAMS AT
    MICROWAVE FREQUENCIES
        CANADIAN JOUR. PHYSICS, VOL. 37, NO. 3, PP. 285-299, MARCH 1959.
MAGALINSKIY, V. B.
    PROPAGATION OF SHOT NOISE IN AN ACCELERATED ELECTRON STREAM WITH
    THERMAL SPREAD
        RADIO ENGINEERING AND ELECTRONIC PHYSICS, VOL. 9, NO. 2, PP. 208-
        215, FEBRUARY 1964.
MANTENA, N. R. AND T. VAN DUZER
    LOW NOISE AND SPACE-CHARGE SMOOTHING IN A CROSSED-FIELD AMPLIFIER
        PROC. IEEE, VOL. 51, NO. 11, PP. 1662-1663, NOVEMBER 1963.
MANTENA, N. R. AND T. VAN DUZER
    CROSSED-FIELD BACKWARD-WAVE AMPLIFIER NOISE-FIGURE STUDIES
        JOUR. ELECTRONICS AND CONTROL, VOL. 17, NO. 5, PP. 497-511,
        NOVEMBER 1964.
MIDDLETON, D.
    THEORY OF PHENOMENOLOGICAL MODELS AND DIRECT MEASUREMENTS OF THE
    FLUCTUATING OUTPUT OF CW MAGNETRONS
        IRE TRANS. ELECTRON DEVICES, VOL. ED-1, NO. 4, PP. 56-90, DECEMBER
        1954.
MIHRAN, T. G. AND B. K. ANDAL
    THE GROWTH OF PEAK VELOCITY, NOISE, AND SIGNALS IN O-TYPE ELECTRON
    BEAMS

IEEE TRANS. ELECTRON DEVICES, VOL. ED-12, NO. 4, PP. 208-216,
APRIL 1965.
MORRISON, J. A.
NOISE PROPAGATION IN DRIFTING MULTIVELOCITY ELECTRON BEAMS
JOUR. APPLIED PHYSICS, VOL. 31, NO. 11, PP. 2066-2067, NOVEMBER
1960.
MUELLER, W. M.
REDUCTION OF BEAM NOISINESS BY MEANS OF POTENTIAL MINIMUM AWAY FROM
THE CATHODE
PROC. IRE, VOL. 49, NO. 3, PP. 642-643, MARCH 1961.
MUELLER, W. M. AND M. R. CURRIE
NOISE PROPAGATION ON UNIFORMLY ACCELERATED MULTIVELOCITY ELECTRON
BEAMS
JOUR. APPLIED PHYSICS, VOL. 30, NO. 12, PP. 1876-1880, DECEMBER
1959.
MUNGALL, A. G.
NOISE IN TRAVELING-WAVE TUBES
IRE TRANS. ELECTRON DEVICES, VOL. ED-2, NO. 2, PP. 12-17, APRIL
1955.
MOTT-SMITH, H. M.
CHANGE OF ELECTRON TEMPERATURE IN AN ELECTRON BEAM
JOUR. APPLIED PHYSICS, VOL. 24, NO. 3, PP. 249-255, MARCH 1953.
NABOKOV, YU. I. AND V. E. AVDEEV
ON THE ANOMALOUS SHOT EFFECT IN DEVICES WITH OXIDE CATHODES
BULLETIN ACAD. SCIENCE USSR, PHYS. SER., VOL. 33, NO. 3,
PP. 421-426, MARCH 1969.
NELSON, J. N. AND B. P. ISRAELSEN
TRAVELING-WAVE TUBE WITH 2.7 DB NOISE FIGURE
PROC. IEEE, VOL. 53, NO. 5, PP. 548-549, MAY 1965.
NICLAS, K. B.
QUANTITATIVE TRANSFORMATION PROCESSES IN A LOW NOISE BEAM FORMING
SYSTEM
ARCHIV FUR ELEKTRO. UBERTRANGUNGSTECHNIK, VOL. 15, NO. 12,
PP. 587-599, DECEMBER 1961.
NICLAS, K. B.
A CONTRIBUTION TO THE THEORY OF THE MINIMUM NOISE FIGURE OF LOW NOISE
TRAVELLING WAVE TUBES
ARCHIV FUR ELEKTR. UBERTRANGUNGSTECHNIK, VOL. 15, NO. 2,
PP. 101-107, FEBRUARY 1961.
NICLAS, K. B.
THEORETICAL AND EXPERIMENTAL INVESTIGATIONS ON LOW-NOISE
TRAVELING-WAVE TUBES IN A REVERSED MAGNETIC FIELD
MICROWAVE JOUR. VOL. 6, NO. 10, PP. 67-75, OCTOBER 1963.
NILSSON, O.
ANOMALOUS NOISE PEAK IN ADLER TUBES
INTERNATIONAL JOUR. ELECTRONICS, VOL. 26, NO. 4, PP. 333-343,
APRIL 1969.
NINOMIYA, K. AND T. OKOSHI
SOME RESULTS FROM THE MEASUREMENT OF THE NOISE PARAMETERS OF AN
ELECTRON BEAM USING A SEALED-OFF TUBE
PROC. IEEE, VOL. 56, NO. 6, PP. 1106-1107, JUNE 1968.
NINOMIYA, K. AND T. OKOSHI
MEASUREMENT OF THE NOISE PARAMETERS S AND TT (PI) OF AN ELECTRON
BEAM USING A SEALED OFF TUBE
ELECTRONICS AND COMMUNICATIONS IN JAPAN, VOL. 52-B, NO. 2, PP. 68-
75, FEBRUARY 1969.
NUDD, G. R. AND W. A. GAMBLING
MEASUREMENTS OF THE NOISE PARAMETERS OF AN ELECTRON BEAM
ELECTRONICS LETTERS, VOL. 2, NO. 3, PP. 81-82, MARCH 1966.
OKOSHI, T.
A NEW METHOD OF MEASURING NOISE PARAMETERS S AND PI OF AN ELECTRON
BEAM
IEEE TRANS. ELECTRON DEVICES, VOL. ED-11, NO. 1, PP. 37-38,
JANUARY 1964.
OZAWA, S. AND T. FUJIOKA
COMPUTER SIMULATION FOR LOW-VELOCITY DRIFTING EFFECTS ON ELECTRON
BEAM NOISE
ELECTRONICS AND COMMUNICATIONS IN JAPAN, VOL. 52-B, NO. 3, PP.
103-107, MARCH 1969.
PARZEN, P.
EFFECT OF THERMAL VELOCITY SPREAD ON THE NOISE FIGURE IN TRAVELING-
WAVE TUBES
JOUR. APPLIED PHYSICS, VOL. 23, NO. 4, PP. 394-406, APRIL 1952.
PETER, R. W.
LOW-NOISE TRAVELING-WAVE AMPLIFIER
RCA REVIEW, VOL. 13, NO. 3, PP. 344-368, SEPTEMBER 1952.
PETER, R. W. AND J. A. RUETZ
INFLUENCE OF SECONDARY ELECTRONS ON NOISE FACTOR AND STABILITY OF
TRAVELING-WAVE TUBES
RCA REVIEW, VOL. 14, NO. 3, PP. 441-452, SEPTEMBER 1953.
PETRIDIS, V. AND P. A. LINDSAY
NEW FORMULATION OF NOISE IN COLLISION-FREE SYSTEMS III
INTERNATIONAL JOUR. ELECTRONICS, VOL. 38, NO. 3, PP. 293-329,
MARCH 1975.
PIERCE, J. R.
NOISE IN RESISTANCES AND ELECTRON STREAMS

BELL SYS. TECH. JOUR., VOL. 27, NO. 1, PP. 158-174, JANUARY 1948.
PIERCE, J. R.
POSSIBLE FLUCTUATIONS IN ELECTRON BEAMS DUE TO IONS
JOUR. APPL. PHYS., VOL. 19, NO. 3, PP. 231-236, MARCH 1948.
PIERCE, J. R.
A NEW METHOD FOR CALCULATING NOISE IN ELECTRON STREAMS
PROC. IRE, VOL. 40, NO. 12, PP. 1675-1680, DECEMBER 1952.
PIERCE, J. R.
A THEOREM CONCERNING NOISE IN ELECTRON STREAMS
JOUR. APPL. PHYS., VOL. 25, NO. 8, PP. 931-933, AUGUST 1954.
PIERCE, J. R.
THE GENERAL SOURCE OF NOISE IN VACUUM TUBES
IRE TRANS. ELECTRON DEVICES, VOL. ED-1, NO. 4, PP. 135-167,
DECEMBER 1954.
PIERCE, J. R.
CALCULATIONS CONCERNING THE NOISINESS OF A DRIFTING STREAM OF
ELECTRONS
PROC. IEE (LONDON), VOL. 105, PT. B, SUPPL. NO. 11, PP. 786-789,
1958.
PIERCE, J. R. AND W. E. DANIELSON
MINIMUM NOISE FIGURE OF TRAVELING WAVE TUBES WITH UNIFORM HELICES
JOUR. APPL. PHYS., VOL. 25, NO. 9, PP. 1161-1165, SEPTEMBER 1954.
POLLACK, M. A.
NOISE AND THE POTENTIAL MINIMUM AT HIGH FREQUENCIES
IRE TRANS. ELECTRON DEVICES, VOL. ED-9, NO. 3, P. 316, MAY 1962.
POLLACK, M. A. AND J. R. WHINNERY
NOISE TRANSPORT IN THE CROSSED-FIELD DIODE
IEEE TRANS. ELECTRON DEVICES, VOL. ED-11, NO. 3, PP. 81-89, MARCH
1964.
RACK, A. J.
EFFECT OF SPACE CHARGE AND TRANSIT TIME ON THE SHOT NOISE IN DIODES
BELL SYSTEM TECH. JOUR., VOL. 17, PP. 592-619, OCTOBER 1938.
RAO, B. V. AND W. A. GAMBLING
NOISE SPECTRA OF BEAM-TYPE MICROWAVE OSCILLATORS
RADIO AND ELECTRONIC ENGINEER, VOL. 35, NO. 3, PP. 165-173,
MARCH 1968.
RIGROD, W. W.
NOISE SPECTRUM OF ELECTRON BEAM IN LONGITUDINAL MAGNETIC FIELD,
I. THE GROWING NOISE PHENOMENA
BELL SYS. TECH. JOUR., VOL. 36, NO. 4, PP. 831-853, JULY 1957.
II. THE U.H.F. NOISE SPECTRUM
BELL SYS. TECH. JOUR., VOL. 36, NO. 4, PP. 855-878, JULY 1957.
RIGROD, W. W.
SPACE-CHARGE WAVE HARMONICS AND NOISE PROPAGATION IN ROTATING
ELECTRON BEAMS
BELL SYSTEM TECH. JOUR., VOL. 38, NO. 1, PP. 119-139, JANUARY 1959.
ROBINSON, F. N. H.
MICROWAVE SHOT NOISE AND AMPLIFIERS
IRE TRANS. ELECTRON DEVICES, VOL. ED-3, NO. 3, PP. 128-133,
JULY 1956.
ROBINSON, F. N. H.
SPACE CHARGE SMOOTHING OF MICROWAVE SHOT NOISE IN ELECTRON BEAMS
PHILOSOPHICAL MAGAZINE, SEVENTH SERIES, VOL. 43, NO. 336, PP. 51-
62, JANUARY 1952.
ROBINSON, F. N. H.
MICROWAVE SHOT NOISE IN ELECTRON BEAMS AND THE MINIMUM NOISE FACTOR
OF TRAVELLING-WAVE TUBES AND KLYSTRONS
JOUR. BRIT. INST. RADIO ENGRS., VOL. 14, NO. 2, PP. 79-86,
FEBRUARY 1954.
ROBINSON, F. N. H.
CURRENT AND VELOCITY FLUCTUATIONS AT THE POTENTIAL MINIMUM
JOUR. ELECTRONICS AND CONTROL, VOL. 5, NO. 2, PP. 152-156, AUGUST
1958.
ROBINSON, F. N. H. AND R. N. FRANKLIN
THE TRANSVERSE ELECTRIC NOISE FROM AN ELECTRON BEAM
JOUR. ELECTRONICS AND CONTROL, VOL. 10, NO. 4, PP. 277-284, APRIL
1961.
ROBINSON, F. N. H. AND H. A. HAUS
ANALYSIS OF NOISE IN ELECTRON BEAMS
JOUR. ELECTRONICS, VOL. 1, NO. 4, PP. 373-384, JANUARY 1956.
ROBINSON, F. N. H. AND R. KOMPFNER
NOISE IN TRAVELING WAVE TUBES
PROC. IRE, VOL. 39, NO. 8, PP. 918-926, AUGUST 1951.
ROCKWELL, R. G.
LOW-NOISE KLYSTRON AMPLIFIERS
IRE TRANS. ELECTRON DEVICES, VOL. ED-6, NO. 4, PP. 428-437,
OCTOBER 1959.
ROWE, H. E.
NOISE ANALYSIS OF A SINGLE-VELOCITY ELECTRON GUN OF FINITE
CROSS SECTION IN AN INFINITE MAGNETIC FIELD
IRE TRANS. ELECTRON DEVICES, NO. PGED-2, PP. 36-46, JANUARY 1953.
RUZICKA, J.
A CONTRIBUTION TOWARD DETERMINING THE NOISE SPECTRUM IN A BACKWARD-
WAVE OSCILLATOR
SLABOPROUDY OBZOR, VOL. 33, NO. 1, PP. 9-13, 1972.

SAITO, S.
    NEW METHOD OF MEASURING THE NOISE PARAMETERS OF AN ELECTRON BEAM
        IRE TRANS. ELECTRON DEVICES, VOL. ED-5, NO. 4, PP. 264-275,
        OCTOBER 1958.
SAITO, S. AND Y. FUJII
    SOME RESULTS FROM THE MEASUREMENT OF THE NOISE PARAMETERS IN ELECTRON
    BEAM
        PROC. IRE, VOL. 50, NO. 7, PP. 1706-1707, JULY 1962.
SAITO, S. AND Y. FUJII
    MEASUREMENT OF MICROWAVE SHOT-NOISE REDUCTION FACTOR BY LASER LIGHT
    INDUCED PHOTOEMISSION
        PROC. IEEE, VOL. 52, NO. 8, P. 980, AUGUST 1964.
SAITO, S., Y. FUJII, AND A. IWAMOTO
    MONTE CARLO CALCULATION AND MEASUREMENT OF SHOT-NOISE REDUCTION FACTOR
        IEEE TRANS. ELECTRON DEVICES, VOL. ED-19, NO. 11, PP. 1190-1198,
        NOVEMBER 1972.
SASAKI, A. AND T. VAN DUZER
    NOISE-FIGURE EXPRESSIONS FOR CROSS-FIELD AMPLIFIERS
        IEEE TRANS. ELECTRON DEVICES, VOL. ED-14, NO. 3, P. 171-172,
        MARCH 1967.
SAYAKHOV, F. L.
    DETERMINING THE OPTIMUM POTENTIAL DISTRIBUTION IN THE GUN OF A
    LOW-NOISE TRAVELLING-WAVE TUBE
        RADIO ENGINEERING AND ELECTRONIC PHYSICS, VOL. 8, NO. 1, PP. 18-24,
        JANUARY-FEBRUARY 1965.
SAYAKHOV, F. L.
    CALCULATION OF THE NOISE FIGURE OF TRAVELING-WAVE TUBE WITH AN
    EXPONENTIAL ELECTRON GUN
        RADIO ENGINEERING AND ELECTRONIC PHYSICS, VOL. 10, NO. 1, PP.
        66-70, JANUARY 1965.
SAYAKHOV, F. L. AND A. I. SHTYROV
    ON THE PHYSICAL MEANING OF NOISE FIGURE MINIMIZATION IN A TRAVELING
    WAVE TUBE
        RADIO ENGINEERING AND ELECTRONIC PHYSICS, VOL. 10, NO. 12, PP.
        1899-1903, DECEMBER 1965.

SHAW, A. W., A. E. SIEGMAN, AND D. A. WATKINS
    REDUCTION OF ELECTRON BEAM NOISINESS BY MEANS OF A LOW-POTENTIAL
    DRIFT REGION
        PROC. IRE, VOL. 41, NO. 2, PP. 334-335, FEBRUARY 1959.
SHENOGIN, A. A.
    LOWERING THE NOISE FACTOR OF TWTS FOR THE 8 MILLIMETER REGION
        RADIO ENGINEERING AND ELECTRONIC PHYSICS, VOL. 15, NO. 5,
        PP. 900-903, MAY 1970.

SHIMODA, K.
    LENGTH OF COHERENT MICROWAVES GENERATED BY AN ELECTRONIC OSCILLATOR
        JOUR. PHYS. SOC. JAPAN, VOL. 8, NO. 1, PP. 131-132, JANUARY-
        FEBRUARY 1953.
SHTYROV, A. I. AND V. S. ANISHCHENKO
    NOISE FIGURE OF A TWT OPERATING IN THE MODE OF CONTINUOUS AND UNIFORM
    CURRENT INTERCEPT BY THE RETARDING DEVICE
        RADIO ENGINEERING AND ELECTRONIC PHYSICS, VOL. 12, NO. 8,
        PP. 1325-1335, AUGUST 1967.
SIDHU, G. S. AND R. P. WADHWA
    NOISE REDUCTION IN KINO LONG GUNS AND ON INCREASING THEIR DYNAMIC
    RANGE
        PROC. IEEE, VOL. 58, NO. 5, PP. 825-826, MAY 1970.
SIEGMAN, A. E. AND S. BLOOM
    AN EQUIVALENT CIRCUIT FOR MICROWAVE NOISE AT THE POTENTIAL MINIMUM
        IRE TRANS. ELECTRON DEVICES, VOL. ED-4, NO. 4, PP. 295-299,
        OCTOBER 1957.
SIEGMAN, A. E. AND D. A. WATKINS
    POTENTIAL MINIMUM NOISE IN THE MICROWAVE DIODE
        IRE TRANS. ELECTRON DEVICES, VOL. ED-4, NO. 1, PP. 82-86,
        JANUARY 1957.
SIEGMAN, A. E., D. A. WATKINS, AND H-C. HSIEH
    DENSITY-FUNCTION CALCULATIONS OF NOISE PROPAGATION ON AN ACCELERATED
    MULTI VELOCITY ELECTRON BEAM
        JOUR. APPLIED PHYSICS, VOL. 28, NO. 10, PP. 1138-1148, OCTOBER 1957
SISODIA, M. L. AND R. P. WADHWA
    NOISE REDUCTION IN CROSSED-FIELD GUNS BY CATHODE TILT
        PROC. IEEE, VOL. 56, NO. 1, PP. 94-95, JANUARY 1968.
SMITH, N. W. W.
    NOISE IN BACKWARD-WAVE OSCILLATORS
        PROC. IEE (LONDON), VOL. 105, PT. B, SUPPL. NO. 11, PP. 800-804,
        1958.
SMITH, N. W. W.
    NOISE REDUCTION IN MICROWAVE TUBES BY GETTER-ION PUMPING
        VACUUM , VOL. 10, NO. 1-2, PP. 106-109, FEBRUARY-APRIL 1960.
SMULLIN, L. D.
    PROPAGATION OF DISTURBANCES IN ONE-DIMENSIONAL ACCELARATED ELECTRON
    STREAMS
        JOUR. APPLIED PHYSICS, VOL. 22, NO. 12, PP. 1496-1498, DECEMBER
        1951.
SMULLIN, L. D. AND C. FRIED
    MICROWAVE NOISE MEASUREMENTS ON ELECTRON BEAMS

IRE TRANS. ELECTRON DEVICES, VOL. ED-1, NO. 4, PP. 168-183,
DECEMBER 1954.
SPROULL, R. L.
EXCESS NOISE IN CAVITY MAGNETRON
JOUR. APPL. PHYS., VOL. 18, NO. 3, PP. 314-320, MARCH 1947.
ST. JOHN, G. E.
MEASUREMENTS OF TRAVELING-WAVE TUBE NOISE FIGURE
IRE TRANS. ELECTRON DEVICES, VOL. ED-1, NO. 4, P. 200, DECEMBER
1954.
TANAKA, K., M. TANAKA, AND O. FUKUMITSU
APPLICATION OF THE BEAM MODE EXPANSION TO THE ANALYSIS OF NOISE
REDUCTION STRUCTURE
IEEE TRANS. MICROWAVE THEORY AND TECHNIQUES, VOL. MTT-23, NO. 7,
PP. 595-598, JULY 1975.
TERPIGOR'YEV, V. G.
EFFECT OF COATING NONUNIFORMITY ON NOISE SPECTRAL DENSITY OF THE
ELECTRON VELOCITY IN A BEAM EMITTED BY AN OXIDE CATHODE
RADIO ENGINEERING AND ELECTRONIC PHYSICS, VOL. 13, NO. 12,
PP. 1936-1941, DECEMBER 1968.
THOMASSEN, K. I. AND D. A. DUNN
EXPERIMENTAL OBSERVATION OF NOISE IN A VOLTAGE-TUNABLE MAGNETRON
OPERATING IN A NOISE MODE
PROC. IEEE, VOL. 53, NO. 2, PP. 202-203, FEBRUARY 1965.
THOMPSON, B. J., D. O. NORTH AND W. A. HARRIS
FLUCTUATIONS IN SPACE CHARGE LIMITED CURRENTS AT MODERATELY HIGH
FREQUENCIES
RCA REV., VOL. 4, NO. 3, PP. 269-285, JANUARY 1940.
RCA REV., VOL. 4, NO. 4, PP. 441-472, APRIL 1940.
RCA REV., VOL. 5, NO. 1, PP. 106-124, JULY 1940.
RCA REV., VOL. 5, NO. 2, PP. 247-260, OCTOBER 1940.
RCA REV., VOL. 5, NO. 3, PP. 371-388, JANUARY 1941.
RCA REV., VOL. 5, NO. 4, PP. 505-524, APRIL 1941.
RCA REV., VOL. 6, NO. 1, PP. 114-124, JULY 1941.
TIEN, P. K.
A DIP IN THE MINIMUM NOISE FIGURE OF BEAM-TYPE MICROWAVE AMPLIFIERS
PROC. IRE, VOL. 44, NO. 7, P. 938, JULY 1956.
TIEN, P. K. AND J. MOSHMAN
MONTE CARLO CALCULATION OF NOISE NEAR THE POTENTIAL MINIMUM OF A
HIGH-FREQUENCY DIODE
JOUR. APPL. PHYS., VOL. 27, NO. 9, PP. 1067-1078, SEPTEMBER 1956.
TREYTL, P. AND R. BAIER
LOW-NOISE KLYSTRONS FOR DOPPLER-RADAR MASTER-OSCILLATOR APPLICATIONS
NACHRICHTENTECH. ZEIT., VOL. 25, NO. 1, PP. 13-15, JANUARY 1971.
TWOMBLY, J. C.
SHOT NOISE AMPLIFICATION IN BEAMS BEYOND CRITICAL PERVEANCE
IRE WESCON CONVENTION RECORD, VOL. 1, PT. 3, PP. 156-162, 1957.
TWOMBLY, J. C. AND J. A. VALENT
ELECTRON BEAM NOISE REDUCTION THROUGH EXPLOITATION OF HIGH
ENERGY-STORAGE SENSITIVITIES NEAR LIMITING PERVEANCE
NACHRICHTENTECH. FACHBER. (NTF), VOL. 35, PP. 347-354, 1968.
VAN DUZER, T.
TRANSFORMATION OF FLUCTUATIONS ALONG ACCELERATING CROSS-FIELD BEAMS
IRE TRANS. ELECTRON DEVICES, VOL. ED-8, NO. 1, PP. 78-86, JANUARY
1961.
VAN DUZER, T.
NOISE IN CROSSED-FIELD ELECTRON BEAMS
IN CROSSED-FIELD MICROWAVE DEVICES, E. OKRESS, EDITOR,
ACADEMIC PRESS, NEW YORK, 1961, PP. 327-357.
VAN DUZER, T.
NOISE-FIGURE CALCULATIONS FOR CROSSED-FIELD FORWARD-WAVE AMPLIFIERS
IEEE TRANS. ELECTRON DEVICES, VOL. ED-10, NO. 6, PP. 370-378,
NOVEMBER 1963.
VAN DUZER, T. AND R. D. HARRIS
A PROPOSED MECHANISM FOR THE BROADBAND NOISE IN LONG CROSSED-FIELD
GUNS
JOUR. APPLIED PHYSICS, VOL. 35, NO. 5, PP. 1642-1643, MAY 1964.
VAN DUZER, T. AND J. R. WHINNERY
HIGH-FREQUENCY BEHAVIOR OF THE CROSSED-FIELD POTENTIAL MINIMUM
IRE TRANS. ELECTRON DEVICES, VOL. ED-8, NO. 4, PP. 331-341,
JULY 1961.
VEHN, R. E. AND R. W. PETER
TRAVELING-WAVE TUBE WITH 1.5 DB NOISE FIGURE IN U.H.F. BAND
PROC. IEEE, VOL. 51, NO. 8, PP. 1140-114, AUGUST 1963.
VIVIAN, W. E.
TRANSPORT NOISE AT MICROWAVE FREQUENCIES THROUGH A SPACE-CHARGE-
LIMITED DIODE
JOUR. APPLIED PHYSICS, VOL. 31, NO. 6, PP. 957-962, JUNE 1960.
VLAARDINGERBROEK, M. T.
COMPARISON OF NOISE IN MICROWAVE TRIODES AND IN ELECTRON BEAMS
NACHRICHTENTECH. FACHBER. (NTF), VOL. 22, PP. 399-402, 1961.
VLAARDINGERBROEK, M. T.
NOISE IN ELECTRON BEAMS AND IN FOUR-TERMINAL NETWORKS
PHILIPS RES. REPTS., VOL. 14, NO. 4, PP. 327-336, AUGUST 1959.
VOKES, J. C. AND T. J. BRIDGES
EXPERIMENTS ON THE NOISE PERFORMANCE OF A D.C. PUMPED QUADRUPOLE
AMPLIFIER

SOLID STATE ELECTRONICS, VOL. 4, NO. 135-141, OCTOBER 1962.
VURAL, B.
  BEAM NOISE REDUCTION IN HIGH MAGNETIC FIELDS
    PROC. IEEE, VOL. 53, NO. 5, PP. 510-511, MAY 1965.
    COMMENTS - B. ZOTTER
    PROC. IEEE, VOL. 53, NO. 12, P. 2160, DECEMBER 1965.
WADE, G., K. AMO, AND D. A. WATKINS
  NOISE IN TRANSVERSE-FIELD TRAVELING-WAVE TUBES
    JOUR. APPLIED PHYSICS, VOL. 25, NO. 12, PP. 1514-1520, DECEMBER
    1954.
WADE, G. AND H. HEFFNER
  GAIN, BANDWIDTH AND NOISE IN A CAVITY TYPE PARAMETRIC AMPLIFIER USING
  AN ELECTRON BEAM
    JOUR. ELECTRONICS AND CONTROL, VOL. 5, NO. 6, PP. 497-509,
    DECEMBER 1958.
WADHWA, R. P.
  DIOCOTRON GAIN REDUCTION AND SPACE CHARGE SMOOTHING IN CROSSED-FIELD
  GUNS
    PROC. IEEE, VOL. 52, NO. 3, PP. 315-316, MARCH 1964.
WADHWA, R. P.
  A MODIFIED SHORT KINO GUN FOR LOW NOISE OPERATION
    JOUR. ELECTRONICS AND CONTROL, VOL. 17, NO. 3, PP. 257-265,
    SEPTEMBER 1964.
WADHWA, R. P.
  TRANSFORMATION OF FLUCTUATIONS IN AN INJECTED-BEAM CROSSED-FIELD
  DEVICE
    JOUR. ELECTRONICS AND CONTROL, VOL. 16, NO. 5, PP. 513-536, MAY
    1964.
WADHWA, R. P.
  NOISE IN MAGNETRON INJECTION GUNS
    PROC. IEEE, VOL. 52, NO. 1, P. 79, JANUARY 1964.
WADHWA, R. P.
  NOISE TRANSPORT IN AN INJECTED-BEAM CROSSED-FIELD DEVICE
    INTERNATIONAL JOUR. ELECTRONICS, VOL. 23, NO. 2, PP. 123-134, 1967.
WADHWA, R. P. AND R. D. HARRIS
  TRANSFORMATION OF FLUCTUATIONS ALONG MAGNETRON INJECTION BEAMS
    IEEE TRANS. ELECTRON DEVICES, VOL. ED-12, NO. 6, PP. 332-343,
    JUNE 1965.
WADHWA, R. P. AND V. K. MISRA
  SOME CONSIDERATIONS FOR REDUCTION OF NOISE AND INSTABILITY
  IMPROVEMENT IN HIGH-POWER CROSS-FIELD DEVICES
    IEEE TRANS. ELECTRON DEVICES, VOL. 16, NO. 12, PP. 977-985,
    DECEMBER 1969.
WADHWA, R. P. AND J. E. ROWE
  MONTE CARLO CALCULATION OF NOISE TRANSPORT IN ELECTRIC AND MAGNETIC
  FIELD
    IEEE TRANS. ELECTRON DEVICES, VOL. ED-10, NO. 6, PP. 378-388,
    NOVEMBER 1963.
WADHWA, R. P. AND J. E. ROWE
  EVALUATION OF THE DC PARAMETERS AND NOISE TRANSPORT IN THE GUN REGION
  OF AN INJECTED-BEAM CROSSED-FIELD DIODE
    IEEE TRANS. ELECTRON DEVICES, VOL. ED-11, NO. 4, PP. 170-180,
    APRIL 1964.
WADHWA, R. P. AND T. VAN DUZER
  A 3.5 DB NOISE FIGURE, S-BAND, MEDIUM-POWER, FORWARD-WAVE, INJECTED-
  BEAM CROSSED-FIELD AMPLIFIER
    PROC. IEEE, VOL. 53, NO. 4, PP. 425-426, APRIL 1965.
WADHWA, R. P. AND T. VAN DUZER
  LOW-NOISE MEASUREMENT ON AN INJECTED-BEAM MEDIUM-POWER CROSSED-FIELD
  AMPLIFIER
    INTERNATIONAL JOUR. ELECTRONICS, VOL. 23, NO. 2, PP. 135-152, 1967.
WATKINS, D. A.
  LOW NOISE TRAVELING-WAVE TUBES FOR X-BAND
    PROC. IRE, VOL. 41, NO. 12, PP. 1741-1746, DECEMBER 1953.
WATKINS, D. A.
  TRAVELING-WAVE TUBE NOISE FIGURE
    PROC. IRE, VOL. 40, NO. 1, PP. 65-70, JANUARY 1952.
WATKINS, D. A.
  THE EFFECT OF VELOCITY DISTRIBUTION IN A MODULATED ELECTRON STREAM
    JOUR. APPL. PHYS., VOL. 23, NO. 5, PP. 568-573, MAY 1952.
WATKINS, D. A.
  NOISE AT THE POTENTIAL MINIMUM IN THE HIGH-FREQUENCY DIODE
    JOUR. APPLIED PHYSICS, VOL. 26, NO. 5, PP. 622-624, MAY 1955.
WEN, C. P. AND J. E. ROWE
  MONTE CARLO CALCULATION OF NOISE TRANSPORT IN A TWO-DIMENSIONAL DIODE
    IEEE TRANS. ELECTRON DEVICES, VOL. ED-11, NO. 3, PP. 90-97,
    MARCH 1964.
WESSEL-BERG, T. AND K. BLOTEKJAER
  MINIMUM NOISE TEMPERATURE OF DC-PUMPED TRANSVERSE-WAVE ELECTRON
  BEAM AMPLIFIERS
    IRE TRANS. ELECTRON DEVICES, VOL. ED-9, NO. 5, PP. 388-399,
    SEPTEMBER 1962.
WESSEL-BERG, T. AND K. BLOTEKJAER
  NOISE REDUCTION SCHEMES IN TRANSVERSE MODULATION TUBES
    SOLID STATE ELECTRONICS, VOL. 4, PP. 142-168, OCTOBER 1962.

WHINNERY, J. R.
   HISTORY AND PROBLEMS OF MICROWAVE TUBE NOISE
      SCIENTIA ELECTRICA (SWITZERLAND), VOL. 5, NO. 4, PP. 133-150,
      DECEMBER 1959.
WHINNERY, J. R.
   HIGH-FREQUENCY EFFECTS OF THE POTENTIAL MINIMUM ON NOISE
      IRE TRANS. ELECTRON DEVICES, VOL. ED-7, NO. 4, PP. 218-230,
      OCTOBER 1960.
WHINNERY, J. R.
   NOISE PHENOMENA IN THE REGION OF THE POTENTIAL MINIMUM
      IRE TRANS. ELECTRON DEVICES, VOL. ED-1, NO. 4, PP. 221-237,
      DECEMBER 1954.
YADAVALLI, S.
   LOW-NOISE ELECTRON GUNS AND THE 6L6 EFFECT
      PROC. IRE, VOL. 49, NO. 6, PP. 1098-1099, JUNE 1961.
YADAVALLI, S. V.
   CROSS-CORRELATION BETWEEN VELOCITY AND CURRENT FLUCTUATIONS IN TUBE
   NOISE
      JOUR. APPLIED PHYSICS, VOL. 26, NO. 5, PP. 605-608, MAY 1955.
YAJIMA, H. AND T. FUJIOKA
   EXPERIMENTAL VERIFICATION OF ELECTRON-BEAM NOISE REDUCTION IN
   LOW-VELOCITY REGION
      ELECTRONICS AND COMMUNICATIONS IN JAPAN, VOL. 52, NO. 8, PP.
      86-92, AUGUST 1969.
YAJIMA, H. AND T. FUJIOKA
   NEW METHOD OF MEASURING NOISE PARAMETERS OF ELECTRON BEAMS USING TWO
   FIXED CAVITIES
      IEEE TRANS. ELECTRON DEVICES, VOL. ED-17, NO. 2, PP. 166-168,
      FEBRUARY 1970.
ZACHARIAS, A. AND L. D. SMULLIN
   NOISE REDUCTION IN ELECTRON BEAMS
      IRE TRANS. ELECTRON DEVICES, VOL. ED-7, NO. 3, PP. 172-173, JULY
      1960.
ZOTTER, B. AND C. M. DE SANTIS
   ULTRA-LOW NOISE FIGURE TRAVELING-WAVE TUBE IN X-BAND
      IEEE TRANS. ELECTRON DEVICES, VOL. ED-14, NO. 12, PP. 858-859,
      DECEMBER 1967.
ZOTTER, B. AND B. VURAL
   BEAM NOISE REDUCTION IN HIGH MAGNETIC FIELD
      PROC. IEEE, VOL. 53, NO. 12, P. 2160, DECEMBER 1965.

****
    4-B. NOISE IN OTHER DEVICES   TUTORIAL REVIEWS AND BIBLIOGRAPHIES
****

****
ELECTRON TUBES
****

BURGESS, R. E., ET AL.
   VALVE AND CIRCUIT NOISE. A SURVEY OF EXISTING KNOWLEDGE AND
   OUTSTANDING PROBLEMS
      RADIO RESEARCH SPECIAL REPORT 20, HIS MAJESTY'S STATIONARY
      OFFICE, LONDON, 1951.
HARVEY, A. F.
   MICROWAVE TUBES   AN INTRODUCTORY REVIEW WITH A BIBLIOGRAPHY
      PROC. IEE (GB), VOL. 107, PT. C, NO. 11, PP. 29-59, MARCH 1960.
SMULLIN, L. D. AND H. A. HAUS, EDITORS
   NOISE IN ELECTRON DEVICES
      TECHNOLOGY PRESS OF M.I.T. AND JOHN WILEY, CAMBRIDGE, MA., 1959.

****
IMAGING DEVICES
****

BARBE, D. F.
   IMAGING DEVICES USING THE CHARGE-COUPLED CONCEPT
      PROC. IEEE, VOL. 63, NO. 1, PP. 38-67, JANUARY 1975.
WEIMER, P. K.
   TELEVISION CAMERA TUBES   A RESEARCH REVIEW
      IN ADVANCES IN ELECTRONICS, VOL. 13, L. MARTON, ED.,
      ACADEMIC PRESS, NEW YORK, 1960, PP. 387-437.

****
INFRARED AND OTHER RADIATION DETECTORS
****

ARAMS, F. R.
   INFRARED-TO-MILLIMETER WAVELENGTH DETECTORS (REPRINT COLLECTION)
      ARTECH HOUSE, DEDHAM, MASS., 1973.
HARVEY, A. F.
   BIBLIOGRAPHY OF MICROWAVE OPTICAL TECHNOLOGY
      PLENUM PUBLISHING CORP., NEW YORK, 1976.

HELSTROM, C. W.
    QUANTUM DETECTION THEORY
        IN PROGRESS IN OPTICS , VOL. 10, E. WOLF, EDITOR
        NORTH HOLLAND, AMSTERDAM, 1972, PP. 291-369.
HUDSON, R. D., JR. AND J. W. HUDSON
    INFRARED DETECTORS
        HALSTED PRESS, DIVISION OF JOHN WILEY, NEW YORK, 1975.
JAKEMAN, E., C. J. OLIVER, AND E. R. PIKE
    OPTICAL HOMODYNE DETECTION
        ADVANCES IN PHYSICS, VOL.24, NO. 3, PP. 349-405, MAY 1975.
JONES, R. C.
    NOISE IN RADIATION DETECTORS
        PROC. IRE, VOL. 47, NO. 9, PP. 1481-1486, SEPTEMBER 1959.
MELCHIOR, H., M. B. FISHER, AND F. R. ARAMS
    PHOTODETECTORS FOR OPTICAL COMMUNICATION SYSTEMS
        PROC. IEEE, VOL. 58, NO. 10, PP. 1466-1486, OCTOBER 1970.
PANKRATOV, N. A. AND V. P. KOROTKOV
    CRYOGENIC SEMICONDUCTOR BOLOMETERS
        SOVIET JOUR. OPTICAL TECHNOLOGY, VOL. 41, NO. 2, PP. 106-123,
        FEBRUARY 1974.
VAN DER ZIEL, A.
    NOISE IN MEASUREMENTS
        JOHN WILEY, NEW YORK, 1976.

****
JOSEPHSON JUNCTION AND OTHER SUPERCONDUCTING DEVICES
****

GALLOP, J. C. AND B. W. PETLEY
    SQUIDS AND THEIR APPLICATIONS
        JOUR. PHYSICS E. SCIENTIFIC INSTRUMENTS, VOL. 9, NO. 6, PP. 417-
        429, JUNE 1976.
KAMPER, R. A.
    REVIEW OF SUPERCONDUCTING ELECTRONICS
        IEEE TRANS. MAGNETICS, VOL. MAG-11, NO. 2, PP. 141-146, MARCH 1975.
NAD, F. YA.
    MILLIMETER AND SUBMILLIMETER RECEIVERS USING JOSEPHSON JUNCTIONS
        INSTRUMENTS AND EXPERIMENTAL TECHNIQUES, VOL. 18, NO. 1, PT. 1,
        PP. 1-15, JANUARY-FEBRUARY 1975.
RICHARDS, P. L., F. AURACHER, AND T. VAN DUZER
    MILLIMETER AND SUBMILLIMETER WAVE DETECTION AND MIXING WITH
    SUPERCONDUCTING WEAK LINKS
        PROC. IEEE, VOL. 61, NO. 1, PP. 36-45, JANUARY 1973.

****
LASERS AND MASERS
****

GRAU, G.
    NOISE AND COHERENCE IN THE OPTICAL SPECTRAL RANGE (IN GERMAN)
        IN LASERS , W. KLEEN AND R. MULLER, EDITORS,
        SPRINGER VERLAG, BERLIN, PP. 459-509.
KLIMONTOVICH, YU. L., A. S. KOVALEV, AND P. S. LANDA
    NATURAL FLUCTUATIONS IN LASERS
        SOVIET PHYSICS - USPEKHI, VOL. 15, NO. 1, PP. 95-113, JULY-AUGUST
        1972.
SIEGMAN, A. E.
    THERMAL NOISE IN MICROWAVE SYSTEMS- PARTS I, II AND III.
        MICROWAVE JOURNAL, VOL. 4, NO. 3, PP. 81-90, MARCH 1961.
        MICROWAVE JOURNAL, VOL. 4, NO. 4, PP. 66-73, APRIL 1961.
        MICROWAVE JOURNAL, VOL. 4, NO. 5, PP. 93-104, MAY 1961.
WEBER, J.
    MASERS
        REVIEWS OF MODERN PHYSICS, VOL. 31, NO. 3, PP. 681-710, JULY 1959.

****
MAGNETIC MATERIALS AND DEVICES
****

MALLINSON, J. C.
    TUTORIAL REVIEW OF MAGNETIC RECORDING
        PROC. IEEE, VOL. 64, NO. 2, PP. 196-208, FEBRUARY 1976.

****
MICROWAVE SEMICONDUCTOR DEVICES
****

ANAND, Y. AND W. J. MORONEY
    MICROWAVE MIXER AND DETECTOR DIODES
        PROC. IEEE, VOL. 59, NO. 8, PP. 1182-1190, AUGUST 1971.
COOKE, H. F.
    MICROWAVE TRANSISTORS   THEORY AND DESIGN

PROC. IEEE, VOL. 59, NO. 8, PP. 1163-1181, AUGUST 1971.
GUPTA, M. S.
   NOISE IN AVALANCHE TRANSIT-TIME DEVICES
      PROC. IEEE, VOL. 59, NO. 12, PP. 1674-1687, DECEMBER 1971.
PUCEL, R. A., H. A. HAUS, AND H. STATZ
   SIGNAL AND NOISE PROPERTIES OF GALLIUM ARSENIDE MICROWAVE
   FIELD-EFFECT TRANSISTORS
      ADVANCES IN ELECTRONICS, VOL. 38, L. MARTON, EDITOR
      ACADEMIC PRESS, NEW YORK, 1975, PP. 195-265.

****
SEMICONDUCTOR JUNCTION AND SPACE-CHARGE-LIMITED DEVICES
****

CHENNETTE, E. R.
   NOISE IN SEMICONDUCTOR DEVICES
      IN ADVANCES IN ELECTRONICS AND ELECTRON PHYSICS , L. MARTON, ED.
      ACADEMIC PRESS, NEW YORK, 1967.
NICOLET, M.-A., H. R. BILGER, AND R. J. J. ZIJLSTRA
   NOISE IN SINGLE AND DOUBLE INJECTION CURRENTS IN SOLIDS. I AND II.
      PHYSICA STATUS SOLIDI, VOL. 70, NO. 1, PP. 9-45, 1 JULY 1975.
      PHYSICA STATUS SOLIDI, VOL. 70, NO. 2, PP. 415-438, 1 AUGUST 1975.
STRUTT, M. J. O.
   NOISE IN SEMICONDUCTORS AND SEMICONDUCTOR DIODES
      SCIENTIA ELECTRICA, VOL. 12, NO. 1, PP. 1-32, 1966.
STRUTT, M. J. O.
   SEMICONDUCTOR DEVICES, VOL. 1 - SEMICONDUCTORS AND SEMICONDUCTOR
   DIODES
      ACADEMIC PRESS, NEW YORK, 1967.
AGAJANIAN, A. H.
   SEMICONDUCTING DEVICES   A BIBLIOGRAPHY OF FABRICATION TECHNOLOGY,
   PROPERTIES, AND APPLICATIONS
      PLENUM PUBLISHING CORP., NEW YORK, 1976.

****
GENERAL
****
CHENETTE, E. R. AND K. M. VAN VLIET
   NOISE IN ELECTRONIC DEVICES
      IN METHODS OF EXPERIMENTAL PHYSICS, VOL. 2, PT. B, L. MARTON, ED.,
      ACADEMIC PRESS, NEW YORK, 1975, PP. 461-500.
HEFFNER, H.
   THE PHYSICAL ASPECTS OF LOW NOISE ELECTRONICS
      SOLID STATE ELECTRONICS, VOL. 4, PP. 3-12, OCTOBER 1962.
HYDE, F. J.
   PHYSICAL BASIS OF ELECTRICAL NOISE
      PHYSICAL ASPECTS OF NOISE IN ELECTRONIC DEVICES (CONFERENCE PROC.)
      INSTITUTE OF PHYSICS AND THE PHYSICAL SOCIETY, LONDON, 1968,
      PP. 1-17.
VAN DER ZIEL, A.
   FLUCTUATION PHENOMENA
      IN ADVANCES IN ELECTRONICS, VOL. 4, L. MARTON, ED., ACADEMIC
      PRESS, NEW YORK, 1952, PP. 110-156.

Part V
Generation of Noise

THE SUBJECT

A noise generator, like a signal generator, is a versatile instrument having many laboratory and field applications. Depending upon the application, the noise generator can be a custom-made, expensive, critical, component or something as simple as a music L.P. record on a turntable [1]. Because noise is omnipresent, a very large number of types of devices can, in principle, be used as noise sources, including passive elements, several types of vacuum and gas-filled electron tubes, various semiconductor devices, radiation detectors like photomultiplier tubes and nuclear particle detectors, and others. All of these devices have indeed been proposed as noise generators in the literature, and their feasibility has been experimentally demonstrated. Not all of these devices have, however, reached the stage of common usage and commercial availability as noise sources because a number of requirements are imposed on noise sources in practice.

To be useful, a noise source should be stable, insensitive to operating conditions, reproducible, inexpensive, simple to operate, and should have a desired power output with known amplitude distribution and power spectral density (usually Gaussian and white, respectively). In practice, therefore, only five of the noise sources have attained the status of "noise generators," are commercially available, and are commonly used. These are as follows:

Source	Mechanism	Frequency Range of Use
1) Resistor	Thermal Noise	Subaudio to Millimeter Waves
2) Gas Discharge Tube	Radiation from Plasma	VHF to Millimeter Waves
3) Saturated Vacuum Diode	Shot Noise	Audio to VHF
4) Avalanche Diode	Avalanche Noise	Audio to Microwave
5) Shift Registers	Pseudorandom Noise	Subaudio to RF

These noise generators differ from each other in many respects, such as the frequency range of utility, the amplitude distribution of noise, the noise power output, controllability and variability of output, temperature and bias sensitivity of output, the internal impedance of the generator and its constancy, and the need for a calibration of output power. This last quality is often used to classify the noise sources into two classes: primary (or absolute) standards in which the noise power output is known in advance, and secondary (or transfer) standards in which the noise output must be separately calibrated against a primary or known noise source. There exist only two primary standards: 1) a resistive load maintained at a known temperature for which the thermal noise can be calculated by Nyquist formula, and 2) a vacuum diode operated in the saturation region of its current–voltage characteristic for which the shot noise output can be calculated by Schottky's formula over a limited range of frequencies. In each case, corrections may be necessary to account for imperfect agreement between the actual noise source and the idealized theoretical model to which the formulas are applicable.

Saturated diodes and gas discharge tubes have been used as noise sources (at frequencies up to UHF and above the UHF region, respectively) for over two decades now. By contrast, the digital and avalanche diode noise sources have come into use (at frequencies up to RF and above the RF range, respectively) only within the last decade. The thermal noise of a resistor has served as a standard of noise power and has been particularly useful at very low noise level, as, for example, are encountered in radio astronomical work. In order to place these and other proposed types noise sources (particularly the proliferating varieties at low frequency) into perspective, a classification scheme for noise sources is desirable. The following scheme is proposed for the present purposes. Only the more commonly used noise sources are explicitly included in the chart which can be extended to include others.

herent noise output from a physical device is used directly or after some signal processing. Thus the fundamental noise sources employ the thermal noise of a resistor or a gaseous plasma, the shot noise due to thermionic emission or avalanche ionization, etc. The synthesized noise sources are more common at low frequencies (particularly audio frequencies and below) where the primary noise output is inadequate in power, contaminated with power frequency or other types of hum, or does not have a flat power spectrum due to the presence of $1/f$ noise. The origin of randomness in the output of a synthesized noise source is still a fundamental noise source. A synthesized noise generator starts with the high-frequency output of one of the primary noise generators, and with the aid of some signal processing, produces low-frequency noise. The necessary signal processing can be carried out by two fundamentally different techniques. In one, the high-frequency noise is sampled at a low frequency, possibly after amplification and clipping, usually resulting in a quantized waveform such as a random telegraph signal. In the other, the analog nature of the signal is preserved by frequency translation of high-frequency noise. The frequency translation is effected in a number of

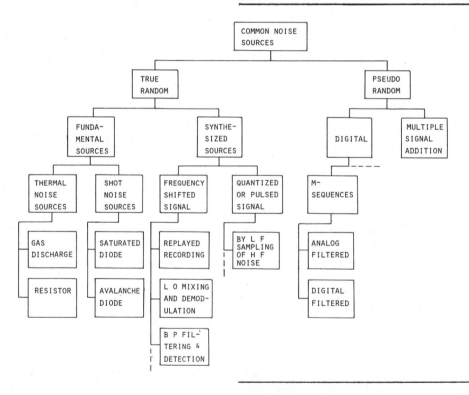

The noise sources are first classified according to the nature of their output power spectrum. A truly random source has no periodic components in its output so that the output power spectrum contains no delta functions, at least in the frequency range of interest. The output of a pseudorandom source, on the other hand, is composed entirely (or primarily) of periodic components, but the separation between delta functions in the frequency spectrum is so small, compared to the frequency range of interest, that the output can be treated as noise.

True random noise sources have been further classified as fundamental or synthesized depending upon whether the in-

well-known ways, such as down conversion by mixing with a local oscillator, detection of the high-frequency noise after narrow-band pass filtering, low-speed replay of a tape or film recording of high-frequency noise, etc.

The pseudorandom noise can be generated, in principle, by both analog and digital techniques, the digital techniques being much more prevalent. Few analog pseudorandom signal sources have been built by directly combining the outputs of a number of low-frequency oscillators, usually of the highly nonsinusoidal type in order to improve the distribution of energy over the frequency band [2]. In the most common digital method, a pseudorandom binary digital signal is generated by a shift register. With an appropriate combination of the signals from some of the flip-flops of the register fed back to the input, the shift register passes through every possible logic state (except the all-zero state). For an n-bit shift register, the binary sequence of maximum length $(2^n - 1)$ bits, called the m-sequence, is thus produced, and serves as a pseudorandom signal with appropriate choice of clock rate and number of bits. The amplitude distribution of this "noise" consists of two delta functions and it can be further filtered, for example by a low-pass filter, to produce a Gaussian (or some other) distribution where necessary. This filtering can be either analog or digital.

The Reprinted Papers

Six different types of noise generators listed in the above classification chart are represented in the five papers reprinted in this part. Only the first of the five is a review paper. The others are not comprehensive review papers surveying the entire literature on the subject, but they all have a significant tutorial value because of their level, approach, or breadth. In two of the five papers, a major part of the discussion is devoted to the measurement of noise power. The close relationship between the development of noise generators and the measurement of noise arises from the fact that one of the important applications of noise generators is to serve as a standard for comparison in the measurement of noise from other sources.

The first paper by Miller, Daywitt, and Arthur is primarily a paper on the precise measurement of noise temperatures, but it contains an extensive review of the literature on the development of thermal, saturated-diode, and gas-discharge noise sources in Section IV. In addition, Section II of the paper gives an idea of why the source impedance and temperature profile are of interest in the design of high precision noise generators. This paper therefore provides the background and motivation for the large number of papers on standard noise generators that are included in the bibliographies.

The second paper by Hart is older than the other four and describes the status of the design of saturated-diode, resistor, and gas-discharge type noise sources. Although the paper is devoted primarily to the work at Philips Laboratories in The Netherlands, it also serves to define the state-of-the-art elsewhere. This state-of-the-art has not advanced appreciably for saturated diodes since the paper was written, at least partly because of the interest in solid-state noise sources. The design of hot and cold passive resistive loads as noise sources has been the subject of considerable work since. The objectives of this work include improving the reproducibility and stability of noise output, reducing the mismatch and reflections, and increasing the frequency range of utility of the noise source. As a result, resistive loads now serve as noise standards up to millimeter wave frequencies. The gas-discharge noise sources have seen improvements in stability, mismatch, and frequency range since then, although to a lesser degree.

Avalanche diodes are a relatively new source of noise and are therefore excluded from the above two survey papers. The third paper by Haitz and Voltmer on avalanche diode noise sources describes the results of an individual effort to determine the noise characteristics of the diodes. While the paper appears to emphasize the experimental verification of theoretical predictions concerning noise in avalanche diodes, it also contains useful information for the design of noise sources. The avalanche diode noise sources have attracted substantial interest because of their large noise output and the other obvious advantages of solid-state devices, such as size, cost, and lifetime. They have also been used as transfer standards of noise.

The last two papers on digital noise sources are categorized under "low-frequency" noise sources. This does not imply a fundamental limitation of the techniques discussed, but only that the digital techniques have so far been used only at low frequencies, typically below 1 MHz. Suggestions have already been made for extending these generators to 1 GHz [3]. No papers are included here on the generation of low-frequency noise by translating the spectrum of a high-frequency noise source. This method is subject to severe stability problems and is not commonly used.

The fourth paper by Sutcliffe and Knott describes, among other things, one possible method of generating a random pulse signal starting from the analog output of a primary noise source. Other, more extensive methods of sampling and shaping primary noise signals are described in some of the papers listed in the bibliography. This paper also contrasts the application of random-pulse secondary noise derived from a primary source and pseudorandom noise of m-sequence type in the measurement of low-frequency noise source spectra. As the paper is not a review paper, it does not discuss the various other methods of sampling a high-frequency noise source, for which the bibliography should be consulted.

The fifth and the final paper by Kramer is a simple and readable description of how the pseudorandom binary sequences of maximum length (m-sequences) are generated and how they approximate a random signal. Once again, this is not a review paper (when this paper was written, there was very little in literature on the subject that could be reviewed). Different algorithms for the generation of maximal-length sequences and the amplitude distribution of the noise generated by this class of noise sources are discussed in other papers included in the bibliography.

The Bibliographies

Seven bibliographies are included in this part.

1) The first, on gas-discharge noise sources, contains papers on the design, construction, performance, and evaluation of these sources, as well as some theoretical and experimental papers on the principle of these sources, i.e., on electron temperature in gaseous plasmas. The subject of fluctuations and oscillations in plasmas is a very extensive one, and is excluded from this bibliography; a review [4] or a book [5] on the subject should be consulted for details. The bibliography deals almost entirely with noise generation at microwave frequencies, although gas-filled electron tubes (thyratrons) have often been used as noise sources at lower frequencies (see, for example, papers by Beecher et al. or Cobine and Curry).

2) The second bibliography is devoted to the use of resistors (passive loads) maintained at suitable temperature as sources of thermal noise. Again, most of the references listed in the bibliography are concerned with microwave and millimeter

wave frequencies, but some (such as those by Knott, Macpherson, or Carsalade *et al*.) dealing with low-frequency noise generation are included. No papers on the "principles" of this noise source are included because the bibliography on thermal noise (in Part III) provides adequate coverage. The thermal noise of a passive device in thermal equilibrium, given by the Nyquist (or the fluctuation-dissipation) theorem serves as a useful noise source at high frequencies, especially in microwave and millimeter wave regions. This type of source has four very important advantages: its noise power output is precisely given by an algebraic expression (Nyquist theorem or fluctuation-dissipation theorem), it can be varied over a significant range by controlling the ambient temperature of the load, its impedance is easier to match over a wider frequency range (compared to gas-discharge sources), and its noise power output is small, which is desirable in the accurate measurement of noise figures of low-noise amplifiers used at microwave frequencies. (This last advantage is also a major limitation of this type of noise sources, precluding its use at other frequency ranges and for many other applications.) The major thrust of the literature on thermal noise sources has been the development of stable, accurately known, and well-matched standard noise generators with precise temperature control. This goal accounts for the close relationship between the subject of noise generation and noise measurement and the measurement orientation in some of the papers included in this bibliography.

3) The third bibliography is a short list of papers on the subject of saturated diodes used as shot noise sources. The age of the papers accurately reflects the state of development of this noise source. The study of shot noise in electron tubes occupied considerable amount of effort and literature more than two decades ago, but a majority of these studies were concerned with the ultimate goal of noise reduction in electron tubes with one or more grids. Those devoted to saturated diodes are nevertheless included in the bibliography to provide the "background" and "principles" of this noise source.

4) The fourth bibliography is on the subject of avalanche diode noise sources. By contrast with all other bibliographies in this part, the listed papers are more recent because this is the most recently developed noise source. Once again, a majority of studies on the noise in avalanche diodes are motivated by the noise limitations of avalanche photodiodes and of microwave avalanche diode amplifiers, oscillators, and mixers; a somewhat dated review [6] of this last field contains further references on that subject. The papers in this bibliography are restricted to those on the design, performance, and applications of avalanche diode noise sources as well as some on the theory and calculation of avalanche noise.

5) The fifth bibliography deals with secondary-noise generators in which the output of a high-frequency primary noise source has been processed to produce a low-frequency random signal. Both frequency-translated and quantized-signal types of generators are included. The papers describe a variety of ways in which this processing can be carried out, and the techniques, actual designs, and measured or calculated performance of such generators.

6) The sixth bibliography pertains to the generation of pseudorandom signals using *m*-sequences. The literature on binary sequences in general is extensive and is not covered in this bibliography; that subject should be approached via some other source, such as the book [7]. The present bibliography is confined to articles relating specifically to the generation of noise by maximum-length pseudorandom sequences. Three of the major goals of work in this area are: 1) development of different algorithms for the generation of *m*-sequences (for example, see Coekin and Wicking), particularly for increasing speeds (i.e., bandwidth of the noise source); 2) evaluation of the different methods of filtering the sequences and the resulting amplitude distributions (see, for example, Matthews); and 3) application of these noise sources in the characterization of systems and the consequences of the pseudorandom character of the signal. Comparisons of pseudorandom and true-random digital noise signals have also been made (for example, papers by Davis and by Sutcliffe and Knott).

7) The final bibliography is formed by merging three short bibliographies containing three kinds of articles: 1) general survey articles dealing with several types of noise sources, 2) noise generators with multiple outputs or for special requirements of power spectrum or amplitude distribution, and 3) miscellaneous noise sources, not covered in any of the earlier bibliographies. These include sources based on photodiodes, photomultiplier tubes, and radioactive materials, as well as electron-beam tubes for very high power noise generation.

REFERENCES

[1] R. Burton and L. D. Hall, "An inexpensive noise source for broadband heteronuclear decoupling experiments," *Canadian J. Chem.*, vol. 48, pp. 2438–2439, 1970.

[2] K. F. Subhani and W. S. Adams, "Low frequency noise sources," *Instruments and Control Syst.*, vol. 43, pp. 113–115, Mar. 1970.

[3] J. A. Coekin and J. R. Wicking, "A new algorithm for the generation of high speed *m*-sequences," *Proc. IREE (Australia)*, vol. 35, pp. 307–309, Oct. 1974.

[4] F. W. Crawford and G. S. Kino, "Oscillations and noise in low-pressure DC discharges," *Proc. IRE*, vol. 49, pp. 1769–1788, Dec. 1961.

[5] G. Bekefi, *Radiation Processes in Plasmas*. New York: Wiley, 1966.
A. G. Sitenko, *Electromagnetic Fluctuations in Plasma*. New York: Academic, 1967.

[6] M. S. Gupta, "Noise in avalanche transit-time devices," *Proc. IEEE*, vol. 59, pp. 1674–1687, Dec. 1971.

[7] S. W. Golomb, *Shift Register Sequences*. San Francisco: Holden-Day, 1967.

Noise Standards, Measurements, and Receiver Noise Definitions

C. K. S. MILLER, W. C. DAYWITT, AND M. G. ARTHUR

Abstract— This paper consists of four sections covering 1) basic principles of noise measurement, 2) the switching radiometer, 3) a survey of noise sources, and 4) concepts of noise factor and noise temperature. The first section presents basic formulas used in analyzing radiometers. The second discusses the switching radiometer, briefly tracing its development and usage in the standards field. The third section surveys the development of hot and cold thermal noise sources, noise diodes, and gas-discharge noise generators. The last section presents and discusses the basic definitions of receiver noise performance.

I. INTRODUCTION

NOISE generally implies any random disturbance which corrupts a desired signal and reduces the certainty with which an observation or measurement may be made. In observing a star, the observer becomes aware of the sensation of the flashing or twinkling. The twinkling is apparently random and is caused by fluctuations in the index of refraction of the atmosphere. In this case the medium fluctuates randomly, perturbing any signal traveling through it, and we have the signal partially obscured by noise. A discriminating look at a continuous sinusoidal signal will reveal a randomness in the amplitude and phase which is in itself a corruption to the signal.

The random quality exhibited by fluctuating or noise-like phenomena requires statistical mathematical methods for its analysis. Generally, one must idealize noise-like characteristics to formulate a mathematical model whose properties approximate those of the natural phenomenon. There are numerous references to which the newcomer can turn for aid [1]–[5].

The intent of this paper is to fill in some of the voids that exist in the literature on the subject of noise measurements. The following sections cover four distinct areas. Section II gives the basic principles of noise measurement, presenting four areas of uncertainty in measuring an effective noise temperature by comparison with a standard. Section III discusses the switching radiometer, briefly tracing its development and usage in the standards field. Section IV surveys the field of reference noise sources. The survey covers the development of heated and cooled resistors and tries to draw from the literature the salient points of consideration in constructing either kind. Noise sources requiring calibration are also considered, and noise diodes are reviewed in the light of attempting to show the need for calibration.

Manuscript received March 7, 1967.
The authors are with the National Bureau of Standards, Boulder, Colo.

Gas-discharge noise sources are covered from their inception to the types of devices currently available commercially. Section V is a discussion of the concepts of noise factor and noise temperature as related to their definitions.

II. BASIC PRINCIPLES OF NOISE MEASUREMENT

Some techniques and formulas for calculating the effects of thermal radiation in communication systems have proved so useful and convenient and of such widespread application that they are worth gathering in one place. These expressions are most easily derived by assuming thermodynamic equilibrium and applying the second law of thermodynamics. However, they have proved to be applicable to many nonequilibrium situations. Their range of applicability includes systems whose components can be specified by some temperature, either effective or physical, and an impedance. The bandwidth of importance is assumed narrow enough to consider the power density constant in frequency. It is to this type of problem that the following noise formulas apply.

Before proceeding to the formulas per se, it is first necessary to review a number of expressions from microwave network analysis [6]. These expressions are, of course, applicable to lower frequency circuits and reflect only a preference for the language of microwave theory.

For immediate convenience, we introduce the circuit in Fig. 1 and restrict ourselves to noiseless components and to a CW signal source.

In Fig. 1, Γ_g, Γ_1, Γ_2, and Γ_L are reflection coefficients, S is the scattering matrix that characterizes the two-port, and P_{1d} and P_{2d} are the net power delivered to the right of reference planes 1 and 2, respectively. Z_M and Z_N are characteristic line impedances which are not necessarily identical. The reflection coefficients and the scattering matrix S are measured at their respective reference planes 1 and 2. Γ_1 is related to the matrix S and Γ_L by the following formula:

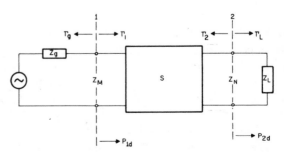

Fig. 1. Circuit schematic used to define M, N, η, and α.

Reprinted from *Proc. IEEE*, vol. 55, pp. 865–877, June 1967.

$$\Gamma_1 = S_{11} + \frac{S_{12}S_{21}\Gamma_L}{1 - S_{22}\Gamma_L}. \qquad (1)$$

A similar expression holds for Γ_2.

There are four derived expressions of interest. They are the mismatch factors M and N associated with the respective reference planes 1 and 2

$$M \equiv \frac{(1 - |\Gamma_g|^2)(1 - |\Gamma_1|^2)}{|1 - \Gamma_g\Gamma_1|^2}, \qquad (2)$$

$$N \equiv \frac{(1 - |\Gamma_L|^2)(1 - |\Gamma_2|^2)}{|1 - \Gamma_L\Gamma_2|^2}; \qquad (3)$$

the efficiency

$$\eta = \frac{P_{2d}}{P_{1d}} = \frac{(Z_M/Z_N)|S_{21}|^2(1 - |\Gamma_L|^2)}{(1 - |\Gamma_1|^2)|1 - S_{22}\Gamma_L|^2}; \qquad (4)$$

and a quantity "α" which is the ratio of the available power at the output to the available power at the input,

$$\alpha \equiv \frac{P_{2a}}{P_{1a}} = \frac{M\eta}{N} = \frac{(Z_M/Z_N)(1 - |\Gamma_g|^2)|S_{21}|^2}{(1 - |\Gamma_2|^2)|1 - S_{11}\Gamma_g|^2}. \qquad (5)$$

Obviously,

$$P_{1d} = MP_{1a} \qquad (6)$$

and

$$P_{2d} = NP_{2a}.$$

Both α and η are measures of the lossiness of the two-port S in that, *if S is lossless,*

$$\eta = 1 = \alpha. \qquad (7)$$

As such, α will turn out to be a direct measure of how much noise the two-port will contribute to the output. Furthermore, α is independent of Γ_L and η is independent of Γ_g.

With the preceding definitions and expressions in mind, it is possible to write very convenient expressions for the relevant noise powers used in radiometer analysis. We assume a constant spectral density across the bandwidth of interest and describe the source in terms of an effective temperature. For example, the available power from an impedance heated to a uniform temperature T is characterized by the physical temperature T itself. That is

$$T \equiv \frac{P_a}{kB} \qquad (8)$$

where k is Boltzmann's constant, B is the bandwidth, P_a the available power, and T will be referred to as the available temperature. It should be kept in mind, however, that T need not be a physical temperature. For example, T may correspond to the available power of an impulse noise source whose physical temperature may be quite different from T. Furthermore, we will adhere to the common but sloppy practice of using the terms "temperature" and "noise power" synonymously.

Figure 2 is typical of many radiometers or portions thereof. In Fig. 2 we have a noise source, either under measurement or being compared to another source of noise power; the source is followed by some two-port, for example an attenuator at ambient temperature T_A or perhaps a section of lossy transmission line. We are interested in either the available power, T, at reference plane 2 or the power delivered, NT, across plane 2 to the receiver. NT consists of two parts: a part that is generated by the noise source and delivered to the receiver, $T_sM\eta = T_sN\alpha$; and a part generated in the two-port itself, $NT_A(1 - \alpha)$. That is

$$\begin{aligned} NT &= M\eta T_S + NT_A(1 - \alpha) \\ &= N[T_S\alpha + T_A(1 - \alpha)]. \end{aligned} \qquad (9)$$

Thus, for example, with proper tuning ($N = 1$), the maximum available power to the receiver is

$$T = T_S\alpha + T_A(1 - \alpha), \qquad (10)$$

and should the component be lossless,

$$T = T_S.$$

It is this last formula for NT that finds its way into so many radiometer applications. To illustrate its use, it will be applied to the comparison of an unknown noise source to a noise standard. In this case the two-port becomes a precision attenuator and, for convenience, we will assume that its input and output are sufficiently isolated from its interior so that S_{11}, S_{22}, and Γ_2 can be considered to be constant as the attenuator is changed from one setting to another. The comparison is described schematically in Fig. 3 [7]. The quantities in Fig. 3 have meanings analogous to those of Figs. 1 and 2. The comparison is made by adjusting α with the standard first connected to the M reference plane (α_S) and then adjusted again when the unknown is connected to this same plane (α_x) so that the receiver samples the same

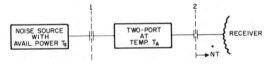

Fig. 2.   Block diagram used to depict the net average noise power, exclusive of receiver noise, delivered to the receiver.

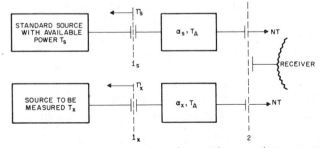

Fig. 3.   Block diagram used to determine an unknown noise temperature T_x by comparison against a known noise temperature T_s.

NT both times. Our assumption of isolation insures that N is the same in both cases and that noise generated in the receiver and reflected back from N remains the same in both cases. Therefore, the effect of receiver noise cancels out and need not appear in the equations. For simplicity, we will assume infinite receiver resolution to changes in NT. The balance equation then reads

$$NT_S\alpha_S + NT_A(1 - \alpha_S) = NT = NT_x\alpha_x + NT_A(1 - \alpha_x),$$

or

$$T_x = (T_S - T_A)\frac{\alpha_S}{\alpha_x} + T_A. \quad (11)$$

Assuming equal characteristic impedances, $Z_{M_S} = Z_{M_x}$ and remembering that Γ_2, S_{11}, and N are constants for the assumed isolation,

$$\frac{\alpha_S}{\alpha_x} = \left|\frac{1 - S_{11}\Gamma_x}{1 - S_{11}\Gamma_S}\right|^2 \left(\frac{1 - |\Gamma_S|^2}{1 - |\Gamma_x|^2}\right)\frac{|S_{21}|^2_S}{|S_{21}|^2_x}. \quad (12)$$

The *phases* of the quantities S_{11}, Γ_x, and Γ_S are seldom if ever measured. Therefore, the first factor in α_S/α_x is uncertain by approximately $2|S_{11}|(|\Gamma_x| + |\Gamma_S|)$ and indicates why the input to the attenuator section is usually tuned ($S_{11} \approx 0$). This first factor causes what is commonly called "mismatch error." Both $|\Gamma_x|^2$ and $|\Gamma_S|^2$ are usually very small and slight inaccuracies in their measurement have little effect on the accuracy of T_x so that the second factor can be considered to be exact. The third ratio can be measured to a high degree of accuracy, but inaccuracy in its measurement will still noticeably affect the accuracy of T_x. The uncertainty in T_x, then, mainly stems from four causes: uncertainty due to mismatch:

$$(\delta T_x)_M \leq \left[1 - \left(\frac{1 + |S_{11}\Gamma_x|}{1 - |S_{11}\Gamma_S|}\right)^2\right]\left(\frac{1 - |\Gamma_S|^2}{1 - |\Gamma_x|^2}\right)\frac{|S_{21}|^2_S}{|S_{21}|^2_x}$$

$$\cdot (T_S - T_A) \quad (13a)$$

$$\approx 2|S_{11}|(|\Gamma_x| + |\Gamma_S|)T_x; \quad (13b)$$

uncertainty in the calibration of the S_{21} ratio:

$$(\delta T_x)_{S_{21}} \leq \left|\frac{1 - S_{11}\Gamma_x}{1 - S_{11}\Gamma_S}\right|^2 \left(\frac{1 - |\Gamma_S|^2}{1 - |\Gamma_x|^2}\right)(T_S - T_A)\delta\left(\frac{|S_{21}|^2_S}{|S_{21}|^2_x}\right) \quad (14a)$$

$$\approx T_x\frac{\delta(|S_{21}|^2_S/|S_{21}|^2_x)}{(|S_{21}|^2_S/|S_{21}|^2_x)}; \quad (14b)$$

uncertainty in the standard temperature T_S:

$$(\delta T_x)_{T_S} \leq \frac{\alpha_S}{\alpha_x}\delta T_S \approx T_x\frac{\delta T_S}{T_S}; \quad (15)$$

and finally, the uncertainty in T_A:

$$(\delta T_x)_{T_A} \leq \left|1 - \frac{\alpha_S}{\alpha_x}\right|\delta T_A \approx \left|1 - \frac{|S_{21}|^2_S}{|S_{21}|^2_x}\right|\delta T_A. \quad (16)$$

III. Discussion of Radiometers for Noise Temperature Comparisons

There are many types of radiometers [8], [9], and they have varied applications. For example, radiometers are used in radio telescopes to chart maps of radio noise sources in outer space, to measure temperatures and temperature variations of stars and sky, to search for icebergs in the oceans, to guide all-weather missiles, to plot temperature profiles of the earth, to study plasmas, and to seek out cancerous areas in human patients. Basically, the radiometer is an instrument to measure noncoherent radio waves.

The radiometer is used as a comparator of noise sources, usually two noise sources, a standard and an unknown. For this purpose the switching radiometer is most frequently used and will be surveyed here. Colvin [8] and Knight [9] give a more general survey including those radiometers excluded in the present discussion.

In 1946, Dicke [10] suggested using a switch in the front end of a radiometer (Fig. 4). Dicke constructed a wheel made of an absorbing material and shaped in such a way that, when rotated by a motor in a slotted section of the waveguide, it produced a nearly square wave modulation with close to equal times in and out of the waveguide. If we assume the wheel and antenna were both nonreflecting, then the effect of the wheel is one of disconnecting the antenna and connecting an equivalent resistance (the absorbing material of the wheel) to the receiver. Note that if the radiation from the wheel and the antenna is the same no change will be noticed on the output meter, while if there is a difference a modulated signal will result and the deflection on the output meter will change.

This scheme was adapted for use in the measurement of standards [11]. The first modification was quite apparent (Fig. 5). The front end of the radiometer was modified so that the switch alternately connected one or the other of the two noise sources, each of which was actually on a separate arm. As depicted, a matching section could be included if so

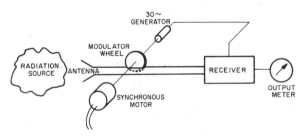

Fig. 4. Schematic diagram of the radiometer used by Dicke.

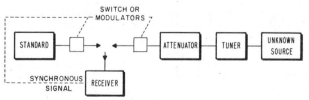

Fig. 5. Block diagram of a switching radiometer used to compare noise sources.

desired. An attenuator was included to lower the noise power level of the unknown, which was usually of considerably higher output than the thermal standard. With subtle differences, and sometimes novel switch designs, this scheme has been employed with a modulator and a synchronous detection signal [12], [13] and with a switch without a synchronous detection signal [14], [15].

The advantage in alternately sampling the two noise sources is that the receiver need only have a stability of the order of time required to complete a full switching cycle. This is extremely valuable since the gain variations and temperature changes encountered are seldom critical for time periods of fractions of a second. The problems caused by receiver instability increase if the switch is of a manual variety and increase even further if the standard and unknown are manually interchanged and connected directly to the receiver [16], [17]. The disadvantages are: 1) the insertion loss of the two paths must be the same or accurately known; 2) asymmetries in the junction of the arms can exist; 3) the switch must be repeatable to the desired degree of accuracy; and 4) the attenuator must have an absolute calibration or the calibration uncertainty in this insertion loss will provide an appreciable error.

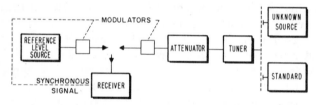

Fig. 6. Block diagram of a modified switching radiometer used for calibration of noise sources.

To improve this type of radiometer, further modifications were made [18]. The modifications (Fig. 6) consist of: 1) using one arm of the previous arrangement as a reference noise source, the only requirement being that it have a stable output over the period of time required to make a measurement, and 2) converting the other arm to a comparison arm, where the comparison of standard and unknown is made. The requirement for the attenuator in the reference arm is that it be stable, and for the attenuator in the comparison arm that it be calibrated for difference attenuation measurements. It should be noted that a difference attenuation calibration is inherently more accurate than an absolute attenuation calibration. In addition, a manually operated switch may be used for ease in connecting standard and unknown to the comparison arm of the radiometer, but the switch must have identical paths (i.e., insertion loss the same) and be repeatable. This modified radiometer has been used by a number of noise metrologists [18], [7], [9], [20], [21]. A further modification of switching the IF rather than the RF signal has also been used successfully [22].

The modified switching radiometer has been analyzed quite extensively and the sources of error are quite well defined [7], [19], [22]. A block diagram (Fig. 7) of the WR90 modified switched radiometer at the National Bureau of Standards is shown. It has a reflectometer incorporated into the comparison arm to simplify tuning the input and measuring the reflection coefficient of the source. To circumvent the problem of matching the attenuator to the transmission line, the attenuator is calibrated with 60 dB of isolation permanently affixed to either side so that the attenuator mismatch effects are included in the calibration.

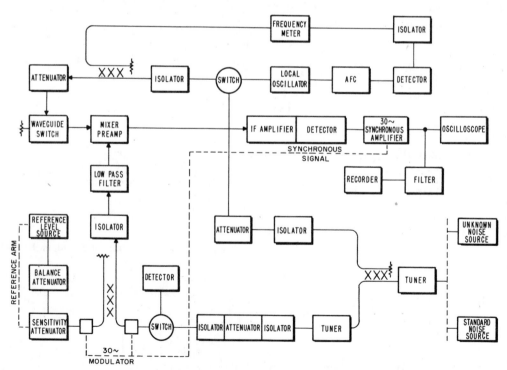

Fig. 7. Block diagram of the NBS modified switching radiometer used to provide microwave noise calibration services.

IV. SURVEY OF REFERENCE NOISE SOURCES

A. The Basic Standard of Noise

Physical measurements of natural phenomena are limited by thermal noise. The noise can be minimized or reduced to the degree that the physical temperature can be reduced. Since thermal noise is directly related to temperature we have the basis of a basic standard. The determination as to whether a thermal noise source can be used as a basic standard lies in whether or not its output is calculable from its dependent parameters.

1) Heated Resistor for Rectangular Waveguides: A heated resistor has been recognized as a thermal noise standard for some time. The difficulty comes in making a practical noise source for use at radio frequencies. A further difficulty arises in applying an analysis to the source which can satisfactorily calculate the noise output.

In 1928, Johnson [23] showed experimentally that a resistor with no externally applied voltage across it has a measurable electrical noise across its terminals. In the same year, Nyquist [24] reported a theoretical calculation of thermal-noise voltage in a resistor. This work has been followed by those who have tried to fabricate practical thermal-noise standards. By present standards some of these attempts may be judged to have been a little crude and somewhat optimistic. Many of these attempts have made contributions to the state of present thermal-noise standard capabilities by the thoroughness of their reporting.

In 1956, Hughes [13] constructed a "hot load" as he called the resistive wedge that he heated to temperatures between 160°C and 230°C. Hughes used three copper-constantan thermocouples (calibrated against a batch of thermometers which were themselves consistent to within 0.5°C) to measure his wedge. The thermocouples were imbedded into the wedge. He reported his error from his heated resistive noise source to be less than 0.5°K. However, Hughes reported no allowances made for the effects of the gradient in the waveguide connecting his noise source to the radiometer.

In the same year, Sutcliffe [14] constructed a resistive wedge which he placed in a steel waveguide to generate thermal noise. He used chromel-alumel thermocouples (calibrated against thermometers to 360°C) and operated the wedge at a nominal 600°C. Sutcliffe did not construct his thermocouples into his wedge so his temperature measurements were taken during simulated rather than operational conditions. He concerned himself with attenuation effects of the gradient in the connecting waveguide but since his measured values in the hot and cold states were so similar he concluded this effect was negligible. Sutcliffe reported that although he could maintain the source temperature to within 1°C, the thermal noise power source was known to only 10°C.

In 1957, Knol [16] constructed a heated resistive noise source using a platinum waveguide and a zinc-titanate based wedge. Knol heated his wedge to 1063°C and reported being able to keep it at the melting point of gold to

within ±1°C. He did not consider the effects of the connecting waveguide; however, he chose platinum to avoid oxidation of the waveguide.

In the same year, Reynard [25] constructed a quartz wedge that was coated with platinum and silicon monoxide. The wedge was placed in a nickel WR 12 waveguide and the operating temperature ranged between 600°C and 700°C. Reynard tried to avoid a waveguide that could become lossy through oxidation. He tried to measure the gradients of the wedge and connecting waveguide as best he could for the size waveguide he was using. He reported that he could maintain the gradient of the load to less than 1°C. There have been other workers who have constructed heated resistors for millimeter wavelengths [26].

In 1958, Sees [27] discussed some considerations and sources of errors to be considered in constructing and using hot-body noise sources. Sees recognized that the waveguide would not only attenuate the noise power generated by the heated resistive wedge but that the waveguide would also generate noise power because of its resistive qualities. Both the attenuation and generation would be functions of the temperature gradient of the waveguide. He correspondingly formulated a method of calculating the resulting macroscopic attenuation and generation effect.

With this background, Estin et al. [18] constructed a hot-body noise source using a silicon-carbide wedge and a gold waveguide. Their oven was a vertical oven and their effort was not completely successful, but it did provide a good beginning for further work at the National Bureau of Standards to develop a hot-body reference noise standard. Later, Wells et al. [7] reported their progress in the development of a hot-body noise source and the National Bureau of Standards announced the availability of a calibration service [28].

Wells et al. used two hot-body noise sources; one was a silicon-carbide wedge in a gold waveguide, the other a zinc-titanate wedge in a platinum-13 percent rhodium waveguide. The shape of the two wedges were different so in effect two entirely different designs were used. They operated the two noise sources over wide temperature ranges. These features showed their standard to be calculable since the differences in configuration did not affect the results, and variations in material and operating temperature could be accounted for as corrections to the noise output to the hot-body noise source. Wells et al. reported having implemented Sees' [27] method of calculating a correction to the noise power output of the resistive wedge due to the resistivity and temperature gradient of the connecting waveguides. They reported the gradient of the wedge to be less than 1°K, but due to the calibration of the thermocouples to ±2.5°K and the temperature effects in the connecting waveguide, they claimed the effective temperature of the load known to an accuracy of ±4.1°K at 1000°C.

It would be remiss not to recognize the work of Birger and Sokov [29], who were apparently unaware of the efforts of all who preceded them with the exception of Sutcliffe. They constructed a wedge of green carborundum on a ceramic base and a connecting waveguide of nickel with internal gold-plating as their thermal noise generator. The operating temperature was a nominal 600°C. They considered the attenuating effects of the connecting waveguide. They constructed a total of four noise generators, one in each waveguide size WR90, WR137, WR187, and WR284, and reported the effective temperature errors not exceeding ±8°C.

Recently, Liedquist [19] reported that the Research Institute of Swedish National Defence (FOA) had purchased a hot-body noise standard from an industrial organization in Japan. The silver-palladium alloy waveguide is gold plated on the inner surface. The reported operating temperature is 1000°K using a single thermocouple for measurement and control. The waveguide necessarily uses a transition from circular to rectangular waveguide. Liedquist reported knowing the wedge temperature to 3°K and the effective temperature to 8°K.

Also, recently, Halford [22] reported the construction of a thermal standard composed of a zinc-titanate wedge in a waveguide composed in part of duralumin and in part of silver plated steel and operated at 400°C. During noise comparisons, temperature measurements are performed by a platinum resistance thermometer located in a channel of the bar which contains the waveguide port. He correlates his thermometer readings with the temperature of the wedge by an auxiliary experiment where he actually measures the wedge temperature with a thermocouple probe. Halford reported knowing the wedge temperature to ±0.5°K and the effective temperature of the source to ±1.5°K.

2) Heated Resistor for Coaxial Waveguide: Halford [22] reported a second effort to build a thermal standard in a coaxial configuration for operation in the 2–4 GHz frequency range. This arrangement uses a carbonyl iron powder bounded by a suitable adhesive cement as the material of the wedge. The wedge is cone shaped with the inner conductor passing through its center and tapered towards the outer conductor. The wedge is operated at 400°C and the temperature measured by a thermometer in an auxiliary channel neighboring the outer conductor. The noise power is coupled to the waveguide by a tunable coaxial to waveguide adapter arrangement. He reported some room for improvement since the coaxial wedge had a local gradient of 3°K.

Zucker et al. [30] built a coaxial heated resistor noise source in 1958. It was designed for use in the 0–1000 MHz frequency range with an operating temperature of 1300°C. The generator was designed with the outer conductor tapered to the inner conductor. The 50-Ω resistive character of the generator was obtained by the deposition of a pyrolytic carbon film on a ceramic base. Temperature measurements of the generator were made by a pyrometer through special ports in the side of the oven. No consideration is recorded by Zucker et al. of the attenuation and generation effects due to the temperature gradient in the connecting coaxial line. They reported that an experienced operator should be able to measure the source temperature to ±3°K.

Gordon-Smith and Lane [20] briefly describe a coaxial heated resistor noise source in a paper published in 1964. They used a 50-Ω metal-oxide resistor having a low temperature coefficient of resistance and an outside diameter of 0.3 cm and length of 1.0 cm. They report that the resistor was heated by immersion into a stirred oil bath maintained at 280°C and that the temperature gradient of the resistor was less than 1°C. Gordon-Smith and Lane considered the attenuation of the coaxial line but not the generation of noise in the line. They reported knowing the effective temperature to ±3.75°K.

Also in 1964 Rusinov and Sorochenko [31] reported the construction of a heated cylindrical carbon resistor for use as a thermal noise standard. The cylindrical carbon resistor is heated from within and is the center conductor as shown in Fig. 8. The outer conductor is composed of two adjoining conical transition sections which are finned for cooling. The resistor is operated at 200–250°C and an accuracy of 3 to 4 percent is claimed.

From the efforts of these authors we recognize that the two major contributors to errors to a heated-resistor noise source are: 1) determining the temperature of the resistor, and 2) evaluating the effects of the connecting transmission line on the output of the resistor.

For determining the effective temperature of the resistor, ideally we would want the resistor to be homogeneous, have a conjugate impedance match, and an infinite thermal conductivity. Similarly, in determining the effects of the transmission line, ideally we would want a stable material with infinite electrical conductivity.

Having less than the ideal resistor, we must measure 1) the gradient in the resistor material to determine an average temperature, and 2) the gradient in the material (from imbedded temperature sensors to the radiating surface) to determine the average temperature of the radiating mass. Unfortunately in physical models the rate of radiation in a transmission line enclosure is not uniform across the radiating surface of the resistor. For one shape of wedge used at the National Bureau of Standards the majority of the power is radiated from a cross section approximately one-third the length of the radiating surface back from the tip of the

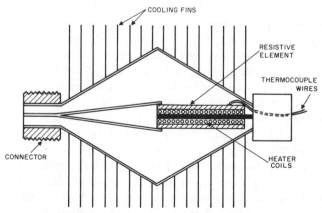

Fig. 8. Schematic diagram of the coaxial thermal noise source used by Rusinov and Sorochenko [31].

wedge. Generally speaking, the lower the temperature gradient, the greater the certainty of the temperature.

Evaluating the effects the transmission line has on the output of the resistor requires 1) knowing the temperature gradient of the transmission line, and 2) knowing the resistive properties of the transmission line as a function of temperature. The attenuation effect of the transmission line can be measured, but for small attenuation values the measurement may be less accurate than calculating the attenuation. This calculation has been programmed for computer solution. To calculate these effects, a material must be selected for the transmission line that will not oxidize at the highest temperatures expected or have uncertain Curie point effects.

The attenuation effect will lower the effective noise temperature while the generation effect will raise the effective noise temperature. These effects can be added because uncorrelated random noise signals are additive. With a heated resistor the attenuation effect is larger than the generation effect; hence, the total effect of the transmission line will be to lower the effective temperature of the resistors. However, considering only the attenuating effect of the connecting transmission line results in a greater error than totally neglecting both attenuation and generation.

3) Cooled Resistor: The noise characteristics of a low-noise system can be more accurately measured by using a cooled resistor as opposed to a heated resistor. Depending on the degree of cooling required, the cooling mechanism will vary. When cryogenic fluids are used, the complexity of the apparatus can compare to that of apparatus used with high temperature resistors.

In 1961, Stelzried [32] constructed a cold resistor in a connecting coaxial line. The coaxial transmission line had both stainless steel conductors coated first with 50 microinches of copper followed by 40 microinches of gold. The resistor was a nominal 50 Ω constructed to give a VSWR less than 1.05 at 960 MHz when immersed in L-He (liquid helium, 4.2°K). He measured the gradient of the coaxial transmission line for differing L-He levels and gives a graph of his calculated equivalent noise temperature for these levels.

In 1962, Mezger and Rother [33] constructed, for the 1–5 GHz frequency range, three configurations where the resistive material was the same and changes were made in the connecting coaxial line. In each case the inner conductor had a 6-mm OD and the outer conductor had a 16-mm ID and was 2 mm thick. One coaxial line was formed from silver-plated brass. A second coaxial line was formed with a 100-mm intermediate section formed from German silver. This section had the wall thickness reduced to 0.5 mm for both conductors (this means the inner conductor was hollow for this section) with the intent of minimizing the heat transfer. The third coaxial line had, in addition to the insulating section, a vented hollow inner conductor to permit the cryogenic fluid to rise to the same level in the inner conductor to that surrounding the outer conductor. Using the first two configurations, the effectiveness of the insulating section using L-N$_2$ (liquid nitrogen) lowers the

noise by 2.5°K. They used the second two configurations to evaluate the effectiveness of cooling as a function of contact surface by changing the depth of L-N$_2$.

In 1964, Eisele [34] constructed a wedge in a section of WR137 waveguide which was cooled to L-He temperatures. The wedge was cut from a single crystal ruby. It was sliced to 0.050 inch, tapered and coated with nichrome and inserted in the connecting waveguide in a manner assuring good thermal contact. For the insulating section, he used a copper-plated stainless steel waveguide. Eisele selected a configuration where the waveguide connection emanated from the bottom of the cryostat. He reported that the noise temperature of the refrigerated termination, excluding transmission line contributions, approached the temperature of the bath to less than 1°K.

In 1965, Stelzried [35] reported having constructed a WR284 waveguide L-He cooled termination. He used a copper-plated stainless steel section as an insulating section and reported an equivalent noise temprature referenced to the connecting flange as 5.0°K.

Also in 1965, Penzias [36] reported having constructed a WR229 waveguide L-He cooled termination as a reference noise source. The waveguide was 90 percent copper brass. The wedge had a pyramid shape and was bathed in L-He. A Mylar septum prevented the L-He from filling the waveguide. Penzias used the same design with L-N$_2$. With the L-N$_2$ he measured a VSWR of 1.05 of the termination compared to 1.04 for L-He. Penzias reported that, for the case of L-N$_2$, the effective equivalent noise output of the device was calculated to be 78.1°K. Penzias measured the temperature gradient of the waveguide by diode thermometers and it appears that he calculated the effects of the connecting waveguide based on Stelzried's [32] earlier work.

In the same year, Jurkus [37] reported having constructed two WR137 waveguide cooled noise sources. The load element, as Jurkus called the resistive material, is made from a resin loaded with iron powder which is shaped in four wedges which fit into the corners of the waveguide. A copper-plated thin-walled 6-inch section of stainless steel waveguide is used as a thermal insulator. The temperature gradient was used to calculate the effect of the connecting waveguide by the method suggested by Wells et al. [7]. Jurkus calculated the effects of the gradient on each of the two noise sources for immersion in L-N$_2$ and L-O$_2$ (liquid oxygen). He also reports on the effect of the depth of immersion in L-N$_2$ on the gradient.

In 1966, Menon and Albaugh [38] reported having constructed three cooled noise sources for use at 2 cm, 9.5 mm, and 3.5 mm. For the thermally isolating section, thin-walled brass and electroformed copper waveguide were used. The resistive element was horizontally positioned through the side of a flask close to the bottom. This configuration was used to reduce the dependence of the effective temperature on the liquid level of the cryogenic fluid (L-N$_2$ in this case).

From the efforts of these authors, we can surmise major difficulties that will need consideration in construction and in determining the errors encountered in using a cooled-resistor noise source. The same two major contributions to errors that were mentioned with heated-resistor noise sources will apply for cooled-resistor noise sources, namely: 1) determining the temperature of the resistor, and 2) evaluating the effects of the connecting transmission line on the output of the resistor.

The same ideal conditions apply as previously stated. With limitations in materials that we can choose for the resistor, we can plan for the most effective thermal contact with the coolant as possible. This means keeping the resistor immersed deep enough or bathed in the coolant. The temperature of the coolant must be known. The boiling point of cryogenic coolants are well defined and they are dependent on pressure. Altitude changes can effect the boiling points of cryogenic fluids and, to a lesser degree, so can local barometric pressure changes. Designs for resistors that provide a reflection coefficient magnitude Γ or VSWR that is a minimum and is independent of temperature are preferred.

The effect the transmission line has on the effective noise temperature of the resistor requires 1) knowing the temperature gradient of the length of transmission line, 2) knowing the resistive properties of the transmission line as a function of temperature, and 3) knowing that the transmission line is free of any condensates. The same comments made previously in regard to heated resistors still apply. The attenuation and generation effects can probably be calculated more accurately than measured for known resistive properties and a measured temperature gradient of the transmission line. The same computer program solves the gradient effects of cooled loads. With cooled resistors, the connecting transmission line has the effect of raising the effective noise temperature output of the resistor, since the generation of noise in the transmission line is greater than the attenuation. Here, too, considering only one effect results in a greater error than in neglecting the total effect.

Keeping the transmission line free of condensates is an additional significant problem. Evacuating, purging, or pressurizing, or some combination of these three, is necessary to keep the transmission line free. For waveguide transmission lines, a window of some kind is usually necessary; polystyrene [35], Teflon [37], or Mylar [34], [36], [38], have been used. A configuration where changing coolant liquid level does not change the effective temperature is preferred. The transmission line connector should be considered as a source of error and has not been identified as such by any of these authors. The new 14-mm and 7-mm precision connectors for coaxial line are improved over the Type N. Still these precision connectors and good flanges (in waveguide) can be serious sources of error to cooled resistors.

B. Noise Sources Needing Calibration or Comparison to a Standard

Noise sources that do not have completely calculable outputs cannot qualify as primary standards. If, however, they are stable, repeatable from one usage to another, and have bandwidths broad enough that they can be used to make noise measurements, then they are valuable in that they may be compared to a standard and so function as an interlaboratory standard. In the case where no absolute standard exists per se, then obviously no comparison can be made. This is a condition that had existed in noise for quite a long time but is now being remedied. The noise sources being used for measuring noise figure, noise factor, and effective input noise temperature need a value and, in the absence of a standard, one of two procedures has been followed. Either a single noise source was isolated and assigned a value and in turn used to assign similar values to other comparable noise sources, or each noise source used had an output that was calculated by an agreed method based on operating conditions and known parameters. The agreed method of calculation needs to be uniformly adapted by all users so that comparable devices have comparable outputs. Both of these methods are still in use in the noise field where standards are lacking. Unfortunately the adoption of these methods can encourage apathy toward the development of standards, and sometimes measurements are attempted with inadequate standards. In this light it may be said that the noise source field is undergoing growing pains.

The noise sources generally used in the field are of two major varieties: a) the noise diode, and b) the gas-discharge noise source. The noise diode is used to around 500 MHz and can be extended in frequency range to higher frequencies with some modifications. The gas-discharge noise source is a comparatively new addition to the noise field and it has been used successfully from 300 MHz and into the millimeter-wave regions. A pleasant discussion of these two noise sources is to be found in Hart [17]. Maxwell and Leon [39] recognized the need for calibrating noise diodes and used a hot-cold resistive standard in 1956. Prinzler [40] used a heated resistor as a standard for his calibration of noise diodes in 1958.

1) The Noise Diode: A noise diode has an output that is generally referred to as shot noise. Shot noise normally implies a fluctuating current, although it could be applied to the fluctuating voltage developed across a conductor through which shot noise current flows. Schottky [41] observed that a flow of electrons from a thermionic cathode was irregular because each emission was a random, independent event. The fluctuating current caused by this irregular flow of electrons he called the "Schroteffekt," from which we get the term "shot noise."

To cover the literature written about the noise diode would be an extensive work in itself. There has been a bibliography [42] made of many of the published papers in noise prior to 1954, so only a few pertinent papers will be referenced.

Tien and Moshman [43], through use of a mathematical model based on the generation of random numbers by the Monte Carlo method, simulated the electron emission at the cathode of a noise diode. Using this, they tried to predict results at low frequencies and high frequencies. They felt the results agreed well with existing theory for low frequencies, but found no correlation at high frequencies. Kosmahl [44] tried to measure the correlation between the induced grid noise current and the fluctuations in the output, but found none.

Most of the noise diodes presently on the market have been described in the literature [45] at one time or another. These devices have their own characteristics as a function of their tube geometry. They are often frequency sensitive and have differing emitter characteristics depending on the material and construction of the emitter. They have noise output characteristics that have some predictability, and it depends on how well the output must be known as to how much faith you should put in the calculated output. In England, a noise generator designed by Harris [46] is employed as an interim noise standard.

The Soviet Union established a thermal noise standard in 1964 in the decimeter bands for the calibration of noise diodes and gas-discharge noise sources [31]. The National Bureau of Standards expects to offer a calibration service at 3 MHz based on a thermal noise standard before the end of 1967.

2) The Gas-Discharge Source: The first suggestion to use a gas-discharge tube as a noise source was made by Mumford [47]. Mumford first used a fluorescent lamp mounted in the *H*-plane of a waveguide. The fluorescent lamp, he reported, gains most of its noise radiation from the mercury gas-discharge in the tube. Mumford used lamps with 10 different types of fluorescent coatings and one germicidal lamp (i.e., no fluorescent coating) to find that the fluorescent coating was not the source of the noise since all of them had comparable noise outputs. He concluded that it was the gas-discharge alone that produced the noise.

The fluorescent lamp noise source appears to have been quite temperature sensitive and some effort was made trying to predict the noise output as a function of mount temperature [48], [49]. The statement has been made that "the noise power, in general, is derived both from the thermal velocities, which are characterized by the electron temperature, and from the dc power, which is characterized by the average current" [50]. The facility of being able to calculate the electron temperature and identifying it with the noise temperature when applied to pure gas discharges provided the basis for suggesting that gas-discharge noise sources be used as microwave noise standards [15]. The trouble is that for pure gas discharges the measured values obtained [51]-[53] did not seem to always support these claims; in fact, the noise temperature is usually less than the electron temperature. These discrepancies could mean that there exist unaccounted for effects [11], [54], or even that the noise temperature as measured is not simply related to the electron temperature. Consequently, gas-discharge noise sources need calibration against a standard to ascertain a numeric value that will classify the noise output.

Johnson and Deremer [53] had trouble extending Mumford's design [47] to 10 GHz, since this design gave a drastically reduced bandwidth of the noise source. They found that with a tube mounted in the *E*-plane of the waveguide at 10° to the axis of the waveguide that the VSWR was much lower than a comparable design using the *H*-plane. Decreasing the angle lowered the VSWR, but lengthened the elliptic waveguide cut to accommodate the tube and also meant the tube itself had to be lengthened. No optimum design was resolved with these limitations. John-

son and Deremer suggested a 10° *E*-plane mount for general use and suggested that a smaller angle could be selected for more exact measurements. In their experiments they varied the gas fill presure from 3 to 30 mm of Hg and tried different gases for gas fill in the tubes. They found gas mixtures minimized problems with fluctuations in the plasma, but also appeared to slightly increase the VSWR. Tubes were designed to have a discharge diameter of 1/4 to 1/3 of the inside width of the waveguide and to operate at discharge currents of 200–250 mA which meant gas fill pressure was to be 20–30 mm of Hg. To eliminate the dependence of noise output on the operating temperature, only inert gases were used.

Originally, Johnson and Deremer designed a series of mounts for use in the 3 to 30 GHz range. This design has been available commercially and has only been modified by manufacturers to satisfy their own specifications. We note that Knol [51] independently built a 10° *E*-plane mount for his own use at about the same time. Others in the field have extended this design for general usage and to increase the upper frequency limits [55], [25], [56], [57].

More recently, it was found that the tube in the typical Johnson and Deremer mount design, if once removed, could not easily be replaced and still repeat the effective noise temperature originally measured [58]. Other weaknesses in current commercial design were noted and an improved mount design was suggested with a 9° insertion angle [58] and a 7° insertion angle [22]. Further, an improvement in tube design is also suggested [22].

To extend the useful range of the gas-discharge noise source to lower frequencies, Johnson and Deremer [53] tried to construct tube shapes that would fill the entire cross section of the coaxial transmission line with a gaseous discharge. They did not regard this work as the most successful but could not pursue the matter due to the termination of the contract. They were aware of a contractural program designed to construct a coaxial helical mount and did not pursue this avenue since it would have been a duplication of effort.

To build a good helical-type coaxial mount we need sufficient coupling between the gas-discharge and the transmission line, and a good impedance match. The helical center conductor is a slow-wave structure as is a lumped-constant low-pass filter. Spencer and Strum [59] believed it was possible to make a coaxial mount on the helix idea that would be usable from 30–5000 MHz. Such a design hinged upon whether the gas tube could be constructed long enough for the necessary length of helix. Spencer and Strum give a lot of design criteria in their paper and state they built a mount having a flat noise output from 200 MHz to 3 GHz. Using the low-pass filter concept they built a coaxial source usable from 50 to 300 MHz. Others have reported having successfully built coaxial helical mounts for gas-discharge tubes [39], [60]-[63].

Alma'ssy and Frigyes [64] present a novel suggestion on building a mount for a gas-discharge noise source. They constructed a coaxial transmission line mount where the inner conductor is replaced by a gas tube having comparable dimensions to the inner conductor. "The filament voltage reaches the tube through a broadband metal extension while the anode is directly connected to the inner conductor." One end may be terminated, thereby simplifying the construction. The device was built, and with a discharge current of 120 mA no difference in output was detected between six tubes. The device was operated between 1800 and 2200 MHz. Another different idea was to construct a gas-discharge tube in two-conductor lines [65].

At present the National Bureau of Standards has the capability to perform calibrations in the frequency range 8.2 to 18.0 GHz, which is covered by two waveguide sizes (WR90 and WR62). The National Bureau of Standards expects to perform calibrations in the 2.60 to 3.95 GHz frequency range (waveguide size WR284) in the immediate future. The remaining frequencies have no standards available to them for calibration. To take up this void Lee and Olsen [66] devised a technique where, by fabricating special tubes, they made transitions between bands.

V. Concepts of Noise Factor and Noise Temperature

A. Introduction

In 1942, North wrote a paper on "The Absolute Sensitivity of Radio Receivers" [67]. This is perhaps the earliest paper that proposed a standardized measurand for the noise performance of a radio receiver. The author suggested a number, N, that he termed "noise factor," as being a fundamental description of the internal noise of a receiver. In 1944, Friis [68] wrote a paper in which he discussed "noise figure" at greater length, and this paper is considered by many to lay the foundation for this quantity. Since 1944, many papers on the subject of the noise performance of amplifiers and receivers have appeared. Such papers have been stimulated by the need to describe this aspect of devices, and by the need to clarify the meaning and application of the noise factor concept.

In recent years, the noise performance of receivers and amplifiers has been described in terms of an effective input noise temperature [69], T_e. For many persons, T_e is a more meaningful measurand than is noise factor, F. However, for a device for which both quantities are defined, a simple relationship exists between the two (23).

The concept of noise temperature comes quite naturally from the relationship between the noise power P_a available from a resistor at a uniform temperature T, which is

$$P_a = kT \text{ watts per hertz.} \qquad (17)$$

For a device having a noise power P_a per unit bandwidth available from one of its ports, a noise temperature of T degrees Kelvin can be associated with that port. Thus the internal noise of an amplifier or receiver, or in fact of any network containing a source of noise, can be expressed in terms of a noise temperature.

B. Noise Factor (Noise Figure)

The noise factor (noise figure), $F(f)$, of a linear two-port, at a specified input frequency f is defined as the ratio of 1) the total noise power per unit bandwidth N_t at a corresponding output frequency available at the output port when the noise temperature of the input termination is standard (290°K), to 2) that portion of 1), N_s, that is engendered at the input frequency of the input termination at the standard noise temperature (290°K) [70]. Thus

$$F(f) = \frac{N_t}{N_s}. \qquad (18)$$

Although this definition is concise, complete, and useful, it must not be misapplied or confused with the more than fifteen additional and different types of noise factor which may be found in the literature.

An alternate definition of noise factor is based upon the concept of signal-to-noise ratio [68]. If S_i/N_i and S_0/N_0 are the signal power-to-noise power ratios at the input port and output port of the linear two-port, respectively, when the input input termination is at 290°K, F is defined by the relationship

$$F = \frac{S_i/N_i}{S_0/N_0}. \qquad (19)$$

Thus, noise factor serves most directly as a measure of the extent to which the noisy two-port degrades the signal-to-noise ratio of the input power.

Equation (19) also shows that, in order to produce a unity output signal-to-noise ratio when the source temperature is 290°K, a signal power S_i' is required where

$$S_i' = FkT_0 \text{ watts per hertz} \qquad (20)$$

and

$$T_0 = 290°\text{K}. \qquad (21)$$

Noise factor, $F(f)$, as defined by the IEEE is a rather idealized concept. It describes the noise performance of a *linear two-port* at a *single* operating frequency. To describe the noise performance over a *band* of frequencies, an average noise factor, $\bar{F}$, is defined [70]. To distinguish the two, the term spot noise factor [70] is sometimes used to identify the single frequency factor. Noise factor is not defined for a nonlinear two-port, nor for a multiport transducer [see Section V-C].

Noise factor is a hyperbolic function of source impedance. It reaches a minimum value at some value of source impedance called the "optimum source impedance." This minimum noise factor is called the optimum noise factor [71], and is independent of source impedance.

Noise factor is based on the temperature of the source impedance being standard, a condition that does not always prevail in practice. Thus, the noise performance of a transducer will differ from that predicted by its noise factor when it operates in a system where the temperature of its input termination is not standard. Furthermore, since the experimental procedures for measuring F can be carried out using source impedance temperatures that are different from 290°K, the quantity that is often measured is not F but, rather, an analogous quantity F'. F' is defined the same way as F except that the noise temperature of the input termination, T_s, is taken to be that which prevails during the measurement. It can then be shown that

$$(F' - 1)T_s = (F - 1)T_0 \qquad (22)$$

so that

$$F = 1 + (F' - 1)\frac{T_s}{T_0}. \qquad (22a)$$

In practice, effort is made to reduce F to its minimum value of unity, thus making the two-port as near as possible to an ideal, noise-free transducer.

C. Effective Input Noise Temperature, T_e

The effective input noise temperature T_e (of a multiport transducer with one port designated as the output port) is defined as the noise temperature in degrees Kelvin which, assigned simultaneously to the specified impedance terminations at the set of frequencies contributing to the output, at all accessible ports except the designated output port of a noise-free equivalent of the transducer, would yield the same available power per unit bandwidth at a specified output frequency at the output port as that of the actual transducer connected to noise-free equivalents of the terminations at all ports except the output port [69].

To help understand this definition, two block diagrams are shown in Fig. 9. In the upper diagram, the input ports of the *noise-free equivalent* of the multiport transducer are terminated in source impedances, each at temperature T_e. The power per unit bandwidth available from the transducer is P_L. In the lower diagram, the input ports of the *noisy* multiport transducer are terminated in source impedances of the same impedance values as before, but each such input termination is noise-free (noise temperature of $0°K$). The power per unit bandwidth available from this transducer is also P_L. The value of T_e necessary to produce the same value of P_L in both diagrams is equal to the value of the effective input noise temperature of the (actual) noisy multiport transducer.

Note that each of the diagrams of Fig. 9 requires the use

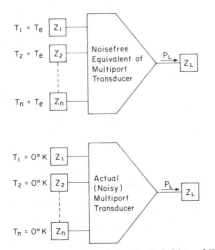

Fig. 9.   Diagrams illustrating the definition of T_e.

of fiction (e.g., noise-free equivalent components). However, the real quantity, T_e, of the real, noisy multiport transducer is established through the use of this fiction. Furthermore, no difficulty is presented to the measurement of T_e through this use.

The definition of T_e pertains to output power at a single frequency. As with noise factor, an average effective input noise temperature, T_e, is defined to describe the noise performance of a multiport transducer over a band of output frequencies [70].

For a linear two-port transducer with a single input and a single output frequency, T_e is related to F [70] by

$$T_e = 290 (F - 1). \qquad (23)$$

Thus the dependence of T_e upon source impedance is also a hyperbolic function. An optimum value of T_e occurs when the input ports are terminated with source impedances that minimize T_e; this optimum T_e is independent of source impedance, and is characteristic of the transducer alone.

T_e is a function of frequency, but is not a function of the temperatures of the input terminations, except that in a nonlinear transducer it may be a function of signal level.

D. Operating Noise Temperature, T_{op}

In a system of transducers, such as may be comprised of one or more sources (e.g., an antenna) and one or more amplifiers, filters, and frequency translators connected in cascade, a useful figure of merit is the operating noise temperature, T_{op}, of a *system* under operating conditions. T_{op} is defined as the temperature in degrees Kelvin given by

$$T_{op} = \frac{N_0}{kG_s} \qquad (24)$$

where N_0 is the output noise power per unit bandwidth at a specified output frequency flowing into the output circuit (under operating conditions), k is Boltzmann's constant, and G_s is the ratio of 1) the signal power delivered at the specified output frequency into the output circuit (under operating conditions), to 2) the signal power available at the corresponding input frequency or frequencies to the system (under operating conditions) at its accessible input terminations [67].

In a linear two-port transducer with a single input and a single output frequency, if the noise power originating in the output termination and reflected at the output port can be neglected, T_{op} is related to the noise temperature of the input termination (source impedance), T_i, and the effective input noise temperature, T_e, by the equation

$$T_{op} = T_i + T_e. \qquad (25)$$

An average operating noise temperature, $\overline{T}_{op}$, is defined [69] when performance over a band of output frequencies is required.

In practice, a system with the smallest T_{op} will provide the best performance from the standpoint of noise degradation of information.

E. Measurement Techniques for $\bar{F}$ and $\bar{T}_e$

Various methods have been devised for measuring the noise factor and/or effective input noise temperature of transducers. A fairly complete description of these methods has been covered previously [70], [72].

REFERENCES

[1] W. R. Bennett, *Electrical Noise*. New York: McGraw-Hill, 1960.
[2] D. K. C. MacDonald, *Noise and Fluctuations: An Introduction*. New York: Wiley, 1962.
[3] A. Van der Ziel, *Noise*. New York: Prentice-Hall, 1954.
[4] J. J. Freeman, "Electrical noise," *Electro-Technology*, pp. 126–144, November 1960.
[5] W. B. Davenport and W. L. Root, *An Introduction to the Theory of Random Signals and Noise*. New York: McGraw-Hill, 1958.
[6] D. M. Kerns and R. W. Beatty, *Basic Theory of Waveguide Junctions and Introductory Microwave Network Analysis*. New York: Pergamon, 1967.
[7] J. S. Wells, W. C. Daywitt, and C. K. S. Miller, "Measurement of effective temperatures of microwave noise sources," *IRE Internat'l Conv. Rec.*, pt. 3, pp. 220–238, March 1962; also in *IEEE Trans. on Instrumentation and Measurement*, vol. IM-13, pp. 17–28, March 1964.
[8] R. S. Colvin, "A study of radio-astronomy receivers," Stanford Radio-Astronomy Inst., Publ. 18A, October 31, 1961.
[9] J. Knight, "Evaluation and analysis of radiometers," thesis, University of Toronto, Canada, September 1961.
[10] R. H. Dicke, "The measurement of thermal radiation at microwave frequencies," *Rev. Sci. Instr.*, vol. 17, pp. 268–275, July 1946.
[11] L. W. Davies and E. Cowcher, "Microwave and metre wave radiation from the positive column of a gas-discharge," *Aust. J. Phys.*, vol. 8, no. 1, pp. 108–128, 1955.
[12] J. J. Freeman, "Noise comparator for microwave," *Television News (Radio-Electronic Engineering Section)*, vol. 49, p. 11, March 1953.
[13] V. A. Hughes, "Absolute calibration of a standard temperature noise source for use with S-band radiometers," *Proc. IEE (London)*, vol. 103, pt. B, pp. 669–672, September 1956.
[14] H. Sutcliffe, "Noise measurements in the 3-cm waveband using a hot source," *Proc. IEE (London)*, vol. 103, pt. B, pp. 673–677, September 1956.
[15] K. W. Olsen, "Reproducible gas-discharge noise sources as possible microwave noise standards," *IRE Trans. on Instrumentation*, vol. I-7, pp. 315–318, December 1958.
[16] K. S. Knol, "A thermal noise standard for microwaves," *Philips Research Repts.*, vol. 12, pp. 123–126, April 1957.
[17] P. A. H. Hart, "Standard noise sources," *Philips Tech. Rev.*, vol. 23, pp. 293–324, July 1962.
[18] A. J. Estin, C. L. Trembath, J. S. Wells, and W. C. Daywitt, "Absolute measurement of temperatures of microwave noise sources," *IRE Trans. on Instrumentation*, vol. I-9, pp. 209–213, September 1960.
[19] L. Liedquist, "Absolutkalibrering av bruskällor inom Mikrovågsområdet" (Absolute calibration of noise sources in the microwave field), *Elteknik*, vol. 9, pt. 3, pp. 43–47, March 1966.
[20] A. C. Gordon-Smith and J. A. Lane, "Standardization of coaxial noise sources," *Electronic Letts.*, vol. 1, pp. 7–8, March 1965.
[21] R. A. Andrews, Jr., "Calibration of gas-discharge noise sources in an industrial standards laboratory," presented at the 20th Ann. ISA Conf., Los Angeles, Calif., October 1965, Preprint 42.1-1-1965.
[22] G. J. Halford, "Noise comparators and standards for S and X bands," *IEEE Trans. on Instrumentation and Measurement*, vol. IM-15, pp. 310–317, December 1966.
[23] J. B. Johnson, "Thermal agitation of electricity in conductors," *Phys. Rev.*, vol. 32, p. 97, July 1928.
[24] H. Nyquist, "Thermal agitation of electric charge in conductors," *Phys. Rev.*, vol. 32, p. 110, July 1928.
[25] A. I. Reynard, "Precision instruments for calibrating radiometers at 4.3 millimeters wavelength," U. S. Naval Research Lab., Rept. 4927, May 1957.
[26a] W. Jasinski and G. Hiller, "Determination of noise temperature of a gas-discharge noise source for four-millimeter waves," *Proc. IRE (Correspondence)*, vol. 49, pp. 807–808, April 1961.
[26b] Q. V. Davis, "A high temperature termination for use at short

[27] J. E. Sees, "Fundamentals in noise source calibrations at microwave frequencies," U. S. Naval Research Lab., Rept. 5051, January 1958.
[28a] "Calibration of microwave noise sources," *NBS Tech. News Bull.*, vol. 47, pp. 31–34, February 1963.
[28b] "Calibration of waveguide noise sources," *Federal Register*, vol. 28, p. 7639, July 26, 1963.
[29] L. A. Birger and I. A. Sokov, "Reference thermal noise generators: high and ultra-high frequency measurements," *Izmeritel. Tekhn.*, pp. 47–50, January 1962. Translation in *Meas. Tech (USSR)*, 1962.
[30] H. Zucker, G. Baskin, S. I. Cohn, J. Lerner, and A. Rosenblum, "Design and development of a standard white noise generator and noise indicating instrument," *IRE Trans. on Instrumentation*, vol. I-7, pp. 279–291, December 1958.
[31] Yu. S. Rusinov and R. L. Sorochenko, "Primary standard of noise radiation in the decimeter band," *Pribory Tekh. Eksper. (USSR)*, pp. 121–122, May–June 1964. Translation in *Instrum. Exper. Tech. (USA)*, pp. 612–613, May–June 1964.
[32] C. T. Stelzried, "A liquid-helium-cooled coaxial termination," *Proc. IRE (Correspondence)*, vol. 49, p. 1224, July 1961.
[33] P. G. Mezger and H. Rother, "Kühlbare Eichwiderstände als Ranschtemparaturnormale" (Coolable calibrating resistance as noise temperature standards), *Frequenz*, vol. 16, no. 10, pp. 386–391, 1962.
[34] K. M. Eisele, "Refrigerated microwave noise sources," *IEEE Trans. on Instrumentation and Measurement*, vol. IM-13, pp. 336–342, December 1964.
[35] C. T. Stelzried, "Temperature calibration of microwave thermal noise sources," *IEEE Trans. on Microwave Theory and Techniques (Correspondence)*, vol. MTT-13, pp. 128–130, January 1965.
[36] A. A. Penzias, "Helium-cooled reference noise source in a 4-kMc waveguide," *Rev. Sci. Instr.*, vol. 36, pp. 68–70, January 1965.
[37] A. Jurkus, "Cold loads as standard noise sources," *Proc. IEEE (Correspondence)*, vol. 53, pp. 176–177, February 1965.
[38] R. C. Menon and N. P. Albaugh, "Cooled loads as calibration noise standards for the mm-wavelength range," *Proc. IEEE (Correspondence)*, vol. 54, pp. 1501–1503, October 1966.
[39] E. Maxwell and B. J. Leon, "Absolute measurement of receiver noise figures at UHF," *IRE Trans. on Microwave Theory and Techniques*, vol. MTT-4, pp. 81–85, April 1956.
[40] H. Prinzler, "A saturated diode noise source for 20 cm wavelengths and its absolute calibration with a heated resistance," *Nachrichtentechnik*, vol. 8, pp. 495–500, November 1958.
[41] W. Schottky, "Uber Spontane Stromschwankungen in Verschiedenen Elektrizitatsleitern," *Ann. Phys. (Leipzig)*, vol. 57, pp. 541–567, December 1918.
[42] P. L. Chessin, "A bibliography of noise," *IRE Trans. on Information Theory*, vol. IT-1, pp. 15–31, September 1955.
[43] P. K. Tien and J. Moshman, "Monte Carlo calculations of the noise near the potential minimum of a HF diode," *J. Appl. Phys.*, vol. 27, pp. 1067–1078, September 1956.
[44] H. Kosmahl, "Correlation factors for noise fluctuations at the potential minimum of a diode (triode)," *Arch. Elekt. Übertragung*, vol. 10, pp. 353–357, August 1956.
[45a] R. W. Slinkman, "Temperature-limited noise diode design," *Sylvania Technol.*, vol. 2, pp. 6–8, October 1949.
[45b] H. Johnson, "A coaxial-line diode noise source for UHF," *RCA Rev.*, vol. 8, pp. 169–185, March 1947.
[45c] H. Groendijk, "A noise diode for ultra-high frequencies," *Philips Tech. Rev.*, vol. 20, pp. 108–110, 1958–1959.
[45d] R. Kompfner et al., "The transmission-line diode as a noise source at centimeter wavelengths," *J. IEE (London)*, vol. 93, pt. IIIA, p. 1436, 1946.
[46] I. A. Harris, "The design of a noise generator for measurements in the frequency range 30–1250 Mc/s," *Proc. IEE (London)*, vol. 108B, November 1961.
[47] W. W. Mumford, "A broad-band microwave noise source," *Bell Sys. Tech. J.*, vol. 28, pp. 608–618, October 1949.
[48] E. L. Chinnock, "A portable, direct-reading microwave noise generator," *Proc. IRE*, vol. 40, pp. 160–164, February 1952.
[49] W. W. Mumford and R. L. Schafersman, "Data on the temperature dependence of X-band fluorescent lamp noise sources," *IRE Trans. on Microwave Theory and Techniques*, vol. MTT-3, pp. 12–17, December 1955.
[50] P. Parzen and L. Goldstein, "Current fluctuations in the direct-

current gas-discharge plasma," *Phys. Rev.*, vol. 82, pp. 724–726, June 1951.

[51] K. S. Knol, "Determination of the electron temperature in gas discharges by noise measurements," *Philips Research Repts.*, vol. 6, pp. 288–302, August 1951.

[52] T. J. Bridges, "A gas-discharge noise source for eight-millimeter waves," *Proc. IRE*, vol. 42, pp. 818–819, May 1954.

[53] H. Johnson and K. R. Deremer, "Gaseous discharge super-high-frequency noise sources," *Proc. IRE*, vol. 39, pp. 908–914, August 1951; also, "Super-high-frequency noise source," US Signal Corps, Project 322C-1, Final Rept., June 1, 1947–March 31, 1951.

[54] H. Prinzler, "Investigations of noise generators in the microwave region," *Acta Tech. Hungar.*, vol. 42, nos. 1–3, pp. 283–292, 1963.

[55] N. Houlding and L. C. Miller, "Discharge tube noise sources," TRE Memo 593, October 1953.

[56] R. Saier, "Rauschgeneratoren für Zentimeter-Wellen," *Frequenz*, vol. 14, pp. 68–70, February 1960.

[57] P. A. H. Hart and G. H. Plantinga, "An experimental noise generator for millimetre waves," *Philips Tech. Rev.*, vol. 22, no. 12, pp. 391–392, 1960–1961.

[58] C. K. S. Miller, W. C. Daywitt, and E. Campbell, "A waveguide noise-tube mount for use as an interlaboratory standard," *Acta IMEKO III*, pp. 371–381, 1964.

[59] W. H. Spencer and P. D. Strum, "Broadband UHF and VHF noise generators," *IRE Trans. on Instrumentation*, vol. PGI-4, pp. 47–50, October 1955.

[60] M. Kollanyi, "Application of gas-discharge tubes as noise sources in the 1700–2300 Mc/s band," *J. Brit. IRE*, vol. 18, pp. 541–548, September 1958.

[61] H. Schittger and D. Weber, "Über einen Gasentladungs-Rauschgenerator mit Verzögerungsleitung," *Nachr.-Techn. Fachber.*, vol. 2, pp. 118–120, 1955.

[62] A. D. Kuz'min and A. N. Khvoshchev, "A broadband noise generator for the decimeter region," *Radiotekhnika*, vol. 13, no. 7, pp. 36–42, 1958.

[63] H. Montague, "Coaxial UHF noise source," U. S. Naval Research Lab., Rept. 4560, August 1955.

[64] G. Alma'ssy and I. Frigyes, "New microwave noise generator for the 2000 Mc/s band," *Periodica Polytech. Elect. Engrg.* (*Hungary*), vol. 4, no. 4, pp. 293–303, 1960.

[65] R. I. Skinner, "Wide-band noise sources using cylindrical gas-discharge tubes in two-conductor lines," *Proc. IEE* (*London*), vol. 103B, pp. 491–496, July 1956.

[66] R. A. Lee and K. W. Olsen, "Absolute values of excess noise ratio traceable to the Bureau of Standards," Bendix, Red Bank, N. J., Engrg. Data Release, Issue 40, File G-19, p. 12, May 1962.

[67] D. O. North, "The absolute sensitivity of radio receivers," *RCA Rev.*, vol. 6, pp. 332–343, January 1942.

[68] H. T. Friis, "Noise figures of radio receivers," *Proc. IRE*, vol. 32, pp. 419–422, July 1944.

[69] "IRE Standards on Electron Tubes: Definition of Terms 1962 (62 IRE 7.S2)," *Proc. IEEE*, vol. 51, pp. 434–435, March 1963.

[70] "Description of the noise performance of amplifiers and receiving systems," *Proc. IEEE*, vol. 51, pp. 436–442, March 1963.

[71] "Representation of noise in linear two-ports," *Proc. IRE*, vol. 48, pp. 69–74, January 1960.

[72] "IRE Standards on Methods of Measuring Noise in Linear Two-ports, 1959," *Proc. IRE*, vol. 48, pp. 60–68, January 1960.

STANDARD NOISE SOURCES

by P. A. H. HART *).

A standard noise source can be a resistor, a saturated diode or a gas discharge. These three types are dealt with in the article below, and it is shown which type is to be preferred in the various frequency ranges. Some standard noise sources specially designed in the Philips Laboratories are discussed.

Introduction

If signals are to be made perceptible — we are concerned here primarily with radio and radar signals — the signal strength must exceed a certain minimum, irrespective of the amplification of the receiving system. The reason for this is the presence of noise. The minimum referred to depends on the particular technique of "information processing" used, and is low in some special techniques such as single-sideband systems and the methods used in radio astronomy.

The noise comes from sources that can be divided into two categories. The first comprises the *external* sources. It is these that cause the receiving aerial to pick up noise in addition to the desired signal. External kinds of noise include atmospherics, thermal noise from the earth and cosmic noise.

The second category comprises the *internal* sources of noise, inside the receiver itself. Their contribution makes the signal-to-noise ratio at the output of the receiver worse than at the input. The noise added by internal sources can be minimized by careful circuitry and the suitable choice of components, but it cannot be entirely eliminated; some noise from resistors, valves and other circuit elements always remains.

The strength of the internal noise is usually measured in a relatively narrow band of frequencies; the average frequency of this band is called "the" frequency at which the noise is measured. The

measurement can be made by comparison with a known noise power delivered by a *standard noise source*. There are various types of standard noise source. Which type is used depends among other things on the frequency at which the measurement is to be made.

A standard noise source that delivers an accurately known noise power is a *resistor of known resistance and temperature*. Requiring no calibration, this noise source is an *absolute* standard.

It is often more convenient to use a *noise diode*, i.e. a diode operated at the saturation current. Because of the shot effect the current fluctuates. A noise diode is a *noise-current generator*. The value of the noise current can be calculated from the direct current flowing in the diode; the noise diode too is therefore an absolute standard, but only in a limited (though wide) range of frequencies. As will presently be shown, this range has both a lower and an upper limit.

At frequencies higher than those at which the noise diode is effective, use can be made of a *gas-discharge noise source*. Noise generators of this type are sub-standards, in the sense that the noise power delivered cannot be exactly calculated but must be determined by calibration. They can be made with highly stable characteristics and are relatively insensitive to fluctuations in supply voltage and ambient temperature. For these reasons it is not necessary to calibrate them individually, unless

*) Philips Research Laboratories, Eindhoven.

Reprinted with permission from *Philips Tech. Rev.*, vol. 23, pp. 293–309, 1961/1962.

extreme precision is required, as for instance in radio-astronomic measurements. Compared with a resistor, gas-discharge noise sources deliver a high noise power. Various designs are possible. As we shall see, in decimetre-wave equipment the gas discharge is coupled to a helix or a Lecher line, and in equipment operating on centimetre or milli-metre wavelengths the discharge tube is mounted in a waveguide. In other designs the gas discharge is in a resonant cavity or in a horn antenna.

Fig. 1 gives a broad indication of the operating ranges of the Philips noise diodes K 81A and 10 P, and of gas-discharge tubes of varying constructions; the boundaries are in fact not as sharp as are shown here.

Measurement of the noise factor with standard noise resistors

An example of the use of resistors as standard noise sources is the measurement of the noise factor of a circuit, an amplifier for instance, that can be treated as a linear four-terminal network.

The noise factor F as defined by the American standard (53 I.R.E. 7 S1) is [1]:

$$F = \frac{N_0 + N_{extra}}{N_0}. \quad \dots \quad (3)$$

Here $N_0 + N_{extra}$ is the total noise power at the output in the narrow frequency band Δf, and N_0 is the share contributed by the thermal noise of an impedance Z_i which is connected externally across

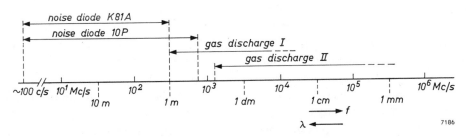

Fig. 1. Rough indication of the frequency and wavelength ranges in which noise diodes and gas discharges can be used as standard noise sources. *Gas discharge I* relates to a discharge tube coupled to a helix (fig. 17a), *gas discharge II* to a discharge tube mounted in a waveguide (fig. 17b and c).

In this article the three main types of noise source — resistor, diode and gas-discharge — will be discussed. We shall consider some special designs, the corrections necessary in certain applications of the noise diode, and gas-discharge noise sources for various wave ranges, including the millimetre band.

The resistor as a noise source

The thermal noise of a resistance R in a relatively narrow frequency band Δf around the frequency f is given by Nyquist's noise theorem:

$$\overline{u^2} = \frac{4\,hf}{e^{hf/kT} - 1}\,R\,\Delta f. \quad \dots \quad (1)$$

Here $\overline{u^2}$ is the mean square noise voltage, k is Boltzmann's constant ($= 1.38 \times 10^{-23}$ joule/°K), T is the absolute temperature of the resistor, and h is Planck's constant ($= 6.6 \times 10^{-34}$ joule-second). If hf is small compared with kT, i.e. if

$$f \ll \frac{k}{h}\,T \approx 2 \times 10^{10}\,T \text{ c/s},$$

formula (1) can be simplified to

$$\overline{u^2} = 4\,k\,T\,R\,\Delta f. \quad \dots \quad (2)$$

the input terminals and is equal to the impedance of the signal source to which the four-terminal network is normally connected (e.g. an antenna); the temperature of Z_i must be 290 °K. Therefore N_{extra} is the portion of the output noise added by the four-terminal network [2].

The noise factor is determined by measuring the output noise power (still in the narrow band Δf) when Z_i is successively at the temperatures T_1 and T_2, which must be known; see *fig. 2*. If only the temperature of Z_i were to change between the two measurements (Z_i itself thus remaining constant) the output noise power when Z_i has the temperature T_1 would be:

$$P_1 = CT_1 + N_{extra},$$

and when Z_i has the temperature T_2:

$$P_2 = CT_2 + N_{extra}.$$

---

[1] See also F. L. H. M. Stumpers and N. van Hurck, An automatic noise figure indicator, Philips tech. Rev. **18**, 141-144, 1956/57.

[2] It would be going too far to deal here with the way in which the noise factor as defined depends on Z_i. In this connection reference may be made to A. G. T. Becking, H. Groendijk and K. S. Knol, The noise factor of four-terminal networks, Philips Res. Repts **10**, 349-357, 1955.

In these two expressions C is a constant. Let the temperature 290 °K, mentioned in the definition, be denoted by T_0; then $N_0 = CT_0$. If we put $P_2/P_1 = a$, it follows from (3) that:

$$F = \frac{\dfrac{T_2}{T_0} + a - 1 - a\dfrac{T_1}{T_0}}{a - 1}. \qquad \ldots (4)$$

For $T_1 = T_0$ this reduces to

$$F = \frac{\dfrac{T_2}{T_0} - 1}{a - 1}. \qquad \ldots \ldots (5)$$

The latter expression is still valid to a good approximation when the difference between T_1 and T_0 is small.

In accordance with the definition of the noise factor we have spoken above of an *impedance* Z_i. Every impedance can be treated as composed of a resistance and a reactance connected in parallel (or in series). As the reactance causes no noise and is not changed during the measurement, it can be regarded as belonging to the four-terminal network. The noise-factor measurement therefore amounts to determining the ratio a of the output noise powers when a *resistor* R of temperature T_1 ($\approx T_0$) and T_2 respectively is connected across the input.

Measurements using a resistor as standard noise source are in principle possible at any frequency. In the range of ultra-high frequencies (decimetre wavelengths and shorter) a resistor is in fact the only absolute standard noise source. Nevertheless, in all frequency ranges — with the sole exception of the audio frequencies [3] — the resistor has been

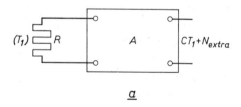

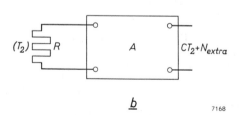

b   7168

Fig. 2. Measurement of the noise factor of a four-terminal network A by means of standard noise resistors. *a*) Connected to the input terminals of A is a resistance R having the temperature T_1. *b*) A resistor R is again connected to the input terminals, but now has the temperature T_2. The noise factor can be calculated from the measured values of the output noise in a narrow frequency band Δf.

superseded as a noise source for routine measurements by the noise diode or the gas discharge. The reasons are of a practical nature:

1) To bring a resistor to the temperature T_2, an oven or a temperature bath is needed, which is a complication.

2) An error is caused by the fact that the resistance changes as a rule with temperature. Two resistors are therefore needed, one of which must have the same resistance at the temperature T_1 as the other at the temperature T_2, which is much higher or lower than T_1.

3) If T_2 is higher than T_1, the temperature difference $T_2 - T_1$ cannot be made very large without damaging the hot resistor. The highest temperature can be achieved with a tungsten filament ($T_2 = 2700$ °K); $T_2 - T_1$ is then about 2400 °K. This means that large noise factors cannot be accurately measured with a hot resistor: N_{extra} is then large compared with CT_1 and CT_2, and therefore $P_2 \approx P_1$, i.e. $a \approx 1$, so that $a - 1$ cannot be determined with great precision.

Apart from its use at audio frequencies, the resistor as a noise source is now mainly used for *calibrating* noise diodes in the decimetre bands and gas-discharge tubes in the centimetre and millimetre bands. An example will be given at the end of this article.

Special designs of resistors as standard noise sources

In noise measurements at very short waves the stray inductance and capacitance of conventional resistors cause an impermissible error. Special resistors are therefore needed, and we shall describe here two that have been designed and constructed in the Philips Research Laboratories at Eindhoven. One serves for calibrating noise diodes in the decimetre bands and is a *cold* resistor ($T_2 = 77$ °K) [4]; the other is used for calibrating gas discharges in the centimetre and millimetre bands, and is a *hot* resistor ($T_2 = 1336$ °K).

a) **A cold resistor for decimetre waves**

The resistor proper (*fig. 3*) consists of a very thin layer of platinum on a hard-glass tube, located at the end of an impedance transformer fitted with shorting plungers. The layer has a resistance of roughly 50 ohm. The resistance, as "seen" from the input of the impedance transformer, is accurately set to 50 ohm by adjusting the plungers until a standing-wave detector, having a 50-ohm characteristic

[3] See e.g. A. van der Ziel, Noise, Prentice Hall, New York 1954, p. 31.
[4] Another cold resistor is described in Philips tech. Rev. **21**, 327 (fig. 13), 1959/60.

impedance, gives a standing-wave ratio of 1. Both the resistor itself and the impedance transformer are immersed in liquid nitrogen (temperature 77 °K). As regards its noise contribution, therefore, the dissipative resistance of the transformer also behaves as a resistor having a temperature of 77 °K. *Fig. 4* shows a photograph of the transformer with the resistor inside.

The second resistor in the measurement can be a conventional type of 50 ohm, kept at room temperature.

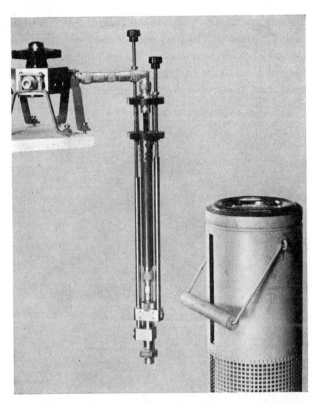

Fig. 4. The impedance transformer represented in fig. 3 (with the noise resistor inside) removed from the nitrogen bath.

b) A hot "resistor" for centimetre and millimetre waves

A waveguide (*fig. 5*) is provided with a matched termination at one end in the form of an absorption wedge, made in this case of the ceramic material "Caslode". The part of the waveguide with the wedge is uniformly heated in an oven to a well-defined high temperature. The temperature is measured with a thermocouple or a pyrometer, which need only be calibrated for this one temperature — which has been chosen as the melting point of gold (1336 °K) [5].

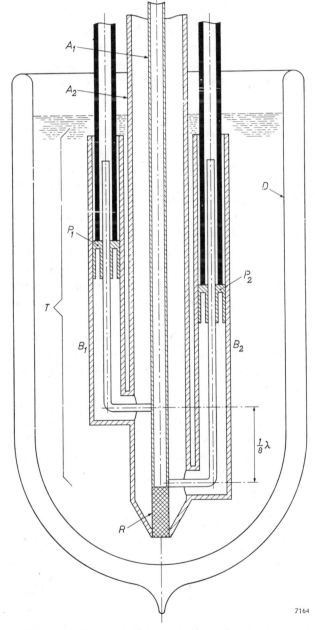

Fig. 3. Construction of a cold standard noise resistor for decimetre waves (not to scale). R is the actual resistor (approx. 50 ohm), being a layer of platinum on glass. T coaxial impedance transformer consisting of an inner conductor A_1 and an outer conductor A_2 with coaxial side branches B_1 and B_2, which are terminated by shorting plungers P_1 and P_2. At the top a coaxial plug (N plug) can be connected. D Dewar vessel filled with liquid nitrogen (temperature $T_2 = 77$ °K).

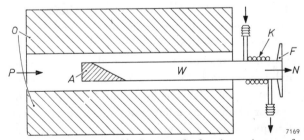

Fig. 5. Cross-section of a hot standard noise resistance for centimetre or millimetre waves. W platinum waveguide. A absorption wedge (the noise source proper). F connecting flange. O electric oven. K pipe carrying cooling water. Arrow N indicates the direction in which the noise leaves the waveguide, arrow P the direction in which the optical pyrometer faces; the latter measures the temperature T_2 of the wedge A.

[5] The method of calibration has been described in: K. S. Knol, A thermal noise standard for microwaves, Philips Res. Repts **12**, 123-126, 1957.

To prevent oxidation, the waveguide is made of (thin) platinum. The part outside the oven is water-cooled to avoid overheating the coupling flange and the components connected to it. *Fig. 6* shows waveguides of this type for wavelengths of 3 cm and 8 mm.

Fig. 6. Hot standard noise sources as in fig. 5, for wavelengths of 3 cm (with oven) and 8 mm (without oven). At *A* can be seen the "Caslode" absorption wedge for the 8 mm waveguide.

The second waveguide used in the measurement contains a matched termination made of wood, at room temperature.

Before concluding this account of standard noise resistors, there are two further points to be noted. One concerns the choice between cold and hot resistors, the other a correction required in certain cases.

We have just discussed a cold resistor for decimetre waves and a hot resistor for centimetre and millimetre waves. In principle the converse is also possible. Compared with a cold resistor a hot resistor has the advantage of delivering a higher noise power, which can increase the accuracy of the measurement (unless the noise figure is low). This is an argument in favour of choosing a hot resistor. Decimetre-wave techniques using Lecher wires or coaxial lines are not so suitable at high temperatures, however, as microwave techniques using waveguides. For this technological reason we decided on a cold resistor for the decimetre bands.

The standard noise resistor which is at the high or low temperature T_2 gives rise to a temperature gradient in the supply line (coaxial cable or wave-

guide). This must be taken into account in two respects. Firstly, the materials employed must be capable of withstanding the temperature gradient. Secondly, the calculated value of the noise generated by the resistor R needs to be corrected if the losses in the supply line are at all significant (e.g. a few percents of those in R), for part of the noise produced by R is lost in the dissipative resistance of the supply line, whilst the latter resistance itself contributes a certain amount of noise.

The diode as a noise source

A diode operated at saturation behaves in a wide range of frequencies like a noise-current generator having an infinite internal resistance. In a relatively narrow band Δf within that range the mean square noise current is given by Schottky's formula:

$$\overline{i^2} = 2 q I_s \Delta f, \qquad \ldots \ldots \quad (6)$$

where q is the charge on the electron ($= 1.60 \times 10^{-19}$ C) and I_s is the saturation current flowing in the diode. I_s depends on the temperature of the filament and therefore the noise current can be given a different value by changing the filament current.

The usual practice for measurements is to connect the noise diode in parallel with a resistor R (*fig. 7a*). In fig. 7*b* the diode is represented by a noise-current source I, and the resistance R by an equal but hypothetically noiseless resistance R^* in series with a noise-voltage source U which accounts for the thermal noise of R. In addition to this thermal noise, given by $4kT_1R\Delta f$ (T_1 being the temperature of R), the resistance R in fig. 7*a* carries the noise voltage generated by the noise current of the diode; this noise is given by $2qI_sR\Delta f$. Since the two noise sources are independent of each other, the total noise is found by simply adding the two contributions mentioned. We now want to find the temperature T_{eq} which a resistor R must have if its thermal noise is to be equal to this total:

$$4 kT_{eq}R\Delta f = 4kT_1R\Delta f + 2qI_sR^2\Delta f.$$

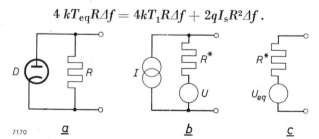

Fig. 7. *a*) Noise diode *D* in parallel with resistor *R*. *b*) Equivalent circuit of (*a*) consisting of a noise-current source *I*, representing the diode noise, shunted across a noiseless resistor *R** which is in series with the noise-voltage source *U* representing the noise from *R*. *c*) Equivalent circuit of (*b*) — likewise of (*a*) — consisting of a noiseless resistor *R** in series with a noise-voltage source U_{eq} which delivers the same noise as a resistor *R* at the temperature T_{eq}.

We can solve this expression for T_{eq}:

$$T_{eq} = \frac{q}{2k} I_s R + T_1 .$$

Insertion of the numerical values of q and k gives $q/2k = 5800$ °K/V $= 20 \times 290$ °K/V $= 20\, T_0$ °K/V, so that the formula for T_{eq} can also be written:

$$T_{eq} = 20\, I_s R T_0 + T_1. \quad \ldots \ldots \quad (7)$$

A noise diode in parallel with a resistor R (fig. 7a) thus constitutes a noise source which is equivalent to a resistor R at the temperature T_{eq} (represented in fig. 7c as a noiseless resistor R^* in series with a noise-voltage source U_{eq}); as can be seen from eq. (7), T_{eq} can be varied by changing the current I_s by means of the filament current of the diode.

Determining the noise factor with a noise diode

The noise factor of a four-terminal network can be determined with a noise diode as follows. The diode with a resistor in parallel is connected to the input terminals of the network, and the noise power P_1 is measured at the output, in the small frequency band Δf, with the diode passing no current; the noise power P_2 is then measured with the diode in operation. Again putting $P_2/P_1 = a$, we find from the definition of the noise factor, using (6) and (7):

$$F = \frac{20\, I_s R - 1 + a + (1-a)\dfrac{T_1}{T_0}}{a - 1} . \quad (8)$$

This formula is very much simpler if the temperature T_1 of the resistor is roughly equal to T_0 ($= 290$ °K) and if a is given the value 2 (i.e. if the diode current I_s is adjusted so that $P_2 = 2P_1$; see [1])). In that case:

$$F = 20\, I_s R . \quad \ldots \ldots \quad (9)$$

Formulae (8) and (9) are valid in a wide frequency range. In this range the diode is an absolute noise standard, since the formulae contain no empirical constants. We shall now consider the limits of this frequency range. We shall see that they are partly of a fundamental nature and partly due to the finite dimensions of the diode.

Lower limit of frequency range

Below a certain frequency a diode shows in addition to the above-mentioned noise, which is due to the shot effect, a kind of noise known as flicker noise [6]). According to one theory, flicker noise is bound up with the "scintillating" character of the emission:

after a certain spot on the cathode has emitted a batch of electrons, some time elapses before the same spot can emit electrons again. Investigations have shown [7]) that a tungsten cathode, as used in noise diodes, exhibits distinct flicker noise only at frequencies of 10 c/s and lower. Impurities in the cathode and traces of gas may raise this limit, however, so that it is safer not to use a diode for noise measurements below, say, 100 or 1000 c/s.

In practice, this limitation is of little significance, noise diodes seldom being used in the audio-frequency range. Noise measurements at audio frequencies are usually done with two different resistors, both at room temperature [8]).

Upper limit of frequency range

At very high frequencies there are two causes of deviations from the Schottky formula (eq. 6): one we shall call the "transformation error" and the other the "transit-time error". The transformation error is attributable to stray capacitances and inductances; the transit-time error is significant at frequencies which are so high that the period of oscillation is not long compared to the time taken by the electrons to travel from cathode to anode. We shall now consider the magnitude of these errors.

The transformation error

Some years ago a short article appeared in this journal [9]) describing a new noise diode (the 10 P type already mentioned) and two circuits making use of this diode for noise measurements in the decimetre wave range; one circuit employed a Lecher system and the other a coaxial system (fig. 8a). To calculate the transformation error of such a circuit, we use the equivalent circuit shown in fig. 9a. Here the current source i represents the noise diode, C_0 the capacitance of the diode, L_0 the inductance of the lead-in wires, Z_1 the impedance of the coupling capacitors, and Z the resistance R and the section of transmission line connected in parallel with it. The other end of this line is fitted with a shorting plunger, which is so adjusted that, at the operating frequency f, the impedance Z is equal to R (i.e. such that the shorted section of line exactly compensates the influence of the various reactances at this frequency). To examine the influence of L_0, C_0 and Z_1 on the noise, we transform this circuit into one in which Z is shunted

[6]) W. Schottky, Small-shot effect and flicker effect, Phys. Rev. **28**, 74-103, 1926.

[7]) J. G. van Wijngaarden, K. M. van Vliet and C. J. van Leeuwen, Low-frequency noise in electron tubes, Physica **18**, 689-704, 1952.

[8]) See page 75 et seq. of the book by Van der Ziel mentioned in footnote [3]).

[9]) H. Groendijk, A noise diode for ultra-high frequencies, Philips tech. Rev. **20**, 108-110, 1958/59.

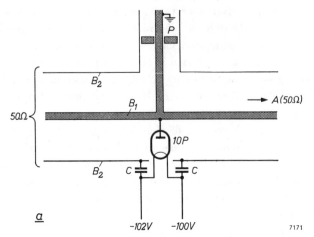

Fig. 8. Type 10 P noise diode in a coaxial system. *a*) Schematic cross-section, *b*) and *c*) side views. The diode projects through a hole in the outer conductor B_2, to which the filament is connected via the coupling capacitors C (each of capacitance $\frac{1}{2}C_1$). The anode is connected to the central conductor B_1. On the right in (*a*) is connected the four-terminal network under measurement; left, a matched termination. The shorting plunger P in the side tube is used to tune out the effects of the diode capacitance and the impedance of the coupling capacitors.

In a new valve holder (*b* and *c*) the capacitors C have a mica dielectric (shown black in the figure). The capacitor plates *1*, *3* and *5* are connected to the outer conductor B_2 of the coaxial system, plates *2* and *4* are connected to one side of the filament, plates *2'* and *4'* to the other side.

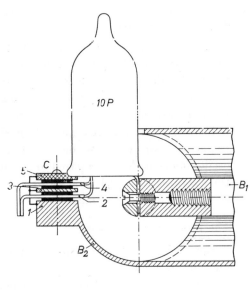

b

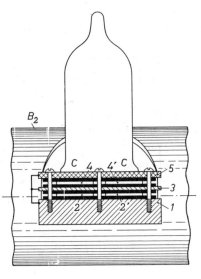

c   7165

across the current source i' (fig. 9*b*); i' is the current which produces across the noiseless resistance R^* (fig. 9*c*) a noise voltage equal to the output noise voltage in fig. 9*a*. It can be calculated that the two circuits are equivalent at the frequency f if the following relation exists between the currents i' and i:

$$i' = \frac{i}{1-(2\pi f)^2 L_0 C_0 + \mathrm{j}\times 2\pi f C_0 Z_1} \quad . \quad . \quad (10)$$

The impedance Z_1 of the two coupling capacitors in parallel can be reasonably approximated by the impedance of a capacitance C_1 and an inductance L_1 in series. Introducing a frequency f_t, defined by $(L_0 + L_1)C_0 = (2\pi f_t)^2$, we find from (10), after taking the mean square, the following expression for the *transformation factor* γ_t:

$$\gamma_t = \frac{\overline{i'^2}}{\overline{i^2}} = \left[1 - \left(\frac{f}{f_t}\right)^2 + \frac{C_0}{C_1}\right]^{-2}.$$

$1-\gamma_t$ is the *transformation error* we wish to find.

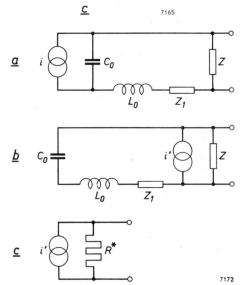

Fig. 9. *a*) Equivalent circuit of a noise diode (noise-current generator i), the capacitance of which is C_0; L_0 is the inductance of the lead-in wires, Z_1 the impedance of the coupling capacitors, and Z the impedance of the resistance R in parallel with a section of transmission line.
b) The current source i in (*a*) has been replaced by a current source i' in parallel with Z.
c) Equivalent circuit of (*b*), consisting of the current source i' in parallel with the noiseless resistance R^*.

There is no objection to raising the capacitance of the coupling capacitors sufficiently for C_1 to be large compared with the capacitance C_0 of the diode. In that case, at least at frequencies f not too close to f_t, we can disregard C_0/C_1, giving:

$$\gamma_t \approx \left[1 - \left(\frac{f}{f_t}\right)^2\right]^{-2} \quad \ldots \ldots \quad (11)$$

Fig. 10 shows a plot of γ_t versus f in accordance with (11), with f_t as parameter. At frequencies a great deal lower than f_t, the transformation factor γ_t is roughly 1, and the error therefore about zero. The value of f_t at which γ_t becomes 1.10 ($f = 0.22\, f_t$) is roughly the upper limit of the frequency range in which the diode can be regarded as an absolute noise standard; at higher frequencies the correction $1 - \gamma_t$ is too inaccurate.

From fig. 10 it can be seen that $1 - \gamma_t$ increases markedly with increasing f_t. The aim is therefore to make $f_t = 1/[2\pi\sqrt{(L_0 + L_1)C_0}]$ as high as possible and thus to minimize L_0, L_1 and C_0. Careful assembly of the diode and the proper choice of coupling capacitors are therefore of considerable importance in this respect. In the 10 P noise diode C_0 has the very low value of 1.8 pF. The values of L_0 and L_1 depend closely on the construction of the valve holder and the coupling capacitors. In the article cited [9] a valve holder was described using ceramic coupling capacitors for coaxial systems. In a later model (fig. 8b and c) these capacitors were replaced by mica capacitors (of 230 pF each, so that $C_1 = 460$ pF). From impedance measurements a value of 2900 Mc/s was found for the frequency f_t of the 10 P diode in this holder; noise measurements [10] at 500 to 1500 Mc/s yielded values from 2600 to 3400 Mc/s, depending on the frequency. Comparison of these

results with the resonance frequency of the diode itself: $1/(2\pi\sqrt{L_0 C_0}) = 3500$ Mc/s, shows that the stray inductance L_1 is satisfactorily low.

The transit-time error

In a diode the electrons take a finite time to travel from the cathode to the anode. In diodes such as the 10 P type, of coaxial cylindrical construction with a cathode diameter of 0.1 mm and an anode diameter of 1 mm, and with an anode voltage V_a which is high enough for operation well within the saturation region, the electron transit time τ is:

$$\tau = \frac{1.08 \times 10^{-9}}{\sqrt{V_a}} \text{ second} \quad \ldots \quad (12)$$

(with V_a in volts). During this time the electron induces a current pulse in the external circuit. Assuming that the current pulses of the different travelling electrons are mutually independent (i.e. that the electron emission is random and that there is no space charge), and moreover that the electrons leave the cathode radially and without an initial velocity, the noise current can be calculated by a suitable summation of the current pulses [11]. We then find that we must add to Schottky's formula (6) a transit-time factor γ_τ:

$$\overline{i^2} = 2\gamma_\tau q I_s \varDelta f. \quad \ldots \ldots \quad (13)$$

For the 10 P diode operating well within the saturation region this factor is given by:

$$\gamma_\tau = 1 - 2.67\,(f\tau)^2,$$

in which the terms in the fourth and higher powers of $f\tau$ are neglected. Using (12) we can then write:

$$\gamma_\tau = 1 - 3.13\,\frac{f^2}{V_a} \times 10^{-18}. \quad \ldots \quad (14)$$

It can be seen from (14) that at low frequencies γ_τ approaches unity, in which case (13) reduces to (6), the original Schottky equation. The use of (6) at higher frequencies gives rise to an error of $1 - \gamma_\tau$, the *transit-time error*.

It should be noted that the error is greater than follows from (14) if the anode voltage is so low that the diode is only barely saturated. One reason is that the assumption of radial emission without initial velocity is then no longer correct (an electron that does not leave the cathode radially is a longer time in transit than one that does, and consequently

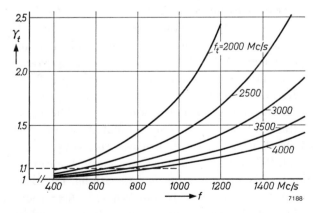

Fig. 10. The transformation factor γ_t, given by (11), as a function of frequency f, with the frequency f_t as parameter.

[10]) See p. 302 under "Experimental determination of $\gamma_t\gamma_\tau$".

[11]) E. Spenke, Die Frequenzabhängigkeit des Schroteffektes, Wiss. Veröff. Siemens-Werke **16**, No. 3, 127-136, 1937. G. Diemer and K. S. Knol, The noise of electronic valves at very high frequencies, I. The diode, Philips tech. Rev. **14**, 153-164, 1952/53.

induces a pulse of different shape). Another reason is that the space charge affects the potential gradient and so changes the transit time of the electrons. The result is that equations (12) and (14) are not valid if V_a is too low.

Some means is therefore needed of ascertaining whether the diode is saturated or not. This cannot be seen clearly enough from a plot of the diode current I_d ($\leq I_s$) versus V_a. A sharper criterion can be derived from the variation with V_a of the factor Γ'^2, which is a measure of the suppression of the space charge. This factor occurs in the formula for the shot noise of a diode at frequencies up to about 100 Mc/s:

$$\overline{i^2} = 2\Gamma'^2 q I_d \Delta f.$$

The factor Γ'^2, which is equal to unity at saturation, quickly drops to a low value if the diode becomes less saturated, making the space-charge effect perceptible. For a given value of V_a a maximum value of I_d can be indicated at which Γ'^2 deviates from unity and saturation is thus no longer present.

Fig. 11 shows Γ'^2 as a function of V_a, with I_d as

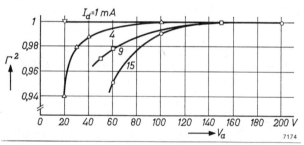

Fig. 11. The factor Γ'^2 of the type 10 P noise diode at 30 Mc/s as a function of anode voltage V_a, for various values of the diode current I_d.

parameter, for the 10 P diode at 30 Mc/s. From this we can determine what the minimum V_a must be, at given values of I_d, if Γ'^2 is not to differ by more than 1%, 2% or 3% from unity; see *fig. 12*. The values of V_a at which the anode dissipation $I_d V_a$ reaches 2 W, which is the maximum permissible value for the 10 P diode, are also shown in this figure; operation in the shaded region is thus not permissible.

Transformation and transit-time errors combined

In general, both errors are present and must be taken into account:

$$\overline{i'^2} = 2\gamma_t\gamma_\tau q I_s \Delta f.$$

The highest frequency, then, at which the Schottky formula is still applicable is that where $\gamma_t\gamma_\tau$ only

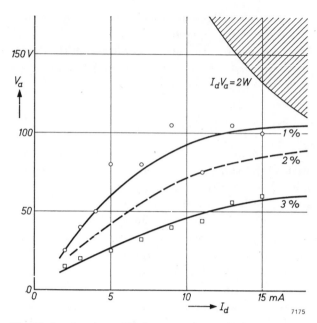

Fig. 12. Anode voltage V_a of type 10 P noise diode as a function of diode current I_d, with Γ'^2 differing from unity by 1%, 2% and 3%, respectively. Curve $I_d V_a = 2$ W indicates the values of V_a at which $I_d V_a = 2$ W.

just permissibly deviates from unity. Having regard to the accuracy required for most noise measurements, the deviation is usually fixed at 10%. If $\gamma_t\gamma_\tau$ is known in a particular frequency range, the Schottky formula can be corrected to enable the diode to be used in that range.

From (11) and (14) it appears that, where f is smaller than f_t, the two γ factors differ from 1 in opposite senses: γ_t is greater than 1 and γ_τ is smaller. The two errors, then, compensate one another to some extent. In usual conditions γ_t differs more from 1 than γ_τ. The compensation can therefore be improved, i.e. the product $\gamma_t\gamma_\tau$ brought closer to unity, by increasing the electron transit time. This can be done by lowering the anode voltage (*fig. 13a*), but not so far that the diode ceases to be saturated; otherwise a space charge would form and γ_τ would then depend on the diode current. The space-charge effect may be greater at high frequencies (of the order of 1 Gc/s) than at lower.

The magnitude of the effect of lowering the anode voltage can only be roughly estimated, because the diode then enters a region where (14) is no longer valid. Calibration is therefore necessary. We can see this plainly by comparing fig. 13a with fig. 13b. Both graphs give γ_τ as a function of frequency for various values of the anode voltage; fig. 13a gives values calculated from (14), and fig. 13b measured values. It is assumed that γ_τ for $V_a = 300$ V varies in accordance with (14); theory and measurements indicate that this will be correct to within about 2%.

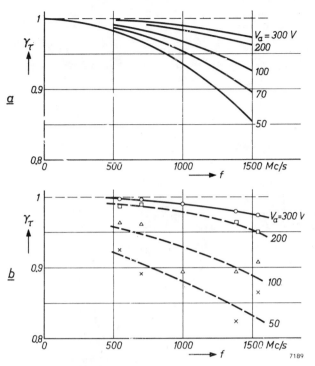

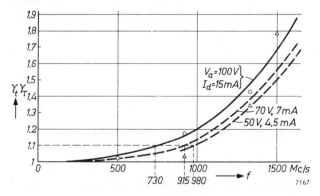

Fig. 13. a) *Calculated* values of the transit-time factor γ_τ of a 10 P noise diode as a function of frequency f, with the anode voltage V_a as parameter.
b) *Measured* values of γ_τ for $V_a = 200$, 100 and 50 V, in relation to the calculated values for $V_a = 300$ V.

The values of γ_τ measured at $V_a = 200$, 100 and 50 V are plotted in relation to the calculated values for $V_a = 300$ V. (As will appear in the next section, γ_τ by itself cannot be measured but the product $\gamma_t \gamma_\tau$ can.) We see from fig. 13a and b that the measured γ_τ differs increasingly from the calculated values as the anode voltage is reduced: the average discrepancy is about 1% at $V_a = 200$ V, about 4% at 100 V, and about 5% at 50 V.

In *fig. 14* the calculated and measured values of $\gamma_t \gamma_\tau$ for the 10 P diode are plotted versus frequency,

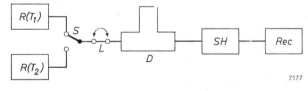

Fig. 14. The curves give the calculated values of $\gamma_t \gamma_\tau$ for a 10 P noise diode with various combinations of V_a and I_d. The experimental points indicated all relate to $V_a = 100$ V; the circles relate to comparison with a cold standard noise source, the triangles to comparison with a gas-discharge noise source as in fig. 17a, which was previously calibrated with a hot standard noise source.

with V_a as parameter. Lowering V_a from 100 to 50 V widens the frequency range from about 730 to about 980 Mc/s, a gain of 35%. Set against this is the great disadvantage that the diode current at $V_a = 50$ V should not be more than 4.5 mA, as otherwise the factor Γ^2 will deviate by more than 1% from unity; see fig. 12. At such a low diode current only small noise factors can be measured satisfactorily, large ones not, at least not without correction.

Experimental determination of $\gamma_t \gamma_\tau$

The quantity $\gamma_t \gamma_\tau$ has been determined experimentally [12] in the following way. A resistor $R(T_1)$ is held at room temperature, and a resistor $R(T_2)$ of roughly the same value at 77 °K (in liquid nitrogen). Both are provided with a variable impedance transformer (e.g. of the type sketched in fig. 3), which has the same temperature as the appertaining resistor. First of all, the transformed resistance of one of the resistors is made as nearly as possible equal (to within about 5%) to 50 ohm, by means of a standing-wave detector. This resistor is then connected via a coaxial switch to a bridge circuit (which need not be calibrated), after which the bridge is balanced. The second resistor is connected to the bridge by turning the switch, and the relevant impedance transformer is adjusted until the bridge is again balanced. This substitutional method makes it possible to equalize the two resistances with an accuracy up to 0.1%.

The next step is to assemble the circuit indicated in *fig. 15*: behind the switch S come a 10 P noise diode in a coaxial holder [13], a superheterodyne

Fig. 15. Measurement of $\gamma_t \gamma_\tau$. The resistors $R(T_1)$ and $R(T_2)$ have as nearly as possible the same value R ($\approx$ 50 ohm) at the respective temperatures T_1 ($\approx$ room temperature) and T_2 (= 77 °K). D noise diode type 10 P in coaxial holder. SH superheterodyne receiver. *Rec* recorder. When switch S is in the position as drawn and the diode is without filament current, the recorder gives a certain deflection. When S is in the other position, the filament current is adjusted so as to produce the same deflection on the recorder. Using eq. (15) it is then possible to calculate $\gamma_t \gamma_\tau$.

At L a coaxial line of 50 ohm characteristic impedance and a quarter wavelength long can be inserted for a second measurement, to reduce the error caused by the fact that $R(T_1)$ and $R(T_2)$ do not have exactly the same values.

---

[12] By W. E. C. Dijkstra of this laboratory.
[13] The switch with resistors thus takes the place of the anode resistor of 50 ohm described in the article mentioned in footnote [9].

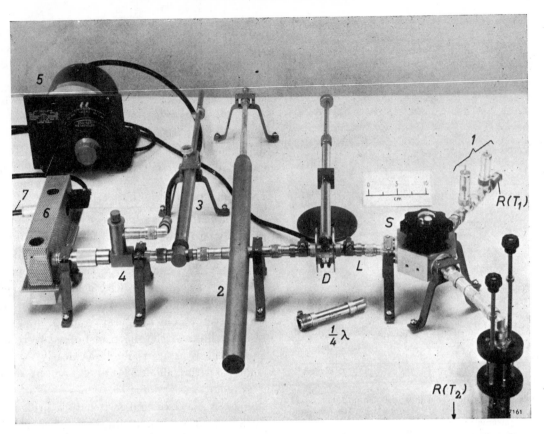

Fig. 16. Part of the equipment for measuring $\gamma_t\gamma_\tau$ at decimetre wavelengths by the method illustrated in fig. 15. $R(T_1)$ standard noise resistor at room temperature, with matching transformer 1. $R(T_2)$ standard noise resistor at 77 °K as in figs 3 and 4 (only the upper part of the impedance transformer can be seen). S coaxial switch. D noise diode type 10 P in holder as shown in fig. 8. At L a section of waveguide a quarter wave in length can be inserted.

The equipment to the left of D is part of the superhet receiver: 2 coaxial resonant cavity, functioning as a filter that only passes a narrow band of frequencies around the measuring frequency f (it therefore does not pass the frequency of the local oscillator, nor the image frequency); 3 impedance transformer for matching the crystal mixer 4 to the resonant cavity 2; 5 local oscillator; 6 first and second stages of the IF amplifier (30 Mc/s); 7 cable to the remaining part of the IF amplifier.

receiver and a recorder. Some of the equipment can be seen in *fig. 16*.

The resistor $R(T_1)$ is now switched in, the diode not yet passing any filament current. The noise of $R(T_1)$, which reaches the receiver through the switch and the diode holder, produces a certain deflection on the recorder. After switching over to $R(T_2)$ the filament current of the diode is adjusted until the recorder shows the same deflection as before. The noise produced by the resistor at 77 °K together with the diode is then equal to the noise of the resistor alone at the temperature T_1. For this condition we easily arrive at the following formula for $\gamma_t\gamma_\tau$:

$$\gamma_t\gamma_\tau = \frac{T_1 - 77}{RT_0 I_s \times 20}. \quad \cdots \quad (15)$$

An important point is that the receiver requires no calibration. Since R occurs in the formula, however, its value must be accurately known.

The effect of a small error ΔR in R can be reduced by a simple expedient: *two* measurements are done as described above, but in the second we insert between the switch and the diode holder, at L in fig. 15, a coaxial line whose characteristic impedance is 50 ohm and which is a quarter wavelength long. In (15) we must then replace R by

$$R' = \frac{50^2}{50 + \Delta R} \approx 50 - \Delta R.$$

The error in R', then, is just as great as in R, but of opposite sign; the average of $\gamma_t\gamma_\tau$ from the two measurements is therefore more accurate than the result of one measurement. A condition, however, is that the characteristic impedance of the inserted line must be more exactly equal to 50 ohms than R, otherwise the improvement is illusory.

Summarizing, it can be said that the 10 P diode, with $V_a = 100$ V and $I_d = 15$ mA, can be used without correction as a standard noise source from about 100 c/s to about 730 Mc/s, the maximum error being 10%. At higher frequencies either correction or compensation is necessary. With wide-band circuits, compensation has the advantage that the noise does not depend on the frequency.

It should also be noted that in certain applications, where for instance the equipment has to be adjusted for minimum noise, only relative noise differences are important; what is needed here, then, is a constant source, which need not be a standard one. For example, the automatic noise-figure indicator mentioned [1] in combination with a 10 P diode has proved very useful for adjusting 4 Gc/s equipment for minimum noise, and this is certainly not the highest frequency at which this is possible.

The gas discharge as a noise source

Mechanism of noise generation by a gas discharge

The positive column of a gas discharge emits electromagnetic radiation having the character of noise. In the centimetre and millimetre wavebands the column of an inert-gas discharge, which is both long and strongly coupled to the measuring circuit, is a particularly suitable noise source for the purposes of measurement.

The positive column consists of ions, electrons and neutral particles. There are roughly just as many positive elementary charges per unit volume as there are negative ones. Such a quasi-neutral mixture is called a *plasma*. The electrons in the plasma have a much higher average velocity than the ions and the neutral particles. This is due to their very low mass (compared with the other particles) as a result of which they are much more accelerated in the electric field and moreover lose very little energy upon elastic collisions with heavy particles.

Owing to the deceleration which an electron suffers upon a collision, a small fraction of its energy is converted into electromagnetic radiation ("Bremsstrahlung"). Since the velocities of the electrons show a random distribution, both in magnitude and direction, the emitted radiation has the character of noise. The power of this radiation depends on the average kinetic energy of the electrons upon collision, that is on the "electron temperature" T_{el}. Let the mass of an electron be m and the mean square velocity be $\overline{v^2}$, then T_{el} is defined as:

$$\tfrac{1}{2}m\,\overline{v^2} = \tfrac{3}{2}\,k\,T_{el}\,.$$

Because the average energy of an electron is much greater than that of the other particles, T_{el} is a high temperature (of the order of 10^4 °K), much higher than the temperature of the gas, which as a rule is not much above room temperature.

It may be asked how the power radiated by the plasma depends on T_{el}. If either the electron

velocities show a Maxwell distribution, or the collision frequency is constant, the radiant power of the plasma is equal to the noise power available from a resistor at the temperature T_{el} [14]), in other words in a small frequency band Δf it is equal to $kT_{el}\,\Delta f$. Experiments have shown [15]) that even though the above conditions are not entirely fulfilled, $kT_{el}\,\Delta f$ can be a good approximation for the power radiated by a plasma in the band Δf. In most cases, then, the power can be assumed to have this value.

Various forms of gas-discharge noise sources

A simple form of gas-discharge noise source consists of a gas-discharge tube mounted in a resonant cavity, with an impedance-matching transformer. This construction is useful for physical research on plasmas, but does not constitute a noise source suitable for a wide range of frequencies. Such a source can be realized, however, in another way. Three examples are given in *fig. 17*; the frequency ranges for which they are suitable will be found in fig. 1 (lines *I* and *II*).

The constructions in fig. 17, although differing considerably from one another, are all based on the same principle. We can make this clear with the aid of the diagram in *fig. 18*. This figure represents the longitudinal cross-section of a waveguide, which has a matched termination at one end in the form of an absorption wedge A, whose temperature is T_1. From B to C extends a homogeneous plasma which, we shall assume, fills the entire cross-section of the waveguide.

Both the wedge A and the plasma emit noise waves. The *wedge* sends out noise waves (power P_A, temperature T_1) from A to D. The waves, after passing through the plasma, where they are attenuated, have the lower power $P_A{}'$ corresponding to a temperature lower than T_1. The *plasma* sends out noise waves to left and right, having powers P_B and P_C, which can be calculated by integrating the emission over the whole plasma column. The wave P_B is completely absorbed in the wedge A, and may therefore be left out of further consideration. The wave leaving the noise generator at D has an equivalent noise temperature T_{eq}, which is the sum of the temperatures corresponding to $P_A{}'$ and P_C (this summation is permissible owing to the fact that

[14]) G. Bekefi and S. C. Brown, Microwave measurements of the radiation temperature of plasmas, J. appl. Phys. **32**, 25-30, 1961 (No. 1).
G. H. Plantinga, The noise temperature of a plasma, Philips Res. Repts **16**, 462-468, 1961 (No. 5).
[15]) K. S. Knol, Determination of the electron temperature in gas discharges by noise measurements, Philips Res. Repts **6**, 288-302, 1951.

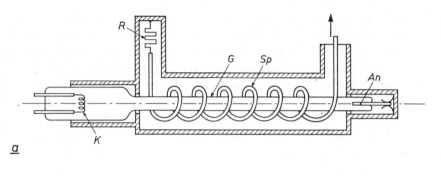

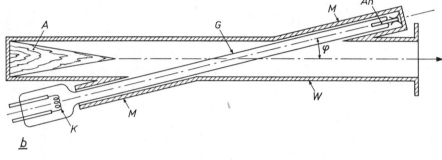

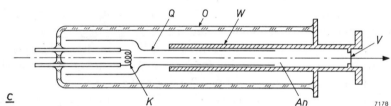

_a_

_b_

_c_

7178

Fig. 17. Three constructions of gas-discharge noise sources. _G_ gas discharge tube with cathode _K_ and anode _An_.

a) Construction for decimetre waves. The discharge tube is coupled via the helix _Sp_ to a system which has a matched termination on the left in the form of a resistor _R_, and on the right goes over into a coaxial system.

b) Construction for centimetre waves, the discharge tube passing obliquely through the waveguide _W_ with side arms _M_. _A_ is an absorption wedge providing a matched termination.

c) Construction with axial gas discharge in a circular waveguide _W_, for millimetre waves. _Q_ thin-walled tube of quartz glass, open at both ends. The glass envelope _O_ is filled with neon. _V_ mica window. The inside wall of the waveguide _W_ just past one end of the tube _Q_ serves as anode (_An_).

the two noise waves are mutually independent). Thus,

$$T_{eq} = T_A{}' + T_C .$$

If _L_ is the attenuation suffered by the power of the waves in passing through the entire plasma column, we can write

$$T_{eq} = \frac{1}{L} T_1 + \left(1 - \frac{1}{L}\right) T_{el} . \qquad . \quad (16)$$

It is assumed here that the reflections at the boundary planes _B_ and _C_ are negligible. (Reflection from _C_ lowers the equivalent temperature T_{eq}; reflection from _B_ lowers or raises T_{eq}, depending on whether the waves are in phase or in anti-phase.) If _L_ is sufficiently large, T_{eq} differs very little from T_{el}, and reflections from _B_ have no further perceptible influence, no more than has the temperature T_1.

Equation (16) can be derived as follows. We imagine that the wedge _A_ is heated to a temperature equal to T_{el}. According to the Nyquist theorem, the total noise power in the band Δf which goes to _D_ must be equal to $kT_{el}\Delta f$, for to the left of _C_ everything is at the temperature T_{el} (the plasma being equivalent to an absorption medium at the temperature T_{el}), and we have assumed that there are no reflections. To the power passing from _C_ to _D_ the wedge _A_ contributes $L^{-1} kT_{el} \Delta f$. The plasma contribution is therefore $(1 - L^{-1})kT_{el}\Delta f$. When the wedge is now cooled to the temperature T_1, its contribution drops to $L^{-1}kT_1\Delta f$, whereas that of the plasma shows no change. The total is thus:

$$kT_{eq}\Delta f = L^{-1} kT_1 \Delta f + (1 - L^{-1})kT_{el}\Delta f ,$$

which leads directly to eq. (16).

Some requirements which must be met by a gas-discharge noise source

In order to build a gas-discharge noise source that will deliver a high and constant noise power in a broad range of frequencies, it is necessary, as we have seen, to start with a plasma of high electron temperature T_{el}, and to introduce this in the circuit

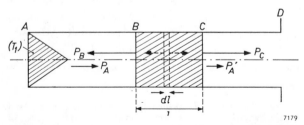

7179

Fig. 18. Illustrating the principle shared by the three designs in fig. 17. _AD_ waveguide terminated at the left by an absorption wedge, with a plasma between _B_ and _C_. The plasma sends out noise waves to left and right, having a power P_B and P_C respectively; the wedge emits noise waves with a power P_A, which are attenuated in the plasma to the power $P_A{}'$.

in such a way that the attenuation L is high and there are no reflections at the boundary plane C (fig. 18).

The electron temperature depends in a complicated manner on the kind of gas used, the gas pressure, the temperature of the column and the discharge current. In inert-gas discharges, T_{el} increases with a decrease in the atomic weight of the gas or its pressure, or the column diameter or the current density [15][16]. Under otherwise identical conditions, therefore, helium gives the highest electron temperature. For various reasons, however — including the short life obtained with helium and the considerable heat generated in a helium discharge — the lightest but one inert gas is preferred, namely neon.

The requirement of a high attenuation L was difficult to fulfil in the millimetre wavebands, and necessitated a new design. The elimination of reflections at C gave the greatest difficulties in the decimetre range. To provide some insight into these problems, we shall consider the complex relative dielectric constant ε_r that can be assigned to a plasma. The imaginary part of ε_r relates to the dissipation, and the ratio of the imaginary to the real part relates to the phase shift:

$$\varepsilon_r = 1 - \frac{\omega_p^2}{\omega^2 + \nu^2} + j\,\frac{\omega_p^2\nu}{\omega(\omega^2 + \nu^2)} \,. \qquad . \,(17)$$

Here ω_p is the plasma frequency, given by $\omega_p = qN^2/\varepsilon_0 m$ (where N is the electron density and ε_0 the dielectric constant of free space), ω is the angular frequency of the wave, and ν the average collision frequency; we assume that ν does not depend on the electron velocity. It may be concluded from (17) that for any given plasma (given ω_p and ν) ε_r is closer to unity the higher is the angular frequency ω, i.e. the shorter the wave. If the remainder of the waveguide in fig. 18 is filled with a non-ionized gas (e.g. air) for which $\varepsilon_r = 1$, there will be no reflection at the boundary planes. However, the absorption in the gas — and hence the attenuation L — shows a marked decrease, for as ω increases, the imaginary part of ε_r approaches zero. A high attenuation L can therefore only be obtained by filling the waveguide with plasma *over a considerable length*.

The construction that most closely approaches this schematic picture is that shown in fig. 17c. Before discussing this, we shall consider the more conventional designs sketched in fig. 17a and b.

A gas-discharge noise source for decimetre waves

Since waveguides for decimetre waves would have to be very large, *coaxial* systems are generally used in this wave range. A suitable noise source can be seen in fig. 17a. A gas-discharge tube G, e.g. a type K 50 A tube, is placed inside a silver-plated helix Sp, which effects the coupling with the gas discharge and at each end passes into the central conductor of a coaxial system. The coupling takes place over the entire length of the helix, and thus has the gradual nature that enables the reflection to be kept at a very low value. One end of the helix is connected to a matched termination (the resistance R of temperature T_1). The helix is dimensioned so that it has a characteristic impedance of 50 ohm, equal to that of the output plug.

Further particulars of the dimensioning will be found in the literature [17].

A gas-discharge noise source for centimetre waves

Fig. 17b shows the commonly used system designed by Johnson and Deremer for waveguides of not all too small dimensions [18]. The gradual transition between the column and the waveguide — necessary for minimizing reflections from the column — is obtained here by passing the gas-discharge tube obliquely through the waveguide (at an angle φ). The tube current (on which ω_p depends, see eq. 17) and the angle φ can be chosen in such a way that the reflection from the column is practically zero. At one end the waveguide has a matched termination in the form of an absorption wedge.

Since the cathode and anode of the tube are outside the waveguide, there is a danger that a considerable part of the noise power will be lost through the side arms M which contain the discharge tube. To avoid such losses, the arms are made so narrow that their lowest cut-off frequency is higher than the operating frequency of the noise generator. This means, of course, that the cross-section of the column must be quite a bit smaller than that of the waveguide; as a result the attenuation L is not maximum, but this is not a serious objection in the centimetre waveband, where the attenuation is still amply sufficient for the purposes for which it is used.

Table I gives some equivalent noise temperatures obtained in our laboratory with noise generators of

[16] A. von Engel and M. Steenbeck, Elektrische Gasentladungen, Part II, Springer, Berlin 1934, p. 85 *et seq.*
F. M. Penning, Electrical discharges in gases, Philips Technical Library 1957, p. 58 *et seq.*

[17] H. Schnittger and D. Weber, Über einen Gasentladungs-Rauschgenerator mit Verzögerungsleitung, Nachr.-techn. Fachber. **2**, 118-120, 1955.
[18] H. Johnson and K. R. Deremer, Gaseous discharge super-high-frequency noise sources, Proc. Inst. Radio Engrs **39**, 908-914, 1951.

Table I. Equivalent noise temperatures obtained with various types of discharge tube in the noise generator shown in fig. 17b (except for the last line, which relates to the construction in fig. 17c).

Freq.	Wave-length	Gas-discharge tubes				Equivalent noise temperature
		type	gas	pressure	current	
Gc/s	cm			torr	mA	°K
4	7.5	K 51 A	Ne		200	23 800
		exper.	Xe	10	150	9 550
6	5	exper.	Ne	8	125	23 400
		exper.	Ar	8	125	14 000
10	3	exper.	He	10	125	28 000
		K 50 A	Ne		125	21 700
		exper.	Ar	8	125	14 000
		exper.	Xe	5.5	125	9 400
34	0.88	exper.	He	40	100	22 700
		exper.	Ne	90	100	20 600
		exper.	Ar	40	75	13 400
		exper.	Kr	20	95	11 200
		exper.	Xe	8	75	9 200
75	0.40	exper.	Ne	100	75	21 000

this type [19]); a few of the discharge tubes employed are shown in *fig. 19*. It can be seen from the table that high noise temperatures T_{eq} are achieved with helium, in accordance with the expected high

---

[19]) For further details of experiments at frequencies from 10 to 75 Gc/s see: P. A. H. Hart and G. H. Plantinga, Millimetrewave noise of a plasma, Proc. 5th Internat. Conf. on ionization phenomena in gases, Munich 1961, pp. 492-499 (North-Holland Publishing Co., Amsterdam 1962).

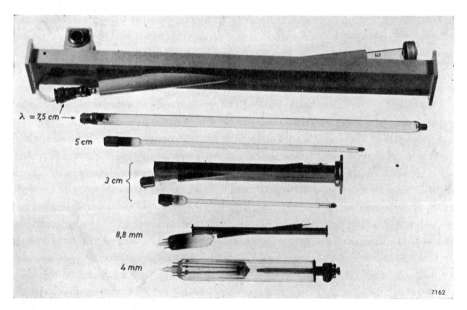

Fig. 19. Experimental gas-discharge noise sources. From top to bottom:
a 7.5 cm waveguide with discharge tube passing obliquely through it,
the 7.5 cm tube separately,
a tube for 5 cm wavelength,
a 3 cm waveguide with tube,
the 3 cm tube separately,
an 8.8 mm waveguide with tube,
a 4 mm tube as described in reference [20]).
The tubes are all shown with the cathode on the left and the anode on the right, the waveguides with the output on the right.

electron temperature T_{el}. We have already mentioned, however, some of the reasons why this gas is nevertheless unsuitable (short life, excessive heat generation). A further reason is that, with helium, T_{el} (and hence T_{eq}) is highly dependent on impurities in the gas. Neon does not have these drawbacks and the noise temperatures obtained are only in a few cases lower than those achieved with helium; this is attributable to the high attenuation L which is possible in neon [19]). Neon is therefore the gas preferred in practice.

It is also important to note that the values of T_{eq} obtained with neon are fairly close to one another, in spite of the markedly divergent conditions (gas pressure, current and frequency), which promises well for its usefulness at even higher frequencies.

A gas-discharge noise source for millimetre waves

The small dimensions of waveguides for waves shorter than about 8 mm make the construction in fig. 17b difficult for practical reasons. For millimetre waves the construction shown in fig. 17c is more suitable; this was described some time ago in this review [20]), and can therefore be dealt with here very briefly.

In a circular waveguide W (fig. 17c) a thin-walled tube Q of quartz glass is introduced. A flared part of the tube contains the oxide cathode K for the gas discharge. The neon is contained (at a pressure of 100 torr) in the tube Q, and also in the waveguide, which is closed at V by a mica window. Since the mean free path in the gas is very short (approx. 6.5 μ), the plasma ends fairly abruptly where the quartz-glass tube ends, and at that position, at An, the anode is formed by the inside of the waveguide; beyond that point, then, the neon is not ionized. The last line of Table I relates to a tube of this type.

From equation (16) we saw that a high attenua-

---

[20]) P. A. H. Hart and G. H. Plantinga, An experimental noise generator for millimetre waves, Philips tech. Rev. **22**, 391-392, 1960/61 (No. 12).

tion L is necessary for matching in a wide range of frequencies. In this construction, the attenuation is obtained through the column, which is here in the axial direction of the waveguide. Since a column can be made arbitrarily long (provided the applied voltage is high enough), the attenuation can in principle be made as large as required by simply making the quartz-glass tube and the waveguide long enough.

The wall of the quartz-glass tube is made very thin (about 0.1 mm); this gives the filling factor of the plasma in the waveguide a high value and limits the losses resulting from noise power leaking away along the tube.

As indicated in the last line of Table I, a noise temperature of 21 000 °K has been achieved at a wavelength of 4 mm.

We shall now briefly consider what the minimum and maximum frequencies are for the noise generator of fig. 17c.

Minimum frequency

The lower the frequency, the larger must be the diameter of the waveguide. This has adverse consequences for the gas discharge. The diameter of a

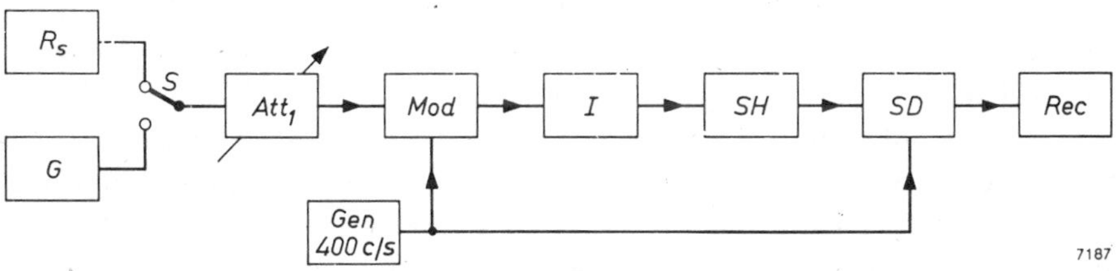

Fig. 20. Calibration of a gas-discharge tube G with a standard noise resistor R_s. S switch. Att_1 calibrated attenuator. Mod modulator (directional isolator) which modulates the noise with a 400 c/s signal delivered by the generator Gen. SH superhet receiver. SD synchronous detector. Rec recorder. I directional isolator which prevents the noise from SH from reaching the modulator (this noise would otherwise be modulated and cause errors).

positive column cannot be widened indefinitely without giving rise to effects such as constriction and striations. The positive column then gradually loses its character and the electron temperature drops. Moreover, as seen from equation (17), as ω decreases the relative dielectric constant differs more and more from unity; this makes it evident that the abrupt ending of the column at the anode will increasingly give rise to reflection.

For these reasons the lowest frequency at which the noise generator in fig. 17c can be used with advantage is in the region of 35 Gc/s.

Maximum frequency

With increasing frequency the attenuation of the column per unit length decreases. To keep L large enough, it is therefore necessary to make the column and the waveguide longer. Lengthening the waveguide increases the losses in the waveguide wall.

There is thus a danger that these losses will finally predominate. The noise would then largely be due to the guide wall, whose temperature is relatively low (400 to 500 °K).

One way of getting around this difficulty is to make the waveguide relatively wide. For a wavelength of 4 mm, for example, we have made the inside diameter of the waveguide 4 mm. Outside the noise generator a gradual transition is then needed to the usual rectangular waveguide.

The upper limiting frequency is probably higher than 300 Gc/s.

Calibration of gas-discharge noise sources

A resistor, and also a noise diode (at least in a wide range of frequencies) can be regarded as absolute noise standards; a gas discharge, however, cannot be so regarded and must therefore be calibrated. To conclude this article, we shall briefly consider this process of calibration.

Our gas-discharge noise sources were calibrated by comparing them with a standard noise source. This consisted of a resistance in the form of an absorption wedge (the matched termination in fig. 17) whose temperature was adjusted as accurately as possible to 1336 °K, the melting point of gold [5]).

Fig. 20 shows a simplified block diagram of the set-up; the major part of the equipment can be seen in *fig. 21*. The part on the right of the switch is a somewhat simplified version of Dicke's noise receiver, as used in radio astronomy [21]). This consists essentially of a modulator *Mod*, which modulates the incoming noise with an audio signal (400 c/s), followed by a superheterodyne system *SH*, a synchronous detector *SD* and a recorder *Rec*. In the synchronous detector the output signal from the superhet receiver is compared with the 400 c/s signal, with the result that the recorder responds only to a signal modulated with 400 c/s, i.e. to the incoming noise.

The modulator is a modified version of a directional isolator using Faraday rotation [22]). The modi-

---

[21]) R. H. Dicke, The measurement of thermal radiation at microwave frequencies, Rev. sci. Instr. **17**, 268-275, 1946. See also: C. A. Muller, A receiver for the radio waves from interstellar hydrogen, II. Design of the receiver, Philips tech. Rev. **17**, 351-361, 1955/56.

[22]) H. G. Beljers, The application of ferroxcube in unidirectional waveguides and its bearing on the principle of reciprocity, Philips tech. Rev. **18**, 158-166, 1956/57.

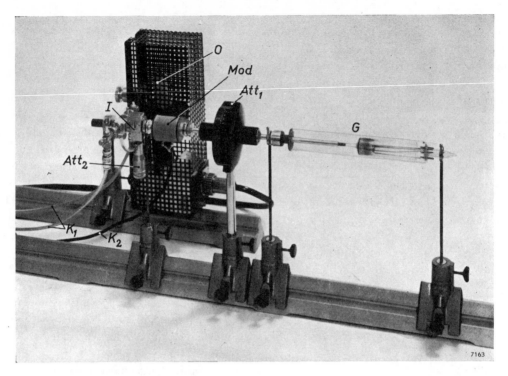

Fig. 21. Part of the equipment for calibrating the gas-discharge tube G (here a 4 mm tube). For Att_1, Mod and I, see fig. 20. O local oscillator (4 mm klystron) of the superhet receiver. Att_2 variable attenuator for adjusting the signal from O to the correct value. The cables K_1 go to the IF amplifier (with push-pull input); cable K_2 comes from the 400 c/s oscillator.

fication consists in the permanent magnet having been replaced by an electromagnet, which is energized by the 400 c/s signal. To prevent the inherent noise of the receiver also being modulated via reflections from the modulator, a second directional isolator is inserted between the modulator and the superhet receiver.

The procedure of calibration is as follows. First of all, the modulator is connected to the standard noise source R_s via a variable attenuator Att_1 set at minimum attenuation; the recorder shows a certain deflection. Next, the attenuator is connected to the gas-discharge noise source G, and the attenuation is adjusted until the recorder shows the same deflection as before. The gas-discharge noise source together with the attenuator now delivers just as much noise as the standard source.

Now the equivalent temperature T_{eq} of the gas-discharge noise source is given by

$$T_{eq} = B\,T_s' - (B-1)\,T_1\,,$$

where B is the inserted attenuation (including the attenuation due to losses in the line), T_1 the temperature of the attenuator, and T_s' the temperature (corrected for line losses) of the standard noise source.

As mentioned at the beginning of this article, gas-discharge noise sources are relatively insensitive to fluctuations in supply voltages and in ambient temperature, so that after calibration they can serve as sub-standards. Because of their reproducibility, it is not necessary to calibrate them individually; it is sufficient to calibrate a few samples.

Summary. Survey of the three main types of standard noise source: resistors, saturated diodes and gas discharges.

Resistors can be used as noise standards at frequencies ranging from the lowest to the ultra-high, in the order of 100 Gc/s (mm waves). For measuring the noise factor of a four-terminal network, two noise resistors are needed, one of which has the same resistance at the temperature T_1 as the other at the temperature T_2. Usually T_1 is made roughly equal to room temperature, and T_2 much higher or much lower. Examples discussed are a cold resistor for a coaxial decimetre-wave system ($T_2 = 77\,°\text{K}$, bath of liquid nitrogen) and a hot resistor, mounted in a waveguide, for centimetre or millimetre waves ($T_2 = 1336\,°\text{K}$, in an oven).

Dealing with the *saturated diode*, the author examines the limits of the useful frequency range. The lower limit is set by flicker noise and lies between 10 and 1000 c/s. The upper limit depends on the extent to which two errors occur: the transformation error and the transit-time error. The correction and mutual compensation of these errors are discussed. The type 10 P noise diode can be used in a new type of holder, without correction, up to 730 Mc/s.

On the subject of *gas-discharge* noise sources, the mechanism of the noise generation is examined. Three noise generators of this kind are reviewed: one for the decimetre band, one for the centimetre band and one for the millimetre band; they differ in the method of coupling the plasma with the waveguide. The article ends with a description of the calibration of this type of noise source.

Noise of a Self-Sustaining Avalanche Discharge in Silicon: Studies at Microwave Frequencies

Roland H. Haitz and Fred W. Voltmer*

Texas Instruments Incorporated, Dallas, Texas

(Received 25 January 1968)

The studies of avalanche noise reported by Haitz are extended to frequencies up to and above the avalanche frequency ω_a. It is found that the open-circuit spectral voltage density is flat within $\pm 5\%$ from less than 100 Hz up to frequencies approaching ω_a. Near ω_a the open-circuit spectral voltage density increases with frequency, goes through a maximum at ω_a and then decreases rapidly. The spectral power density is similarly flat but fails to exhibit an expected maximum at ω_a. For $\omega \gg \omega_a$ the spectral power density decreases with ω^{-4}. The experimental results are in good agreement with Hines' theory of avalanche noise for $\omega \ll \omega_a$ and $\omega \gg \omega_a$. The discrepancies observed near ω_a are thought to be caused by an oversimplification in Hines' theory which neglects internal diode losses. Compared with conventional noise sources such as temperature-limited diodes or gas discharge tubes, avalanche diodes have several advantages: larger noise output, larger bandwidth, lower $1/f$ noise, low power consumption, small size, and low weight.

I. INTRODUCTION

Recent interest in avalanche devices such as microwave oscillators and avalanche photodetectors has triggered extensive studies of the noise properties of a stable-burning, self-sustaining avalanche discharge. Several papers on the theory of avalanche noise have been published within the last two years. Hines[1] discussed the avalanche noise in terms of a simplified junction model using equal ionization rates for electrons and holes. Gummel and Blue[2] developed a more general theory for arbitrary junction profiles and realistic ionization coefficients $\alpha \neq \beta$. Their theory reduces to Hines' theory for the corresponding simplifying assumptions. In a different approach based on the multiplication of carriers injected into the junction both Tager[3-5] and McIntyre[6] calculated the low-frequency noise of a self-sustaining avalanche discharge by extending the carrier-multiplication factor to infinity.

The noise theory of Hines[1] has several advantages. It yields fairly simple analytical solutions for the spectral noise density at frequencies below and above the avalanche frequency ω_a, while McIntyre[6] gives only a low-frequency solution. The theory of Gummel and Blue[2] yields only a computer solution in its most general case. The analytical solution of Gummel and Blue's theory for a simplified but still more general model than in the case of Hines' theory is still too complicated for most practical purposes. It is, therefore, of considerable interest to determine the validity and accuracy of Hines'[1] theory.

There is also an interesting discrepancy between the theories of Hines[1] and Tager.[3] While Hines' theory predicts that the spectral power density decreases with ω^{-4} for $\omega \gg \omega_a$ Tager's theory predicts only an ω^{-2} cutoff. Furthermore, in Tager's theory the avalanche frequency depends on the carrier-multiplication factor M, which is a sensitive function of the diode leakage current for large values of M. Tager's theory, therefore, predicts a strong temperature dependence of the avalanche frequency, which is expected to be in the range of 10^6–10^8 Hz, while Hines' values for $\omega_a/2\pi$ are in the range of 10^9–10^{10} Hz and practically temperature independent. Avalanche-noise measurements at microwave frequencies are expected to clarify this discrepancy.

The experimental studies of avalanche noise at Texas Instruments were divided into two sections. In the first part[7] we investigated the validity and accuracy of Hines' low-frequency solution of avalanche noise. There it was found that Hines' theory very accurately describes the functional relationship between the open-circuit spectral voltage density and breakdown voltage and avalanche current. In spite of its crude simplifications Hines' low-frequency solution is accurate within better than a factor of two.

It is the subject of this paper to investigate the validity and accuracy of Hines' theory at microwave frequencies, particularly around and above the avalanche frequency ω_a. Preliminary results of this work were reported in an earlier publication.[8]

* Summer Development Program participant from University of California, Berkeley, Calif.

[1] M. E. Hines, IEEE Trans. Electron Devices **ED-13,** 158 (1966).

[2] H. K. Gummel and J. L. Blue, IEEE Trans. Electron Devices **ED-14,** 569 (1967).

[3] A. S. Tager, Fiz. Tverd. Tela **6,** 2418 (1964) [Sov. Phys.—Solid State **6,** 1919 (1965)].

[4] V. M. Val'd-Perlov, A. V. Krasilov, and A. S. Tager, Radiotekhn. i Elektron. **11,** 2008 (1966) [Radio Eng. Electron. (USSR) **11,** 1764 (1966)].

[5] A. S. Tager, Usp. Fiz. Nauk **90,** 631 (1966) [Sov. Phys.—Usp. **9,** 892 (1967)].

[6] R. J. McIntyre, IEEE Trans. Electron Devices **ED-13,** 164 (1966).

[7] R. H. Haitz, J. Appl. Phys. **38,** 2935 (1967).

[8] R. H. Haitz and F. W. Voltmer, Appl. Phys. Letters **9,** 381 (1966).

Reprinted with permission from *J. Appl. Phys.*, vol. 39, pp. 3379–3384, June 1968.

327

II. HINES' THEORY OF AVALANCHE NOISE

Hines' theory of avalanche noise leads to a fairly simple expression for the open-circuit spectral voltage density[7,9]

$$\langle u^2 \rangle / \Delta f = (q\kappa^2/3m^2)(\tau_{tr}^2/\tau_z^2)(V_b^2/I)[1/(1-\omega^2/\omega_a^2)^2], \quad (1)$$

where the avalanche frequency ω_a is given by[9]

$$\omega_a^2 = 3mv_d I/\kappa\epsilon A V_b. \quad (2)$$

Here q denotes the electron charge; $m\approx5$ is a constant describing the power dependence of the ionization coefficient on field; $\kappa\approx0.5$ is defined by $V_a=\kappa V_b$, where V_a is the voltage across the avalanche region and V_b the breakdown voltage; $\tau_{tr}=W/v_d$ is the transit time of carriers through the space-charge layer of width W at the saturated drift velocity v_d; τ_z is the mean time between two ionization events of a carrier in the avalanche zone; I is the avalanche current; and A denotes the breakdown area. Since τ_{tr}/τ_z is constant for all practical purposes we can express the first two terms of Eq. (1) by a constant a^2:

$$a^2 = (q\kappa^2/3m^2)(\tau_{tr}^2/\tau_z^2). \quad (3)$$

From noise measurements at low frequencies it is found that $a^2=3.3\times10^{-20}$ A/Hz, in good agreement with Eq. (3).

From Eq. (1) it is seen that the open-circuit spectral voltage density is frequency independent up to frequencies approaching ω_a. Near ω_a the open-circuit noise voltage increases rapidly with frequency. For an ideal diode with no internal losses the open-circuit spectral voltage density would reach infinity at ω_a. Above ω_a the noise voltage rapidly decreases with frequency according to the ω^{-4} relation of Eq. (1).

The avalanche frequency ω_a is usually located in the microwave range where noise measurements have to be made as power measurements. In a simple circuit containing no reactive elements one obtains the following expression for the spectral power density transferred to the load R_L:

$$w_n(\omega) = a^2(V_b^2/I)[1/(1-\omega^2/\omega_a^2)^2]$$
$$\times [R_L/(R_{sc}+R_{sp}+R_L)^2]. \quad (4)$$

Here R_{sc} denotes the space-charge resistance which is given by Eq. (5) for small transit angles $\theta\approx\omega\tau_{tr}$:

$$R_{sc} = (0.7W)^2/2\epsilon v_d A(1-\omega^2/\omega_a^2) = R_{sc}^0/(1-\omega^2/\omega_a^2). \quad (5)$$

The term R_{sc}^0 is the low-frequency space-charge resist-

---

ance. The spreading resistance R_{sp} can be approximated by

$$R_{sp} = \rho/2D, \quad (6)$$

with ρ denoting the resistivity of the lowly doped side of the p–n junction, usually the bulk resistivity, and D the diameter of the breakdown area A.

According to Eq. (4) the spectral power density is independent of frequency for $\omega\ll\omega_a$. A small frequency dependence, however, arises from the thermal resistance of the diode which is neglected in Eq. (4). This frequency dependence is of the order of 1 dB/decade or less.

For frequencies approaching ω_a, R_{sc} is the dominating contribution to the diode resistance. If the external load resistance R_L is tuned for maximum power transfer ($R_L=R_{sc}\gg R_{sp}$) then $w_n(\omega)$ is given by

$$w_n(\omega) = a^2 V_b^2/4I R_{sc}^0(1-\omega^2/\omega_a^2). \quad (7)$$

For frequencies $\omega\gg\omega_a$ the spreading resistance R_{sp} is the dominating contribution to the diode resistance. Tuning R_L for maximum power transfer leads to

$$w_n(\omega) = a^2 V_b^2/4I R_{sp}(1-\omega^2/\omega_a^2)^2. \quad (8)$$

For $\omega\gg\omega_a$, Eq. (8) predicts an ω^{-4} dependence of noise power on frequency.

III. EXPERIMENTAL RESULTS AND DISCUSSION

A. Diodes

During the studies of low-frequency avalanche noise[7] it became evident that special precautions have to be taken in order to prevent the generation of excessive noise resulting from nonuniform breakdown. The combination of both a guard ring to prevent edge breakdown and a small breakdown area to reduce material nonuniformities have led to satisfactory results.

The small p^+n guard-ring diodes manufactured for the avalanche-noise studies at microwave frequencies are essentially the same as the diodes used for low-frequency noise studies. The only difference is their reduced guard-ring size to cut down the junction capacitance. For a diode with a breakdown region of diameter D (equal to the inner diameter of the guard ring) the external diameter of the guard ring is equal to $D+42\ \mu$. The guard-ring width of 21 μ is partly due to the width of the guard-ring mask (10 μ) and partly due to underdiffusion ($2\times5.5\ \mu$). The diodes are manufactured by a double boron diffusion into n-type Si. The back contact is made by alloying AuSb. The front contact consists of an Al dot of diameter $D+11\ \mu$. The diodes are mounted in a microwave pill package.

Diodes with five different breakdown diameters are obtained from each slice: $D=13, 36, 63, 89$, and 113 μ. The small 13-μ diodes are of no use for these noise studies because of (1) their large diode resistance,

---

[9] As pointed out by C. A. Lee and R. L. Batdorf at the 1964 Solid State Device Res. Conf., Boulder, Colorado, and by Gummel and Blue[2] the numerical constant in Eq. (2) has to be 3 instead of 2. As a result, the numerical constant in the denominator of Eq. (1) has also to be replaced by 3.

(2) their relatively large uncertainty in breakdown area, and (3) their relatively large guard-ring capacitance. The 36-μ diodes are best suitable for avalanche-noise studies because their series resistance at microwave frequencies is fairly close to the characteristic impedance of a 50-Ω coaxial line. The larger diodes have a correspondingly smaller impedance and the impedance matching required to maximize the noise power transfer leads to correspondingly larger errors. Diodes with breakdown voltages of 9.5 V, 21.5 V, and 44 V were produced. The 21-V diodes were studied most extensively.

B. Measuring Techniques

The open-circuit spectral voltage-density measurements are carried out by biasing the diode through a 3-kΩ bias resistor which is large compared with the internal diode resistance. For measurements at frequencies below 10 kHz the avalanche noise is amplified by a preamplifier with low $1/f$ noise and then measured with a wave analyzer. For measurements between 10 kHz and 18 MHz the avalanche noise is measured directly with another wave analyzer. Between 18 MHz and 200 MHz the avalanche noise is down converted to 1 MHz and then measured with a wave analyzer. The open-circuit spectral voltage density for frequencies above 200 MHz is obtained from noise power and impedance measurements discussed in the remainder of this subsection.

The spectral noise power-density measurements are made with the coaxial system shown in Fig. 1. The diode is biased through a bias tee. The double stub tuner is used to tune for maximum noise power output

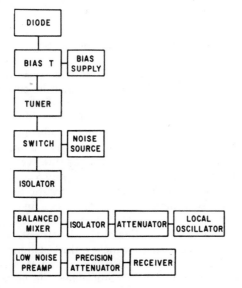

FIG. 1. Block diagram of circuit used to measure spectral noise power density.

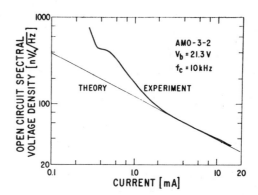

FIG. 2. Open-circuit spectral voltage density at low frequencies as a function of current.

by matching the diode impedance to the load. The mixer converts the avalanche noise to an i.f. frequency of 30 MHz. The mixer output is amplified and measured with a 30-MHz precision receiver. The avalanche noise is compared with the output of a gas discharge noise standard.

The diode impedance is measured using the standard standing-wave technique. A coaxial slotted line is inserted between bias tee and diode. A shorted package is used as reference short. The injected signal is kept low enough so that the peak ac current does not exceed 50% of the dc current.

C. Results and Discussion

The low-frequency open-circuit spectral voltage density shows the typical current dependence of small guard-ring diodes.[7] The results are plotted in Fig. 2. At high currents $(\langle u^2 \rangle / \Delta f)^{1/2}$ decreases inversely with $I^{1/2}$ as predicted by Eq. (1). The straight line of Fig. 2 is calculated from Eqs. (1) and (3) using a value of $a^2 = 3.3 \times 10^{-20}$ A/Hz. This value of a^2 was found earlier[7] to be characteristic for a large variety of p^+n, n^+p and $n^+p\pi p^+$ diodes with breakdown voltages between 6 V and 90 V. The agreement between these measurements and earlier low-frequency noise studies[7] is excellent.

At currents below 2.5 mA the measured noise is larger than the noise predicted by Eq. (1). This discrepancy, which is typical for avalanche diodes at low-current densities, is not serious. It is caused by extremely small nonuniformities of the breakdown voltage.[10] As for the diode of Fig. 2, the breakdown uniformity of all diodes used in this noise study is checked by a measurement of the low-frequency noise-current characteristic. Only diodes with good agreement between calculated and measured noise at currents in the 3–20 mA range are used.

[10] R. H. Haitz, *Physics of Failure in Electronics*, Vol. V, pp. 447–456 (1967), Rome Air Development Center, USAF.

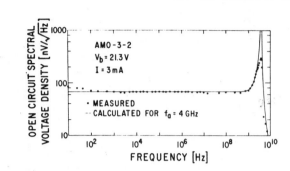

FIG. 3. Open-circuit spectral voltage density at a current of 3 mA as a function of frequency.

The open-circuit spectral voltage density, plotted in Fig. 3 as a function of frequency, exhibits the expected dependence on frequency. For $\omega \ll \omega_a$ the noise voltage is independent of frequency. The variation of $(\langle u^2 \rangle/\Delta f)^{1/2}$ between 100 Hz and 1 GHz is less than $\pm 5\%$. Below 100 Hz the noise increases slightly. This increase is believed to be caused either by a $1/f$-type noise contribution superimposed on the avalanche noise or by 60-Hz pickup passing through the filter tails of the wave analyzer. Above 1 GHz the noise voltage increases rapidly with frequency reaching a maximum approximately four times above the low-frequency noise level. The solid line in Fig. 3 is calculated from Eqs. (1) and (3) for $a^2 = 3.3 \times 10^{-20}$ A/Hz and $\omega_a/2\pi = f_a = 4.0$ GHz. The discrepancy between experimental and calculated noise near f_a is at least partly caused by simplifications in Hines' theory which completely neglects internal diode losses. It has to be mentioned here that the experimental points above 1 GHz are calculated from admittance and noise-power measurements which will be reported in the remainder of this section.

The admittance measurements needed to convert noise-power measurements to open-circuit spectral-voltage density data are plotted in Fig. 4 for diode AMO-3-2. At frequencies below 800 MHz the diode presents a fairly constant conductance of approximately 15 mS. The conductance drops rapidly above 1 GHz reaching a minimum of 3.3 mS near 4.0 GHz. The susceptance is negligibly small up to 2 GHz. Above 2 GHz the capacitive susceptance increases rapidly. The capacitive susceptance at frequencies below the conductance minimum (4.0 GHz) is not expected from Gilden and Hines'[11] avalanche-diode theory, which predicts an inductive admittance for $\omega \lesssim \omega_a$. Realizing, however, that the guard-ring area, and consequently the guard-ring capacitance, is four times bigger than the breakdown area one finds a quantitative agreement between measured and expected susceptance.

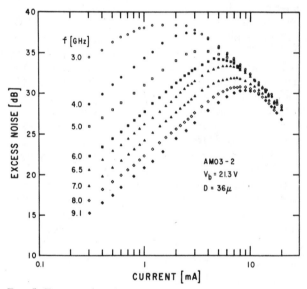

FIG. 5. Excess noise above kT as a function of current for frequencies between 3.0 and 9.1 GHz.

The essential experimental results on spectral power density $w_n(\omega)$, as a function of avalanche current and frequency are presented in Fig. 5 for diode AMO-3-2. $w_n(\omega)$ is expressed in terms of excess noise above thermal noise. To interpret these results in terms of Hines' noise theory we will first give a qualitative interpretation and then go into quantitative details.

From a qualitative inspection of Fig. 5 it is evident that all curves have several characteristic features in common. With increasing current $w_n(\omega)$ first increases approximately linearly, then goes through a maximum and finally decreases linearly with current. Furthermore, with increasing frequency the maximum of the spectral power density decreases and shifts to higher currents. Finally, for currents above the noise maxi-

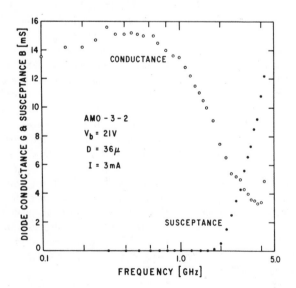

FIG. 4. Diode conductance and susceptance at a current of 3 mA as a function of frequency.

[11] M. Gilden and M. E. Hines, IEEE Trans. Electron Devices ED-13, 169 (1966).

mum, $w_n(\omega)$ is practically frequency-independent while for currents below the noise maximum it is very strongly dependent on frequency.

The qualitative interpretation is continued by comparing the above-mentioned characteristic features with the functional relations predicted by Eqs. (7) and (8). As a first step it is assumed that at the current at which $w_n(\omega)$ reaches a maximum, I_{max}, the avalanche frequency equals the measurement frequency: $\omega_a = \omega(I_{max})$. Under this assumption, which will find its justification later in the quantitative interpretation, it is evident that ω_a increases with current as predicted by Eq. (2). Furthermore, one can conclude that $\omega \lessgtr \omega_a$ for $I \gtrless I_{max}$, or in other words, to the right of the maxima $\omega < \omega_a$ and to the left $\omega > \omega_a$. For $\omega < \omega_a$ Eq. (7) predicts a frequency dependence in the immediate vicinity of ω_a, but frequency-independent noise for $\omega \ll \omega_a$. While the data of Fig. 5 do not confirm the first prediction they agree very well for $\omega \ll \omega_a$. For $\omega > \omega_a$ Eq. (8) predicts strongly decreasing noise with increasing frequency turning into an ω^{-4} dependence of $w_n(\omega)$ for $\omega \gg \omega_a$. This predicted strong frequency dependence of $w_n(\omega)$ for $\omega > \omega_a$ is obviously met for currents to the left of the noise maxima of Fig. 5.

Summarizing the qualitative interpretation it can be said that Eqs. (7) and (8) together with Eq. (2) fairly well predict the functional relationship between spectral power density and current and frequency.

The quantitative interpretation is discussed with reference to Fig. 6, where $w_n(\omega)$ is plotted as a function of frequency. Noise power-density data from Fig. 5 and data from additional measurements are plotted for three current levels: 0.5, 3, and 15 mA. In keeping with the admittance measurements of Fig. 4 and the noise data of Fig. 6 the avalanche frequency, $f_a = \omega_a/2\pi$, is chosen as 4.0 GHz for $I = 3$ mA. Using the $\omega_a - I$ relationship of Eq. (2) one obtains: $f_a = 1.8$ GHz for $I = 0.5$ mA and $f_a = 9.0$ GHz for $I = 15$ mA. The spectral power density $w_n(\omega)$ is calculated from Eq. (7) for $\omega < \omega_a$ and from Eq. (8) for $\omega > \omega_a$. The resulting curves

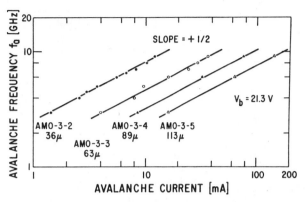

FIG. 7. Experimental values of avalanche frequency f_a as a function of current for diodes with diameters of 36, 63, 89, and 113 μ. The solid lines have a slope of $\frac{1}{2}$.

for $\omega < \omega_a$ fit the experimental data quite well for $\omega \ll \omega_a$ (low-frequency approach). In the vicinity of ω_a the experimental results do not increase as expected. This discrepancy is very likely caused by the fact that Hines' theory does not take internal losses into account. This simplification may be especially critical in the case of guard-ring diodes where the guard ring presents a substantial capacitive load. The spectral power density calculated from Eq. (8) for $\omega > \omega_a$ using $R_{sp} = 20$ Ω is approximately 5 dB above the experimental results. Since the experimental points for $I = 0.5$ mA exhibit the expected ω^{-4} dependence over a range of 20 dB we have used a larger resistance value in order to fit the curves to the experimental points. A value of $R_s = 67$ Ω (equal to the low-frequency diode resistance) is then used to calculate the two curves in Fig. 6 for $\omega > \omega_a$.

To conclude the quantitative interpretation we will investigate the current and area dependence of ω_a predicted by Eq. (2). The avalanche frequency $f_a = \omega_a/2\pi$ is determined experimentally from $w_n(\omega) - I$ plots similar to that of Fig. 6 by extrapolating the linear sections below and above I_{max}. The intersection of the two lines yields a point for the $f_a - I$ curves of Fig. 7. The solid lines used to fit the experimental points have a slope of $\frac{1}{2}$ as predicted by Eq. (2). From these results it can be concluded that Eq. (2) yields the proper current dependence of ω_a.

The dependence of ω_a on area A is investigated by similar $f_a - I$ determinations obtained on larger diodes from the same slice. The three curves for diodes with diameters of 63, 89 and 113 μ are also plotted in Fig. 7. The resulting relationship between f_a and breakdown diameter is plotted in Fig. 8 for $I = 10$ mA. The solid curve in Fig. 8 is calculated from Eq. (2) for $m = 5$, $v_d = 10^7$ cm/sec, $\kappa = 0.5$, $\epsilon = 10^{-12}$ A·sec/V·cm and $V_b = 21.3$ V. For the two larger diodes the agreement between experimental points and calculated curve is quite good. However, with decreasing diameter the experimental points seem to drop further and further

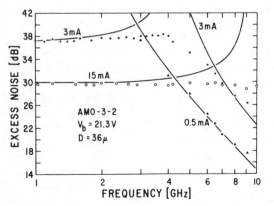

FIG. 6. Excess noise above kT as a function of frequency for currents of 0.5, 3, and 15 mA.

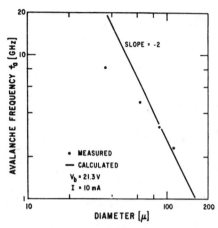

FIG. 8. Avalanche frequency f_a as a function of breakdown diameter. The line with a slope of -2 is calculated from Eq. (2).

below the calculated curve. This discrepancy for diodes with decreasing breakdown diameter is very likely caused by the increasing importance of parasitic capacitance from the guard ring and package.

The 20-dB drop in noise power following an ω^{-4} relationship and the good agreement between measured values of ω_a and the values calculated from Hines' theory strongly indicate that Tager's[3] approach of calculating the noise of a self-sustaining avalanche discharge using a carrier multiplication model is invalid.

IV. APPLICATION AS NOISE SOURCE

The experimental results presented in Figs. 3 and 6 and similar investigations by Constant et al.[12] and Dalman et al.[13] suggest the application of avalanche diodes as noise sources. Actually, such an avalanche-noise source is expected to have several advantages over conventional noise sources such as a temperature-limited diode or a gas-discharge tube. The following comments hold for a 20-V, 50-Ω diode in a coaxial system:

High Noise Output: Noise figures of 40 dB up to 3 GHz, 30 dB up to 10 GHz, and 25 dB up to 18 GHz can be obtained readily.

Large Bandwidth: A broadband coaxial noise source covering the range from 100 MHz to 12 GHz with a noise figure of 30 dB and an output variation of ± 0.5 dB over the entire frequency range can be designed by slightly modifying the AMO-3-2 diode. A broadband coaxial source from 100 MHz to 18 GHz with a noise figure of 25 dB is also feasible.

Low 1/f Noise: Uniform open-circuit spectral voltage density down to frequencies of less than 10 Hz.

Low Power Consumption: An input power of 50 mW is sufficient to drive a diode generating noise 40 dB

[12] E. Constant, R. Gabillard, A. Hautducoeur, and A. Chadelas, Compt. Rend. **262B**, 16 (1966).
[13] G. C. Dalman and L. F. Eastman, presented at the 1967 Intern. Electron Device Meeting, Washington, D. C.

above kT up to 3 GHz, 300 mW for 30 dB up to 12 GHz and 800 mW for 25 dB up to 18 GHz. The increase in power with frequency is caused by the current dependence of ω_a.

Variable Noise Output: Since the noise output is a linear function of I^{-1} the noise output level can be readily varied thus greatly simplifying a large fraction of conventional noise measurements.

Pulsed Operation: The diodes can be readily pulsed into breakdown with subnanosecond rise and fall times. No voltage spikes are required to trigger the avalanche discharge.

High Reliability: The mean time between failure of such guard-ring avalanche diodes is expected to greatly exceed the failure time of conventional gas discharge tubes or temperature-limited noise diodes.

Large Operating Temperature Range: The avalanche noise diodes are expected to operate satisfactorily over a range from less than $-200°C$ to more than $+200°C$ with only a small temperature dependence of the noise output ($<10^{-2}/°C$).

Small Size and Low Weight: Size and weight of the complete noise source will be determined by coaxial holder and power supply. Holder size can readily be kept below 10 cm³. Compared with a conventional gas-discharge tube the size of the power supply can be greatly reduced because of the small power consumption of an avalanche noise diode.

The above cited advantages of avalanche noise sources make this device the ideal noise source for all applications with critical restrictions in size, weight, power consumption, reliability, output level and bandwidth. It is believed that the frequency range of specially designed avalanche noise diodes can be extended far into the millimeter wave region.

V. CONCLUSIONS

The studies of avalanche noise reported here and in an earlier publication[7] lead to the conclusion that Hines' noise theory,[1] in spite of its simplifications, yields a fairly accurate description of avalanche noise. For frequencies $\omega \ll \omega_a$ Eqs. (1) and (4), which are based on Hines' theory, describe the open-circuit spectral voltage density and the spectral power density quite well. For $\omega \gg \omega_a$ Eq. (4) predicts the proper ω^{-4} dependence of the spectral power density but predicts a value approximately 5 dB above the experimental points. Around ω_a Hines' theory is most inaccurate since it neglects internal diode losses. Part of the inaccuracy near ω_a and above is also thought to be caused by the relatively large guard-ring capacitance of the diodes under investigation. Small mesa diodes recently of interest as avalanche transit time oscillators should be more suitable for accurate noise measurements around and above ω_a.

Standard L.F. Noise Sources using Digital Techniques and their Application to the Measurement of Noise Spectra

By

Professor H. SUTCLIFFE,
M.A., Ph.D., C.Eng., F.I.E.E.[†]

and

K. F. KNOTT, B.Eng., Ph.D.[†]

Reprinted from the Proceedings of the I.E.R.E. Conference on 'Digital Methods of Measurement' held at the University of Kent at Canterbury on 23rd to 25th July 1969.

Circuits of a random noise source and a pseudo-random noise source, both using digital techniques, are described and an account is given of their auto-correlation functions and power spectra. Their value as standard signal sources in noise power measurements is discussed and conclusions are reached about their relative merits.

1. Introduction

Within the general field of electronic instrumentation there is a trend towards the replacement of analogue circuits by their more precise digital versions, and the provision of precise noise sources is no exception. Especially at low frequencies the construction of a random noise source of known spectral intensity can be achieved more effectively by employing digital techniques than by attempting to exploit physical effects such as thermionic shot noise. In Section 2 of this paper a standard random low-frequency noise source based on digital techniques is described and in Section 3 a rather similar circuit employing pseudo-random sequences is discussed briefly. A comparison of the performances of these two circuits, together with some comments on experimental techniques concerning the measurement of noise spectra, appears in Section 4.

2. A Random Low-frequency Standard Noise Source

Of all the qualities needed of a standard noise source the accuracy of its spectral intensity is the most important. The distribution of amplitude is of little significance for measurements of noise spectra, since filtering action during the measurements has the effect of producing a normal distribution. This statement is justified by considering the response of a frequency selective filter in the time domain when the input function is wide-band in the frequency domain. The filter output at any instant may be regarded as the sum of a large number of independent effects, a situation typical of the normal distribution.

It is appropriate, therefore, to use as a basis for the noise source a random waveform which can be generated by precise processes, and the *random telegraph signal* is a natural choice. This waveform

---

† Department of Electrical Engineering, University of Salford, Salford M5 4WT.

was described in the early literature on random fluctuations, possibly because the ease of deducing its auto-correlation function makes it a good example for demonstrating the Wiener-Khintchine theorem.[1] This theorem is given in equations (1) and (2) in its practical form, that is for real sinusoids of positive frequency:

$$v(t)v(t-\tau) = R(\tau)$$

$$= \int_0^\infty G(f) \cos 2\pi\tau f . df \quad \text{volts}^2 \quad ......(1)$$

$$G(f) = \int_0^\infty 4R(\tau) \cos 2\pi\tau f . d\tau \quad \text{volts}^2/\text{Hz} \quad ...(2)$$

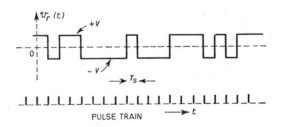

Fig. 1. Random telegraph signal.

The random telegraph wave $v_r(t)$ is shown in Fig. 1, together with a train of clock pulses with which it is associated. On the arrival of a clock pulse at intervals T_s a random choice is imposed on $v_r(t)$, whether to acquire the value $+V$ or $-V$ for the duration of the next interval T_s. The autocorrelation function $R_r(\tau)$ of $v_r(t)$ is derived by simple reasoning and is shown in Fig. 2 together with the corresponding spectral intensity $G_r(f)$ derived from equation (2). For frequencies small compared with $f_s = 1/T_s$ the value of $G_r(f)$ is $2V^2 T_s$. In this expression V is the amplitude of $v_r(t)$ and can be defined with great precision by simple circuits. T_s is also precise since it is simply the period of a regular pulse train. The

Reprinted with permission from *The Radio and Electron. Engineer*, vol. 40, pp. 132–136, Sept. 1970.

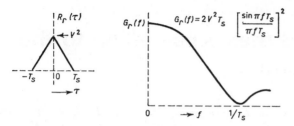

Fig. 2. Auto-correlation function of $v_r(t)$ and its spectral intensity.

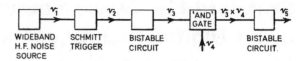

Fig. 3. Block diagram of circuit for generating random voltages.
- v_1 wideband h.f. noise
- v_2 h.f. random pulse train
- v_3 h.f. random square wave
- v_4 pulse train of interval T_s
- v_5 output (becomes $v_r(t)$ after clipping)

remaining feature open to question in the system is the method of realizing the random choice $\pm V$ at the pulse instant. The method favoured by the authors is illustrated in Fig. 3.

In Fig. 3 the inclusion of bistable C ensures that v_3 has equal probability of '1 or 0'. The broadband h.f. waveform v_1 ensures that there is no correlation in waveform v_3 between adjacent pulse intervals. Thus if the minimum setting of T_s is 100 μs, the bandwidth of v_1 is adequately large if it is in the region of 1 MHz. An exact analysis of this aspect of the design presents difficulties and would provide an interesting problem for theorists.

3. A Pseudo-random Noise Source

In recent years there has been an abundance of publications on the subject of pseudo-random binary signals generated as '*m*-sequences' or maximum length sequences.[2] The conventional method of generating these is shown in Fig. 4.

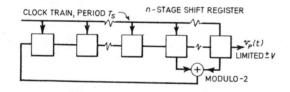

Fig. 4. Generation of p.r.b.s. as maximum length sequences.

Circuits of this type include the following features among their properties. The output waveform $v_p(t)$ often has a similar appearance, when viewed on an oscilloscope, to the random waveform $v_r(t)$ described in the previous Section. The difference is that waveform $v_p(t)$ is periodic and is repeated for every (2^n-1) intervals between clock pulses. This number is called the sequence length L, thus if the clock pulse interval is T_s, the fundamental period of $v_p(t)$ is $T_s \times L$ or $T_s(2^n-1)$. The auto-correlation function of $v_p(t)$ is shown in Fig. 5, together with the power spectrum $G_p(f)$. It is interesting to note that if L is large, then

$G_p(f)$ approximates closely to a continuous power spectrum $S_p(f)$ such that:

$$S_p(f) = 2V^2 T_s \left[\frac{\sin \pi f T_s}{\pi f T_s} \right]^2 \quad \text{volts}^2/\text{Hz}$$

This expression is similar to that of $G_r(f)$ in the previous section.

Instruments embodying these two types of noise source have been constructed and used extensively in noise spectrum investigations by the authors. Comments on the performance of the two types will be made in the next Section.

4. Application to Noise Spectrum Measurement

The first part of this paper described the different properties of true random and pseudo-random pulse trains. Let us now consider the different behaviour of the two types of waveform in l.f. noise spectra measurement. For a complete understanding of their application in this field it is perhaps useful to describe briefly a typical system used for measuring l.f. noise spectra. For the greatest accuracy it is preferable to compare the unknown source of noise with a calibrated noise generator. The block diagram of Fig. 6 illustrates such a system. The noise to be measured is amplified, filtered and detected. The change in

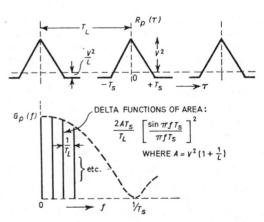

Fig. 5. Auto-correlation function of $v_p(t)$ and its power spectrum.

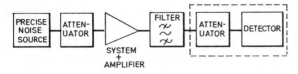

Fig. 6. Noise measurement system.

the detector reading when a known amount of additional noise is fed into the system then gives the value of the unknown noise. The detector may be a true r.m.s. meter but distinct advantages are gained if a transit-counting detector[3] is used instead. The main advantages of this type of detector are that the period of integration may be set to an arbitrary length and that the reading is immune from drift in the datum level of the signal. However, this type of detector gives a reading which is proportional to the mean magnitude of the signal. For noise which has a Gaussian amplitude probability density function there is a constant relationship between mean magnitude and r.m.s. values. For almost all types of noise found in practice the amplitude probability density function approximates closely to Gaussian after band-pass filtering so that this is not a serious restriction. A further advantage of this type of detector is that the output is in digital form and hence the accuracy of the reading can be high. The factors governing the use of the two types of digital noise generators for narrowband noise spectra will now be considered.

If a true random generator is used for calibrating the system then the detector output must be integrated over a period of time which satisfies the usual standard deviation law

$$\sigma \ll 1$$

where

$$\sigma = \left(\frac{1}{BT}\right)^{\frac{1}{2}}$$

In this expression B is the bandwidth of signal, T is the time of observation and σ is the fractional standard deviation of the observed noise power. For example, if a measurement were being made to an accuracy of 10% over a bandwidth of 0·1 Hz centred at 1 Hz, the observation time would be 1000 seconds.

If a pseudo-random generator is used for calibrating the system there may be a saving in calibration time depending on the number of discrete lines required in the frequency spectrum of the system. The spacing of the lines depends on the sampling frequency and the length of the sequence of pulses. To avoid errors due to the shape of envelope of the power spectrum of the generator, the sampling frequency should be set to about 20 times the frequency of the noise measurement so that for a fixed

frequency of measurement the lower limit of sampling frequency is fixed. This, therefore, means that the time for one sequence is inversely proportional to the spacing of the lines in the power spectrum and since the calibration time need be only one sequence period it follows that it also is inversely proportional to the line spacing.

There will thus be a saving of time if the line spacing can be increased, but this raises the question of the effect on the apparent power spectral density of reducing the number of discrete lines. This effect will depend on the shape of the frequency response of band-pass filters. Consider the ideal rectangular band-pass response as shown in Fig. 7(a), then if the centre frequency is varied slightly a serious error could be introduced in calculating the power in the bandwidth. Whereas, if a filter with a less sharp cut-off characteristic as in Fig. 7(b) were used, there would be no abrupt changes in the power contained in the bandwidth. Figure 7(b) is perhaps more typical, since it is usual in l.f. noise spectra measurements to use only two single-tuned circuits in cascade for the filter.

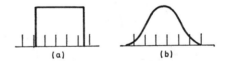

Fig. 7. Possible situations in bandpass systems.

A further question needs to be considered, namely the errors introduced by the departure of the amplitude probability density function from Gaussian when the number of lines in the spectrum is small. This will conceivably introduce errors if a transit counting or other mean magnitude method is used. It is also of interest to discover whether errors introduced in this manner are more serious than those that would be obtained from the effect described in the previous paragraph even if a true r.m.s. meter were used.

It has not yet been possible to find a satisfactory analytical solution to the problem. To obtain a measurement of the likely errors involved the following experiments were carried out using a true random noise generator and a pseudo-random generator of equal power spectral density as calculated from the sampling frequency and amplitude of the pulses. The two generators were compared by feeding them into a Bruel and Kjaer audio frequency analyser type 2107 and measuring the output with a transit counting detector. In the first experiment the setting of the analyser was kept constant at 500 Hz centre frequency and 21% bandwidth and the sampling frequency of

335

the noise generators was kept constant at 10 kHz. The sequence length of the pseudo-random generator was varied to give from 2·5 lines to 80 lines in this bandwidth. The reading of the detector was taken over one sequence period for the pseudo-random generator and for the random generator the reading was averaged over a length of time, such that the fractional standard deviation was less than 0·01. The readings agreed to within 2% for number of lines as low as 5.

In the second experiment the frequency of the analyser was varied gradually to ± 10% of the initial value and also the bandwidth was varied from 21% to 12%. The response of the detector was plotted for the pseudo-random generator set to give 2·5 to 90 lines in the bandwidth. The graph of Fig. 8 illustrates the results obtained for 5 lines in 3 dB bandwidths of 21% and 12%. Also on this graph is the response of the detector for true random noise.

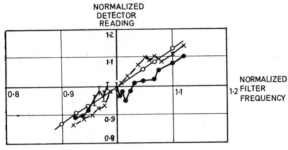

o TRUE RANDOM GENERATOR
● PSEUDO-RANDOM 5·7 LINES IN 12% B.W.
(ꟷ DENOTES READING WHICH FLUCTUATES FROM SEQUENCE TO SEQUENCE)
× PSEUDO-RANDOM 5·0 LINES IN 21% B.W.

Fig. 8. Variation in detector reading for ± 10% variation in filter frequency.

Considering the results for the pseudo-random generator it is seen that there are sharp fluctuations in the detector reading for quite small changes in frequency. Another thing to note is that when using narrow bandwidth and few lines one can encounter certain conditions under which the reading varies between sequences. This last result is rather unexpected since all input sequences are identical. A possible explanation of the result is that when there are only a few lines in the signal spectrum their phase relation is critical. If there were any slight changes in the phase response of the filter with time the reading might be affected.

Curves such as those in Fig. 8 were plotted for each value of bandwidth and for numbers of lines between 2·5 and 90. To obtain a comparison between the results the response of the detector to true random

noise was taken as a reference. The greatest deviation in the detector response encountered in the ± 10% frequency range was then measured for each value of the number of lines. In Fig. 9 this deviation, expressed in noise power, is plotted as a function of the number of lines for the two values of bandwidth used. Also shown in this figure are experimental points obtained using a true r.m.s. meter.

It is seen from Fig. 9 that firstly there is little difference between the r.m.s. meter and the transit-counting detector for numbers of lines greater than 10. Secondly that for numbers of lines above 20 the deviation between the pseudo- and true-random generators is small. For a maximum deviation of 1% the number of lines required appears to be about 100.

For a 10% measurement 10 lines would be adequate. In this case the time taken for one sequence, assuming the bandwidth is 10% and the sampling frequency is 20 times the frequency of the measurement, is

$$T = 100/f_0$$

where

$$f_0 = \text{centre frequency}$$

For example, in a measurement over a bandwidth of 0·1 Hz the calibration time would be reduced from the previous 1000 s to 100 s.

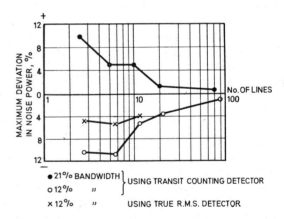

● 21% BANDWIDTH ⎫
o 12%   ,,   ⎬ USING TRANSIT COUNTING DETECTOR
× 12%   ,,   USING TRUE R.M.S. DETECTOR

Fig. 9. Deviation between pseudo- and true-random generators for ± 10% change in filter frequency.

Although the reduction in calibration time appears to be a big advantage for l.f. measurements there is one point to bear in mind if a pseudo-random generator is used. To obtain this reduction in calibration time the level of the pseudo-random signal fed into the system must be much larger than the inherent noise in it. This means that one must be careful to ensure that the signal at all points in the system is within the linear range of the constituent parts of the system. This restriction does not apply when a true random signal is used since this can be fed into the system at the same level as the inherent noise.

In conclusion, the discussion has shown that the pseudo-random noise source has theoretical advantages, in that the time of calibration of systems may be reduced. This advantage is partly offset by the added complexity of the experimental procedures.

5. Acknowledgments

The authors would like to express their thanks to their colleague Mr. H. Dunderdale for his collaboration and advice on pseudo-random sources.

6. References

1. Rice, S. O., 'Mathematical analysis of random noise', *Bell Syst. Tech. J.*, **23**, p. 282, 1944 and **24**, p. 46, 1945.
2. Kramer, C., 'A low frequency pseudo-random noise generator', *Electronic Engineering*, **37**, p. 465, July 1965.
3. Sutcliffe, H., 'Mean detector for slow fluctuations', *Electronics Letters*, **4**, No. 6, p. 97, March 1968.

Manuscript first received by the Institution on 19th May 1969 and in final form on 14th July 1970. (Paper No. 1341/CC87).

A Low-Frequency Pseudo-Random Noise Generator

By C. Kramer*

A circuit is described which generates noise in the sub-audio range by means of a feedback shift register. The noise is nearly random, with a binomial amplitude distribution. The spectrum is flat in the region of interest. The statistical characteristics are highly stable and predictable.

A NOISE generator in the sub-audio frequency range may sometimes be needed. The most common applications of such an instrument are testing of control systems, vibration testing, and the solution of problems involving noise by means of an analogue computer. It is well known that white noise in this frequency region cannot be generated by simply amplifying the output of a noisy device, as all amplifiers, transistors as well as valves, have a noise spectrum which varies inversely with the frequency ("$1/f$ noise").

Generally, therefore, three different types of low-frequency generator are constructed. In the first type a narrow-band noise signal is heterodyned to the low-frequency range. In a typical realization[1] a 2kc/s noise signal, generated by a thyratron followed by a 2kc/s selective amplifier, is mixed with a 2kc/s sine wave. As the noise amplitude will be directly influenced by the noise properties of the thyratron and by the amplification factors of the amplifiers, stability will always be a problem in noise generators of this kind.

In a second type[2] the noise, generated by a physical source, is converted into a random step function by means of low-frequency sampling. This type of noise generator suffers from the same stability difficulties. These can be partially alleviated by feedback control of the noise amplitude, as the authors suggest, but this affects the noise spectrum in such a way that the lowest frequencies are attenuated.

In the third type[3] the original noise is first clipped and then, by means of sampling, converted into a random binary sequence (also called random telegraph wave). After filtering in a low-pass filter this gives a low-frequency noise signal, in accordance with a theorem by S. O. Rice[4]. Though this type of noise generator is less liable to stability problems than the others, it is still not fully independent of the properties of the noise source.

In the present case it was decided to generate a random binary sequence, not with the aid of a physical noise source, but by a mathematical method. For this new type of noise generator use is made of the properties of a so-called feedback shift register.

The Feedback Shift Register as a Pseudo-Random Sequence Generator

A feedback shift register is a binary circuit, the properties of which were first described by D. A. Huffmann[5]. It consists of a number of flip-flops and 'exclusive OR' gates (Fig. 1). Upon application of a trigger pulse the information contained by the flip-flop chain is shifted one step to the right. At the same time a bit, obtained by a parity check on the content of a number of the flip-flops, is shifted into the first flip-flop. In this way a

cyclic code will circulate through the shift register. If the number and positions of the feedback connexions are properly chosen, this code will contain all possible combinations of N bits, except that in which all bits are positive. For instance the circuit shown in Fig. 1 generates the code shown in Fig. 3. These maximal length shift regis-

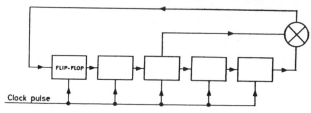

Fig. 1. Feedback shift register

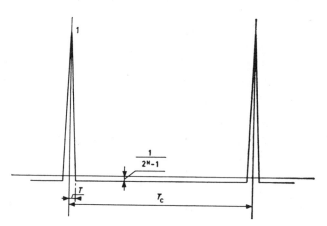

Fig. 2. Autocorrelation function of shift register code

ter codes have a length of $2^N - 1$ bits, where N is the number of flip-flops.

Now these codes have the property that, when such a code is multiplied by a time-shifted replica of itself, the result is again the same code with a different time shift. This property can be used to calculate their auto-correlation function. Because the number of negative bits exceeds the number of positive bits by one, the auto-correlation function $1/T_c \int_0^{T_c} f(t) f(t-\tau)dt$ will be

$-T/T_c = -1/2^N - 1$ for all values of $\tau \neq 0, T_c, 2T_c° \ldots$ For $\tau = 0, T_c \ldots$ the autocorrelation function is 1. So the autocorrelation function has the form sketched in Fig. 2.

This autocorrelation function is the same as that of a real random binary sequence, except for two differences: this sequence is not purely random but periodic, and the autocorrelation between the peaks is not exactly zero but

---

* *N.V. Philips' Gloeilampenfabrieken Eindhoven-Netherlands*

Reprinted with permission from *Electron. Engineering*, vol. 37, pp. 465–467, July 1965.

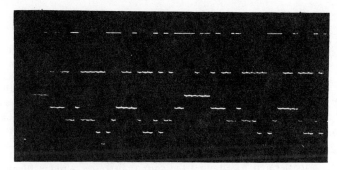

Fig. 3. 31-bit code and step signal

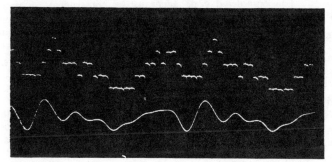

Fig. 4. Filtered code and step signal

has a small negative value. Both differences, however, can be made as small as desired by making N sufficiently large. For these reasons a maximal length shift register code can be designated as a pseudo-random binary sequence. The number N which is required, depends on the problem for which the noise generator is applied. Generally it can be stated that the period $(2^N-1)T_c$ of the code has to be large compared with the response time of the tested circuit.

Pseudo-Random Step Function

Now this binary sequence might be converted into a noiselike signal by applying it to a low-pass filter. The structure of the codes, however, suggests another, more elegant method of conversion. Because all possible combinations of positive and negative bits, except all bits positive, occur once in a cycle in the content of the shift-register, the number of positive bits will have a binomial probability distribution, with the one known exception. So when a signal is constructed, proportional in amplitude to the number of positive bits in the shift register, this

signal will have a binomial amplitude distribution. As is well-known, this distribution is the best possible approximation to a Gaussian distribution when the number of possible states is finite. Now this pseudo-random step function can be generated very simply by adding together the contents of all the shift register stages by means of a network of N equal resistors. As an example, the code and the random step function generated by the circuit in Fig. 1 are shown in Fig. 3. The random step function has the form 0 1 2 3 3 3 3 2 1 1 2 1 2 2 3 3 4 4 4 4 4 3 3 3 2 2 3 2 2 2 1 etc. The distribution function is 1, 5, 10, 10, 5 which is nearly binomial.

The random step method is not too different from the low-pass filter method. In fact the action of a low-pass filter can be simulated by adding together a number of delayed versions of the same signal. As these delayed signals are already present in the shift register, it is rather obvious to make use of them. That the shift register-adder combination acts as a low-pass filter is shown in Fig. 4, where a filtered binary sequence and the step signal are compared. It can also be shown by calculating the spectrum of the step signal.

This could be done in a straightforward manner, but it is most easily solved by means of an analogy. It will be noted that the autocorrelation function, as shown in Fig. 2, is the same as that of a positive pulse of width T, repeated after a period T_c. The autocorrelation function of the step signal is therefore the same as that of the pulse after passing the shift register-adder combination. This of course is a pulse of width NT. Now, because the spectrum is fully determined by the autocorrelation function by means of the Wiener-Khintchine relation, the power spec-

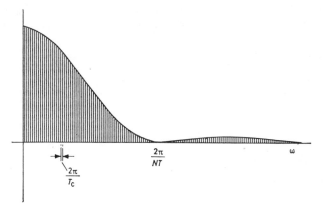

Fig. 6. Noise output from the generator

Fig. 5. Spectrum of the step signal

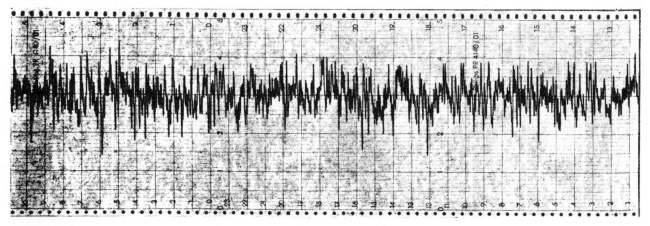

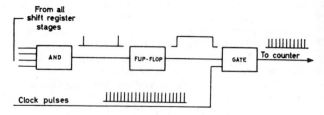

Fig. 7. Circuit for checking the code length

trum of the step signal is the same as that of a pulse of width NT, repeated with a period T_c. This spectrum is a line spectrum with an envelope $\left\{\dfrac{\sin \omega NT/2}{\omega NT/2}\right\}^2$ and a line spacing of $2\pi/T_c$ (Fig. 5).

Realization and Results

A noise generator, based on the principles described above, has been constructed, with a shift register of 16 stages and a clock frequency of 100c/s. The length of the generated code is $2^{16} - 1 = 65535$, so that the generated pseudo-noise has a period of nearly eleven minutes. This noise is shown in Fig. 6. The bandwidth of the noise to the first zero is $1/NT = 100/16$ c/s $= 6\cdot25$c/s. Higher frequencies are filtered out by means of a simple low-pass filter to smooth the signal.

A difficulty which occurs when one tries to build such a code generator is that it has to be determined whether indeed the code has the maximum length. This can be determined by the circuit given in Fig. 7.

The AND gate determines the moment that all shift register stages are in the negative state, which occurs once in every period. By the output of the AND gate a flip-flop is triggered, so this will give off a ' 1 ' output on alternate periods. The output of the flip-flop controls a gate which directs clock-pulses to a counter. Thus the counter directly indicates the length of the code. By this means it was determined that in the 16 flip-flop code generator a code of the maximum length 65535 was indeed generated when the inputs to the parity check network were taken from the flip-flops 1, 3, 12 and 16. The determination could be accelerated by running the shift register on a much higher clock frequency.

The completed code generator is constructed of standard digital circuit blocks, mounted on printed wiring boards. The generator has been applied to the solution of a problem on a non-linear noisy oscillator by an analogue computer, and in this application it gave satisfactory results.

Conclusion

In the foregoing it has been shown that a nearly random noiselike signal with exactly known and stable characteristics can be generated by standard digital techniques. Even with only a limited number of components a very good approximation to real noise can be realized. The distribution function is exactly binomial.

REFERENCES

1. BELL, D. A., ROSIE, A. M. A Low-frequency Noise Generator with Gaussian Distribution. *Electronic Technol. 37*, 241 (1960).
2. DIAMANTIDES, N. D., MCCRAY, C. E. Generating Random Forcing Functions for Control System Simulation. *Electronics. 34*, 60 (18 August, 1961).
3. SHIVA SHANKAR, T. N. A low-frequency Random Noise Generator. *J. Inst. Tel. Engrs. New Delhi. 9*, 63 (1963).
4. RICE, S. O. A Mathematical Analysis of Random Noise. *Bell Syst. Tech. J. 23*, 282 (1944).
5. HUFFMAN, D. A. The Synthesis of Linear Sequential Coding Networks. Proc. 3rd London Symp. on Inf. Theory.

Bibliography for Part V

****
  5-A. GAS-DISCHARGE NOISE SOURCES
****

ALMA'SSY, G. AND I. FRIGYES
    NEW MICROWAVE NOISE GENERATOR FOR THE 2000 MC/S BAND
        PERIODICA POLYTECHNICA ELECTRICAL ENGG., VOL. 4, NO. 4, PP. 293-
        303, 1960.
APUSHKINSKII, G. P.
    CERTAIN SINGULARITIES OF THE NOISE SPECTRUM OF A GAS-DISCHARGE TUBE
        RADIO ENGINEERING AND ELECTRONIC PHYSICS, VOL. 4, NO. 12,
        PP. 220-222, DECEMBER 1959.
BARKER, K. & F. A. BENSON.
    SPECTRAL-NOISE CHARACTERISTICS OF GLOW DISCHARGES IN THE FREQUENCY
    RANGE 1 KC/S - 8 MC/S.
        PROC. IEE. (LONDON), VOL. 113, NO. 6, PP. 937-942, JUNE 1966.
BEAM, W. R. AND R. D. HUGHES
    MICROWAVE NOISE SOURCE MODULATOR AND POWER SUPPLY
        IRE TRANS. ELECTRON DEVICES, VOL. ED-4, NO. 2, PP. 185-186,
        APRIL 1957.
BEECHER, D. E., R. R. BENNETT, AND H. LOW
    STABILIZED NOISE SOURCE FOR AIR-WEAPONS DESIGN
        ELECTRONICS, VOL. 27, NO. 7, PP. 163-165, JULY 1954.
BEKEFI, G. AND S. C. BROWN
    MICROWAVE MEASUREMENTS OF THE RADIATION TEMPERATURE OF PLASMAS
        JOUR. APPLIED PHYSICS, VOL. 32, NO. 1, PP. 25-30, JANUARY 1960.
BEKEFI, G., J. L. HIRSCHFIELD, AND S. C. BROWN
    INCOHERENT MICROWAVE RADIATION FROM PLASMAS
        PHYSICAL REV., VOL. 116, NO. 5, PP. 1051-1056, DECEMBER 1, 1959.
BEKEFI, G., J. H. HIRSCHFIELD, AND S. C. BROWN
    KIRCHHOFF'S RADIATION LAW FOR PLASMAS WITH NON-MAXWELLIAN DISTRIBUTIONS
        PHYSICS OF FLUIDS, VOL. 4, NO. 2, PP. 173-176, FEBRUARY 1961.
BENJAMIN, S. K., P. M. LEVY AND S. KING
    LOW FREQUENCY SPECTRUM OUTPUT OF A NOISE GENERATOR
        PROC. NATIONAL ELECTRONICS CONF., VOL. 13, PP. 106-114, 1957.
BENSON, F. A.
    GAS-FILLED VOLTAGE STABILIZERS
        ELECTRONIC AND RADIO ENGINEER, VOL. 34, NO. 1, PP. 16-20, JANUARY
        1957.
BERGMANN, S. M.
    SPECTRAL DISTRIBUTION OF THERMAL NOISE IN A GAS DISCHARGE
        IRE TRANS. MICROWAVE THEORY AND TECHNIQUES, VOL. MTT-5, NO. 4,
        PP. 237-238, OCTOBER 1957.
BOSE, K. K. AND P. L. DASGUPTA
    GRID CONTROL OF GAS TUBE NOISE
        INDIAN JOUR. PHYSICS, VOL. 36, NO. 7, PP. 364-368, JULY 1962.
BRIDGES, T. J.
    A GAS-DISCHARGE NOISE SOURCE FOR EIGHT-MILLIMETER WAVES
        PROC. IRE, VOL. 42, NO. 5, PP. 818-819, MAY 1954.
CHINNOCK, E. F.
    A PROTABLE, DIRECT-READING MICROWAVE NOISE GENERATOR
        PROC. IRE, VOL. 40, NO. 2, PP. 160-164, FEBRUARY 1952.
COBINE, J. D. AND J. R. CURRY
    ELECTRICAL NOISE GENERATORS
        PROC. IRE, VOL. 35, NO. 9, PP. 875-879, SEPTEMBER 1947.
COLLINGS, E. W.
    NOISE AND ELECTRON TEMPERATURES OF SOME COLD CATHODE ARGON DISCHARGES
        JOUR. APPL. PHYS. VOL. 29, NO. 8, PP. 1215-1219, AUGUST 1958.
CRAWFORD, F. W. AND G. S. KINO
    OSCILLATIONS AND NOISE IN LOW-PRESSURE D.C. DISCHARGES
        PROC. IRE, VOL. 49, NO. 12, PP. 1769-1788, DECEMBER 1961.
CRAWFORD, F. W. AND G. N. OETZEL
    NOISE GENERATION AND TRANSMISSION ALONG A PLASMA COLUMN
        PHYS. LETTERS, VOL. 13, NO. 2, PP. 119-120, NOVEMBER 15, 1964.
DAVIES, L. W. AND E. COWCHER
    MICROWAVE AND METRE-WAVE RADIATION FROM THE POSITIVE COLUMN OF A
    GAS DISCHARGE
        AUSTRALIAN JOUR. PHYSICS, VOL. 8, NO. 1, PP. 108-128, MARCH 1955.
DENSON, C. I. AND G. J. HALFORD
    PLASMA NOISE SOURCES OF IMPROVED ACCURACY.
        IEEE TRANS. MICROWAVE THEORY TECH., VOL. MTT-16, NO. 9, PP. 653-
        61, SEPT. 1968.
EASLEY, M. A.
    PROBE TECHNIQUE FOR THE MEASUREMENT OF ELECTRON TEMPERATURE
        JOUR. APPL. PHYS., VOL. 22, NO. 5, PP. 590-593, MAY 1951.
EASLEY, M. A. AND W. W. MUMFORD
    ELECTRON TEMPERATURE VS. NOISE TEMPERATURE IN LOW PRESSURE MERCURY-
    ARGON DISCHARGES
        JOUR. APPL. PHYS., VOL. 22, NO. 6, PP. 846-847, JUNE 1951.
GORDON-SMITH, A. C. AND J. A. LANE
    MEASUREMENTS ON GAS DISCHARGE NOISE SOURCES AT CENTIMETRE WAVELENGTHS

        PROC. IEE (GB), VOL. 105B, NO. 24, PP. 545-547, NOVEMBER 1958.
GORDON-SMITH, A. C. AND J. A. LANE
    STANDARDIZATION OF COAXIAL NOISE SOURCES
        ELECTRONICS LETTERS, VOL. 1, NO. 1, PP. 7-8, MARCH 1965.
GUENTZLER, R. E.
    THE INFLUENCE OF CATAPHORESIS UPON THE NOISE TEMPERATURE OF F8T5 LAMPS
        IEEE TRANS. MICROWAVE THEORY AND TECHNIQUES, VOL. MTT-18, NO. 7,
        PP. 393-400, JULY 1970.
GUENTZLER, R. E.
    NOISE TEMPERATURE DATA ON CATAPHORETICALLY PUMPED F13T5 LAMPS
        IEEE TRANS. MICROWAVE THEORY AND TECHNIQUES, VOL. MTT-19, NO. 3,
        PP. 339-341, MARCH 1971.
GUENTZLER, R. E.
    NOISE TEMPERATURE DATA ON LOW PRESSURE ARGON DISCHARGES
        IEEE TRANS ELECTRON DEVICES, VOL. ED-19, NO. 2, PP. 160-163,
        FEBRUARY 1972.
GUENTZLER, R. E.
    COMMENTS ON 'MEASURED NOISE TEMPERATURE VERSUS THEORETICAL ELECTRON
    TEMPERATURE FOR GAS DISCHARGE NOISE SOURCES'.
        IEEE TRANS. MICROWAVE THEORY AND TECH., VOL. MTT-22, NO. 4, PP.
        469-470, APRIL 1974.
GUENTZLER, R. E.
    ADDITIONAL INFORMATION ON THE NOISE-TEMPERATURE BEHAVOIR OF F8T5
    LAMPS
        IEEE TRANS. MICROWAVE THEORY AND TECHNIQUES, VOL. MTT-24, NO. 8,
        PP. 538-539, AUGUST 1976.
HALFORD, G. J. AND E. G. ROBUS
    NOISE SOURCE CALIBRATION IN THE DECIMETRE BAND
        RADIO AND ELECTRONIC ENGINEER, VOL. 34, NO. 4 , PP. 227-234,
        APRIL 1968.
HART, P. A. H.
    GAS DISCHARGE AS A SOURCE OF INCOHERENT RADIATION AT MILLIMETRE AND
    SUB-MILLIMETRE WAVELENGTHS
        PHILIPS RES. REPTS., VOL. 22, NO. 2, PP. 77-109, APRIL 1967.
HART, P. A. H. AND G. H. PLANTINGA
    AN EXPERIMENTAL NOISE GENERATOR FOR MILLIMETRE WAVES
        PHILIPS TECHNICAL REV., VOL. 22, NO. 12, PP. 391-392, SEPTEMBER 1961.
HELFF, E.
    NOISE MEASUREMENTS OF GLOW DISCHARGES (IN GERMAN)
        WISS. Z. TECH. UNIV. DRESDEN, VOL. 13, NO. 1, PP. 217-22, 1964.
HEYMANN, P.
    THE USE OF LOW CURRENT POSITIVE COLUMNS IN NOISE GENERATORS OF
    ELEVATED TEMPERATURE
        NACHRICHTENTECHNIK (GERMANY), VOL. 19, NO. 9, PP. 321-4, SEPT 1969
HUGHES, V. A.
    ABSOLUTE CALIBRATION OF A STANDARD TEMPERATURE NOISE SOURCE FOR USE
    WITH S-BAND RADIOMETERS
        PROC. IEE (LONDON), VOL. 103, PT. B, PP. 669-672, SEPTEMBER 1956.
JASINSKI, W. AND G. HILLER
    DETERMINATION OF NOISE TEMPERATURE OF A GAS-DISCHARGE NOISE SOURCE
    FOR FOUR-MILLIMETER WAVES
        PROC. IRE, VOL. 49, NO. 4, PP. 807-808, APRIL 1961.
JOHNSON, H. AND K. R. DEREMER
    GASEOUS DISCHARGE SUPER HIGH FREQUENCY NOISE SOURCES
        PROC. IRE, VOL. 39, NO. 8, PP. 908-914, AUGUST 1951.
JONES, R. C. AND W. J. GRAHAM
    NOISE GENERATION IN HIGH-POWER MICROWAVE GAS DISCHARGES
        ELECTRONIC DESIGN, VOL. 5, NO. 10, PP. 144 - 145, MAY 15, 1957
KANDYBA, V. V., G. L. IOSELSON & V. A. LANDER
    ON THE DEVELOPMENT OF A UHF STANDARD OF HIGH PLASMA TEMPERATURE
        METROLOGIA (GERMANY), VOL. 9, NO. 1, PP. 1-7, JAN 1973
KAZANSKAYA, A. YA. & A. B. YASTREBOV.
    RADIATION OF MICROWAVE NOISE OF ANOMALOUS INTENSITY FORM A GAS
    DISCHARGE PLASMA.
        RADIO ENGINEERING AND ELECTRONIC PHYSICS, VOL. 12, NO. 4,
        PP. 688-689, APRIL 1967.
KNOL, K. S.
    DETERMINATION OF THE ELECTRON TEMPERATURE IN GAS DISCHARGES BY NOISE
    MEASUREMENTS
        PHILIPS RES. REPTS., VOL. 6, NO. 4, PP. 288-302, AUGUST 1951.
KOJIMA, S. AND K. TAKAYAMA
    NOISE TEMPERATURE OF A D.C. GAS DISCHARGE PLASMA
        PHYSICAL REV., VOL. 80, NO. 5, P. 907, DECEMBER 1, 1950.
KOLLANYI, M.
    APPLICATION OF GAS DISCHARGE TUBES AS NOISE SOURCES IN THE 1700-2300
    MC BAND
        JOUR. BRIT. INST. RADIO ENGRS., VOL. 18, NO. 9, PP. 541-548,
        SEPTEMBER 1958.
KONONENKO, K. I., S. P. MOVCHAN & A. I. YATSENKO.
    A METHOD FOR DECREASING THE NOISE OF THE GASEOUS DISCHARGE PLASMA
        RADIO ENGINEERING AND ELECTRONIC PHYSICS, VOL. 10, NO. 12,
        PP. 1950-1953, DECEMBER 1965.
KUHN, N. J. AND M. R. NEGRETE
    GAS DISCHARGE NOISE SOURCES IN PULSED OPERATION
        IRE INTERNATIONAL CONVENTION RECORD, PT. 3, PP. 166-173, 1961.

KUZ'MIN, A. D. AND A. N. KHVOSHCHEV
    A BROADBAND NOISE GENERATOR FOR THE DECIMETRE WAVEBAND
        RADIOENGINEERING, VOL. 13, NO. 7, PP. 47-55, 1958.
MARTIN, H. AND H. A. WOODS
    LOW-FREQUENCY NOISE SPECTRA OF HOT-FILAMENT LOW-PRESSURE DISCHARGE
    TUBES
        PROC. PHYSICAL SOC. (LONDON), SEC. B, VOL. 65, PT. 4,
        PP. 281-286, APRIL 1952.
MELTSIN, A. L. AND V. I. OBEREMKO
    NOISE CHARACTERISTICS OF A GAS-DISCHARGE CELL
        OPTICS AND SPECTROSCOPY, VOL. 33, NO. 5, PP. 543-544, NOVEMBER 1972
MESSENGER, R. A. AND A. VAN DER ZIEL
    HIGH FREQUENCY GAS DISCHARGE NOISE
        PHYSICA, VOL. 34, NO. 4, PP. 534-540, 1967.
        ERRATUM - PHYSICA, VOL. 37, NO. 1, P. 138, 1967.
MIKHALEV, L. A.
    THE THEORY OF A COAXIAL GAS-DISCHARGE OSCILLATOR LOADED BY A HELICAL
    LINE
        RADIO ENGINEERING AND ELECTRONIC PHYSICS, VOL. 17, NO. 6,
        PP. 973-975, JUNE 1972.
MUMFORD, W. W.
    A BROADBAND MICROWAVE NOISE SOURCE
        BELL SYS. TECH. JOUR., VOL. 28, NO. 4, PP. 608-618, OCTOBER 1949.
MUMFORD, W. W. AND R. L. SCHAFERSMAN
    DATA ON THE TEMPERATURE DEPENDENCE OF X-BAND FLUORESCENT LAMP NOISE
    SOURCES
        IRE TRANS. MICROWAVE THEORY AND TECHNIQUES, VOL. MTT-3, NO. 6,
        PP. 12-17, DECEMBER 1955.
OLSON, K. W.
    REPRODUCIBLE GAS-DISCHARGE NOISE SOURCES AS POSSIBLE MICROWAVE NOISE
    STANDARDS
        IRE TRANS. INSTRUMENTATION, VOL. I-7, NO. 3-4, PP. 315-318,
        DECEMBER 1958.
OLSON, K. W.
    MEASURED NOISE TEMPERATURE VERSUS THEORETICAL ELECTRON TEMPERATURE
    FOR GAS DISCHARGE NOISE SOURCES
        IEEE TRANS. MICROWAVE THEORY AND TECHNIQUES, VOL. MTT-16, NO. 9,
        PP. 640-645, SEPTEMBER 1968.
OLSON, K. W.
    BAND TRANSITION NOISE MEASUREMENTS OF GAS DISCHARGE NOISE SOURCES
        IEEE TRANS. MICROWAVE THEORY AND TECHNIQUES, VOL. MTT-16, NO. 9,
        P. 797, SEPTEMBER 1968.
OLSON, K. W.
    APPARENT FREQUENCY DEPENDENCE OF MICROWAVE POWER RADIATED FROM
    GAS-DISCHARGE NOISE SOURCES
        IEEE TRANS. MICROWAVE THEORY AND TECHNIQUES, VOL. MTT-19, NO. 9,
        PP. 776-778, SEPTEMBER 1971.
PAK, T. S.
    NOISE FROM HOT-CATHODE DISCHARGES
        PROC. PHYSICAL SOC. (LONDON), SEC. B, VOL. 68, PT. 5,
        PP. 292-296, MAY 1955.
PARZEN, P. AND L. GOLDSTEIN
    CURRENT FLUCTUATIONS IN DC GAS DISCHARGE PLASMA
        PHYS. REV., VOL. 79, NO. 1, PP. 190-191, JULY 1, 1950.
PARZEN, P. AND L. GOLDSTEIN
    CURRENT FLUCTUATIONS IN THE DIRECT CURRENT GAS DISCHARGE PLASMA
        PHYS. REV. VOL. 82, NO. 5, PP. 724-726, JUNE 1951.
PETROSYAN, G. G.
    INVESTIGATION RADIATION TEMPERATURE OF GAS DISCHARGE NOISE GENERATORS
    IN THE CENTIMETER RANGE
        MEAS. TECH. (USA), NO. 4, APRIL 1969.
PLANTINGA, G. H.
    THE NOISE TEMPERATURE OF A PLASMA
        PHILIPS RESEARCH REPTS., VOL. 16, NO. 5, PP. 462-468, OCTOBER 1961.
PRINZLER, H.
    ELECTRON- AND GAS-TEMPERATURE IN THE POSITIVE COLUMN OF A GAS
    DISCHARGE AT ATMOSPHERIC PRESSURE (IN GERMAN)
        ANNALEN DER PHYSIK, SER. 7, VOL. 8, NO. 1-2, PP. 42-59, 1961.
RAJA RAO, B. V. AND S SINHA
    LOW FREQUENCY NOISE GENERATION USING A THYRATRON NOISE SOURCE
        J. INSTN. ENG. (INDIA) ELECTRON. TELECOMMUN. ENG. DIV., VOL. 50,
        NO. 9, PT ET3, PP. 138-42, MAY 1970
SAIER, R.
    NOISE GENERATOR FOR CENTIMETER WAVES
        FREQUENZ, VOL. 14, NO. 2, PP. 68-70, FEBRUARY 1960.
SCHNITGER, H.
    GAS-DISCHARGE NOISE TUBES IN THE RANGE OF HIGH DISCHARGE ADMITTANCES
    (IN GERMAN)
        NACHRICHTENTECHNISCHE ZEITSCHRIFT, VOL. 10, NO. 5, PP. 236-240,
        MAY 1957.
SKINNER, R. I.
    WIDE-BAND NOISE SOURCES USING CYLINDRICAL GAS DISCHARGE TUBES IN TWO-
    CONDUCTOR LINES
        PROC. IEE (GB), VOL. 103, PT. B, NO. 10, PP. 491-496, JULY 1956.
SMITH, K. L.
    NOISE - CONFUSION IN MORE WAYS THAN ONE. 3. MICROWAVE NOISE

GENERATORS AND AERIAL TEMPERATURE
        WIRELESS WORLD, VOL. 81, NO. 1473, PP. 235-237, MAY 1975.
SPENCER, W. H. AND P. D. STRUM
    BROADBAND UHF AND VHF NOISE GENERATOR
        IRE TRANS. INSTRUMENTATION, VOL. PGI-4, PP. 47-50, OCTOBER 1955.
STEWARD, K. W. F.
    HELIX COUPLED GAS TUBE NOISE SOURCES
        THE MARCONI REV., VOL. 21, NO. 129, PP. 43-55, 2ND QUARTER 1958.
VAN DER ZIEL, A.
    NOISE IN GAS DISCHARGES
        JOUR. APPLIED PHYSICS, VOL. 24, NO. 2, PP. 223-224, FEBRUARY 1953.
WHITE, W. D. AND J. G. GREENE
    ON THE EFFECTIVE NOISE TEMPERATURE OF GAS DISCHARGE NOISE GENERATORS
        PROC. IRE, VOL. 44, NO. 7, P. 939, JULY 1956.

****
    5-B. TEMPERATURE LIMITED DIODE NOISE SOURCES
****

BREZGUNOV, V. N. AND V. A. UDAL'TSOV
    STABILIZED DIODE NOISE GENERATOR WITH CONTINUOUS AND MODULATED REGIMES
        INSTRUMENTS AND EXPERIMENTAL TECHNIQUES, NO. 6, PP. 1385-1389,
        NOVEMBER-DECEMBER 1968.
DIEMER, G. AND K. S. KNOL
    THE NOISE OF ELECTRONIC VALVES AT VERY HIGH FREQUENCIES, I. THE DIODE
        PHILIPS TECHNICAL REV., VOL. 14, NO. 6, PP. 153-164, DECEMBER 1952.
GROENDIJK, H.
    A NOISE DIODE FOR ULTRA-HIGH FREQUENCIES
        PHILIPS TECHNICAL REV., VOL. 20, NO. 4, PP. 108-110, OCTOBER 1958.
HARRIS, I. A.

    THE DESIGN OF A NOISE GENERATOR FOR MEASUREMENTS IN THE FREQUENCY
    RANGE 30-1250 MC/S
        PROC. IEE (LONDON), VOL. 108, PT. B, PP. 651-658, NOVEMBER 1961.
JOHNSON, H.
    A COAXIAL LINE NOISE DIODE FOR UHF
        RCA REV., VOL. 8, NO. 1, PP. 169-185, MARCH 1947.
KOMPFNER, R., J. HATTON, E. E. SCHNEIDER AND L. A. G. DRESEL
    THE TRANSMISSION LINE DIODE AS A NOISE SOURCE AT CENTIMETRE
    WAVELENGTHS
        JOUR. IEE, VOL. 93, PT. IIIA, NO. 9, PP. 1436-1442, MARCH-MAY 1946.
MOFFATT, J.
    A DIODE NOISE GENERATOR
        JOUR. IEE (GB), VOL. 93, PT. IIIA, NO. 8, PP. 1335-1337, MARCH-MAY
        1946.
PRINZLER, H.
    A SATURATED DIODE NOISE SOURCE FOR 20 CM WAVELENGTHS AND ITS
    ABSOLUTE CALIBRATION WITH A HEATED RESISTANCE
        NACHRICHTENTECHNIK, VOL. 8, NO. 11, PP. 495-500, NOVEMBER 1958.
SEDMAK, G.
    THE STUDY AND CONSTRUCTION OF A SECONDARY STANDARD OF NOISE USING A
    CONTROLLABLE, STABILIZED SATURATED DIODE (IN ITALIAN)
        ALTA FREQUENZA, VOL. 36, NO. 10, PP. 959-963, OCTOBER 1967.
SLINKMAN, R. W.

    TEMPERATURE-LIMITED NOISE DIODE DESIGN
        SYLVANIA TECHNOLOGIST, VOL. 2, NO. 4, PP. 6-8, OCTOBER 1949.
SMITH, K. L.
    NOISE - CONFUSION IN MORE WAYS THAN ONE. 2 . NOISE TEMPERATURE AND
    NOISE GENERATORS
        WIRELESS WORLD, VOL. 81, NO. 1472, PP. 169-173, APRIL 1975.
VOGEL, M. AND J. P. HATCH
    ELECTRON EMISSION COOLS NOISE-DIODE CATHODE
        PROC. IEEE, VOL. 54, NO. 10, PP. 1492-1494, OCTOBER 1966.

****
    5-C. RESISTIVE LOADS AS NOISE SOURCES
****

ANON.
    NOISE SOURCES IN FOUR WAVEGUIDE SIZES
    NBS TECHNICAL NEWS BULLETIN, VOL. 53, NO. 1, P. 16, JANUARY 1969.
BIRGER, L. A. AND I. A. SOKOV
    REFERENCE THERMAL NOISE GENERATORS   HIGH AND ULTRA-HIGH FREQUENCY
    MEASUREMENTS
        MEASUREMENT TECHNIQUES, NO. 1, PP. 62-65, JANUARY 1962.
BLUNDELL, D. J.
    UNITED KINGDOM NATIONAL STANDARD OF MICROWAVE NOISE AT 4.1 GHZ AND
    77 K
        JOUR. PHYSICS E, VOL. 8, NO. 11, PP. 925-929, NOVEMBER 1975.
BLUNDELL, D. J., E. W. HOUGHTON AND M. W. SINCLAIR
    MICROWAVE NOISE STANDARDS IN THE UNITED KINGDOM
        IEEE TRANS. INSTRUMENTATION AND MEASUREMENTS, VOL. IM-21, NO. 4,
        PP. 484-488, NOVEMBER 1972.

344

BRUN, P.
    WHITE-NOISE GENERATOR OPERATING IN THE AUDIO-FREQUENCY RANGE
ANNALES DES TELECOMMUNICATIONS, VOL. 18, NO. 3-4, PP. 42-45, MARCH-APRIL
    1963.
CARSALADE, H., J. C. HOFFMANN AND G. LACAZE
    NOISE GENERATOR IN AUDIO FREQUENCY CAPABLE OF STANDARDIZATION AT
    ABSOLUTE VALUE (IN FRENCH)
        ONDE ELECT., VOL. 50, NO. 515, PP. 148-51, FEBRUARY 1970.
COLLINGS, E. W.
    A FILAMENT NOISE SOURCE FOR 3 GC/S
        PROC. IEE (LONDON), VOL. 106C, NO. 9, PP. 97-101, MARCH 1959.
DAGLISH, H. N. AND J. W. CARTER.
    COOLED TERMINATIONS FOR USE AS 4 GC/S MICROWAVE STANDARD NOISE
    SOURCES.
        PROC. INSTN. ELECT. ENGRS. (GB), VOL. 112, NO. 4, PP. 705-7, APRIL
        1965.
DAVIS, Q. V.
    A HIGH TEMPERATURE TERMINATION FOR USE AT SHORT MILLIMETER WAVELENGTHS
        JOUR. SCIENTIFIC INSTRUMENTS, VOL. 40, NO. 11, PP. 524-525,
        NOVEMBER 1963.
DAYWITT, W. C.
    A REFERENCE NOISE STANDARD FOR MILLIMETER WAVES
        IEEE TRANS. MICROWAVE THEORY AND TECHNIQUES, VOL. MTT-21, NO. 12,
        PP. 845-847, DECEMBER 1973.
DAYWITT, W. C., W. J. FOOTE, AND E. CAMPBELL
    WR15 THERMAL NOISE STANDARD
        NBS TECHNICAL NOTE 615, NATIONAL BUREAU OF STANDARDS, BOULDER,
        COLO., MARCH 1972.
EISELE, K. M.
    REFRIGERATED MICROWAVE NOISE SOURCES
        IEEE TRANS. INSTRUMENTATION AND MEASUREMENT, VOL. IM-13, NO. 4,
        PP. 336-342, DECEMBER 1964.
ESTIN, A. J., C. L. TREMBATH, J. S. WELLS, AND C. W. DAYWITT
    ABSOLUTE MEASUREMENT OF TEMPERATURES OF MICROWAVE NOISE SOURCES
        IRE TRANS. INSTRUMENTATION AND MEASUREMENT, VOL. IM-9, NO. ,
        PP. 209-213, SEPTEMBER 1960.
HALFORD, G. J.
    NOISE COMPARATORS AND STANDARDS FOR S AND X BANDS
        IEEE TRANS. INSTRUMENTATION AND MEASUREMENT, VOL. IM-15, NO. 4,
        PP. 310-317, DECEMBER 1966.
HALFORD, G. J. AND E. G. ROBUS
    NOISE SOURCE CALIBRATION IN THE DECIMETER BAND
        RADIO AND ELECTRONIC ENGINEERING, VOL. 35, NO. 4, PP. 227-235,
        APRIL 1968.
HOLLWAY, D. L. AND P. I. SOMLO
    A STABLE BROAD-BAND VARIABLE NOISE SOURCE FOR MICROWAVE RADIOMETRY
        ELECTRONICS LETTERS, VOL. 4, NO. 2, PP. 24-25, JANUARY 26, 1968.
HORNBOSTEL, D. H.
    HOT AND COLD BODY REFERENCE NOISE GENERATORS FROM O TO 40 GHZ.
        IEEE TRANS. INSTRUMENTATION AND MEASUREMENT, VOL. IM-23, NO. 2,
        PP. 120-131, JUNE 1974.
HOUGHTON, E. W. AND M. W. SINCLAIR.
    COOLED THERMAL REFERENCE NOISE STANDARDS AND COMPARATOR SYSTEMS FOR
    CALIBRATING COOLED NOISE SOURCES.
        1969 EUROPEAN MICROWAVE CONFERENCE, LONDON, ENGLAND, 8-12 SEPT.
        1969, PP. 410-15.
HOWELL, T. F. AND C. FIELD
    COAXIAL TERMINATIONS FOR USE AS LOW-TEMPERATURE NOISE-REFERENCE
    SOURCES
        ELECTRONICS LETTERS, VOL. 2, NO. 6, PP. 198-199, JUNE 1966.
JORDAN, P. R., JR.
    A NOVEL MICROWAVE NOISE TEMPERATURE GENERATOR HAVING AN OUTPUT
    TEMPERATURE OF FROM BELOW 40 TO ABOVE 370K
        REV. SCI. INSTR., VOL. 41, NO. 11, PP. 1649-1651, NOVEMBER 1970.
JUNG, H.
    A NEW STANDARD FOR EXTREMELY SMALL NOISE POWERS IN THE MICROWAVE
    RANGE (IN GERMAN)
        HOCHFREQUENZTECH. UND ELEKT. AKUST., VOL. 64, NO. 2, PP. 50-52,
        SEPTEMBER 1956.
JURKUS, A.
    COLD LOADS AS STANDARD NOISE SOURCES
        PROC. IEEE, VOL. 53, NO. 2, PP. 176-177, FEBRUARY 1965.
KEISER, J. J. PRIESE & G. K. SHCMIDT
    THERMAL NOISE STANDARDS AND COMMERCIAL NOISE GENERATORS IN THE
    WAVELENGTH RANGE FROM 3 CM TO 38 CM (IN GERMAN)
        14TH INTERNATIONAL SCIENTIFIC COLLOQUIM, MICROWAVE TECHNIQUE,
        ILLMENAU, GERMANY, 29 SEP.-3 OCT. 1969, PP. 161-180.
KIRSTETTER, B.
    WIDEBAND NOISE STANDARDS FOR THE TEMPERATURE RANGE OF 6 K TO 950 K
    (IN GERMAN)
        Z. ANGEW. PHYS., VOL. 22, NO. 2, PP. 127-133, JANUARY 1967.
KNOL, K. S.
    A THERMAL NOISE STANDARD FOR MICROWAVES
        PHILIPS RESEARCH REPTS., VOL. 12, PP. 123-126, APRIL 1957.
KNOTT, K. F.
    PRECISE WIDEBAND GAUSSIAN NOISE SOURCES USING JNFET AMPLIFIERS AND

RESISTOR NOISE
J. PHYS. E (GB), VOL. 5, NO. 3, PP. 254-7, MAR 1972

MACPHERSON, A. C.
LOW FREQUENCY STANDARD NOISE SOURCE
REV. SCIENTIFIC INSTRUMENTS, VOL. 35, NO. , PP. 386-387, 1962.

MENON, R. C., N. P. ALBOUGHAND J. W. DOZIER
COOL LOADS AS CALIBRATION NOISE STANDARDS FOR THE MILLIMETER WAVE-
LENGTH RANGE
PROC. IEEE, VOL. 54, PP. 1501-1503, OCTOBER 1966.

MEZGER, P. G. AND H. ROTHER
COOLABLE CALIBRATING RESISTANCE AS NOISE TEMPERATURE STANDARDS (IN
GERMAN)
FREQUENZ, VOL. 16, NO. 10, PP. 386-391, OCTOBER 1962.

MUKAIHATA, T.
APPLICATIONS AND ANALYSIS OF NOISE GENERATION IN N-CASCADED
MISMATCHED TWO-PORT NETWORKS
IEEE TRANS. MICROWAVE THEORY AND TECHNIQUES, VOL. MTT-16, NO. 9,
PP. 699-708, SEPTEMBER 1968.

MUKAIHATA, T. AND P. S. ROBERTS
MAINTENANCE-FREE CRYOGENIC NOISE STANDARD FOR CONTINUOUS ANTENNA
OPERATION INDEPENDENT OF ELEVATION ANGLE
IEEE TRANS. INSTRUMENTATION AND MEASUREMENT, VOL. IM-19, NO. 4,
PP. 403-407, NOVEMBER 1970.

MUKAIHATA, T., B. L. WALSH, M. F. BOTTJER, AND E. B. ROBERTS
SUBTLE DIFFERENCES IN SYSTEM NOISE MEASUREMENTS AND CALIBRATION OF
NOISE STANDARDS
IRE TRANS. MICROWAVE THEORY AND TECHNIQUES, VOL. MTT-10, NO. 6,
PP. 506-516, NOVEMBER 1962.

PENZIAS, A. A.
HELIUM COOLED REFERENCE NOISE SOURCE IN A 4-KMC WAVEGUIDE
REV. SCI. INSTR., VOL. 36, NO. 1, PP. 68-70, JANUARY 1965.

PETROSYA, G. G.
GOVERNMENT STANDARDS UNITS OF SPECTRAL DENSITY OF POWER OF NOISE
RADIATION IN RANGE 2.6 - 17.4 GHZ
MEASUREMENT TECHNIQUES, VOL. 14, NO. 6, PP. 886-890, DECEMBER 1971.

PRIESE, J.
THERMAL NOISE GENERATOR FOR UHF WITH DIRECT INDICATION OF THE NOISE
TEMPERATURE (IN GERMAN)
HOCHFREQUENZTECHNIK UND ELEKTROAKUSTIK, VOL. 77, NO. 5-6,
PP. 209-214, DECEMBER 1968.

RUSINOV, YU. S. AND R. L. SOROCHENKO
PRIMARY STANDARD OF NOISE RADIATION IN THE DECIMETER BAND
INSTRUM. EXPER. TECH., NO. 3, PP. 612-613, MAY-JUNE 1964.

SOMLO, P. I. AND D. L. HOLLWAY
THE AUSTRALIAN NATIONAL LABORATORY X-BAND RADIOMETER FOR CALIBRATION
OF NOISE SOURCES
IEEE TRANS. MICROWAVE THEORY AND TECHNIQUES, VOL. MTT-16, NO. 9,
PP. 664-669, SEPTEMEBR 1968.

STELZRIED, C. T.
A LIQUID-HELIUM-COOLED COAXIAL TERMINATION
PROC. IRE, VOL. 49, NO. 7, P. 1224, JULY 1961.

STELZRIED, C. T.
IMPROVEMENT IN THE PERFORMANCE OF AN AUTOMATIC NOISE FIGURE METER
WITH A LIQUID-NITROGEN-COOLED TERMINATION
PROC. IRE, VOL. 49, NO. 12, P. 1963, DECEMBER 1961.

STELZRIED, C. T.
TEMPERATURE CALIBRATION AND MICROWAVE THERMAL NOISE SOURCES
IEEE TRANS. MICROWAVE THEORY AND TECHNIQUES, VOL. MTT-13, NO. 1,
PP. 128-130, JANUARY 1965.

STELZRIED, C. T.
MICROWAVE THERMAL NOISE STANDARDS
IEEE TRANS. MICROWAVE THEORY AND TECHNIQUES, VOL. MTT-16, NO. 9,
PP. 646-655, SEPTEMBER 1968.

SUTCLIFFE, H.
NOISE MEASUREMENTS IN THE 3-CM WAVEBAND USING A HOT SOURCE
PROC. IEE (LONDON), VOL. 103, PT. B, PP. 673-677, SEPTEMBER 1956.

TREMBATH, C. L., D. F. WAIT, G. F. ENGEN AND W. J. FOOTE
A LOW-TEMPERATURE MICROWAVE NOISE STANDARD
IEEE TRANS. MICROWAVE THEORY AND TECHNIQUES, VOL. MTT-16, NO. 9,
PP. 709-714, SEPTEMBER 1968.

TREMBATH, C. L.
LIQUID-NITROGEN COOLED MICROWAVE NOISE STANDARD
REV. SCIENTIFIC INSTRUM., VOL. 42, NO. 8, PP. 1261-1262, AUGUST
1971.

WAIT, D. F.
THE PRECISION MEASUREMENT OF NOISE TEMPERATURE OF MISMATCHED NOISE
GENERATORS
IEEE TRANS. MICROWAVE THEORY AND TECHNIQUES, VOL. MTT-18, NO. 10,
PP. 715-724, OCTOBER 1970.

WAIT, D. F. AND T. NEMOTO
THE MEASUREMENT OF THE NOISE TEMPERATURE OF A MISMATCHED NOISE SOURCE
IEEE TRANS. MICROWAVE THEORY AND TECHNIQUES, VOL. MTT-16, NO. 9,
PP. 670-675, SEPTEMBER 1968.

WELLS, J. S., W. C. DAYWITT, AND C. K. S. MILLER
MEASUREMENT OF EFFECTIVE TEMPERATURES OF MICROWAVE NOISE SOURCES
1962 IRE INTERNATIONAL CONVENTION RECORD, PT. 3, PP. 220-238. ALSO

IEEE TRANS. INSTRUMENTATION AND MEASUREMENT, VOL. IM-13, NO. 1,
PP. 17-28, MARCH 1964.
YOKOSHIMA, I.
AN ESTIMATION METHOD FOR NOISE TEMPERATURE OF MICROWAVE STANDARD
NOISE SOURCES
ELECTRONICS AND COMMUNICATIONS IN JAPAN, VOL. 52B, NO. 8, PP. 76-
85, AUGUST 1969.
YOKOSHIMA, I.
DIRECT MEASUREMENT TECHNIQUES OF TRANSMISSION LINE CORRECTIONS FOR
THERMAL NOISE STANDARDS
IEEE TRANS. INSTRUMENTATION AND MEASUREMENTS, VOL. IM-25, NO. 2,
PP. 138-145, JUNE 1976.
ZUCKER, H., Y. BASKIN, S. I. COHN, I. LERNER AND A. ROSENBLUM
DESIGN AND DEVELOPMENT OF A STANDARD WHITE NOISE GENERATOR AND NOISE
INDICATING INSTRUMENT
IRE TRANS. INSTRUMENTATION, VOL. I-7, NO. 3-4, PP. 279-291,
DECEMBER 1958.

```
****************************************************************************
```
5-D. AVALANCHE DIODES AS NOISE SOURCES
```
****************************************************************************
```

ALADINSKII, V. K.
THEORY OF MICROPLASMA PHENOMENA IN P-N JUNCTIONS
SOVIET PHYSICS - SEMICONDUCTORS, VOL. 6, NO. 10, PP. 1731-1736,
APRIL 1973.
ALADINSKIY, V. K., V. I. DASHKIN, A. S. SUSHCHIK, AND A. M. TIMERBULATOV
AVALANCHE CURRENT FLUCTUATIONS IN AN ARTIFICIAL MICROPLASMA IN SI
RADIO ENGINEERING AND ELECTRONIC PHYSICS, VOL. 18, NO. 2, PP. 244-
250, FEBRUARY 1973.
BAREISHA, L. I.
AVALANCHE TRANSIT DIODE NOISE GENERATOR
RADIO ENGINEERING AND ELECTRONIC PHYSICS, VOL. 14, NO. 1,
PP. 104-110, JANUARY 1969.
BAREYSHA, L. I., A. I. MEL'NIKOV, T. V. SPRICHEVA, A. S. TAGER,
G. M. FEDOROVA, AND F. M. SHAPIRO
AVALANCHE-TRANSIT DIODE NOISE GENERATOR
RADIO ENGINEERING AND ELECTRONIC PHYSICS, VOL. 14, NO. 1,
PP. 87-92, JANUARY 1969.
CHADELAS, A., E. CONSTANT, AND A. HAUDUCOEUR
USE OF AVALANCHE DIODES AS A SOURCE OF NOISE WITH ELECTRONIC CONTROL
(IN FRENCH)
ONDE ELECTRIQUE, VOL. 48, NO. 496-497, PP. 733-734, JULY 1968.
CHADHA, K. C.
AVALANCHE DIODES AS NOISE SOURCES.
J. INSTN. TELECOMM. ENGRS. (INDIA), VOL. 13, NO. 11, PP. 418-427,
NOVEMBER 1967.
CHAMPLIN, K. S.
MICROPLASMA FLUCTUATIONS IN SILICON
JOUR. APPLIED PHYSICS, VOL. 30, NO. 7, PP. 1039-1050, JULY 1959.
CHASEK, N.
AVALANCHE DIODES PERMIT IN-SERVICE MEASUREMENTS OF CRITICAL
PARAMETERS IN MICROWAVE EQUIPMENT
ELECTRONICS, VOL. 43, NO. , PP. 87-91, JUNE 19, 1970.
CONSTANT, E., R. GABILLARD, A. HAUTDUCOEUR, AND A. CHADELAS
THE USE OF VARACTOR DIODES IN THE AVALANCHE ZONE AS INTENSE SOURCES
OF WHITE NOISE (IN FRENCH)
COMPTES RENDUS ACAD. SCI. B, VOL. 262, NO. 1, PP. 16-18, JANUARY 3,
1966.
CONSTANT, E., B. KRAMER, AND L. RACZY
AVALANCHE NOISE FROM SEMICONDUCTOR JUNCTIONS IN THE RADIO RANGE AND
BEYOND (IN FRENCH)
COMPTES RENDUS ACAD. SCI. B, VOL. 265, NO. 7, PP. 385-388,
16 AUGUST 1967.
COOK, E. J.
HIGH-FREQUENCY COAXIAL LINE CIRCUIT FOR AN AVALANCHE DIODE NOISE
GENERATOR
U. S. PATENT 3594657, 11 APRIL 1969. PUBLISHED 20 JULY 1971 AS
U. S. A. 815465.
GOLOVKO, A. G. AND T. D. SHERMERGOR
DEPENDENCE OF THE LOW-FREQUENCY NOISE ON THE CURRENT AT THE
BEGINNING OF BREAKDOWN IN SILICON P-N(+) JUNCTIONS
SOVIET PHYSICS - SEMICONDUCTORS, VOL. 8, NO. 7, PP. 877-878,
JANUARY 1975.
GUMMEL, H. K. AND J. L. BLUE
A SMALL-SIGNAL THEORY OF AVALANCHE NOISE IN IMPATT DIODES
IEEE TRANS. ELECTRON DEVICES, VOL. ED-14, NO. 9, PP. 569-580,
SEPTEMBER 1967.
GUPTA, M. S.
NOISE IN AVALANCHE TRANSIT-TIME DEVICES
PROC. IEEE, VOL. 59, NO. 12, PP. 1674-1687, DECEMBER 1971.
GUPTA, M. S.
A SMALL-SIGNAL AND NOISE EQUIVALENT CIRCUIT FOR IMPATT DIODES
IEEE TRANS. MICROWAVE THEORY AND TECHNIQUES, VOL. MTT-21, NO. 9,
PP. 591-594, SEPTEMBER 1973.

HAITZ, R. H.
    MODEL FOR THE ELECTRICAL BEHAVIOR OF A MICROPLASMA
        JOUR. APPLIED PHYSICS, VOL. 35, NO. 5, PP. 1370-1376, MAY 1964.
HAITZ, R.H.
    CONTROLLED NOISE GENERATOR WITH AVALANCHE DIODES. I. LOW PULSE RATE
    DESIGN.
        IEEE TRANS ELECTRON DEVICES, VOL ED-12, NO 4, PP 198-207,
        APRIL 1965.
HAITZ, R. H.
    MECHANISMS CONTRIBUTING TO THE NOISE PULSE RATE OF AVALANCHE DIODES.
        JOUR. APPLIED PHYSICS, VOL. 36, NO. 10, PP. 3123-3131, OCTOBER 1965
HAITZ, R. H.
    CONTROLLED NOISE GENERATION WITH AVALANCHE DIODES II. HIGH PULSE RATE
    DESIGN.
        IEEE TRANS ELECTRON DEVICES, VOL. ED-13, NO. 3, PP. 342-346, MARCH
        1966.
HAITZ, R. H.
    NOISE OF A SELF-SUSTAINING AVALANCHE DISCHARGE IN SILICON   LOW
    FREQUENCY NOISE STUDIES
        JOUR. APPLIED PHYSICS, VOL. 38, NO. 6, PP. 2935-2946, JUNE 1967.
HAITZ, R. H. AND F. W. VOLTMER
    NOISE STUDIES IN UNIFORM AVALANCHE DIODES
        APPLIED PHYSICS LETTERS, VOL. 9, NO. , PP. 381-383, NOVEMBER 1966.
HAITZ, R. H. AND F. W. VOLTMER
    NOISE OF A SELF-SUSTAINING AVALANCHE DISCHARGE IN SILICON   STUDIES
    AT MICROWAVE FREQUENCIES
        JOUR. APPLIED PHYSICS, VOL. 39, NO. 6, PP. 3379-3384, JUNE 1968.
HINES, M. E.
    NOISE THEORY FOR THE READ TYPE AVALANCHE DIODE
        IEEE TRANS. ELECTRON DEVICES, VOL. ED-13, NO. 1, PP. 158-163,
        JANUARY 1966.
KANDA, M.
    AN IMPROVED SOLID-STATE NOISE SOURCE
        IEEE TRANS. MICROWAVE THEORY AND TECHNIQUES, VOL. MTT-24, NO. 12,
        PP. 990-995, DECEMBER 1976.
KEEN, N. J.
    AVALANCHE DIODES AS TRANSFER NOISE STANDARDS FOR MICROWAVE RADIOMETERS
        RADIO AND ELECTRONIC ENGINEER, VOL. 41, NO. 3, PP. 133-136, MARCH
        1971.
KEEN, N. J.
    AVALANCHE DIODE NOISE SOURCES AT SHORT CENTIMETER AND MILLIMETER
    WAVELENGTHS
        IEEE TRANS. MICROWAVE THEORY AND TECHNIQUES, VOL. MTT-24, NO. 3,
        PP. 153-155, MARCH 1976.
LECOY, G., R. ALABEDRA, AND B. BARBAN
    NOISE STUDIES IN INTERNAL FIELD EMISSION DIODES
        SOLID-STATE ELECTRONICS, VOL. 15, NO. 12, PP. 1273-1276, DECEMBER
        1972.
MARTINACHE, J-M., A. SEMICHON, E. CONSTANT, AND A. VANOVERSCHELDE
    ON THE BACKGROUND NOISE FROM A METAL-SEMICONDUCTOR BARRIER AVALANCHE
    DIODE
        COMPTES RENDUS ACAD. SCI., VOL. B269, NO. 14, PP. 644-647,
        OCTOBER 6, 1969.
MCINTYRE, R.J.
    MULTIPLICATION NOISE IN UNIFORM AVALANCHE DIODES.
        IEEE TRANS. ELECTRON DEVICES, VOL. ED-13, NO. 1, PP. 164-168,
        JANUARY 1966.
MINDEN, H. T.
    NOISE BEHAVIOR OF AVALANCHING SILICON DIODES
        PROC. IEEE, VOL. 54, NO. 8, PP. 1124-1125, AUGUST 1966.
MONCH, W.
    ON THE PHYSICS OF AVALANCHE BREAKDOWN IN SEMICONDUCTORS
        PHYSICA STATUS SOLIDI, VOL. 36, NO. , PP. 9-48, 1969.
MOUTHAAN, K.
    LOW-FREQUENCY MULTIPLICATION NOISE IN AVALANCHE TRANSIT-TIME DIODES
        PHILIPS RESEARCH REPTS., VOL. 26, NO. , PP. 298-325, AUGUST 1971.
NAQVI, I. M.
    EXPERIMENTAL OBSERVATION OF THE DEPENDENCE OF AVALANCHE NOISE ON
    CARRIER IONIZATION COEFFICIENTS
        PROC. IEEE, VOL. 60, NO. 12, PP. 1555-1556, DECEMBER 1972.
NAQVI, I. M.
    EFFECT OF TIME DEPENDENCE OF MULTIPLICATION PROCESS ON AVALANCHE NOISE
        SOLID-STATE ELECTRONICS, VOL. 16, NO. 1, PP. 19-28, JANUARY 1973.
PENFIELD, H.
    WHY NOT THE AVALANCHE DIODE AS AN RF NOISE SOURCE
        ELECTRONIC DESIGN, VOL. 12, PP. 32-35, APRIL 12, 1965.
RINGO, J. A. AND P. O. LAURITZEN
    LOW-FREQUENCY WHITE NOISE IN REFERENCE DIODES
        SOLID-STATE ELECTRONICS, VOL. 15, NO. 6, PP. 625-634, JUNE 1972.
RINGO, J. A. AND P. O. LAURITZEN
    1/F NOISE IN UNIFORM AVALANCHE DIODES
        SOLID-STATE ELECTRONICS, VOL. 16, NO. 3, PP. 327-328, MARCH 1973.
ROBERTS, J. A. AND W. GOSLING
    MEASUREMENTS ON A SOLID-STATE NOISE SOURCE
        RADIO AND ELECTRONIC ENGINEER, VOL. 40, NO. 6, PP. 323-324,
        DECEMBER 1970.

ROSE, D. J.
    MICROPLASMA IN SILICON
        PHYSICAL REV., VOL. 105, NO. 2, PP. 413-418, JANUARY 15, 1957.
SHERR, S. AND S. KING
    AVALANCHE NOISE IN P-N JUNCTIONS
        SEMICONDUCTOR PRODUCTS, VOL. 2, NO. 5, PP. 21-25, MAY 1959.
SINIGAGLIA, G. AND G. TOMASSETTI
    A SOLID STATE WIDE BAND NOISE GENERATOR
        ELECTRONIC ENGINEERING, VOL. 45, NO. 548, P. 15, OCTOBER 1973.
SOMLO, P. I.
    ZENER DIODE NOISE GENERATORS
        ELECTRONICS LETTERS, VOL. 11, NO. 14, P. 290, JULY 10, 1975.
SUSANS, D. E.
    NOISE CALIBRATOR FOR V.H.F. AND U.H.F. FIELD-STRENGTH-MEASURING
    RECEIVERS
        ELECTRONICS LETTERS, VOL. 3, NO. 8, PP. 354-355, AUGUST 1967.
SUSANS, D. E.
    SEMICONDUCTOR-DIODE V.H.F. AND U.H.F. NOISE SOURCES
        ELECTRONICS LETTERS, VOL. 4, NO. 4, PP. 72-73, FEBRUARY 23, 1968.
TAGER, A.S.
    CURRENT FLUCTUATIONS IN A SEMICONDUCTOR(DIELECTRIC) UNDER THE
    CONDITIONS OF IMPACT IONIZATION AND AVALANCHE BREAKDOWN.
        SOVIET PHYSICS - SOLID STATE, VOL. 6, NO. 8, PP. 1919-1925,
        FEBRUARY 1965.
VAL'D PERLOV, V. M., A. V. KRASILOV, AND A. S. TAGER
    THE AVALANCHING TRANSIT-TIME DIODE - A NEW SEMICONDUCTOR MICROWAVE
    DEVICE
        RADIO ENGINEERING AND ELECTRONIC PHYSICS, VOL. 11, NO. 11,
        PP. 1764-1779, NOVEMBER 1966.
VOLLMAN, E.
    EFFECT OF DOPING PROFILE ON AVALANCHE NOISE OF SILICON IMPATT DIODES
        ELECTRONICS LETTERS, VOL. 9, NO. 25, PP. 602-603, DECEMBER 13, 1973
WIERICH, R. L.
    COMPUTER SIMULATION OF AVALANCHE NOISE
        ELECTRONICS LETTERS, VOL. 8, NO. 3, PP. 58 - 59, 10 FEBRUARY 1972.
ZAKHODYAKIN, A. I.
    CONTROLLABLE NOISE GENERATOR BASED ON AN AVALANCHE-DRIFT DIODE
        INSTRUMENTS AND EXPERIMENTAL TECHNIQUES, VOL. 16, NO. 1, PT. 1,
        P. 154, JANUARY - FEBRUARY 1973.
ZAKHODYAKIN, A. I.
    NOISE-VOLTAGE GENERATOR WITH AMPLITUDE MODULATION
        INSTRUMENTS AND EXPERIEMNTAL TECHNIQUES, VOL. 16, NO. 4, PT. 1,
        PP. 1137-1138, JULY-AUGUST 1973.

****
    5-E. FREQUENCY-TRANSLATED NOISE SOURCES
****

BELL, D. A. AND A. M. ROSIE
    A LOW-FREQUENCY NOISE GENERATOR WITH GAUSSIAN DISTRIBUTION
        ELECTRONIC TECHNOLOGY, VOL. 37, NO. 6, PP. 241-245, JUNE 1960.
BENNETT, R. R. AND A. S. FULTON
    THE GENERATION AND MEASUREMENT OF LOW FREQUENCY RANDOM NOISE
        JOUR. APPL. PHYS., VOL. 22, NO. 9, PP. 1187-1191, SEPTEMBER 1951.
BONDAR, V. A., V. P. BONDAREN, A. K. KONDAKOV, AND I. SHIPUNOV
    LOW-FREQUENCY AND VERY-LOW-FREQUENCY NOISE GENERATOR
        INSTRUMENTS AND EXPERIMENTAL TECHNIQUES, NO. 3, PP. 946-948,
        MAY-JUNE 1970.
BOYES, J. D.
    BINARY NOISE SOURCES INCORPORATING MODULO-N DIVIDERS
        IEEE TRANS. COMPUTERS, VOL. C-23, NO. 5, PP. 550-552, MAY 1974.
BRODERICK, P.
    THE ANALYSIS AND DEMONSTRATION OF A METHOD OF GENERATING V.L.F. NOISE
        RADIO ELECTRONIC ENGRG (GB), VOL. 30, NO. 1, PP. 46-52, JULY 1965.
COHN, C. E.
    THE PERFORMANCE OF RANDOM-BIT GENERATORS
        SIMULATION, VOL. 17, NO. 6, PP. 234-236, DECEMBER 1971.
COHN, C.E.,
    CORRELATIONS IN A RANDOM-PULSE GENERATOR
        SIMULATION, VOL. 23, NO. 4, P. 128, OCTOBER 1974.
DIAMANTIDES, N. D. AND C. E. MC CRAY
    GENERATING RANDOM FORCING FUNCTIONS FOR CONTROL SYSTEM SIMULATION
        ELECTRONICS, VOL. 34, NO. 33, PP. 60-63, AUGUST 18, 1961.
DOUCE, J. L. AND J. M. SHACKLETON
    L. F. RANDOM SIGNAL GENERATOR
        ELECTRONIC AND RADIO ENGINEERING, VOL. 35, NO. 8, PP. 295-297,
        AUGUST 1958.
FRENCH, A. S.
    SYNTHESIS OF LOW-FREQUENCY NOISE FOR USE IN BIOLOGICAL EXPERIMENTS
        IEEE TRANS. BIO-MEDICAL ENGINEERING, VOL. BME-21, NO. 3, PP. 251 -
        252, MAY 1974.
GLADKIKH, G. A., V. G. PANOV, I. P. PAKHOMOV, AND P. D. CHICHIK
    INFRALOW-FREQUENCY NOISE GENERATOR
        INSTRUMENTS AND EXPERIMENTAL TECHNIQUES, VOL. 14, NO. 3, PT. 1,
        PP. 797 - 798, MAY - JUNE 1971.

HATA, S., H. SHIBATA AND T. TAKAI
    TRANSISTORISED ULTRALOW-FREQUENCY WHITE NOISE GENERATOR
        ELECTRICAL ENGINEERING IN JAPAN, VOL. 89, NO.5, PP. 61-68, MAY 1969
KABALEVSKII, A. N.
    STATISTICAL DESIGN OF A RANDOM-FUNCTION GENERATOR OF THE GSF-2 TYPE
        AUTOMATION AND REMOTE CONTROL, VOL. 29, NO. 12, PP. 1952-1960,
        DECEMBER 1968.
KNOTT, K. F.
    PRECISE WIDEBAND GAUSSIAN NOISE SOURCES USING JN FET AMPLIFIERS
    AND RESISTOR NOISE
        J. PHYSICS E, VOL. 5, NO. 3, PP. 254 - 257, MARCH 1972.
NIKIFORUK, P. N., R. PRONOVOST & G. SQUIRES.
    A LOW FREQUENCY RANDOM SIGNAL GENERATOR.
        ELECTRONIC ENGINEERING, VOL. 38, NO. 456, PP. 100-101, FEBRUARY
        1966.
SHIVA SHANKAR, T. N.
    A LOW-FREQUENCY RANDOM NOISE GENERATOR
        JOUR. INSTITUTION OF TELECOMMUNICATION ENGINEERS (INDIA), VOL. 9,
        NO. 1, PP. 63-68, JANUARY 1963.
SLATER, N. T.
    A LOW FREQUENCY NOISE GENERATOR
        ELECTRONIC ENGINEERING, VOL. 32, NO. 390, PP. 473-475, AUGUST 1960.
SUTCLIFFE, H.
    NOISE-SPECTRUM MEASUREMENT AT SUBAUDIO FREQUENCIES
        PROC. IEE (LONDON), VOL. 112, NO. 2, PP. 301-309, FEBRUARY 1965.
SUTCLIFFE, H. AND G. H. TOMLINSON
    A LOW- FREQUENCY GAUSSIAN WHITE-NOISE GENERATOR
        INTERNATIONAL JOUR. CONTROL, VOL. 8, NO. 5, PP. 457-471, NOVEMBER
        1968.
TAIT, D. A. G. AND M. SKINNER
    A RANDOM SIGNAL GENERATOR
        ELECTRONIC ENGINEERING , VOL. 38, NO. 455, PP. 2 - 7, JANUARY 1966.
WEST, J. C. AND G. T. ROBERTS
    A LOW FREQUENCY RANDOM NOISE SIGNAL GENERATOR
        JOUR. SCI. INSTRUM., VOL. 34, NO. 11, PP. 447-450, NOVEMBER 1957.
WINTER, D. F.
    A GAUSSIAN NOISE GENERATOR FOR FREQUENCIES DC IN TO 0.001 CYCLES PER
    SECOND
        1954 IRE CONVENTION RECORD, PT. 4, PP. 23-29, MARCH 1954.
YEOWART, N. S.
    LOW-FREQUENCY NOISE GENERATOR
        ELECTRONIC ENGINEERING, VOL. 40, NO. 482, PP. 212-214, APRIL 1968.

****
    5-F. PSEUDO-RANDOM NOISE SOURCES
****

ANDERSON, G. C., B. F. FINNIE AND G. T. ROBERTS
    PSEUDO-RANDOM AND RANDOM TEST SIGNALS
        HEWLETT-PACKARD JOUR., VOL. 19, NO. 1, PP. 2-17, SEPTEMBER 1967.
ARDICHVILI, P.
    A BINARY AND GAUSSIAN NOISE GENERATOR (IN FRENCH)
        ELECTRONIQUE INDUSTR., NO. 122, PP. 187-90, 1969.
BALL, J. R., A. H. SPITTLE, AND H. T. LIU
    HIGH-SPEED M-SEQUENCE GENERATION   A FURTHER NOTE
        ELECTRONICS LETTERS, VOL. 11, NO. 5, PP. 107-108, 6 MARCH 1975.
BEASTALL, H. R.
    AF WHITE-NOISE GENERATOR
        ELECTRON. AUSTRALIA, VOL. 34, NO. 4, PP. 63-65, JULY 1972.
BEASTALL, H. R.
    WHITE-NOISE GENERATOR
        WIRELESS WORLD, VOL. 78, NO. 1437, PP. 127 - 128, MARCH 1972.
    COMMENTS BY D. E. WADDINGTON
        WIRELESS WORLD, VOL. 78, PP. 264-265, JUNE 1972.
BRIGGS, P. A. N. AND K. R. GODFREY
    PSEUDORANDOM SIGNALS FOR THE DYNAMIC ANALYSIS OF MULTIVARIABLE SYSTEMS
        PROC. IEE (LONDON), VOL. 113, NO. 7, PP. 1259-1267, 1966.
CHEKRIZOV, V. G.
    A GENERATOR OF PSEUDORANDOM SIGNALS WITH UNIFORM SPECTRUM
        INSTRUMENTS AND EXPERIMENTAL TECHNIQUES, VOL. 18, NO. 3, PT. 1,
        PP. 806-807, MAY-JUNE 1975.
COEKIN, J. A. AND J. R. WICKING
    GENERATING AND CORRELATING PSEUDO-RANDOM BINARY SEQUENCES AT 1
    GIGABIT/SECOND
        CONFERENCE ON DIGITAL INSTRUMENTATION, LONDON, ENGLAND, 12-14 NOV
        1973, PP. 139-144
COEKIN, J. A. AND J. R. WICKING
    A NEW ALGORITHM FOR THE GENERATION OF HIGH SPEED M-SEQUENCES
        PROC. INST. RADIO AND ELECTRONICS ENGRS. (AUSTRALIA), VOL. 35,
        NO. 10, PP. 307 - 309, OCTOBER 1974.
CROVINI, L. AND A. ACTIS
    A PRECISE VARIABLE LEVEL BINARY NOISE GENERATOR
        ALTA FREQUENZA, VOL. 44, NO. 10, PP. 617-621 (327E-331E), OCTOBER
        1975.

CUMMINGS, I. G.
    AUTOCORRELATION FUNCTION AND SPECTRUM OF A FILTERED PSEUDORANDOM
    BINARY SEQUENCE
        PROC. IEE (LONDON), VOL. 114, NO. 9, PP. 1360-1362, SEPTEMBER 1967
DARNELL, M.
    SYNTHESIS OF PSEUDORANDOM SIGNALS DERIVED FROM P-LEVEL M-SEQUENCES
        ELECTRONICS LETTERS, VOL. 2, NO. 11, PP. 428-430, NOVEMBER 1966.
DAVIES, A. C.
    PROBABILITY DISTRIBUTION OF PSEUDORANDOM WAVEFORMS OBTAINED FROM
    M SEQUENCES
        ELECTRONICS LETTERS, VOL. 3, NO. 3, PP. 115-117, MARCH 1967.
DAVIES, A. C.
    DIGITAL FILTERING OF BINARY SEQUENCES
        ELECTRONICS LETTERS, VOL. 3, NO. 7, PP. 318-319, JULY 1967.
DAVIES, A. C.
    PROBABILITY DISTRIBUTIONS OF NOISELIKE WAVEFORMS GENERATED BY A
    DIGITAL TECHNIQUE
        ELECTRONICS LETTERS, VOL. 4, NO. 19, PP. 421-423, SEPTEMBER 20,
        1968.
DAVIES, A. C.
    PROBABILITY-DENSITY FUNCTIONS OF DIGITALLY FILTERED M-SEQUENCES
        ELECTRONICS LETTERS, VOL. 5, NO. 10, PP. 2½2-224, MAY 15, 1969.
DAVIO, M.
    RANDOM AND PSEUDORANDOM NUMBER GENERATORS
        ELECTRONIC ENGINEERING VOL. 39, NO. 475, PP. 558 - 559, SEPTEMBER
        1967.
DEUTSCH, S.
    A PSEUDO-RANDOM NOISE GENERATOR
        IEEE TRANS. INSTRUMENTATION AND MEASUREMENTS, VOL IM-16, NO 1, PP
        23-32, MARCH 1967.
EVERETT, D.
    PERIODIC DIGITAL SEQUENCES WITH PSEUDONOISE PROPERTIES.
        GEC J. SCI. TECHNOL. , VOL. 33, NO. 3, PP. 115-126, 1966.
FETH, L. L.
    A PSEUDORANDOM NOISE GENERATOR FOR USE IN AUDITORY RESEARCH
        BEHAVIOR RESEARCH METHODS AND INSTRUMENTATION, VOL. 2, NO. 4,
        P. 169, 1970.
GARDINER, A. B.
    LOGIC P.R.B.S. DELAY CALCULATOR AND DELAYED-VERSION GENERATOR WITH
    AUTOMATIC DELAY-CHANGING FACILITY
        ELECTRONICS LETTERS, VOL. 1, NO. 5, PP. 123-125, JULY 1965.
GILSON, R. P.
    SOME RESULTS OF AMPLITUDE DISTRIBUTION EXPERIMENTS ON SHIFT REGISTER
    GENERATED PSEUDO-RANDOM NOISE.
        IEEE TRANS. ELECTRONIC COMPUTERS, VOL. EC-15, NO. 6, PP. 926-927,
        DECEMBER 1966.
GOLOVKO, A. G., I. YA. KOZYR' AND V. M. PISCEV
    INFRA-LOW-FREQUENCY NOISE GENERATOR FOR AN INSTRUMENT THAT MEASURES
    THE NOISE OF SEMICONDUCTOR DEVICES.
        INSTRUM. AND EXP. TECH. (USA), VOL 16, NO. 4, PT. 1, PP. 1139-
        41, JULY-AUG 1973.
HAMPTON, R. L. T.
    A HYBRID ANALOG-DIGITAL PSEUDO RANDOM NOISE GENERATOR
        SIMULATION, VOL. 4, NO. 3, PP. 179-187, MARCH 1965.
HAMPTON, R. L. T.
    EXPERIMENTS USING PSEUDO-RANDOM NOISE
        SIMULATION, VOL. 4, NO. 4, PP. 246-254, APRIL 1965.
HAMPTON, R., G. A. KORN, AND B. MITCHELL
    HYBRID ANALOG-DIGITAL RANDOM-NOISE GENERATION
        IEEE TRANS. ELECTRONIC COMPUTERS, VOL. EC-12, NO. 4, PP. 412-413,
        AUGUST 1963.
HARTLEY, M. G.
    DEVELOPMENT, DESIGN AND TEST PROCEDURE FOR RANDOM GENERATORS USING
    CHAIN CODES
        PROC. IEE (LONDON), VOL. 116, NO. 1, PP. 22-26, JANUARY 1969.
HARVEY, J. T.
    HIGH-SPEED M-SEQUENCE GENERATION
        ELECTRONICS LETTERS, VOL. 10, NO. 23, PP. 480-481, 14 NOVEMBER
        1974.
HENNE, W.
    A REPRODUCIBLE PRECISION NOISE GENERATOR. PT. 1 (IN GERMAN)
        ARCH. TECH. MESSEN., NO. 385, PP. 37-42, FEBRUARY 1968.
HENNE, W.
    A REPRODUCIBLE PRECISION NOISE GENERATOR. PT. 2 (IN GERMAN)
        ARCH. TECH. MESSEN., NO. 386, PP. 59-62, MARCH 1968.
HSIAO, M. Y.
    GENERATING PSEUDO NOISE SEQUENCE
        IBM TECHNICAL DISCLOSURE BULLETIN, VOL. 11, NO. 4, PP. 393-394,
        SEPTEMBER 1968.
HURD, W. J.
    A WIDEBAND GAUSSIAN NOISE GENERATOR UTILIZING SIMULTANEOUSLY
    GENERATED PN-SEQUENCES
        PROC. OF THE 5TH HAWAII INTERNATIONAL CONFERENCE ON SYSTEM
        SCIENCE, HONOLULU, HAWAII, 11-13 JAN 1972, PP. 168-70
IKAI, T., H. KOSAKO & Y. KOJIMA
    TERNARY PSEUDO-RANDOM NOISE GENERATOR.

BULL. UNIV. OSAKA PREFECTURE A (JAPAN), VOL. 17, NO. 1, PP. 113-28
1968.
IRELAND, B. AND J. E. MARSHALL
    MATRIX METHOD OF DETERMINING SHIFT REGISTER CONNECTION FOR DELAYED
    PSEUDORANDOM BINARY SEQUENCES
        ELECTRONICS LETTERS, VOL. 4, NO. 15, PP. 309-310, JULY 26, 1968.
IRELAND, B. AND J. E. MARSHALL
    MATRIX METHOD OF DETERMINING SHIFT REGISTER CONNECTION FOR DELAYED
    PSEUDORANDOM BINARY SEQUENCES
        ELECTRONICS LETTERS, VOL. 4, NO. 21, PP. 467-468, OCTOBER 18, 1968.
JORDAN, H. F. AND D. C. M. WOOD
    ON THE DISTRIBUTION OF SUMS OF SUCCESSIVE BITS OF SHIFT-REGISTER
    SEQUENCES
        IEEE TRANS. COMPUTERS, VOL. C-22, NO. 4, PP. 400-408, APRIL 1973.
KEELE, D. B.
    THE DESIGN AND USE OF A SIMPLE PSEUDORANDOM PINK-NOISE GENERATOR
        JOUR. AUDIO ENGINEERING SOC., VOL. 21, NO. 1, PP. 33-41,
        JANUARY/FEBRUARY 1973.
    FURTHER ON SIMPLE PSEUDORANDOM PINK NOISE GENERATOR
        JOUR. AUDIO ENGINEERING SOC., VOL. 21, NO. 3, P. 198, APRIL 1973.
KOCK, K.
    GENERATOR FOR PSEUDORANDOM NOISE SIGNALS (IN GERMAN)
        ELECTRONIK, VOL. 17, NO. 3, PP. 69-72, MARCH 1968.
KRAMER, C.
    A LOW FREQUENCY PSEUDO-RANDOM NOISE GENERATOR
        ELECTRONIC ENGINEERING, VOL. 37, NO. 449, PP. 465-467, JULY 1965.
LATAWIEC, K. J.
    NEW METHOD OF GENERATION OF SHIFTED LINEAR PSEUDO-RANDOM BINARY
    SEQUENCES
        PROC. IEE (LONDON), VOL. 121, NO. 8, PP. 905-906, AUGUST 1974.
    DISCUSSIONS IN
        PROC. IEE (LONDON), VOL. 122, NO. 4, P. 448, APRIL 1975.
        PROC. IEE (LONDON), VOL. 123, NO. 2, P. 182, FEBRUARY 1976.
LEMPEL, A. AND W. L. EASTMAN
    HIGH SPEED GENERATION OF MAXIMAL LENGTH SEQUENCES
        IEEE TRANS. COMPUTERS, VOL. C-20, NO. 2, PP. 227-229, FEBRUARY 1971.
LINDHOLM, J. H.
    AN ANALYSIS OF THE PSEUDO-RANDOMNESS PROPERTIES OF SUBSEQUENCES OF
    LONG M-SEQUENCES
        IEEE TRANS. INFORMATION THEORY, VOL. IT-14, NO. 4, PP. 569-576,
        JULY 1968.
LIPSON, E. D., K. W. FOSTER, AND M. P. WALSH
    A VERSATILE PSEUDO-RANDOM NOISE GENERATOR
  ·   IEEE TRANS. INSTRUMENTATION AND MEASUREMENTS, VOL. IM-25, NO. 2,
        PP. 112-116, JUNE 1976.
MATTHEWS, S. B.
    GENERATION OF PSEUDORANDOM NOISE HAVING A GAUSSIAN SPECTRAL
    DENSITY
        IEEE TRANS COMPUTERS, VOL C-17, NO 4, PP 312-385, APRIL 1968.
MEYER, F.
    GIGABIT/S M-SEQUENCE GENERATION
        ELECTRONICS LETTERS, VOL. 12, NO. 4, P. 353, JULY 8, 1976.
MIMAKI, T.
    A STATIONARY HIGH-LEVEL GAUSSIAN NOISE GENERATOR
        JOUR. PHYS. E - SCIENTIFIC INSTR., VOL. 5, NO. 3, PP. 208-211,
        MARCH 1972.
MORGAN, D.P. AND J.G. SUTHERLAND,
    GENERATION OF PSEUDONOISE SEQUENCES USING SURFACE ACOUSTIC WAVES,
        IEEE TRANS. MICROWAVES THEORY TECHNIQUES, VOL MTT-21, NO 4, P. 306,
        APRIL 1973
MORGAN, D. P.
    GENERATION OF PSEUDONOISE SEQUENCES USING SURFACE ACOUSTIC WAVES
        IEEE TRANS. MICROWAVE THEORY AND TECHNIQUES, VOL. MTT-21, NO. 4,
        P. 306, APRIL 1973, AND IEEE TRANS. SONICS AND ULTRASONICS, VOL.
        SU-20, NO 2, P 224, APRIL 1973.
NAKAMURA , S.
    A METHOD OF GENERATING A RANDOM SIGNAL USING OPERATIONAL AMPLIFIERS
        PROC. IEEE, VOL. 62, NO. 5, PP. 651-653, MAY 1974.
NEUVO, Y. AND W. H. KU
    ANALYSIS AND DIGITAL REALIZATION OF A PSEUDORANDOM GAUSSIAN AND
    IMPULSIVE NOISE SOURCE
        IEEE TRANS. COMMUNICATIONS, VOL. COM-23, NO. 9, PP. 849-858,
        SEPTEMBER 1975.
PANGRATZ, H. AND H. WEINRICHTER
    PSEUDORANDOM GENERATOR FOR SIMULATION OF A POISSON PROCESS (IN GERMAN)
        NACHRICHTENTECHNISCHE ZEITSCHRIFT, VOL. 27, NO. 9, PP. 325 -
        SEPTEMBER 1974.
PAUMARD, A.
    NOISE GENERATOR IN THE FIELD OF LOW FREQUENCY (IN FRENCH)
        ONDE ELECT., VOL. 40, PP. 1095-1098, OCTOBER 1966.
PETTERNELLA, M. & A. RUBERTI.
    A LOW-FREQUENCY WHITE NOISE GENERATOR WITH ARTIFICIAL SOURCE.
        PROC. 4TH INTERNATIONAL ANALOGUE COMPUTATION MEETINGS, BRIGHTON,
        1964, PP. 289-300. ALSO IN
        ALTA FREQUENZA, VOL. 33, NO. 6, PP. 376-86, JUNE 1964. (IN
        ITALIAN).

ROBERTS, G., B. FINNIE, AND G. ANDERSON
    SYNTHESIZING LOW-FREQUENCY NOISE
        INSTRUMENT. CONTROL SYSTEMS, VOL. 40, NO. 8, PP. 120-124, AUGUST
        1967.
ROBERTS, P. D. AND R. H. DAVIS
    STATISTICAL PROPERTIES OF SMOOTHED MAXIMUM-LENGTH LINEAR BINARY
    SEQUENCES
        PROC. IEE (LONDON), VOL. 113, NO. 1, PP. 190-196, JANUARY 1966.
ROWE, I. H. AND I. M. KERR
    A BROAD-SPECTRUM PSEUDORANDOM GAUSSIAN NOISE GENERATOR
        11TH JOINT AUTOMATIC CONTROL CONF. OF THE AMERICAN AUTOMATIC
        CONTROL COUNCIL, ATLANTA, GA., USA, 22-26 JUNE 1970, PP. 733-739,
        ALSO AS
        IEEE TRANS. AUTOMATIC CONTROL, VOL. AC-15, NO. 5, PP. 529-535,
        OCTOBER 1970.
SIMPSON, H. R.
    STATISTICAL PROPERTIES OF A CLASS OF PSEUDORANDOM SEQUENCES
        PROC. IEE, VOL. 113, NO. 12, PP. 2075 - 2080, DECEMBER 1966.
SINHA, S. AND B. V. RAJA RAO
    A PSEUDO-RANDOM NOISE GENERATOR
        J. INSTN. ENG. (INDIA) ELECTRON TELECOMMUN. ENG. DIV., VOL. 50,
        NO. 9, PT. ET 3, PP. 143-6, MAY 1970
SMITH, B. M.
    AN ASYMMETRICAL PROPERTY OF BINARY PSEUDORANDOM NOISE GENERATORS.
        PROC. IEEE, VOL. 54, NO. 5, PP. 793-794, MAY 1966.
SMITH, K. D. AND J. C. HAMILTON.
    THE LOGICAL DESIGN OF A DIGITAL PSEUDORANDOM NOISE GENERATOR.
        IEEE TRANS. NUCLEAR SCI., VOL. NS-13, NO. 1, PP. 371-81, FEB. 1966
SUTCLIFFE, H. AND K. F. KNOTT
    STANDARD L. F. NOISE SOURCES USING DIGITAL TECHNIQUES AND THEIR
    APPLICATION TO THE MEASUREMENT OF NOISE SPECTRA
        RADIO AND ELECTRONIC ENGINEER, VOL. 40, NO. 3, PP. 132 - 136,
        SEPTEMBER 1970.
SUZUKI, Y., P.S. PAK AND K. FUJII,
    PSEUDO GAUSSIAN NOISE GENERATOR OF HYBRID TYPE
        TECHNOL. REP. OSAKA UNIV. (JAPAN), VOL. 22, NO. 1053-1089, PP.623-
        633, OCTOBER 1972.
SZAJNOWSKI, W. J.
    LOW-COST PSEUDORANDOM GAUSSIAN NOISE GENERATOR
        ELECTRONIC ENGINEERING, VOL. 48, NO. 579, P. 22, MAY 1976.
TOMLINSON, G. H. AND P. GALVIN
    ANALYSIS OF SKEWING IN AMPLITUDE DISTRIBUTIONS OF FILTERED
    M-SEQUENCES
        PROC. IEE, VOL. 121, NO. 12, PP. 1475-1479, DECEMBER 1974.
TOMLINSON, G. H. AND P. GALVIN
    ELIMINATION OF SKEWING IN THE AMPLITUDE DISTRIBUTIONS OF LONG
    M-SEQUENCES SUBJECTED TO LOWPASS FILTERING
        ELECTRONICS LETTERS, VOL. 11, NO. 4, PP. 77-78, 20 FEBRUARY 1975.
TOMLINSON, G. H. AND P. GALVIN
    GENERATION OF GAUSSIAN SIGNALS FROM SUMMED M- SEQUENCES
        ELECTRONICS LETTERS, VOL. 11, NO. 21, PP. 521-522, OCT. 16, 1975.
TOMLINSON, G. H. AND P. GALVIN
    DESIGN CRITERION FOR GENERATION OF GAUSSIAN SIGNALS FROM SMOOTHED
    M-SEQUENCES
        ELECTRONICS LETTERS, VOL. 12, NO. 14, PP. 349-350, JULY 8, 1976.
TRICHARD, CL.
    PSEUDORANDOM NOISE GENERATOR (IN FRENCH)
        ELECTRONIQUE, NO. 140, PP. 61-63, JANUARY-FEBRUARY 1971.
VORONIN, A. A.
    THE SPECTRA OF PSEUDORANDOM BINARY SEQUENCES
        TELECOMMUNICATIONS AND RADIO ENGINEERING, PT. 1 - TELECOMM.,
        NO. 2, PP. 60-62, FEBRUARY 1965.
WAINBERG, S. AND J. K. WOLF
    SUBSEQUENCES OF PSEUDORANDOM SEQUENCES
        IEEE TRANS. COMMUNICATIONS, VOL. COM-18, P. 606-612, 1970.
WANKE, E.
    NEW CRITERION FOR FEED BACK-SHIFT REGISTER CONNECTION.
        ELECTRONICS LETTERS, VOL. 1, NO. 10, P. 290, DEC. 1965.
WONHAM, W. M. AND A. T. FULLER
    PROBABILITY DENSITIES OF THE SMOOTHED RANDOM TELEGRAPH SIGNAL
        JOUR. ELECTRONICS AND CONTROL, VOL. 4, NO. 6, PP. 567-576, JUNE
        1958.
YUEN, W.L.,
    A MAPPING PROPERTY OF PSEUDORANDOM SEQUENCE,
        INTERN. J. CONTROL, VOL. 17, NO 6, 1217-1223, JUNE 1973

```
***************************************************************
    5-G.   OTHER PAPERS ON NOISE SOURCES
***************************************************************

***************************************************************
        REVIEW PAPERS ON NOISE SOURCES
***************************************************************
```

BENNETT, W. R.
    EQUIPMENT FOR GENERATING NOISE

ELECTRONICS, VOL. 29, NO. 4, PP. 134-137, APRIL 1956.
CASTRIOTA, L. J.
    NOISE GENERATION
        IN HANDBOOK OF MICROWAVE MEASUREMENTS   VOL. 1, 2ND EDITION,
        M. WIND AND H. RAPAPORT, EDITORS, POLYTECHNIC INSTITUTE OF BROOKLYN
        NEW YORK, 1955.
HART, P. A. H.
    STANDARD NOISE SOURCES
        PHILIPS TECH. REV., VOL. 23, NO. 10, PP. 293-309, JULY 25, 1962.
KIRSCHNER, U.
    NOISE GENERATORS AND THEIR APPLICATION (IN GERMAN)
        MESS. U. PRUFEN., VOL. 5, NO. 4, PP. 237-240, APRIL 1969.
MICHEL, H. ST. AND H. PRINZLER
    NOISE GENERATORS IN THE MICROWAVE RANGE. PT. I. (IN GERMAN)
        NACHRICHTENTECHNIK, VOL. 15, NO. 1, PP. 33-37, JANUARY 1965.
MICHEL, H. ST. AND H. PRINZLER
    NOISE GENERATORS IN THE MICROWAVE RANGE. PT. 2. (IN GERMAN)
        NACHRICHTENTECHNIK, VOL. 15, NO. 2, PP. 65-67, FEBRUARY 1965.
MILLER, C. K. S., W. C. DAYWITT, AND M. G. ARTHUR
    NOISE STANDARDS, MEASUREMENTS, AND RECEIVER NOISE DEFINITIONS
        PROC. IEEE, VOL. 55, NO. 6, PP. 865-876, JUNE 1967.
NAGLE, J. J.
    NOISE AND NOISE GENERATORS I
        CQ RADIO AMAT. J., VOL. 28, NO. 5, PP. 46-47, 81-2, MAY 1972.
NAGLE, J. J.
    NOISE AND NOISE GENERATORS II
        CQ RADIO AMAT. JOUR., VOL. 28, NO. 6, PP. 28-30, 81, JUNE 1972.

****
NOISE SOURCES WITH SPECIFIED SPECTRA, DISTRIBUTIONS, OR CORRELATIONS
****

ALEIXANDRE, V., J. L. GARCIA AND L. A. BAILON,
    A NOISE GENERATOR FOR DIFFERENT TYPES OF DISTRIBUTIONS
        INTERN. JOUR. ELECTRONICS, VOL. 33, NO. 2, PP. 133-142, AUGUST 1972
BOULTON, P. I. AND R. J. KAVANAGH
    A METHOD OF PRODUCING MULTIPLE NONCORRELATED RANDOM SIGNALS FROM
    A SINGLE GAUSSIAN NOISE SOURCE
        IEEE TRANS. APPLICATIONS AND INDUSTRY, VOL. 82, NO. 65, PP. 46-52,
        MARCH 1963.
BROWN, J. L. JR.
    GENERATING UNCORRELATED RANDOM OUTPUTS BY NONLINEAR PROCESSING OF A
    SINGLE NOISE SOURCE.
        IEEE TRANS. APPLICATIONS AND INDUSTRY, VOL. 83, NO. 75, PP. 408-
        410, NOVEMBER 1964.
BRYAN, R. E.
    GENERATION OF NONSTATIONARY RANDOM PROCESSES
        SIMULATION, VOL. 4, NO. 1, PP. 42-48, JANUARY 1965.
DIAMANTIDES, N. D.
    ANALOGUE COMPUTER GENERATION OF PROBABILITY DISTRIBUTIONS FOR
    OPERATIONS RESEARCH
        TRANS. AMER. INST. ELECTRICAL ENGRS., PT. 1 - COMMUNICATIONS AND
        ELECTRONICS, VOL. 75, NO. 3, PP. 86-91, MARCH 1956.
GELB, A. AND P. PALOSKY.
    GENRATING DISCRETE COLORED NOISE FROM DISCRETE WHITE NOISE.
        IEEE TRANS. AUTOMATIC CONTROL, VOL. AC-11, NO. 1, PP. 148-149,
        JANUARY 1966.
GUJAR, U. G. AND R. J. KAVANAGH
    GENERATION OF RANDOM SIGNALS WITH SPECIFIED PROBABILITY DENSITY
    FUNCTIONS AND POWER DENSITY SPECTRA
        IEEE TRANS. AUTOMATIC CONTROL, VOL. AC-13, NO. 6, PP. 716-719,
        DECEMBER 1968.
KHRISTYUK, V. A. AND B. M. CHERNOVOI
    A GENERATOR PRODUCING RANDOM QUANTITIES HAVING A STIPULATED
    DISTRIBUTION LAW
        INSTRUMENTS AND EXPERIMENTAL TECHNIQUES, VOL. 17, NO. 2, PT. 1, PP.
        430 - 43, MARCH - APRIL 1974.
KROSCHEL, K.
    GENERATION OF RANDOM PROCESS WITH PRESCRIBED DENSITY AND AUTO-
    CORRELATION FUNCTION.
        PROC. OF THE 5TH COLLOQUIM ON MICROWAVE COMM. VOL. I. MICROWAVE
        COMM. SYSTEMS AND COMM. SYS. THEORY, P. ST-11/181-9, 24-30 JUNE
        1974.
YAKOVLEV, V. P.
    OBTAINING A SET OF NON-CORRELATED RANDOM PROCESSES VIA NON-LINEAR
    TRANSFORMATIONS OF A SINGLE RANDOM SIGNAL
        AUTOMATION AND REMOTE CONTROL, VOL. 26, NO. 6, PP. 1086-1092,
        JUNE 1965.
ZAKHODYAKIN, A. I.
    NOISE-VOLTAGE GENERATOR WITH AMPLITUDE MODULATION.
        INSTRUM. AND EXP. TECH., VOL. 16, NO. 4, PT. 1, PP. 1137-8,
        JULY-AUG 1973.

```
*****************************************************************************
                     OTHER MISCELLANEOUS NOISE SOURCES
*****************************************************************************

ARNAUD, J., M. PROGENT, W. SOBOTKA AND L. TEYSSIER
    HIGH POWER NOISE SOURCES (IN FRENCH)
        NACHRICHTENTECH. FACHBER. (NTF) GERMANY, VOL. 35, PP. 333-336, 1968
BURTON, R. AND L. D. HALL
    AN INEXPENSIVE NOISE SOURCE FOR BROAD-BAND HETERONUCLEAR DECOUPLING
    EXPERIMENTS
        CANADIAN JOUR. CHEMISTRY, VOL. 48, NO.  , PP. 2438-2439, 1970.
COLEMAN, J. T.
    DARK NOISE GENERATION BY SUPER POWER TUBES
        IRE TRANS. RADIO FREQUENCY INTERFERENCE, VOL. RFI-4, NO. 3, PP.
        44-48, OCTOBER 1962.
DAVYDOV, YU. T.
    PHOTODIODE AS A NOISE OSCILLATOR
        RADIO ENGINEERING AND ELECTRONIC PHYSICS, VOL. 12, NO. 12,
        PP. 2088-2090, DECEMBER 1967.
DIAMOND, J.M.
    CALIBRATION OF AN AUDIO-FREQUENCY NOISE GENERATOR,
        IEEE TRANS. AUDIO AND ELECTROACOUSTICS, VOL-AU 14, NO 2, PP 96-100,
        JUNE 1966.
FARAN, J. J., JR.
    RANDOM NOISE GENERATORS
        GR EXPERIMENTER, VOL. 42, NO. 1, PP. 3-13, JANUARY 1968.
HENRY, R. F.
    RANDOM NOISE GENERATION BY HYBRID COMPUTER
        IEEE TRANS. ELECTRONIC COMPUTERS, VOL. EC-16, NO. 6, PP. 872-873,
        DECEMBER 1967.
MANELIS, J. B.
    GENERATING RANDOM NOISE WITH RADIO-ACTIVE SOURCES
        ELECTRONICS, VOL. 34, NO. 36, PP. 66-69, SEPTEMBER 8, 1961.
MARTYNICHEV, A. K. AND E. M. BELOV
    CONTROLLED NOISE GENERATION
        INSTRUMENTS AND EXPERIMENTAL TECHNIQUES, NO. 1, PP. 130-131,
        JANUARY-FEBRUARY 1968.
MC DOWELL, H. L. AND G. K. FARNEY
    CROSSED-FIELD NOISE GENERATION DEVICES
        IN  MICROWAVE POWER ENGINEERING , VOL. 1, E. C. OKRESS, ED.,
        ACADEMIC PRESS, NEW YORK, 1968, PP. 75-83.
OLIVER, W. AND A. J. SPENCER.
    CHANNEL WHITE NOISE GENERATOR.
        MARCONI INSTRUM. (GB), VOL. 10, NO. 2, PP. 23-25, AUG. 1965.
OSEPCHUK, J. M.
    NON-LINEAR SINGLE-CHARGE THEORY OF CROSSED-FIELD INTERACTION & ITS
    APPLICATION TO CFA'S & NOISE GENERATORS
        NACHRICHTENTECH. FACHBER (NTF) GERMANY, VOL. 35, PP. 294-302, 1968
PALIN, J. AND G. GOUREVITCH
    AN IMPROVED NARROW-BAND NOISE SOURCE
        EEG CLIN. NEUROPHYSIOL. (NETHERLANDS), VOL. 29, NO.5, PP. 523-4, NOV 1970
ROSS, G. F.
    NOISE PROPERTIES OF BEAM SWITCHING TUBES
        ELECTRONIC INDUSTR., VOL. 20, NO. 7, PP. 96-99, JULY 1961.
SHIBATA, H., T. TAKAI, AND S. HATA
    ULTRA LOW FREQUENCY WHITE NOISE GENERATOR
        BULLETIN UNIVERSITY OF OSAKA PREFECTUER A, VOL. 15, NO. 2,
        PP. 99-108, 1966.
SHIVA SHANKAR, T. N.
    A LOW FREQUENCY RANDOM NOISE GENERATOR
        JOUR. INSTITUTION OF TELECOMM. ENGINEERS (INDIA), VOL. 9, NO. 1,
        PP. 63-68, JANUARY 1963.
SIZ'MIN, A. M.
    BROADBAND NOISE GENERATORS BASED ON INTERMITTENT PHOTOELECTRON
    MULTIPLIERS
        INSTRUM. & EXP. TECH (USA), NO. 4, PP. 1184-6, JUL-AUG, 1970
SUBHANI, K. F. AND W. S. ADAMS
    LOW FREQUENCY NOISE SOURCES
        INSTRUMENTS AND CONTROL SYSTEMS, VOL. 43, NO. 3, PP. 113-115,
        MARCH 1970.
TIPTON, R. B.
    A STUDY OF THE STATIONARITY OF LABORATORY RANDOM NOISE GENERATORS
        IEEE TRANS. INSTRUMENTATION AND MEASUREMENTS, VOL. IM-15, NO. 1/2,
        PP. 20 - 24, MARCH/JUNE 1966.
VOLCHKOV, N. M.
    FORMATION OF RANDOM PULSE SEQUENCE WITH A GIVEN LAW OF DISTRIBUTION
    OF INTERVALS
        INSTRUMENTS AND EXPERIMENTAL TECHNIQUES, NO. 2, PP. 382 - 383,
        MARCH - APRIL 1969.
WHITE, G. M.
    ELECTRONIC PROBABILITY GENERATOR
        REV. SCI. INSTR., VOL. 30, NO. 9, PP. 825-829, SEPTEMBER 1959.
YAKUTIS, A. J.
    SOLID-STATE RF NOISE SOURCE
        PROC. IEEE, VOL. 56, NO. 2, P. 228, FEBRUARY 1968.
```

Author Index

Subject Index

Editor's Biography

Madhu-Sudan Gupta (S'68–M'72) received the M.S. and Ph.D. degrees from the University of Michigan, Ann Arbor, in 1968 and 1972, respectively.

From 1968 to 1972 he carried out research on large-signal and noise characteristics of IMPATT diodes at the Electron Physics Laboratory, Department of Electrical and Computer Engineering, University of Michigan, Ann Arbor. He was an Assistant Professor of Electrical Engineering at Queen's University, Kingston, Ont., Canada, during the year 1972–1973. He is presently an Assistant Professor in the Department of Electrical Engineering, Massachusetts Institute of Technology, Cambridge, and is engaged in research on semi-conductor microwave devices and noise at the M.I.T. Research Laboratory of Electronics.

Dr. Gupta is a member of Eta Kappa Nu, Sigma Xi, and Phi Kappa Phi, and is a Registered Professional Engineer in the Province of Ontario.